# BNi Building News

# GENERAL CONSTRUCTION 1997 COSTBOOK

## SEVENTH EDITION

BNi Building News

Los Angeles • Anaheim

Boston • Washington, D.C.

**BNi Building News**

**EDITOR-IN-CHIEF**
*William D. Mahoney, P.E.*

**TECHNICAL SERVICES**
*Edward B. Wetherill, CSI*
*Anthony Jackson*
*Alison Kermode*
*Ramon Lopez*

**DESIGN**
*Robert O. Wright*

**COVER DESIGN**
*Hannus Design Associates*

**BNI Publications, Inc.**

**LOS ANGELES**
10801 NATIONAL BLVD.
LOS ANGELES, CA 90064

**ANAHEIM**
1612 S. CLEMENTINE STREET
ANAHEIM, CA 92802

**BOSTON**
629 HIGHLAND AVENUE
NEEDHAM HEIGHTS, MA 02194

**WASHINGTON, D.C.**
502 MAPLE AVENUE WEST
VIENNA, VA 22180

**1-800-873-6397**

**ISBN 1557011591**

BNi® Building News

# PREFACE

For the past 50 years, Building News has been dedicated to providing construction professionals with timely and reliable information. Based on this experience, our staff has researched and compiled thousands of up-to-the-minute costs for the **Building News 1997 Costbooks**. This book is an essential reference for contractors, engineers, architects, facilities managers — any construction professional who must provide an estimate on any type of building project.

Whether working up a preliminary estimate or submitting a formal bid, the costs listed here can quickly and easily be tailored to your needs. All costs are based on national averages, while a table of modifiers is provided for regional adjustments. Overhead and profit are included in all costs.

Complete man-hour tables follow the unit costs to provide data on typical durations of specific tasks. This information can be used to schedule projects as well as to determine specific labor costs based on local labor rates.

All data is categorized according to the MASTERFORMAT of the Construction Specifications Institute (CSI). This industry standard provides an all-inclusive checklist to ensure that no element of a project is overlooked. In addition, to make specific items even easier to locate, there is a complete alphabetical index.

This costbook contains an appendix with reference charts and tables taken from an array of sources. Text explains the costs in certain categories and provides helpful pointers that should be taken into account with every estimate.

A section on square foot costs provides an overview of project costs for different building types — commercial, residential, etc. — with summaries of the actual projects. Square foot costs are invaluable for making budget estimates and checking prices when time is a factor.

The "Features in this Book" section presents a clear overview of the many features of this book. Included is an explanation of the data, sample page layout and discussion of how to best use the information in the book.

Of course, all buildings and construction projects are unique. The costs provided in this book are based on averages from well-managed projects with good labor productivity under normal working conditions (eight hours a day). Other circumstances affecting costs such as overtime, unusual working conditions, savings from buying bulk quantities for large projects, and unusual or hidden costs must be factored in as they arise.

The data provided in this book is for estimating purposes only. Check all applicable federal, state and local codes and regulations for specific requirements.

BNi® Building News

# TABLE OF CONTENTS

# CSI MASTERFORMAT

All data in the Costbook pages and Man-Hour tables is organized according to the CSI MASTERFORMAT — the industry standard numbering/classification system. The data is divided into the 16 divisions as shown below. The five digit numbers within each division correspond to the MASTERFORMAT Broadscope designations. Each section is further broken down into MASTERFORMAT Mediumscope designations. These numbers, in most cases, are the same as those used in architectural and engineering specifications.

The construction estimating information in this book is divided into two main sections: Costbook Pages and Man-Hour Tables. Each is organized to the 16 divisions of the CSI MASTERFORMAT as shown below. In addition, there are extensive Construction Reference Tables, Geographic Cost Modifiers, Square Foot Tables and a detailed Index. Sample pages with graphic explanations are included before the Costbook pages, Man-Hour Tables and Square Foot Tables. These explanations, along with the discussions below, will provide a good understanding of what is included in this book and how it can best be used for construction estimating.

| GENERAL REQUIREMENTS | *Division 1* |
|---|---|
| Summary of Work | 01010 |
| Allowances | 01020 |
| Measurement and Payment | 01025 |
| Alternates/Alternatives | 01030 |
| Modification Procedures | 01035 |
| Coordination | 01040 |
| Field Engineering | 01050 |
| Regulatory Requirements | 01060 |
| Identification Systems | 01070 |
| References | 01090 |
| Special Project Procedures | 01100 |
| Project Meetings | 01200 |
| Submittals | 01300 |
| Quality Control | 01400 |
| Construction Facilities & Temporary Controls | 01500 |
| Material and Equipment | 01600 |
| Facility Startup/Commissioning | 01650 |
| Contract Closeout | 01700 |
| Maintenance | 01800 |

| SITEWORK | *Division 2* |
|---|---|
| Subsurface Investigation | 02010 |
| Demolition | 02050 |
| Site Preparation | 02100 |
| Dewatering | 02140 |
| Shoring and Underpinning | 02150 |
| Excavation Support Systems | 02160 |
| Cofferdams | 02170 |
| Earthwork | 02200 |
| Tunneling | 02300 |
| Piles and Caissons | 02350 |
| Railroad Work | 02450 |
| Marine Work | 02480 |
| Paving and Surfacing | 02500 |
| Utility Piping Materials | 02600 |
| Water Distribution | 02660 |
| Fuel and Steam Distribution | 02680 |
| Sewerage and Drainage | 02700 |
| Restoration of Underground Pipe | 02760 |
| Ponds and Reservoirs | 02770 |
| Power and Communications | 02780 |
| Site Improvements | 02800 |
| Landscaping | 02900 |

| CONCRETE | *Division 3* |
|---|---|
| Concrete Formwork | 03100 |
| Concrete Reinforcement | 03200 |
| Concrete Accessories | 03250 |
| Cast-In-Place Concrete | 03300 |
| Concrete Curing | 03370 |
| Precast Concrete | 03400 |
| Cementitious Decks and Toppings | 03500 |
| Grout | 03600 |
| Concrete Restoration and Cleaning | 03700 |
| Mass Concrete | 03800 |

| MASONRY | *Division 4* |
|---|---|
| Mortar and Masonry Grout | 04100 |
| Masonry Accessories | 04150 |
| Unit Masonry | 04200 |
| Stone | 04400 |
| Masonry Restoration and Cleaning | 04500 |
| Refractories | 04550 |
| Corrosion Resistant Masonry | 04600 |
| Simulated Masonry | 04700 |

| METALS | *Division 5* |
|---|---|
| Metal Materials | 05010 |
| Metal Coatings | 05030 |
| Metal Fastening | 05050 |
| Structural Metal Framing | 05100 |
| Metal Joists | 05200 |
| Metal Decking | 05300 |
| Cold Formed Metal Framing | 05400 |

## WOOD AND PLASTICS *Division 6*



## THERMAL AND MOISTURE PROTECTION *Division 7*



## DOORS AND WINDOWS *Division 8*



## FINISHES *Division 9*



## SPECIALTIES *Division 10*



## EQUIPMENT *Division 11*

## FURNISHINGS *Division 12*



## SPECIAL CONSTRUCTION *Division 13*



## CONVEYING SYSTEMS *Division 14*



## MECHANICAL *Division 15*



## ELECTRICAL *Division 16*

# FEATURES IN THIS BOOK

The construction estimating information in this book is divided into two main sections: Costbook Pages and Man-Hour Tables. Each section is organized according to the 16 divisions of the CSI MASTERFORMAT as shown on the previous pages. In addition, there are extensive Supporting Construction Reference tables, Geographic Costs Modifiers, Square Foot tables and a detailed Index.

Sample pages with graphic explanations are included before the Costbook pages and Man-Hour tables. These explanations, along with the discussions below, will provide a good understanding of what is included in this book and how it can best be used in construction estimating.

**Material Costs**

The material costs used in this book represent national averages for prices that a contractor would expect to pay plus an allowance for freight (if applicable), handling and storage. These costs reflect neither the lowest or highest prices, but rather a typical average cost over time. Periodic fluctuations in availability and in certain commodities (eg. copper, lumber) can significantly affect local material pricing. In the final estimating and bidding stages of a project when the highest degree of accuracy is required, it is best to check local, current prices.

**Labor Costs**

Labor costs include the basic wage, plus commonly applicable taxes, insurance and markups for overhead and profit. The labor rates used here to develop the costs are typical average prevailing wage rates. Rates for different trades are used where appropriate for each type of work.

Taxes and insurance which are most often applied to labor rates include employer-paid Social Security/Medicare taxes (FICA), Worker's Compensation insurance, state and federal unemployment taxes, and business insurance. Fixed government rates as well as average allowances are included in the labor costs. However, most of these items vary significantly from state to state and within states. For more specific data, local agencies and sources should be consulted.

**Equipment Costs**

Costs for various types and pieces of equipment are included in Division 1 - General Requirements and can be included in an estimate when required either as a total "Equipment" category or with specific appropriate trades. Costs for equipment are included when appropriate in the installation of costs in the Costbook pages.

**Overhead And Profit**

Included in the labor costs are allowances for overhead and profit for the contractor/employer whose workers are performing the specific tasks. No cost allowances or fees are included for management of subcontractors by the general contractor or construction manager. These costs, where appropriate, must be added to the costs as listed in the book.

The allowance for overhead is included to account for office overhead, the contractors' typical costs of doing business. These costs normally include in-house office staff salaries and benefits, office rent and operating expenses, professional fees, vehicle costs and other operating costs which are not directly applicable to specific jobs. It should be noted for this book that office overhead as included should be distinguished form project overhead, the General Requirements (CSI Division 1) which are specific to particular projects. Project overhead should be included on an item by item basis for each job.

Depending on the trade, an allowance of 10-15 percent is incorporated into the labor/installation costs to account for typical profit of the installing contractor. See Division 1, General

Requirements, for a more detailed review of typical profit allowances.

**Adjustments to Costs**

The costs as presented in this book attempt to represent national averages. Costs, however, vary among regions, states and even between adjacent localities.

In order to more closely approximate the probable costs for specific locations throughout the U.S., a table of Geographic Cost Modifiers is provided. These adjustment factors are used to modify costs obtained from this book to help account for regional variations of construction costs. Whenever local current costs are known, whether material or equipment prices or labor rates, they should be used if more accuracy is required.

**Man-Hour Tables**

The man-hour data used to develop the labor costs are listed in the second main section of this book, the "Man-Hour Tables." These productivities represent typical installation labor for thousands of construction items. The data takes into account all activities involved in normal construction under commonly experienced working conditions such as site movement, material handling, start-up, etc. As with the Costbook pages, these items are listed according to the CSI MASTERFORMAT.

**Square Foot Tables**

Included as an additional reference are Square Foot Tables which list hundreds of actual projects for dozens of building types, each with associated building size and total square foot building cost. This data provides an overview of construction costs by building type. These costs are for actual projects. The variations within similar building types may be due, among other factors, to size, location, quality and specified components, material and processes. Depending upon all such factors, specific building costs can vary significantly and may not necessarily fall within the range of costs as presented.

**Editors' Note:** The **Building News 1997 Costbooks** are intended to provide accurate, reliable, average costs and typical productivities for thousands of common construction components. The data is developed and compiled from various industry sources, including government, manufacturers, suppliers and working professionals. The intent of the information is to provide assistance and guidelines to construction professionals in estimating. The user should be aware that local conditions, material and labor availability and cost variations, economic considerations, weather, local codes and regulations, etc., all affect the actual cost of construction. These and other such factors must be considered and incorporated into any and all construction estimates.

# Sample Costbook Page

In order to best use the information in this book, please review this sample page and read the "Features In This Book" section.

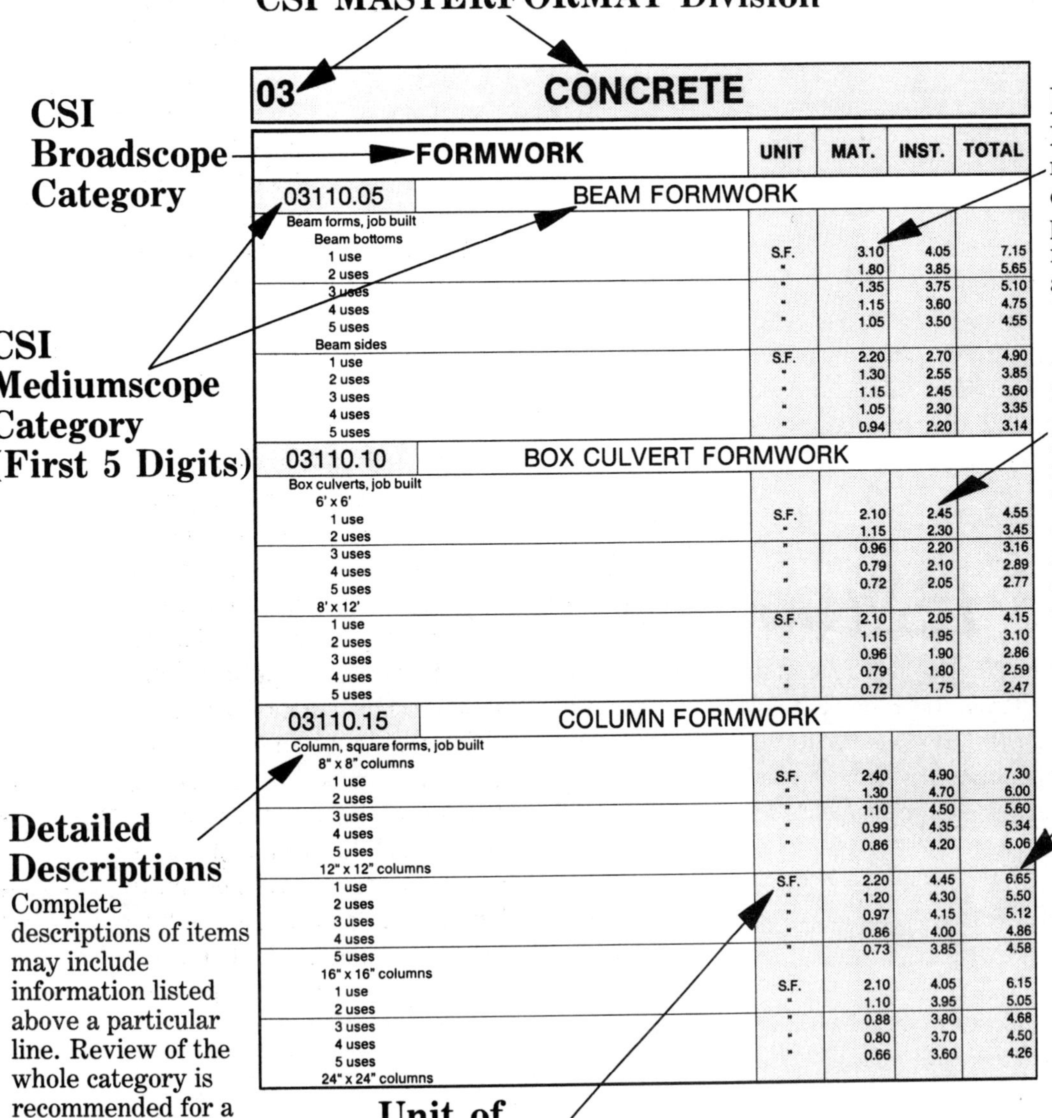


| 03 | CONCRETE | | | |
|---|---|---|---|---|
| **FORMWORK** | **UNIT** | **MAT.** | **INST.** | **TOTAL** |
| **03110.05** BEAM FORMWORK | | | | |
| Beam forms, job built | | | | |
| Beam bottoms | | | | |
| 1 use | S.F. | 3.10 | 4.05 | 7.15 |
| 2 uses | " | 1.80 | 3.85 | 5.65 |
| 3 uses | " | 1.35 | 3.75 | 5.10 |
| 4 uses | " | 1.15 | 3.60 | 4.75 |
| 5 uses | " | 1.05 | 3.50 | 4.55 |
| Beam sides | | | | |
| 1 use | S.F. | 2.20 | 2.70 | 4.90 |
| 2 uses | " | 1.30 | 2.55 | 3.85 |
| 3 uses | " | 1.15 | 2.45 | 3.60 |
| 4 uses | " | 1.05 | 2.30 | 3.35 |
| 5 uses | " | 0.94 | 2.20 | 3.14 |
| **03110.10** BOX CULVERT FORMWORK | | | | |
| Box culverts, job built | | | | |
| 6' x 6' | | | | |
| 1 use | S.F. | 2.10 | 2.45 | 4.55 |
| 2 uses | " | 1.15 | 2.30 | 3.45 |
| 3 uses | " | 0.96 | 2.20 | 3.16 |
| 4 uses | " | 0.79 | 2.10 | 2.89 |
| 5 uses | " | 0.72 | 2.05 | 2.77 |
| 8' x 12' | | | | |
| 1 use | S.F. | 2.10 | 2.05 | 4.15 |
| 2 uses | " | 1.15 | 1.95 | 3.10 |
| 3 uses | " | 0.96 | 1.90 | 2.86 |
| 4 uses | " | 0.79 | 1.80 | 2.59 |
| 5 uses | " | 0.72 | 1.75 | 2.47 |
| **03110.15** COLUMN FORMWORK | | | | |
| Column, square forms, job built | | | | |
| 8" x 8" columns | | | | |
| 1 use | S.F. | 2.40 | 4.90 | 7.30 |
| 2 uses | " | 1.30 | 4.70 | 6.00 |
| 3 uses | " | 1.10 | 4.50 | 5.60 |
| 4 uses | " | 0.99 | 4.35 | 5.34 |
| 5 uses | " | 0.86 | 4.20 | 5.06 |
| 12" x 12" columns | | | | |
| 1 use | S.F. | 2.20 | 4.45 | 6.65 |
| 2 uses | " | 1.20 | 4.30 | 5.50 |
| 3 uses | " | 0.97 | 4.15 | 5.12 |
| 4 uses | " | 0.86 | 4.00 | 4.86 |
| 5 uses | " | 0.73 | 3.85 | 4.58 |
| 16" x 16" columns | | | | |
| 1 use | S.F. | 2.10 | 4.05 | 6.15 |
| 2 uses | " | 1.10 | 3.95 | 5.05 |
| 3 uses | " | 0.88 | 3.80 | 4.68 |
| 4 uses | " | 0.80 | 3.70 | 4.50 |
| 5 uses | " | 0.66 | 3.60 | 4.26 |
| 24" x 24" columns | | | | |

### Detailed Descriptions

Complete descriptions of items may include information listed above a particular line. Review of the whole category is recommended for a complete description.

### Unit of Measurement

Each item (and cost) is defined in terms of the common estimating unit. All costs are listed in dollars per unit.

### Material Cost

Material costs represent average contractor prices plus an allowance for freight, handling and storage.

### Installation Cost

Installation costs include basic wage rates, markups for taxes, insurance overhead and profit and also include equipment costs where appropriate.

### Total Cost

The total cost is the sum of material and installation costs. This total represents typical contractors' costs including overhead and profit, but does not include markups for the general contractor or construction management fees.

BNi® Building News

# 01 GENERAL

| REQUIREMENTS | UNIT | MAT. | INST. | TOTAL |
|---|---|---|---|---|
| **01020.10 ALLOWANCES** | | | | |
| Overhead | | | | |
| $20,000 project | | | | |
| Minimum | PCT. | | | 15.00 |
| Average | " | | | 20.00 |
| Maximum | " | | | 40.00 |
| $100,000 project | | | | |
| Minimum | PCT. | | | 12.00 |
| Average | " | | | 15.00 |
| Maximum | " | | | 25.00 |
| $500,000 project | | | | |
| Minimum | PCT. | | | 10.00 |
| Average | " | | | 12.00 |
| Maximum | " | | | 20.00 |
| $1,000,000 project | | | | |
| Minimum | PCT. | | | 6.00 |
| Average | " | | | 10.00 |
| Maximum | " | | | 12.00 |
| $10,000,000 project | | | | |
| Minimum | PCT. | | | 1.50 |
| Average | " | | | 5.00 |
| Maximum | " | | | 8.00 |
| Profit | | | | |
| $20,000 project | | | | |
| Minimum | PCT. | | | 10.00 |
| Average | " | | | 15.00 |
| Maximum | " | | | 25.00 |
| $100,000 project | | | | |
| Minimum | PCT. | | | 10.00 |
| Average | " | | | 12.00 |
| Maximum | " | | | 20.00 |
| $500,000 project | | | | |
| Minimum | PCT. | | | 5.00 |
| Average | " | | | 10.00 |
| Maximum | " | | | 15.00 |
| $1,000,000 project | | | | |
| Minimum | PCT. | | | 3.00 |
| Average | " | | | 8.00 |
| Maximum | " | | | 15.00 |
| Professional fees | | | | |
| Architectural | | | | |
| $100,000 project | | | | |
| Minimum | PCT. | | | 5.00 |
| Average | " | | | 10.00 |
| Maximum | " | | | 20.00 |
| $500,000 project | | | | |
| Minimum | PCT. | | | 5.00 |
| Average | " | | | 8.00 |
| Maximum | " | | | 12.00 |
| $1,000,000 project | | | | |
| Minimum | PCT. | | | 3.50 |
| Average | " | | | 7.00 |
| Maximum | " | | | 10.00 |

# 01 GENERAL

| REQUIREMENTS | UNIT | MAT. | INST. | TOTAL |
|---|---|---|---|---|
| **01020.10 ALLOWANCES** | | | | |
| Structural engineering | | | | |
| Minimum | PCT. | | | 2.00 |
| Average | " | | | 3.00 |
| Maximum | " | | | 5.00 |
| Mechanical engineering | | | | |
| Minimum | PCT. | | | 4.00 |
| Average | " | | | 5.00 |
| Maximum | " | | | 15.00 |
| Electrical engineering | | | | |
| Minimum | PCT. | | | 3.00 |
| Average | " | | | 5.00 |
| Maximum | " | | | 12.00 |
| Taxes | | | | |
| Sales tax | | | | |
| Minimum | PCT. | | | 4.00 |
| Average | " | | | 5.00 |
| Maximum | " | | | 8.25 |
| Unemployment | | | | |
| Minimum | PCT. | | | 2.00 |
| Average | " | | | 5.00 |
| Maximum | " | | | 7.00 |
| Social security (FICA) | " | | | 7.85 |
| **01050.10 FIELD STAFF** | | | | |
| Superintendent | | | | |
| Minimum | YEAR | | | 44,400 |
| Average | " | | | 72,150 |
| Maximum | " | | | 116,550 |
| Field engineer | | | | |
| Minimum | YEAR | | | 38,850 |
| Average | " | | | 57,720 |
| Maximum | " | | | 94,350 |
| Foreman | | | | |
| Minimum | YEAR | | | 27,750 |
| Average | " | | | 44,400 |
| Maximum | " | | | 61,050 |
| Bookkeeper/timekeeper | | | | |
| Minimum | YEAR | | | 13,320 |
| Average | " | | | 21,090 |
| Maximum | " | | | 31,080 |
| Watchman | | | | |
| Minimum | YEAR | | | 11,100 |
| Average | " | | | 13,875 |
| Maximum | " | | | 19,980 |
| **01310.10 SCHEDULING** | | | | |
| Scheduling for | | | | |
| $100,000 project | | | | |
| Minimum | PCT. | | | 1.00 |
| Average | " | | | 2.00 |
| Maximum | " | | | 4.00 |
| $500,000 project | | | | |

# 01 GENERAL

| REQUIREMENTS | UNIT | MAT. | INST. | TOTAL |
|---|---|---|---|---|
| **01310.10 SCHEDULING** | | | | |
| Minimum | PCT. | | | 0.50 |
| Average | " | | | 1.00 |
| Maximum | " | | | 2.00 |
| $1,000,000 project | | | | |
| Minimum | PCT. | | | 0.30 |
| Average | " | | | 0.80 |
| Maximum | " | | | 1.50 |
| Scheduling software, not including computer | | | | |
| Minimum | EA. | | | 504.55 |
| Average | " | | | 2,523 |
| Maximum | " | | | 40,364 |
| **01330.10 SURVEYING** | | | | |
| Surveying | | | | |
| Small crew | DAY | 0.00 | 480.00 | 480.00 |
| Average crew | " | 0.00 | 730.00 | 730.00 |
| Large crew | " | 0.00 | 970.00 | 970.00 |
| Lot lines and boundaries | | | | |
| Minimum | ACRE | 0.00 | 350.00 | 350.00 |
| Average | " | 0.00 | 730.00 | 730.00 |
| Maximum | " | 0.00 | 1,210 | 1,210 |
| **01380.10 JOB REQUIREMENTS** | | | | |
| Job photographs, small jobs | | | | |
| Minimum | EA. | | | 55.50 |
| Average | " | | | 111.00 |
| Maximum | " | | | 222.00 |
| Large projects | | | | |
| Minimum | EA. | | | 388.50 |
| Average | " | | | 555.00 |
| Maximum | " | | | 1,110 |
| **01410.10 TESTING** | | | | |
| Testing concrete, per test | | | | |
| Minimum | EA. | | | 12.21 |
| Average | " | | | 22.20 |
| Maximum | " | | | 44.40 |
| Soil, per test | | | | |
| Minimum | EA. | | | 22.20 |
| Average | " | | | 66.60 |
| Maximum | " | | | 177.60 |
| Welding, per test | | | | |
| Minimum | EA. | | | 11.10 |
| Average | " | | | 22.20 |
| Maximum | " | | | 83.25 |
| **01500.10 TEMPORARY FACILITIES** | | | | |
| Barricades, temporary | | | | |
| Highway | | | | |
| Concrete | L.F. | 7.75 | 2.40 | 10.15 |

# 01 GENERAL

| REQUIREMENTS | UNIT | MAT. | INST. | TOTAL |
|---|---|---|---|---|
| **01500.10 TEMPORARY FACILITIES** | | | | |
| Wood | L.F. | 2.20 | 0.97 | 3.17 |
| Steel | " | 2.45 | 0.81 | 3.26 |
| Pedestrian barricades | | | | |
| Plywood | S.F. | 1.45 | 0.81 | 2.26 |
| Chain link fence | " | 1.90 | 0.81 | 2.71 |
| Trailers, general office type, per month | | | | |
| Minimum | EA. | | | 166.50 |
| Average | " | | | 244.20 |
| Maximum | " | | | 555.00 |
| Crew change trailers, per month | | | | |
| Minimum | EA. | | | 66.60 |
| Average | " | | | 77.70 |
| Maximum | " | | | 116.55 |
| **01505.10 MOBILIZATION** | | | | |
| Equipment mobilization | | | | |
| Bulldozer | | | | |
| Minimum | EA. | | | 111.00 |
| Average | " | | | 222.00 |
| Maximum | " | | | 366.30 |
| Backhoe/front-end loader | | | | |
| Minimum | EA. | | | 55.50 |
| Average | " | | | 111.00 |
| Maximum | " | | | 244.20 |
| Crane, crawler type | | | | |
| Minimum | EA. | | | 1,110 |
| Average | " | | | 2,775 |
| Maximum | " | | | 6,105 |
| Truck crane | | | | |
| Minimum | EA. | | | 277.50 |
| Average | " | | | 416.25 |
| Maximum | " | | | 721.50 |
| Pile driving rig | | | | |
| Minimum | EA. | | | 5,550 |
| Average | " | | | 11,100 |
| Maximum | " | | | 19,980 |
| **01525.10 CONSTRUCTION AIDS** | | | | |
| Scaffolding/staging, rent per month | | | | |
| Measured by lineal feet of base | | | | |
| 10' high | L.F. | | | 7.77 |
| 20' high | " | | | 13.32 |
| 30' high | " | | | 19.43 |
| 40' high | " | | | 22.20 |
| 50' high | " | | | 26.64 |
| Measured by square foot of surface | | | | |
| Minimum | S.F. | | | 0.33 |
| Average | " | | | 0.56 |
| Maximum | " | | | 1.05 |
| Safety nets, heavy duty, per job | | | | |

# 01 GENERAL

| REQUIREMENTS | UNIT | MAT. | INST. | TOTAL |
|---|---|---|---|---|
| **01525.10 CONSTRUCTION AIDS** | | | | |
| Minimum | S.F. | | | 0.22 |
| Average | " | | | 0.28 |
| Maximum | " | | | 0.61 |
| Tarpaulins, fabric, per job | | | | |
| Minimum | S.F. | | | 0.17 |
| Average | " | | | 0.28 |
| Maximum | " | | | 0.72 |
| **01570.10 SIGNS** | | | | |
| Construction signs, temporary | | | | |
| Signs, 2' x 4' | | | | |
| Minimum | EA. | | | 22.20 |
| Average | " | | | 53.28 |
| Maximum | " | | | 188.70 |
| Signs, 4' x 8' | | | | |
| Minimum | EA. | | | 46.62 |
| Average | " | | | 122.10 |
| Maximum | " | | | 521.70 |
| Signs, 8' x 8' | | | | |
| Minimum | EA. | | | 59.94 |
| Average | " | | | 188.70 |
| Maximum | " | | | 1,943 |
| **01600.10 EQUIPMENT** | | | | |
| Air compressor | | | | |
| 60 cfm | | | | |
| By day | EA. | | | 57.72 |
| By week | " | | | 173.16 |
| By month | " | | | 521.70 |
| 300 cfm | | | | |
| By day | EA. | | | 122.10 |
| By week | " | | | 366.30 |
| By month | " | | | 1,099 |
| 600 cfm | | | | |
| By day | EA. | | | 333.00 |
| By week | " | | | 999.00 |
| By month | " | | | 2,997 |
| Air tools, per compressor, per day | | | | |
| Minimum | EA. | | | 22.20 |
| Average | " | | | 27.75 |
| Maximum | " | | | 38.85 |
| Generators, 5 kw | | | | |
| By day | EA. | | | 49.95 |
| By week | " | | | 149.85 |
| By month | " | | | 449.55 |
| Heaters, salamander type, per week | | | | |
| Minimum | EA. | | | 66.60 |
| Average | " | | | 94.35 |
| Maximum | " | | | 199.80 |
| Pumps, submersible | | | | |

# 01 GENERAL

| REQUIREMENTS | UNIT | MAT. | INST. | TOTAL |
|---|---|---|---|---|
| **01600.10** EQUIPMENT | | | | |
| 50 gpm | | | | |
| By day | EA. | | | 44.40 |
| By week | " | | | 133.20 |
| By month | " | | | 399.60 |
| 100 gpm | | | | |
| By day | EA. | | | 55.50 |
| By week | " | | | 166.50 |
| By month | " | | | 499.50 |
| 500 gpm | | | | |
| By day | EA. | | | 88.80 |
| By week | " | | | 266.40 |
| By month | " | | | 799.20 |
| Diaphragm pump, by week | | | | |
| Minimum | EA. | | | 66.60 |
| Average | " | | | 88.80 |
| Maximum | " | | | 133.20 |
| Pickup truck | | | | |
| By day | EA. | | | 77.70 |
| By week | " | | | 233.10 |
| By month | " | | | 699.30 |
| Dump truck | | | | |
| 6 cy truck | | | | |
| By day | EA. | | | 222.00 |
| By week | " | | | 666.00 |
| By month | " | | | 1,998 |
| 10 cy truck | | | | |
| By day | EA. | | | 277.50 |
| By week | " | | | 832.50 |
| By month | " | | | 2,498 |
| 16 cy truck | | | | |
| By day | EA. | | | 444.00 |
| By week | " | | | 1,332 |
| By month | " | | | 3,996 |
| Backhoe, track mounted | | | | |
| 1/2 cy capacity | | | | |
| By day | EA. | | | 455.10 |
| By week | " | | | 1,388 |
| By month | " | | | 4,107 |
| 1 cy capacity | | | | |
| By day | EA. | | | 721.50 |
| By week | " | | | 2,165 |
| By month | " | | | 6,494 |
| 2 cy capacity | | | | |
| By day | EA. | | | 1,221 |
| By week | " | | | 3,663 |
| By month | " | | | 10,989 |
| 3 cy capacity | | | | |
| By day | EA. | | | 2,331 |
| By week | " | | | 6,993 |
| By month | " | | | 20,979 |
| Backhoe/loader, rubber tired | | | | |
| 1/2 cy capacity | | | | |

# 01 GENERAL

| REQUIREMENTS | UNIT | MAT. | INST. | TOTAL |
|---|---|---|---|---|
| **01600.10 EQUIPMENT** | | | | |
| By day | EA. | | | 277.50 |
| By week | " | | | 832.50 |
| By month | " | | | 2,498 |
| 3/4 cy capacity | | | | |
| By day | EA. | | | 333.00 |
| By week | " | | | 999.00 |
| By month | " | | | 2,997 |
| Bulldozer | | | | |
| 75 hp | | | | |
| By day | EA. | | | 388.50 |
| By week | " | | | 1,166 |
| By month | " | | | 3,497 |
| 200 hp | | | | |
| By day | EA. | | | 1,110 |
| By week | " | | | 3,330 |
| By month | " | | | 9,990 |
| 400 hp | | | | |
| By day | EA. | | | 1,665 |
| By week | " | | | 4,995 |
| By month | " | | | 14,985 |
| Cranes, crawler type | | | | |
| 15 ton capacity | | | | |
| By day | EA. | | | 499.50 |
| By week | " | | | 1,499 |
| By month | " | | | 4,496 |
| 25 ton capacity | | | | |
| By day | EA. | | | 610.50 |
| By week | " | | | 1,832 |
| By month | " | | | 5,495 |
| 50 ton capacity | | | | |
| By day | EA. | | | 1,110 |
| By week | " | | | 3,330 |
| By month | " | | | 9,990 |
| 100 ton capacity | | | | |
| By day | EA. | | | 1,665 |
| By week | " | | | 4,995 |
| By month | " | | | 14,985 |
| Truck mounted, hydraulic | | | | |
| 15 ton capacity | | | | |
| By day | EA. | | | 471.75 |
| By week | " | | | 1,415 |
| By month | " | | | 4,246 |
| Loader, rubber tired | | | | |
| 1 cy capacity | | | | |
| By day | EA. | | | 333.00 |
| By week | " | | | 999.00 |
| By month | " | | | 2,997 |
| 2 cy capacity | | | | |
| By day | EA. | | | 499.50 |
| By week | " | | | 1,943 |
| By month | " | | | 5,828 |
| 3 cy capacity | | | | |

# 01 GENERAL

| REQUIREMENTS | UNIT | MAT. | INST. | TOTAL |
|---|---|---|---|---|
| **01600.10 EQUIPMENT** | | | | |
| By day | EA. | | | 888.00 |
| By week | " | | | 2,664 |
| By month | " | | | 7,992 |
| **01740.10 BONDS** | | | | |
| Performance bonds | | | | |
| Minimum | PCT. | | | 0.60 |
| Average | " | | | 1.90 |
| Maximum | " | | | 3.00 |

## SOIL TESTS

| 02010.10 SOIL BORING | UNIT | MAT. | INST. | TOTAL |
|---|---|---|---|---|
| Borings, uncased, stable earth | | | | |
| 2-1/2" dia. | | | | |
| Minimum | L.F. | 0.00 | 10.90 | 10.90 |
| Average | " | 0.00 | 16.35 | 16.35 |
| Maximum | " | 0.00 | 26.20 | 26.20 |
| 4" dia. | | | | |
| Minimum | L.F. | 0.00 | 11.90 | 11.90 |
| Average | " | 0.00 | 18.70 | 18.70 |
| Maximum | " | 0.00 | 32.70 | 32.70 |
| Cased, including samples | | | | |
| 2-1/2" dia. | | | | |
| Minimum | L.F. | 0.00 | 13.10 | 13.10 |
| Average | " | 0.00 | 21.80 | 21.80 |
| Maximum | " | 0.00 | 43.60 | 43.60 |
| 4" dia. | | | | |
| Minumum | L.F. | 0.00 | 26.20 | 26.20 |
| Average | " | 0.00 | 37.40 | 37.40 |
| Maximum | " | 0.00 | 52.50 | 52.50 |
| Drilling in rock | | | | |
| No sampling | | | | |
| Minimum | L.F. | 0.00 | 23.75 | 23.75 |
| Average | " | 0.00 | 34.40 | 34.40 |
| Maximum | " | 0.00 | 46.70 | 46.70 |
| With casing and sampling | | | | |
| Minimum | L.F. | 0.00 | 32.70 | 32.70 |
| Average | " | 0.00 | 43.60 | 43.60 |
| Maximum | " | 0.00 | 65.50 | 65.50 |
| Test pits | | | | |
| Light soil | | | | |
| Minimum | EA. | 0.00 | 160.00 | 160.00 |
| Average | " | 0.00 | 220.00 | 220.00 |
| Maximum | " | 0.00 | 440.00 | 440.00 |
| Heavy soil | | | | |
| Minimum | EA. | 0.00 | 260.00 | 260.00 |
| Average | " | 0.00 | 330.00 | 330.00 |
| Maximum | " | 0.00 | 650.00 | 650.00 |

## DEMOLITION

| 02060.10 BUILDING DEMOLITION | UNIT | MAT. | INST. | TOTAL |
|---|---|---|---|---|
| Building, complete with disposal | | | | |
| Wood frame | C.F. | 0.00 | 0.18 | 0.18 |
| Concrete | " | 0.00 | 0.27 | 0.27 |
| Steel frame | " | 0.00 | 0.36 | 0.36 |

## DEMOLITION

### 02060.10 BUILDING DEMOLITION

| DEMOLITION | UNIT | MAT. | INST. | TOTAL |
|---|---|---|---|---|
| Partition removal | | | | |
| Concrete block partitions | | | | |
| 4" thick | S.F. | 0.00 | 1.20 | 1.20 |
| 8" thick | " | 0.00 | 1.60 | 1.60 |
| 12" thick | " | 0.00 | 2.20 | 2.20 |
| Brick masonry partitions | | | | |
| 4" thick | S.F. | 0.00 | 1.20 | 1.20 |
| 8" thick | " | 0.00 | 1.50 | 1.50 |
| 12" thick | " | 0.00 | 2.00 | 2.00 |
| 16" thick | " | 0.00 | 3.00 | 3.00 |
| Cast in place concrete partitions | | | | |
| Unreinforced | | | | |
| 6" thick | S.F. | 0.00 | 8.70 | 8.70 |
| 8" thick | " | 0.00 | 9.35 | 9.35 |
| 10" thick | " | 0.00 | 10.90 | 10.90 |
| 12" thick | " | 0.00 | 13.10 | 13.10 |
| Reinforced | | | | |
| 6" thick | S.F. | 0.00 | 10.05 | 10.05 |
| 8" thick | " | 0.00 | 13.10 | 13.10 |
| 10" thick | " | 0.00 | 14.55 | 14.55 |
| 12" thick | " | 0.00 | 17.45 | 17.45 |
| Terra cotta | | | | |
| To 6" thick | S.F. | 0.00 | 1.20 | 1.20 |
| Stud partitions | | | | |
| Metal or wood, with drywall both sides | S.F. | 0.00 | 1.20 | 1.20 |
| Metal studs, both sides, lath and plaster | " | 0.00 | 1.60 | 1.60 |
| Concrete, elevated slabs, mesh reinforcing | | | | |
| Under 5 cf | C.F. | 0.00 | 24.20 | 24.20 |
| Over 5 cf | " | 0.00 | 20.15 | 20.15 |
| Bar reinforcing | | | | |
| Under 5 cf | C.F. | 0.00 | 40.30 | 40.30 |
| Over 5 cf | " | 0.00 | 30.30 | 30.30 |
| Walls, concrete, bar reinforcing | | | | |
| Small jobs | C.F. | 0.00 | 16.15 | 16.15 |
| Large jobs | " | 0.00 | 13.45 | 13.45 |
| Brick walls, not including toothing | | | | |
| 4" thick | S.F. | 0.00 | 1.20 | 1.20 |
| 8" thick | " | 0.00 | 1.50 | 1.50 |
| 12" thick | " | 0.00 | 2.00 | 2.00 |
| 16" thick | " | 0.00 | 3.00 | 3.00 |
| Concrete block walls, not including toothing | | | | |
| 4" thick | S.F. | 0.00 | 1.35 | 1.35 |
| 6" thick | " | 0.00 | 1.40 | 1.40 |
| 8" thick | " | 0.00 | 1.50 | 1.50 |
| 10" thick | " | 0.00 | 1.75 | 1.75 |
| 12" thick | " | 0.00 | 2.00 | 2.00 |
| Rubbish handling | | | | |
| Load in dumpster or truck | | | | |
| Minimum | C.F. | 0.00 | 0.54 | 0.54 |
| Maximum | " | 0.00 | 0.81 | 0.81 |
| For use of elevators, add | | | | |
| Minimum | C.F. | 0.00 | 0.12 | 0.12 |

# 02 SITEWORK

| DEMOLITION | UNIT | MAT. | INST. | TOTAL |
|---|---|---|---|---|
| **02060.10 BUILDING DEMOLITION** | | | | |
| Maximum | C.F. | 0.00 | 0.24 | 0.24 |
| Rubbish hauling | | | | |
| Hand loaded on trucks, 2 mile trip | C.Y. | 0.00 | 19.35 | 19.35 |
| Machine loaded on trucks, 2 mile trip | " | 0.00 | 13.10 | 13.10 |

| HIGHWAY DEMOLITION | UNIT | MAT. | INST. | TOTAL |
|---|---|---|---|---|
| **02065.10 PAVEMENT DEMOLITION** | | | | |
| Bituminous pavement, up to 3" thick | | | | |
| On streets | S.Y. | 0.00 | 5.25 | 5.25 |
| On pipe trench | " | 0.00 | 6.55 | 6.55 |
| Concrete pavement, 6" thick | | | | |
| No reinforcement | S.Y. | 0.00 | 8.70 | 8.70 |
| With wire mesh | " | 0.00 | 13.10 | 13.10 |
| With rebars | " | 0.00 | 16.35 | 16.35 |
| 9" thick | | | | |
| No reinforcement | S.Y. | 0.00 | 10.90 | 10.90 |
| With wire mesh | " | 0.00 | 16.35 | 16.35 |
| With rebars | " | 0.00 | 21.80 | 21.80 |
| 12" thick | | | | |
| No reinforcement | S.Y. | 0.00 | 13.10 | 13.10 |
| With wire mesh | " | 0.00 | 18.70 | 18.70 |
| With rebars | " | 0.00 | 26.20 | 26.20 |
| Sidewalk, 4" thick, with disposal | " | 0.00 | 4.35 | 4.35 |
| Removal of pavement markings by waterblasting | S.F. | 0.00 | 0.12 | 0.12 |
| **02065.15 SAW CUTTING PAVEMENT** | | | | |
| Pavement, bituminous | | | | |
| 2" thick | L.F. | 0.00 | 0.97 | 0.97 |
| 3" thick | " | 0.00 | 1.20 | 1.20 |
| 4" thick | " | 0.00 | 1.50 | 1.50 |
| 5" thick | " | 0.00 | 1.60 | 1.60 |
| 6" thick | " | 0.00 | 1.75 | 1.75 |
| Concrete pavement, with wire mesh | | | | |
| 4" thick | L.F. | 0.00 | 1.85 | 1.85 |
| 5" thick | " | 0.00 | 2.00 | 2.00 |
| 6" thick | " | 0.00 | 2.20 | 2.20 |
| 8" thick | " | 0.00 | 2.40 | 2.40 |
| 10" thick | " | 0.00 | 2.70 | 2.70 |
| Plain concrete, unreinforced | | | | |
| 4" thick | L.F. | 0.00 | 1.60 | 1.60 |
| 5" thick | " | 0.00 | 1.85 | 1.85 |
| 6" thick | " | 0.00 | 2.00 | 2.00 |

# 02 SITEWORK

| HIGHWAY DEMOLITION | UNIT | MAT. | INST. | TOTAL |
|---|---|---|---|---|
| **02065.15 SAW CUTTING PAVEMENT** | | | | |
| 8" thick | L.F. | 0.00 | 2.20 | 2.20 |
| 10" thick | " | 0.00 | 2.40 | 2.40 |
| **02065.80 CURB & GUTTER** | | | | |
| Removal, plain concrete curb | L.F. | 0.00 | 3.25 | 3.25 |
| Plain concrete curb and 2' gutter | " | 0.00 | 4.50 | 4.50 |
| **02065.85 GUARDRAILS** | | | | |
| Remove standard guardrail | | | | |
| Steel | L.F. | 0.00 | 4.35 | 4.35 |
| Wood | " | 0.00 | 3.35 | 3.35 |
| **02075.80 CORE DRILLING** | | | | |
| Concrete | | | | |
| 6" thick | | | | |
| 3" dia. | EA. | 0.00 | 22.15 | 22.15 |
| 4" dia. | " | 0.00 | 25.80 | 25.80 |
| 6" dia. | " | 0.00 | 31.00 | 31.00 |
| 8" dia. | " | 0.00 | 51.50 | 51.50 |
| 8" thick | | | | |
| 3" dia. | EA. | 0.00 | 31.00 | 31.00 |
| 4" dia. | " | 0.00 | 38.80 | 38.80 |
| 6" dia. | " | 0.00 | 44.30 | 44.30 |
| 8" dia. | " | 0.00 | 62.00 | 62.00 |
| 10" thick | | | | |
| 3" dia. | EA. | 0.00 | 38.80 | 38.80 |
| 4" dia. | " | 0.00 | 44.30 | 44.30 |
| 6" dia. | " | 0.00 | 51.50 | 51.50 |
| 8" dia. | " | 0.00 | 77.50 | 77.50 |
| 12" thick | | | | |
| 3" dia. | EA. | 0.00 | 51.50 | 51.50 |
| 4" dia. | " | 0.00 | 62.00 | 62.00 |
| 6" dia. | " | 0.00 | 77.50 | 77.50 |
| 8" dia. | " | 0.00 | 100.00 | 100.00 |

| HAZARDOUS WASTE | UNIT | MAT. | INST. | TOTAL |
|---|---|---|---|---|
| **02080.10 ASBESTOS REMOVAL** | | | | |
| Enclosure using wood studs & poly, install & remove | S.F. | 0.74 | 0.60 | 1.35 |
| Trailer (change room) | DAY | | | 68.95 |
| Disposal suits (4 suits per man day) | " | | | 28.53 |
| Type C respirator mask, includes hose & filters, per man | " | | | 14.26 |

# 02 SITEWORK

## HAZARDOUS WASTE

| | UNIT | MAT. | INST. | TOTAL |
|---|---|---|---|---|
| **02080.10 ASBESTOS REMOVAL** | | | | |
| Respirator mask & filter, light contamination | DAY | | | 5.58 |
| Air monitoring test, 12 tests per day | | | | |
| Off job testing | DAY | | | 737.04 |
| On the job testing | " | | | 974.80 |
| Asbestos vacuum with attachments | EA. | | | 428.02 |
| Hydraspray piston pump | " | | | 546.90 |
| Negative air pressure system | " | | | 499.28 |
| Grade D breathing air equipment | " | | | 1,249 |
| Glove bag, 44" x 60" x 6 mil plastic | " | | | 3.80 |
| 40 CY asbestos dumpster | | | | |
| Weekly rental | EA. | | | 475.08 |
| Pick up/delivery | " | | | 225.88 |
| Asbestos dump fee | " | | | 118.77 |

## SITE DEMOLITION

| | UNIT | MAT. | INST. | TOTAL |
|---|---|---|---|---|
| **02105.10 CATCH BASINS/MANHOLES** | | | | |
| Abandon catch basin or manhole (fill with sand) | EA. | 0.00 | 260.00 | 260.00 |
| Remove and reset frame and cover | " | 0.00 | 130.00 | 130.00 |
| Remove catch basin, to 10' deep | | | | |
| Masonry | EA. | 0.00 | 330.00 | 330.00 |
| Concrete | " | 0.00 | 440.00 | 440.00 |
| **02105.20 FENCES** | | | | |
| Remove fencing | | | | |
| Chain link, 8' high | | | | |
| For disposal | L.F. | 0.00 | 1.20 | 1.20 |
| For reuse | " | 0.00 | 3.00 | 3.00 |
| Wood | | | | |
| 4' high | S.F. | 0.00 | 0.81 | 0.81 |
| 6' high | " | 0.00 | 0.97 | 0.97 |
| 8' high | " | 0.00 | 1.20 | 1.20 |
| Masonry | | | | |
| 8" thick | | | | |
| 4' high | S.F. | 0.00 | 2.40 | 2.40 |
| 6' high | " | 0.00 | 3.00 | 3.00 |
| 8' high | " | 0.00 | 3.45 | 3.45 |
| 12" thick | | | | |
| 4' high | S.F. | 0.00 | 4.05 | 4.05 |
| 6' high | " | 0.00 | 4.85 | 4.85 |
| 8' high | " | 0.00 | 6.05 | 6.05 |
| 12' high | " | 0.00 | 8.05 | 8.05 |

# 02 SITEWORK

| SITE DEMOLITION | UNIT | MAT. | INST. | TOTAL |
|---|---|---|---|---|
| **02105.30 HYDRANTS** | | | | |
| Remove fire hydrant | EA. | 0.00 | 220.00 | 220.00 |
| Remove and reset fire hydrant | " | 0.00 | 650.00 | 650.00 |
| **02105.42 DRAINAGE PIPING** | | | | |
| Remove drainage pipe, not including excavation | | | | |
| 12" dia. | L.F. | 0.00 | 5.45 | 5.45 |
| 18" dia. | " | 0.00 | 6.90 | 6.90 |
| 24" dia. | " | 0.00 | 8.70 | 8.70 |
| 36" dia. | " | 0.00 | 10.90 | 10.90 |
| **02105.43 GAS PIPING** | | | | |
| Remove welded steel pipe, not including excavation | | | | |
| 4" dia. | L.F. | 0.00 | 8.15 | 8.15 |
| 5" dia. | " | 0.00 | 13.10 | 13.10 |
| 6" dia. | " | 0.00 | 16.35 | 16.35 |
| 8" dia. | " | 0.00 | 26.20 | 26.20 |
| 10" dia. | " | 0.00 | 32.70 | 32.70 |
| **02105.45 SANITARY PIPING** | | | | |
| Remove sewer pipe, not including excavation | | | | |
| 4" dia. | L.F. | 0.00 | 5.25 | 5.25 |
| 6" dia. | " | 0.00 | 5.95 | 5.95 |
| 8" dia. | " | 0.00 | 6.55 | 6.55 |
| 10" dia. | " | 0.00 | 6.90 | 6.90 |
| 12" dia. | " | 0.00 | 7.25 | 7.25 |
| 15" dia. | " | 0.00 | 7.70 | 7.70 |
| 18" dia. | " | 0.00 | 8.70 | 8.70 |
| 24" dia. | " | 0.00 | 10.90 | 10.90 |
| 30" dia. | " | 0.00 | 13.10 | 13.10 |
| 36" dia. | " | 0.00 | 16.35 | 16.35 |
| **02105.48 WATER PIPING** | | | | |
| Remove water pipe, not including excavation | | | | |
| 4" dia. | L.F. | 0.00 | 5.95 | 5.95 |
| 6" dia. | " | 0.00 | 6.25 | 6.25 |
| 8" dia. | " | 0.00 | 6.90 | 6.90 |
| 10" dia. | " | 0.00 | 7.25 | 7.25 |
| 12" dia. | " | 0.00 | 7.70 | 7.70 |
| 14" dia. | " | 0.00 | 8.15 | 8.15 |
| 16" dia. | " | 0.00 | 8.70 | 8.70 |
| 18" dia. | " | 0.00 | 9.35 | 9.35 |
| 20" dia. | " | 0.00 | 10.05 | 10.05 |
| Remove valves | | | | |
| 6" | EA. | 0.00 | 65.50 | 65.50 |
| 10" | " | 0.00 | 72.50 | 72.50 |
| 14" | " | 0.00 | 81.50 | 81.50 |
| 18" | " | 0.00 | 110.00 | 110.00 |

| SITE DEMOLITION | UNIT | MAT. | INST. | TOTAL |
|---|---|---|---|---|
| **02105.60 UNDERGROUND TANKS** | | | | |
| Remove underground storage tank, and backfill | | | | |
| 50 to 250 gals | EA. | 0.00 | 440.00 | 440.00 |
| 600 gals | " | 0.00 | 440.00 | 440.00 |
| 1000 gals | " | 0.00 | 650.00 | 650.00 |
| 4000 gals | " | 0.00 | 1,050 | 1,050 |
| 5000 gals | " | 0.00 | 1,050 | 1,050 |
| 10,000 gals | " | 0.00 | 1,740 | 1,740 |
| 12,000 gals | " | 0.00 | 2,180 | 2,180 |
| 15,000 gals | " | 0.00 | 2,620 | 2,620 |
| 20,000 gals | " | 0.00 | 3,270 | 3,270 |
| **02105.66 SEPTIC TANKS** | | | | |
| Remove septic tank | | | | |
| 1000 gals | EA. | 0.00 | 110.00 | 110.00 |
| 2000 gals | " | 0.00 | 130.00 | 130.00 |
| 5000 gals | " | 0.00 | 160.00 | 160.00 |
| 15,000 gals | " | 0.00 | 1,310 | 1,310 |
| 25,000 gals | " | 0.00 | 1,740 | 1,740 |
| 40,000 gals | " | 0.00 | 2,620 | 2,620 |
| **02105.80 WALLS, EXTERIOR** | | | | |
| Concrete wall | | | | |
| Light reinforcing | | | | |
| 6" thick | S.F. | 0.00 | 6.55 | 6.55 |
| 8" thick | " | 0.00 | 6.90 | 6.90 |
| 10" thick | " | 0.00 | 7.25 | 7.25 |
| 12" thick | " | 0.00 | 8.15 | 8.15 |
| Medium reinforcing | | | | |
| 6" thick | S.F. | 0.00 | 6.90 | 6.90 |
| 8" thick | " | 0.00 | 7.25 | 7.25 |
| 10" thick | " | 0.00 | 8.15 | 8.15 |
| 12" thick | " | 0.00 | 9.35 | 9.35 |
| Heavy reinforcing | | | | |
| 6" thick | S.F. | 0.00 | 7.70 | 7.70 |
| 8" thick | " | 0.00 | 8.15 | 8.15 |
| 10" thick | " | 0.00 | 9.35 | 9.35 |
| 12" thick | " | 0.00 | 10.90 | 10.90 |
| Masonry | | | | |
| No reinforcing | | | | |
| 8" thick | S.F. | 0.00 | 2.90 | 2.90 |
| 12" thick | " | 0.00 | 3.25 | 3.25 |
| 16" thick | " | 0.00 | 3.75 | 3.75 |
| Horizontal reinforcing | | | | |
| 8" thick | S.F. | 0.00 | 3.25 | 3.25 |
| 12" thick | " | 0.00 | 3.55 | 3.55 |
| 16" thick | " | 0.00 | 4.20 | 4.20 |
| Vertical reinforcing | | | | |
| 8" thick | S.F. | 0.00 | 4.20 | 4.20 |
| 12" thick | " | 0.00 | 4.85 | 4.85 |
| 16" thick | " | 0.00 | 5.95 | 5.95 |
| Remove concrete headwall | | | | |

# 02 SITEWORK

| SITE DEMOLITION | UNIT | MAT. | INST. | TOTAL |
|---|---|---|---|---|
| **02105.80 WALLS, EXTERIOR** | | | | |
| 15" pipe | EA. | 0.00 | 93.50 | 93.50 |
| 18" pipe | " | 0.00 | 110.00 | 110.00 |
| 24" pipe | " | 0.00 | 120.00 | 120.00 |
| 30" pipe | " | 0.00 | 130.00 | 130.00 |
| 36" pipe | " | 0.00 | 150.00 | 150.00 |
| 48" pipe | " | 0.00 | 190.00 | 190.00 |
| 60" pipe | " | 0.00 | 260.00 | 260.00 |
| **02110.10 CLEARING AND GRUBBING** | | | | |
| Clear wooded area | | | | |
| Light density | ACRE | 0.00 | 3,270 | 3,270 |
| Medium density | " | 0.00 | 4,360 | 4,360 |
| Heavy density | " | 0.00 | 5,230 | 5,230 |
| **02110.50 TREE CUTTING & CLEARING** | | | | |
| Cut trees and clear out stumps | | | | |
| 9" to 12" dia. | EA. | 0.00 | 260.00 | 260.00 |
| To 24" dia. | " | 0.00 | 330.00 | 330.00 |
| 24" dia. and up | " | 0.00 | 440.00 | 440.00 |
| Loading and trucking | | | | |
| For machine load, per load, round trip | | | | |
| 1 mile | EA. | 0.00 | 52.50 | 52.50 |
| 3 mile | " | 0.00 | 59.50 | 59.50 |
| 5 mile | " | 0.00 | 65.50 | 65.50 |
| 10 mile | " | 0.00 | 87.00 | 87.00 |
| 20 mile | " | 0.00 | 130.00 | 130.00 |
| Hand loaded, round trip | | | | |
| 1 mile | EA. | 0.00 | 120.00 | 120.00 |
| 3 mile | " | 0.00 | 140.00 | 140.00 |
| 5 mile | " | 0.00 | 160.00 | 160.00 |
| 10 mile | " | 0.00 | 190.00 | 190.00 |
| 20 mile | " | 0.00 | 240.00 | 240.00 |
| Tree trimming for pole line construction | | | | |
| Light cutting | L.F. | 0.00 | 0.65 | 0.65 |
| Medium cutting | " | 0.00 | 0.87 | 0.87 |
| Heavy cutting | " | 0.00 | 1.30 | 1.30 |

| DEWATERING | UNIT | MAT. | INST. | TOTAL |
|---|---|---|---|---|
| **02144.10 WELLPOINT SYSTEMS** | | | | |
| Pumping, gas driven, 50' hose | | | | |
| 3" header pipe | DAY | 0.00 | 480.00 | 480.00 |

# 02 SITEWORK

## DEWATERING

| | UNIT | MAT. | INST. | TOTAL |
|---|---|---|---|---|
| **02144.10 WELLPOINT SYSTEMS** | | | | |
| 6" header pipe | DAY | 0.00 | 610.00 | 610.00 |
| Wellpoint system per job | | | | |
| 6" header pipe | L.F. | 1.05 | 1.95 | 3.00 |
| 8" header pipe | " | 1.35 | 2.40 | 3.75 |
| 10" header pipe | " | 1.75 | 3.25 | 5.00 |
| Jetting wellpoint system | | | | |
| 14' long | EA. | 33.80 | 32.30 | 66.10 |
| 18' long | " | 40.80 | 40.40 | 81.20 |
| Sand filter for wellpoints | L.F. | 1.75 | 0.81 | 2.56 |
| Replacement of wellpoint components | EA. | 0.00 | 9.70 | 9.70 |

## SHORING AND UNDERPINNING

| | UNIT | MAT. | INST. | TOTAL |
|---|---|---|---|---|
| **02162.10 TRENCH SHEETING** | | | | |
| Closed timber, including pull and salvage, excavation | | | | |
| 8' deep | S.F. | 2.05 | 4.35 | 6.40 |
| 10' deep | " | 2.05 | 4.55 | 6.60 |
| 12' deep | " | 2.05 | 4.80 | 6.85 |
| 14' deep | " | 2.05 | 5.10 | 7.15 |
| 16' deep | " | 2.05 | 5.40 | 7.45 |
| 18' deep | " | 2.05 | 6.65 | 8.70 |
| 20' deep | " | 2.05 | 6.20 | 8.25 |
| **02170.10 COFFERDAMS** | | | | |
| Cofferdam, steel, driven from shore | | | | |
| 15' deep | S.F. | 9.95 | 10.90 | 20.85 |
| 20' deep | " | 9.95 | 10.15 | 20.10 |
| 25' deep | " | 9.95 | 9.50 | 19.45 |
| 30' deep | " | 9.95 | 8.95 | 18.90 |
| 40' deep | " | 9.95 | 8.45 | 18.40 |
| Driven from barge | | | | |
| 20' deep | S.F. | 9.95 | 11.70 | 21.65 |
| 30' deep | " | 9.95 | 10.90 | 20.85 |
| 40' deep | " | 9.95 | 10.15 | 20.10 |
| 50' deep | " | 9.95 | 9.50 | 19.45 |

| EARTHWORK | UNIT | MAT. | INST. | TOTAL |
|---|---|---|---|---|
| **02210.10 HAULING MATERIAL** | | | | |
| Haul material by 10 cy dump truck, round trip distance | | | | |
| 1 mile | C.Y. | 0.00 | 2.70 | 2.70 |
| 2 mile | " | 0.00 | 3.25 | 3.25 |
| 5 mile | " | 0.00 | 4.40 | 4.40 |
| 10 mile | " | 0.00 | 4.85 | 4.85 |
| 20 mile | " | 0.00 | 5.40 | 5.40 |
| 30 mile | " | 0.00 | 6.45 | 6.45 |
| Site grading, cut & fill, sandy clay, 200' haul, 75 hp dozer | " | 0.00 | 1.95 | 1.95 |
| Spread topsoil by equipment on site | " | 0.00 | 2.15 | 2.15 |
| Site grading (cut and fill to 6") less than 1 acre | | | | |
| 75 hp dozer | C.Y. | 0.00 | 3.25 | 3.25 |
| 1.5 cy backhoe/loader | " | 0.00 | 4.85 | 4.85 |
| **02210.30 BULK EXCAVATION** | | | | |
| Excavation, by small dozer | | | | |
| Large areas | C.Y. | 0.00 | 0.97 | 0.97 |
| Small areas | " | 0.00 | 1.60 | 1.60 |
| Trim banks | " | 0.00 | 2.40 | 2.40 |
| Drag line | | | | |
| 1-1/2 cy bucket | | | | |
| Sand or gravel | C.Y. | 0.00 | 2.20 | 2.20 |
| Light clay | " | 0.00 | 2.90 | 2.90 |
| Heavy clay | " | 0.00 | 3.25 | 3.25 |
| Unclassified | " | 0.00 | 3.50 | 3.50 |
| 2 cy bucket | | | | |
| Sand or gravel | C.Y. | 0.00 | 2.00 | 2.00 |
| Light clay | " | 0.00 | 2.60 | 2.60 |
| Heavy clay | " | 0.00 | 2.90 | 2.90 |
| Unclassified | " | 0.00 | 3.10 | 3.10 |
| 2-1/2 cy bucket | | | | |
| Sand or gravel | C.Y. | 0.00 | 1.85 | 1.85 |
| Light clay | " | 0.00 | 2.40 | 2.40 |
| Heavy clay | " | 0.00 | 2.60 | 2.60 |
| Unclassified | " | 0.00 | 2.75 | 2.75 |
| 3 cy bucket | | | | |
| Sand or gravel | C.Y. | 0.00 | 1.65 | 1.65 |
| Light clay | " | 0.00 | 2.20 | 2.20 |
| Heavy clay | " | 0.00 | 2.40 | 2.40 |
| Unclassified | " | 0.00 | 2.50 | 2.50 |
| Hydraulic excavator | | | | |
| 1 cy capacity | | | | |
| Light material | C.Y. | 0.00 | 2.20 | 2.20 |
| Medium material | " | 0.00 | 2.60 | 2.60 |
| Wet material | " | 0.00 | 3.25 | 3.25 |
| Blasted rock | " | 0.00 | 3.75 | 3.75 |
| 1-1/2 cy capacity | | | | |
| Light material | C.Y. | 0.00 | 0.87 | 0.87 |
| Medium material | " | 0.00 | 1.15 | 1.15 |
| Wet material | " | 0.00 | 1.40 | 1.40 |
| Blasted rock | " | 0.00 | 1.75 | 1.75 |
| 2 cy capacity | | | | |
| Light material | C.Y. | 0.00 | 0.77 | 0.77 |

| EARTHWORK | UNIT | MAT. | INST. | TOTAL |
|---|---|---|---|---|
| **02210.30 BULK EXCAVATION** | | | | |
| Medium material | C.Y. | 0.00 | 0.99 | 0.99 |
| Wet material | " | 0.00 | 1.15 | 1.15 |
| Blasted rock | " | 0.00 | 1.40 | 1.40 |
| Wheel mounted front-end loader | | | | |
| 7/8 cy capacity | | | | |
| Light material | C.Y. | 0.00 | 1.75 | 1.75 |
| Medium material | " | 0.00 | 2.00 | 2.00 |
| Wet material | " | 0.00 | 2.30 | 2.30 |
| Blasted rock | " | 0.00 | 2.80 | 2.80 |
| 1-1/2 cy capacity | | | | |
| Light material | C.Y. | 0.00 | 0.99 | 0.99 |
| Medium material | " | 0.00 | 1.05 | 1.05 |
| Wet material | " | 0.00 | 1.15 | 1.15 |
| Blasted rock | " | 0.00 | 1.25 | 1.25 |
| 2-1/2 cy capacity | | | | |
| Light material | C.Y. | 0.00 | 0.82 | 0.82 |
| Medium material | " | 0.00 | 0.87 | 0.87 |
| Wet material | " | 0.00 | 0.93 | 0.93 |
| Blasted rock | " | 0.00 | 0.99 | 0.99 |
| 3-1/2 cy capacity | | | | |
| Light material | C.Y. | 0.00 | 0.77 | 0.77 |
| Medium material | " | 0.00 | 0.82 | 0.82 |
| Wet material | " | 0.00 | 0.87 | 0.87 |
| Blasted rock | " | 0.00 | 0.93 | 0.93 |
| 6 cy capacity | | | | |
| Light material | C.Y. | 0.00 | 0.46 | 0.46 |
| Medium material | " | 0.00 | 0.50 | 0.50 |
| Wet material | " | 0.00 | 0.53 | 0.53 |
| Blasted rock | " | 0.00 | 0.58 | 0.58 |
| Track mounted front-end loader | | | | |
| 1-1/2 cy capacity | | | | |
| Light material | C.Y. | 0.00 | 1.15 | 1.15 |
| Medium material | " | 0.00 | 1.25 | 1.25 |
| Wet material | " | 0.00 | 1.40 | 1.40 |
| Blasted rock | " | 0.00 | 1.55 | 1.55 |
| 2-3/4 cy capacity | | | | |
| Light material | C.Y. | 0.00 | 0.69 | 0.69 |
| Medium material | " | 0.00 | 0.77 | 0.77 |
| Wet material | " | 0.00 | 0.87 | 0.87 |
| Blasted rock | " | 0.00 | 0.99 | 0.99 |
| **02220.10 BORROW** | | | | |
| Borrow fill, F.O.B. at pit | | | | |
| Sand, haul to site, round trip | | | | |
| 10 mile | C.Y. | 5.35 | 6.95 | 12.30 |
| 20 mile | " | 5.35 | 11.55 | 16.90 |
| 30 mile | " | 5.35 | 17.35 | 22.70 |
| Place borrow fill and compact | | | | |
| Less than 1 in 4 slope | C.Y. | 5.35 | 3.45 | 8.80 |
| Greater than 1 in 4 slope | " | 5.35 | 4.65 | 10.00 |

| EARTHWORK | UNIT | MAT. | INST. | TOTAL |
|---|---|---|---|---|
| **02220.20 GRAVEL AND STONE** | | | | |
| F.O.B. PLANT | | | | |
| No. 21 crusher run stone | C.Y. | | | 19.36 |
| No. 26 crusher run stone | " | | | 18.83 |
| No. 57 stone | " | | | 19.90 |
| No. 67 gravel | " | | | 9.68 |
| No. 68 stone | " | | | 19.36 |
| No. 78 stone | " | | | 19.36 |
| No. 78 gravel, (pea gravel) | " | | | 11.29 |
| No. 357 or B-3 stone | " | | | 19.90 |
| Structural & foundation backfill | | | | |
| No. 21 crusher run stone | TON | | | 17.20 |
| No. 26 crusher run stone | " | | | 17.75 |
| No. 57 stone | " | | | 19.36 |
| No. 67 gravel | " | | | 8.60 |
| No. 68 stone | " | | | 17.20 |
| No. 78 stone | " | | | 17.20 |
| No. 78 gravel, (pea gravel) | " | | | 13.99 |
| No. 357 or B-3 stone | " | | | 19.90 |
| **02220.40 BUILDING EXCAVATION** | | | | |
| Structural excavation, unclassified earth | | | | |
| 3/8 cy backhoe | C.Y. | 0.00 | 9.25 | 9.25 |
| 3/4 cy backhoe | " | 0.00 | 6.95 | 6.95 |
| 1 cy backhoe | " | 0.00 | 5.80 | 5.80 |
| Foundation backfill and compaction by machine | " | 0.00 | 13.90 | 13.90 |
| **02220.50 UTILITY EXCAVATION** | | | | |
| Trencher, sandy clay, 8" wide trench | | | | |
| 18" deep | L.F. | 0.00 | 1.10 | 1.10 |
| 24" deep | " | 0.00 | 1.20 | 1.20 |
| 36" deep | " | 0.00 | 1.40 | 1.40 |
| Trench backfill, 95% compaction | | | | |
| Tamp by hand | C.Y. | 0.00 | 15.10 | 15.10 |
| Vibratory compaction | " | 0.00 | 12.10 | 12.10 |
| Trench backfilling, with borrow sand, place & compact | " | 4.75 | 12.10 | 16.85 |
| **02220.60 TRENCHING** | | | | |
| Trenching and continuous footing excavation | | | | |
| By gradall | | | | |
| 1 cy capacity | | | | |
| Light soil | C.Y. | 0.00 | 2.00 | 2.00 |
| Medium soil | " | 0.00 | 2.15 | 2.15 |
| Heavy/wet soil | " | 0.00 | 2.30 | 2.30 |
| Loose rock | " | 0.00 | 2.50 | 2.50 |
| Blasted rock | " | 0.00 | 2.65 | 2.65 |
| By hydraulic excavator | | | | |
| 1/2 cy capacity | | | | |
| Light soil | C.Y. | 0.00 | 2.30 | 2.30 |
| Medium soil | " | 0.00 | 2.50 | 2.50 |
| Heavy/wet soil | " | 0.00 | 2.80 | 2.80 |
| Loose rock | " | 0.00 | 3.10 | 3.10 |

| EARTHWORK | UNIT | MAT. | INST. | TOTAL |
|---|---|---|---|---|
| **02220.60 TRENCHING** | | | | |
| Blasted rock | C.Y. | 0.00 | 3.45 | 3.45 |
| 1 cy capacity | | | | |
| Light soil | C.Y. | 0.00 | 1.65 | 1.65 |
| Medium soil | " | 0.00 | 1.75 | 1.75 |
| Heavy/wet soil | " | 0.00 | 1.85 | 1.85 |
| Loose rock | " | 0.00 | 2.00 | 2.00 |
| Blasted rock | " | 0.00 | 2.15 | 2.15 |
| 1-1/2 cy capacity | | | | |
| Light soil | C.Y. | 0.00 | 1.45 | 1.45 |
| Medium soil | " | 0.00 | 1.55 | 1.55 |
| Heavy/wet soil | " | 0.00 | 1.65 | 1.65 |
| Loose rock | " | 0.00 | 1.75 | 1.75 |
| Blasted rock | " | 0.00 | 1.85 | 1.85 |
| 2 cy capacity | | | | |
| Light soil | C.Y. | 0.00 | 1.40 | 1.40 |
| Medium soil | " | 0.00 | 1.45 | 1.45 |
| Heavy/wet soil | " | 0.00 | 1.55 | 1.55 |
| Loose rock | " | 0.00 | 1.65 | 1.65 |
| Blasted rock | " | 0.00 | 1.75 | 1.75 |
| 2-1/2 cy capacity | | | | |
| Light soil | C.Y. | 0.00 | 1.25 | 1.25 |
| Medium soil | " | 0.00 | 1.30 | 1.30 |
| Heavy/wet soil | " | 0.00 | 1.40 | 1.40 |
| Loose rock | " | 0.00 | 1.45 | 1.45 |
| Blasted rock | " | 0.00 | 1.55 | 1.55 |
| Trencher, chain, 1' wide to 4' deep | | | | |
| Light soil | C.Y. | 0.00 | 1.20 | 1.20 |
| Medium soil | " | 0.00 | 1.40 | 1.40 |
| Heavy soil | " | 0.00 | 1.60 | 1.60 |
| Hand excavation | | | | |
| Bulk, wheeled 100' | | | | |
| Normal soil | C.Y. | 0.00 | 26.90 | 26.90 |
| Sand or gravel | " | 0.00 | 24.20 | 24.20 |
| Medium clay | " | 0.00 | 34.60 | 34.60 |
| Heavy clay | " | 0.00 | 48.40 | 48.40 |
| Loose rock | " | 0.00 | 60.50 | 60.50 |
| Trenches, up to 2' deep | | | | |
| Normal soil | C.Y. | 0.00 | 30.30 | 30.30 |
| Sand or gravel | " | 0.00 | 26.90 | 26.90 |
| Medium clay | " | 0.00 | 40.30 | 40.30 |
| Heavy clay | " | 0.00 | 60.50 | 60.50 |
| Loose rock | " | 0.00 | 80.50 | 80.50 |
| Trenches, to 6' deep | | | | |
| Normal soil | C.Y. | 0.00 | 34.60 | 34.60 |
| Sand or gravel | " | 0.00 | 30.30 | 30.30 |
| Medium clay | " | 0.00 | 48.40 | 48.40 |
| Heavy clay | " | 0.00 | 80.50 | 80.50 |
| Loose rock | " | 0.00 | 120.00 | 120.00 |
| Backfill trenches | | | | |
| With compaction | | | | |
| By hand | C.Y. | 0.00 | 20.15 | 20.15 |
| By 60 hp tracked dozer | " | 0.00 | 1.20 | 1.20 |

## EARTHWORK

| EARTHWORK | UNIT | MAT. | INST. | TOTAL |
|---|---|---|---|---|
| **02220.60 TRENCHING** | | | | |
| By 200 hp tracked dozer | C.Y. | 0.00 | 0.77 | 0.77 |
| By small front-end loader | " | 0.00 | 1.40 | 1.40 |
| Spread dumped fill or gravel, no compaction | | | | |
| 6" layers | S.Y. | 0.00 | 0.81 | 0.81 |
| 12" layers | " | 0.00 | 0.97 | 0.97 |
| Compaction in 6" layers | | | | |
| By hand with air tamper | S.Y. | 0.00 | 0.62 | 0.62 |
| Backfill trenches, sand bedding, no compaction | | | | |
| By hand | C.Y. | 6.10 | 20.15 | 26.25 |
| By small front-end loader | " | 6.10 | 2.00 | 8.10 |
| **02220.70 ROADWAY EXCAVATION** | | | | |
| Roadway excavation | | | | |
| 1/4 mile haul | C.Y. | 0.00 | 1.40 | 1.40 |
| 2 mile haul | " | 0.00 | 2.30 | 2.30 |
| 5 mile haul | " | 0.00 | 3.45 | 3.45 |
| Excavation of open ditches | " | 0.00 | 0.99 | 0.99 |
| Trim banks, swales or ditches | S.Y. | 0.00 | 1.15 | 1.15 |
| Bulk swale excavation by dragline | | | | |
| Small jobs | C.Y. | 0.00 | 3.25 | 3.25 |
| Large jobs | " | 0.00 | 1.85 | 1.85 |
| Spread base course | " | 0.00 | 1.75 | 1.75 |
| Roll and compact | " | 0.00 | 2.30 | 2.30 |
| **02220.71 BASE COURSE** | | | | |
| Base course, crushed stone | | | | |
| 3" thick | S.Y. | 2.20 | 0.35 | 2.55 |
| 4" thick | " | 3.00 | 0.38 | 3.38 |
| 6" thick | " | 4.35 | 0.41 | 4.76 |
| 8" thick | " | 5.60 | 0.46 | 6.06 |
| 10" thick | " | 6.95 | 0.50 | 7.45 |
| 12" thick | " | 7.85 | 0.58 | 8.43 |
| Base course, bank run gravel | | | | |
| 4" deep | S.Y. | 1.05 | 0.36 | 1.42 |
| 6" deep | " | 1.55 | 0.40 | 1.95 |
| 8" deep | " | 2.10 | 0.43 | 2.53 |
| 10" deep | " | 2.60 | 0.46 | 3.06 |
| 12" deep | " | 3.10 | 0.53 | 3.63 |
| Prepare and roll sub base | | | | |
| Minimum | S.Y. | 0.00 | 0.35 | 0.35 |
| Average | " | 0.00 | 0.43 | 0.43 |
| Maximum | " | 0.00 | 0.58 | 0.58 |
| **02220.90 HAND EXCAVATION** | | | | |
| Excavation | | | | |
| To 2' deep | | | | |
| Normal soil | C.Y. | 0.00 | 26.90 | 26.90 |
| Sand and gravel | " | 0.00 | 24.20 | 24.20 |
| Medium clay | " | 0.00 | 30.30 | 30.30 |
| Heavy clay | " | 0.00 | 34.60 | 34.60 |

# 02 SITEWORK

| EARTHWORK | UNIT | MAT. | INST. | TOTAL |
|---|---|---|---|---|
| **02220.90 HAND EXCAVATION** | | | | |
| Loose rock | C.Y. | 0.00 | 40.30 | 40.30 |
| To 6' deep | | | | |
| Normal soil | C.Y. | 0.00 | 34.60 | 34.60 |
| Sand and gravel | " | 0.00 | 30.30 | 30.30 |
| Medium clay | " | 0.00 | 40.30 | 40.30 |
| Heavy clay | " | 0.00 | 48.40 | 48.40 |
| Loose rock | " | 0.00 | 60.50 | 60.50 |
| Backfilling foundation without compaction, 6" lifts | " | 0.00 | 15.10 | 15.10 |
| Compaction of backfill around structures or in trench | | | | |
| By hand with air tamper | C.Y. | 0.00 | 17.30 | 17.30 |
| By hand with vibrating plate tamper | " | 0.00 | 16.15 | 16.15 |
| 1 ton roller | " | 0.00 | 24.20 | 24.20 |
| Miscellaneous hand labor | | | | |
| Trim slopes, sides of excavation | S.F. | 0.00 | 0.04 | 0.04 |
| Trim bottom of excavation | " | 0.00 | 0.05 | 0.05 |
| Excavation around obstructions and services | C.Y. | 0.00 | 80.50 | 80.50 |
| **02240.05 SOIL STABILIZATION** | | | | |
| Straw bale secured with rebar | L.F. | 1.20 | 0.81 | 2.01 |
| Filter barrier, 18" high filter fabric | " | 1.20 | 2.40 | 3.60 |
| Sediment fence, 36" fabric with 6" mesh | " | 2.65 | 3.00 | 5.65 |
| Soil stabilization with tar paper, burlap, straw and stakes | S.F. | 0.23 | 0.04 | 0.27 |
| **02240.30 GEOTEXTILE** | | | | |
| Filter cloth, light reinforcement | | | | |
| Woven | | | | |
| 12'-6" wide x 50' long | S.F. | 0.02 | 0.04 | 0.06 |
| Various lengths | " | 0.03 | 0.04 | 0.07 |
| Non-woven | | | | |
| 14'-8" wide x 430' long | S.F. | 0.12 | 0.04 | 0.16 |
| Various lengths | " | 0.18 | 0.04 | 0.22 |
| **02270.10 SLOPE PROTECTION** | | | | |
| Gabions, stone filled | | | | |
| 6" deep | S.Y. | 15.85 | 12.10 | 27.95 |
| 9" deep | " | 19.80 | 13.85 | 33.65 |
| 12" deep | " | 26.00 | 16.15 | 42.15 |
| 18" deep | " | 33.40 | 19.35 | 52.75 |
| 36" deep | " | 59.00 | 32.30 | 91.30 |
| **02270.40 RIPRAP** | | | | |
| Riprap | | | | |
| Crushed stone blanket, max size 2-1/2" | TON | 18.10 | 36.10 | 54.20 |
| Stone, quarry run, 300 lb. stones | " | 22.65 | 33.40 | 56.05 |
| 400 lb. stones | " | 22.65 | 31.00 | 53.65 |
| 500 lb. stones | " | 22.65 | 28.90 | 51.55 |
| 750 lb. stones | " | 22.65 | 27.10 | 49.75 |
| Dry concrete riprap in bags 3" thick, 80 lb. per bag | BAG | 3.60 | 1.80 | 5.40 |

# 02 SITEWORK

| EARTHWORK | UNIT | MAT. | INST. | TOTAL |
|---|---|---|---|---|
| **02280.20 SOIL TREATMENT** | | | | |
| Soil treatment, termite control pretreatment | | | | |
| Under slabs | S.F. | 0.08 | 0.13 | 0.21 |
| By walls | " | 0.08 | 0.16 | 0.24 |
| **02290.30 WEED CONTROL** | | | | |
| Weed control, bromicil, 15 lb./acre, wettable powder | ACRE | 180.00 | 120.00 | 300.00 |
| Vegetation control, by application of plant killer | S.Y. | 0.02 | 0.10 | 0.12 |
| Weed killer, lawns and fields | " | 0.12 | 0.05 | 0.17 |

| TUNNELING | UNIT | MAT. | INST. | TOTAL |
|---|---|---|---|---|
| **02300.10 PIPE JACKING** | | | | |
| Pipe casing, horizontal jacking | | | | |
| 18" dia. | L.F. | 50.00 | 48.20 | 98.20 |
| 21" dia. | " | 59.50 | 51.50 | 111.00 |
| 24" dia. | " | 63.00 | 54.00 | 117.00 |
| 27" dia. | " | 69.00 | 54.00 | 123.00 |
| 30" dia. | " | 76.00 | 57.00 | 133.00 |
| 36" dia. | " | 86.50 | 62.00 | 148.50 |
| 42" dia. | " | 120.00 | 68.00 | 188.00 |
| 48" dia. | " | 150.00 | 72.50 | 222.50 |

| PILES AND CAISSONS | UNIT | MAT. | INST. | TOTAL |
|---|---|---|---|---|
| **02360.50 PRESTRESSED PILING** | | | | |
| Prestressed concrete piling, less than 60' long | | | | |
| 10" sq. | L.F. | 7.45 | 3.15 | 10.60 |
| 12" sq. | " | 10.85 | 3.30 | 14.15 |
| 14" sq. | " | 11.30 | 3.40 | 14.70 |
| 16" sq. | " | 13.95 | 3.45 | 17.40 |
| 18" sq. | " | 19.25 | 3.70 | 22.95 |
| 20" sq. | " | 26.00 | 3.80 | 29.80 |
| 24" sq. | " | 31.10 | 3.90 | 35.00 |

## PILES AND CAISSONS

| PILES AND CAISSONS | UNIT | MAT. | INST. | TOTAL |
|---|---|---|---|---|
| **02360.50 PRESTRESSED PILING** | | | | |
| More than 60' long | | | | |
| 12" sq. | L.F. | 10.85 | 2.70 | 13.55 |
| 14" sq. | " | 11.80 | 2.75 | 14.55 |
| 16" sq. | " | 14.70 | 2.80 | 17.50 |
| 18" sq. | " | 19.25 | 2.85 | 22.10 |
| 20" sq. | " | 26.00 | 2.95 | 28.95 |
| 24" sq. | " | 31.10 | 3.00 | 34.10 |
| Straight cylinder, less than 60' long | | | | |
| 12" dia. | L.F. | 10.10 | 3.45 | 13.55 |
| 14" dia. | " | 13.70 | 3.55 | 17.25 |
| 16" dia. | " | 16.65 | 3.65 | 20.30 |
| 18" dia. | " | 19.05 | 3.70 | 22.75 |
| 20" dia. | " | 24.95 | 3.80 | 28.75 |
| 24" dia. | " | 30.90 | 3.90 | 34.80 |
| More than 60' long | | | | |
| 12" dia. | L.F. | 10.10 | 2.75 | 12.85 |
| 14" dia. | " | 13.70 | 2.80 | 16.50 |
| 16" dia. | " | 16.65 | 2.85 | 19.50 |
| 18" dia. | " | 19.05 | 2.95 | 22.00 |
| 20" dia. | " | 24.95 | 3.00 | 27.95 |
| 24" dia. | " | 31.50 | 3.05 | 34.55 |
| Concrete sheet piling | | | | |
| 12" thick x 20' long | S.F. | 11.55 | 7.60 | 19.15 |
| 25' long | " | 11.55 | 6.90 | 18.45 |
| 30' long | " | 11.55 | 6.35 | 17.90 |
| 35' long | " | 11.55 | 5.85 | 17.40 |
| 40' long | " | 11.55 | 5.45 | 17.00 |
| 16" thick x 40' long | " | 15.80 | 4.25 | 20.05 |
| 45' long | " | 15.80 | 4.00 | 19.80 |
| 50' long | " | 15.80 | 3.80 | 19.60 |
| 55' long | " | 15.80 | 3.65 | 19.45 |
| 60' long | " | 15.80 | 3.45 | 19.25 |
| **02360.60 STEEL PILES** | | | | |
| H-section piles | | | | |
| 8x8 | | | | |
| 36 lb/ft | | | | |
| 30' long | L.F. | 11.30 | 6.35 | 17.65 |
| 40' long | " | 11.30 | 5.10 | 16.40 |
| 50' long | " | 11.30 | 4.25 | 15.55 |
| 10x10 | | | | |
| 42 lb/ft | | | | |
| 30' long | L.F. | 12.75 | 6.35 | 19.10 |
| 40' long | " | 12.75 | 5.10 | 17.85 |
| 50' long | " | 12.75 | 4.25 | 17.00 |
| 57 lb/ft | | | | |
| 30' long | L.F. | 16.40 | 6.35 | 22.75 |
| 40' long | " | 16.40 | 5.10 | 21.50 |
| 50' long | " | 16.40 | 4.25 | 20.65 |
| 12x12 | | | | |
| 53 lb/ft | | | | |
| 30' long | L.F. | 15.30 | 6.90 | 22.20 |

| PILES AND CAISSONS | UNIT | MAT. | INST. | TOTAL |
|---|---|---|---|---|
| **02360.60 STEEL PILES** | | | | |
| 40' long | L.F. | 15.30 | 5.45 | 20.75 |
| 50' long | " | 15.30 | 4.25 | 19.55 |
| 74 lb/ft | | | | |
| 30' long | L.F. | 20.95 | 6.90 | 27.85 |
| 40' long | " | 20.95 | 5.45 | 26.40 |
| 50' long | " | 20.95 | 4.25 | 25.20 |
| 14x14 | | | | |
| 73 lb/ft | | | | |
| 40' long | L.F. | 20.95 | 6.90 | 27.85 |
| 50' long | " | 20.95 | 5.45 | 26.40 |
| 60' long | " | 20.95 | 4.25 | 25.20 |
| 89 lb/ft | | | | |
| 40' long | L.F. | 26.60 | 6.90 | 33.50 |
| 50' long | " | 26.60 | 5.45 | 32.05 |
| 60' long | " | 26.60 | 4.25 | 30.85 |
| 102 lb/ft | | | | |
| 40' long | L.F. | 30.60 | 6.90 | 37.50 |
| 50' long | " | 30.60 | 5.45 | 36.05 |
| 60' long | " | 30.60 | 4.25 | 34.85 |
| 117 lb/ft | | | | |
| 40' long | L.F. | 35.10 | 7.25 | 42.35 |
| 50' long | " | 35.10 | 5.65 | 40.75 |
| 60' long | " | 35.10 | 4.35 | 39.45 |
| Splice | | | | |
| 8" | EA. | 59.00 | 40.30 | 99.30 |
| 10" | " | 65.00 | 48.40 | 113.40 |
| 12" | " | 88.50 | 48.40 | 136.90 |
| 14" | " | 110.00 | 60.50 | 170.50 |
| Driving cap | | | | |
| 8" | EA. | 33.00 | 24.20 | 57.20 |
| 10" | " | 33.00 | 30.30 | 63.30 |
| 12" | " | 33.00 | 30.30 | 63.30 |
| 14" | " | 33.00 | 34.60 | 67.60 |
| Standard point | | | | |
| 8" | EA. | 43.60 | 24.20 | 67.80 |
| 10" | " | 54.00 | 30.30 | 84.30 |
| 12" | " | 76.50 | 34.60 | 111.10 |
| 14" | " | 100.00 | 40.30 | 140.30 |
| Heavy duty point | | | | |
| 8" | EA. | 45.90 | 26.90 | 72.80 |
| 10" | " | 56.50 | 34.60 | 91.10 |
| 12" | " | 79.00 | 40.30 | 119.30 |
| 14" | " | 100.00 | 48.40 | 148.40 |
| Tapered friction piles, with fluted steel casing, up to 50' | | | | |
| With 4000 psi concrete no reinforcing | | | | |
| 12" dia. | L.F. | 13.80 | 3.80 | 17.60 |
| 14" dia. | " | 15.90 | 3.90 | 19.80 |
| 16" dia. | " | 19.05 | 4.00 | 23.05 |
| 18" dia. | " | 21.20 | 4.50 | 25.70 |

## PILES AND CAISSONS

| | UNIT | MAT. | INST. | TOTAL |
|---|---|---|---|---|
| **02360.65 STEEL PIPE PILES** | | | | |
| Concrete filled, 3000# concrete, up to 40' | | | | |
| 8" dia. | L.F. | 11.90 | 5.45 | 17.35 |
| 10" dia. | " | 14.65 | 5.65 | 20.30 |
| 12" dia. | " | 18.35 | 5.85 | 24.20 |
| 14" dia. | " | 20.20 | 6.10 | 26.30 |
| 16" dia. | " | 22.95 | 6.35 | 29.30 |
| 18" dia. | " | 25.60 | 6.60 | 32.20 |
| Pipe piles, non-filled | | | | |
| 8" dia. | L.F. | 10.10 | 4.25 | 14.35 |
| 10" dia. | " | 11.90 | 4.35 | 16.25 |
| 12" dia. | " | 14.65 | 4.50 | 19.15 |
| 14" dia. | " | 15.60 | 4.75 | 20.35 |
| 16" dia. | " | 18.35 | 4.90 | 23.25 |
| 18" dia. | " | 20.20 | 5.10 | 25.30 |
| Splice | | | | |
| 8" dia. | EA. | 34.60 | 48.40 | 83.00 |
| 10" dia. | " | 36.70 | 48.40 | 85.10 |
| 12" dia. | " | 41.30 | 60.50 | 101.80 |
| 14" dia. | " | 45.80 | 60.50 | 106.30 |
| 16" dia. | " | 61.00 | 80.50 | 141.50 |
| 18" dia. | " | 81.50 | 80.50 | 162.00 |
| Standard point | | | | |
| 8" dia. | EA. | 45.30 | 48.40 | 93.70 |
| 10" dia. | " | 51.00 | 48.40 | 99.40 |
| 12" dia. | " | 79.50 | 60.50 | 140.00 |
| 14" dia. | " | 120.00 | 60.50 | 180.50 |
| 16" dia. | " | 150.00 | 80.50 | 230.50 |
| 18" dia. | " | 170.00 | 80.50 | 250.50 |
| Heavy duty point | | | | |
| 8" dia. | EA. | 56.50 | 60.50 | 117.00 |
| 10" dia. | " | 68.00 | 60.50 | 128.50 |
| 12" dia. | " | 90.50 | 80.50 | 171.00 |
| 14" dia. | " | 140.00 | 80.50 | 220.50 |
| 16" dia. | " | 170.00 | 97.00 | 267.00 |
| 18" dia. | " | 220.00 | 97.00 | 317.00 |
| **02360.70 STEEL SHEET PILING** | | | | |
| Steel sheet piling,12" wide | | | | |
| 20' long | S.F. | 9.25 | 7.60 | 16.85 |
| 35' long | " | 9.25 | 5.45 | 14.70 |
| 50' long | " | 9.25 | 3.80 | 13.05 |
| Over 50' long | " | 9.25 | 3.45 | 12.70 |
| **02360.80 WOOD AND TIMBER PILES** | | | | |
| Treated wood piles, 12" butt, 8" tip | | | | |
| 25' long | L.F. | 5.90 | 7.60 | 13.50 |
| 30' long | " | 6.45 | 6.35 | 12.80 |
| 35' long | " | 6.45 | 5.45 | 11.90 |
| 40' long | " | 6.45 | 4.75 | 11.20 |
| 12" butt, 7" tip | | | | |

# 02 SITEWORK

| PILES AND CAISSONS | UNIT | MAT. | INST. | TOTAL |
|---|---|---|---|---|
| **02360.80 WOOD AND TIMBER PILES** | | | | |
| 40' long | L.F. | 7.00 | 4.75 | 11.75 |
| 45' long | " | 7.00 | 4.25 | 11.25 |
| 50' long | " | 8.05 | 3.80 | 11.85 |
| 55' long | " | 8.05 | 3.45 | 11.50 |
| 60' long | " | 8.05 | 3.15 | 11.20 |
| **02360.90 PILE TESTING** | | | | |
| Pile test | | | | |
| 50 ton to 100 ton | EA. | | | 13,586 |
| To 200 ton | " | | | 20,380 |
| To 300 ton | " | | | 22,644 |
| To 400 ton | " | | | 33,966 |
| To 600 ton | " | | | 39,627 |
| **02380.10 CAISSONS** | | | | |
| Caisson, including 3000# concrete, in stable ground | | | | |
| 18" dia. | L.F. | 10.60 | 15.25 | 25.85 |
| 24" dia. | " | 17.65 | 15.85 | 33.50 |
| 30" dia. | " | 28.30 | 19.05 | 47.35 |
| 36" dia. | " | 38.90 | 21.75 | 60.65 |
| 48" dia. | " | 67.00 | 25.40 | 92.40 |
| 60" dia. | " | 110.00 | 34.60 | 144.60 |
| 72" dia. | " | 180.00 | 42.30 | 222.30 |
| 84" dia. | " | 220.00 | 54.50 | 274.50 |
| Wet ground, casing required but pulled | | | | |
| 18" dia. | L.F. | 10.60 | 19.05 | 29.65 |
| 24" dia. | " | 17.65 | 21.15 | 38.80 |
| 30" dia. | " | 28.30 | 23.80 | 52.10 |
| 36" dia. | " | 38.90 | 25.40 | 64.30 |
| 48" dia. | " | 67.00 | 31.70 | 98.70 |
| 60" dia. | " | 110.00 | 42.30 | 152.30 |
| 72" dia. | " | 160.00 | 63.50 | 223.50 |
| 84" dia. | " | 220.00 | 95.00 | 315.00 |
| Soft rock | | | | |
| 18" dia. | L.F. | 10.60 | 54.50 | 65.10 |
| 24" dia. | " | 17.65 | 95.00 | 112.65 |
| 30" dia. | " | 28.30 | 130.00 | 158.30 |
| 36" dia. | " | 38.90 | 190.00 | 228.90 |
| 48" dia. | " | 67.00 | 250.00 | 317.00 |
| 60" dia. | " | 110.00 | 380.00 | 490.00 |
| 72" dia. | " | 180.00 | 420.00 | 600.00 |
| 84" dia. | " | 220.00 | 480.00 | 700.00 |

## RAILROAD WORK

### 02450.10 RAILROAD WORK

| RAILROAD WORK | UNIT | MAT. | INST. | TOTAL |
|---|---|---|---|---|
| Rail | | | | |
| 90 lb | L.F. | 16.35 | 0.52 | 16.87 |
| 100 lb | " | 18.65 | 0.52 | 19.17 |
| 115 lb | " | 21.00 | 0.52 | 21.52 |
| 132 lb | " | 23.30 | 0.52 | 23.82 |
| Rail relay | | | | |
| 90 lb | L.F. | 6.55 | 0.52 | 7.07 |
| 100 lb | " | 7.25 | 0.52 | 7.77 |
| 115 lb | " | 8.75 | 0.52 | 9.27 |
| 132 lb | " | 10.80 | 0.52 | 11.32 |
| New angle bars, per pair | | | | |
| 90 lb | EA. | 60.50 | 0.65 | 61.15 |
| 100 lb | " | 65.50 | 0.65 | 66.15 |
| 115 lb | " | 84.00 | 0.65 | 84.65 |
| 132 lb | " | 97.00 | 0.65 | 97.65 |
| Angle bar relay | | | | |
| 90 lb | EA. | 25.10 | 0.65 | 25.75 |
| 100 lb | " | 25.50 | 0.65 | 26.15 |
| 115 lb | " | 26.80 | 0.65 | 27.45 |
| 132 lb | " | 28.60 | 0.65 | 29.25 |
| New tie plates | | | | |
| 90 lb | EA. | 8.15 | 0.47 | 8.62 |
| 100 lb | " | 8.50 | 0.47 | 8.97 |
| 115 lb | " | 9.30 | 0.47 | 9.77 |
| 132 lb | " | 9.90 | 0.47 | 10.37 |
| Tie plate relay | | | | |
| 90 lb | EA. | 2.55 | 0.47 | 3.02 |
| 100 lb | " | 3.60 | 0.47 | 4.07 |
| 115 lb | " | 3.60 | 0.47 | 4.07 |
| 132 lb | " | 4.45 | 0.47 | 4.92 |
| Track accessories | | | | |
| Wooden cross ties, 8' | EA. | 32.70 | 3.25 | 35.95 |
| Concrete cross ties, 8' | " | 200.00 | 6.55 | 206.55 |
| Tie plugs, 5" | " | 12.85 | 0.33 | 13.18 |
| Track bolts and nuts, 1" | " | 3.15 | 0.33 | 3.48 |
| Lockwashers, 1" | " | 0.81 | 0.22 | 1.03 |
| Track spikes, 6" | " | 0.75 | 1.30 | 2.05 |
| Wooden switch ties | B.F. | 1.15 | 0.33 | 1.48 |
| Rail anchors | EA. | 3.20 | 1.20 | 4.40 |
| Ballast | TON | 9.20 | 6.55 | 15.75 |
| Gauge rods | EA. | 25.10 | 5.25 | 30.35 |
| Compromise splice bars | " | 310.00 | 8.70 | 318.70 |
| Turnout | | | | |
| 90 lb | EA. | 8,750 | 1,310 | 10,060 |
| 100 lb | " | 9,210 | 1,310 | 10,520 |
| 110 lb | " | 9,910 | 1,310 | 11,220 |
| 115 lb | " | 10,150 | 1,310 | 11,460 |
| 132 lb | " | 11,080 | 1,310 | 12,390 |
| Turnout relay | | | | |
| 90 lb | EA. | 5,600 | 1,310 | 6,910 |
| 100 lb | " | 6,180 | 1,310 | 7,490 |
| 110 lb | " | 6,420 | 1,310 | 7,730 |

| RAILROAD WORK | UNIT | MAT. | INST. | TOTAL |
|---|---|---|---|---|
| **02450.10 RAILROAD WORK** | | | | |
| 115 lb | EA. | 6,760 | 1,310 | 8,070 |
| 132 lb | " | 7,350 | 1,310 | 8,660 |
| Railroad track in place, complete | | | | |
| New rail | | | | |
| 90 lb | L.F. | 110.00 | 13.10 | 123.10 |
| 100 lb | " | 110.00 | 13.10 | 123.10 |
| 110 lb | " | 100.00 | 13.10 | 113.10 |
| 115 lb | " | 120.00 | 13.10 | 133.10 |
| 132 lb | " | 120.00 | 13.10 | 133.10 |
| Rail relay | | | | |
| 90 lb | L.F. | 61.00 | 13.10 | 74.10 |
| 100 lb | " | 67.50 | 13.10 | 80.60 |
| 110 lb | " | 67.50 | 13.10 | 80.60 |
| 115 lb | " | 73.50 | 13.10 | 86.60 |
| 132 lb | " | 76.00 | 13.10 | 89.10 |
| No. 8 turnout | | | | |
| 90 lb | EA. | 21,920 | 1,740 | 23,660 |
| 100 lb | " | 24,370 | 1,740 | 26,110 |
| 110 lb | " | 26,710 | 1,740 | 28,450 |
| 115 lb | " | 27,290 | 1,740 | 29,030 |
| 132 lb | " | 27,870 | 1,740 | 29,610 |
| No. 8 turnout relay | | | | |
| 90 lb | EA. | 18,190 | 1,740 | 19,930 |
| 100 lb | " | 18,190 | 1,740 | 19,930 |
| 110 lb | " | 18,190 | 1,740 | 19,930 |
| 115 lb | " | 20,640 | 1,740 | 22,380 |
| 132 lb | " | 20,640 | 1,740 | 22,380 |
| Railroad crossings, asphalt, based on 8" thick x 20' | | | | |
| Including track and approach | | | | |
| 12' roadway | EA. | 530.00 | 330.00 | 860.00 |
| 15' roadway | " | 620.00 | 370.00 | 990.00 |
| 18' roadway | " | 720.00 | 440.00 | 1,160 |
| 21' roadway | " | 800.00 | 520.00 | 1,320 |
| 24' roadway | " | 910.00 | 650.00 | 1,560 |
| Precast concrete inserts | | | | |
| 12' roadway | EA. | 760.00 | 130.00 | 890.00 |
| 15' roadway | " | 910.00 | 160.00 | 1,070 |
| 18' roadway | " | 1,080 | 220.00 | 1,300 |
| 21' roadway | " | 1,400 | 260.00 | 1,660 |
| 24' roadway | " | 1,690 | 290.00 | 1,980 |
| Molded rubber, with headers | | | | |
| 12' roadway | EA. | 4,550 | 130.00 | 4,680 |
| 15' roadway | " | 5,710 | 160.00 | 5,870 |
| 18' roadway | " | 6,760 | 220.00 | 6,980 |
| 21' roadway | " | 7,460 | 260.00 | 7,720 |
| 24' roadway | " | 9,100 | 290.00 | 9,390 |

| PAVING AND SURFACING | UNIT | MAT. | INST. | TOTAL |
|---|---|---|---|---|
| **02510.20 ASPHALT SURFACES** | | | | |
| Asphalt wearing surface, for flexible pavement | | | | |
| 1" thick | S.Y. | 1.80 | 1.25 | 3.05 |
| 1-1/2" thick | " | 2.70 | 1.50 | 4.20 |
| 2" thick | " | 3.60 | 1.90 | 5.50 |
| 3" thick | " | 5.30 | 2.55 | 7.85 |
| Bituminous sidewalk, no base | | | | |
| 2" thick | S.Y. | 3.85 | 1.55 | 5.40 |
| 3" thick | " | 5.65 | 1.65 | 7.30 |
| **02520.10 CONCRETE PAVING** | | | | |
| Concrete paving, reinforced, 5000 psi concrete | | | | |
| 6" thick | S.Y. | 15.95 | 11.90 | 27.85 |
| 7" thick | " | 17.85 | 12.70 | 30.55 |
| 8" thick | " | 20.20 | 13.60 | 33.80 |
| 9" thick | " | 22.00 | 14.65 | 36.65 |
| 10" thick | " | 24.40 | 15.85 | 40.25 |
| 11" thick | " | 26.90 | 17.30 | 44.20 |
| 12" thick | " | 28.80 | 19.05 | 47.85 |
| 15" thick | " | 36.00 | 23.80 | 59.80 |
| Concrete paving, for pipe trench, reinforced | | | | |
| 7" thick | S.Y. | 33.30 | 13.10 | 46.40 |
| 8" thick | " | 38.00 | 14.55 | 52.55 |
| 9" thick | " | 41.60 | 16.35 | 57.95 |
| 10" thick | " | 46.40 | 18.70 | 65.10 |
| Fibrous concrete | | | | |
| 5" thick | S.Y. | 15.90 | 14.65 | 30.55 |
| 8" thick | " | 20.30 | 15.85 | 36.15 |
| Roller compacted concrete, (RCC), place and compact | | | | |
| 8" thick | S.Y. | 20.80 | 19.05 | 39.85 |
| 12" thick | " | 30.30 | 23.80 | 54.10 |
| Steel edge forms up to | | | | |
| 12" deep | L.F. | 0.54 | 0.81 | 1.35 |
| 15" deep | " | 0.67 | 0.97 | 1.64 |
| Paving finishes | | | | |
| Belt dragged | S.Y. | 0.00 | 1.20 | 1.20 |
| Curing | " | 0.26 | 0.24 | 0.50 |
| **02545.10 ASPHALT REPAIR** | | | | |
| Coal tar emulsion seal coat, rubber additive, fuel resistant | S.Y. | 0.80 | 0.35 | 1.15 |
| Bituminous surface treatment, single | " | 0.74 | 0.24 | 0.98 |
| Double | " | 0.99 | 0.03 | 1.02 |
| Bituminous prime coat | " | 0.50 | 0.03 | 0.53 |
| Tack coat | " | 0.24 | 0.02 | 0.26 |
| Crack sealing, concrete paving | L.F. | 0.41 | 0.16 | 0.57 |
| Bituminous paving for pipe trench, 4" thick | S.Y. | 4.75 | 8.70 | 13.45 |
| Polypropylene, nonwoven paving fabric | " | 0.62 | 0.12 | 0.74 |
| Rubberized asphalt | " | 0.99 | 2.20 | 3.19 |
| Asphalt slurry seal | " | 2.40 | 1.40 | 3.80 |

## PAVING AND SURFACING

### 02580.10 PAVEMENT MARKINGS

| PAVING AND SURFACING | UNIT | MAT. | INST. | TOTAL |
|---|---|---|---|---|
| Pavement line marking, paint | | | | |
| 4" wide | L.F. | 0.04 | 0.06 | 0.10 |
| 6" wide | " | 0.07 | 0.13 | 0.20 |
| 8" wide | " | 0.08 | 0.20 | 0.28 |
| Reflective paint, 4" wide | " | 0.20 | 0.20 | 0.40 |
| Airfield markings, retro-reflective | | | | |
| White | L.F. | 0.33 | 0.20 | 0.53 |
| Yellow | " | 0.37 | 0.20 | 0.57 |
| Preformed tape, 4" wide | | | | |
| Inlaid reflective | L.F. | 0.99 | 0.04 | 1.02 |
| Reflective paint | " | 0.68 | 0.06 | 0.74 |
| Thermoplastic | | | | |
| White | L.F. | 0.31 | 0.12 | 0.43 |
| Yellow | " | 0.32 | 0.12 | 0.44 |
| 12" wide, thermoplastic, white | " | 0.85 | 0.35 | 1.20 |
| Directional arrows, reflective preformed tape | EA. | 71.50 | 24.20 | 95.70 |
| Messages, reflective preformed tape (per letter) | " | 34.50 | 12.10 | 46.60 |
| Handicap symbol, preformed tape | " | 60.50 | 24.20 | 84.70 |
| Parking stall painting | " | 0.93 | 4.85 | 5.78 |

## UTILITIES

### 02605.30 MANHOLES

| UTILITIES | UNIT | MAT. | INST. | TOTAL |
|---|---|---|---|---|
| Precast sections, 48" dia. | | | | |
| Base section | EA. | 120.00 | 110.00 | 230.00 |
| 1'0" riser | " | 33.30 | 87.00 | 120.30 |
| 1'4" riser | " | 40.40 | 93.50 | 133.90 |
| 2'8" riser | " | 60.00 | 100.00 | 160.00 |
| 4'0" riser | " | 110.00 | 110.00 | 220.00 |
| 2'8" cone top | " | 72.50 | 130.00 | 202.50 |
| Precast manholes, 48" dia. | | | | |
| 4' deep | EA. | 270.00 | 260.00 | 530.00 |
| 6' deep | " | 420.00 | 330.00 | 750.00 |
| 7' deep | " | 480.00 | 370.00 | 850.00 |
| 8' deep | " | 540.00 | 440.00 | 980.00 |
| 10' deep | " | 600.00 | 520.00 | 1,120 |
| Cast-in-place, 48" dia., with frame and cover | | | | |
| 5' deep | EA. | 390.00 | 650.00 | 1,040 |
| 6' deep | " | 520.00 | 750.00 | 1,270 |
| 8' deep | " | 750.00 | 870.00 | 1,620 |
| 10' deep | " | 880.00 | 1,050 | 1,930 |
| Brick manholes, 48" dia. with cover, 8" thick | | | | |
| 4' deep | EA. | 490.00 | 300.00 | 790.00 |

## UTILITIES

| UTILITIES | UNIT | MAT. | INST. | TOTAL |
|---|---|---|---|---|
| **02605.30 MANHOLES** | | | | |
| 6' deep | EA. | 620.00 | 330.00 | 950.00 |
| 8' deep | " | 790.00 | 380.00 | 1,170 |
| 10' deep | " | 990.00 | 430.00 | 1,420 |
| 12' deep | " | 1,230 | 500.00 | 1,730 |
| 14' deep | " | 1,500 | 600.00 | 2,100 |
| Inverts for manholes | | | | |
| Single channel | EA. | 79.50 | 120.00 | 199.50 |
| Triple channel | " | 90.50 | 150.00 | 240.50 |
| Frames and covers, 24" diameter | | | | |
| 300 lb | EA. | 250.00 | 24.20 | 274.20 |
| 400 lb | " | 260.00 | 26.90 | 286.90 |
| 500 lb | " | 300.00 | 34.60 | 334.60 |
| Watertight, 350 lb | " | 310.00 | 80.50 | 390.50 |
| For heavy equipment, 1200 lb | " | 690.00 | 120.00 | 810.00 |
| Steps for manholes | | | | |
| 7" x 9" | EA. | 8.60 | 4.85 | 13.45 |
| 8" x 9" | " | 11.05 | 5.40 | 16.45 |
| Curb inlet, 4' throat, cast in place | | | | |
| 12"-30" pipe | EA. | 250.00 | 650.00 | 900.00 |
| 36"-48" pipe | " | 270.00 | 750.00 | 1,020 |
| Raise exist frame and cover, when repaving | " | 0.00 | 260.00 | 260.00 |
| **02610.10 CAST IRON FLANGED PIPE** | | | | |
| Cast iron flanged sections | | | | |
| 4" pipe, with one bolt set | | | | |
| 3' section | EA. | 120.00 | 11.90 | 131.90 |
| 4' section | " | 120.00 | 13.10 | 133.10 |
| 5' section | " | 140.00 | 14.55 | 154.55 |
| 6' section | " | 150.00 | 16.35 | 166.35 |
| 8' section | " | 170.00 | 18.70 | 188.70 |
| 10' section | " | 190.00 | 26.20 | 216.20 |
| 12' section | " | 210.00 | 43.60 | 253.60 |
| 15' section | " | 240.00 | 65.50 | 305.50 |
| 18' section | " | 270.00 | 87.00 | 357.00 |
| 6" pipe, with one bolt set | | | | |
| 3' section | EA. | 150.00 | 13.10 | 163.10 |
| 4' section | " | 160.00 | 15.40 | 175.40 |
| 5' section | " | 170.00 | 17.45 | 187.45 |
| 6' section | " | 190.00 | 20.10 | 210.10 |
| 8' section | " | 210.00 | 29.10 | 239.10 |
| 10' section | " | 240.00 | 32.70 | 272.70 |
| 12' section | " | 260.00 | 43.60 | 303.60 |
| 15' section | " | 290.00 | 65.50 | 355.50 |
| 18' section | " | 360.00 | 93.50 | 453.50 |
| 8" pipe, with one bolt set | | | | |
| 3' section | EA. | 200.00 | 16.35 | 216.35 |
| 4' section | " | 220.00 | 18.70 | 238.70 |
| 5' section | " | 240.00 | 21.80 | 261.80 |
| 6' section | " | 250.00 | 26.20 | 276.20 |
| 8' section | " | 280.00 | 37.40 | 317.40 |
| 10' section | " | 320.00 | 43.60 | 363.60 |

| UTILITIES | UNIT | MAT. | INST. | TOTAL |
|---|---|---|---|---|
| **02610.10 CAST IRON FLANGED PIPE** | | | | |
| 12' section | EA. | 340.00 | 65.50 | 405.50 |
| 15' section | " | 400.00 | 87.00 | 487.00 |
| 18' section | " | 470.00 | 110.00 | 580.00 |
| 10" pipe, with one bolt set | | | | |
| 3' section | EA. | 250.00 | 16.75 | 266.75 |
| 4' section | " | 290.00 | 19.25 | 309.25 |
| 5' section | " | 310.00 | 22.55 | 332.55 |
| 6' section | " | 340.00 | 27.20 | 367.20 |
| 8' section | " | 370.00 | 39.60 | 409.60 |
| 10' section | " | 410.00 | 46.70 | 456.70 |
| 12' section | " | 460.00 | 72.50 | 532.50 |
| 15' section | " | 510.00 | 93.50 | 603.50 |
| 18' section | " | 590.00 | 130.00 | 720.00 |
| 12" pipe, with one bolt set | | | | |
| 3' section | EA. | 340.00 | 18.15 | 358.15 |
| 4' section | " | 360.00 | 21.10 | 381.10 |
| 5' section | " | 400.00 | 25.10 | 425.10 |
| 6' section | " | 420.00 | 29.70 | 449.70 |
| 8' section | " | 480.00 | 43.60 | 523.60 |
| 10' section | " | 530.00 | 50.50 | 580.50 |
| 12' section | " | 590.00 | 81.50 | 671.50 |
| 15' section | " | 670.00 | 110.00 | 780.00 |
| 18' section | " | 710.00 | 150.00 | 860.00 |
| **02610.11 CAST IRON FITTINGS** | | | | |
| Mechanical joint, with 2 bolt kits | | | | |
| 90 deg bend | | | | |
| 4" | EA. | 79.50 | 16.15 | 95.65 |
| 6" | " | 98.50 | 18.60 | 117.10 |
| 8" | " | 120.00 | 24.20 | 144.20 |
| 10" | " | 190.00 | 34.60 | 224.60 |
| 12" | " | 280.00 | 48.40 | 328.40 |
| 14" | " | 400.00 | 60.50 | 460.50 |
| 16" | " | 450.00 | 80.50 | 530.50 |
| 45 deg bend | | | | |
| 4" | EA. | 68.00 | 16.15 | 84.15 |
| 6" | " | 90.50 | 18.60 | 109.10 |
| 8" | " | 120.00 | 24.20 | 144.20 |
| 10" | " | 160.00 | 34.60 | 194.60 |
| 12" | " | 220.00 | 48.40 | 268.40 |
| 14" | " | 260.00 | 60.50 | 320.50 |
| 16" | " | 340.00 | 80.50 | 420.50 |
| Tee, with 3 bolt kits | | | | |
| 4" x 4" | EA. | 120.00 | 24.20 | 144.20 |
| 6" x 6" | " | 170.00 | 30.30 | 200.30 |
| 8" x 8" | " | 230.00 | 40.30 | 270.30 |
| 10" x 10" | " | 340.00 | 60.50 | 400.50 |
| 12" x 12" | " | 450.00 | 80.50 | 530.50 |
| Wye, with 3 bolt kits | | | | |
| 6" x 6" | EA. | 230.00 | 30.30 | 260.30 |
| 8" x 8" | " | 340.00 | 40.30 | 380.30 |

| UTILITIES | UNIT | MAT. | INST. | TOTAL |
|---|---|---|---|---|
| **02610.11 CAST IRON FITTINGS** | | | | |
| 10" x 10" | EA. | 450.00 | 60.50 | 510.50 |
| 12" x 12" | " | 620.00 | 80.50 | 700.50 |
| Reducer, with 2 bolt kits | | | | |
| 6" x 4" | EA. | 85.00 | 30.30 | 115.30 |
| 8" x 6" | " | 110.00 | 40.30 | 150.30 |
| 10" x 8" | " | 200.00 | 60.50 | 260.50 |
| 12" x 10" | " | 250.00 | 80.50 | 330.50 |
| Flanged, 90 deg bend, 125 lb. | | | | |
| 4" | EA. | 51.00 | 20.15 | 71.15 |
| 6" | " | 70.00 | 24.20 | 94.20 |
| 8" | " | 98.50 | 30.30 | 128.80 |
| 10" | " | 180.00 | 40.30 | 220.30 |
| 12" | " | 240.00 | 60.50 | 300.50 |
| 14" | " | 490.00 | 80.50 | 570.50 |
| 16" | " | 720.00 | 80.50 | 800.50 |
| Tee | | | | |
| 4" | EA. | 88.50 | 30.30 | 118.80 |
| 6" | " | 120.00 | 34.60 | 154.60 |
| 8" | " | 190.00 | 40.30 | 230.30 |
| 10" | " | 350.00 | 48.40 | 398.40 |
| 12" | " | 480.00 | 60.50 | 540.50 |
| 14" | " | 1,050 | 80.50 | 1,131 |
| 16" | " | 1,590 | 120.00 | 1,710 |
| **02610.13 GATE VALVES** | | | | |
| Gate valve, (AWWA) mechanical joint, with adjustable box | | | | |
| 4" valve | EA. | 460.00 | 43.60 | 503.60 |
| 6" valve | " | 520.00 | 52.50 | 572.50 |
| 8" valve | " | 690.00 | 65.50 | 755.50 |
| 10" valve | " | 1,040 | 77.00 | 1,117 |
| 12" valve | " | 1,390 | 93.50 | 1,484 |
| 14" valve | " | 3,460 | 110.00 | 3,570 |
| 16" valve | " | 4,620 | 120.00 | 4,740 |
| 18" valve | " | 5,770 | 130.00 | 5,900 |
| Flanged, with box, post indicator (AWWA) | | | | |
| 4" valve | EA. | 420.00 | 52.50 | 472.50 |
| 6" valve | " | 490.00 | 59.50 | 549.50 |
| 8" valve | " | 690.00 | 72.50 | 762.50 |
| 10" valve | " | 1,040 | 87.00 | 1,127 |
| 12" valve | " | 1,500 | 110.00 | 1,610 |
| 14" valve | " | 3,460 | 130.00 | 3,590 |
| 16" valve | " | 4,620 | 160.00 | 4,780 |
| **02610.15 WATER METERS** | | | | |
| Water meter, displacement type | | | | |
| 1" | EA. | 160.00 | 34.20 | 194.20 |
| 1-1/2" | " | 580.00 | 38.00 | 618.00 |
| 2" | " | 870.00 | 42.70 | 912.70 |

# 02 SITEWORK

| UTILITIES | UNIT | MAT. | INST. | TOTAL |
|---|---|---|---|---|
| **02610.17 CORPORATION STOPS** | | | | |
| Stop for flared copper service pipe | | | | |
| 3/4" | EA. | 18.50 | 17.10 | 35.60 |
| 1" | " | 25.40 | 19.00 | 44.40 |
| 1-1/4" | " | 67.00 | 22.75 | 89.75 |
| 1-1/2" | " | 81.00 | 28.50 | 109.50 |
| 2" | " | 120.00 | 34.20 | 154.20 |
| **02610.40 DUCTILE IRON PIPE** | | | | |
| Ductile iron pipe, cement lined, slip-on joints | | | | |
| 4" | L.F. | 6.45 | 3.65 | 10.10 |
| 6" | " | 8.05 | 3.85 | 11.90 |
| 8" | " | 10.55 | 4.10 | 14.65 |
| 10" | " | 14.50 | 4.35 | 18.85 |
| 12" | " | 17.90 | 5.25 | 23.15 |
| 14" | " | 22.60 | 6.55 | 29.15 |
| 16" | " | 27.70 | 7.25 | 34.95 |
| 18" | " | 31.10 | 8.15 | 39.25 |
| 20" | " | 35.10 | 9.35 | 44.45 |
| Mechanical joint pipe | | | | |
| 4" | L.F. | 7.95 | 5.05 | 13.00 |
| 6" | " | 9.45 | 5.45 | 14.90 |
| 8" | " | 12.45 | 5.95 | 18.40 |
| 10" | " | 16.30 | 6.55 | 22.85 |
| 12" | " | 20.70 | 8.70 | 29.40 |
| 14" | " | 26.00 | 10.05 | 36.05 |
| 16" | " | 28.50 | 11.90 | 40.40 |
| 18" | " | 32.20 | 13.10 | 45.30 |
| 20" | " | 37.00 | 14.55 | 51.55 |
| Fittings, mechanical joint | | | | |
| 90 degree elbow | | | | |
| 4" | EA. | 98.00 | 16.15 | 114.15 |
| 6" | " | 130.00 | 18.60 | 148.60 |
| 8" | " | 180.00 | 24.20 | 204.20 |
| 10" | " | 270.00 | 34.60 | 304.60 |
| 12" | " | 360.00 | 48.40 | 408.40 |
| 14" | " | 550.00 | 60.50 | 610.50 |
| 16" | " | 690.00 | 80.50 | 770.50 |
| 18" | " | 1,040 | 97.00 | 1,137 |
| 20" | " | 1,150 | 120.00 | 1,270 |
| 45 degree elbow | | | | |
| 4" | EA. | 100.00 | 16.15 | 116.15 |
| 6" | " | 120.00 | 18.60 | 138.60 |
| 8" | " | 160.00 | 24.20 | 184.20 |
| 10" | " | 230.00 | 34.60 | 264.60 |
| 12" | " | 330.00 | 48.40 | 378.40 |
| 14" | " | 460.00 | 60.50 | 520.50 |
| 16" | " | 550.00 | 80.50 | 630.50 |
| 18" | " | 810.00 | 120.00 | 930.00 |
| 20" | " | 960.00 | 120.00 | 1,080 |
| Tee | | | | |
| 4"x3" | EA. | 140.00 | 30.30 | 170.30 |
| 4"x4" | " | 150.00 | 30.30 | 180.30 |

# 02 SITEWORK

| UTILITIES | UNIT | MAT. | INST. | TOTAL |
|---|---|---|---|---|
| **02610.40 DUCTILE IRON PIPE** | | | | |
| 6"x3" | EA. | 170.00 | 34.60 | 204.60 |
| 6"x4" | " | 180.00 | 34.60 | 214.60 |
| 6"x6" | " | 200.00 | 34.60 | 234.60 |
| 8"x4" | " | 240.00 | 40.30 | 280.30 |
| 8"x6" | " | 280.00 | 40.30 | 320.30 |
| 8"x8" | " | 260.00 | 40.30 | 300.30 |
| 10"x4" | " | 320.00 | 48.40 | 368.40 |
| 10"x6" | " | 360.00 | 48.40 | 408.40 |
| 10"x8" | " | 370.00 | 48.40 | 418.40 |
| 10"x10" | " | 420.00 | 48.40 | 468.40 |
| 12"x4" | " | 510.00 | 60.50 | 570.50 |
| 12"x6" | " | 390.00 | 60.50 | 450.50 |
| 12"x8" | " | 430.00 | 60.50 | 490.50 |
| 12"x10" | " | 490.00 | 60.50 | 550.50 |
| 12"x12" | " | 520.00 | 64.50 | 584.50 |
| 14"x4" | " | 630.00 | 69.00 | 699.00 |
| 14"x6" | " | 680.00 | 69.00 | 749.00 |
| 14"x8" | " | 690.00 | 69.00 | 759.00 |
| 14"x10" | " | 710.00 | 69.00 | 779.00 |
| 14"x12" | " | 730.00 | 74.50 | 804.50 |
| 14"x14" | " | 720.00 | 74.50 | 794.50 |
| 16"x4" | " | 820.00 | 80.50 | 900.50 |
| 16"x6" | " | 840.00 | 80.50 | 920.50 |
| 16"x8" | " | 740.00 | 80.50 | 820.50 |
| 16"x10" | " | 760.00 | 80.50 | 840.50 |
| 16"x12" | " | 740.00 | 80.50 | 820.50 |
| 16"x14" | " | 780.00 | 80.50 | 860.50 |
| 16"x16" | " | 800.00 | 80.50 | 880.50 |
| 18"x6" | " | 950.00 | 88.00 | 1,038 |
| 18"x8" | " | 970.00 | 88.00 | 1,058 |
| 18"x10" | " | 990.00 | 88.00 | 1,078 |
| 18"x12" | " | 1,000 | 88.00 | 1,088 |
| 18"x14" | " | 1,120 | 88.00 | 1,208 |
| 18"x16" | " | 1,110 | 88.00 | 1,198 |
| 18"x18" | " | 1,260 | 88.00 | 1,348 |
| 20"x6" | " | 1,210 | 97.00 | 1,307 |
| 20"x8" | " | 1,220 | 97.00 | 1,317 |
| 20"x10" | " | 1,250 | 97.00 | 1,347 |
| 20"x12" | " | 1,270 | 97.00 | 1,367 |
| 20"x14" | " | 1,310 | 97.00 | 1,407 |
| 20"x16" | " | 1,520 | 97.00 | 1,617 |
| 20"x18" | " | 1,590 | 97.00 | 1,687 |
| 20"x20" | " | 1,630 | 97.00 | 1,727 |
| Cross | | | | |
| 4"x3" | EA. | 170.00 | 40.30 | 210.30 |
| 4"x4" | " | 190.00 | 40.30 | 230.30 |
| 6"x3" | " | 200.00 | 48.40 | 248.40 |
| 6"x4" | " | 210.00 | 48.40 | 258.40 |
| 6"x6" | " | 230.00 | 48.40 | 278.40 |
| 8"x4" | " | 270.00 | 54.00 | 324.00 |
| 8"x6" | " | 290.00 | 54.00 | 344.00 |
| 8"x8" | " | 320.00 | 54.00 | 374.00 |

| UTILITIES | UNIT | MAT. | INST. | TOTAL |
|---|---|---|---|---|
| **02610.40 DUCTILE IRON PIPE** | | | | |
| 10"x4" | EA. | 370.00 | 60.50 | 430.50 |
| 10"x6" | " | 390.00 | 60.50 | 450.50 |
| 10"x8" | " | 430.00 | 60.50 | 490.50 |
| 10"x10" | " | 510.00 | 60.50 | 570.50 |
| 12"x4" | " | 470.00 | 69.00 | 539.00 |
| 12"x6" | " | 530.00 | 69.00 | 599.00 |
| 12"x8" | " | 520.00 | 69.00 | 589.00 |
| 12"x10" | " | 600.00 | 74.50 | 674.50 |
| 12"x12" | " | 650.00 | 74.50 | 724.50 |
| 14"x4" | " | 610.00 | 80.50 | 690.50 |
| 14"x6" | " | 690.00 | 80.50 | 770.50 |
| 14"x8" | " | 720.00 | 80.50 | 800.50 |
| 14"x10" | " | 770.00 | 80.50 | 850.50 |
| 14"x12" | " | 830.00 | 88.00 | 918.00 |
| 14"x14" | " | 910.00 | 88.00 | 998.00 |
| 16"x4" | " | 800.00 | 97.00 | 897.00 |
| 16"x6" | " | 820.00 | 97.00 | 917.00 |
| 16"x8" | " | 870.00 | 97.00 | 967.00 |
| 16"x10" | " | 920.00 | 97.00 | 1,017 |
| 16"x12" | " | 960.00 | 97.00 | 1,057 |
| 16"x14" | " | 1,050 | 97.00 | 1,147 |
| 16"x16" | " | 1,110 | 97.00 | 1,207 |
| 18"x6" | " | 1,030 | 110.00 | 1,140 |
| 18"x8" | " | 1,060 | 110.00 | 1,170 |
| 18"x10" | " | 1,110 | 110.00 | 1,220 |
| 18"x12" | " | 1,170 | 110.00 | 1,280 |
| 18"x14" | " | 1,390 | 110.00 | 1,500 |
| 18"x16" | " | 1,480 | 110.00 | 1,590 |
| 18"x18" | " | 1,560 | 110.00 | 1,670 |
| 20"x6" | " | 1,240 | 120.00 | 1,360 |
| 20"x8" | " | 1,270 | 120.00 | 1,390 |
| 20"x10" | " | 1,330 | 120.00 | 1,450 |
| 20"x12" | " | 1,390 | 120.00 | 1,510 |
| 20"x14" | " | 1,460 | 120.00 | 1,580 |
| 20"x16" | " | 1,690 | 120.00 | 1,810 |
| 20"x18" | " | 1,800 | 120.00 | 1,920 |
| 20"x20" | " | 1,910 | 120.00 | 2,030 |
| **02610.60 PLASTIC PIPE** | | | | |
| PVC, class 150 pipe | | | | |
| 4" dia. | L.F. | 3.05 | 3.25 | 6.30 |
| 6" dia. | " | 5.75 | 3.55 | 9.30 |
| 8" dia. | " | 9.15 | 3.75 | 12.90 |
| 10" dia. | " | 13.00 | 4.10 | 17.10 |
| 12" dia. | " | 19.25 | 4.35 | 23.60 |
| Schedule 40 pipe | | | | |
| 1-1/2" dia. | L.F. | 1.70 | 1.40 | 3.10 |
| 2" dia. | " | 2.00 | 1.50 | 3.50 |
| 2-1/2" dia. | " | 2.20 | 1.60 | 3.80 |
| 3" dia. | " | 3.15 | 1.75 | 4.90 |
| 4" dia. | " | 4.55 | 2.00 | 6.55 |

| UTILITIES | UNIT | MAT. | INST. | TOTAL |
|---|---|---|---|---|
| **02610.60 PLASTIC PIPE** | | | | |
| 6" dia. | L.F. | 7.65 | 2.40 | 10.05 |
| 90 degree elbows | | | | |
| 1" | EA. | 1.15 | 4.05 | 5.20 |
| 1-1/2" | " | 1.90 | 4.05 | 5.95 |
| 2" | " | 2.15 | 4.40 | 6.55 |
| 2-1/2" | " | 5.45 | 4.85 | 10.30 |
| 3" | " | 6.55 | 5.40 | 11.95 |
| 4" | " | 18.10 | 6.05 | 24.15 |
| 6" | " | 40.80 | 8.05 | 48.85 |
| 45 degree elbows | | | | |
| 1" | EA. | 1.25 | 4.05 | 5.30 |
| 1-1/2" | " | 2.05 | 4.05 | 6.10 |
| 2" | " | 2.70 | 4.40 | 7.10 |
| 2-1/2" | " | 5.55 | 4.85 | 10.40 |
| 3" | " | 6.75 | 5.40 | 12.15 |
| 4" | " | 19.80 | 6.05 | 25.85 |
| 6" | " | 43.00 | 8.05 | 51.05 |
| Tees | | | | |
| 1" | EA. | 1.50 | 4.85 | 6.35 |
| 1-1/2" | " | 2.15 | 4.85 | 7.00 |
| 2" | " | 2.95 | 5.40 | 8.35 |
| 2-1/2" | " | 5.30 | 6.05 | 11.35 |
| 3" | " | 7.25 | 6.90 | 14.15 |
| 4" | " | 20.40 | 8.05 | 28.45 |
| 6" | " | 40.80 | 9.70 | 50.50 |
| Couplings | | | | |
| 1" | EA. | 0.68 | 4.05 | 4.73 |
| 1-1/2" | " | 0.79 | 4.05 | 4.84 |
| 2" | " | 1.35 | 4.40 | 5.75 |
| 2-1/2" | " | 2.70 | 4.85 | 7.55 |
| 3" | " | 5.10 | 5.40 | 10.50 |
| 4" | " | 11.30 | 6.05 | 17.35 |
| 6" | " | 14.70 | 8.05 | 22.75 |
| Drainage pipe | | | | |
| PVC schedule 80 | | | | |
| 1" dia. | L.F. | 1.05 | 1.40 | 2.45 |
| 1-1/2" dia. | " | 1.25 | 1.40 | 2.65 |
| ABS, 2" dia. | " | 1.60 | 1.50 | 3.10 |
| 2-1/2" dia. | " | 2.30 | 1.60 | 3.90 |
| 3" dia. | " | 2.70 | 1.75 | 4.45 |
| 4" dia. | " | 3.70 | 2.00 | 5.70 |
| 6" dia. | " | 6.15 | 2.40 | 8.55 |
| 8" dia. | " | 8.20 | 3.45 | 11.65 |
| 10" dia. | " | 10.95 | 4.10 | 15.05 |
| 12" dia. | " | 17.90 | 4.35 | 22.25 |
| 90 degree elbows | | | | |
| 1" | EA. | 1.85 | 4.05 | 5.90 |
| 1-1/2" | " | 2.30 | 4.05 | 6.35 |
| 2" | " | 2.80 | 4.40 | 7.20 |
| 2-1/2" | " | 6.65 | 4.85 | 11.50 |
| 3" | " | 6.85 | 5.40 | 12.25 |
| 4" | " | 12.10 | 6.05 | 18.15 |

| UTILITIES | UNIT | MAT. | INST. | TOTAL |
|---|---|---|---|---|
| **02610.60 PLASTIC PIPE** | | | | |
| 6" | EA. | 26.60 | 8.05 | 34.65 |
| 45 degree elbows | | | | |
| 1" | EA. | 3.00 | 4.05 | 7.05 |
| 1-1/2" | " | 3.80 | 4.05 | 7.85 |
| 2" | " | 4.75 | 4.40 | 9.15 |
| 2-1/2" | " | 8.90 | 4.85 | 13.75 |
| 3" | " | 9.45 | 5.40 | 14.85 |
| 4" | " | 17.90 | 6.05 | 23.95 |
| 6" | " | 41.60 | 8.05 | 49.65 |
| Tees | | | | |
| 1" | EA. | 1.95 | 4.85 | 6.80 |
| 1-1/2" | " | 6.25 | 4.85 | 11.10 |
| 2" | " | 7.60 | 5.40 | 13.00 |
| 2-1/2" | " | 8.90 | 6.05 | 14.95 |
| 3" | " | 9.70 | 6.90 | 16.60 |
| 4" | " | 18.50 | 8.05 | 26.55 |
| 6" | " | 36.40 | 9.70 | 46.10 |
| Couplings | | | | |
| 1" | EA. | 1.60 | 4.05 | 5.65 |
| 1-1/2" | " | 2.80 | 4.05 | 6.85 |
| 2" | " | 4.10 | 4.40 | 8.50 |
| 2-1/2" | " | 8.65 | 4.85 | 13.50 |
| 3" | " | 8.90 | 5.40 | 14.30 |
| 4" | " | 9.25 | 6.05 | 15.30 |
| 6" | " | 15.60 | 8.05 | 23.65 |
| Pressure pipe | | | | |
| PVC, class 200 pipe | | | | |
| 3/4" | L.F. | 1.00 | 1.20 | 2.20 |
| 1" | " | 1.15 | 1.25 | 2.40 |
| 1-1/4" | " | 1.25 | 1.35 | 2.60 |
| 1-1/2" | " | 1.55 | 1.40 | 2.95 |
| 2" | " | 1.65 | 1.50 | 3.15 |
| 2-1/2" | " | 2.40 | 1.60 | 4.00 |
| 3" | " | 3.05 | 1.75 | 4.80 |
| 4" | " | 5.30 | 2.00 | 7.30 |
| 6" | " | 10.05 | 2.40 | 12.45 |
| 8" | " | 15.25 | 3.75 | 19.00 |
| 90 degree elbows | | | | |
| 3/4" | EA. | 0.51 | 4.05 | 4.56 |
| 1" | " | 0.63 | 4.05 | 4.68 |
| 1-1/4" | " | 1.00 | 4.05 | 5.05 |
| 1-1/2" | " | 1.20 | 4.05 | 5.25 |
| 2" | " | 2.40 | 4.40 | 6.80 |
| 2-1/2" | " | 3.50 | 4.85 | 8.35 |
| 3" | " | 7.90 | 5.40 | 13.30 |
| 4" | " | 20.30 | 6.05 | 26.35 |
| 6" | " | 38.70 | 8.05 | 46.75 |
| 8" | " | 76.00 | 12.10 | 88.10 |
| 45 degree elbows | | | | |
| 3/4" | EA. | 0.75 | 4.05 | 4.80 |
| 1" | " | 0.95 | 4.05 | 5.00 |
| 1-1/4" | " | 1.40 | 4.05 | 5.45 |

| UTILITIES | UNIT | MAT. | INST. | TOTAL |
|---|---|---|---|---|
| **02610.60 PLASTIC PIPE** | | | | |
| 1-1/2" | EA. | 1.70 | 4.05 | 5.75 |
| 2" | " | 2.35 | 4.40 | 6.75 |
| 2-1/2" | " | 3.85 | 4.85 | 8.70 |
| 3" | " | 8.90 | 5.40 | 14.30 |
| 4" | " | 16.65 | 6.05 | 22.70 |
| 6" | " | 41.00 | 8.05 | 49.05 |
| 8" | " | 83.00 | 12.10 | 95.10 |
| Tees | | | | |
| 3/4" | EA. | 0.63 | 4.85 | 5.48 |
| 1" | " | 0.82 | 4.85 | 5.67 |
| 1-1/4" | " | 1.20 | 4.85 | 6.05 |
| 1-1/2" | " | 1.65 | 4.85 | 6.50 |
| 2" | " | 2.40 | 5.40 | 7.80 |
| 2-1/2" | " | 3.80 | 6.05 | 9.85 |
| 3" | " | 11.70 | 6.90 | 18.60 |
| 4" | " | 16.75 | 8.05 | 24.80 |
| 6" | " | 58.50 | 9.70 | 68.20 |
| 8" | " | 120.00 | 13.45 | 133.45 |
| Couplings | | | | |
| 3/4" | EA. | 0.39 | 4.05 | 4.44 |
| 1" | " | 0.58 | 4.05 | 4.63 |
| 1-1/4" | " | 0.75 | 4.05 | 4.80 |
| 1-1/2" | " | 0.82 | 4.05 | 4.87 |
| 2" | " | 1.15 | 4.40 | 5.55 |
| 2-1/2" | " | 2.40 | 4.85 | 7.25 |
| 3" | " | 3.80 | 5.40 | 9.20 |
| 4" | " | 5.35 | 5.40 | 10.75 |
| 6" | " | 16.75 | 6.05 | 22.80 |
| 8" | " | 30.00 | 8.05 | 38.05 |
| **02610.90 VITRIFIED CLAY PIPE** | | | | |
| Vitrified clay pipe, extra strength | | | | |
| 6" dia. | L.F. | 2.90 | 5.95 | 8.85 |
| 8" dia. | " | 3.45 | 6.25 | 9.70 |
| 10" dia. | " | 5.30 | 6.55 | 11.85 |
| 12" dia. | " | 7.60 | 8.70 | 16.30 |
| 15" dia. | " | 13.85 | 13.10 | 26.95 |
| 18" dia. | " | 20.80 | 14.55 | 35.35 |
| 24" dia. | " | 38.10 | 18.70 | 56.80 |
| 30" dia. | " | 64.50 | 26.20 | 90.70 |
| 36" dia. | " | 92.50 | 37.40 | 129.90 |
| **02630.10 TAPPING SADDLES & SLEEVES** | | | | |
| Tapping saddle, tap size to 2" | | | | |
| 4" saddle | EA. | 39.30 | 12.10 | 51.40 |
| 6" saddle | " | 46.20 | 15.10 | 61.30 |
| 8" saddle | " | 53.00 | 20.15 | 73.15 |
| 10" saddle | " | 62.50 | 24.20 | 86.70 |
| 12" saddle | " | 74.00 | 34.60 | 108.60 |
| 14" saddle | " | 83.00 | 48.40 | 131.40 |
| Tapping sleeve | | | | |

| UTILITIES | UNIT | MAT. | INST. | TOTAL |
|---|---|---|---|---|
| **02630.10 TAPPING SADDLES & SLEEVES** | | | | |
| 4x4 | EA. | 420.00 | 16.15 | 436.15 |
| 6x4 | " | 530.00 | 18.60 | 548.60 |
| 6x6 | " | 540.00 | 18.60 | 558.60 |
| 8x4 | " | 550.00 | 24.20 | 574.20 |
| 8x6 | " | 580.00 | 24.20 | 604.20 |
| 10x4 | " | 890.00 | 52.50 | 942.50 |
| 10x6 | " | 1,270 | 52.50 | 1,323 |
| 10x8 | " | 1,330 | 52.50 | 1,383 |
| 10x10 | " | 1,350 | 54.50 | 1,405 |
| 12x4 | " | 1,360 | 54.50 | 1,415 |
| 12x6 | " | 1,390 | 59.50 | 1,450 |
| 12x8 | " | 1,420 | 65.50 | 1,486 |
| 12x10 | " | 1,500 | 72.50 | 1,573 |
| 12x12 | " | 1,560 | 81.50 | 1,642 |
| Tapping valve, mechanical joint | | | | |
| 4" valve | EA. | 390.00 | 160.00 | 550.00 |
| 6" valve | " | 460.00 | 220.00 | 680.00 |
| 8" valve | " | 690.00 | 330.00 | 1,020 |
| 10" valve | " | 1,090 | 440.00 | 1,530 |
| 12" valve | " | 1,850 | 650.00 | 2,500 |
| Tap hole in pipe | | | | |
| 4" hole | EA. | 0.00 | 30.30 | 30.30 |
| 6" hole | " | 0.00 | 48.40 | 48.40 |
| 8" hole | " | 0.00 | 80.50 | 80.50 |
| 10" hole | " | 0.00 | 97.00 | 97.00 |
| 12" hole | " | 0.00 | 120.00 | 120.00 |
| **02640.15 VALVE BOXES** | | | | |
| Valve box, adjustable, for valves up to 20" | | | | |
| 3' deep | EA. | 98.00 | 8.05 | 106.05 |
| 4' deep | " | 98.00 | 9.70 | 107.70 |
| 5' deep | " | 98.00 | 12.10 | 110.10 |
| **02640.19 THRUST BLOCKS** | | | | |
| Thrust block, 3000# concrete | | | | |
| 1/4 c.y. | EA. | 64.50 | 51.50 | 116.00 |
| 1/2 c.y. | " | 90.00 | 62.00 | 152.00 |
| 3/4 c.y. | " | 100.00 | 100.00 | 200.00 |
| 1 c.y. | " | 130.00 | 210.00 | 340.00 |
| **02645.10 FIRE HYDRANTS** | | | | |
| Standard, 3 way post, 6" mechanical joint | | | | |
| 2' deep | EA. | 960.00 | 440.00 | 1,400 |
| 4' deep | " | 1,030 | 520.00 | 1,550 |
| 6' deep | " | 1,140 | 650.00 | 1,790 |
| 8' deep | " | 1,280 | 750.00 | 2,030 |

| UTILITIES | UNIT | MAT. | INST. | TOTAL |
|---|---|---|---|---|
| **02665.10 CHILLED WATER SYSTEMS** | | | | |
| Chilled water pipe, 2" thick insulation, w/casing | | | | |
| Align and tack weld on sleepers | | | | |
| 1-1/2" dia. | L.F. | 10.70 | 1.20 | 11.90 |
| 3" dia. | " | 17.20 | 1.85 | 19.05 |
| 4" dia. | " | 19.80 | 2.60 | 22.40 |
| 6" dia. | " | 22.60 | 3.25 | 25.85 |
| 8" dia. | " | 31.80 | 3.75 | 35.55 |
| 10" dia. | " | 40.60 | 4.35 | 44.95 |
| 12" dia. | " | 48.30 | 5.25 | 53.55 |
| 14" dia. | " | 65.50 | 5.70 | 71.20 |
| 16" dia. | " | 83.50 | 6.55 | 90.05 |
| Align and tack weld on trench bottom | | | | |
| 18" dia. | L.F. | 86.00 | 7.25 | 93.25 |
| 20" dia. | " | 110.00 | 8.15 | 118.15 |
| Preinsulated fittings | | | | |
| Align and tack weld on sleepers | | | | |
| Elbows | | | | |
| 1-1/2" | EA. | 220.00 | 21.35 | 241.35 |
| 3" | " | 280.00 | 34.20 | 314.20 |
| 4" | " | 360.00 | 42.70 | 402.70 |
| 6" | " | 510.00 | 57.00 | 567.00 |
| 8" | " | 720.00 | 68.50 | 788.50 |
| Tees | | | | |
| 1-1/2" | EA. | 340.00 | 22.75 | 362.75 |
| 3" | " | 480.00 | 38.00 | 518.00 |
| 4" | " | 360.00 | 48.80 | 408.80 |
| 6" | " | 790.00 | 68.50 | 858.50 |
| 8" | " | 1,050 | 85.50 | 1,136 |
| Reducers | | | | |
| 3" | EA. | 360.00 | 28.50 | 388.50 |
| 4" | " | 580.00 | 34.20 | 614.20 |
| 6" | " | 780.00 | 42.70 | 822.70 |
| 8" | " | 810.00 | 57.00 | 867.00 |
| Anchors, not including concrete | | | | |
| 4" | EA. | 160.00 | 42.70 | 202.70 |
| 6" | " | 230.00 | 42.70 | 272.70 |
| Align and tack weld on trench bottom | | | | |
| Elbows | | | | |
| 10" | EA. | 940.00 | 81.50 | 1,022 |
| 12" | " | 1,130 | 93.50 | 1,224 |
| 14" | " | 1,410 | 100.00 | 1,510 |
| 16" | " | 1,530 | 110.00 | 1,640 |
| 18" | " | 1,710 | 120.00 | 1,830 |
| 20" | " | 2,060 | 130.00 | 2,190 |
| Tees | | | | |
| 10" | EA. | 1,530 | 81.50 | 1,612 |
| 12" | " | 1,940 | 93.50 | 2,034 |
| 14" | " | 2,060 | 100.00 | 2,160 |
| 16" | " | 2,180 | 110.00 | 2,290 |
| 18" | " | 2,470 | 120.00 | 2,590 |
| 20" | " | 2,830 | 130.00 | 2,960 |
| Reducers | | | | |

| UTILITIES | UNIT | MAT. | INST. | TOTAL |
|---|---|---|---|---|
| **02665.10 CHILLED WATER SYSTEMS** | | | | |
| 10" | EA. | 1,170 | 54.50 | 1,225 |
| 12" | " | 1,410 | 59.50 | 1,470 |
| 14" | " | 1,530 | 65.50 | 1,596 |
| 16" | " | 1,880 | 72.50 | 1,953 |
| 18" | " | 2,060 | 81.50 | 2,142 |
| 20" | " | 2,180 | 93.50 | 2,274 |
| Anchors, not including concrete | | | | |
| 10" | EA. | 330.00 | 54.50 | 384.50 |
| 12" | " | 360.00 | 59.50 | 419.50 |
| 14" | " | 410.00 | 65.50 | 475.50 |
| 16" | " | 580.00 | 72.50 | 652.50 |
| 18" | " | 810.00 | 81.50 | 891.50 |
| 20" | " | 1,100 | 93.50 | 1,194 |
| **02670.10 WELLS** | | | | |
| Domestic water, drilled and cased | | | | |
| 4" dia. | L.F. | 9.80 | 130.00 | 139.80 |
| 6" dia. | " | 9.95 | 150.00 | 159.95 |
| 8" dia. | " | 12.70 | 190.00 | 202.70 |
| **02685.10 GAS DISTRIBUTION** | | | | |
| Gas distribution lines | | | | |
| Polyethylene, 60 psi coils | | | | |
| 1-1/4" dia. | L.F. | 0.81 | 2.30 | 3.11 |
| 1-1/2" dia. | " | 1.10 | 2.45 | 3.55 |
| 2" dia. | " | 1.40 | 2.85 | 4.25 |
| 3" dia. | " | 2.70 | 3.40 | 6.10 |
| 30' pipe lengths | | | | |
| 3" dia. | L.F. | 2.90 | 3.80 | 6.70 |
| 4" dia. | " | 4.55 | 4.25 | 8.80 |
| 6" dia. | " | 7.25 | 5.70 | 12.95 |
| 8" dia. | " | 13.40 | 6.85 | 20.25 |
| Steel, schedule 40, plain end | | | | |
| 1" dia. | L.F. | 3.05 | 2.85 | 5.90 |
| 2" dia. | " | 4.20 | 3.10 | 7.30 |
| 3" dia. | " | 7.80 | 3.40 | 11.20 |
| 4" dia. | " | 10.70 | 8.70 | 19.40 |
| 5" dia. | " | 21.00 | 9.35 | 30.35 |
| 6" dia. | " | 23.90 | 10.90 | 34.80 |
| 8" dia. | " | 35.00 | 11.90 | 46.90 |
| Natural gas meters, direct digital reading, threaded | | | | |
| 250 cfh @ 5 lbs | EA. | 110.00 | 68.50 | 178.50 |
| 425 cfh @ 10 lbs | " | 260.00 | 68.50 | 328.50 |
| 800 cfh @ 20 lbs | " | 370.00 | 85.50 | 455.50 |
| 1000 cfh @ 25 lbs | " | 1,100 | 85.50 | 1,186 |
| 1,400 cfh @ 100 lbs | " | 2,450 | 110.00 | 2,560 |
| 2,300 cfh @ 100 lbs | " | 3,500 | 170.00 | 3,670 |
| 5,000 cfh @ 100 lbs | " | 5,130 | 340.00 | 5,470 |
| Gas pressure regulators | | | | |
| Threaded | | | | |
| 3/4" | EA. | 45.50 | 42.70 | 88.20 |

# 02 SITEWORK

| UTILITIES | UNIT | MAT. | INST. | TOTAL |
|---|---|---|---|---|
| **02685.10 GAS DISTRIBUTION** | | | | |
| 1" | EA. | 47.80 | 57.00 | 104.80 |
| 1-1/4" | " | 50.00 | 57.00 | 107.00 |
| 1-1/2" | " | 330.00 | 57.00 | 387.00 |
| 2" | " | 340.00 | 68.50 | 408.50 |
| Flanged | | | | |
| 3" | EA. | 960.00 | 85.50 | 1,046 |
| 4" | " | 1,340 | 110.00 | 1,450 |
| **02690.10 STORAGE TANKS** | | | | |
| Oil storage tank, underground | | | | |
| Steel | | | | |
| 500 gals | EA. | 510.00 | 160.00 | 670.00 |
| 1,000 gals | " | 890.00 | 220.00 | 1,110 |
| 4,000 gals | " | 2,490 | 440.00 | 2,930 |
| 5,000 gals | " | 3,340 | 650.00 | 3,990 |
| 10,000 gals | " | 4,760 | 1,310 | 6,070 |
| Fiberglass, double wall | | | | |
| 550 gals | EA. | 1,920 | 220.00 | 2,140 |
| 1,000 gals | " | 2,490 | 220.00 | 2,710 |
| 2,000 gals | " | 2,890 | 330.00 | 3,220 |
| 4,000 gals | " | 3,850 | 650.00 | 4,500 |
| 6,000 gals | " | 4,420 | 870.00 | 5,290 |
| 8,000 gals | " | 6,000 | 1,310 | 7,310 |
| 10,000 gals | " | 7,590 | 1,630 | 9,220 |
| 12,000 gals | " | 9,060 | 2,180 | 11,240 |
| 15,000 gals | " | 10,640 | 2,910 | 13,550 |
| 20,000 gals | " | 13,590 | 3,270 | 16,860 |
| Above ground | | | | |
| Steel | | | | |
| 275 gals | EA. | 200.00 | 130.00 | 330.00 |
| 500 gals | " | 670.00 | 220.00 | 890.00 |
| 1,000 gals | " | 1,000 | 260.00 | 1,260 |
| 1,500 gals | " | 1,640 | 330.00 | 1,970 |
| 2,000 gals | " | 2,090 | 440.00 | 2,530 |
| 5,000 gals | " | 5,550 | 650.00 | 6,200 |
| Fill cap | " | 27.20 | 34.20 | 61.40 |
| Vent cap | " | 27.20 | 34.20 | 61.40 |
| Level indicator | " | 62.50 | 34.20 | 96.70 |
| **02695.80 STEAM METERS** | | | | |
| In-line turbine, direct reading, 300 lb, flanged | | | | |
| 2" | EA. | 3,030 | 42.70 | 3,073 |
| 3" | " | 3,270 | 57.00 | 3,327 |
| 4" | " | 3,620 | 68.50 | 3,689 |
| Threaded, 2" | | | | |
| 5" line | EA. | 5,710 | 340.00 | 6,050 |
| 6" line | " | 5,830 | 340.00 | 6,170 |
| 8" line | " | 5,890 | 340.00 | 6,230 |
| 10" line | " | 6,300 | 340.00 | 6,640 |
| 12" line | " | 6,410 | 340.00 | 6,750 |

# 02 SITEWORK

| UTILITIES | UNIT | MAT. | INST. | TOTAL |
|---|---|---|---|---|
| **02695.80 STEAM METERS** | | | | |
| 14" line | EA. | 6,650 | 340.00 | 6,990 |
| 16" line | " | 7,000 | 340.00 | 7,340 |

| SEWERAGE AND DRAINAGE | UNIT | MAT. | INST. | TOTAL |
|---|---|---|---|---|
| **02720.10 CATCH BASINS** | | | | |
| Standard concrete catch basin | | | | |
| Cast in place, 3'8" x 3'8", 6" thick wall | | | | |
| 2' deep | EA. | 200.00 | 330.00 | 530.00 |
| 3' deep | " | 270.00 | 330.00 | 600.00 |
| 4' deep | " | 360.00 | 440.00 | 800.00 |
| 5' deep | " | 420.00 | 440.00 | 860.00 |
| 6' deep | " | 470.00 | 520.00 | 990.00 |
| 4'x4', 8" thick wall, cast in place | | | | |
| 2' deep | EA. | 210.00 | 330.00 | 540.00 |
| 3' deep | " | 300.00 | 330.00 | 630.00 |
| 4' deep | " | 400.00 | 440.00 | 840.00 |
| 5' deep | " | 460.00 | 440.00 | 900.00 |
| 6' deep | " | 510.00 | 520.00 | 1,030 |
| Frames and covers, cast iron | | | | |
| Round | | | | |
| 24" dia. | EA. | 180.00 | 60.50 | 240.50 |
| 26" dia. | " | 200.00 | 60.50 | 260.50 |
| 28" dia. | " | 240.00 | 60.50 | 300.50 |
| Rectangular | | | | |
| 23"x23" | EA. | 160.00 | 60.50 | 220.50 |
| 27"x20" | " | 190.00 | 60.50 | 250.50 |
| 24"x24" | " | 180.00 | 60.50 | 240.50 |
| 26"x26" | " | 200.00 | 60.50 | 260.50 |
| Curb inlet frames and covers | | | | |
| 27"x27" | EA. | 380.00 | 60.50 | 440.50 |
| 24"x36" | " | 230.00 | 60.50 | 290.50 |
| 24"x25" | " | 210.00 | 60.50 | 270.50 |
| 24"x22" | " | 180.00 | 60.50 | 240.50 |
| 20"x22" | " | 240.00 | 60.50 | 300.50 |
| Airfield catch basin frame and grating, galvanized | | | | |
| 2'x4' | EA. | 450.00 | 60.50 | 510.50 |
| 2'x2' | " | 310.00 | 60.50 | 370.50 |
| **02720.40 STORM DRAINAGE** | | | | |
| Concrete pipe | | | | |
| Plain, bell and spigot joint, class II | | | | |
| 6" pipe | L.F. | 3.40 | 5.95 | 9.35 |
| 8" pipe | " | 3.75 | 6.55 | 10.30 |

## SEWERAGE AND DRAINAGE

### 02720.40 STORM DRAINAGE

| SEWERAGE AND DRAINAGE | UNIT | MAT. | INST. | TOTAL |
|---|---|---|---|---|
| 10" pipe | L.F. | 3.85 | 6.90 | 10.75 |
| 12" pipe | " | 4.55 | 7.25 | 11.80 |
| 15" pipe | " | 5.50 | 7.70 | 13.20 |
| 18" pipe | " | 7.45 | 8.15 | 15.60 |
| 21" pipe | " | 9.35 | 8.70 | 18.05 |
| 24" pipe | " | 11.65 | 9.35 | 21.00 |
| Reinforced, class III, tongue and groove joint | | | | |
| 12" pipe | L.F. | 7.95 | 7.25 | 15.20 |
| 15" pipe | " | 8.15 | 7.70 | 15.85 |
| 18" pipe | " | 10.30 | 8.15 | 18.45 |
| 21" pipe | " | 13.60 | 8.70 | 22.30 |
| 24" pipe | " | 17.00 | 9.35 | 26.35 |
| 27" pipe | " | 20.40 | 10.05 | 30.45 |
| 30" pipe | " | 22.65 | 10.90 | 33.55 |
| 36" pipe | " | 34.00 | 11.90 | 45.90 |
| 42" pipe | " | 44.20 | 13.10 | 57.30 |
| 48" pipe | " | 59.00 | 14.55 | 73.55 |
| 54" pipe | " | 77.00 | 16.35 | 93.35 |
| 60" pipe | " | 90.50 | 18.70 | 109.20 |
| 66" pipe | " | 100.00 | 21.80 | 121.80 |
| 72" pipe | " | 140.00 | 26.20 | 166.20 |
| Flared end-section, concrete | | | | |
| 12" pipe | L.F. | 40.40 | 7.25 | 47.65 |
| 15" pipe | " | 47.60 | 7.70 | 55.30 |
| 18" pipe | " | 53.00 | 8.15 | 61.15 |
| 24" pipe | " | 65.50 | 9.35 | 74.85 |
| 30" pipe | " | 80.50 | 10.90 | 91.40 |
| 36" pipe | " | 110.00 | 11.90 | 121.90 |
| 42" pipe | " | 120.00 | 13.10 | 133.10 |
| 48" pipe | " | 130.00 | 14.55 | 144.55 |
| 54" pipe | " | 140.00 | 16.35 | 156.35 |
| Corrugated metal pipe, coated, paved invert | | | | |
| 16 ga. | | | | |
| 8" pipe | L.F. | 5.90 | 4.35 | 10.25 |
| 10" pipe | " | 7.95 | 4.50 | 12.45 |
| 12" pipe | " | 9.15 | 4.65 | 13.80 |
| 15" pipe | " | 11.30 | 5.05 | 16.35 |
| 18" pipe | " | 13.70 | 5.45 | 19.15 |
| 21" pipe | " | 16.15 | 5.95 | 22.10 |
| 24" pipe | " | 19.65 | 6.55 | 26.20 |
| 30" pipe | " | 26.60 | 7.25 | 33.85 |
| 36" pipe | " | 37.00 | 8.15 | 45.15 |
| 12 ga., 48" pipe | " | 55.50 | 9.35 | 64.85 |
| 10 ga. | | | | |
| 60" pipe | L.F. | 81.00 | 10.90 | 91.90 |
| 72" pipe | " | 110.00 | 13.10 | 123.10 |
| Galvanized or aluminum, plain | | | | |
| 16 ga. | | | | |
| 8" pipe | L.F. | 5.20 | 4.35 | 9.55 |
| 10" pipe | " | 6.95 | 4.50 | 11.45 |
| 12" pipe | " | 8.10 | 4.65 | 12.75 |
| 15" pipe | " | 10.60 | 5.05 | 15.65 |

| SEWERAGE AND DRAINAGE | UNIT | MAT. | INST. | TOTAL |
|---|---|---|---|---|
| **02720.40 STORM DRAINAGE** | | | | |
| 18" pipe | L.F. | 12.70 | 5.45 | 18.15 |
| 24" pipe | " | 17.90 | 6.55 | 24.45 |
| 30" pipe | " | 25.40 | 7.25 | 32.65 |
| 36" pipe | " | 32.30 | 8.15 | 40.45 |
| 12 ga., 48" pipe | " | 53.00 | 9.35 | 62.35 |
| 10 ga., 60" pipe | " | 75.00 | 10.90 | 85.90 |
| Galvanized or aluminum, coated oval arch | | | | |
| 16 ga. | | | | |
| 17" x 13" | L.F. | 15.70 | 5.95 | 21.65 |
| 21" x 15" | " | 19.05 | 6.55 | 25.60 |
| 14 ga. | | | | |
| 28" x 20" | L.F. | 29.70 | 7.25 | 36.95 |
| 35" x 24" | " | 38.10 | 9.35 | 47.45 |
| 12 ga. | | | | |
| 42" x 29" | L.F. | 52.00 | 10.90 | 62.90 |
| 57" x 38" | " | 69.50 | 13.10 | 82.60 |
| 64" x 43" | " | 81.00 | 13.75 | 94.75 |
| Oval arch culverts, plain | | | | |
| 16 ga. | | | | |
| 17" x 13" | L.F. | 9.20 | 5.95 | 15.15 |
| 21" x 15" | " | 11.00 | 6.55 | 17.55 |
| 14 ga. | | | | |
| 28" x 20" | L.F. | 24.75 | 7.25 | 32.00 |
| 35" x 24" | " | 30.90 | 9.35 | 40.25 |
| 12 ga. | | | | |
| 57" x 38" | L.F. | 46.20 | 10.90 | 57.10 |
| 64" x 43" | " | 52.00 | 13.10 | 65.10 |
| 71" x 47" | " | 81.00 | 13.75 | 94.75 |
| Nestable corrugated metal pipe | | | | |
| 16 ga. | | | | |
| 10" pipe | L.F. | 7.85 | 4.50 | 12.35 |
| 12" pipe | " | 10.45 | 4.65 | 15.10 |
| 15" pipe | " | 13.10 | 5.05 | 18.15 |
| 18" pipe | " | 15.45 | 5.45 | 20.90 |
| 24" pipe | " | 21.40 | 6.55 | 27.95 |
| 30" pipe | " | 25.00 | 7.25 | 32.25 |
| 14 ga., 36" pipe | " | 29.70 | 8.15 | 37.85 |
| Headwalls, cast in place, 30 deg wingwall | | | | |
| 12" pipe | EA. | 250.00 | 77.00 | 327.00 |
| 15" pipe | " | 310.00 | 77.00 | 387.00 |
| 18" pipe | " | 380.00 | 88.00 | 468.00 |
| 24" pipe | " | 570.00 | 88.00 | 658.00 |
| 30" pipe | " | 680.00 | 100.00 | 780.00 |
| 36" pipe | " | 740.00 | 150.00 | 890.00 |
| 42" pipe | " | 920.00 | 150.00 | 1,070 |
| 48" pipe | " | 980.00 | 210.00 | 1,190 |
| 54" pipe | " | 1,110 | 260.00 | 1,370 |
| 60" pipe | " | 1,330 | 310.00 | 1,640 |
| 4" cleanout for storm drain | | | | |
| 4" pipe | EA. | 350.00 | 30.30 | 380.30 |
| 6" pipe | " | 420.00 | 30.30 | 450.30 |
| 8" pipe | " | 580.00 | 30.30 | 610.30 |

## SEWERAGE AND DRAINAGE

| SEWERAGE AND DRAINAGE | UNIT | MAT. | INST. | TOTAL |
|---|---|---|---|---|
| **02720.40 STORM DRAINAGE** | | | | |
| Connect new drain line | | | | |
| To existing manhole | EA. | 75.00 | 80.50 | 155.50 |
| To new manhole | " | 63.50 | 48.40 | 111.90 |
| **02720.70 UNDERDRAIN** | | | | |
| Drain tile, clay | | | | |
| 6" pipe | L.F. | 2.20 | 2.90 | 5.10 |
| 8" pipe | " | 4.40 | 3.05 | 7.45 |
| 12" pipe | " | 7.65 | 3.25 | 10.90 |
| Porous concrete, standard strength | | | | |
| 6" pipe | L.F. | 2.85 | 2.90 | 5.75 |
| 8" pipe | " | 3.15 | 3.05 | 6.20 |
| 12" pipe | " | 6.45 | 3.25 | 9.70 |
| 15" pipe | " | 7.80 | 3.65 | 11.45 |
| 18" pipe | " | 10.20 | 4.35 | 14.55 |
| Corrugated metal pipe, perforated type | | | | |
| 6" pipe | L.F. | 5.55 | 3.25 | 8.80 |
| 8" pipe | " | 6.45 | 3.45 | 9.90 |
| 10" pipe | " | 7.65 | 3.65 | 11.30 |
| 12" pipe | " | 10.75 | 3.85 | 14.60 |
| 18" pipe | " | 13.85 | 4.10 | 17.95 |
| Perforated clay pipe | | | | |
| 6" pipe | L.F. | 2.15 | 3.75 | 5.90 |
| 8" pipe | " | 3.05 | 3.85 | 6.90 |
| 12" pipe | " | 6.70 | 3.95 | 10.65 |
| Drain tile, concrete | | | | |
| 6" pipe | L.F. | 2.05 | 2.90 | 4.95 |
| 8" pipe | " | 3.15 | 3.05 | 6.20 |
| 12" pipe | " | 6.35 | 3.25 | 9.60 |
| Perforated rigid PVC underdrain pipe | | | | |
| 4" pipe | L.F. | 0.89 | 2.20 | 3.09 |
| 6" pipe | " | 1.65 | 2.60 | 4.25 |
| 8" pipe | " | 2.95 | 2.90 | 5.85 |
| 10" pipe | " | 3.65 | 3.25 | 6.90 |
| 12" pipe | " | 6.20 | 3.75 | 9.95 |
| Underslab drainage, crushed stone | | | | |
| 3" thick | S.F. | 0.17 | 0.44 | 0.61 |
| 4" thick | " | 0.21 | 0.50 | 0.71 |
| 6" thick | " | 0.23 | 0.55 | 0.78 |
| 8" thick | " | 0.28 | 0.57 | 0.85 |
| Plastic filter fabric for drain lines | " | 0.04 | 0.24 | 0.28 |
| Gravel fill in trench, crushed or bank run, 1/2" to 3/4" | C.Y. | 16.40 | 32.70 | 49.10 |
| **02730.10 SANITARY SEWERS** | | | | |
| Clay | | | | |
| 6" pipe | L.F. | 3.85 | 4.35 | 8.20 |
| 8" pipe | " | 4.75 | 4.65 | 9.40 |
| 10" pipe | " | 6.90 | 5.05 | 11.95 |
| 12" pipe | " | 9.10 | 5.45 | 14.55 |
| PVC | | | | |

# 02 SITEWORK

| SEWERAGE AND DRAINAGE | UNIT | MAT. | INST. | TOTAL |
|---|---|---|---|---|
| **02730.10 SANITARY SEWERS** | | | | |
| 4" pipe | L.F. | 0.88 | 3.25 | 4.13 |
| 6" pipe | " | 1.80 | 3.45 | 5.25 |
| 8" pipe | " | 2.45 | 3.65 | 6.10 |
| 10" pipe | " | 3.95 | 3.85 | 7.80 |
| 12" pipe | " | 6.00 | 4.10 | 10.10 |
| Cleanout | | | | |
| 4" pipe | EA. | 4.45 | 30.30 | 34.75 |
| 6" pipe | " | 9.45 | 30.30 | 39.75 |
| 8" pipe | " | 28.80 | 30.30 | 59.10 |
| Connect new sewer line | | | | |
| To existing manhole | EA. | 52.50 | 80.50 | 133.00 |
| To new manhole | " | 35.00 | 48.40 | 83.40 |
| **02740.10 DRAINAGE FIELDS** | | | | |
| Perforated PVC pipe, for drain field | | | | |
| 4" pipe | L.F. | 0.90 | 2.90 | 3.80 |
| 6" pipe | " | 1.70 | 3.10 | 4.80 |
| **02740.50 SEPTIC TANKS** | | | | |
| Septic tank, precast concrete | | | | |
| 1000 gals | EA. | 520.00 | 220.00 | 740.00 |
| 2000 gals | " | 980.00 | 330.00 | 1,310 |
| 5000 gals | " | 4,270 | 650.00 | 4,920 |
| 25,000 gals | " | 17,320 | 2,620 | 19,940 |
| 40,000 gals | " | 24,830 | 4,360 | 29,190 |
| Leaching pit, precast concrete, 72" diameter | | | | |
| 3' deep | EA. | 380.00 | 160.00 | 540.00 |
| 6' deep | " | 490.00 | 190.00 | 680.00 |
| 8' deep | " | 600.00 | 220.00 | 820.00 |
| **02760.10 PIPELINE RESTORATION** | | | | |
| Relining existing water main | | | | |
| 6" dia. | L.F. | 4.75 | 19.05 | 23.80 |
| 8" dia. | " | 5.35 | 20.05 | 25.40 |
| 10" dia. | " | 6.00 | 21.15 | 27.15 |
| 12" dia. | " | 6.55 | 22.40 | 28.95 |
| 14" dia. | " | 7.00 | 23.80 | 30.80 |
| 16" dia. | " | 7.60 | 25.40 | 33.00 |
| 18" dia. | " | 8.20 | 27.20 | 35.40 |
| 20" dia. | " | 9.05 | 29.30 | 38.35 |
| 24" dia. | " | 9.60 | 31.70 | 41.30 |
| 36" dia. | " | 10.45 | 38.10 | 48.55 |
| 48" dia. | " | 11.75 | 42.30 | 54.05 |
| 72" dia. | " | 14.85 | 47.60 | 62.45 |
| Replacing in line gate valves | | | | |
| 6" valve | EA. | 480.00 | 250.00 | 730.00 |
| 8" valve | " | 740.00 | 320.00 | 1,060 |
| 10" valve | " | 1,120 | 380.00 | 1,500 |
| 12" valve | " | 1,940 | 480.00 | 2,420 |
| 16" valve | " | 4,400 | 540.00 | 4,940 |
| 18" valve | " | 6,660 | 630.00 | 7,290 |

## SEWERAGE AND DRAINAGE

### 02760.10 PIPELINE RESTORATION

| SEWERAGE AND DRAINAGE | UNIT | MAT. | INST. | TOTAL |
|---|---|---|---|---|
| 20" valve | EA. | 9,150 | 760.00 | 9,910 |
| 24" valve | " | 13,080 | 950.00 | 14,030 |
| 36" valve | " | 35,660 | 1,270 | 36,930 |

## SITE IMPROVEMENTS

### 02810.40 LAWN IRRIGATION

| SITE IMPROVEMENTS | UNIT | MAT. | INST. | TOTAL |
|---|---|---|---|---|
| Residential system, complete | | | | |
| Minimum | ACRE | | | 10,729 |
| Maximum | " | | | 19,825 |
| Commercial system, complete | | | | |
| Minimum | ACRE | | | 15,743 |
| Maximum | " | | | 24,490 |
| Components | | | | |
| Pop-up head | EA. | 9.30 | 15.10 | 24.40 |
| Impact head | " | 9.30 | 15.10 | 24.40 |
| Rotary impact head | " | 25.70 | 30.30 | 56.00 |
| Shrub head | " | 2.45 | 15.10 | 17.55 |
| Hose bibb | " | 4.55 | 30.30 | 34.85 |

### 02830.10 CHAIN LINK FENCE

| SITE IMPROVEMENTS | UNIT | MAT. | INST. | TOTAL |
|---|---|---|---|---|
| Chain link fence, 9 ga., galvanized, with posts 10' o.c. | | | | |
| 4' high | L.F. | 4.15 | 1.75 | 5.90 |
| 5' high | " | 5.55 | 2.20 | 7.75 |
| 6' high | " | 6.30 | 3.00 | 9.30 |
| 7' high | " | 7.15 | 3.70 | 10.85 |
| 8' high | " | 8.25 | 4.85 | 13.10 |
| For barbed wire with hangers, add | | | | |
| 3 strand | L.F. | 1.55 | 1.20 | 2.75 |
| 6 strand | " | 2.65 | 2.00 | 4.65 |
| Corner or gate post, 3" post | | | | |
| 4' high | EA. | 48.50 | 8.05 | 56.55 |
| 5' high | " | 53.50 | 8.95 | 62.45 |
| 6' high | " | 59.50 | 10.50 | 70.00 |
| 7' high | " | 71.50 | 12.10 | 83.60 |
| 8' high | " | 75.00 | 13.45 | 88.45 |
| 4" post | | | | |
| 4' high | EA. | 83.00 | 8.95 | 91.95 |
| 5' high | " | 98.00 | 10.50 | 108.50 |
| 6' high | " | 110.00 | 12.10 | 122.10 |
| 7' high | " | 120.00 | 13.45 | 133.45 |
| 8' high | " | 130.00 | 15.10 | 145.10 |
| Gate with gate posts, galvanized, 3' wide | | | | |

| SITE IMPROVEMENTS | UNIT | MAT. | INST. | TOTAL |
|---|---|---|---|---|
| **02830.10 CHAIN LINK FENCE** | | | | |
| 4' high | EA. | 53.50 | 60.50 | 114.00 |
| 5' high | " | 69.00 | 80.50 | 149.50 |
| 6' high | " | 82.50 | 80.50 | 163.00 |
| 7' high | " | 99.00 | 120.00 | 219.00 |
| 8' high | " | 110.00 | 120.00 | 230.00 |
| Fabric, galvanized chain link, 2" mesh, 9 ga. | | | | |
| 4' high | L.F. | 1.90 | 0.81 | 2.71 |
| 5' high | " | 2.30 | 0.97 | 3.27 |
| 6' high | " | 2.40 | 1.20 | 3.60 |
| 8' high | " | 3.70 | 1.60 | 5.30 |
| Line post, no rail fitting, galvanized, 2-1/2" dia. | | | | |
| 4' high | EA. | 14.30 | 6.90 | 21.20 |
| 5' high | " | 15.60 | 7.55 | 23.15 |
| 6' high | " | 17.10 | 8.05 | 25.15 |
| 7' high | " | 19.40 | 9.70 | 29.10 |
| 8' high | " | 21.65 | 12.10 | 33.75 |
| 1-7/8" H beam | | | | |
| 4' high | EA. | 19.65 | 6.90 | 26.55 |
| 5' high | " | 21.95 | 7.55 | 29.50 |
| 6' high | " | 26.30 | 8.05 | 34.35 |
| 7' high | " | 29.80 | 9.70 | 39.50 |
| 8' high | " | 32.20 | 12.10 | 44.30 |
| 2-1/4" H beam | | | | |
| 4' high | EA. | 14.30 | 6.90 | 21.20 |
| 5' high | " | 17.85 | 7.55 | 25.40 |
| 6' high | " | 20.50 | 8.05 | 28.55 |
| 7' high | " | 23.80 | 9.70 | 33.50 |
| 8' high | " | 27.60 | 12.10 | 39.70 |
| Vinyl coated, 9 ga., with posts 10' o.c. | | | | |
| 4' high | L.F. | 4.50 | 1.75 | 6.25 |
| 5' high | " | 5.35 | 2.20 | 7.55 |
| 6' high | " | 6.40 | 3.00 | 9.40 |
| 7' high | " | 7.00 | 3.70 | 10.70 |
| 8' high | " | 7.95 | 4.85 | 12.80 |
| For barbed wire w/hangers, add | | | | |
| 3 strand | L.F. | 1.70 | 1.20 | 2.90 |
| 6 Strand | " | 2.70 | 2.00 | 4.70 |
| Corner, or gate post, 4' high | | | | |
| 3" dia. | EA. | 59.50 | 8.05 | 67.55 |
| 4" dia. | " | 91.50 | 8.05 | 99.55 |
| 6" dia. | " | 110.00 | 9.70 | 119.70 |
| Gate, with posts, 3' wide | | | | |
| 4' high | EA. | 65.00 | 60.50 | 125.50 |
| 5' high | " | 77.00 | 80.50 | 157.50 |
| 6' high | " | 89.00 | 80.50 | 169.50 |
| 7' high | " | 100.00 | 120.00 | 220.00 |
| 8' high | " | 120.00 | 120.00 | 240.00 |
| Line post, no rail fitting, 2-1/2" dia. | | | | |
| 4' high | EA. | 25.10 | 6.90 | 32.00 |
| 5' high | " | 34.10 | 7.55 | 41.65 |
| 6' high | " | 41.00 | 8.05 | 49.05 |
| 7' high | " | 48.30 | 9.70 | 58.00 |

## SITE IMPROVEMENTS

### 02830.10 CHAIN LINK FENCE

| | UNIT | MAT. | INST. | TOTAL |
|---|---|---|---|---|
| 8' high | EA. | 53.00 | 12.10 | 65.10 |
| Corner post, no top rail fitting, 4" dia. | | | | |
| 4' high | EA. | 100.00 | 8.05 | 108.05 |
| 5' high | " | 110.00 | 8.95 | 118.95 |
| 6' high | " | 130.00 | 10.50 | 140.50 |
| 7' high | " | 140.00 | 12.10 | 152.10 |
| 8' high | " | 150.00 | 13.45 | 163.45 |
| Fabric, vinyl, chain link, 2" mesh, 9 ga. | | | | |
| 4' high | L.F. | 3.45 | 0.81 | 4.26 |
| 5' high | " | 4.15 | 0.97 | 5.12 |
| 6' high | " | 4.85 | 1.20 | 6.05 |
| 8' high | " | 6.55 | 1.60 | 8.15 |
| Swing gates, galvanized, 4' high | | | | |
| Single gate | | | | |
| 3' wide | EA. | 120.00 | 60.50 | 180.50 |
| 4' wide | " | 140.00 | 60.50 | 200.50 |
| Double gate | | | | |
| 10' wide | EA. | 320.00 | 97.00 | 417.00 |
| 12' wide | " | 350.00 | 97.00 | 447.00 |
| 14' wide | " | 360.00 | 97.00 | 457.00 |
| 16' wide | " | 400.00 | 97.00 | 497.00 |
| 18' wide | " | 430.00 | 140.00 | 570.00 |
| 20' wide | " | 450.00 | 140.00 | 590.00 |
| 22' wide | " | 500.00 | 140.00 | 640.00 |
| 24' wide | " | 510.00 | 160.00 | 670.00 |
| 26' wide | " | 540.00 | 160.00 | 700.00 |
| 28' wide | " | 570.00 | 190.00 | 760.00 |
| 30' wide | " | 610.00 | 190.00 | 800.00 |
| 5' high | | | | |
| Single gate | | | | |
| 3' wide | EA. | 130.00 | 80.50 | 210.50 |
| 4' wide | " | 150.00 | 80.50 | 230.50 |
| Double gate | | | | |
| 10' wide | EA. | 350.00 | 120.00 | 470.00 |
| 12' wide | " | 370.00 | 120.00 | 490.00 |
| 14' wide | " | 630.00 | 120.00 | 750.00 |
| 16' wide | " | 420.00 | 120.00 | 540.00 |
| 18' wide | " | 430.00 | 140.00 | 570.00 |
| 20' wide | " | 490.00 | 140.00 | 630.00 |
| 22' wide | " | 510.00 | 140.00 | 650.00 |
| 24' wide | " | 530.00 | 160.00 | 690.00 |
| 26' wide | " | 550.00 | 160.00 | 710.00 |
| 28' wide | " | 620.00 | 190.00 | 810.00 |
| 30' wide | " | 640.00 | 190.00 | 830.00 |
| 6' high | | | | |
| Single gate | | | | |
| 3' wide | EA. | 150.00 | 80.50 | 230.50 |
| 4' wide | " | 160.00 | 80.50 | 240.50 |
| Double gate | | | | |
| 10' wide | EA. | 360.00 | 120.00 | 480.00 |
| 12' wide | " | 400.00 | 120.00 | 520.00 |
| 14' wide | " | 430.00 | 120.00 | 550.00 |

| SITE IMPROVEMENTS | UNIT | MAT. | INST. | TOTAL |
|---|---|---|---|---|
| **02830.10 CHAIN LINK FENCE** | | | | |
| 16' wide | EA. | 460.00 | 120.00 | 580.00 |
| 18' wide | " | 490.00 | 140.00 | 630.00 |
| 20' wide | " | 510.00 | 140.00 | 650.00 |
| 22' wide | " | 550.00 | 140.00 | 690.00 |
| 24' wide | " | 590.00 | 160.00 | 750.00 |
| 26' wide | " | 610.00 | 160.00 | 770.00 |
| 28' wide | " | 660.00 | 190.00 | 850.00 |
| 30' wide | " | 690.00 | 190.00 | 880.00 |
| 7' high | | | | |
| Single gate | | | | |
| 3' wide | EA. | 170.00 | 120.00 | 290.00 |
| 4' wide | " | 180.00 | 120.00 | 300.00 |
| Double gate | | | | |
| 10' wide | EA. | 430.00 | 160.00 | 590.00 |
| 12' wide | " | 470.00 | 160.00 | 630.00 |
| 14' wide | " | 500.00 | 160.00 | 660.00 |
| 16' wide | " | 540.00 | 160.00 | 700.00 |
| 18' wide | " | 570.00 | 190.00 | 760.00 |
| 20' wide | " | 610.00 | 190.00 | 800.00 |
| 22' wide | " | 640.00 | 190.00 | 830.00 |
| 24' wide | " | 690.00 | 240.00 | 930.00 |
| 26' wide | " | 730.00 | 240.00 | 970.00 |
| 28' wide | " | 770.00 | 300.00 | 1,070 |
| 30' wide | " | 880.00 | 300.00 | 1,180 |
| 8' high | | | | |
| Single gate | | | | |
| 3' wide | EA. | 180.00 | 120.00 | 300.00 |
| 4' wide | " | 190.00 | 120.00 | 310.00 |
| Double gate | | | | |
| 10' wide | EA. | 470.00 | 160.00 | 630.00 |
| 12' wide | " | 500.00 | 160.00 | 660.00 |
| 14' wide | " | 530.00 | 160.00 | 690.00 |
| 16' wide | " | 580.00 | 160.00 | 740.00 |
| 18' wide | " | 610.00 | 190.00 | 800.00 |
| 20' wide | " | 630.00 | 190.00 | 820.00 |
| 22' wide | " | 660.00 | 190.00 | 850.00 |
| 24' wide | " | 740.00 | 240.00 | 980.00 |
| 26' wide | " | 760.00 | 240.00 | 1,000 |
| 28' wide | " | 810.00 | 300.00 | 1,110 |
| 30' wide | " | 910.00 | 300.00 | 1,210 |
| Vinyl coated swing gates, 4' high | | | | |
| Single gate | | | | |
| 3' wide | EA. | 190.00 | 60.50 | 250.50 |
| 4' wide | " | 210.00 | 60.50 | 270.50 |
| Double gate | | | | |
| 10' wide | EA. | 450.00 | 97.00 | 547.00 |
| 12' wide | " | 570.00 | 97.00 | 667.00 |
| 14' wide | " | 610.00 | 97.00 | 707.00 |
| 16' wide | " | 680.00 | 97.00 | 777.00 |
| 18' wide | " | 760.00 | 140.00 | 900.00 |
| 20' wide | " | 900.00 | 140.00 | 1,040 |
| 22' wide | " | 1,000 | 140.00 | 1,140 |

| SITE IMPROVEMENTS | UNIT | MAT. | INST. | TOTAL |
|---|---|---|---|---|
| **02830.10 CHAIN LINK FENCE** | | | | |
| 24' wide | EA. | 1,070 | 160.00 | 1,230 |
| 26' wide | " | 1,130 | 160.00 | 1,290 |
| 28' wide | " | 1,310 | 190.00 | 1,500 |
| 30' wide | " | 1,370 | 190.00 | 1,560 |
| 5' high | | | | |
| Single gate | | | | |
| 3' wide | EA. | 210.00 | 80.50 | 290.50 |
| 4' wide | " | 230.00 | 80.50 | 310.50 |
| Double gate | | | | |
| 10' wide | EA. | 610.00 | 120.00 | 730.00 |
| 12' wide | " | 650.00 | 120.00 | 770.00 |
| 14' wide | " | 730.00 | 120.00 | 850.00 |
| 16' wide | " | 760.00 | 120.00 | 880.00 |
| 18' wide | " | 920.00 | 140.00 | 1,060 |
| 20' wide | " | 1,000 | 140.00 | 1,140 |
| 22' wide | " | 1,100 | 140.00 | 1,240 |
| 24' wide | " | 1,250 | 160.00 | 1,410 |
| 26' wide | " | 1,310 | 160.00 | 1,470 |
| 28' wide | " | 1,440 | 190.00 | 1,630 |
| 30' wide | " | 1,560 | 190.00 | 1,750 |
| 6' high | | | | |
| Single gate | | | | |
| 3' wide | EA. | 220.00 | 80.50 | 300.50 |
| 4' wide | " | 230.00 | 80.50 | 310.50 |
| Double gate | | | | |
| 10' wide | EA. | 550.00 | 120.00 | 670.00 |
| 12' wide | " | 620.00 | 120.00 | 740.00 |
| 14' wide | " | 720.00 | 120.00 | 840.00 |
| 16' wide | " | 830.00 | 120.00 | 950.00 |
| 18' wide | " | 920.00 | 140.00 | 1,060 |
| 20' wide | " | 1,000 | 140.00 | 1,140 |
| 22' wide | " | 1,090 | 140.00 | 1,230 |
| 24' wide | " | 1,210 | 160.00 | 1,370 |
| 26' wide | " | 1,280 | 160.00 | 1,440 |
| 28' wide | " | 1,430 | 190.00 | 1,620 |
| 30' wide | " | 1,570 | 190.00 | 1,760 |
| 7' high | | | | |
| Single gate | | | | |
| 3' wide | EA. | 240.00 | 120.00 | 360.00 |
| 4' wide | " | 310.00 | 120.00 | 430.00 |
| Double gate | | | | |
| 10' wide | EA. | 610.00 | 160.00 | 770.00 |
| 12' wide | " | 730.00 | 160.00 | 890.00 |
| 14' wide | " | 820.00 | 160.00 | 980.00 |
| 16' wide | " | 920.00 | 160.00 | 1,080 |
| 18' wide | " | 1,040 | 190.00 | 1,230 |
| 20' wide | " | 1,190 | 190.00 | 1,380 |
| 22' wide | " | 1,320 | 190.00 | 1,510 |
| 24' wide | " | 1,440 | 240.00 | 1,680 |
| 26' wide | " | 1,560 | 240.00 | 1,800 |
| 28' wide | " | 1,690 | 300.00 | 1,990 |
| 30' wide | " | 1,810 | 300.00 | 2,110 |

## SITE IMPROVEMENTS

| SITE IMPROVEMENTS | UNIT | MAT. | INST. | TOTAL |
|---|---|---|---|---|
| **02830.10 CHAIN LINK FENCE** | | | | |
| 8' high | | | | |
| Single gate | | | | |
| 3' wide | EA. | 260.00 | 120.00 | 380.00 |
| 4' wide | " | 310.00 | 120.00 | 430.00 |
| Double gate | | | | |
| 10' wide | EA. | 620.00 | 160.00 | 780.00 |
| 12' wide | " | 730.00 | 160.00 | 890.00 |
| 14' wide | " | 830.00 | 160.00 | 990.00 |
| 16' wide | " | 590.00 | 160.00 | 750.00 |
| 18' wide | " | 1,050 | 190.00 | 1,240 |
| 20' wide | " | 1,220 | 190.00 | 1,410 |
| 22' wide | " | 1,400 | 190.00 | 1,590 |
| 24' wide | " | 1,490 | 240.00 | 1,730 |
| 28' wide | " | 1,730 | 240.00 | 1,970 |
| 30' wide | " | 1,860 | 300.00 | 2,160 |
| Motor operator for gates, no wiring | " | | | 3,580 |
| Drilling fence post holes | | | | |
| In soil | | | | |
| By hand | EA. | 0.00 | 12.10 | 12.10 |
| By machine auger | " | 0.00 | 7.75 | 7.75 |
| In rock | | | | |
| By jackhammer | EA. | 0.00 | 100.00 | 100.00 |
| By rock drill | " | 0.00 | 31.00 | 31.00 |
| Aluminum privacy slats, installed vertically | S.F. | 0.64 | 0.60 | 1.25 |
| Post hole, dig by hand | EA. | 0.00 | 16.15 | 16.15 |
| Set fence post in concrete | " | 6.35 | 12.10 | 18.45 |
| **02830.70 RECREATIONAL COURTS** | | | | |
| Walls, galvanized steel | | | | |
| 8' high | L.F. | 8.30 | 4.85 | 13.15 |
| 10' high | " | 9.85 | 5.40 | 15.25 |
| 12' high | " | 11.30 | 6.35 | 17.65 |
| Vinyl coated | | | | |
| 8' high | L.F. | 7.95 | 4.85 | 12.80 |
| 10' high | " | 9.75 | 5.40 | 15.15 |
| 12' high | " | 10.85 | 6.35 | 17.20 |
| Gates, galvanized steel | | | | |
| Single, 3' transom | | | | |
| 3'x7' | EA. | 200.00 | 120.00 | 320.00 |
| 4'x7' | " | 210.00 | 140.00 | 350.00 |
| 5'x7' | " | 280.00 | 160.00 | 440.00 |
| 6'x7' | " | 310.00 | 190.00 | 500.00 |
| Double, 3' transom | | | | |
| 10'x7' | EA. | 470.00 | 480.00 | 950.00 |
| 12'x7' | " | 610.00 | 540.00 | 1,150 |
| 14'x7' | " | 730.00 | 610.00 | 1,340 |
| Double, no transom | | | | |
| 10'x10' | EA. | 510.00 | 400.00 | 910.00 |
| 12'x10' | " | 610.00 | 480.00 | 1,090 |
| 14'x10' | " | 700.00 | 540.00 | 1,240 |
| Vinyl coated | | | | |
| Single, 3' transom | | | | |

| SITE IMPROVEMENTS | UNIT | MAT. | INST. | TOTAL |
|---|---|---|---|---|
| **02830.70 RECREATIONAL COURTS** | | | | |
| 3'x7' | EA. | 380.00 | 120.00 | 500.00 |
| 4'x7' | " | 420.00 | 140.00 | 560.00 |
| 5'x7' | " | 420.00 | 160.00 | 580.00 |
| 6'x7' | " | 430.00 | 190.00 | 620.00 |
| Double, 3' | | | | |
| 10'x7' | EA. | 1,130 | 480.00 | 1,610 |
| 12'x7' | " | 1,150 | 540.00 | 1,690 |
| 14'x7' | " | 1,250 | 610.00 | 1,860 |
| Double, no transom | | | | |
| 10'x10' | EA. | 1,120 | 400.00 | 1,520 |
| 12'x10' | " | 1,140 | 480.00 | 1,620 |
| 14'x10' | " | 1,250 | 540.00 | 1,790 |
| Baseball backstop, regulation | | | | |
| Galvanized | EA. | | | 4,851 |
| Vinyl coated | " | | | 6,755 |
| Softball backstop, regulation | | | | |
| 14' high | | | | |
| Galvanized | EA. | | | 4,504 |
| Vinyl coated | " | | | 6,698 |
| 18' high | | | | |
| Galvanized | EA. | | | 5,312 |
| Vinyl coated | " | | | 7,680 |
| 20' high | | | | |
| Galvanized | EA. | | | 6,294 |
| Vinyl coated | " | | | 9,065 |
| 22' high | | | | |
| Galvanized | EA. | | | 7,276 |
| Vinyl coated | " | | | 10,625 |
| 24' high | | | | |
| Galvanized | EA. | | | 8,846 |
| Vinyl coated | " | | | 14,551 |
| Wire and miscellaneous metal fences | | | | |
| Chicken wire, post 4' o.c. | | | | |
| 2" mesh | | | | |
| 4' high | L.F. | 0.93 | 1.20 | 2.13 |
| 6' high | " | 1.05 | 1.60 | 2.65 |
| Galvanized steel | | | | |
| 12 gauge, 2" by 4" mesh, posts 5' o.c. | | | | |
| 3' high | L.F. | 1.50 | 1.20 | 2.70 |
| 5' high | " | 2.15 | 1.50 | 3.65 |
| 14 gauge, 1" by 2" mesh, posts 5' o.c. | | | | |
| 3' high | L.F. | 1.25 | 1.20 | 2.45 |
| 5' high | " | 2.05 | 1.50 | 3.55 |
| **02840.30 GUARDRAILS** | | | | |
| Pipe bollard, steel pipe, concrete filled, painted | | | | |
| 6" dia. | EA. | 100.00 | 20.15 | 120.15 |
| 8" dia. | " | 150.00 | 30.30 | 180.30 |
| 12" dia. | " | 240.00 | 80.50 | 320.50 |
| Corrugated steel, guardrail, galvanized | L.F. | 17.05 | 2.20 | 19.25 |
| End section, wrap around or flared | EA. | 48.50 | 24.20 | 72.70 |

| SITE IMPROVEMENTS | UNIT | MAT. | INST. | TOTAL |
|---|---|---|---|---|
| **02840.30 GUARDRAILS** | | | | |
| Timber guardrail, 4" x 8" | L.F. | 20.20 | 1.65 | 21.85 |
| Guard rail, 3 cables, 3/4" dia. | | | | |
| Steel posts | L.F. | 9.10 | 6.55 | 15.65 |
| Wood posts | " | 9.80 | 5.25 | 15.05 |
| Steel box beam | | | | |
| 6" x 6" | L.F. | 38.70 | 7.25 | 45.95 |
| 6" x 8" | " | | | |
| Concrete posts | EA. | 22.50 | 12.10 | 34.60 |
| Barrel type impact barrier | " | 320.00 | 24.20 | 344.20 |
| Light shield, 6' high | L.F. | 20.80 | 4.85 | 25.65 |
| **02840.40 PARKING BARRIERS** | | | | |
| Timber, treated, 8' long | | | | |
| 4" x 4" | EA. | 11.90 | 20.15 | 32.05 |
| 6" x 6" | " | 22.00 | 24.20 | 46.20 |
| Precast concrete, 4' long, with dowels | | | | |
| 12" x 6" | EA. | 21.40 | 12.10 | 33.50 |
| 12" x 8" | " | 24.95 | 13.45 | 38.40 |
| **02840.60 SIGNAGE** | | | | |
| Traffic signs | | | | |
| Reflectorized signs per OSHA standards, including post | | | | |
| Stop, 24"x24" | EA. | 40.80 | 16.15 | 56.95 |
| Yield, 30" triangle | " | 23.80 | 16.15 | 39.95 |
| Speed limit, 12"x18" | " | 27.20 | 16.15 | 43.35 |
| Directional, 12"x18" | " | 37.40 | 16.15 | 53.55 |
| Exit, 12"x18" | " | 37.40 | 16.15 | 53.55 |
| Entry, 12"x18" | " | 37.40 | 16.15 | 53.55 |
| Warning, 24"x24" | " | 47.60 | 16.15 | 63.75 |
| Informational, 12"x18" | " | 17.00 | 16.15 | 33.15 |
| Handicap parking, 12"x18" | " | 18.10 | 16.15 | 34.25 |
| **02860.40 RECREATIONAL FACILITIES** | | | | |
| Bleachers, outdoor, portable, per seat | | | | |
| 10 tiers | | | | |
| Minimum | EA. | 21.00 | 8.15 | 29.15 |
| Maximum | " | 40.80 | 10.90 | 51.70 |
| 20 tiers | | | | |
| Minimum | EA. | 25.70 | 7.70 | 33.40 |
| Maximum | " | 49.00 | 10.05 | 59.05 |
| Grandstands, fixed, wood seat, steel frame, per seat | | | | |
| 15 tiers | | | | |
| Minimum | EA. | 36.20 | 13.10 | 49.30 |
| Maximum | " | 62.00 | 21.80 | 83.80 |
| 30 tiers | | | | |
| Minimum | EA. | 37.30 | 11.90 | 49.20 |
| Maximum | " | 79.50 | 18.70 | 98.20 |
| Seats | | | | |
| Seat backs only | | | | |
| Fiberglass | EA. | 22.15 | 2.40 | 24.55 |
| Steel and wood seat | " | 30.30 | 2.40 | 32.70 |

## SITE IMPROVEMENTS

### 02860.40 RECREATIONAL FACILITIES

| SITE IMPROVEMENTS | UNIT | MAT. | INST. | TOTAL |
|---|---|---|---|---|
| Seat restoration, fiberglass on wood | | | | |
| Seats | EA. | 15.15 | 4.85 | 20.00 |
| Plain bench, no backs | " | 9.30 | 2.00 | 11.30 |
| Benches | | | | |
| Park, precast concrete with backs | | | | |
| 4' long | EA. | 580.00 | 80.50 | 660.50 |
| 8' long | " | 1,280 | 120.00 | 1,400 |
| Fiberglass, with backs | | | | |
| 4' long | EA. | 490.00 | 60.50 | 550.50 |
| 8' long | " | 930.00 | 80.50 | 1,011 |
| Wood, with backs and fiberglass supports | | | | |
| 4' long | EA. | 250.00 | 60.50 | 310.50 |
| 8' long | " | 260.00 | 80.50 | 340.50 |
| Steel frame, 6' long | | | | |
| All steel | EA. | 230.00 | 60.50 | 290.50 |
| Hardwood boards | " | 160.00 | 60.50 | 220.50 |
| Players bench (no back), steel frame, fir seat, 10' long | " | 180.00 | 80.50 | 260.50 |
| Backstops | | | | |
| Handball or squash court, outdoor | | | | |
| Wood | EA. | | | 19,825 |
| Masonry | " | | | 20,408 |
| Soccer goal posts | PAIR | | | 1,633 |
| Running track | | | | |
| Gravel and cinders over stone base | S.Y. | 5.70 | 3.25 | 8.95 |
| Rubber-cork base resilient pavement | " | 8.05 | 26.20 | 34.25 |
| For colored surfaces, add | " | 5.05 | 2.60 | 7.65 |
| Colored rubberized asphalt | " | 10.80 | 32.70 | 43.50 |
| Artificial resilient mat over asphalt | " | 25.40 | 65.50 | 90.90 |
| Tennis courts | | | | |
| Bituminous pavement, 2-1/2" thick | S.Y. | 8.15 | 8.15 | 16.30 |
| Colored sealer, acrylic emulsion | | | | |
| 3 coats | S.Y. | 4.25 | 1.60 | 5.85 |
| For 2 color seal coating, add | " | 0.78 | 0.24 | 1.02 |
| For preparing old courts, add | " | 1.80 | 0.16 | 1.96 |
| Net, nylon, 42' long | EA. | 250.00 | 30.30 | 280.30 |
| Paint markings on asphalt, 2 coats | " | 64.00 | 240.00 | 304.00 |
| Complete court with fence, etc., bituminous | | | | |
| Minimum | EA. | | | 10,029 |
| Average | " | | | 17,259 |
| Maximum | " | | | 24,490 |
| Clay court | | | | |
| Minimum | EA. | | | 10,437 |
| Average | " | | | 14,227 |
| Maximum | " | | | 21,574 |
| Playground equipment | | | | |
| Basketball backboard | | | | |
| Minimum | EA. | 440.00 | 60.50 | 500.50 |
| Maximum | " | 840.00 | 69.00 | 909.00 |
| Bike rack, 10' long | " | 340.00 | 48.40 | 388.40 |
| Golf shelter, fiberglass | " | 1,690 | 60.50 | 1,751 |
| Ground socket for movable posts | | | | |
| Minimum | EA. | 76.00 | 15.10 | 91.10 |

# 02 SITEWORK

## SITE IMPROVEMENTS

### 02860.40 RECREATIONAL FACILITIES

| | UNIT | MAT. | INST. | TOTAL |
|---|---|---|---|---|
| Maximum | EA. | 150.00 | 15.10 | 165.10 |
| Horizontal monkey ladder, 14' long | " | 500.00 | 40.30 | 540.30 |
| Posts, tether ball | " | 240.00 | 12.10 | 252.10 |
| Multiple purpose, 10' long | " | 250.00 | 24.20 | 274.20 |
| See-saw, steel | | | | |
| Minimum | EA. | 410.00 | 97.00 | 507.00 |
| Average | " | 760.00 | 120.00 | 880.00 |
| Maximum | " | 1,170 | 160.00 | 1,330 |
| Slide | | | | |
| Minimum | EA. | 740.00 | 190.00 | 930.00 |
| Maximum | " | 1,280 | 220.00 | 1,500 |
| Swings, plain seats | | | | |
| 8' high | | | | |
| Minimum | EA. | 550.00 | 160.00 | 710.00 |
| Maximum | " | 1,040 | 190.00 | 1,230 |
| 12' high | | | | |
| Minimum | EA. | 840.00 | 190.00 | 1,030 |
| Maximum | " | 1,520 | 270.00 | 1,790 |

### 02870.10 PREFABRICATED PLANTERS

| | UNIT | MAT. | INST. | TOTAL |
|---|---|---|---|---|
| Concrete precast, circular | | | | |
| 24" dia., 18" high | EA. | 190.00 | 24.20 | 214.20 |
| 42" dia., 30" high | " | 250.00 | 30.30 | 280.30 |
| Fiberglass, circular | | | | |
| 36" dia., 27" high | EA. | 350.00 | 12.10 | 362.10 |
| 60" dia., 39" high | " | 820.00 | 13.45 | 833.45 |
| Tapered, circular | | | | |
| 24" dia., 36" high | EA. | 270.00 | 11.00 | 281.00 |
| 40" dia., 36" high | " | 450.00 | 12.10 | 462.10 |
| Square | | | | |
| 2' by 2', 17" high | EA. | 230.00 | 11.00 | 241.00 |
| 4' by 4', 39" high | " | 820.00 | 13.45 | 833.45 |
| Rectangular | | | | |
| 4' by 1', 18" high | EA. | 260.00 | 12.10 | 272.10 |

## LANDSCAPING

### 02910.10 SHRUB & TREE MAINTENANCE

| | UNIT | MAT. | INST. | TOTAL |
|---|---|---|---|---|
| Moving shrubs on site | | | | |
| 12" ball | EA. | 0.00 | 30.30 | 30.30 |
| 24" ball | " | 0.00 | 40.30 | 40.30 |
| 3' high | " | 0.00 | 24.20 | 24.20 |
| 4' high | " | 0.00 | 26.90 | 26.90 |
| 5' high | " | 0.00 | 30.30 | 30.30 |

| LANDSCAPING | UNIT | MAT. | INST. | TOTAL |
|---|---|---|---|---|
| **02910.10 SHRUB & TREE MAINTENANCE** | | | | |
| 18" spread | EA. | 0.00 | 34.60 | 34.60 |
| 30" spread | " | 0.00 | 40.30 | 40.30 |
| Moving trees on site | | | | |
| 24" ball | EA. | 0.00 | 65.50 | 65.50 |
| 48" ball | " | 0.00 | 87.00 | 87.00 |
| Trees | | | | |
| 3' high | EA. | 0.00 | 26.20 | 26.20 |
| 6' high | " | 0.00 | 29.10 | 29.10 |
| 8' high | " | 0.00 | 32.70 | 32.70 |
| 10' high | " | 0.00 | 43.60 | 43.60 |
| Palm trees | | | | |
| 7' high | EA. | 0.00 | 32.70 | 32.70 |
| 10' high | " | 0.00 | 43.60 | 43.60 |
| 20' high | " | 0.00 | 130.00 | 130.00 |
| 40' high | " | 0.00 | 260.00 | 260.00 |
| Guying trees | | | | |
| 4" dia. | EA. | 6.25 | 12.10 | 18.35 |
| 8" dia. | " | 6.25 | 15.10 | 21.35 |
| **02920.10 TOPSOIL** | | | | |
| Spread topsoil, with equipment | | | | |
| Minimum | C.Y. | 0.00 | 6.95 | 6.95 |
| Maximum | " | 0.00 | 8.70 | 8.70 |
| By hand | | | | |
| Minimum | C.Y. | 0.00 | 24.20 | 24.20 |
| Maximum | " | 0.00 | 30.30 | 30.30 |
| Area preparation for seeding (grade, rake and clean) | | | | |
| Square yard | S.Y. | 0.00 | 0.19 | 0.19 |
| By acre | ACRE | 0.00 | 970.00 | 970.00 |
| Remove topsoil and stockpile on site | | | | |
| 4" deep | C.Y. | 0.00 | 5.80 | 5.80 |
| 6" deep | " | 0.00 | 5.35 | 5.35 |
| Spreading topsoil from stock pile | | | | |
| By loader | C.Y. | 0.00 | 6.30 | 6.30 |
| By hand | " | 0.00 | 69.50 | 69.50 |
| Top dress by hand | S.Y. | 0.00 | 0.69 | 0.69 |
| Place imported top soil | | | | |
| By loader | | | | |
| 4" deep | S.Y. | 0.00 | 0.69 | 0.69 |
| 6" deep | " | 0.00 | 0.77 | 0.77 |
| By hand | | | | |
| 4" deep | S.Y. | 0.00 | 2.70 | 2.70 |
| 6" deep | " | 0.00 | 3.00 | 3.00 |
| Plant bed preparation, 18" deep | | | | |
| With backhoe/loader | S.Y. | 0.00 | 1.75 | 1.75 |
| By hand | " | 0.00 | 4.05 | 4.05 |
| **02930.30 SEEDING** | | | | |
| Mechanical seeding, 175 lb/acre | | | | |
| By square yard | S.Y. | 0.46 | 0.06 | 0.52 |
| By acre | ACRE | 1,830 | 310.00 | 2,140 |
| 450 lb/acre | | | | |

# 02 SITEWORK

| LANDSCAPING | UNIT | MAT. | INST. | TOTAL |
|---|---|---|---|---|
| **02930.30 SEEDING** | | | | |
| By square yard | S.Y. | 0.73 | 0.08 | 0.81 |
| By acre | ACRE | 2,830 | 390.00 | 3,220 |
| Seeding by hand, 10 lb per 100 s.y. | | | | |
| By square yard | S.Y. | 0.48 | 0.08 | 0.56 |
| By acre | ACRE | 1,920 | 400.00 | 2,320 |
| Reseed disturbed areas | S.F. | 0.42 | 0.12 | 0.54 |
| **02950.10 PLANTS** | | | | |
| Euonymus coloratus, 18" (purple wintercreeper) | EA. | 1.40 | 4.05 | 5.45 |
| Hedera Helix, 2-1/4" pot (English ivy) | " | 0.57 | 4.05 | 4.62 |
| Liriope muscari, 2" clumps | " | 2.50 | 2.40 | 4.90 |
| Santolina, 12" | " | 2.85 | 2.40 | 5.25 |
| Vinca major or minor, 3" pot | " | 0.46 | 2.40 | 2.86 |
| Cortaderia argentia, 2 gallon (pampas grass) | " | 9.05 | 2.40 | 11.45 |
| Ophiopogan japonicus, 1 quart (4" pot) | " | 2.50 | 2.40 | 4.90 |
| Ajuga reptans, 2-3/4" pot (carpet bugle) | " | 0.46 | 2.40 | 2.86 |
| Pachysandra terminalis, 2-3/4" pot (Japanese spurge) | " | 0.62 | 2.40 | 3.02 |
| **02950.30 SHRUBS** | | | | |
| Juniperus conferia litoralis, 18"-24" (Shore Juniper) | EA. | 20.95 | 9.70 | 30.65 |
| Horizontalis plumosa, 18"-24" (Andorra Juniper) | " | 22.20 | 9.70 | 31.90 |
| Sabina tamar-iscfolia-tamarix juniper, 18"-24" | " | 22.20 | 9.70 | 31.90 |
| Chin San Hose, 18"-24" (San Hose Juniper) | " | 22.20 | 9.70 | 31.90 |
| Sargenti, 18"-24" (Sargent's Juniper) | " | 20.95 | 9.70 | 30.65 |
| Nandina domestica, 18"-24" (Heavenly Bamboo) | " | 14.05 | 9.70 | 23.75 |
| Raphiolepis Indica Springtime, 18"-24" Indian Hawthorn | " | 15.10 | 9.70 | 24.80 |
| Osmanthus Heterophyllus Gulftide, 18"-24" (Osmanthus) | " | 16.15 | 9.70 | 25.85 |
| Ilex Cornuta Burfordi Nana, 18"-24" (Dwarf Burford Holly) | " | 18.40 | 9.70 | 28.10 |
| Glabra, 18"-24" (Inkberry Holly) | " | 17.35 | 9.70 | 27.05 |
| Azalea, Indica types, 18"-24" | " | 19.55 | 9.70 | 29.25 |
| Kurume types, 18"-24" | " | 21.85 | 9.70 | 31.55 |
| Berberis Julianae, 18"-24" (Wintergreen Barberry) | " | 12.80 | 9.70 | 22.50 |
| Pieris Japonica Japanese, 18"-24" (Japanese Pieris) | " | 12.80 | 9.70 | 22.50 |
| Ilex Cornuta Rotunda, 18"-24" (Dwarf Chinese Holly) | " | 15.15 | 9.70 | 24.85 |
| Juniperus Horizontalis Plumosa, 24"-30" (Andorra Juniper) | " | 14.00 | 12.10 | 26.10 |
| Rhodopendrow Hybrids, 24"-30" | " | 37.30 | 12.10 | 49.40 |
| Aucuba Japonica Varigata, 24"-30" (Gold Dust Aucuba) | " | 12.75 | 12.10 | 24.85 |
| Ilex Crenata Willow Leaf, 24"-30" (Japanese Holly) | " | 14.00 | 12.10 | 26.10 |
| Cleyera Japonica, 30"-36" (Japanese Cleyera) | " | 16.35 | 15.10 | 31.45 |
| Pittosporum Tobira, 30"-36" | " | 18.70 | 15.10 | 33.80 |
| Prumus Laurocerasus, 30"-36" | " | 35.10 | 15.10 | 50.20 |
| Ilex Cornuta Burfordi, 30"-36" (Burford Holly) | " | 18.70 | 15.10 | 33.80 |
| Abelia Grandiflora, 24"-36" (Yew Podocarpus) | " | 12.80 | 12.10 | 24.90 |
| Podocarpos Macrophylla, 24"-36" (Yew Podocarpus) | " | 20.95 | 12.10 | 33.05 |
| Pyracantha Coccinea Lalandi, 3'-4' (Firethorn) | " | 11.60 | 15.10 | 26.70 |
| Photinia Frazieri, 3'-4' (Red Photinia) | " | 18.95 | 15.10 | 34.05 |
| Forsythia Suspensa, 3'-4' (Weeping Forsythia) | " | 11.60 | 15.10 | 26.70 |
| Camellia Japonica, 3'-4' (Common Camellia) | " | 21.05 | 15.10 | 36.15 |
| Juniperus Chin Torulosa, 3'-4' (Hollywood Juniper) | " | 22.35 | 15.10 | 37.45 |
| Cupressocyparis Leylandi, 3'-4' | " | 18.80 | 15.10 | 33.90 |
| Ilex Opaca Fosteri, 5'-6' (Foster's Holly) | " | 76.00 | 20.15 | 96.15 |

# 02 SITEWORK

| LANDSCAPING | UNIT | MAT. | INST. | TOTAL |
|---|---|---|---|---|
| **02950.30 SHRUBS** | | | | |
| Opaca, 5'-6' (American Holly) | EA. | 110.00 | 20.15 | 130.15 |
| Nyrica Cerifera, 4'-5' (Southern Wax Myrtles) | " | 23.80 | 17.30 | 41.10 |
| Ligustrum Japonicum, 4'-5' (Japanese Privet) | " | 18.70 | 17.30 | 36.00 |
| **02950.60 TREES** | | | | |
| Cornus Florida, 5'-6' (White flowering Dogwood) | EA. | 52.50 | 20.15 | 72.65 |
| Prunus Serrulata Kwanzan, 6'-8' (Kwanzan Cherry) | " | 58.50 | 24.20 | 82.70 |
| Caroliniana, 6'-8' (Carolina Cherry Laurel) | " | 69.00 | 24.20 | 93.20 |
| Cercis Canadensis, 6'-8' (Eastern Redbud) | " | 47.60 | 24.20 | 71.80 |
| Koelreuteria Paniculata, 8'-10' (Goldenrain tree) | " | 81.50 | 30.30 | 111.80 |
| Acer Platanoides, 1-3/4"-2" (11'-13') (Norway Maple) | " | 110.00 | 40.30 | 150.30 |
| Rubrum, 1-3/4"-2" (11'-13') (Red Maple) | " | 81.50 | 40.30 | 121.80 |
| Saccharum, 1-3/4"-2" (Sugar Maple) | " | 150.00 | 40.30 | 190.30 |
| Fraxinus Pennsylvanica, 1-3/4"-2" Laneolata-Green Ash | " | 70.00 | 40.30 | 110.30 |
| Celtis Occidentalis, 1-3/4"-2" (American Hackberry) | " | 110.00 | 40.30 | 150.30 |
| Glenditsia Triacantos Inermis, 2" | " | 96.00 | 40.30 | 136.30 |
| Prunus Cerasifera 'Thundercloud', 6'-8' | " | 55.50 | 24.20 | 79.70 |
| Yeodensis, 6'-8' (Yoshino Cherry) | " | 59.00 | 24.20 | 83.20 |
| Lagerstroemia Indica, 8'-10' (Crapemyrtle) | " | 93.50 | 30.30 | 123.80 |
| Crataegus Phaenopyrum, 8'-10' Washington Hawthorn | " | 150.00 | 30.30 | 180.30 |
| Quercus Borealis, 1-3/4"-2" (Northern Red Oak) | " | 87.00 | 40.30 | 127.30 |
| Quercus Acutissima, 1-3/4"-2" (8'-10') (Sawtooth Oak) | " | 81.50 | 40.30 | 121.80 |
| Saliz Babylonica, 1-3/4"-2" (Weeping Willow) | " | 40.80 | 40.30 | 81.10 |
| Tilia Cordata Greenspire, 1-3/4"-2" (10'-12') | " | 180.00 | 40.30 | 220.30 |
| Malus, 2"-2-1/2" (8'-10') (Flowering Crabapple) | " | 87.00 | 40.30 | 127.30 |
| Platanus Occidentalis, (12'-14') | " | 140.00 | 48.40 | 188.40 |
| Pyrus Calleryana Bradford, 2"-2-1/2" (Bradford Pear) | " | 100.00 | 40.30 | 140.30 |
| Quercus Palustris, 2"-2-1/2" (12'-14') (Pin Oak) | " | 120.00 | 40.30 | 160.30 |
| Phellos, 2-1/2"-3" (Willow Oak) | " | 130.00 | 48.40 | 178.40 |
| Nigra, 2"-2-1/2" (Water Oak) | " | 110.00 | 40.30 | 150.30 |
| Magnolia Soulangeana, 4'-5' (Saucer Magnolia) | " | 63.50 | 20.15 | 83.65 |
| Grandiflora, 6'-8' (Southern Magnolia) | " | 87.00 | 24.20 | 111.20 |
| Cedrus Deodara, 10'-12' (Deodare Cedar) | " | 140.00 | 40.30 | 180.30 |
| Ginkgo Biloba, 10'-12' (2"-2-1/2") (Maidenhair Tree) | " | 140.00 | 40.30 | 180.30 |
| Pinus Thunbergi, 5'-6' (Japanese Black Pine) | " | 53.00 | 20.15 | 73.15 |
| Strobus, 6'-8' (White Pine) | " | 59.00 | 24.20 | 83.20 |
| Taeda, 6'-8' (Loblolly Pine) | " | 49.80 | 24.20 | 74.00 |
| Quercus Virginiana, 2"-2-1/2" (live oak) | " | 130.00 | 48.40 | 178.40 |
| **02970.10 FERTILIZING** | | | | |
| Fertilizing (23#/1000 sf) | | | | |
| By square yard | S.Y. | 0.02 | 0.08 | 0.10 |
| By acre | ACRE | 89.50 | 390.00 | 479.50 |
| Liming (70#/1000 sf) | | | | |
| By square yard | S.Y. | 0.02 | 0.10 | 0.12 |
| By acre | ACRE | 89.50 | 520.00 | 609.50 |

| LANDSCAPING | UNIT | MAT. | INST. | TOTAL |
|---|---|---|---|---|
| **02980.10 LANDSCAPE ACCESSORIES** | | | | |
| Steel edging, 3/16" x 4" | L.F. | 0.37 | 0.30 | 0.67 |
| Landscaping stepping stones, 15"x15", white | EA. | 3.55 | 1.20 | 4.75 |
| Wood chip mulch | C.Y. | 30.00 | 16.15 | 46.15 |
| 2" thick | S.Y. | 1.85 | 0.48 | 2.33 |
| 4" thick | " | 3.45 | 0.69 | 4.14 |
| 6" thick | " | 5.20 | 0.88 | 6.08 |
| Gravel mulch, 3/4" stone | C.Y. | 23.10 | 24.20 | 47.30 |
| White marble chips, 1" deep | S.F. | 0.43 | 0.24 | 0.67 |
| Peat moss | | | | |
| 2" thick | S.Y. | 2.15 | 0.54 | 2.69 |
| 4" thick | " | 4.15 | 0.81 | 4.96 |
| 6" thick | " | 6.35 | 1.00 | 7.35 |
| Landscaping timbers, treated lumber | | | | |
| 4" x 4" | L.F. | 0.93 | 0.81 | 1.74 |
| 6" x 6" | " | 1.85 | 0.86 | 2.71 |
| 8" x 8" | " | 3.00 | 1.00 | 4.00 |

# 03 CONCRETE

| FORMWORK | UNIT | MAT. | INST. | TOTAL |
|---|---|---|---|---|
| **03110.05 BEAM FORMWORK** | | | | |
| Beam forms, job built | | | | |
| Beam bottoms | | | | |
| 1 use | S.F. | 3.35 | 5.15 | 8.50 |
| 2 uses | " | 1.95 | 4.90 | 6.85 |
| 3 uses | " | 1.50 | 4.75 | 6.25 |
| 4 uses | " | 1.25 | 4.55 | 5.80 |
| 5 uses | " | 1.15 | 4.40 | 5.55 |
| Beam sides | | | | |
| 1 use | S.F. | 2.40 | 3.45 | 5.85 |
| 2 uses | " | 1.40 | 3.25 | 4.65 |
| 3 uses | " | 1.25 | 3.10 | 4.35 |
| 4 uses | " | 1.15 | 2.95 | 4.10 |
| 5 uses | " | 1.00 | 2.80 | 3.80 |
| **03110.10 BOX CULVERT FORMWORK** | | | | |
| Box culverts, job built | | | | |
| 6' x 6' | | | | |
| 1 use | S.F. | 2.25 | 3.10 | 5.35 |
| 2 uses | " | 1.25 | 2.95 | 4.20 |
| 3 uses | " | 1.05 | 2.80 | 3.85 |
| 4 uses | " | 0.85 | 2.70 | 3.55 |
| 5 uses | " | 0.77 | 2.55 | 3.32 |
| 8' x 12' | | | | |
| 1 use | S.F. | 2.25 | 2.55 | 4.80 |
| 2 uses | " | 1.25 | 2.45 | 3.70 |
| 3 uses | " | 1.05 | 2.40 | 3.45 |
| 4 uses | " | 0.85 | 2.30 | 3.15 |
| 5 uses | " | 0.77 | 2.20 | 2.97 |
| **03110.15 COLUMN FORMWORK** | | | | |
| Column, square forms, job built | | | | |
| 8" x 8" columns | | | | |
| 1 use | S.F. | 2.60 | 6.20 | 8.80 |
| 2 uses | " | 1.40 | 5.95 | 7.35 |
| 3 uses | " | 1.20 | 5.70 | 6.90 |
| 4 uses | " | 1.10 | 5.50 | 6.60 |
| 5 uses | " | 0.93 | 5.30 | 6.23 |
| 12" x 12" columns | | | | |
| 1 use | S.F. | 2.40 | 5.60 | 8.00 |
| 2 uses | " | 1.30 | 5.40 | 6.70 |
| 3 uses | " | 1.05 | 5.25 | 6.30 |
| 4 uses | " | 0.93 | 5.05 | 5.98 |
| 5 uses | " | 0.78 | 4.90 | 5.68 |
| 16" x 16" columns | | | | |
| 1 use | S.F. | 2.25 | 5.15 | 7.40 |
| 2 uses | " | 1.20 | 5.00 | 6.20 |
| 3 uses | " | 0.95 | 4.85 | 5.80 |
| 4 uses | " | 0.88 | 4.70 | 5.58 |
| 5 uses | " | 0.71 | 4.55 | 5.26 |
| 24" x 24" columns | | | | |

| FORMWORK | UNIT | MAT. | INST. | TOTAL |
|---|---|---|---|---|
| **03110.15 COLUMN FORMWORK** | | | | |
| 1 use | S.F. | 2.25 | 4.75 | 7.00 |
| 2 uses | " | 1.05 | 4.60 | 5.65 |
| 3 uses | " | 0.85 | 4.50 | 5.35 |
| 4 uses | " | 0.71 | 4.35 | 5.06 |
| 5 uses | " | 0.65 | 4.25 | 4.90 |
| 36" x 36" columns | | | | |
| 1 use | S.F. | 2.25 | 4.40 | 6.65 |
| 2 uses | " | 1.10 | 4.30 | 5.40 |
| 3 uses | " | 0.90 | 4.15 | 5.05 |
| 4 uses | " | 0.77 | 4.05 | 4.82 |
| 5 uses | " | 0.71 | 3.95 | 4.66 |
| Round fiber forms, 1 use | | | | |
| 10" dia. | L.F. | 3.10 | 6.20 | 9.30 |
| 12" dia. | " | 3.80 | 6.30 | 10.10 |
| 14" dia. | " | 5.00 | 6.55 | 11.55 |
| 16" dia. | " | 6.55 | 6.85 | 13.40 |
| 18" dia. | " | 10.70 | 7.35 | 18.05 |
| 24" dia. | " | 13.10 | 7.90 | 21.00 |
| 30" dia. | " | 19.60 | 8.60 | 28.20 |
| 36" dia. | " | 24.40 | 9.35 | 33.75 |
| 42" dia. | " | 44.60 | 10.30 | 54.90 |
| **03110.18 CURB FORMWORK** | | | | |
| Curb forms | | | | |
| Straight, 6" high | | | | |
| 1 use | L.F. | 1.55 | 3.10 | 4.65 |
| 2 uses | " | 0.93 | 2.95 | 3.88 |
| 3 uses | " | 0.69 | 2.80 | 3.49 |
| 4 uses | " | 0.63 | 2.70 | 3.33 |
| 5 uses | " | 0.57 | 2.55 | 3.12 |
| Curved, 6" high | | | | |
| 1 use | L.F. | 1.65 | 3.85 | 5.50 |
| 2 uses | " | 1.05 | 3.65 | 4.70 |
| 3 uses | " | 0.80 | 3.45 | 4.25 |
| 4 uses | " | 0.74 | 3.30 | 4.04 |
| 5 uses | " | 0.69 | 3.15 | 3.84 |
| **03110.20 ELEVATED SLAB FORMWORK** | | | | |
| Elevated slab formwork | | | | |
| Slab, with drop panels | | | | |
| 1 use | S.F. | 2.65 | 2.45 | 5.10 |
| 2 uses | " | 1.50 | 2.40 | 3.90 |
| 3 uses | " | 1.15 | 2.30 | 3.45 |
| 4 uses | " | 1.05 | 2.20 | 3.25 |
| 5 uses | " | 0.93 | 2.15 | 3.08 |
| Floor slab, hung from steel beams | | | | |
| 1 use | S.F. | 2.10 | 2.40 | 4.50 |
| 2 uses | " | 1.15 | 2.30 | 3.45 |
| 3 uses | " | 1.05 | 2.20 | 3.25 |
| 4 uses | " | 0.91 | 2.15 | 3.06 |

| FORMWORK | UNIT | MAT. | INST. | TOTAL |
|---|---|---|---|---|
| **03110.20 ELEVATED SLAB FORMWORK** | | | | |
| 5 uses | S.F. | 0.78 | 2.05 | 2.83 |
| Floor slab, with pans or domes | | | | |
| 1 use | S.F. | 3.50 | 2.80 | 6.30 |
| 2 uses | " | 2.35 | 2.70 | 5.05 |
| 3 uses | " | 2.10 | 2.55 | 4.65 |
| 4 uses | " | 2.00 | 2.45 | 4.45 |
| 5 uses | " | 1.75 | 2.40 | 4.15 |
| Equipment curbs, 12" high | | | | |
| 1 use | L.F. | 2.00 | 3.10 | 5.10 |
| 2 uses | " | 1.30 | 2.95 | 4.25 |
| 3 uses | " | 1.10 | 2.80 | 3.90 |
| 4 uses | " | 1.00 | 2.70 | 3.70 |
| 5 uses | " | 0.83 | 2.55 | 3.38 |
| **03110.25 EQUIPMENT PAD FORMWORK** | | | | |
| Equipment pad, job built | | | | |
| 1 use | S.F. | 2.50 | 3.85 | 6.35 |
| 2 uses | " | 1.50 | 3.65 | 5.15 |
| 3 uses | " | 1.20 | 3.45 | 4.65 |
| 4 uses | " | 0.92 | 3.25 | 4.17 |
| 5 uses | " | 0.77 | 3.10 | 3.87 |
| **03110.35 FOOTING FORMWORK** | | | | |
| Wall footings, job built, continuous | | | | |
| 1 use | S.F. | 1.20 | 3.10 | 4.30 |
| 2 uses | " | 0.83 | 2.95 | 3.78 |
| 3 uses | " | 0.69 | 2.80 | 3.49 |
| 4 uses | " | 0.61 | 2.70 | 3.31 |
| 5 uses | " | 0.53 | 2.55 | 3.08 |
| Column footings, spread | | | | |
| 1 use | S.F. | 1.25 | 3.85 | 5.10 |
| 2 uses | " | 0.92 | 3.65 | 4.57 |
| 3 uses | " | 0.65 | 3.45 | 4.10 |
| 4 uses | " | 0.54 | 3.25 | 3.79 |
| 5 uses | " | 0.50 | 3.10 | 3.60 |
| **03110.50 GRADE BEAM FORMWORK** | | | | |
| Grade beams, job built | | | | |
| 1 use | S.F. | 1.95 | 3.10 | 5.05 |
| 2 uses | " | 1.10 | 2.95 | 4.05 |
| 3 uses | " | 0.85 | 2.80 | 3.65 |
| 4 uses | " | 0.71 | 2.70 | 3.41 |
| 5 uses | " | 0.60 | 2.55 | 3.15 |
| **03110.53 PILE CAP FORMWORK** | | | | |
| Pile cap forms, job built | | | | |
| Square | | | | |
| 1 use | S.F. | 2.15 | 3.85 | 6.00 |
| 2 uses | " | 1.25 | 3.65 | 4.90 |
| 3 uses | " | 0.98 | 3.45 | 4.43 |

# 03 CONCRETE

| FORMWORK | UNIT | MAT. | INST. | TOTAL |
|---|---|---|---|---|
| **03110.53 PILE CAP FORMWORK** | | | | |
| 4 uses | S.F. | 0.85 | 3.25 | 4.10 |
| 5 uses | " | 0.71 | 3.10 | 3.81 |
| Triangular | | | | |
| 1 use | S.F. | 2.25 | 4.40 | 6.65 |
| 2 uses | " | 1.50 | 4.10 | 5.60 |
| 3 uses | " | 1.20 | 3.85 | 5.05 |
| 4 uses | " | 0.98 | 3.65 | 4.63 |
| 5 uses | " | 0.78 | 3.45 | 4.23 |
| **03110.55 SLAB/MAT FORMWORK** | | | | |
| Mat foundations, job built | | | | |
| 1 use | S.F. | 1.85 | 3.85 | 5.70 |
| 2 uses | " | 1.05 | 3.65 | 4.70 |
| 3 uses | " | 0.78 | 3.45 | 4.23 |
| 4 uses | " | 0.65 | 3.25 | 3.90 |
| 5 uses | " | 0.52 | 3.10 | 3.62 |
| Edge forms | | | | |
| 6" high | | | | |
| 1 use | L.F. | 1.85 | 2.80 | 4.65 |
| 2 uses | " | 1.05 | 2.70 | 3.75 |
| 3 uses | " | 0.78 | 2.55 | 3.33 |
| 4 uses | " | 0.65 | 2.45 | 3.10 |
| 5 uses | " | 0.52 | 2.40 | 2.92 |
| 12" high | | | | |
| 1 use | L.F. | 1.70 | 3.10 | 4.80 |
| 2 uses | " | 0.98 | 2.95 | 3.93 |
| 3 uses | " | 0.73 | 2.80 | 3.53 |
| 4 uses | " | 0.60 | 2.70 | 3.30 |
| 5 uses | " | 0.49 | 2.55 | 3.04 |
| Formwork for openings | | | | |
| 1 use | S.F. | 2.50 | 6.20 | 8.70 |
| 2 uses | " | 1.40 | 5.60 | 7.00 |
| 3 uses | " | 1.20 | 5.15 | 6.35 |
| 4 uses | " | 0.92 | 4.75 | 5.67 |
| 5 uses | " | 0.78 | 4.40 | 5.18 |
| **03110.60 STAIR FORMWORK** | | | | |
| Stairway forms, job built | | | | |
| 1 use | S.F. | 3.25 | 6.20 | 9.45 |
| 2 uses | " | 1.85 | 5.60 | 7.45 |
| 3 uses | " | 1.40 | 5.15 | 6.55 |
| 4 uses | " | 1.30 | 4.75 | 6.05 |
| 5 uses | " | 1.10 | 4.40 | 5.50 |
| Stairs, elevated | | | | |
| 1 use | S.F. | 3.95 | 6.20 | 10.15 |
| 2 uses | " | 2.10 | 5.15 | 7.25 |
| 3 uses | " | 1.85 | 4.40 | 6.25 |
| 4 uses | " | 1.60 | 4.10 | 5.70 |
| 5 uses | " | 1.30 | 3.85 | 5.15 |

| FORMWORK | UNIT | MAT. | INST. | TOTAL |
|---|---|---|---|---|
| **03110.65 WALL FORMWORK** | | | | |
| Wall forms, exterior, job built | | | | |
| Up to 8' high wall | | | | |
| 1 use | S.F. | 2.15 | 3.10 | 5.25 |
| 2 uses | " | 1.20 | 2.95 | 4.15 |
| 3 uses | " | 1.05 | 2.80 | 3.85 |
| 4 uses | " | 0.91 | 2.70 | 3.61 |
| 5 uses | " | 0.79 | 2.55 | 3.34 |
| Over 8' high wall | | | | |
| 1 use | S.F. | 2.40 | 3.85 | 6.25 |
| 2 uses | " | 1.35 | 3.65 | 5.00 |
| 3 uses | " | 1.25 | 3.45 | 4.70 |
| 4 uses | " | 1.15 | 3.25 | 4.40 |
| 5 uses | " | 0.98 | 3.10 | 4.08 |
| Over 16' high wall | | | | |
| 1 use | S.F. | 2.50 | 4.40 | 6.90 |
| 2 uses | " | 1.50 | 4.10 | 5.60 |
| 3 uses | " | 1.35 | 3.85 | 5.20 |
| 4 uses | " | 1.25 | 3.65 | 4.90 |
| 5 uses | " | 1.15 | 3.45 | 4.60 |
| Radial wall forms | | | | |
| 1 use | S.F. | 2.40 | 4.75 | 7.15 |
| 2 uses | " | 1.40 | 4.40 | 5.80 |
| 3 uses | " | 1.30 | 4.10 | 5.40 |
| 4 uses | " | 1.20 | 3.85 | 5.05 |
| 5 uses | " | 1.10 | 3.65 | 4.75 |
| Retaining wall forms | | | | |
| 1 use | S.F. | 2.00 | 3.45 | 5.45 |
| 2 uses | " | 1.10 | 3.25 | 4.35 |
| 3 uses | " | 0.93 | 3.10 | 4.03 |
| 4 uses | " | 0.80 | 2.95 | 3.75 |
| 5 uses | " | 0.68 | 2.80 | 3.48 |
| Radial retaining wall forms | | | | |
| 1 use | S.F. | 2.25 | 5.15 | 7.40 |
| 2 uses | " | 1.40 | 4.75 | 6.15 |
| 3 uses | " | 1.20 | 4.40 | 5.60 |
| 4 uses | " | 1.15 | 4.10 | 5.25 |
| 5 uses | " | 0.98 | 3.85 | 4.83 |
| Column pier and pilaster | | | | |
| 1 use | S.F. | 2.40 | 6.20 | 8.60 |
| 2 uses | " | 1.40 | 5.60 | 7.00 |
| 3 uses | " | 1.30 | 5.15 | 6.45 |
| 4 uses | " | 1.20 | 4.75 | 5.95 |
| 5 uses | " | 1.10 | 4.40 | 5.50 |
| Interior wall forms | | | | |
| Up to 8' high | | | | |
| 1 use | S.F. | 2.15 | 2.80 | 4.95 |
| 2 uses | " | 1.20 | 2.70 | 3.90 |
| 3 uses | " | 1.10 | 2.55 | 3.65 |
| 4 uses | " | 0.92 | 2.45 | 3.37 |
| 5 uses | " | 0.78 | 2.40 | 3.18 |
| Over 8' high | | | | |
| 1 use | S.F. | 2.40 | 3.45 | 5.85 |

# 03 CONCRETE

| FORMWORK | UNIT | MAT. | INST. | TOTAL |
|---|---|---|---|---|
| **03110.65 WALL FORMWORK** | | | | |
| 2 uses | S.F. | 1.35 | 3.25 | 4.60 |
| 3 uses | " | 1.25 | 3.10 | 4.35 |
| 4 uses | " | 1.15 | 2.95 | 4.10 |
| 5 uses | " | 0.99 | 2.80 | 3.79 |
| Over 16' high | | | | |
| 1 use | S.F. | 2.50 | 3.85 | 6.35 |
| 2 uses | " | 1.50 | 3.65 | 5.15 |
| 3 uses | " | 1.35 | 3.45 | 4.80 |
| 4 uses | " | 1.25 | 3.25 | 4.50 |
| 5 uses | " | 1.15 | 3.10 | 4.25 |
| Radial wall forms | | | | |
| 1 use | S.F. | 2.40 | 4.10 | 6.50 |
| 2 uses | " | 1.40 | 3.85 | 5.25 |
| 3 uses | " | 1.30 | 3.65 | 4.95 |
| 4 uses | " | 1.20 | 3.45 | 4.65 |
| 5 uses | " | 1.10 | 3.25 | 4.35 |
| Curved wall forms, 24" sections | | | | |
| 1 use | S.F. | 2.40 | 6.20 | 8.60 |
| 2 uses | " | 1.40 | 5.60 | 7.00 |
| 3 uses | " | 1.30 | 5.15 | 6.45 |
| 4 uses | " | 1.20 | 4.75 | 5.95 |
| 5 uses | " | 1.10 | 4.40 | 5.50 |
| PVC form liner, per side, smooth finish | | | | |
| 1 use | S.F. | 3.90 | 2.55 | 6.45 |
| 2 uses | " | 2.10 | 2.45 | 4.55 |
| 3 uses | " | 1.80 | 2.40 | 4.20 |
| 4 uses | " | 1.40 | 2.20 | 3.60 |
| 5 uses | " | 1.10 | 2.05 | 3.15 |
| **03110.90 MISCELLANEOUS FORMWORK** | | | | |
| Keyway forms (5 uses) | | | | |
| 2 x 4 | L.F. | 0.14 | 1.55 | 1.69 |
| 2 x 6 | " | 0.21 | 1.70 | 1.91 |
| Bulkheads | | | | |
| Walls, with keyways | | | | |
| 2 piece | L.F. | 2.25 | 2.80 | 5.05 |
| 3 piece | " | 2.85 | 3.10 | 5.95 |
| Elevated slab, with keyway | | | | |
| 2 piece | L.F. | 2.70 | 2.55 | 5.25 |
| 3 piece | " | 3.80 | 2.80 | 6.60 |
| Ground slab, with keyway | | | | |
| 2 piece | L.F. | 2.80 | 2.20 | 5.00 |
| 3 piece | " | 3.75 | 2.40 | 6.15 |
| Chamfer strips | | | | |
| Wood | | | | |
| 1/2" wide | L.F. | 0.19 | 0.69 | 0.88 |
| 3/4" wide | " | 0.23 | 0.69 | 0.92 |
| 1" wide | " | 0.29 | 0.69 | 0.98 |
| PVC | | | | |
| 1/2" wide | L.F. | 0.57 | 0.69 | 1.26 |
| 3/4" wide | " | 0.62 | 0.69 | 1.31 |
| 1" wide | " | 0.80 | 0.69 | 1.49 |

# 03 CONCRETE

## FORMWORK

| 03110.90 MISCELLANEOUS FORMWORK | UNIT | MAT. | INST. | TOTAL |
|---|---|---|---|---|
| Radius | | | | |
| 1" | L.F. | 0.78 | 0.73 | 1.51 |
| 1-1/2" | " | 1.40 | 0.73 | 2.13 |
| Reglets | | | | |
| Galvanized steel, 24 ga. | L.F. | 0.91 | 1.25 | 2.16 |
| Metal formwork | | | | |
| Straight edge forms | | | | |
| 4" high | L.F. | 0.12 | 1.95 | 2.07 |
| 6" high | " | 0.13 | 2.05 | 2.18 |
| 8" high | " | 0.16 | 2.20 | 2.36 |
| 12" high | " | 0.20 | 2.40 | 2.60 |
| 16" high | " | 0.24 | 2.55 | 2.79 |
| Curb form, S-shape | | | | |
| 12" x | | | | |
| 1'-6" | L.F. | 0.24 | 4.40 | 4.64 |
| 2' | " | 0.30 | 4.10 | 4.40 |
| 2'-6" | " | 0.33 | 3.85 | 4.18 |
| 3' | " | 0.38 | 3.45 | 3.83 |

## REINFORCEMENT

| 03210.05 BEAM REINFORCING | UNIT | MAT. | INST. | TOTAL |
|---|---|---|---|---|
| Beam-girders | | | | |
| #3 - #4 | TON | 660.00 | 870.00 | 1,530 |
| #5 - #6 | " | 590.00 | 700.00 | 1,290 |
| #7 - #8 | " | 570.00 | 580.00 | 1,150 |
| #9 - #10 | " | 570.00 | 500.00 | 1,070 |
| #11 - #12 | " | 570.00 | 470.00 | 1,040 |
| #13 - #14 | " | 570.00 | 440.00 | 1,010 |
| Galvanized | | | | |
| #3 - #4 | TON | 1,160 | 870.00 | 2,030 |
| #5 - #6 | " | 1,100 | 700.00 | 1,800 |
| #7 - #8 | " | 1,060 | 580.00 | 1,640 |
| #9 - #10 | " | 1,060 | 500.00 | 1,560 |
| #11 - #12 | " | 1,060 | 470.00 | 1,530 |
| #13 - #14 | " | 1,060 | 440.00 | 1,500 |
| Bond beams | | | | |
| #3 - #4 | TON | 660.00 | 1,170 | 1,830 |
| #5 - #6 | " | 590.00 | 870.00 | 1,460 |
| #7 - #8 | " | 570.00 | 780.00 | 1,350 |
| Galvanized | | | | |
| #3 - #4 | TON | 1,160 | 1,170 | 2,330 |
| #5 - #6 | " | 1,100 | 870.00 | 1,970 |
| #7 - #8 | " | 1,060 | 780.00 | 1,840 |

# 03 CONCRETE

| REINFORCEMENT | UNIT | MAT. | INST. | TOTAL |
|---|---|---|---|---|
| **03210.10 BOX CULVERT REINFORCING** | | | | |
| Box culverts | | | | |
| #3 - #4 | TON | 660.00 | 440.00 | 1,100 |
| #5 - #6 | " | 590.00 | 390.00 | 980.00 |
| #7 - #8 | " | 570.00 | 350.00 | 920.00 |
| #9 - #10 | " | 570.00 | 320.00 | 890.00 |
| #11 - #12 | " | 570.00 | 290.00 | 860.00 |
| Galvanized | | | | |
| #3 - #4 | TON | 1,160 | 440.00 | 1,600 |
| #5 - #6 | " | 1,100 | 390.00 | 1,490 |
| #7 - #8 | " | 1,060 | 350.00 | 1,410 |
| #9 - #10 | " | 1,060 | 320.00 | 1,380 |
| #11 - #12 | " | 1,060 | 290.00 | 1,350 |
| **03210.15 COLUMN REINFORCING** | | | | |
| Columns | | | | |
| #3 - #4 | TON | 660.00 | 1,000 | 1,660 |
| #5 - #6 | " | 590.00 | 780.00 | 1,370 |
| #7 - #8 | " | 570.00 | 700.00 | 1,270 |
| #9 - #10 | " | 570.00 | 640.00 | 1,210 |
| #11 - #12 | " | 570.00 | 580.00 | 1,150 |
| #13 - #14 | " | 570.00 | 540.00 | 1,110 |
| #15 - #16 | " | 570.00 | 500.00 | 1,070 |
| Galvanized | | | | |
| #3 - #4 | TON | 1,160 | 1,000 | 2,160 |
| #5 - #6 | " | 1,100 | 780.00 | 1,880 |
| #7 - #8 | " | 1,060 | 700.00 | 1,760 |
| #9 - #10 | " | 1,060 | 640.00 | 1,700 |
| #11 - #12 | " | 1,060 | 580.00 | 1,640 |
| #13 - #14 | " | 1,060 | 540.00 | 1,600 |
| #15 - #16 | " | 1,060 | 500.00 | 1,560 |
| Spirals | | | | |
| 8" to 24" dia. | TON | 1,130 | 870.00 | 2,000 |
| 24" to 48" dia. | " | 1,130 | 780.00 | 1,910 |
| 48" to 84" dia. | " | 1,250 | 700.00 | 1,950 |
| **03210.20 ELEVATED SLAB REINFORCING** | | | | |
| Elevated slab | | | | |
| #3 - #4 | TON | 660.00 | 440.00 | 1,100 |
| #5 - #6 | " | 590.00 | 390.00 | 980.00 |
| #7 - #8 | " | 570.00 | 350.00 | 920.00 |
| #9 - #10 | " | 570.00 | 320.00 | 890.00 |
| #11 - #12 | " | 570.00 | 290.00 | 860.00 |
| Galvanized | | | | |
| #3 - #4 | TON | 1,160 | 440.00 | 1,600 |
| #5 - #6 | " | 1,100 | 390.00 | 1,490 |
| #7 - #8 | " | 1,060 | 350.00 | 1,410 |
| #9 - #10 | " | 1,060 | 320.00 | 1,380 |
| #11 - #12 | " | 1,060 | 290.00 | 1,350 |

# 03 CONCRETE

| REINFORCEMENT | UNIT | MAT. | INST. | TOTAL |
|---|---|---|---|---|
| **03210.25 EQUIP. PAD REINFORCING** | | | | |
| Equipment pad | | | | |
| #3 - #4 | TON | 660.00 | 700.00 | 1,360 |
| #5 - #6 | " | 590.00 | 640.00 | 1,230 |
| #7 - #8 | " | 570.00 | 580.00 | 1,150 |
| #9 - #10 | " | 570.00 | 540.00 | 1,110 |
| #11 - #12 | " | 570.00 | 500.00 | 1,070 |
| **03210.35 FOOTING REINFORCING** | | | | |
| Footings | | | | |
| Grade 50 | | | | |
| #3 - #4 | TON | 660.00 | 580.00 | 1,240 |
| #5 - #6 | " | 590.00 | 500.00 | 1,090 |
| #7 - #8 | " | 570.00 | 440.00 | 1,010 |
| #9 - #10 | " | 570.00 | 390.00 | 960.00 |
| Grade 60 | | | | |
| #3 - #4 | TON | 680.00 | 580.00 | 1,260 |
| #5 - #6 | " | 620.00 | 500.00 | 1,120 |
| #7 - #8 | " | 590.00 | 440.00 | 1,030 |
| #9 - #10 | " | 590.00 | 390.00 | 980.00 |
| Grade 70 | | | | |
| #3 - #4 | TON | 700.00 | 580.00 | 1,280 |
| #5 - #6 | " | 640.00 | 500.00 | 1,140 |
| #7 - #8 | " | 620.00 | 440.00 | 1,060 |
| #9 - #10 | " | 620.00 | 390.00 | 1,010 |
| #11- #12 | " | 620.00 | 350.00 | 970.00 |
| Straight dowels, 24" long | | | | |
| 1" dia. (#8) | EA. | 2.05 | 3.50 | 5.55 |
| 3/4" dia. (#6) | " | 1.80 | 3.50 | 5.30 |
| 5/8" dia. (#5) | " | 1.60 | 2.90 | 4.50 |
| 1/2" dia. (#4) | " | 1.15 | 2.50 | 3.65 |
| **03210.45 FOUNDATION REINFORCING** | | | | |
| Foundations | | | | |
| #3 - #4 | TON | 660.00 | 580.00 | 1,240 |
| #5 - #6 | " | 590.00 | 500.00 | 1,090 |
| #7 - #8 | " | 570.00 | 440.00 | 1,010 |
| #9 - #10 | " | 570.00 | 390.00 | 960.00 |
| #11 - #12 | " | 570.00 | 350.00 | 920.00 |
| Galvanized | | | | |
| #3 - #4 | TON | 1,160 | 580.00 | 1,740 |
| #5 - #6 | " | 1,100 | 500.00 | 1,600 |
| #7 - #8 | " | 1,060 | 440.00 | 1,500 |
| #9 - #10 | " | 1,060 | 390.00 | 1,450 |
| #11 - #12 | " | 1,060 | 350.00 | 1,410 |
| **03210.50 GRADE BEAM REINFORCING** | | | | |
| Grade beams | | | | |
| #3 - #4 | TON | 660.00 | 540.00 | 1,200 |
| #5 - #6 | " | 590.00 | 470.00 | 1,060 |
| #7 - #8 | " | 570.00 | 410.00 | 980.00 |
| #9 - #10 | " | 570.00 | 370.00 | 940.00 |
| #11 - #12 | " | 570.00 | 330.00 | 900.00 |

| REINFORCEMENT | UNIT | MAT. | INST. | TOTAL |
|---|---|---|---|---|
| **03210.50 GRADE BEAM REINFORCING** | | | | |
| Galvanized | | | | |
| #3 - #4 | TON | 1,160 | 540.00 | 1,700 |
| #5 - #6 | " | 1,100 | 470.00 | 1,570 |
| #7 - #8 | " | 1,060 | 410.00 | 1,470 |
| #9 - #10 | " | 1,060 | 370.00 | 1,430 |
| #11 - #12 | " | 1,060 | 330.00 | 1,390 |
| **03210.53 PILE CAP REINFORCING** | | | | |
| Pile caps | | | | |
| #3 - #4 | TON | 660.00 | 870.00 | 1,530 |
| #5 - #6 | " | 590.00 | 780.00 | 1,370 |
| #7 - #8 | " | 570.00 | 700.00 | 1,270 |
| #9 - #10 | " | 570.00 | 640.00 | 1,210 |
| #11 - #12 | " | 570.00 | 580.00 | 1,150 |
| Galvanized | | | | |
| #3 - #4 | TON | 1,160 | 870.00 | 2,030 |
| #5 - #6 | " | 1,100 | 780.00 | 1,880 |
| #7 - #8 | " | 1,060 | 700.00 | 1,760 |
| #9 - #10 | " | 1,060 | 640.00 | 1,700 |
| #11 - #12 | " | 1,060 | 580.00 | 1,640 |
| **03210.55 SLAB/MAT REINFORCING** | | | | |
| Bars, slabs | | | | |
| #3 - #4 | TON | 660.00 | 580.00 | 1,240 |
| #5 - #6 | " | 590.00 | 500.00 | 1,090 |
| #7 - #8 | " | 570.00 | 440.00 | 1,010 |
| #9 - #10 | " | 570.00 | 390.00 | 960.00 |
| #11 - #12 | " | 570.00 | 350.00 | 920.00 |
| Galvanized | | | | |
| #3 - #4 | TON | 1,160 | 580.00 | 1,740 |
| #5 - #6 | " | 1,100 | 500.00 | 1,600 |
| #7 - #8 | " | 1,060 | 440.00 | 1,500 |
| #9 - #10 | " | 1,060 | 390.00 | 1,450 |
| #11 - #12 | " | 1,060 | 350.00 | 1,410 |
| Wire mesh, slabs | | | | |
| Galvanized | | | | |
| 4x4 | | | | |
| W1.4xW1.4 | S.F. | 0.17 | 0.23 | 0.40 |
| W2.0xW2.0 | " | 0.22 | 0.25 | 0.47 |
| W2.9xW2.9 | " | 0.32 | 0.27 | 0.59 |
| W4.0xW4.0 | " | 0.48 | 0.29 | 0.77 |
| 6x6 | | | | |
| W1.4xW1.4 | S.F. | 0.16 | 0.17 | 0.34 |
| W2.0xW2.0 | " | 0.22 | 0.19 | 0.41 |
| W2.9xW2.9 | " | 0.31 | 0.21 | 0.52 |
| W4.0xW4.0 | " | 0.33 | 0.23 | 0.56 |
| Standard | | | | |
| 2x2 | | | | |
| W.9xW.9 | S.F. | 0.18 | 0.23 | 0.41 |
| 4x4 | | | | |
| W1.4xW1.4 | S.F. | 0.12 | 0.23 | 0.35 |
| W2.0xW2.0 | " | 0.16 | 0.25 | 0.41 |

| REINFORCEMENT | UNIT | MAT. | INST. | TOTAL |
|---|---|---|---|---|
| **03210.55 SLAB/MAT REINFORCING** | | | | |
| W2.9xW2.9 | S.F. | 0.21 | 0.27 | 0.48 |
| W4.0xW4.0 | " | 0.33 | 0.29 | 0.62 |
| 6x6 | | | | |
| W1.4xW1.4 | S.F. | 0.08 | 0.17 | 0.26 |
| W2.0xW2.0 | " | 0.11 | 0.19 | 0.30 |
| W2.9xW2.9 | " | 0.16 | 0.21 | 0.37 |
| W4.0xW4.0 | " | 0.22 | 0.23 | 0.45 |
| **03210.60 STAIR REINFORCING** | | | | |
| Stairs | | | | |
| #3 - #4 | TON | 660.00 | 700.00 | 1,360 |
| #5 - #6 | " | 590.00 | 580.00 | 1,170 |
| #7 - #8 | " | 570.00 | 500.00 | 1,070 |
| #9 - #10 | " | 570.00 | 440.00 | 1,010 |
| Galvanized | | | | |
| #3 - #4 | TON | 1,160 | 700.00 | 1,860 |
| #5 - #6 | " | 1,100 | 580.00 | 1,680 |
| #7 - #8 | " | 1,060 | 500.00 | 1,560 |
| #9 - #10 | " | 1,060 | 440.00 | 1,500 |
| **03210.65 WALL REINFORCING** | | | | |
| Walls | | | | |
| #3 - #4 | TON | 660.00 | 500.00 | 1,160 |
| #5 - #6 | " | 590.00 | 440.00 | 1,030 |
| #7 - #8 | " | 570.00 | 390.00 | 960.00 |
| #9 - #10 | " | 570.00 | 350.00 | 920.00 |
| Galvanized | | | | |
| #3 - #4 | TON | 1,160 | 500.00 | 1,660 |
| #5 - #6 | " | 1,100 | 440.00 | 1,540 |
| #7 - #8 | " | 1,060 | 390.00 | 1,450 |
| #9 - #10 | " | 1,060 | 350.00 | 1,410 |
| Masonry wall (horizontal) | | | | |
| #3 - #4 | TON | 660.00 | 1,400 | 2,060 |
| #5 - #6 | " | 590.00 | 1,170 | 1,760 |
| Galvanized | | | | |
| #3 - #4 | TON | 1,160 | 1,400 | 2,560 |
| #5 - #6 | " | 1,100 | 1,170 | 2,270 |
| Masonry wall (vertical) | | | | |
| #3 - #4 | TON | 660.00 | 1,750 | 2,410 |
| #5 - #6 | " | 590.00 | 1,400 | 1,990 |
| Galvanized | | | | |
| #3 - #4 | TON | 1,160 | 1,750 | 2,910 |
| #5 - #6 | " | 1,100 | 1,400 | 2,500 |

# 03 CONCRETE

| ACCESSORIES | UNIT | MAT. | INST. | TOTAL |
|---|---|---|---|---|
| **03250.40 CONCRETE ACCESSORIES** | | | | |
| Expansion joint, poured | | | | |
| Asphalt | | | | |
| 1/2" x 1" | L.F. | 0.36 | 0.48 | 0.84 |
| 1" x 2" | " | 1.10 | 0.53 | 1.63 |
| Liquid neoprene, cold applied | | | | |
| 1/2" x 1" | L.F. | 1.35 | 0.49 | 1.84 |
| 1" x 2" | " | 5.40 | 0.54 | 5.94 |
| Polyurethane, 2 parts | | | | |
| 1/2" x 1" | L.F. | 1.25 | 0.81 | 2.06 |
| 1" x 2" | " | 5.20 | 0.88 | 6.08 |
| Rubberized asphalt, cold | | | | |
| 1/2" x 1" | L.F. | 0.30 | 0.48 | 0.78 |
| 1" x 2" | " | 0.97 | 0.53 | 1.50 |
| Hot, fuel resistant | | | | |
| 1/2" x 1" | L.F. | 0.57 | 0.48 | 1.05 |
| 1" x 2" | " | 2.85 | 0.53 | 3.38 |
| Expansion joint, premolded, in slabs | | | | |
| Asphalt | | | | |
| 1/2" x 6" | L.F. | 0.39 | 0.60 | 1.00 |
| 1" x 12" | " | 0.68 | 0.81 | 1.49 |
| Cork | | | | |
| 1/2" x 6" | L.F. | 0.91 | 0.60 | 1.51 |
| 1" x 12" | " | 3.60 | 0.81 | 4.41 |
| Neoprene sponge | | | | |
| 1/2" x 6" | L.F. | 1.25 | 0.60 | 1.86 |
| 1" x 12" | " | 4.65 | 0.81 | 5.46 |
| Polyethylene foam | | | | |
| 1/2" x 6" | L.F. | 0.43 | 0.60 | 1.03 |
| 1" x 12" | " | 2.25 | 0.81 | 3.06 |
| Polyurethane foam | | | | |
| 1/2" x 6" | L.F. | 0.61 | 0.60 | 1.22 |
| 1" x 12" | " | 1.35 | 0.81 | 2.16 |
| Polyvinyl chloride foam | | | | |
| 1/2" x 6" | L.F. | 1.35 | 0.60 | 1.96 |
| 1" x 12" | " | 4.85 | 0.81 | 5.66 |
| Rubber, gray sponge | | | | |
| 1/2" x 6" | L.F. | 2.15 | 0.60 | 2.75 |
| 1" x 12" | " | 9.05 | 0.81 | 9.86 |
| Asphalt felt control joints or bond breaker, screed joints | | | | |
| 4" slab | L.F. | 0.57 | 0.48 | 1.05 |
| 6" slab | " | 0.73 | 0.54 | 1.27 |
| 8" slab | " | 0.97 | 0.60 | 1.58 |
| 10" slab | " | 1.35 | 0.69 | 2.04 |
| Keyed cold expansion and control joints, 24 ga. | | | | |
| 4" slab | L.F. | 0.51 | 1.50 | 2.01 |
| 5" slab | " | 0.62 | 1.50 | 2.12 |
| 6" slab | " | 0.73 | 1.60 | 2.33 |
| 8" slab | " | 0.89 | 1.75 | 2.64 |
| 10" slab | " | 1.00 | 1.85 | 2.85 |
| Waterstops | | | | |
| Polyvinyl chloride | | | | |
| Ribbed | | | | |

| ACCESSORIES | UNIT | MAT. | INST. | TOTAL |
|---|---|---|---|---|
| **03250.40 CONCRETE ACCESSORIES** | | | | |
| 3/16" thick x | | | | |
| 4" wide | L.F. | 0.85 | 1.20 | 2.05 |
| 6" wide | " | 1.05 | 1.35 | 2.40 |
| 1/2" thick x | | | | |
| 9" wide | L.F. | 2.85 | 1.50 | 4.35 |
| Ribbed with center bulb | | | | |
| 3/16" thick x 9" wide | L.F. | 2.50 | 1.50 | 4.00 |
| 3/8" thick x 9" wide | " | 3.05 | 1.50 | 4.55 |
| Dumbbell type, 3/8" thick x 6" wide | " | 3.05 | 1.35 | 4.40 |
| Plain, 3/8" thick x 9" wide | " | 3.70 | 1.50 | 5.20 |
| Center bulb, 3/8" thick x 9" wide | " | 5.10 | 1.50 | 6.60 |
| Rubber | | | | |
| Flat dumbbell | | | | |
| 3/8" thick x | | | | |
| 6" wide | L.F. | 4.90 | 1.35 | 6.25 |
| 9" wide | " | 7.80 | 1.50 | 9.30 |
| Center bulb | | | | |
| 3/8" thick x | | | | |
| 6" wide | L.F. | 5.20 | 1.35 | 6.55 |
| 9" wide | " | 9.30 | 1.50 | 10.80 |
| Vapor barrier | | | | |
| 4 mil polyethylene | S.F. | 0.02 | 0.08 | 0.10 |
| 6 mil polyethylene | " | 0.03 | 0.08 | 0.11 |
| Gravel porous fill, under floor slabs, 3/4" stone | C.Y. | 14.15 | 40.30 | 54.45 |
| Reinforcing accessories | | | | |
| Beam bolsters | | | | |
| 1-1/2" high, plain | L.F. | 0.32 | 0.35 | 0.67 |
| Galvanized | " | 0.40 | 0.35 | 0.75 |
| 3" high | | | | |
| Plain | L.F. | 0.78 | 0.44 | 1.22 |
| Galvanized | " | 0.83 | 0.44 | 1.27 |
| Slab bolsters | | | | |
| 1" high | | | | |
| Plain | L.F. | 0.22 | 0.17 | 0.40 |
| Galvanized | " | 0.30 | 0.17 | 0.47 |
| 2" high | | | | |
| Plain | L.F. | 0.30 | 0.19 | 0.49 |
| Galvanized | " | 0.34 | 0.19 | 0.53 |
| Chairs, high chairs | | | | |
| 3" high | | | | |
| Plain | EA. | 0.36 | 0.87 | 1.23 |
| Galvanized | " | 0.47 | 0.87 | 1.34 |
| 5" high | | | | |
| Plain | EA. | 0.47 | 0.92 | 1.39 |
| Galvanized | " | 0.58 | 0.92 | 1.50 |
| 8" high | | | | |
| Plain | EA. | 0.95 | 1.00 | 1.95 |
| Galvanized | " | 1.25 | 1.00 | 2.25 |
| 12" high | | | | |
| Plain | EA. | 2.05 | 1.15 | 3.20 |
| Galvanized | " | 2.40 | 1.15 | 3.55 |
| Continuous, high chair | | | | |

# 03 CONCRETE

## ACCESSORIES

| 03250.40 CONCRETE ACCESSORIES | UNIT | MAT. | INST. | TOTAL |
|---|---|---|---|---|
| 3" high | | | | |
| Plain | L.F. | 0.41 | 0.23 | 0.64 |
| Galvanized | " | 0.51 | 0.23 | 0.74 |
| 5" high | | | | |
| Plain | L.F. | 0.63 | 0.25 | 0.88 |
| Galvanized | " | 0.83 | 0.25 | 1.08 |
| 8" high | | | | |
| Plain | L.F. | 0.85 | 0.27 | 1.12 |
| Galvanized | " | 1.10 | 0.27 | 1.37 |
| 12" high | | | | |
| Plain | L.F. | 1.85 | 0.29 | 2.14 |
| Galvanized | " | 2.15 | 0.29 | 2.44 |

## CAST-IN-PLACE CONCRETE

| 03300.10 CONCRETE ADMIXTURES | UNIT | MAT. | INST. | TOTAL |
|---|---|---|---|---|
| Concrete admixtures | | | | |
| Water reducing admixture | GAL | | | 10.10 |
| Set retarder | " | | | 17.87 |
| Air entraining agent | " | | | 5.99 |

| 03350.10 CONCRETE FINISHES | UNIT | MAT. | INST. | TOTAL |
|---|---|---|---|---|
| Floor finishes | | | | |
| Broom | S.F. | 0.00 | 0.35 | 0.35 |
| Screed | " | 0.00 | 0.30 | 0.30 |
| Darby | " | 0.00 | 0.30 | 0.30 |
| Steel float | " | 0.00 | 0.40 | 0.40 |
| Granolithic topping | | | | |
| 1/2" thick | S.F. | 0.20 | 1.10 | 1.30 |
| 1" thick | " | 0.37 | 1.20 | 1.57 |
| 2" thick | " | 0.65 | 1.35 | 2.00 |
| Wall finishes | | | | |
| Burlap rub, with cement paste | S.F. | 0.06 | 0.40 | 0.46 |
| Float finish | " | 0.08 | 0.60 | 0.69 |
| Etch with acid | " | 0.26 | 0.40 | 0.66 |
| Sandblast | | | | |
| Minimum | S.F. | 0.06 | 0.62 | 0.68 |
| Maximum | " | 0.30 | 0.62 | 0.92 |
| Bush hammer | | | | |
| Green concrete | S.F. | 0.00 | 1.20 | 1.20 |
| Cured concrete | " | 0.00 | 1.85 | 1.85 |
| Break ties and patch holes | " | 0.00 | 0.48 | 0.48 |
| Carborundum | | | | |
| Dry rub | S.F. | 0.00 | 0.81 | 0.81 |

# 03 CONCRETE

| CAST-IN-PLACE CONCRETE | UNIT | MAT. | INST. | TOTAL |
|---|---|---|---|---|
| **03350.10 CONCRETE FINISHES** | | | | |
| Wet rub | S.F. | 0.00 | 1.20 | 1.20 |
| Floor hardeners | | | | |
| Metallic | | | | |
| Light service | S.F. | 0.20 | 0.30 | 0.50 |
| Heavy service | " | 0.65 | 0.40 | 1.05 |
| Non-metallic | | | | |
| Light service | S.F. | 0.10 | 0.30 | 0.40 |
| Heavy service | " | 0.46 | 0.40 | 0.86 |
| Rusticated concrete finish | | | | |
| Beveled edge | L.F. | 0.23 | 1.35 | 1.58 |
| Square edge | " | 0.29 | 1.75 | 2.04 |
| Solid board concrete finish | | | | |
| Standard | S.F. | 0.60 | 2.00 | 2.60 |
| Rustic | " | 0.53 | 2.40 | 2.93 |
| **03360.10 PNEUMATIC CONCRETE** | | | | |
| Pneumatic applied concrete (gunite) | | | | |
| 2" thick | S.F. | 2.95 | 1.65 | 4.60 |
| 3" thick | " | 3.80 | 2.20 | 6.00 |
| 4" thick | " | 4.50 | 2.60 | 7.10 |
| Finish surface | | | | |
| Minimum | S.F. | 0.00 | 1.55 | 1.55 |
| Maximum | " | 0.00 | 3.10 | 3.10 |
| **03370.10 CURING CONCRETE** | | | | |
| Sprayed membrane | | | | |
| Slabs | S.F. | 0.04 | 0.05 | 0.09 |
| Walls | " | 0.07 | 0.06 | 0.13 |
| Curing paper | | | | |
| Slabs | S.F. | 0.07 | 0.06 | 0.13 |
| Walls | " | 0.07 | 0.07 | 0.14 |
| Burlap | | | | |
| 7.5 oz. | S.F. | 0.06 | 0.08 | 0.14 |
| 12 oz. | " | 0.08 | 0.09 | 0.17 |

| PLACING CONCRETE | UNIT | MAT. | INST. | TOTAL |
|---|---|---|---|---|
| **03380.05 BEAM CONCRETE** | | | | |
| Beams and girders | | | | |
| 2500# or 3000# concrete | | | | |
| By crane | C.Y. | 64.50 | 46.20 | 110.70 |
| By pump | " | 64.50 | 42.00 | 106.50 |

## PLACING CONCRETE

| PLACING CONCRETE | UNIT | MAT. | INST. | TOTAL |
|---|---|---|---|---|
| **03380.05 BEAM CONCRETE** | | | | |
| By hand buggy | C.Y. | 64.50 | 24.20 | 88.70 |
| 3500# or 4000# concrete | | | | |
| By crane | C.Y. | 69.00 | 46.20 | 115.20 |
| By pump | " | 69.00 | 42.00 | 111.00 |
| By hand buggy | " | 69.00 | 24.20 | 93.20 |
| 5000# concrete | | | | |
| By crane | C.Y. | 73.50 | 46.20 | 119.70 |
| By pump | " | 73.50 | 42.00 | 115.50 |
| By hand buggy | " | 73.50 | 24.20 | 97.70 |
| Bond beam, 3000# concrete | | | | |
| By pump | | | | |
| 8" high | | | | |
| 4" wide | L.F. | 0.17 | 0.93 | 1.10 |
| 6" wide | " | 0.41 | 1.05 | 1.46 |
| 8" wide | " | 0.53 | 1.15 | 1.68 |
| 10" wide | " | 0.71 | 1.30 | 2.01 |
| 12" wide | " | 0.95 | 1.45 | 2.40 |
| 16" high | | | | |
| 8" wide | L.F. | 1.30 | 1.45 | 2.75 |
| 10" wide | " | 1.75 | 1.65 | 3.40 |
| 12" wide | " | 2.30 | 1.95 | 4.25 |
| By crane | | | | |
| 8" high | | | | |
| 4" wide | L.F. | 0.20 | 1.00 | 1.20 |
| 6" wide | " | 0.41 | 1.10 | 1.51 |
| 8" wide | " | 0.53 | 1.15 | 1.68 |
| 10" wide | " | 0.71 | 1.30 | 2.01 |
| 12" wide | " | 0.95 | 1.45 | 2.40 |
| 16" high | | | | |
| 8" wide | L.F. | 1.30 | 1.45 | 2.75 |
| 10" wide | " | 1.75 | 1.55 | 3.30 |
| 12" wide | " | 2.30 | 1.80 | 4.10 |
| **03380.15 COLUMN CONCRETE** | | | | |
| Columns | | | | |
| 2500# or 3000# concrete | | | | |
| By crane | C.Y. | 64.50 | 42.00 | 106.50 |
| By pump | " | 64.50 | 38.50 | 103.00 |
| 3500# or 4000# concrete | | | | |
| By crane | C.Y. | 69.00 | 42.00 | 111.00 |
| By pump | " | 69.00 | 38.50 | 107.50 |
| 5000# concrete | | | | |
| By crane | C.Y. | 73.50 | 42.00 | 115.50 |
| By pump | " | 73.50 | 38.50 | 112.00 |
| **03380.20 ELEVATED SLAB CONCRETE** | | | | |
| Elevated slab | | | | |
| 2500# or 3000# concrete | | | | |
| By crane | C.Y. | 64.50 | 23.10 | 87.60 |
| By pump | " | 64.50 | 17.80 | 82.30 |
| By hand buggy | " | 64.50 | 24.20 | 88.70 |
| 3500# or 4000# concrete | | | | |

# 03 CONCRETE

| PLACING CONCRETE | UNIT | MAT. | INST. | TOTAL |
| --- | --- | --- | --- | --- |
| **03380.20 ELEVATED SLAB CONCRETE** | | | | |
| By crane | C.Y. | 69.00 | 23.10 | 92.10 |
| By pump | " | 69.00 | 17.80 | 86.80 |
| By hand buggy | " | 69.00 | 24.20 | 93.20 |
| 5000# concrete | | | | |
| By crane | C.Y. | 73.50 | 23.10 | 96.60 |
| By pump | " | 73.50 | 17.80 | 91.30 |
| By hand buggy | " | 73.50 | 24.20 | 97.70 |
| Topping | | | | |
| 2500# or 3000# concrete | | | | |
| By crane | C.Y. | 64.50 | 23.10 | 87.60 |
| By pump | " | 64.50 | 17.80 | 82.30 |
| By hand buggy | " | 64.50 | 24.20 | 88.70 |
| 3500# or 4000# concrete | | | | |
| By crane | C.Y. | 69.00 | 23.10 | 92.10 |
| By pump | " | 69.00 | 17.80 | 86.80 |
| By hand buggy | " | 69.00 | 24.20 | 93.20 |
| 5000# concrete | | | | |
| By crane | C.Y. | 73.50 | 23.10 | 96.60 |
| By pump | " | 73.50 | 17.80 | 91.30 |
| By hand buggy | " | 73.50 | 24.20 | 97.70 |
| **03380.25 EQUIPMENT PAD CONCRETE** | | | | |
| Equipment pad | | | | |
| 2500# or 3000# concrete | | | | |
| By chute | C.Y. | 64.50 | 8.05 | 72.55 |
| By pump | " | 64.50 | 33.00 | 97.50 |
| By crane | " | 64.50 | 38.50 | 103.00 |
| 3500# or 4000# concrete | | | | |
| By chute | C.Y. | 69.00 | 8.05 | 77.05 |
| By pump | " | 69.00 | 33.00 | 102.00 |
| By crane | " | 69.00 | 38.50 | 107.50 |
| 5000# concrete | | | | |
| By chute | C.Y. | 73.50 | 8.05 | 81.55 |
| By pump | " | 73.50 | 33.00 | 106.50 |
| By crane | " | 73.50 | 38.50 | 112.00 |
| **03380.35 FOOTING CONCRETE** | | | | |
| Continuous footing | | | | |
| 2500# or 3000# concrete | | | | |
| By chute | C.Y. | 64.50 | 8.05 | 72.55 |
| By pump | " | 64.50 | 28.90 | 93.40 |
| By crane | " | 64.50 | 33.00 | 97.50 |
| 3500# or 4000# concrete | | | | |
| By chute | C.Y. | 69.00 | 8.05 | 77.05 |
| By pump | " | 69.00 | 28.90 | 97.90 |
| By crane | " | 69.00 | 33.00 | 102.00 |
| 5000# concrete | | | | |
| By chute | C.Y. | 73.50 | 8.05 | 81.55 |
| By pump | " | 73.50 | 28.90 | 102.40 |
| By crane | " | 73.50 | 33.00 | 106.50 |
| Spread footing | | | | |

## PLACING CONCRETE

| | UNIT | MAT. | INST. | TOTAL |
|---|---|---|---|---|
| **03380.35 FOOTING CONCRETE** | | | | |
| 2500# or 3000# concrete | | | | |
| Under 5 cy | | | | |
| By chute | C.Y. | 64.50 | 8.05 | 72.55 |
| By pump | " | 64.50 | 30.80 | 95.30 |
| By crane | " | 64.50 | 35.60 | 100.10 |
| Over 5 cy | | | | |
| By chute | C.Y. | 64.50 | 6.05 | 70.55 |
| By pump | " | 64.50 | 27.20 | 91.70 |
| By crane | " | 64.50 | 30.80 | 95.30 |
| 3500# or 4000# concrete | | | | |
| Under 5 c.y. | | | | |
| By chute | C.Y. | 69.00 | 8.05 | 77.05 |
| By pump | " | 69.00 | 30.80 | 99.80 |
| By crane | " | 69.00 | 35.60 | 104.60 |
| Over 5 c.y. | | | | |
| By chute | C.Y. | 69.00 | 6.05 | 75.05 |
| By pump | " | 69.00 | 27.20 | 96.20 |
| By crane | " | 69.00 | 30.80 | 99.80 |
| 5000# concrete | | | | |
| Under 5 c.y. | | | | |
| By chute | C.Y. | 73.50 | 8.05 | 81.55 |
| By pump | " | 73.50 | 30.80 | 104.30 |
| By crane | " | 73.50 | 35.60 | 109.10 |
| Over 5 c.y. | | | | |
| By chute | C.Y. | 73.50 | 6.05 | 79.55 |
| By pump | " | 73.50 | 27.20 | 100.70 |
| By crane | " | 73.50 | 30.80 | 104.30 |
| **03380.50 GRADE BEAM CONCRETE** | | | | |
| Grade beam | | | | |
| 2500# or 3000# concrete | | | | |
| By chute | C.Y. | 64.50 | 8.05 | 72.55 |
| By crane | " | 64.50 | 33.00 | 97.50 |
| By pump | " | 64.50 | 28.90 | 93.40 |
| By hand buggy | " | 64.50 | 24.20 | 88.70 |
| 3500# or 4000# concrete | | | | |
| By chute | C.Y. | 69.00 | 8.05 | 77.05 |
| By crane | " | 69.00 | 33.00 | 102.00 |
| By pump | " | 69.00 | 28.90 | 97.90 |
| By hand buggy | " | 69.00 | 24.20 | 93.20 |
| 5000# concrete | | | | |
| By chute | C.Y. | 73.50 | 8.05 | 81.55 |
| By crane | " | 73.50 | 33.00 | 106.50 |
| By pump | " | 73.50 | 28.90 | 102.40 |
| By hand buggy | " | 73.50 | 24.20 | 97.70 |
| **03380.53 PILE CAP CONCRETE** | | | | |
| Pile cap | | | | |
| 2500# or 3000 concrete | | | | |
| By chute | C.Y. | 64.50 | 8.05 | 72.55 |
| By crane | " | 64.50 | 38.50 | 103.00 |

| PLACING CONCRETE | UNIT | MAT. | INST. | TOTAL |
|---|---|---|---|---|
| **03380.53 PILE CAP CONCRETE** | | | | |
| By pump | C.Y. | 64.50 | 33.00 | 97.50 |
| By hand buggy | " | 64.50 | 24.20 | 88.70 |
| 3500# or 4000# concrete | | | | |
| By chute | C.Y. | 69.00 | 8.05 | 77.05 |
| By crane | " | 69.00 | 38.50 | 107.50 |
| By pump | " | 69.00 | 33.00 | 102.00 |
| By hand buggy | " | 69.00 | 24.20 | 93.20 |
| 5000# concrete | | | | |
| By chute | C.Y. | 73.50 | 8.05 | 81.55 |
| By crane | " | 73.50 | 38.50 | 112.00 |
| By pump | " | 73.50 | 33.00 | 106.50 |
| By hand buggy | " | 73.50 | 24.20 | 97.70 |
| **03380.55 SLAB/MAT CONCRETE** | | | | |
| Slab on grade | | | | |
| 2500# or 3000# concrete | | | | |
| By chute | C.Y. | 64.50 | 6.05 | 70.55 |
| By crane | " | 64.50 | 19.25 | 83.75 |
| By pump | " | 64.50 | 16.50 | 81.00 |
| By hand buggy | " | 64.50 | 16.15 | 80.65 |
| 3500# or 4000# concrete | | | | |
| By chute | C.Y. | 69.00 | 6.05 | 75.05 |
| By crane | " | 69.00 | 19.25 | 88.25 |
| By pump | " | 69.00 | 16.50 | 85.50 |
| By hand buggy | " | 69.00 | 16.15 | 85.15 |
| 5000# concrete | | | | |
| By chute | C.Y. | 73.50 | 6.05 | 79.55 |
| By crane | " | 73.50 | 19.25 | 92.75 |
| By pump | " | 73.50 | 16.50 | 90.00 |
| By hand buggy | " | 73.50 | 16.15 | 89.65 |
| Foundation mat | | | | |
| 2500# or 3000# concrete, over 20 cy | | | | |
| By chute | C.Y. | 64.50 | 4.85 | 69.35 |
| By crane | " | 64.50 | 16.50 | 81.00 |
| By pump | " | 64.50 | 14.45 | 78.95 |
| By hand buggy | " | 64.50 | 12.10 | 76.60 |
| **03380.58 SIDEWALKS** | | | | |
| Walks, cast in place with wire mesh, base not incl. | | | | |
| 4" thick | S.F. | 0.88 | 0.81 | 1.69 |
| 5" thick | " | 1.20 | 0.97 | 2.17 |
| 6" thick | " | 1.40 | 1.20 | 2.60 |
| **03380.60 STAIR CONCRETE** | | | | |
| Stairs | | | | |
| 2500# or 3000# concrete | | | | |
| By chute | C.Y. | 64.50 | 8.05 | 72.55 |
| By crane | " | 64.50 | 38.50 | 103.00 |
| By pump | " | 64.50 | 33.00 | 97.50 |
| By hand buggy | " | 64.50 | 24.20 | 88.70 |
| 3500# or 4000# concrete | | | | |
| By chute | C.Y. | 69.00 | 8.05 | 77.05 |

# 03 CONCRETE

## PLACING CONCRETE

| PLACING CONCRETE | UNIT | MAT. | INST. | TOTAL |
|---|---|---|---|---|
| **03380.60 STAIR CONCRETE** | | | | |
| By crane | C.Y. | 69.00 | 38.50 | 107.50 |
| By pump | " | 69.00 | 33.00 | 102.00 |
| By hand buggy | " | 69.00 | 24.20 | 93.20 |
| 5000# concrete | | | | |
| By chute | C.Y. | 73.50 | 8.05 | 81.55 |
| By crane | " | 73.50 | 38.50 | 112.00 |
| By pump | " | 73.50 | 33.00 | 106.50 |
| By hand buggy | " | 73.50 | 24.20 | 97.70 |
| **03380.65 WALL CONCRETE** | | | | |
| Walls | | | | |
| 2500# or 3000# concrete | | | | |
| To 4' | | | | |
| By chute | C.Y. | 64.50 | 6.90 | 71.40 |
| By crane | " | 64.50 | 38.50 | 103.00 |
| By pump | " | 64.50 | 35.60 | 100.10 |
| To 8' | | | | |
| By crane | C.Y. | 64.50 | 42.00 | 106.50 |
| By pump | " | 64.50 | 38.50 | 103.00 |
| To 16' | | | | |
| By crane | C.Y. | 64.50 | 46.20 | 110.70 |
| By pump | " | 64.50 | 42.00 | 106.50 |
| Over 16' | | | | |
| By crane | C.Y. | 64.50 | 51.50 | 116.00 |
| By pump | " | 64.50 | 46.20 | 110.70 |
| 3500# or 4000# concrete | | | | |
| To 4' | | | | |
| By chute | C.Y. | 69.00 | 6.90 | 75.90 |
| By crane | " | 69.00 | 38.50 | 107.50 |
| By pump | " | 69.00 | 35.60 | 104.60 |
| To 8' | | | | |
| By crane | C.Y. | 69.00 | 42.00 | 111.00 |
| By pump | " | 69.00 | 38.50 | 107.50 |
| To 16' | | | | |
| By crane | C.Y. | 69.00 | 46.20 | 115.20 |
| By pump | " | 69.00 | 42.00 | 111.00 |
| Over 16' | | | | |
| By crane | C.Y. | 69.00 | 51.50 | 120.50 |
| By pump | " | 69.00 | 46.20 | 115.20 |
| 5000# concrete | | | | |
| To 4' | | | | |
| By chute | C.Y. | 73.50 | 6.90 | 80.40 |
| By crane | " | 73.50 | 38.50 | 112.00 |
| By pump | " | 73.50 | 35.60 | 109.10 |
| To 8' | | | | |
| By crane | C.Y. | 73.50 | 42.00 | 115.50 |
| By pump | " | 73.50 | 38.50 | 112.00 |
| To 16' | | | | |
| By crane | C.Y. | 73.50 | 46.20 | 119.70 |
| By pump | " | 73.50 | 42.00 | 115.50 |
| Filled block (CMU) | | | | |
| 3000# concrete, by pump | | | | |

# 03 CONCRETE

## PLACING CONCRETE

| 03380.65 WALL CONCRETE | UNIT | MAT. | INST. | TOTAL |
|---|---|---|---|---|
| 4" wide | S.F. | 0.23 | 1.65 | 1.88 |
| 6" wide | " | 0.53 | 1.95 | 2.48 |
| 8" wide | " | 0.83 | 2.30 | 3.13 |
| 10" wide | " | 1.10 | 2.70 | 3.80 |
| 12" wide | " | 1.40 | 3.30 | 4.70 |
| Pilasters, 3000# concrete | C.F. | 3.25 | 46.20 | 49.45 |
| Wall cavity, 2" thick, 3000# concrete | S.F. | 0.60 | 1.55 | 2.15 |

## PRECAST CONCRETE

| 03400.10 PRECAST BEAMS | UNIT | MAT. | INST. | TOTAL |
|---|---|---|---|---|
| Prestressed, double tee, 24" deep, 8' wide | | | | |
| 35' span | | | | |
| 115 psf | S.F. | 5.85 | 0.64 | 6.49 |
| 140 psf | " | 6.20 | 0.64 | 6.84 |
| 40' span | | | | |
| 80 psf | S.F. | 5.60 | 0.68 | 6.28 |
| 143 psf | " | 5.95 | 0.68 | 6.63 |
| 45' span | | | | |
| 50 psf | S.F. | 5.35 | 0.59 | 5.94 |
| 70 psf | " | 5.70 | 0.59 | 6.29 |
| 100 psf | " | 5.85 | 0.59 | 6.44 |
| 130 psf | " | 6.45 | 0.59 | 7.04 |
| 50' span | | | | |
| 75 psf | S.F. | 5.35 | 0.53 | 5.88 |
| 100 psf | " | 5.85 | 0.53 | 6.38 |
| Precast beams, girders and joists | | | | |
| 1000 lb/lf live load | | | | |
| 10' span | L.F. | 49.30 | 12.70 | 62.00 |
| 20' span | " | 52.00 | 7.60 | 59.60 |
| 30' span | " | 65.50 | 6.35 | 71.85 |
| 3000 lb/lf live load | | | | |
| 10' span | L.F. | 51.50 | 12.70 | 64.20 |
| 20' span | " | 58.50 | 7.60 | 66.10 |
| 30' span | " | 77.50 | 6.35 | 83.85 |
| 5000 lb/lf live load | | | | |
| 10' span | L.F. | 53.00 | 12.70 | 65.70 |
| 20' span | " | 69.00 | 7.60 | 76.60 |
| 30' span | " | 87.00 | 6.35 | 93.35 |

| 03400.20 PRECAST COLUMNS | UNIT | MAT. | INST. | TOTAL |
|---|---|---|---|---|
| Prestressed concrete columns | | | | |
| 10" x 10" | | | | |

# 03 CONCRETE

| PRECAST CONCRETE | UNIT | MAT. | INST. | TOTAL |
|---|---|---|---|---|
| **03400.20 PRECAST COLUMNS** | | | | |
| 10' long | EA. | 140.00 | 76.00 | 216.00 |
| 15' long | " | 210.00 | 79.50 | 289.50 |
| 20' long | " | 290.00 | 84.50 | 374.50 |
| 25' long | " | 370.00 | 90.50 | 460.50 |
| 30' long | " | 440.00 | 95.00 | 535.00 |
| 12" x 12" | | | | |
| 20' long | EA. | 390.00 | 95.00 | 485.00 |
| 25' long | " | 490.00 | 100.00 | 590.00 |
| 30' long | " | 610.00 | 110.00 | 720.00 |
| 16" x 16" | | | | |
| 20' long | EA. | 700.00 | 95.00 | 795.00 |
| 25' long | " | 890.00 | 100.00 | 990.00 |
| 30' long | " | 1,060 | 110.00 | 1,170 |
| 20" x 20" | | | | |
| 20' long | EA. | 1,090 | 100.00 | 1,190 |
| 25' long | " | 1,490 | 110.00 | 1,600 |
| 30' long | " | 1,750 | 110.00 | 1,860 |
| 24" x 24" | | | | |
| 20' long | EA. | 1,660 | 110.00 | 1,770 |
| 25' long | " | 2,020 | 110.00 | 2,130 |
| 30' long | " | 2,500 | 120.00 | 2,620 |
| 28" x 28" | | | | |
| 20' long | EA. | 2,260 | 120.00 | 2,380 |
| 25' long | " | 2,730 | 130.00 | 2,860 |
| 30' long | " | 3,390 | 140.00 | 3,530 |
| 32" x 32" | | | | |
| 20' long | EA. | 2,850 | 130.00 | 2,980 |
| 25' long | " | 3,690 | 140.00 | 3,830 |
| 30' long | " | 4,220 | 150.00 | 4,370 |
| 36" x 36" | | | | |
| 20' long | EA. | 3,570 | 140.00 | 3,710 |
| 25' long | " | 4,460 | 150.00 | 4,610 |
| 30' long | " | 5,350 | 160.00 | 5,510 |
| **03400.30 PRECAST SLABS** | | | | |
| Prestressed flat slab | | | | |
| 6" thick, 4' wide | | | | |
| 20' span | | | | |
| 80 psf | S.F. | 9.40 | 1.60 | 11.00 |
| 110 psf | " | 9.45 | 1.60 | 11.05 |
| 25' span | | | | |
| 80 psf | S.F. | 9.85 | 1.50 | 11.35 |
| Cored slab | | | | |
| 6" thick, 4' wide | | | | |
| 20' span | | | | |
| 80 psf | S.F. | 4.15 | 1.60 | 5.75 |
| 100 psf | " | 4.20 | 1.60 | 5.80 |
| 130 psf | " | 4.25 | 1.60 | 5.85 |
| 8" thick, 4' wide | | | | |
| 25' span | | | | |
| 70 psf | S.F. | 4.65 | 1.50 | 6.15 |

# 03 CONCRETE

| PRECAST CONCRETE | UNIT | MAT. | INST. | TOTAL |
|---|---|---|---|---|
| **03400.30 PRECAST SLABS** | | | | |
| 125 psf | S.F. | 4.75 | 1.50 | 6.25 |
| 170 psf | " | 4.80 | 1.50 | 6.30 |
| 30' span | | | | |
| 70 psf | S.F. | 4.65 | 1.25 | 5.90 |
| 90 psf | " | 4.95 | 1.25 | 6.20 |
| 35' span | | | | |
| 70 psf | S.F. | 4.85 | 1.20 | 6.05 |
| 10" thick, 4' wide | | | | |
| 30' span | | | | |
| 75 psf | S.F. | 4.85 | 1.25 | 6.10 |
| 100 psf | " | 5.00 | 1.25 | 6.25 |
| 130 psf | " | 5.10 | 1.25 | 6.35 |
| 35' span | | | | |
| 60 psf | S.F. | 5.00 | 1.20 | 6.20 |
| 80 psf | " | 5.10 | 1.20 | 6.30 |
| 120 psf | " | 5.35 | 1.20 | 6.55 |
| 40' span | | | | |
| 65 psf | S.F. | 5.35 | 0.95 | 6.30 |
| Slabs, roof and floor members, 4' wide | | | | |
| 6" thick, 25' span | S.F. | 4.25 | 1.50 | 5.75 |
| 8" thick, 30' span | " | 4.95 | 1.15 | 6.10 |
| 10" thick, 40' span | " | 6.05 | 1.05 | 7.10 |
| Tee members | | | | |
| Multiple tee, roof and floor | | | | |
| Minimum | S.F. | 5.60 | 0.95 | 6.55 |
| Maximum | " | 7.00 | 1.90 | 8.90 |
| Double tee wall member | | | | |
| Minimum | S.F. | 5.10 | 1.10 | 6.20 |
| Maximum | " | 6.50 | 2.10 | 8.60 |
| Single tee | | | | |
| Short span, roof members | | | | |
| Minimum | S.F. | 5.75 | 1.15 | 6.90 |
| Maximum | " | 7.15 | 2.40 | 9.55 |
| Long span, roof members | | | | |
| Minimum | S.F. | 7.30 | 0.95 | 8.25 |
| Maximum | " | 8.70 | 1.90 | 10.60 |
| **03400.40 PRECAST WALLS** | | | | |
| Wall panel, 8' x 20' | | | | |
| Gray cement | | | | |
| Liner finish | | | | |
| 4" wall | S.F. | 7.25 | 1.10 | 8.35 |
| 5" wall | " | 7.85 | 1.10 | 8.95 |
| 6" wall | " | 8.75 | 1.15 | 9.90 |
| 8" wall | " | 9.20 | 1.20 | 10.40 |
| Sandblast finish | | | | |
| 4" wall | S.F. | 8.25 | 1.10 | 9.35 |
| 5" wall | " | 9.05 | 1.10 | 10.15 |
| 6" wall | " | 10.00 | 1.15 | 11.15 |
| 8" wall | " | 10.40 | 1.20 | 11.60 |
| White cement | | | | |
| Liner finish | | | | |

| PRECAST CONCRETE | UNIT | MAT. | INST. | TOTAL |
|---|---|---|---|---|
| **03400.40 PRECAST WALLS** | | | | |
| 4" wall | S.F. | 8.75 | 1.10 | 9.85 |
| 5" wall | " | 9.30 | 1.10 | 10.40 |
| 6" wall | " | 10.15 | 1.15 | 11.30 |
| 8" wall | " | 10.80 | 1.20 | 12.00 |
| Sandblast finish | | | | |
| 4" wall | S.F. | 9.35 | 1.10 | 10.45 |
| 5" wall | " | 9.95 | 1.10 | 11.05 |
| 6" wall | " | 10.75 | 1.15 | 11.90 |
| 8" wall | " | 11.30 | 1.20 | 12.50 |
| Double tee wall panel, 24" deep | | | | |
| Gray cement | | | | |
| Liner finish | S.F. | 4.85 | 1.25 | 6.10 |
| Sandblast finish | " | 6.10 | 1.25 | 7.35 |
| White cement | | | | |
| Form liner finish | S.F. | 6.85 | 1.25 | 8.10 |
| Sandblast finish | " | 8.80 | 1.25 | 10.05 |
| Partition panels | | | | |
| 4" wall | S.F. | 7.30 | 1.25 | 8.55 |
| 5" wall | " | 7.90 | 1.25 | 9.15 |
| 6" wall | " | 8.70 | 1.25 | 9.95 |
| 8" wall | " | 9.40 | 1.25 | 10.65 |
| Cladding panels | | | | |
| 4" wall | S.F. | 7.25 | 1.35 | 8.60 |
| 5" wall | " | 7.90 | 1.35 | 9.25 |
| 6" wall | " | 8.75 | 1.35 | 10.10 |
| 8" wall | " | 9.30 | 1.35 | 10.65 |
| Sandwich panel, 2.5" cladding panel, 2" insulation | | | | |
| 5" wall | S.F. | 10.70 | 1.35 | 12.05 |
| 6" wall | " | 11.25 | 1.35 | 12.60 |
| 8" wall | " | 11.85 | 1.35 | 13.20 |
| **03400.90 PRECAST SPECIALTIES** | | | | |
| Precast concrete, coping, 4' to 8' long | | | | |
| 12" wide | L.F. | 4.85 | 3.25 | 8.10 |
| 10" wide | " | 4.35 | 3.75 | 8.10 |
| Splash block, 30"x12"x4" | EA. | 7.25 | 21.80 | 29.05 |
| Stair unit, per riser | " | 47.50 | 21.80 | 69.30 |
| Sun screen and trellis, 8' long, 12" high | | | | |
| 4" thick blades | EA. | 54.50 | 16.35 | 70.85 |
| 5" thick blades | " | 64.00 | 16.35 | 80.35 |
| 6" thick blades | " | 79.50 | 17.45 | 96.95 |
| 8" thick blades | " | 100.00 | 17.45 | 117.45 |
| Bearing pads for precast members, 2" wide strips | | | | |
| 1/8" thick | L.F. | 0.17 | 0.10 | 0.27 |
| 1/4" thick | " | 0.23 | 0.10 | 0.33 |
| 1/2" thick | " | 0.24 | 0.10 | 0.34 |
| 3/4" thick | " | 0.51 | 0.11 | 0.62 |
| 1" thick | " | 0.53 | 0.12 | 0.65 |
| 1-1/2" thick | " | 0.65 | 0.12 | 0.77 |

## CEMENTITOUS TOPPINGS

### 03550.10 CONCRETE TOPPINGS

| CEMENTITOUS TOPPINGS | UNIT | MAT. | INST. | TOTAL |
|---|---|---|---|---|
| Gypsum fill | | | | |
| 2" thick | S.F. | 0.97 | 0.24 | 1.21 |
| 2-1/2" thick | " | 1.10 | 0.24 | 1.34 |
| 3" thick | " | 1.35 | 0.25 | 1.60 |
| 3-1/2" thick | " | 1.55 | 0.26 | 1.81 |
| 4" thick | " | 1.80 | 0.29 | 2.09 |
| Formboard | | | | |
| Mineral fiber board | | | | |
| 1" thick | S.F. | 0.92 | 0.60 | 1.52 |
| 1-1/2" thick | " | 2.40 | 0.69 | 3.09 |
| Cement fiber board | | | | |
| 1" thick | S.F. | 0.71 | 0.81 | 1.52 |
| 1-1/2" thick | " | 0.92 | 0.93 | 1.85 |
| Glass fiber board | | | | |
| 1" thick | S.F. | 1.10 | 0.60 | 1.71 |
| 1-1/2" thick | " | 1.45 | 0.69 | 2.14 |
| Poured deck | | | | |
| Vermiculite or perlite | | | | |
| 1 to 4 mix | C.Y. | 120.00 | 38.50 | 158.50 |
| 1 to 6 mix | " | 110.00 | 35.60 | 145.60 |
| Vermiculite or perlite | | | | |
| 2" thick | | | | |
| 1 to 4 mix | S.F. | 0.97 | 0.24 | 1.21 |
| 1 to 6 mix | " | 0.71 | 0.22 | 0.93 |
| 3" thick | | | | |
| 1 to 4 mix | S.F. | 1.35 | 0.36 | 1.71 |
| 1 to 6 mix | " | 1.05 | 0.33 | 1.38 |
| Concrete plank, lightweight | | | | |
| 2" thick | S.F. | 5.05 | 1.90 | 6.95 |
| 2-1/2" thick | " | 5.20 | 1.90 | 7.10 |
| 3-1/2" thick | " | 5.50 | 2.10 | 7.60 |
| 4" thick | " | 5.65 | 2.10 | 7.75 |
| Channel slab, lightweight, straight | | | | |
| 2-3/4" thick | S.F. | 3.90 | 1.90 | 5.80 |
| 3-1/2" thick | " | 4.10 | 1.90 | 6.00 |
| 3-3/4" thick | " | 4.35 | 1.90 | 6.25 |
| 4-3/4" thick | " | 5.50 | 2.10 | 7.60 |
| Gypsum plank | | | | |
| 2" thick | S.F. | 1.85 | 1.90 | 3.75 |
| 3" thick | " | 1.90 | 1.90 | 3.80 |
| Cement fiber, T and G planks | | | | |
| 1" thick | S.F. | 1.00 | 1.75 | 2.75 |
| 1-1/2" thick | " | 1.05 | 1.75 | 2.80 |
| 2" thick | " | 1.20 | 1.90 | 3.10 |
| 2-1/2" thick | " | 1.30 | 1.90 | 3.20 |
| 3" thick | " | 1.70 | 1.90 | 3.60 |
| 3-1/2" thick | " | 1.95 | 2.10 | 4.05 |
| 4" thick | " | 2.20 | 2.10 | 4.30 |

| GROUT | UNIT | MAT. | INST. | TOTAL |
|---|---|---|---|---|
| **03600.10 GROUTING** | | | | |
| Grouting for bases | | | | |
| Nonshrink | | | | |
| Metallic grout | | | | |
| 1" deep | S.F. | 4.30 | 6.20 | 10.50 |
| 2" deep | " | 8.20 | 6.90 | 15.10 |
| Non-metallic grout | | | | |
| 1" deep | S.F. | 3.15 | 6.20 | 9.35 |
| 2" deep | " | 6.10 | 6.90 | 13.00 |
| Fluid type | | | | |
| Non-metallic | | | | |
| 1" deep | S.F. | 3.30 | 6.20 | 9.50 |
| 2" deep | " | 6.00 | 6.90 | 12.90 |
| Grouting for joints | | | | |
| Portland cement grout (1 cement to 3 sand, by volume) | | | | |
| 1/2" joint thickness | | | | |
| 6" wide joints | L.F. | 0.07 | 1.05 | 1.12 |
| 8" wide joints | " | 0.10 | 1.25 | 1.35 |
| 1" joint thickness | | | | |
| 4" wide joints | L.F. | 0.10 | 0.97 | 1.07 |
| 6" wide joints | " | 0.16 | 1.05 | 1.21 |
| 8" wide joints | " | 0.19 | 1.30 | 1.49 |
| Nonshrink, nonmetallic grout | | | | |
| 1/2" joint thickness | | | | |
| 4" wide joint | L.F. | 0.47 | 0.89 | 1.36 |
| 6" wide joint | " | 0.67 | 1.05 | 1.72 |
| 8" wide joint | " | 0.88 | 1.25 | 2.13 |
| 1" joint thickness | | | | |
| 4" wide joint | L.F. | 0.88 | 0.97 | 1.85 |
| 6" wide joint | " | 1.35 | 1.05 | 2.40 |
| 8" wide joint | " | 1.80 | 1.30 | 3.10 |

| CONCRETE RESTORATION | UNIT | MAT. | INST. | TOTAL |
|---|---|---|---|---|
| **03730.10 CONCRETE REPAIR** | | | | |
| Epoxy grout floor patch, 1/4" thick | S.F. | 3.80 | 2.40 | 6.20 |
| Grout, epoxy, 2 component system | C.F. | | | 190.25 |
| Epoxy sand | BAG | | | 12.79 |
| Epoxy modifier | GAL | | | 82.03 |
| Epoxy gel grout | S.F. | 1.90 | 24.20 | 26.10 |
| Injection valve, 1 way, threaded plastic | EA. | 5.25 | 4.85 | 10.10 |
| Grout crack seal, 2 component | C.F. | 440.00 | 24.20 | 464.20 |
| Grout, non shrink | " | 45.20 | 24.20 | 69.40 |
| Concrete, epoxy modified | | | | |
| Sand mix | C.F. | 71.50 | 9.70 | 81.20 |

## CONCRETE RESTORATION

### 03730.10 CONCRETE REPAIR

| | UNIT | MAT. | INST. | TOTAL |
|---|---|---|---|---|
| Gravel mix | C.F. | 54.50 | 8.95 | 63.45 |
| Concrete repair | | | | |
| Soffit repair | | | | |
| 16" wide | L.F. | 2.25 | 4.85 | 7.10 |
| 18" wide | " | 2.40 | 5.05 | 7.45 |
| 24" wide | " | 2.80 | 5.40 | 8.20 |
| 30" wide | " | 3.20 | 5.75 | 8.95 |
| 32" wide | " | 3.45 | 6.05 | 9.50 |
| Edge repair | | | | |
| 2" spall | L.F. | 1.05 | 6.05 | 7.10 |
| 3" spall | " | 1.05 | 6.35 | 7.40 |
| 4" spall | " | 1.15 | 6.55 | 7.70 |
| 6" spall | " | 1.20 | 6.70 | 7.90 |
| 8" spall | " | 1.25 | 7.10 | 8.35 |
| 9" spall | " | 1.30 | 8.05 | 9.35 |
| Crack repair, 1/8" crack | " | 2.10 | 2.40 | 4.50 |
| Reinforcing steel repair | | | | |
| 1 bar, 4 ft | | | | |
| #4 bar | L.F. | 0.30 | 4.35 | 4.65 |
| #5 bar | " | 0.42 | 4.35 | 4.77 |
| #6 bar | " | 0.53 | 4.65 | 5.18 |
| #8 bar | " | 0.93 | 4.65 | 5.58 |
| #9 bar | " | 1.20 | 5.00 | 6.20 |
| #11 bar | " | 1.85 | 5.00 | 6.85 |
| Form fabric, nylon | | | | |
| 18" diameter | L.F. | | | 7.96 |
| 20" diameter | " | | | 8.09 |
| 24" diameter | " | | | 13.31 |
| 30" diameter | " | | | 13.68 |
| 36" diameter | " | | | 15.68 |
| Pile repairs | | | | |
| Polyethylene wrap | | | | |
| 30 mil thick | | | | |
| 60" wide | S.F. | 8.55 | 8.05 | 16.60 |
| 72" wide | " | 9.30 | 9.70 | 19.00 |
| 60 mil thick | | | | |
| 60" wide | S.F. | 10.30 | 8.05 | 18.35 |
| 80" wide | " | 11.90 | 11.00 | 22.90 |
| Pile spall, average repair 3' | | | | |
| 18" x 18" | EA. | 26.80 | 20.15 | 46.95 |
| 20" x 20" | " | 35.60 | 24.20 | 59.80 |

## MORTAR AND GROUT

| 04100.10 MASONRY GROUT | UNIT | MAT. | INST. | TOTAL |
|---|---|---|---|---|
| Grout, non shrink, non-metallic, trowelable | C.F. | 7.40 | 0.87 | 8.27 |
| Grout door frame, hollow metal | | | | |
| Single | EA. | 10.75 | 32.70 | 43.45 |
| Double | " | 16.10 | 34.40 | 50.50 |
| Grout-filled concrete block (CMU) | | | | |
| 4" wide | S.F. | 0.31 | 1.10 | 1.41 |
| 6" wide | " | 0.64 | 1.20 | 1.84 |
| 8" wide | " | 1.10 | 1.30 | 2.40 |
| 12" wide | " | 1.60 | 1.40 | 3.00 |
| Grout-filled individual CMU cells | | | | |
| 4" wide | L.F. | 0.16 | 0.65 | 0.81 |
| 6" wide | " | 0.33 | 0.65 | 0.98 |
| 8" wide | " | 0.44 | 0.65 | 1.09 |
| 10" wide | " | 0.59 | 0.75 | 1.34 |
| 12" wide | " | 0.67 | 0.75 | 1.42 |
| Bond beams or lintels, 8" deep | | | | |
| 6" thick | L.F. | 0.67 | 1.05 | 1.72 |
| 8" thick | " | 0.88 | 1.15 | 2.03 |
| 10" thick | " | 1.10 | 1.30 | 2.40 |
| 12" thick | " | 1.30 | 1.45 | 2.75 |
| Cavity walls | | | | |
| 2" thick | S.F. | 0.74 | 1.55 | 2.29 |
| 3" thick | " | 1.10 | 1.55 | 2.65 |
| 4" thick | " | 1.45 | 1.65 | 3.10 |
| 6" thick | " | 2.20 | 1.95 | 4.15 |

| 04150.10 MASONRY ACCESSORIES | UNIT | MAT. | INST. | TOTAL |
|---|---|---|---|---|
| Foundation vents | EA. | 18.50 | 12.00 | 30.50 |
| Bar reinforcing | | | | |
| Horizontal | | | | |
| #3 - #4 | Lb. | 0.38 | 1.20 | 1.58 |
| #5 - #6 | " | 0.36 | 1.00 | 1.36 |
| Vertical | | | | |
| #3 - #4 | Lb. | 0.38 | 1.50 | 1.88 |
| #5 - #6 | " | 0.36 | 1.20 | 1.56 |
| Horizontal joint reinforcing | | | | |
| Truss type | | | | |
| 4" wide, 6" wall | L.F. | 0.14 | 0.12 | 0.26 |
| 6" wide, 8" wall | " | 0.14 | 0.13 | 0.27 |
| 8" wide, 10" wall | " | 0.18 | 0.13 | 0.31 |
| 10" wide, 12" wall | " | 0.18 | 0.14 | 0.32 |
| 12" wide, 14" wall | " | 0.21 | 0.14 | 0.35 |
| Ladder type | | | | |
| 4" wide, 6" wall | L.F. | 0.11 | 0.12 | 0.23 |
| 6" wide, 8" wall | " | 0.12 | 0.13 | 0.25 |
| 8" wide, 10" wall | " | 0.13 | 0.13 | 0.26 |
| 10" wide, 12" wall | " | 0.14 | 0.13 | 0.27 |
| Rectangular wall ties | | | | |
| 3/16" dia., galvanized | | | | |
| 2" x 6" | EA. | 0.16 | 0.50 | 0.66 |
| 2" x 8" | " | 0.17 | 0.50 | 0.67 |

| MORTAR AND GROUT | UNIT | MAT. | INST. | TOTAL |
|---|---|---|---|---|
| **04150.10 MASONRY ACCESSORIES** | | | | |
| 2" x 10" | EA. | 0.19 | 0.50 | 0.69 |
| 2" x 12" | " | 0.22 | 0.50 | 0.72 |
| 4" x 6" | " | 0.18 | 0.60 | 0.78 |
| 4" x 8" | " | 0.20 | 0.60 | 0.80 |
| 4" x 10" | " | 0.27 | 0.60 | 0.87 |
| 4" x 12" | " | 0.30 | 0.60 | 0.90 |
| 1/4" dia., galvanized | | | | |
| 2" x 6" | EA. | 0.30 | 0.50 | 0.80 |
| 2" x 8" | " | 0.32 | 0.50 | 0.82 |
| 2" x 10" | " | 0.37 | 0.50 | 0.87 |
| 2" x 12" | " | 0.41 | 0.50 | 0.91 |
| 4" x 6" | " | 0.33 | 0.60 | 0.93 |
| 4" x 8" | " | 0.37 | 0.60 | 0.97 |
| 4" x 10" | " | 0.41 | 0.60 | 1.01 |
| 4" x 12" | " | 0.43 | 0.60 | 1.03 |
| "Z" type wall ties, galvanized | | | | |
| 6" long | | | | |
| 1/8" dia. | EA. | 0.16 | 0.50 | 0.66 |
| 3/16" dia. | " | 0.17 | 0.50 | 0.67 |
| 1/4" dia. | " | 0.18 | 0.50 | 0.68 |
| 8" long | | | | |
| 1/8" dia. | EA. | 0.17 | 0.50 | 0.67 |
| 3/16" dia. | " | 0.18 | 0.50 | 0.68 |
| 1/4" dia. | " | 0.19 | 0.50 | 0.69 |
| 10" long | | | | |
| 1/8" dia. | EA. | 0.18 | 0.50 | 0.68 |
| 3/16" dia. | " | 0.20 | 0.50 | 0.70 |
| 1/4" dia. | " | 0.23 | 0.50 | 0.73 |
| Dovetail anchor slots | | | | |
| Galvanized steel, filled | | | | |
| 24 ga. | L.F. | 0.41 | 0.75 | 1.16 |
| 20 ga. | " | 0.52 | 0.75 | 1.27 |
| 16 oz. copper, foam filled | " | 1.05 | 0.75 | 1.80 |
| Dovetail anchors | | | | |
| 16 ga. | | | | |
| 3-1/2" long | EA. | 0.12 | 0.50 | 0.62 |
| 5-1/2" long | " | 0.14 | 0.50 | 0.64 |
| 12 ga. | | | | |
| 3-1/2" long | EA. | 0.16 | 0.50 | 0.66 |
| 5-1/2" long | " | 0.32 | 0.50 | 0.82 |
| Dovetail, triangular galvanized ties, 12 ga. | | | | |
| 3" x 3" | EA. | 0.30 | 0.50 | 0.80 |
| 5" x 5" | " | 0.32 | 0.50 | 0.82 |
| 7" x 7" | " | 0.36 | 0.50 | 0.86 |
| 7" x 9" | " | 0.38 | 0.50 | 0.88 |
| Brick anchors | | | | |
| Corrugated, 3-1/2" long | | | | |
| 16 ga. | EA. | 0.12 | 0.50 | 0.62 |
| 12 ga. | " | 0.20 | 0.50 | 0.70 |
| Non-corrugated, 3-1/2" long | | | | |
| 16 ga. | EA. | 0.16 | 0.50 | 0.66 |
| 12 ga. | " | 0.29 | 0.50 | 0.79 |

# 04 MASONRY

| MORTAR AND GROUT | UNIT | MAT. | INST. | TOTAL |
|---|---|---|---|---|
| **04150.10 MASONRY ACCESSORIES** | | | | |
| Cavity wall anchors, corrugated, galvanized | | | | |
| 5" long | | | | |
| 16 ga. | EA. | 0.36 | 0.50 | 0.86 |
| 12 ga. | " | 0.52 | 0.50 | 1.02 |
| 7" long | | | | |
| 28 ga. | EA. | 0.39 | 0.50 | 0.89 |
| 24 ga. | " | 0.49 | 0.50 | 0.99 |
| 22 ga. | " | 0.50 | 0.50 | 1.00 |
| 16 ga. | " | 0.57 | 0.50 | 1.07 |
| Mesh ties, 16 ga., 3" wide | | | | |
| 8" long | EA. | 0.47 | 0.50 | 0.97 |
| 12" long | " | 0.52 | 0.50 | 1.02 |
| 20" long | " | 0.72 | 0.50 | 1.22 |
| 24" long | " | 0.80 | 0.50 | 1.30 |
| **04150.20 MASONRY CONTROL JOINTS** | | | | |
| Control joint, cross shaped PVC | L.F. | 2.60 | 0.75 | 3.35 |
| Closed cell joint filler | | | | |
| 1/2" | L.F. | 0.39 | 0.75 | 1.14 |
| 3/4" | " | 0.73 | 0.75 | 1.48 |
| Rubber, for | | | | |
| 4" wall | L.F. | 2.90 | 0.75 | 3.65 |
| 6" wall | " | 4.60 | 0.79 | 5.39 |
| 8" wall | " | 5.20 | 0.83 | 6.03 |
| PVC, for | | | | |
| 4" wall | L.F. | 2.25 | 0.75 | 3.00 |
| 6" wall | " | 2.80 | 0.79 | 3.59 |
| 8" wall | " | 3.10 | 0.83 | 3.93 |
| **04150.50 MASONRY FLASHING** | | | | |
| Through-wall flashing | | | | |
| 5 oz. coated copper | S.F. | 2.90 | 2.50 | 5.40 |
| 0.030" elastomeric | " | 0.59 | 2.00 | 2.59 |

| UNIT MASONRY | UNIT | MAT. | INST. | TOTAL |
|---|---|---|---|---|
| **04210.10 BRICK MASONRY** | | | | |
| Standard size brick, running bond | | | | |
| Face brick, red (6.4/sf) | | | | |
| Veneer | S.F. | 2.05 | 5.00 | 7.05 |
| Cavity wall | " | 2.05 | 4.30 | 6.35 |
| 9" solid wall | " | 4.30 | 8.55 | 12.85 |
| Common brick (6.4/sf) | | | | |

# 04 MASONRY

| UNIT MASONRY | UNIT | MAT. | INST. | TOTAL |
|---|---|---|---|---|
| **04210.10 BRICK MASONRY** | | | | |
| Select common for veneers | S.F. | 2.40 | 5.00 | 7.40 |
| Back-up | | | | |
| 4" thick | S.F. | 2.15 | 3.75 | 5.90 |
| 8" thick | " | 4.30 | 6.00 | 10.30 |
| Firewall | | | | |
| 12" thick | S.F. | 5.90 | 10.00 | 15.90 |
| 16" thick | " | 7.80 | 13.65 | 21.45 |
| Glazed brick (7.4/sf) | | | | |
| Veneer | S.F. | 8.50 | 5.45 | 13.95 |
| Buff or gray face brick (6.4/sf) | | | | |
| Veneer | S.F. | 2.40 | 5.00 | 7.40 |
| Cavity wall | " | 2.40 | 4.30 | 6.70 |
| Jumbo or oversize brick (3/sf) | | | | |
| 4" veneer | S.F. | 4.30 | 3.00 | 7.30 |
| 4" back-up | " | 4.30 | 2.50 | 6.80 |
| 8" back-up | " | 8.25 | 4.30 | 12.55 |
| 12" firewall | " | 12.35 | 7.50 | 19.85 |
| 16" firewall | " | 16.40 | 10.00 | 26.40 |
| Norman brick, red face, (4.5/sf) | | | | |
| 4" veneer | S.F. | 3.60 | 3.75 | 7.35 |
| Cavity wall | " | 3.50 | 3.35 | 6.85 |
| Chimney, standard brick, including flue | | | | |
| 16" x 16" | L.F. | 15.50 | 30.00 | 45.50 |
| 16" x 20" | " | 18.35 | 30.00 | 48.35 |
| 16" x 24" | " | 21.75 | 30.00 | 51.75 |
| 20" x 20" | " | 23.80 | 37.50 | 61.30 |
| 20" x 24" | " | 30.60 | 37.50 | 68.10 |
| 20" x 32" | " | 38.50 | 42.90 | 81.40 |
| Window sill, face brick on edge | " | 2.30 | 7.50 | 9.80 |
| **04210.20 STRUCTURAL TILE** | | | | |
| Structural glazed tile | | | | |
| 6T series, 5-1/2" x 12" | | | | |
| Glazed on one side | | | | |
| 2" thick | S.F. | 4.55 | 3.00 | 7.55 |
| 4" thick | " | 5.65 | 3.00 | 8.65 |
| 6" thick | " | 7.95 | 3.35 | 11.30 |
| 8" thick | " | 10.20 | 3.75 | 13.95 |
| Glazed on two sides | | | | |
| 4" thick | S.F. | 8.50 | 3.75 | 12.25 |
| 6" thick | " | 11.00 | 4.30 | 15.30 |
| Special shapes | | | | |
| Group 1 | S.F. | 4.40 | 6.00 | 10.40 |
| Group 2 | " | 6.10 | 6.00 | 12.10 |
| Group 3 | " | 7.60 | 6.00 | 13.60 |
| Group 4 | " | 16.40 | 6.00 | 22.40 |
| Group 5 | " | 19.80 | 6.00 | 25.80 |
| Fire rated | | | | |
| 4" thick, 1 hr rating | S.F. | 11.90 | 3.00 | 14.90 |
| 6" thick, 2 hr rating | " | 15.30 | 3.35 | 18.65 |
| 8W series, 8" x 16" | | | | |

# 04 MASONRY

| UNIT MASONRY | UNIT | MAT. | INST. | TOTAL |
|---|---|---|---|---|
| **04210.20 STRUCTURAL TILE** | | | | |
| Glazed on one side | | | | |
| 2" thick | S.F. | 4.85 | 2.00 | 6.85 |
| 4" thick | " | 4.95 | 2.00 | 6.95 |
| 6" thick | " | 6.95 | 2.30 | 9.25 |
| 8" thick | " | 8.25 | 2.30 | 10.55 |
| Glazed on two sides | | | | |
| 4" thick | S.F. | 8.70 | 2.50 | 11.20 |
| 6" thick | " | 11.00 | 3.00 | 14.00 |
| 8" thick | " | 14.70 | 3.00 | 17.70 |
| Special shapes | | | | |
| Group 1 | S.F. | 7.20 | 4.30 | 11.50 |
| Group 2 | " | 9.05 | 4.30 | 13.35 |
| Group 3 | " | 9.95 | 4.30 | 14.25 |
| Group 4 | " | 15.85 | 4.30 | 20.15 |
| Group 5 | " | 44.20 | 4.30 | 48.50 |
| Fire rated | | | | |
| 4" thick, 1 hr rating | S.F. | 12.25 | 4.30 | 16.55 |
| 6" thick, 2 hr rating | " | 15.85 | 4.30 | 20.15 |
| **04210.60 PAVERS, MASONRY** | | | | |
| Brick walk laid on sand, sand joints | | | | |
| Laid flat, (4.5 per sf) | S.F. | 2.60 | 3.35 | 5.95 |
| Laid on edge, (7.2 per sf) | " | 4.05 | 5.00 | 9.05 |
| Precast concrete patio blocks | | | | |
| 2" thick | | | | |
| Natural | S.F. | 1.55 | 1.00 | 2.55 |
| Colors | " | 2.15 | 1.00 | 3.15 |
| Exposed aggregates, local aggregate | | | | |
| Natural | S.F. | 1.70 | 1.00 | 2.70 |
| Colors | " | 1.90 | 1.00 | 2.90 |
| Granite or limestone aggregate | " | 1.60 | 1.00 | 2.60 |
| White tumblestone aggregate | " | 2.25 | 1.00 | 3.25 |
| Stone pavers, set in mortar | | | | |
| Bluestone | | | | |
| 1" thick | | | | |
| Irregular | S.F. | 2.00 | 7.50 | 9.50 |
| Snapped rectangular | " | 3.10 | 6.00 | 9.10 |
| 1-1/2" thick, random rectangular | " | 3.30 | 7.50 | 10.80 |
| 2" thick, random rectangular | " | 3.75 | 8.55 | 12.30 |
| Slate | | | | |
| Natural cleft | | | | |
| Irregular, 3/4" thick | S.F. | 1.80 | 8.55 | 10.35 |
| Random rectangular | | | | |
| 1-1/4" thick | S.F. | 4.30 | 7.50 | 11.80 |
| 1-1/2" thick | " | 4.75 | 8.35 | 13.10 |
| Granite blocks | | | | |
| 3" thick, 3" to 6" wide | | | | |
| 4" to 12" long | S.F. | 4.30 | 10.00 | 14.30 |
| 6" to 15" long | " | 2.70 | 8.55 | 11.25 |
| Crushed stone, white marble, 3" thick | " | 0.85 | 0.48 | 1.33 |

| UNIT MASONRY | UNIT | MAT. | INST. | TOTAL |
|---|---|---|---|---|
| **04220.10 CONCRETE MASONRY UNITS** | | | | |
| Hollow, load bearing | | | | |
| 4" | S.F. | 1.20 | 2.20 | 3.40 |
| 6" | " | 1.50 | 2.30 | 3.80 |
| 8" | " | 1.95 | 2.50 | 4.45 |
| 10" | " | 2.60 | 2.75 | 5.35 |
| 12" | " | 2.75 | 3.00 | 5.75 |
| Solid, load bearing | | | | |
| 4" | S.F. | 1.60 | 2.20 | 3.80 |
| 6" | " | 1.95 | 2.30 | 4.25 |
| 8" | " | 2.25 | 2.50 | 4.75 |
| 10" | " | 3.05 | 2.75 | 5.80 |
| 12" | " | 3.20 | 3.00 | 6.20 |
| Back-up block, 8" x 16" | | | | |
| 2" | S.F. | 0.79 | 1.70 | 2.49 |
| 4" | " | 1.15 | 1.75 | 2.90 |
| 6" | " | 1.45 | 1.85 | 3.30 |
| 8" | " | 1.65 | 2.00 | 3.65 |
| 10" | " | 2.25 | 2.15 | 4.40 |
| 12" | " | 2.35 | 2.30 | 4.65 |
| Foundation wall, 8" x 16" | | | | |
| 6" | S.F. | 1.55 | 2.15 | 3.70 |
| 8" | " | 1.70 | 2.30 | 4.00 |
| 10" | " | 2.60 | 2.50 | 5.10 |
| 12" | " | 2.90 | 2.75 | 5.65 |
| Solid | | | | |
| 6" | S.F. | 2.15 | 2.30 | 4.45 |
| 8" | " | 2.55 | 2.50 | 5.05 |
| 10" | " | 3.25 | 2.75 | 6.00 |
| 12" | " | 3.95 | 3.00 | 6.95 |
| Exterior, styrofoam inserts, standard weight, 8" x 16" | | | | |
| 6" | S.F. | 2.20 | 2.30 | 4.50 |
| 8" | " | 2.45 | 2.50 | 4.95 |
| 10" | " | 3.45 | 2.75 | 6.20 |
| 12" | " | 3.65 | 3.00 | 6.65 |
| Lightweight | | | | |
| 6" | S.F. | 2.35 | 2.30 | 4.65 |
| 8" | " | 2.85 | 2.50 | 5.35 |
| 10" | " | 3.60 | 2.75 | 6.35 |
| 12" | " | 3.80 | 3.00 | 6.80 |
| Acoustical slotted block | | | | |
| 4" | S.F. | 2.30 | 2.75 | 5.05 |
| 6" | " | 2.95 | 2.75 | 5.70 |
| 8" | " | 4.75 | 3.00 | 7.75 |
| Filled cavities | | | | |
| 4" | S.F. | 3.10 | 3.35 | 6.45 |
| 6" | " | 3.80 | 3.55 | 7.35 |
| 8" | " | 5.70 | 3.75 | 9.45 |
| Hollow, split face | | | | |
| 4" | S.F. | 2.90 | 2.20 | 5.10 |
| 6" | " | 3.25 | 2.30 | 5.55 |
| 8" | " | 4.05 | 2.50 | 6.55 |
| 10" | " | 4.40 | 2.75 | 7.15 |

| UNIT MASONRY | UNIT | MAT. | INST. | TOTAL |
|---|---|---|---|---|
| **04220.10 CONCRETE MASONRY UNITS** | | | | |
| 12" | S.F. | 4.90 | 3.00 | 7.90 |
| Split rib profile | | | | |
| 4" | S.F. | 2.10 | 2.75 | 4.85 |
| 6" | " | 2.60 | 2.75 | 5.35 |
| 8" | " | 3.45 | 3.00 | 6.45 |
| 10" | " | 3.75 | 3.00 | 6.75 |
| 12" | " | 4.15 | 3.00 | 7.15 |
| High strength block, 3500 psi | | | | |
| 2" | S.F. | 1.15 | 2.20 | 3.35 |
| 4" | " | 1.45 | 2.30 | 3.75 |
| 6" | " | 1.70 | 2.30 | 4.00 |
| 8" | " | 1.95 | 2.50 | 4.45 |
| 10" | " | 2.25 | 2.75 | 5.00 |
| 12" | " | 2.70 | 3.00 | 5.70 |
| Solar screen concrete block | | | | |
| 4" thick | | | | |
| 6" x 6" | S.F. | 3.95 | 6.65 | 10.60 |
| 8" x 8" | " | 3.25 | 6.00 | 9.25 |
| 12" x 12" | " | 2.50 | 4.60 | 7.10 |
| 8" thick | | | | |
| 8" x 16" | S.F. | 2.70 | 4.30 | 7.00 |
| Glazed block | | | | |
| Cove base, glazed 1 side, 2" | L.F. | 5.55 | 3.35 | 8.90 |
| 4" | " | 5.95 | 3.35 | 9.30 |
| 6" | " | 6.45 | 3.75 | 10.20 |
| 8" | " | 7.15 | 3.75 | 10.90 |
| Single face | | | | |
| 2" | S.F. | 5.75 | 2.50 | 8.25 |
| 4" | " | 6.25 | 2.50 | 8.75 |
| 6" | " | 6.80 | 2.75 | 9.55 |
| 8" | " | 7.50 | 3.00 | 10.50 |
| 10" | " | 8.15 | 3.35 | 11.50 |
| 12" | " | 9.05 | 3.55 | 12.60 |
| Double face | | | | |
| 4" | S.F. | 9.45 | 3.15 | 12.60 |
| 6" | " | 10.10 | 3.35 | 13.45 |
| 8" | " | 10.70 | 3.75 | 14.45 |
| Corner or bullnose | | | | |
| 2" | EA. | 7.50 | 3.75 | 11.25 |
| 4" | " | 9.80 | 4.30 | 14.10 |
| 6" | " | 11.55 | 4.30 | 15.85 |
| 8" | " | 12.10 | 5.00 | 17.10 |
| 10" | " | 13.30 | 5.45 | 18.75 |
| 12" | " | 14.45 | 6.00 | 20.45 |
| Gypsum unit masonry | | | | |
| Partition blocks (12"x30") | | | | |
| Solid | | | | |
| 2" | S.F. | 0.69 | 1.20 | 1.89 |
| Hollow | | | | |
| 3" | S.F. | 0.70 | 1.20 | 1.90 |
| 4" | " | 0.79 | 1.25 | 2.04 |
| 6" | " | 0.85 | 1.35 | 2.20 |

# 04 MASONRY

| UNIT MASONRY | UNIT | MAT. | INST. | TOTAL |
|---|---|---|---|---|
| **04220.10 CONCRETE MASONRY UNITS** | | | | |
| Vertical reinforcing | | | | |
| 4' o.c., add 5% to labor | | | | |
| 2'8" o.c., add 15% to labor | | | | |
| Interior partitions, add 10% to labor | | | | |
| **04220.90 BOND BEAMS & LINTELS** | | | | |
| Bond beam, no grout or reinforcement | | | | |
| 8" x 16" x | | | | |
| 4" thick | L.F. | 1.50 | 2.30 | 3.80 |
| 6" thick | " | 1.85 | 2.40 | 4.25 |
| 8" thick | " | 2.30 | 2.50 | 4.80 |
| 10" thick | " | 2.40 | 2.60 | 5.00 |
| 12" thick | " | 2.40 | 2.75 | 5.15 |
| Beam lintel, no grout or reinforcement | | | | |
| 8" x 16" x | | | | |
| 10" thick | L.F. | 4.20 | 3.00 | 7.20 |
| 12" thick | " | 4.20 | 3.35 | 7.55 |
| Precast masonry lintel | | | | |
| 6 lf, 8" high x | | | | |
| 4" thick | L.F. | 3.85 | 5.00 | 8.85 |
| 6" thick | " | 5.05 | 5.00 | 10.05 |
| 8" thick | " | 5.65 | 5.45 | 11.10 |
| 10" thick | " | 6.85 | 5.45 | 12.30 |
| 10 lf, 8" high x | | | | |
| 4" thick | L.F. | 5.00 | 3.00 | 8.00 |
| 6" thick | " | 6.20 | 3.00 | 9.20 |
| 8" thick | " | 6.90 | 3.35 | 10.25 |
| 10" thick | " | 9.30 | 3.35 | 12.65 |
| Steel angles and plates | | | | |
| Minimum | Lb. | 0.32 | 0.43 | 0.75 |
| Maximum | " | 0.36 | 0.75 | 1.11 |
| Various size angle lintels | | | | |
| 1/4" stock | | | | |
| 3" x 3" | L.F. | 1.65 | 1.85 | 3.50 |
| 3" x 3-1/2" | " | 1.75 | 1.85 | 3.60 |
| 3/8" stock | | | | |
| 3" x 4" | L.F. | 2.90 | 1.85 | 4.75 |
| 3-1/2" x 4" | " | 3.05 | 1.85 | 4.90 |
| 4" x 4" | " | 3.15 | 1.85 | 5.00 |
| 5" x 3-1/2" | " | 3.50 | 1.85 | 5.35 |
| 6" x 3-1/2" | " | 3.75 | 1.85 | 5.60 |
| 1/2" stock | | | | |
| 6" x 4" | L.F. | 5.25 | 1.85 | 7.10 |
| **04240.10 CLAY TILE** | | | | |
| Hollow clay tile, for back-up, 12" x 12" | | | | |
| Scored face | | | | |
| Load bearing | | | | |
| 4" thick | S.F. | 3.00 | 2.15 | 5.15 |
| 6" thick | " | 3.70 | 2.20 | 5.90 |
| 8" thick | " | 4.55 | 2.30 | 6.85 |

## UNIT MASONRY

| UNIT MASONRY | UNIT | MAT. | INST. | TOTAL |
|---|---|---|---|---|
| **04240.10 CLAY TILE** | | | | |
| 10" thick | S.F. | 5.20 | 2.40 | 7.60 |
| 12" thick | " | 6.35 | 2.50 | 8.85 |
| Non-load bearing | | | | |
| 3" thick | S.F. | 2.10 | 2.05 | 4.15 |
| 4" thick | " | 2.65 | 2.15 | 4.80 |
| 6" thick | " | 3.10 | 2.20 | 5.30 |
| 8" thick | " | 4.25 | 2.30 | 6.55 |
| 12" thick | " | 5.55 | 2.50 | 8.05 |
| Partition, 12" x 12" | | | | |
| In walls | | | | |
| 3" thick | S.F. | 2.10 | 2.50 | 4.60 |
| 4" thick | " | 2.65 | 2.50 | 5.15 |
| 6" thick | " | 3.10 | 2.60 | 5.70 |
| 8" thick | " | 4.25 | 2.75 | 7.00 |
| 10" thick | " | 5.45 | 2.85 | 8.30 |
| 12" thick | " | 5.55 | 3.00 | 8.55 |
| Clay tile floors | | | | |
| 4" thick | S.F. | 2.65 | 1.65 | 4.30 |
| 6" thick | " | 3.70 | 1.75 | 5.45 |
| 8" thick | " | 4.60 | 1.85 | 6.45 |
| 10" thick | " | 5.10 | 2.00 | 7.10 |
| 12" thick | " | 5.65 | 2.15 | 7.80 |
| Terra cotta | | | | |
| Coping, 10" or 12" wide, 3" thick | L.F. | 16.15 | 6.00 | 22.15 |
| **04270.10 GLASS BLOCK** | | | | |
| Glass block, 4" thick | | | | |
| 6" x 6" | S.F. | 26.80 | 10.00 | 36.80 |
| 8" x 8" | " | 23.30 | 7.50 | 30.80 |
| 12" x 12" | " | 21.00 | 6.00 | 27.00 |
| **04295.10 PARGING/MASONRY PLASTER** | | | | |
| Parging | | | | |
| 1/2" thick | S.F. | 0.40 | 2.00 | 2.40 |
| 3/4" thick | " | 0.51 | 2.50 | 3.01 |
| 1" thick | " | 0.65 | 3.00 | 3.65 |

## STONE

| STONE | UNIT | MAT. | INST. | TOTAL |
|---|---|---|---|---|
| **04400.10 STONE** | | | | |
| Rubble stone | | | | |
| Walls set in mortar | | | | |
| 8" thick | S.F. | 9.05 | 7.50 | 16.55 |
| 12" thick | " | 10.95 | 12.00 | 22.95 |

# 04 MASONRY

| STONE | UNIT | MAT. | INST. | TOTAL |
|---|---|---|---|---|
| **04400.10 STONE** | | | | |
| 18" thick | S.F. | 14.55 | 15.00 | 29.55 |
| 24" thick | " | 18.20 | 20.00 | 38.20 |
| Dry set wall | | | | |
| 8" thick | S.F. | 10.20 | 5.00 | 15.20 |
| 12" thick | " | 11.55 | 7.50 | 19.05 |
| 18" thick | " | 15.95 | 10.00 | 25.95 |
| 24" thick | " | 19.45 | 12.00 | 31.45 |
| Cut stone | | | | |
| Imported marble | | | | |
| Facing panels | | | | |
| 3/4" thick | S.F. | 24.25 | 12.00 | 36.25 |
| 1-1/2" thick | " | 36.40 | 13.65 | 50.05 |
| 2-1/4" thick | " | 42.40 | 16.65 | 59.05 |
| Base | | | | |
| 1" thick | | | | |
| 4" high | L.F. | 10.95 | 15.00 | 25.95 |
| 6" high | " | 13.45 | 15.00 | 28.45 |
| Columns, solid | | | | |
| Plain faced | C.F. | 60.00 | 200.00 | 260.00 |
| Fluted | " | 180.00 | 200.00 | 380.00 |
| Flooring, travertine, minimum | S.F. | 7.95 | 4.60 | 12.55 |
| Average | " | 10.35 | 6.00 | 16.35 |
| Maximum | " | 13.85 | 6.65 | 20.50 |
| Domestic marble | | | | |
| Facing panels | | | | |
| 7/8" thick | S.F. | 21.95 | 12.00 | 33.95 |
| 1-1/2" thick | " | 32.90 | 13.65 | 46.55 |
| 2-1/4" thick | " | 39.80 | 16.65 | 56.45 |
| Stairs | | | | |
| 12" treads | L.F. | 19.65 | 15.00 | 34.65 |
| 6" risers | " | 14.55 | 10.00 | 24.55 |
| Thresholds, 7/8" thick, 3' long, 4" to 6" wide | | | | |
| Plain | EA. | 18.70 | 25.00 | 43.70 |
| Beveled | " | 20.70 | 25.00 | 45.70 |
| Window sill | | | | |
| 6" wide, 2" thick | L.F. | 9.40 | 12.00 | 21.40 |
| Stools | | | | |
| 5" wide, 7/8" thick | L.F. | 13.85 | 12.00 | 25.85 |
| Limestone panels up to 12' x 5', smooth finish | | | | |
| 2" thick | S.F. | 16.75 | 5.25 | 22.00 |
| 3" thick | " | 20.80 | 5.25 | 26.05 |
| 4" thick | " | 27.10 | 5.25 | 32.35 |
| Miscellaneous limestone items | | | | |
| Steps, 14" wide, 6" deep | L.F. | 46.80 | 20.00 | 66.80 |
| Coping, smooth finish | C.F. | 64.50 | 10.00 | 74.50 |
| Sills, lintels, jambs, smooth finish | " | 64.50 | 12.00 | 76.50 |
| Granite veneer facing panels, polished | | | | |
| 7/8" thick | | | | |
| Black | S.F. | 26.00 | 12.00 | 38.00 |
| Gray | " | 22.60 | 12.00 | 34.60 |
| Base | | | | |
| 4" high | L.F. | 12.10 | 6.00 | 18.10 |

# 04 MASONRY

| STONE | UNIT | MAT. | INST. | TOTAL |
|---|---|---|---|---|
| **04400.10 STONE** | | | | |
| 6" high | L.F. | 14.55 | 6.65 | 21.20 |
| Curbing, straight, 6" x 16" | " | 13.40 | 21.80 | 35.20 |
| Radius curbs, radius over 5' | " | 16.40 | 29.10 | 45.50 |
| Ashlar veneer | | | | |
| 4" thick, random | S.F. | 37.00 | 12.00 | 49.00 |
| Pavers, 4" x 4" split | | | | |
| Gray | S.F. | 21.85 | 6.00 | 27.85 |
| Pink | " | 21.35 | 6.00 | 27.35 |
| Black | " | 21.15 | 6.00 | 27.15 |
| Slate, panels | | | | |
| 1" thick | S.F. | 11.55 | 12.00 | 23.55 |
| 2" thick | " | 19.10 | 13.65 | 32.75 |
| Sills or stools | | | | |
| 1" thick | | | | |
| 6" wide | L.F. | 7.25 | 12.00 | 19.25 |
| 10" wide | " | 11.80 | 13.05 | 24.85 |
| 2" thick | | | | |
| 6" wide | L.F. | 11.85 | 13.65 | 25.50 |
| 10" wide | " | 19.65 | 15.00 | 34.65 |

| MASONRY RESTORATION | UNIT | MAT. | INST. | TOTAL |
|---|---|---|---|---|
| **04520.10 RESTORATION AND CLEANING** | | | | |
| Masonry cleaning | | | | |
| Washing brick | | | | |
| Smooth surface | S.F. | 0.29 | 0.50 | 0.79 |
| Rough surface | " | 0.39 | 0.67 | 1.06 |
| Steam clean masonry | | | | |
| Smooth face | | | | |
| Minimum | S.F. | 0.00 | 0.39 | 0.39 |
| Maximum | " | 0.00 | 0.56 | 0.56 |
| Rough face | | | | |
| Minimum | S.F. | 0.00 | 0.52 | 0.52 |
| Maximum | " | 0.00 | 0.78 | 0.78 |
| Sandblast masonry | | | | |
| Minimum | S.F. | 0.33 | 0.62 | 0.95 |
| Maximum | " | 0.51 | 1.05 | 1.56 |
| Pointing masonry | | | | |
| Brick | S.F. | 0.88 | 1.20 | 2.08 |
| Concrete block | " | 0.39 | 0.86 | 1.25 |
| Cut and repoint | | | | |
| Brick | | | | |
| Minimum | S.F. | 0.24 | 1.50 | 1.74 |

# 04 MASONRY

| MASONRY RESTORATION | UNIT | MAT. | INST. | TOTAL |
|---|---|---|---|---|
| **04520.10 RESTORATION AND CLEANING** | | | | |
| Maximum | S.F. | 0.41 | 3.00 | 3.41 |
| Stone work | L.F. | 1.10 | 2.30 | 3.40 |
| Cut and recaulk | | | | |
| Oil base caulks | L.F. | 0.85 | 2.00 | 2.85 |
| Butyl caulks | " | 0.75 | 2.00 | 2.75 |
| Polysulfides and acrylics | " | 1.45 | 2.00 | 3.45 |
| Silicones | " | 1.70 | 2.00 | 3.70 |
| Cement and sand grout on walls, to 1/8" thick | | | | |
| Minimum | S.F. | 0.38 | 1.20 | 1.58 |
| Maximum | " | 0.99 | 1.50 | 2.49 |
| Brick removal and replacement | | | | |
| Minimum | EA. | 0.39 | 3.75 | 4.14 |
| Average | " | 0.51 | 5.00 | 5.51 |
| Maximum | " | 0.99 | 15.00 | 15.99 |
| **04550.10 REFRACTORIES** | | | | |
| Flue liners | | | | |
| Rectangular | | | | |
| 8" x 12" | L.F. | 5.20 | 5.00 | 10.20 |
| 12" x 12" | " | 7.60 | 5.45 | 13.05 |
| 12" x 18" | " | 11.45 | 6.00 | 17.45 |
| 16" x 16" | " | 12.70 | 6.65 | 19.35 |
| 18" x 18" | " | 16.75 | 7.15 | 23.90 |
| 20" x 20" | " | 26.60 | 7.50 | 34.10 |
| 24" x 24" | " | 35.80 | 8.55 | 44.35 |
| Round | | | | |
| 18" dia. | L.F. | 25.40 | 7.15 | 32.55 |
| 24" dia. | " | 41.60 | 8.55 | 50.15 |

# 05 METALS

| METAL FASTENING | UNIT | MAT. | INST. | TOTAL |
|---|---|---|---|---|
| **05050.10 STRUCTURAL WELDING** | | | | |
| Welding | | | | |
| Single pass | | | | |
| 1/8" | L.F. | 0.38 | 1.70 | 2.08 |
| 3/16" | " | 0.72 | 2.25 | 2.97 |
| 1/4" | " | 0.99 | 2.80 | 3.79 |
| Miscellaneous steel shapes | | | | |
| Plain | Lb. | 0.72 | 0.07 | 0.79 |
| Galvanized | " | 1.10 | 0.11 | 1.21 |
| Plates | | | | |
| Plain | Lb. | 0.67 | 0.09 | 0.76 |
| Galvanized | " | 1.00 | 0.14 | 1.14 |
| **05050.90 METAL ANCHORS** | | | | |
| Anchor bolts | | | | |
| 3/8" x | | | | |
| 8" long | EA. | | | 3.12 |
| 10" long | " | | | 3.35 |
| 12" long | " | | | 3.57 |
| 1/2" x | | | | |
| 8" long | EA. | | | 4.27 |
| 10" long | " | | | 4.40 |
| 12" long | " | | | 4.73 |
| 18" long | " | | | 5.08 |
| 5/8" x | | | | |
| 8" long | EA. | | | 3.93 |
| 10" long | " | | | 4.27 |
| 12" long | " | | | 4.51 |
| 18" long | " | | | 4.73 |
| 24" long | " | | | 5.08 |
| 3/4" x | | | | |
| 8" long | EA. | | | 5.61 |
| 12" long | " | | | 6.18 |
| 18" long | " | | | 8.08 |
| 24" long | " | | | 10.97 |
| 7/8" x | | | | |
| 8" long | EA. | | | 5.50 |
| 12" long | " | | | 5.95 |
| 18" long | " | | | 9.07 |
| 24" long | " | | | 11.32 |
| 1" x | | | | |
| 12" long | EA. | | | 9.07 |
| 18" long | " | | | 12.35 |
| 24" long | " | | | 15.36 |
| 36" long | " | | | 23.44 |
| Expansion shield | | | | |
| 1/4" | EA. | | | 0.52 |
| 3/8" | " | | | 0.80 |
| 1/2" | " | | | 1.32 |
| 5/8" | " | | | 2.03 |
| 3/4" | " | | | 3.19 |
| 1" | " | | | 3.63 |

# 05 METALS

| METAL FASTENING | UNIT | MAT. | INST. | TOTAL |
|---|---|---|---|---|
| **05050.90 METAL ANCHORS** | | | | |
| Non-drilling anchor | | | | |
| 1/4" | EA. | | | 0.41 |
| 3/8" | " | | | 0.52 |
| 1/2" | " | | | 0.80 |
| 5/8" | " | | | 1.32 |
| 3/4" | " | | | 2.25 |
| Self-drilling anchor | | | | |
| 1/4" | EA. | | | 0.52 |
| 5/16" | " | | | 0.69 |
| 3/8" | " | | | 0.80 |
| 1/2" | " | | | 1.15 |
| 5/8" | " | | | 2.03 |
| 3/4" | " | | | 3.57 |
| 7/8" | " | | | 5.77 |
| Add 25% for galvanized anchor bolts | | | | |
| Channel door frame, with anchors | Lb. | 1.10 | 0.38 | 1.48 |
| Corner guard angle, with anchors | " | 0.99 | 0.56 | 1.55 |
| **05050.95 METAL LINTELS** | | | | |
| Lintels, steel | | | | |
| Plain | Lb. | 0.70 | 0.84 | 1.54 |
| Galvanized | " | 1.00 | 0.84 | 1.85 |
| **05120.10 STRUCTURAL STEEL** | | | | |
| Beams and girders, A-36 | | | | |
| Welded | TON | 1,180 | 380.00 | 1,560 |
| Bolted | " | 1,150 | 350.00 | 1,500 |
| Columns | | | | |
| Pipe | | | | |
| 6" dia. | Lb. | 0.79 | 0.38 | 1.17 |
| 12" dia. | " | 0.68 | 0.32 | 1.00 |
| Purlins and girts | | | | |
| Welded | TON | 1,470 | 630.00 | 2,100 |
| Bolted | " | 1,450 | 540.00 | 1,990 |
| Column base plates | | | | |
| Up to 150 lb each | Lb. | 0.85 | 0.23 | 1.08 |
| Over 150 lb each | " | 0.70 | 0.28 | 0.98 |
| Structural pipe | | | | |
| 3" to 5" o.d. | TON | 1,590 | 760.00 | 2,350 |
| 6" to 12" o.d. | " | 1,470 | 540.00 | 2,010 |
| Structural tube | | | | |
| 6" square | | | | |
| Light sections | TON | 1,840 | 760.00 | 2,600 |
| Heavy sections | " | 1,720 | 540.00 | 2,260 |
| 6" wide rectangular | | | | |
| Light sections | TON | 1,840 | 630.00 | 2,470 |
| Heavy sections | " | 1,720 | 480.00 | 2,200 |
| Greater than 6" wide rectangular | | | | |
| Light sections | TON | 2,020 | 630.00 | 2,650 |
| Heavy sections | " | 1,900 | 480.00 | 2,380 |

## METAL FASTENING

| METAL FASTENING | UNIT | MAT. | INST. | TOTAL |
|---|---|---|---|---|
| **05120.10 STRUCTURAL STEEL** | | | | |
| Miscellaneous structural shapes | | | | |
| Steel angle | TON | 1,310 | 950.00 | 2,260 |
| Steel plate | " | 1,490 | 630.00 | 2,120 |
| Trusses, field welded | | | | |
| 60 lb/lf | TON | 1,900 | 480.00 | 2,380 |
| 100 lb/lf | " | 1,660 | 380.00 | 2,040 |
| 150 lb/lf | " | 1,570 | 320.00 | 1,890 |
| Bolted | | | | |
| 60 lb/lf | TON | 1,880 | 420.00 | 2,300 |
| 100 lb/lf | " | 1,640 | 350.00 | 1,990 |
| 150 lb/lf | " | 1,560 | 290.00 | 1,850 |
| Add for galvanizing | " | | | 356.64 |
| **05200.10 METAL JOISTS** | | | | |
| Joist | | | | |
| DLH series | TON | 830.00 | 250.00 | 1,080 |
| K series | " | 850.00 | 250.00 | 1,100 |
| LH series | " | 820.00 | 250.00 | 1,070 |
| **05300.10 METAL DECKING** | | | | |
| Roof, 1-1/2" deep, non-composite | | | | |
| 16 ga. | | | | |
| Primed | S.F. | 1.65 | 0.64 | 2.29 |
| Galvanized | " | 1.75 | 0.64 | 2.38 |
| 18 ga. | | | | |
| Primed | S.F. | 1.20 | 0.64 | 1.84 |
| Galvanized | " | 1.35 | 0.64 | 1.99 |
| 20 ga. | | | | |
| Primed | S.F. | 0.98 | 0.64 | 1.62 |
| Galvanized | " | 1.10 | 0.64 | 1.74 |
| 22 ga. | | | | |
| Primed | S.F. | 0.74 | 0.64 | 1.38 |
| Galvanized | " | 0.92 | 0.64 | 1.55 |
| Open type decking, galvanized | | | | |
| 1-1/2" deep | | | | |
| 18 ga. | S.F. | 1.70 | 0.64 | 2.34 |
| 20 ga. | " | 1.35 | 0.64 | 1.99 |
| 22 ga. | " | 1.10 | 0.64 | 1.74 |
| 3" deep | | | | |
| 16 ga. | S.F. | 4.00 | 0.69 | 4.69 |
| 18 ga. | " | 3.40 | 0.69 | 4.09 |
| 20 ga. | " | 2.95 | 0.69 | 3.64 |
| 22 ga. | " | 1.95 | 0.69 | 2.64 |
| 4-1/2" deep | | | | |
| 16 ga. | S.F. | 4.50 | 0.76 | 5.26 |
| 18 ga. | " | 3.90 | 0.76 | 4.66 |
| 6" deep | | | | |
| 16 ga. | S.F. | 4.80 | 0.85 | 5.65 |
| 18 ga. | " | 4.00 | 0.85 | 4.85 |
| 7-1/2" deep | | | | |
| 16 ga. | S.F. | 5.15 | 0.89 | 6.04 |
| 18 ga. | " | 4.25 | 0.89 | 5.13 |

## METAL FASTENING

| 05300.10 METAL DECKING | UNIT | MAT. | INST. | TOTAL |
|---|---|---|---|---|
| Cellular type | | | | |
| 1-1/2" deep, galvanized | | | | |
| 18-18 ga. | S.F. | 4.00 | 0.76 | 4.76 |
| 22-18 ga. | " | 3.65 | 0.76 | 4.41 |
| 3" deep, galvanized | | | | |
| 16-16 ga. | S.F. | 5.70 | 0.85 | 6.55 |
| 18-16 ga. | " | 5.35 | 0.85 | 6.20 |
| 18-18 ga. | " | 4.90 | 0.85 | 5.75 |
| 20-18 ga. | " | 4.35 | 0.85 | 5.20 |
| 4-1/2" deep, galvanized | | | | |
| 16-16 ga. | S.F. | 8.60 | 0.91 | 9.51 |
| 18-16 ga. | " | 8.30 | 0.91 | 9.21 |
| 18-18 ga. | " | 7.55 | 0.91 | 8.46 |
| 20-18 ga. | " | 6.85 | 0.91 | 7.76 |
| Composite deck, non-cellular, galvanized | | | | |
| 1-1/2" deep | | | | |
| 18 ga. | S.F. | 1.25 | 0.69 | 1.94 |
| 20 ga. | " | 1.10 | 0.69 | 1.79 |
| 22 ga. | " | 0.90 | 0.69 | 1.59 |
| 3" deep | | | | |
| 18 ga. | S.F. | 1.70 | 0.73 | 2.43 |
| 20 ga. | " | 1.10 | 0.73 | 1.83 |
| 22 ga. | " | 0.90 | 0.73 | 1.63 |
| Slab form floor deck | | | | |
| 9/16" deep | | | | |
| 28 ga. | S.F. | 0.48 | 0.69 | 1.17 |
| 1-5/16" deep | | | | |
| 24 ga. | S.F. | 0.75 | 0.72 | 1.47 |
| 22 ga. | " | 0.87 | 0.72 | 1.59 |

## COLD FORMED FRAMING

| 05410.10 METAL FRAMING | UNIT | MAT. | INST. | TOTAL |
|---|---|---|---|---|
| Furring channel, galvanized | | | | |
| Beams and columns, 3/4" | | | | |
| 12" o.c. | S.F. | 0.34 | 3.40 | 3.74 |
| 16" o.c. | " | 0.29 | 3.05 | 3.34 |
| Walls, 3/4" | | | | |
| 12" o.c. | S.F. | 0.34 | 1.70 | 2.04 |
| 16" o.c. | " | 0.29 | 1.40 | 1.69 |
| 24" o.c. | " | 0.22 | 1.15 | 1.37 |
| 1-1/2" | | | | |
| 12" o.c. | S.F. | 0.47 | 1.70 | 2.17 |

# 05 METALS

## COLD FORMED FRAMING

| 05410.10 METAL FRAMING | UNIT | MAT. | INST. | TOTAL |
|---|---|---|---|---|
| 16" o.c. | S.F. | 0.38 | 1.40 | 1.78 |
| 24" o.c. | " | 0.31 | 1.15 | 1.46 |
| Stud, load bearing | | | | |
| 16" o.c. | | | | |
| 16 ga. | | | | |
| 2-1/2" | S.F. | 1.10 | 1.50 | 2.60 |
| 3-5/8" | " | 1.30 | 1.50 | 2.80 |
| 4" | " | 1.40 | 1.50 | 2.90 |
| 6" | " | 1.80 | 1.70 | 3.50 |
| 18 ga. | | | | |
| 2-1/2" | S.F. | 0.91 | 1.50 | 2.41 |
| 3-5/8" | " | 1.15 | 1.50 | 2.65 |
| 4" | " | 1.20 | 1.50 | 2.70 |
| 6" | " | 1.50 | 1.70 | 3.20 |
| 8" | " | 1.65 | 1.70 | 3.35 |
| 20 ga. | | | | |
| 2-1/2" | S.F. | 0.73 | 1.50 | 2.23 |
| 3-5/8" | " | 0.85 | 1.50 | 2.35 |
| 4" | " | 0.97 | 1.50 | 2.47 |
| 6" | " | 1.10 | 1.70 | 2.80 |
| 8" | " | 1.20 | 1.70 | 2.90 |
| 24" o.c. | | | | |
| 16 ga. | | | | |
| 2-1/2" | S.F. | 0.79 | 1.30 | 2.09 |
| 3-5/8" | " | 1.00 | 1.30 | 2.30 |
| 4" | " | 1.15 | 1.30 | 2.45 |
| 6" | " | 1.55 | 1.40 | 2.95 |
| 8" | " | 1.65 | 1.40 | 3.05 |
| 18 ga. | | | | |
| 2-1/2" | S.F. | 0.73 | 1.30 | 2.03 |
| 3-5/8" | " | 0.91 | 1.30 | 2.21 |
| 4" | " | 1.00 | 1.30 | 2.30 |
| 6" | " | 1.20 | 1.40 | 2.60 |
| 8" | " | 1.30 | 1.40 | 2.70 |
| 20 ga. | | | | |
| 2-1/2" | S.F. | 0.54 | 1.30 | 1.84 |
| 3-5/8" | " | 0.65 | 1.30 | 1.95 |
| 4" | " | 0.77 | 1.30 | 2.07 |
| 6" | " | 0.89 | 1.40 | 2.29 |
| 8" | " | 1.00 | 1.40 | 2.40 |

## METAL FABRICATIONS

| METAL FABRICATIONS | UNIT | MAT. | INST. | TOTAL |
|---|---|---|---|---|
| **05510.10 STAIRS** | | | | |
| Stock unit, steel, complete, per riser | | | | |
| Tread | | | | |
| 3'-6" wide | EA. | 81.00 | 42.30 | 123.30 |
| 4' wide | " | 93.50 | 48.30 | 141.80 |
| 5' wide | " | 110.00 | 56.50 | 166.50 |
| Metal pan stair, cement filled, per riser | | | | |
| 3'-6" wide | EA. | 86.50 | 33.80 | 120.30 |
| 4' wide | " | 98.00 | 37.60 | 135.60 |
| 5' wide | " | 110.00 | 42.30 | 152.30 |
| Landing, steel pan | S.F. | 34.60 | 8.45 | 43.05 |
| Cast iron tread, steel stringers, stock units, per riser | | | | |
| Tread | | | | |
| 3'-6" wide | EA. | 140.00 | 42.30 | 182.30 |
| 4' wide | " | 160.00 | 48.30 | 208.30 |
| 5' wide | " | 190.00 | 56.50 | 246.50 |
| Stair treads, abrasive, 12" x 3'-6" | | | | |
| Cast iron | | | | |
| 3/8" | EA. | 81.00 | 16.90 | 97.90 |
| 1/2" | " | 92.50 | 16.90 | 109.40 |
| Cast aluminum | | | | |
| 5/16" | EA. | 92.50 | 16.90 | 109.40 |
| 3/8" | " | 100.00 | 16.90 | 116.90 |
| 1/2" | " | 130.00 | 16.90 | 146.90 |
| **05515.10 LADDERS** | | | | |
| Ladder, 18" wide | | | | |
| With cage | L.F. | 55.50 | 22.55 | 78.05 |
| Without cage | " | 32.30 | 16.90 | 49.20 |
| **05520.10 RAILINGS** | | | | |
| Railing, pipe | | | | |
| 1-1/4" diameter, welded steel | | | | |
| 2-rail | | | | |
| Primed | L.F. | 13.85 | 6.75 | 20.60 |
| Galvanized | " | 16.75 | 6.75 | 23.50 |
| 3-rail | | | | |
| Primed | L.F. | 17.35 | 8.45 | 25.80 |
| Galvanized | " | 21.95 | 8.45 | 30.40 |
| Wall mounted, single rail, welded steel | | | | |
| Primed | L.F. | 5.20 | 5.20 | 10.40 |
| Galvanized | " | 7.25 | 5.20 | 12.45 |
| 1-1/2" diameter, welded steel | | | | |
| 2-rail | | | | |
| Primed | L.F. | 15.00 | 6.75 | 21.75 |
| Galvanized | " | 18.50 | 6.75 | 25.25 |
| 3-rail | | | | |
| Primed | L.F. | 18.50 | 8.45 | 26.95 |
| Galvanized | " | 23.10 | 8.45 | 31.55 |
| Wall mounted, single rail, welded steel | | | | |
| Primed | L.F. | 6.05 | 5.20 | 11.25 |

# 05 METALS

## METAL FABRICATIONS

| 05520.10 RAILINGS | UNIT | MAT. | INST. | TOTAL |
|---|---|---|---|---|
| Galvanized | L.F. | 7.50 | 5.20 | 12.70 |
| 2" diameter, welded steel | | | | |
| 2-rail | | | | |
| Primed | L.F. | 18.50 | 7.50 | 26.00 |
| Galvanized | " | 23.10 | 7.50 | 30.60 |
| 3-rail | | | | |
| Primed | L.F. | 23.10 | 9.65 | 32.75 |
| Galvanized | " | 27.70 | 9.65 | 37.35 |
| Wall mounted, single rail, welded steel | | | | |
| Primed | L.F. | 6.60 | 5.65 | 12.25 |
| Galvanized | " | 7.80 | 5.65 | 13.45 |

| 05530.10 METAL GRATING | UNIT | MAT. | INST. | TOTAL |
|---|---|---|---|---|
| Floor plate, checkered, steel | | | | |
| 1/4" | | | | |
| Primed | S.F. | 5.20 | 0.48 | 5.68 |
| Galvanized | " | 6.95 | 0.48 | 7.43 |
| 3/8" | | | | |
| Primed | S.F. | 6.95 | 0.52 | 7.47 |
| Galvanized | " | 9.25 | 0.52 | 9.77 |
| Aluminum grating, pressure-locked bearing bars | | | | |
| 3/4" x 1/8" | S.F. | 9.80 | 0.84 | 10.65 |
| 1" x 1/8" | " | 11.25 | 0.84 | 12.10 |
| 1-1/4" x 1/8" | " | 12.10 | 0.84 | 12.95 |
| 1-1/4" x 3/16" | " | 15.60 | 0.84 | 16.45 |
| 1-1/2" x 1/8" | " | 13.30 | 0.84 | 14.15 |
| 1-3/4" x 3/16" | " | 18.20 | 0.84 | 19.05 |
| Miscellaneous expenses | | | | |
| Cutting | | | | |
| Minimum | L.F. | 0.00 | 2.25 | 2.25 |
| Maximum | " | 0.00 | 3.40 | 3.40 |
| Banding | | | | |
| Minimum | L.F. | 0.00 | 5.65 | 5.65 |
| Maximum | " | 0.00 | 6.75 | 6.75 |
| Toe plates | | | | |
| Minimum | L.F. | 0.00 | 6.75 | 6.75 |
| Maximum | " | 0.00 | 8.45 | 8.45 |
| Steel grating, primed | | | | |
| 3/4" x 1/8" | S.F. | 4.90 | 1.15 | 6.05 |
| 1" x 1/8" | " | 5.10 | 1.15 | 6.25 |
| 1-1/4" x 1/8" | " | 5.45 | 1.15 | 6.60 |
| 1-1/4" x 3/16" | " | 6.65 | 1.15 | 7.80 |
| 1-1/2" x 1/8" | " | 6.05 | 1.15 | 7.20 |
| 1-3/4" x 3/16" | " | 8.65 | 1.15 | 9.80 |
| Galvanized | | | | |
| 3/4" x 1/8" | S.F. | 6.05 | 1.15 | 7.20 |
| 1" x 1/8" | " | 6.45 | 1.15 | 7.60 |
| 1-1/4" x 1/8" | " | 7.05 | 1.15 | 8.20 |
| 1-1/4" x 3/16" | " | 8.40 | 1.15 | 9.55 |
| 1-1/2" x 1/8" | " | 7.50 | 1.15 | 8.65 |
| 1-3/4" x 3/16" | " | 10.95 | 1.15 | 12.10 |

## METAL FABRICATIONS

| | UNIT | MAT. | INST. | TOTAL |
|---|---|---|---|---|
| **05530.10 METAL GRATING** | | | | |
| Miscellaneous expenses | | | | |
| Cutting | | | | |
| Minimum | L.F. | 0.00 | 2.40 | 2.40 |
| Maximum | " | 0.00 | 3.75 | 3.75 |
| Banding | | | | |
| Minimum | L.F. | 0.00 | 6.15 | 6.15 |
| Maximum | " | 0.00 | 7.50 | 7.50 |
| Toe plates | | | | |
| Minimum | L.F. | 0.00 | 7.50 | 7.50 |
| Maximum | " | 0.00 | 9.65 | 9.65 |
| **05540.10 CASTINGS** | | | | |
| Miscellaneous castings | | | | |
| Light sections | Lb. | 1.00 | 0.68 | 1.68 |
| Heavy sections | " | 0.85 | 0.48 | 1.33 |
| Manhole covers and frames | | | | |
| Regular, city type | | | | |
| 18" dia. | | | | |
| 100 lb | EA. | 210.00 | 67.50 | 277.50 |
| 24" dia. | | | | |
| 200 lb | EA. | 220.00 | 67.50 | 287.50 |
| 300 lb | " | 220.00 | 75.00 | 295.00 |
| 400 lb | " | 230.00 | 75.00 | 305.00 |
| 26" dia., 475 lb | " | 280.00 | 84.50 | 364.50 |
| 30" dia., 600 lb | " | 330.00 | 96.50 | 426.50 |
| 8" square, 75 lb | " | 93.00 | 13.50 | 106.50 |
| 24" square | | | | |
| 126 lb | EA. | 200.00 | 67.50 | 267.50 |
| 500 lb | " | 330.00 | 84.50 | 414.50 |
| Watertight type | | | | |
| 20" dia., 200 lb | EA. | 190.00 | 84.50 | 274.50 |
| 24" dia., 350 lb | " | 320.00 | 110.00 | 430.00 |
| Steps, cast iron | | | | |
| 7" x 9" | EA. | 8.95 | 6.75 | 15.70 |
| 8" x 9" | " | 10.70 | 7.50 | 18.20 |
| Manhole covers and frames, aluminum | | | | |
| 12" x 12" | EA. | 42.80 | 13.50 | 56.30 |
| 18" x 18" | " | 44.40 | 13.50 | 57.90 |
| 24" x 24" | " | 48.70 | 16.90 | 65.60 |
| Corner protection | | | | |
| Steel angle guard with anchors | | | | |
| 2" x 2" x 3/16" | L.F. | 9.30 | 4.85 | 14.15 |
| 2" x 3" x 1/4" | " | 10.55 | 4.85 | 15.40 |
| 3" x 3" x 5/16" | " | 12.25 | 4.85 | 17.10 |
| 3" x 4" x 5/16" | " | 14.70 | 5.20 | 19.90 |
| 4" x 4" x 5/16" | " | 15.30 | 5.20 | 20.50 |

| MISC. FABRICATIONS | UNIT | MAT. | INST. | TOTAL |
|---|---|---|---|---|
| **05580.10 METAL SPECIALTIES** | | | | |
| Kick plate | | | | |
| 4" high x 1/4" thick | | | | |
| Primed | L.F. | 4.40 | 6.75 | 11.15 |
| Galvanized | " | 4.95 | 6.75 | 11.70 |
| 6" high x 1/4" thick | | | | |
| Primed | L.F. | 4.80 | 7.50 | 12.30 |
| Galvanized | " | 5.65 | 7.50 | 13.15 |
| **05700.10 ORNAMENTAL METAL** | | | | |
| Railings, vertical square bars, 6" o.c., with shaped top rails | | | | |
| Steel | L.F. | 41.60 | 16.90 | 58.50 |
| Aluminum | " | 68.00 | 16.90 | 84.90 |
| Bronze | " | 110.00 | 22.55 | 132.55 |
| Stainless steel | " | 100.00 | 22.55 | 122.55 |
| Laminated metal or wood handrails with metal supports | | | | |
| 2-1/2" round or oval shape | L.F. | 93.50 | 16.90 | 110.40 |
| Grilles and louvers | | | | |
| Fixed type louvers | | | | |
| 4 through 10 sf | S.F. | 16.75 | 5.65 | 22.40 |
| Over 10 sf | " | 20.20 | 4.20 | 24.40 |
| Movable type louvers | | | | |
| 4 through 10 sf | S.F. | 20.20 | 5.65 | 25.85 |
| Over 10 sf | " | 22.80 | 4.20 | 27.00 |
| Aluminum louvers | | | | |
| Residential use, fixed type, with screen | | | | |
| 8" x 8" | EA. | 11.10 | 16.90 | 28.00 |
| 12" x 12" | " | 12.25 | 16.90 | 29.15 |
| 12" x 18" | " | 14.65 | 16.90 | 31.55 |
| 14" x 24" | " | 21.10 | 16.90 | 38.00 |
| 18" x 24" | " | 23.70 | 16.90 | 40.60 |
| 30" x 24" | " | 32.30 | 18.80 | 51.10 |
| **05800.10 EXPANSION CONTROL** | | | | |
| Expansion joints with covers, floor assembly type | | | | |
| With 1" space | | | | |
| Aluminum | L.F. | 16.05 | 5.65 | 21.70 |
| Bronze | " | 33.50 | 5.65 | 39.15 |
| Stainless steel | " | 26.00 | 5.65 | 31.65 |
| With 2" space | | | | |
| Aluminum | L.F. | 19.85 | 5.65 | 25.50 |
| Bronze | " | 35.20 | 5.65 | 40.85 |
| Stainless steel | " | 31.00 | 5.65 | 36.65 |
| Ceiling and wall assembly type | | | | |
| With 1" space | | | | |
| Aluminum | L.F. | 10.30 | 6.75 | 17.05 |
| Bronze | " | 35.20 | 6.75 | 41.95 |
| Stainless steel | " | 31.80 | 6.75 | 38.55 |
| With 2" space | | | | |
| Aluminum | L.F. | 11.30 | 6.75 | 18.05 |
| Bronze | " | 38.10 | 6.75 | 44.85 |
| Stainless steel | " | 33.50 | 6.75 | 40.25 |

## MISC. FABRICATIONS

| 05800.10 EXPANSION CONTROL | UNIT | MAT. | INST. | TOTAL |
|---|---|---|---|---|
| Exterior roof and wall, aluminum | | | | |
| Roof to roof | | | | |
| With 1" space | L.F. | 30.00 | 5.65 | 35.65 |
| With 2" space | " | 32.40 | 5.65 | 38.05 |
| Roof to wall | | | | |
| With 1" space | L.F. | 23.30 | 6.15 | 29.45 |
| With 2" space | " | 27.70 | 6.15 | 33.85 |
| Flat wall to wall | | | | |
| With 1" space | L.F. | 11.30 | 5.65 | 16.95 |
| With 2" space | " | 12.35 | 5.65 | 18.00 |
| Corner to flat wall | | | | |
| With 1" space | L.F. | 13.65 | 6.75 | 20.40 |
| With 2" in space | " | 13.85 | 6.75 | 20.60 |

## FASTENERS AND ADHESIVES

### 06050.10 ACCESSORIES

| | UNIT | MAT. | INST. | TOTAL |
|---|---|---|---|---|
| Column/post base, cast aluminum | | | | |
| 4" x 4" | EA. | 1.90 | 7.70 | 9.60 |
| 6" x 6" | " | 4.20 | 7.70 | 11.90 |
| Bridging, metal, per pair | | | | |
| 12" o.c. | EA. | 0.16 | 3.10 | 3.26 |
| 16" o.c. | " | 0.16 | 2.80 | 2.96 |
| Anchors | | | | |
| Bolts, threaded two ends, with nuts and washers | | | | |
| 1/2" dia. | | | | |
| 4" long | EA. | 0.93 | 1.95 | 2.88 |
| 7-1/2" long | " | 1.15 | 1.95 | 3.10 |
| 3/4" dia. | | | | |
| 7-1/2" long | EA. | 1.80 | 1.95 | 3.75 |
| 15" long | " | 2.35 | 1.95 | 4.30 |
| Framing anchors | | | | |
| 10 gauge | EA. | 0.51 | 2.55 | 3.06 |
| Bolts, carriage | | | | |
| 1/4 x 4 | EA. | 0.94 | 3.10 | 4.04 |
| 5/16 x 6 | " | 1.00 | 3.25 | 4.25 |
| 3/8 x 6 | " | 1.15 | 3.25 | 4.40 |
| 1/2 x 6 | " | 1.30 | 3.25 | 4.55 |
| Joist and beam hangers | | | | |
| 18 ga. | | | | |
| 2 x 4 | EA. | 0.47 | 3.10 | 3.57 |
| 2 x 6 | " | 0.59 | 3.10 | 3.69 |
| 2 x 8 | " | 0.70 | 3.10 | 3.80 |
| 2 x 10 | " | 0.81 | 3.45 | 4.26 |
| 2 x 12 | " | 0.93 | 3.85 | 4.78 |
| 16 ga. | | | | |
| 3 x 6 | EA. | 1.05 | 3.45 | 4.50 |
| 3 x 8 | " | 1.15 | 3.45 | 4.60 |
| 3 x 10 | " | 1.45 | 3.65 | 5.10 |
| 3 x 12 | " | 1.75 | 4.10 | 5.85 |
| 3 x 14 | " | 2.10 | 4.40 | 6.50 |
| 4 x 6 | " | 1.30 | 3.45 | 4.75 |
| 4 x 8 | " | 1.40 | 3.45 | 4.85 |
| 4 x 10 | " | 1.85 | 3.65 | 5.50 |
| 4 x 12 | " | 2.10 | 4.10 | 6.20 |
| 4 x 14 | " | 2.20 | 4.40 | 6.60 |
| Rafter anchors, 18 ga., 1-1/2" wide | | | | |
| 5-1/4" long | EA. | 0.52 | 2.55 | 3.07 |
| 10-3/4" long | " | 0.79 | 2.55 | 3.34 |
| Shear plates | | | | |
| 2-5/8" dia. | EA. | 1.45 | 2.40 | 3.85 |
| 4" dia. | " | 3.05 | 2.55 | 5.60 |
| Sill anchors | | | | |
| Embedded in concrete | EA. | 1.30 | 3.10 | 4.40 |
| Split rings | | | | |
| 2-1/2" dia. | EA. | 0.88 | 3.45 | 4.33 |
| 4" dia. | " | 1.70 | 3.85 | 5.55 |
| Strap ties, 14 ga., 1-3/8" wide | | | | |
| 12" long | EA. | 1.05 | 2.55 | 3.60 |

# 06 WOOD AND PLASTICS

## FASTENERS AND ADHESIVES

| FASTENERS AND ADHESIVES | UNIT | MAT. | INST. | TOTAL |
|---|---|---|---|---|
| **06050.10 ACCESSORIES** | | | | |
| 18" long | EA. | 1.15 | 2.80 | 3.95 |
| 24" long | " | 1.45 | 3.10 | 4.55 |
| 36" long | " | 2.20 | 3.45 | 5.65 |
| Toothed rings | | | | |
| 2-5/8" dia. | EA. | 0.79 | 5.15 | 5.94 |
| 4" dia. | " | 0.93 | 6.20 | 7.13 |

## ROUGH CARPENTRY

| ROUGH CARPENTRY | UNIT | MAT. | INST. | TOTAL |
|---|---|---|---|---|
| **06110.10 BLOCKING** | | | | |
| Steel construction | | | | |
| Walls | | | | |
| 2x4 | L.F. | 0.49 | 2.05 | 2.54 |
| 2x6 | " | 0.68 | 2.40 | 3.08 |
| 2x8 | " | 0.98 | 2.55 | 3.53 |
| 2x10 | " | 1.35 | 2.80 | 4.15 |
| 2x12 | " | 1.80 | 3.10 | 4.90 |
| Ceilings | | | | |
| 2x4 | L.F. | 0.49 | 2.40 | 2.89 |
| 2x6 | " | 0.68 | 2.80 | 3.48 |
| 2x8 | " | 0.98 | 3.10 | 4.08 |
| 2x10 | " | 1.35 | 3.45 | 4.80 |
| 2x12 | " | 1.80 | 3.85 | 5.65 |
| Wood construction | | | | |
| Walls | | | | |
| 2x4 | L.F. | 0.49 | 1.70 | 2.19 |
| 2x6 | " | 0.68 | 1.95 | 2.63 |
| 2x8 | " | 0.98 | 2.05 | 3.03 |
| 2x10 | " | 1.35 | 2.20 | 3.55 |
| 2x12 | " | 1.80 | 2.40 | 4.20 |
| Ceilings | | | | |
| 2x4 | L.F. | 0.49 | 1.95 | 2.44 |
| 2x6 | " | 0.68 | 2.20 | 2.88 |
| 2x8 | " | 0.98 | 2.40 | 3.38 |
| 2x10 | " | 1.35 | 2.55 | 3.90 |
| 2x12 | " | 1.80 | 2.80 | 4.60 |
| **06110.20 CEILING FRAMING** | | | | |
| Ceiling joists | | | | |
| 12" o.c. | | | | |
| 2x4 | S.F. | 0.58 | 0.73 | 1.31 |
| 2x6 | " | 0.85 | 0.77 | 1.62 |
| 2x8 | " | 1.20 | 0.81 | 2.01 |
| 2x10 | " | 1.80 | 0.86 | 2.66 |

## ROUGH CARPENTRY

| | UNIT | MAT. | INST. | TOTAL |
|---|---|---|---|---|
| **06110.20 CEILING FRAMING** | | | | |
| 2x12 | S.F. | 2.40 | 0.91 | 3.31 |
| 16" o.c. | | | | |
| 2x4 | S.F. | 0.49 | 0.59 | 1.08 |
| 2x6 | " | 0.73 | 0.62 | 1.35 |
| 2x8 | " | 1.00 | 0.64 | 1.64 |
| 2x10 | " | 1.50 | 0.67 | 2.17 |
| 2x12 | " | 1.95 | 0.70 | 2.65 |
| 24" o.c. | | | | |
| 2x4 | S.F. | 0.39 | 0.49 | 0.88 |
| 2x6 | " | 0.58 | 0.52 | 1.10 |
| 2x8 | " | 0.79 | 0.54 | 1.33 |
| 2x10 | " | 1.20 | 0.57 | 1.77 |
| 2x12 | " | 1.60 | 0.60 | 2.21 |
| Headers and nailers | | | | |
| 2x4 | L.F. | 0.49 | 1.00 | 1.49 |
| 2x6 | " | 0.68 | 1.05 | 1.73 |
| 2x8 | " | 0.98 | 1.10 | 2.08 |
| 2x10 | " | 1.35 | 1.20 | 2.55 |
| 2x12 | " | 1.80 | 1.30 | 3.10 |
| Sister joists for ceilings | | | | |
| 2x4 | L.F. | 0.49 | 2.20 | 2.69 |
| 2x6 | " | 0.68 | 2.55 | 3.23 |
| 2x8 | " | 0.98 | 3.10 | 4.08 |
| 2x10 | " | 1.35 | 3.85 | 5.20 |
| 2x12 | " | 1.80 | 5.15 | 6.95 |
| **06110.30 FLOOR FRAMING** | | | | |
| Floor joists | | | | |
| 12" o.c. | | | | |
| 2x6 | S.F. | 0.85 | 0.62 | 1.47 |
| 2x8 | " | 1.20 | 0.63 | 1.83 |
| 2x10 | " | 1.80 | 0.64 | 2.44 |
| 2x12 | " | 2.40 | 0.67 | 3.07 |
| 2x14 | " | 2.65 | 0.64 | 3.29 |
| 3x6 | " | 1.95 | 0.66 | 2.61 |
| 3x8 | " | 2.55 | 0.67 | 3.22 |
| 3x10 | " | 3.20 | 0.70 | 3.90 |
| 3x12 | " | 3.80 | 0.73 | 4.54 |
| 3x14 | " | 4.35 | 0.77 | 5.12 |
| 4x6 | " | 2.55 | 0.64 | 3.19 |
| 4x8 | " | 3.25 | 0.67 | 3.92 |
| 4x10 | " | 4.20 | 0.70 | 4.90 |
| 4x12 | " | 5.10 | 0.73 | 5.84 |
| 4x14 | " | 5.95 | 0.77 | 6.72 |
| 16" o.c. | | | | |
| 2x6 | S.F. | 0.73 | 0.52 | 1.25 |
| 2x8 | " | 1.00 | 0.52 | 1.52 |
| 2x10 | " | 1.50 | 0.53 | 2.03 |
| 2x12 | " | 1.95 | 0.55 | 2.50 |
| 2x14 | " | 2.40 | 0.57 | 2.97 |
| 3x6 | " | 1.65 | 0.53 | 2.18 |

## ROUGH CARPENTRY

| ROUGH CARPENTRY | UNIT | MAT. | INST. | TOTAL |
|---|---|---|---|---|
| **06110.30 FLOOR FRAMING** | | | | |
| 3x8 | S.F. | 2.15 | 0.55 | 2.70 |
| 3x10 | " | 2.65 | 0.57 | 3.22 |
| 3x12 | " | 3.15 | 0.59 | 3.74 |
| 3x14 | " | 3.75 | 0.62 | 4.37 |
| 4x6 | " | 2.10 | 0.53 | 2.63 |
| 4x8 | " | 2.85 | 0.55 | 3.40 |
| 4x10 | " | 3.55 | 0.57 | 4.12 |
| 4x12 | " | 4.20 | 0.59 | 4.79 |
| 4x14 | " | 4.95 | 0.62 | 5.57 |
| Sister joists for floors | | | | |
| 2x4 | L.F. | 0.49 | 1.95 | 2.44 |
| 2x6 | " | 0.68 | 2.20 | 2.88 |
| 2x8 | " | 0.98 | 2.55 | 3.53 |
| 2x10 | " | 1.35 | 3.10 | 4.45 |
| 2x12 | " | 1.80 | 3.85 | 5.65 |
| 3x6 | " | 1.65 | 3.10 | 4.75 |
| 3x8 | " | 2.15 | 3.45 | 5.60 |
| 3x10 | " | 2.65 | 3.85 | 6.50 |
| 3x12 | " | 3.15 | 4.40 | 7.55 |
| 4x6 | " | 2.05 | 3.10 | 5.15 |
| 4x8 | " | 2.85 | 3.45 | 6.30 |
| 4x10 | " | 3.55 | 3.85 | 7.40 |
| 4x12 | " | 4.20 | 4.40 | 8.60 |
| **06110.40 FURRING** | | | | |
| Furring, wood strips | | | | |
| Walls | | | | |
| On masonry or concrete walls | | | | |
| 1x2 furring | | | | |
| 12" o.c. | S.F. | 0.20 | 0.96 | 1.17 |
| 16" o.c. | " | 0.17 | 0.88 | 1.05 |
| 24" o.c. | " | 0.13 | 0.81 | 0.94 |
| 1x3 furring | | | | |
| 12" o.c. | S.F. | 0.28 | 0.96 | 1.25 |
| 16" o.c. | " | 0.23 | 0.88 | 1.11 |
| 24" o.c. | " | 0.18 | 0.81 | 0.99 |
| On wood walls | | | | |
| 1x2 furring | | | | |
| 12" o.c. | S.F. | 0.20 | 0.69 | 0.89 |
| 16" o.c. | " | 0.17 | 0.62 | 0.79 |
| 24" o.c. | " | 0.13 | 0.56 | 0.69 |
| 1x3 furring | | | | |
| 12" o.c. | S.F. | 0.28 | 0.69 | 0.97 |
| 16" o.c. | " | 0.23 | 0.62 | 0.85 |
| 24" o.c. | " | 0.18 | 0.56 | 0.74 |
| Ceilings | | | | |
| On masonry or concrete ceilings | | | | |
| 1x2 furring | | | | |
| 12" o.c. | S.F. | 0.20 | 1.70 | 1.90 |
| 16" o.c. | " | 0.17 | 1.55 | 1.72 |
| 24" o.c. | " | 0.13 | 1.40 | 1.53 |

# 06 WOOD AND PLASTICS

## ROUGH CARPENTRY

| ROUGH CARPENTRY | UNIT | MAT. | INST. | TOTAL |
|---|---|---|---|---|
| **06110.40 FURRING** | | | | |
| 1x3 furring | | | | |
| 12" o.c. | S.F. | 0.28 | 1.70 | 1.98 |
| 16" o.c. | " | 0.23 | 1.55 | 1.78 |
| 24" o.c. | " | 0.18 | 1.40 | 1.58 |
| On wood ceilings | | | | |
| 1x2 furring | | | | |
| 12" o.c. | S.F. | 0.20 | 1.15 | 1.35 |
| 16" o.c. | " | 0.17 | 1.05 | 1.22 |
| 24" o.c. | " | 0.13 | 0.94 | 1.07 |
| 1x3 | | | | |
| 12" o.c. | S.F. | 0.28 | 1.15 | 1.43 |
| 16" o.c. | " | 0.23 | 1.05 | 1.28 |
| 24" o.c. | " | 0.18 | 0.94 | 1.12 |
| **06110.50 ROOF FRAMING** | | | | |
| Roof framing | | | | |
| Rafters, gable end | | | | |
| 0-2 pitch (flat to 2-in-12) | | | | |
| 12" o.c. | | | | |
| 2x4 | S.F. | 0.58 | 0.64 | 1.22 |
| 2x6 | " | 0.87 | 0.67 | 1.54 |
| 2x8 | " | 1.20 | 0.70 | 1.90 |
| 2x10 | " | 1.80 | 0.73 | 2.54 |
| 2x12 | " | 2.40 | 0.77 | 3.17 |
| 16" o.c. | | | | |
| 2x6 | S.F. | 0.73 | 0.55 | 1.28 |
| 2x8 | " | 1.00 | 0.57 | 1.57 |
| 2x10 | " | 1.50 | 0.59 | 2.09 |
| 2x12 | " | 1.95 | 0.62 | 2.57 |
| 24" o.c. | | | | |
| 2x6 | S.F. | 0.58 | 0.47 | 1.05 |
| 2x8 | " | 0.79 | 0.48 | 1.27 |
| 2x10 | " | 1.20 | 0.50 | 1.70 |
| 2x12 | " | 1.60 | 0.52 | 2.12 |
| 4-6 pitch (4-in-12 to 6-in-12) | | | | |
| 12" o.c. | | | | |
| 2x4 | S.F. | 0.61 | 0.67 | 1.28 |
| 2x6 | " | 0.92 | 0.70 | 1.62 |
| 2x8 | " | 1.30 | 0.73 | 2.04 |
| 2x10 | " | 1.95 | 0.77 | 2.72 |
| 2x12 | " | 2.60 | 0.81 | 3.41 |
| 16" o.c. | | | | |
| 2x6 | S.F. | 0.77 | 0.57 | 1.34 |
| 2x8 | " | 1.10 | 0.59 | 1.69 |
| 2x10 | " | 1.60 | 0.62 | 2.22 |
| 2x12 | " | 2.15 | 0.64 | 2.79 |
| 24" o.c. | | | | |
| 2x6 | S.F. | 0.61 | 0.48 | 1.09 |
| 2x8 | " | 0.85 | 0.50 | 1.35 |
| 2x10 | " | 1.30 | 0.53 | 1.83 |
| 2x12 | " | 1.70 | 0.59 | 2.29 |

| ROUGH CARPENTRY | UNIT | MAT. | INST. | TOTAL |
|---|---|---|---|---|
| **06110.50 ROOF FRAMING** | | | | |
| 8-12 pitch (8-in-12 to 12-in-12) | | | | |
| 12" o.c. | | | | |
| 2x4 | S.F. | 0.64 | 0.70 | 1.34 |
| 2x6 | " | 0.99 | 0.73 | 1.73 |
| 2x8 | " | 1.35 | 0.77 | 2.12 |
| 2x10 | " | 2.05 | 0.81 | 2.86 |
| 2x12 | " | 2.75 | 0.86 | 3.61 |
| 16" o.c. | | | | |
| 2x6 | S.F. | 0.82 | 0.59 | 1.41 |
| 2x8 | " | 1.15 | 0.62 | 1.77 |
| 2x10 | " | 1.70 | 0.64 | 2.34 |
| 2x12 | " | 2.30 | 0.67 | 2.97 |
| 24" o.c. | | | | |
| 2x6 | S.F. | 0.64 | 0.50 | 1.14 |
| 2x8 | " | 0.90 | 0.52 | 1.42 |
| 2x10 | " | 1.35 | 0.53 | 1.88 |
| 2x12 | " | 1.80 | 0.55 | 2.35 |
| Ridge boards | | | | |
| 2x6 | L.F. | 0.68 | 1.55 | 2.23 |
| 2x8 | " | 0.98 | 1.70 | 2.68 |
| 2x10 | " | 1.35 | 1.95 | 3.30 |
| 2x12 | " | 1.80 | 2.20 | 4.00 |
| Hip rafters | | | | |
| 2x6 | L.F. | 0.68 | 1.10 | 1.78 |
| 2x8 | " | 0.98 | 1.15 | 2.13 |
| 2x10 | " | 1.35 | 1.20 | 2.55 |
| 2x12 | " | 1.80 | 1.25 | 3.05 |
| Jack rafters | | | | |
| 4-6 pitch (4-in-12 to 6-in-12) | | | | |
| 16" o.c. | | | | |
| 2x6 | S.F. | 0.78 | 0.91 | 1.69 |
| 2x8 | " | 1.10 | 0.94 | 2.04 |
| 2x10 | " | 1.60 | 1.00 | 2.60 |
| 2x12 | " | 2.20 | 1.05 | 3.25 |
| 24" o.c. | | | | |
| 2x6 | S.F. | 0.61 | 0.70 | 1.31 |
| 2x8 | " | 0.85 | 0.72 | 1.57 |
| 2x10 | " | 1.30 | 0.75 | 2.05 |
| 2x12 | " | 1.75 | 0.77 | 2.52 |
| 8-12 pitch (8-in-12 to 12-in-12) | | | | |
| 16" o.c. | | | | |
| 2x6 | S.F. | 1.20 | 0.96 | 2.17 |
| 2x8 | " | 1.70 | 1.00 | 2.70 |
| 2x10 | " | 2.55 | 1.05 | 3.60 |
| 2x12 | " | 3.40 | 1.05 | 4.45 |
| 24" o.c. | | | | |
| 2x6 | S.F. | 0.99 | 0.73 | 1.73 |
| 2x8 | " | 1.35 | 0.75 | 2.10 |
| 2x10 | " | 2.05 | 0.77 | 2.82 |
| 2x12 | " | 2.75 | 0.79 | 3.54 |
| Sister rafters | | | | |
| 2x4 | L.F. | 0.49 | 2.20 | 2.69 |

## ROUGH CARPENTRY

| ROUGH CARPENTRY | UNIT | MAT. | INST. | TOTAL |
|---|---|---|---|---|
| **06110.50 ROOF FRAMING** | | | | |
| 2x6 | L.F. | 0.68 | 2.55 | 3.23 |
| 2x8 | " | 0.98 | 3.10 | 4.08 |
| 2x10 | " | 1.35 | 3.85 | 5.20 |
| 2x12 | " | 1.80 | 5.15 | 6.95 |
| Fascia boards | | | | |
| 2x4 | L.F. | 0.49 | 1.55 | 2.04 |
| 2x6 | " | 0.68 | 1.55 | 2.23 |
| 2x8 | " | 0.98 | 1.70 | 2.68 |
| 2x10 | " | 1.35 | 1.70 | 3.05 |
| 2x12 | " | 1.80 | 1.95 | 3.75 |
| Cant strips | | | | |
| Fiber | | | | |
| 3x3 | L.F. | 0.23 | 0.88 | 1.11 |
| 4x4 | " | 0.31 | 0.94 | 1.25 |
| Wood | | | | |
| 3x3 | L.F. | 1.30 | 0.94 | 2.24 |
| **06110.60 SLEEPERS** | | | | |
| Sleepers, over concrete | | | | |
| 12" o.c. | | | | |
| 1x2 | S.F. | 0.20 | 0.70 | 0.90 |
| 1x3 | " | 0.28 | 0.73 | 1.01 |
| 2x4 | " | 0.58 | 0.86 | 1.44 |
| 2x6 | " | 0.85 | 0.91 | 1.76 |
| 16" o.c. | | | | |
| 1x2 | S.F. | 0.17 | 0.62 | 0.79 |
| 1x3 | " | 0.23 | 0.62 | 0.85 |
| 2x4 | " | 0.49 | 0.73 | 1.23 |
| 2x6 | " | 0.73 | 0.77 | 1.50 |
| **06110.65 SOFFITS** | | | | |
| Soffit framing | | | | |
| 2x3 | L.F. | 0.38 | 2.20 | 2.58 |
| 2x4 | " | 0.49 | 2.40 | 2.89 |
| 2x6 | " | 0.68 | 2.55 | 3.23 |
| 2x8 | " | 0.98 | 2.80 | 3.78 |
| **06110.70 WALL FRAMING** | | | | |
| Framing wall, studs | | | | |
| 12" o.c. | | | | |
| 2x3 | S.F. | 0.42 | 0.57 | 0.99 |
| 2x4 | " | 0.58 | 0.57 | 1.15 |
| 2x6 | " | 0.85 | 0.62 | 1.47 |
| 2x8 | " | 1.20 | 0.64 | 1.84 |
| 16" o.c. | | | | |
| 2x3 | S.F. | 0.36 | 0.48 | 0.84 |
| 2x4 | " | 0.49 | 0.48 | 0.97 |
| 2x6 | " | 0.73 | 0.52 | 1.25 |
| 2x8 | " | 1.00 | 0.53 | 1.53 |
| 24" o.c. | | | | |

# 06 WOOD AND PLASTICS

| ROUGH CARPENTRY | UNIT | MAT. | INST. | TOTAL |
|---|---|---|---|---|
| **06110.70 WALL FRAMING** | | | | |
| 2x3 | S.F. | 0.27 | 0.42 | 0.69 |
| 2x4 | " | 0.39 | 0.42 | 0.81 |
| 2x6 | " | 0.58 | 0.44 | 1.02 |
| 2x8 | " | 0.79 | 0.45 | 1.24 |
| Plates, top or bottom | | | | |
| 2x3 | L.F. | 0.38 | 0.91 | 1.29 |
| 2x4 | " | 0.49 | 0.96 | 1.46 |
| 2x6 | " | 0.68 | 1.05 | 1.73 |
| 2x8 | " | 0.98 | 1.10 | 2.08 |
| Headers, door or window | | | | |
| 2x6 | | | | |
| Single | | | | |
| 3' long | EA. | 2.20 | 15.45 | 17.65 |
| 6' long | " | 4.35 | 19.30 | 23.65 |
| Double | | | | |
| 3' long | EA. | 4.35 | 17.15 | 21.50 |
| 6' long | " | 8.75 | 22.05 | 30.80 |
| 2x8 | | | | |
| Single | | | | |
| 4' long | EA. | 4.05 | 19.30 | 23.35 |
| 8' long | " | 8.15 | 23.75 | 31.90 |
| Double | | | | |
| 4' long | EA. | 8.15 | 22.05 | 30.20 |
| 8' long | " | 16.25 | 28.10 | 44.35 |
| 2x10 | | | | |
| Single | | | | |
| 5' long | EA. | 7.55 | 23.75 | 31.30 |
| 10' long | " | 15.10 | 30.90 | 46.00 |
| Double | | | | |
| 5' long | EA. | 15.10 | 25.70 | 40.80 |
| 10' long | " | 30.20 | 30.90 | 61.10 |
| 2x12 | | | | |
| Single | | | | |
| 6' long | EA. | 11.70 | 23.75 | 35.45 |
| 12' long | " | 23.45 | 30.90 | 54.35 |
| Double | | | | |
| 6' long | EA. | 23.45 | 28.10 | 51.55 |
| 12' long | " | 46.90 | 34.30 | 81.20 |
| **06115.10 FLOOR SHEATHING** | | | | |
| Sub-flooring, plywood, CDX | | | | |
| 1/2" thick | S.F. | 0.46 | 0.39 | 0.85 |
| 5/8" thick | " | 0.54 | 0.44 | 0.98 |
| 3/4" thick | " | 0.64 | 0.52 | 1.16 |
| Structural plywood | | | | |
| 1/2" thick | S.F. | 0.50 | 0.39 | 0.89 |
| 5/8" thick | " | 0.60 | 0.44 | 1.04 |
| 3/4" thick | " | 0.71 | 0.47 | 1.19 |
| Board type subflooring | | | | |
| 1x6 | | | | |
| Minimum | S.F. | 0.74 | 0.69 | 1.43 |
| Maximum | " | 1.00 | 0.77 | 1.77 |

## ROUGH CARPENTRY

| ROUGH CARPENTRY | UNIT | MAT. | INST. | TOTAL |
|---|---|---|---|---|
| **06115.10 FLOOR SHEATHING** | | | | |
| 1x8 | | | | |
| Minimum | S.F. | 0.90 | 0.65 | 1.55 |
| Maximum | " | 1.00 | 0.73 | 1.73 |
| 1x10 | | | | |
| Minimum | S.F. | 1.05 | 0.62 | 1.67 |
| Maximum | " | 1.10 | 0.69 | 1.79 |
| Underlayment | | | | |
| Hardboard, 1/4" tempered | S.F. | 0.37 | 0.39 | 0.76 |
| Plywood, CDX | | | | |
| 3/8" thick | S.F. | 0.40 | 0.39 | 0.79 |
| 1/2" thick | " | 0.46 | 0.41 | 0.87 |
| 5/8" thick | " | 0.54 | 0.44 | 0.98 |
| 3/4" thick | " | 0.64 | 0.47 | 1.12 |
| **06115.20 ROOF SHEATHING** | | | | |
| Sheathing | | | | |
| Plywood, CDX | | | | |
| 3/8" thick | S.F. | 0.40 | 0.40 | 0.80 |
| 1/2" thick | " | 0.46 | 0.41 | 0.87 |
| 5/8" thick | " | 0.54 | 0.44 | 0.98 |
| 3/4" thick | " | 0.64 | 0.47 | 1.12 |
| Structural plywood | | | | |
| 3/8" thick | S.F. | 0.43 | 0.40 | 0.83 |
| 1/2" thick | " | 0.50 | 0.41 | 0.91 |
| 5/8" thick | " | 0.60 | 0.44 | 1.04 |
| 3/4" thick | " | 0.71 | 0.47 | 1.19 |
| **06115.30 WALL SHEATHING** | | | | |
| Sheathing | | | | |
| Plywood, CDX | | | | |
| 3/8" thick | S.F. | 0.40 | 0.46 | 0.86 |
| 1/2" thick | " | 0.46 | 0.47 | 0.94 |
| 5/8" thick | " | 0.54 | 0.52 | 1.05 |
| 3/4" thick | " | 0.64 | 0.56 | 1.20 |
| Waferboard | | | | |
| 3/8" thick | S.F. | 0.34 | 0.46 | 0.80 |
| 1/2" thick | " | 0.42 | 0.47 | 0.90 |
| 5/8" thick | " | 0.51 | 0.52 | 1.02 |
| 3/4" thick | " | 0.57 | 0.56 | 1.13 |
| Structural plywood | | | | |
| 3/8" thick | S.F. | 0.43 | 0.46 | 0.89 |
| 1/2" thick | " | 0.50 | 0.47 | 0.97 |
| 5/8" thick | " | 0.60 | 0.52 | 1.12 |
| 3/4" thick | " | 0.71 | 0.56 | 1.27 |
| Gypsum, 1/2" thick | " | 0.37 | 0.47 | 0.84 |
| Asphalt impregnated fiberboard, 1/2" thick | " | 0.39 | 0.47 | 0.86 |

## ROUGH CARPENTRY

| ROUGH CARPENTRY | UNIT | MAT. | INST. | TOTAL |
|---|---|---|---|---|
| **06125.10 WOOD DECKING** | | | | |
| Decking, T&G solid | | | | |
| Cedar | | | | |
| 3" thick | S.F. | 5.65 | 0.77 | 6.42 |
| 4" thick | " | 6.85 | 0.82 | 7.67 |
| Fir | | | | |
| 3" thick | S.F. | 2.40 | 0.77 | 3.17 |
| 4" thick | " | 2.90 | 0.82 | 3.72 |
| Southern yellow pine | | | | |
| 3" thick | S.F. | 2.30 | 0.88 | 3.18 |
| 4" thick | " | 2.50 | 0.95 | 3.45 |
| White pine | | | | |
| 3" thick | S.F. | 2.95 | 0.77 | 3.72 |
| 4" thick | " | 3.85 | 0.82 | 4.67 |
| **06130.10 HEAVY TIMBER** | | | | |
| Mill framed structures | | | | |
| Beams to 20' long | | | | |
| Douglas fir | | | | |
| 6x8 | L.F. | 4.40 | 3.85 | 8.25 |
| 6x10 | " | 5.20 | 4.00 | 9.20 |
| 6x12 | " | 6.25 | 4.30 | 10.55 |
| 6x14 | " | 7.45 | 4.45 | 11.90 |
| 6x16 | " | 8.15 | 4.60 | 12.75 |
| 8x10 | " | 6.90 | 4.00 | 10.90 |
| 8x12 | " | 8.15 | 4.30 | 12.45 |
| 8x14 | " | 9.40 | 4.45 | 13.85 |
| 8x16 | " | 10.70 | 4.60 | 15.30 |
| Southern yellow pine | | | | |
| 6x8 | L.F. | 3.50 | 3.85 | 7.35 |
| 6x10 | " | 4.25 | 4.00 | 8.25 |
| 6x12 | " | 5.45 | 4.30 | 9.75 |
| 6x14 | " | 6.25 | 4.45 | 10.70 |
| 6x16 | " | 6.90 | 4.60 | 11.50 |
| 8x10 | " | 5.75 | 4.00 | 9.75 |
| 8x12 | " | 7.00 | 4.30 | 11.30 |
| 8x14 | " | 8.05 | 4.45 | 12.50 |
| 8x16 | " | 9.30 | 4.60 | 13.90 |
| Columns to 12' high | | | | |
| Douglas fir | | | | |
| 6x6 | L.F. | 3.15 | 5.80 | 8.95 |
| 8x8 | " | 5.45 | 5.80 | 11.25 |
| 10x10 | " | 9.50 | 6.40 | 15.90 |
| 12x12 | " | 11.80 | 6.40 | 18.20 |
| Southern yellow pine | | | | |
| 6x6 | L.F. | 2.70 | 5.80 | 8.50 |
| 8x8 | " | 4.60 | 5.80 | 10.40 |
| 10x10 | " | 7.15 | 6.40 | 13.55 |
| 12x12 | " | 9.95 | 6.40 | 16.35 |
| Posts, treated | | | | |
| 4x4 | L.F. | 1.50 | 1.25 | 2.75 |
| 6x6 | " | 3.45 | 1.55 | 5.00 |

# 06 WOOD AND PLASTICS

## ROUGH CARPENTRY

### 06190.20 WOOD TRUSSES

| ROUGH CARPENTRY | UNIT | MAT. | INST. | TOTAL |
|---|---|---|---|---|
| Truss, fink, 2x4 members | | | | |
| 3-in-12 slope | | | | |
| 24' span | EA. | 47.50 | 33.00 | 80.50 |
| 26' span | " | 53.50 | 33.00 | 86.50 |
| 28' span | " | 56.00 | 35.00 | 91.00 |
| 30' span | " | 60.50 | 35.00 | 95.50 |
| 34' span | " | 71.50 | 37.30 | 108.80 |
| 38' span | " | 84.50 | 37.30 | 121.80 |
| 5-in-12 slope | | | | |
| 24' span | EA. | 48.70 | 34.00 | 82.70 |
| 28' span | " | 58.50 | 35.00 | 93.50 |
| 30' span | " | 65.50 | 36.10 | 101.60 |
| 32' span | " | 81.00 | 36.10 | 117.10 |
| 40' span | " | 98.50 | 38.50 | 137.00 |
| Gable, 2x4 members | | | | |
| 5-in-12 slope | | | | |
| 24' span | EA. | 68.00 | 34.00 | 102.00 |
| 26' span | " | 72.50 | 34.00 | 106.50 |
| 28' span | " | 83.00 | 35.00 | 118.00 |
| 30' span | " | 88.00 | 36.10 | 124.10 |
| 32' span | " | 90.50 | 36.10 | 126.60 |
| 36' span | " | 97.50 | 37.30 | 134.80 |
| 40' span | " | 110.00 | 38.50 | 148.50 |
| King post type, 2x4 members | | | | |
| 4-in-12 slope | | | | |
| 16' span | EA. | 38.00 | 31.20 | 69.20 |
| 18' span | " | 39.20 | 32.10 | 71.30 |
| 24' span | " | 42.80 | 34.00 | 76.80 |
| 26' span | " | 47.60 | 34.00 | 81.60 |
| 30' span | " | 63.00 | 36.10 | 99.10 |
| 34' span | " | 68.00 | 36.10 | 104.10 |
| 38' span | " | 85.50 | 37.30 | 122.80 |
| 42' span | " | 100.00 | 39.90 | 139.90 |

## FINISH CARPENTRY

### 06200.10 FINISH CARPENTRY

| FINISH CARPENTRY | UNIT | MAT. | INST. | TOTAL |
|---|---|---|---|---|
| Mouldings and trim | | | | |
| Apron, flat | | | | |
| 9/16 x 2 | L.F. | 0.81 | 1.55 | 2.36 |
| 9/16 x 3-1/2 | " | 1.45 | 1.65 | 3.10 |
| Base | | | | |
| Colonial | | | | |

| FINISH CARPENTRY | UNIT | MAT. | INST. | TOTAL |
|---|---|---|---|---|
| **06200.10 FINISH CARPENTRY** | | | | |
| 7/16 x 2-1/4 | L.F. | 0.93 | 1.55 | 2.48 |
| 7/16 x 3 | " | 1.10 | 1.55 | 2.65 |
| 7/16 x 3-1/4 | " | 1.20 | 1.55 | 2.75 |
| 9/16 x 3 | " | 1.20 | 1.65 | 2.85 |
| 9/16 x 3-1/4 | " | 1.30 | 1.65 | 2.95 |
| 11/16 x 2-1/4 | " | 1.10 | 1.70 | 2.80 |
| Ranch | | | | |
| 7/16 x 2-1/4 | L.F. | 0.97 | 1.55 | 2.52 |
| 7/16 x 3-1/4 | " | 1.30 | 1.55 | 2.85 |
| 9/16 x 2-1/4 | " | 1.05 | 1.65 | 2.70 |
| 9/16 x 3 | " | 1.25 | 1.65 | 2.90 |
| 9/16 x 3-1/4 | " | 1.30 | 1.65 | 2.95 |
| Casing | | | | |
| 11/16 x 2-1/2 | L.F. | 1.10 | 1.40 | 2.50 |
| 11/16 x 3-1/2 | " | 1.35 | 1.45 | 2.80 |
| Chair rail | | | | |
| 9/16 x 2-1/2 | L.F. | 1.00 | 1.55 | 2.55 |
| 9/16 x 3-1/2 | " | 1.50 | 1.55 | 3.05 |
| Closet pole | | | | |
| 1-1/8" dia. | L.F. | 1.15 | 2.05 | 3.20 |
| 1-5/8" dia. | " | 1.35 | 2.05 | 3.40 |
| Cove | | | | |
| 9/16 x 1-3/4 | L.F. | 0.79 | 1.55 | 2.34 |
| 11/16 x 2-3/4 | " | 1.10 | 1.55 | 2.65 |
| Crown | | | | |
| 9/16 x 1-5/8 | L.F. | 1.15 | 2.05 | 3.20 |
| 9/16 x 2-5/8 | " | 1.30 | 2.40 | 3.70 |
| 11/16 x 3-5/8 | " | 1.55 | 2.55 | 4.10 |
| 11/16 x 4-1/4 | " | 2.15 | 2.80 | 4.95 |
| 11/16 x 5-1/4 | " | 2.80 | 3.10 | 5.90 |
| Drip cap | | | | |
| 1-1/16 x 1-5/8 | L.F. | 1.10 | 1.55 | 2.65 |
| Glass bead | | | | |
| 3/8 x 3/8 | L.F. | 0.31 | 1.95 | 2.26 |
| 1/2 x 9/16 | " | 0.39 | 1.95 | 2.34 |
| 5/8 x 5/8 | " | 0.41 | 1.95 | 2.36 |
| 3/4 x 3/4 | " | 0.51 | 1.95 | 2.46 |
| Half round | | | | |
| 1/2 | L.F. | 0.38 | 1.25 | 1.63 |
| 5/8 | " | 0.51 | 1.25 | 1.76 |
| 3/4 | " | 0.64 | 1.25 | 1.89 |
| Lattice | | | | |
| 1/4 x 7/8 | L.F. | 0.30 | 1.25 | 1.55 |
| 1/4 x 1-1/8 | " | 0.32 | 1.25 | 1.57 |
| 1/4 x 1-3/8 | " | 0.34 | 1.25 | 1.59 |
| 1/4 x 1-3/4 | " | 0.39 | 1.25 | 1.64 |
| 1/4 x 2 | " | 0.46 | 1.25 | 1.71 |
| Ogee molding | | | | |
| 5/8 x 3/4 | L.F. | 0.64 | 1.55 | 2.19 |
| 11/16 x 1-1/8 | " | 0.97 | 1.55 | 2.52 |
| 11/16 x 1-3/8 | " | 1.15 | 1.55 | 2.70 |
| Parting bead | | | | |

| FINISH CARPENTRY | UNIT | MAT. | INST. | TOTAL |
|---|---|---|---|---|
| **06200.10 FINISH CARPENTRY** | | | | |
| 3/8 x 7/8 | L.F. | 0.49 | 1.95 | 2.44 |
| Quarter round | | | | |
| 1/4 x 1/4 | L.F. | 0.16 | 1.25 | 1.41 |
| 3/8 x 3/8 | " | 0.23 | 1.25 | 1.48 |
| 1/2 x 1/2 | " | 0.31 | 1.25 | 1.56 |
| 11/16 x 11/16 | " | 0.33 | 1.35 | 1.68 |
| 3/4 x 3/4 | " | 0.37 | 1.35 | 1.72 |
| 1-1/16 x 1-1/16 | " | 0.47 | 1.40 | 1.87 |
| Railings, balusters | | | | |
| 1-1/8 x 1-1/8 | L.F. | 1.70 | 3.10 | 4.80 |
| 1-1/2 x 1-1/2 | " | 2.00 | 2.80 | 4.80 |
| Screen moldings | | | | |
| 1/4 x 3/4 | L.F. | 0.40 | 2.55 | 2.95 |
| 5/8 x 5/16 | " | 0.50 | 2.55 | 3.05 |
| Shoe | | | | |
| 7/16 x 11/16 | L.F. | 0.50 | 1.25 | 1.75 |
| Sash beads | | | | |
| 1/2 x 3/4 | L.F. | 0.63 | 2.55 | 3.18 |
| 1/2 x 7/8 | " | 0.74 | 2.55 | 3.29 |
| 1/2 x 1-1/8 | " | 0.80 | 2.80 | 3.60 |
| 5/8 x 7/8 | " | 0.80 | 2.80 | 3.60 |
| Stop | | | | |
| 5/8 x 1-5/8 | | | | |
| Colonial | L.F. | 0.39 | 1.95 | 2.34 |
| Ranch | " | 0.38 | 1.95 | 2.33 |
| Stools | | | | |
| 11/16 x 2-1/4 | L.F. | 1.80 | 3.45 | 5.25 |
| 11/16 x 2-1/2 | " | 1.90 | 3.45 | 5.35 |
| 11/16 x 5-1/4 | " | 2.05 | 3.85 | 5.90 |
| Exterior trim, casing, select pine, 1x3 | " | 1.35 | 1.55 | 2.90 |
| Douglas fir | | | | |
| 1x3 | L.F. | 0.71 | 1.55 | 2.26 |
| 1x4 | " | 0.97 | 1.55 | 2.52 |
| 1x6 | " | 1.15 | 1.70 | 2.85 |
| 1x8 | " | 1.70 | 1.95 | 3.65 |
| Cornices, white pine, #2 or better | | | | |
| 1x2 | L.F. | 0.43 | 1.55 | 1.98 |
| 1x4 | " | 0.67 | 1.55 | 2.22 |
| 1x6 | " | 1.05 | 1.70 | 2.75 |
| 1x8 | " | 1.15 | 1.80 | 2.95 |
| 1x10 | " | 1.60 | 1.95 | 3.55 |
| 1x12 | " | 2.00 | 2.05 | 4.05 |
| Shelving, pine | | | | |
| 1x8 | L.F. | 1.40 | 2.40 | 3.80 |
| 1x10 | " | 1.70 | 2.45 | 4.15 |
| 1x12 | " | 2.20 | 2.55 | 4.75 |
| Plywood shelf, 3/4", with edge band, 12" wide | " | 1.50 | 3.10 | 4.60 |
| Adjustable shelf, and rod, 12" wide | | | | |
| 3' to 4' long | EA. | 10.50 | 7.70 | 18.20 |
| 5' to 8' long | " | 19.80 | 10.30 | 30.10 |
| Prefinished wood shelves with brackets and supports | | | | |
| 8" wide | | | | |

## FINISH CARPENTRY

| FINISH CARPENTRY | UNIT | MAT. | INST. | TOTAL |
|---|---|---|---|---|
| **06200.10 FINISH CARPENTRY** | | | | |
| 3' long | EA. | 29.20 | 7.70 | 36.90 |
| 4' long | " | 33.80 | 7.70 | 41.50 |
| 6' long | " | 50.00 | 7.70 | 57.70 |
| 10" wide | | | | |
| 3' long | EA. | 32.70 | 7.70 | 40.40 |
| 4' long | " | 47.80 | 7.70 | 55.50 |
| 6' long | " | 56.00 | 7.70 | 63.70 |
| **06220.10 MILLWORK** | | | | |
| Countertop, laminated plastic | | | | |
| 25" x 7/8" thick | | | | |
| Minimum | L.F. | 13.85 | 7.70 | 21.55 |
| Average | " | 21.35 | 10.30 | 31.65 |
| Maximum | " | 33.50 | 12.35 | 45.85 |
| 25" x 1-1/4" thick | | | | |
| Minimum | L.F. | 15.00 | 10.30 | 25.30 |
| Average | " | 22.50 | 12.35 | 34.85 |
| Maximum | " | 37.00 | 15.45 | 52.45 |
| Add for cutouts | EA. | 0.00 | 19.30 | 19.30 |
| Backsplash, 4" high, 7/8" thick | L.F. | 12.70 | 6.20 | 18.90 |
| Plywood, sanded, A-C | | | | |
| 1/4" thick | S.F. | 0.67 | 1.05 | 1.72 |
| 3/8" thick | " | 0.72 | 1.10 | 1.82 |
| 1/2" thick | " | 0.81 | 1.20 | 2.01 |
| A-D | | | | |
| 1/4" thick | S.F. | 0.60 | 1.05 | 1.65 |
| 3/8" thick | " | 0.69 | 1.10 | 1.79 |
| 1/2" thick | " | 0.78 | 1.20 | 1.98 |
| Base cabinets, 34-1/2" high, 24" deep, hardwood, no tops | | | | |
| Minimum | L.F. | 40.40 | 12.35 | 52.75 |
| Average | " | 57.50 | 15.45 | 72.95 |
| Maximum | " | 100.00 | 20.60 | 120.60 |
| Wall cabinets | | | | |
| Minimum | L.F. | 28.90 | 10.30 | 39.20 |
| Average | " | 46.20 | 12.35 | 58.55 |
| Maximum | " | 89.00 | 15.45 | 104.45 |

## WOOD TREATMENT

| WOOD TREATMENT | UNIT | MAT. | INST. | TOTAL |
|---|---|---|---|---|
| **06300.10 WOOD TREATMENT** | | | | |
| Creosote preservative treatment | | | | |
| 8 lb/cf | B.F. | | | 0.24 |

## WOOD TREATMENT

| 06300.10 WOOD TREATMENT | UNIT | MAT. | INST. | TOTAL |
|---|---|---|---|---|
| 10 lb/cf | B.F. | | | 0.32 |
| Salt preservative treatment | | | | |
| Oil borne | | | | |
| Minimum | B.F. | | | 0.18 |
| Maximum | " | | | 0.28 |
| Water borne | | | | |
| Minimum | B.F. | | | 0.14 |
| Maximum | " | | | 0.22 |
| Fire retardant treatment | | | | |
| Minimum | B.F. | | | 0.31 |
| Maximum | " | | | 0.38 |
| Kiln dried, softwood, add to framing costs | | | | |
| 1" thick | B.F. | | | 0.10 |
| 2" thick | " | | | 0.16 |
| 3" thick | " | | | 0.21 |
| 4" thick | " | | | 0.26 |

## ARCHITECTURAL WOODWORK

| 06420.10 PANEL WORK | UNIT | MAT. | INST. | TOTAL |
|---|---|---|---|---|
| Hardboard, tempered, 1/4" thick | | | | |
| Natural faced | S.F. | 0.42 | 0.77 | 1.19 |
| Plastic faced | " | 0.71 | 0.88 | 1.59 |
| Pegboard, natural | " | 0.49 | 0.77 | 1.26 |
| Plastic faced | " | 0.64 | 0.88 | 1.52 |
| Untempered, 1/4" thick | | | | |
| Natural faced | S.F. | 0.38 | 0.77 | 1.15 |
| Plastic faced | " | 0.69 | 0.88 | 1.57 |
| Pegboard, natural | " | 0.40 | 0.77 | 1.17 |
| Plastic faced | " | 0.63 | 0.88 | 1.51 |
| Plywood unfinished, 1/4" thick | | | | |
| Birch | | | | |
| Natural | S.F. | 0.97 | 1.05 | 2.02 |
| Select | " | 1.25 | 1.05 | 2.30 |
| Knotty pine | " | 1.25 | 1.05 | 2.30 |
| Cedar (closet lining) | | | | |
| Standard boards T&G | S.F. | 1.15 | 1.05 | 2.20 |
| Particle board | " | 0.93 | 1.05 | 1.98 |
| Plywood, prefinished, 1/4" thick, premium grade | | | | |
| Birch veneer | S.F. | 1.70 | 1.25 | 2.95 |
| Cherry veneer | " | 1.90 | 1.25 | 3.15 |
| Chestnut veneer | " | 3.40 | 1.25 | 4.65 |
| Lauan veneer | " | 0.61 | 1.25 | 1.86 |
| Mahogany veneer | " | 1.90 | 1.25 | 3.15 |

# 06 WOOD AND PLASTICS

| ARCHITECTURAL WOODWORK | UNIT | MAT. | INST. | TOTAL |
|---|---|---|---|---|
| **06420.10 PANEL WORK** | | | | |
| Oak veneer (red) | S.F. | 1.90 | 1.25 | 3.15 |
| Pecan veneer | " | 2.25 | 1.25 | 3.50 |
| Rosewood veneer | " | 3.15 | 1.25 | 4.40 |
| Teak veneer | " | 4.05 | 1.25 | 5.30 |
| Walnut veneer | " | 2.85 | 1.25 | 4.10 |
| **06430.10 STAIRWORK** | | | | |
| Risers, 1x8, 42" wide | | | | |
| White oak | EA. | 13.85 | 15.45 | 29.30 |
| Pine | " | 9.80 | 15.45 | 25.25 |
| Treads, 1-1/16" x 9-1/2" x 42" | | | | |
| White oak | EA. | 19.65 | 19.30 | 38.95 |
| **06440.10 COLUMNS** | | | | |
| Column, hollow, round wood | | | | |
| 12" diameter | | | | |
| 10' high | EA. | 380.00 | 43.60 | 423.60 |
| 12' high | " | 470.00 | 46.70 | 516.70 |
| 14' high | " | 550.00 | 52.50 | 602.50 |
| 16' high | " | 640.00 | 65.50 | 705.50 |
| 24" diameter | | | | |
| 16' high | EA. | 1,920 | 65.50 | 1,986 |
| 18' high | " | 2,220 | 69.00 | 2,289 |
| 20' high | " | 2,680 | 69.00 | 2,749 |
| 22' high | " | 2,800 | 72.50 | 2,873 |
| 24' high | " | 3,030 | 72.50 | 3,103 |

## MOISTURE PROTECTION

### 07100.10 WATERPROOFING

| MOISTURE PROTECTION | UNIT | MAT. | INST. | TOTAL |
|---|---|---|---|---|
| Membrane waterproofing, elastomeric | | | | |
| Butyl | | | | |
| 1/32" thick | S.F. | 0.57 | 0.97 | 1.54 |
| 1/16" thick | " | 0.74 | 1.00 | 1.74 |
| Butyl with nylon | | | | |
| 1/32" thick | S.F. | 0.68 | 0.97 | 1.65 |
| 1/16" thick | " | 0.80 | 1.00 | 1.80 |
| Neoprene | | | | |
| 1/32" thick | S.F. | 0.93 | 0.97 | 1.90 |
| 1/16" thick | " | 1.55 | 1.00 | 2.55 |
| Neoprene with nylon | | | | |
| 1/32" thick | S.F. | 1.00 | 0.97 | 1.97 |
| 1/16" thick | " | 1.65 | 1.00 | 2.65 |
| Bituminous membrane waterproofing, asphalt felt, 15 lb. | | | | |
| One ply | S.F. | 0.33 | 0.60 | 0.94 |
| Two ply | " | 0.39 | 0.73 | 1.12 |
| Three ply | " | 0.50 | 0.86 | 1.36 |
| Four ply | " | 0.56 | 1.00 | 1.56 |
| Five ply | " | 0.61 | 1.25 | 1.86 |
| Modified asphalt membrane waterproofing, fibrous asphalt | | | | |
| One ply | S.F. | 0.78 | 1.00 | 1.78 |
| Two ply | " | 1.15 | 1.20 | 2.35 |
| Three ply | " | 1.30 | 1.35 | 2.65 |
| Four ply | " | 1.60 | 1.60 | 3.20 |
| Five ply | " | 1.85 | 1.95 | 3.80 |
| Asphalt coated protective board | | | | |
| 1/8" thick | S.F. | 0.43 | 0.60 | 1.03 |
| 1/4" thick | " | 0.61 | 0.60 | 1.22 |
| 3/8" thick | " | 0.68 | 0.60 | 1.28 |
| 1/2" thick | " | 0.81 | 0.64 | 1.45 |
| Cement protective board | | | | |
| 3/8" thick | S.F. | 1.50 | 0.81 | 2.31 |
| 1/2" thick | " | 2.10 | 0.81 | 2.91 |
| Fluid applied, neoprene | | | | |
| 50 mil | S.F. | 1.80 | 0.81 | 2.61 |
| 90 mil | " | 2.50 | 0.81 | 3.31 |
| Tab extended polyurethane | | | | |
| .050" thick | S.F. | 0.98 | 0.60 | 1.59 |
| Fluid applied rubber based polyurethane | | | | |
| 6 mil | S.F. | 0.52 | 0.76 | 1.28 |
| 15 mil | " | 0.97 | 0.60 | 1.58 |
| Bentonite waterproofing, panels | | | | |
| 3/16" thick | S.F. | 0.88 | 0.60 | 1.49 |
| 1/4" thick | " | 0.99 | 0.60 | 1.60 |
| 5/8" thick | " | 1.50 | 0.64 | 2.14 |
| Granular admixtures, trowel on, 3/8" thick | " | 1.20 | 0.60 | 1.80 |
| Metallic oxide waterproofing, iron compound, troweled | | | | |
| 5/8" thick | S.F. | 0.90 | 0.60 | 1.50 |
| 3/4" thick | " | 1.10 | 0.69 | 1.79 |

# 07 THERMAL AND MOISTURE

## MOISTURE PROTECTION

| MOISTURE PROTECTION | UNIT | MAT. | INST. | TOTAL |
|---|---|---|---|---|
| **07150.10 DAMPROOFING** | | | | |
| Silicone dampproofing, sprayed on | | | | |
| Concrete surface | | | | |
| 1 coat | S.F. | 0.36 | 0.13 | 0.49 |
| 2 coats | " | 0.59 | 0.19 | 0.78 |
| Concrete block | | | | |
| 1 coat | S.F. | 0.36 | 0.16 | 0.52 |
| 2 coats | " | 0.59 | 0.22 | 0.81 |
| Brick | | | | |
| 1 coat | S.F. | 0.41 | 0.19 | 0.60 |
| 2 coats | " | 0.65 | 0.24 | 0.89 |
| **07160.10 BITUMINOUS DAMPPROOFING** | | | | |
| Building paper, asphalt felt | | | | |
| 15 lb | S.F. | 0.08 | 0.97 | 1.05 |
| 30 lb | " | 0.14 | 1.00 | 1.14 |
| Asphalt dampproofing, troweled, cold, primer plus | | | | |
| 1 coat | S.F. | 0.50 | 0.81 | 1.31 |
| 2 coats | " | 0.74 | 1.20 | 1.94 |
| 3 coats | " | 1.10 | 1.50 | 2.60 |
| Fibrous asphalt dampproofing, hot troweled, primer plus | | | | |
| 1 coat | S.F. | 0.78 | 0.97 | 1.75 |
| 2 coats | " | 0.97 | 1.35 | 2.32 |
| 3 coats | " | 1.05 | 1.75 | 2.80 |
| Asphaltic paint dampproofing, per coat | | | | |
| Brush on | S.F. | 0.14 | 0.35 | 0.49 |
| Spray on | " | 0.24 | 0.27 | 0.51 |
| **07190.10 VAPOR BARRIERS** | | | | |
| Vapor barrier, polyethylene | | | | |
| 2 mil | S.F. | 0.01 | 0.12 | 0.13 |
| 6 mil | " | 0.02 | 0.12 | 0.14 |
| 8 mil | " | 0.03 | 0.13 | 0.16 |
| 10 mil | " | 0.04 | 0.13 | 0.17 |

## INSULATION

| INSULATION | UNIT | MAT. | INST. | TOTAL |
|---|---|---|---|---|
| **07210.10 BATT INSULATION** | | | | |
| Ceiling, fiberglass, unfaced | | | | |
| 3-1/2" thick, R11 | S.F. | 0.26 | 0.28 | 0.55 |
| 6" thick, R19 | " | 0.43 | 0.32 | 0.75 |
| 9" thick, R30 | " | 0.67 | 0.37 | 1.04 |
| Suspended ceiling, unfaced | | | | |
| 3-1/2" thick, R11 | S.F. | 0.26 | 0.27 | 0.53 |

| INSULATION | UNIT | MAT. | INST. | TOTAL |
|---|---|---|---|---|
| **07210.10 BATT INSULATION** | | | | |
| 6" thick, R19 | S.F. | 0.43 | 0.30 | 0.73 |
| 9" thick, R30 | " | 0.67 | 0.35 | 1.02 |
| Crawl space, unfaced | | | | |
| 3-1/2" thick, R11 | S.F. | 0.26 | 0.37 | 0.63 |
| 6" thick, R19 | " | 0.43 | 0.40 | 0.83 |
| 9" thick, R30 | " | 0.67 | 0.44 | 1.11 |
| Wall, fiberglass | | | | |
| Paper backed | | | | |
| 2" thick, R7 | S.F. | 0.17 | 0.26 | 0.42 |
| 3" thick, R8 | " | 0.18 | 0.27 | 0.45 |
| 4" thick, R11 | " | 0.30 | 0.28 | 0.58 |
| 6" thick, R19 | " | 0.44 | 0.30 | 0.74 |
| Foil backed, 1 side | | | | |
| 2" thick, R7 | S.F. | 0.38 | 0.26 | 0.64 |
| 3" thick, R11 | " | 0.40 | 0.27 | 0.67 |
| 4" thick, R14 | " | 0.44 | 0.28 | 0.72 |
| 6" thick, R21 | " | 0.47 | 0.30 | 0.77 |
| Foil backed, 2 sides | | | | |
| 2" thick, R7 | S.F. | 0.44 | 0.28 | 0.72 |
| 3" thick, R11 | " | 0.54 | 0.30 | 0.84 |
| 4" thick, R14 | " | 0.64 | 0.32 | 0.96 |
| 6" thick, R21 | " | 0.69 | 0.35 | 1.04 |
| Unfaced | | | | |
| 2" thick, R7 | S.F. | 0.18 | 0.26 | 0.44 |
| 3" thick, R9 | " | 0.21 | 0.27 | 0.48 |
| 4" thick, R11 | " | 0.26 | 0.28 | 0.55 |
| 6" thick, R19 | " | 0.43 | 0.30 | 0.73 |
| Mineral wool batts | | | | |
| Paper backed | | | | |
| 2" thick, R6 | S.F. | 0.37 | 0.26 | 0.63 |
| 4" thick, R12 | " | 0.44 | 0.27 | 0.71 |
| 6" thick, R19 | " | 0.51 | 0.30 | 0.81 |
| Fasteners, self adhering, attached to ceiling deck | | | | |
| 2-1/2" long | EA. | 0.11 | 0.40 | 0.51 |
| 4-1/2" long | " | 0.13 | 0.44 | 0.57 |
| Capped, self-locking washers for fastening insulation | " | 0.11 | 0.24 | 0.35 |
| **07210.20 BOARD INSULATION** | | | | |
| Insulation, rigid | | | | |
| Fiberglass, roof | | | | |
| 0.75" thick, R2.78 | S.F. | 0.36 | 0.22 | 0.58 |
| 1.06" thick, R4.17 | " | 0.52 | 0.23 | 0.75 |
| 1.31" thick, R5.26 | " | 0.72 | 0.24 | 0.96 |
| 1.63" thick, R6.67 | " | 0.90 | 0.26 | 1.16 |
| 2.25" thick, R8.33 | " | 0.88 | 0.27 | 1.15 |
| Composite board, roof | | | | |
| 1-1/2" thick, R6.67 | S.F. | 0.72 | 0.24 | 0.96 |
| 1-5/8" thick, R7.69 | " | 0.78 | 0.26 | 1.03 |
| 2" thick, R10.0 | " | 1.40 | 0.27 | 1.67 |
| 2-1/4" thick, R12.50 | " | 1.55 | 0.28 | 1.84 |
| 2-1/2" thick, R14.29 | " | 1.70 | 0.30 | 2.00 |
| 2-3/4" thick, R16.67 | " | 1.85 | 0.32 | 2.17 |

| INSULATION | UNIT | MAT. | INST. | TOTAL |
|---|---|---|---|---|
| **07210.20 BOARD INSULATION** | | | | |
| 3-1/4" thick, R20.00 | S.F. | 2.40 | 0.35 | 2.75 |
| Perlite board, roof | | | | |
| 1.00" thick, R2.78 | S.F. | 0.47 | 0.20 | 0.67 |
| 1.50" thick, R4.17 | " | 0.72 | 0.21 | 0.93 |
| 2.00" thick, R5.92 | " | 0.89 | 0.22 | 1.11 |
| 2.50" thick, R6.67 | " | 1.10 | 0.23 | 1.33 |
| 3.00" thick, R8.33 | " | 1.40 | 0.24 | 1.64 |
| 4.00" thick, R10.00 | " | 1.55 | 0.26 | 1.80 |
| 5.25" thick, R14.29 | " | 1.70 | 0.27 | 1.97 |
| Rigid urethane | | | | |
| Roof | | | | |
| 1" thick, R6.67 | S.F. | 0.63 | 0.20 | 0.83 |
| 1.20" thick, R8.33 | " | 0.71 | 0.21 | 0.92 |
| 1.50" thick, R11.11 | " | 0.87 | 0.21 | 1.08 |
| 2" thick, R14.29 | " | 1.15 | 0.22 | 1.37 |
| 2.25" thick, R16.67 | " | 1.45 | 0.23 | 1.68 |
| Wall | | | | |
| 1" thick, R6.67 | S.F. | 0.63 | 0.26 | 0.89 |
| 1.5" thick, R11.11 | " | 0.87 | 0.27 | 1.14 |
| 2" thick, R14.29 | " | 1.15 | 0.28 | 1.44 |
| Polystyrene | | | | |
| Roof | | | | |
| 1.0" thick, R4.17 | S.F. | 0.60 | 0.20 | 0.80 |
| 1.5" thick, R6.26 | " | 0.99 | 0.21 | 1.20 |
| 2.0" thick, R8.33 | " | 1.10 | 0.22 | 1.32 |
| Wall | | | | |
| 1.0" thick, R4.17 | S.F. | 0.60 | 0.26 | 0.85 |
| 1.5" thick, R6.26 | " | 0.99 | 0.27 | 1.26 |
| 2.0" thick, R8.33 | " | 1.10 | 0.28 | 1.39 |
| Rigid board insulation, deck | | | | |
| Mineral fiberboard | | | | |
| 1" thick, R3.0 | S.F. | 0.36 | 0.20 | 0.56 |
| 2" thick, R5.26 | " | 0.80 | 0.22 | 1.02 |
| Fiberglass | | | | |
| 1" thick, R4.3 | S.F. | 0.62 | 0.20 | 0.82 |
| 2" thick, R8.5 | " | 0.92 | 0.22 | 1.14 |
| Polystyrene | | | | |
| 1" thick, R5.4 | S.F. | 0.60 | 0.20 | 0.80 |
| 2" thick, R10.8 | " | 1.10 | 0.22 | 1.32 |
| Urethane | | | | |
| .75" thick, R5.4 | S.F. | 0.58 | 0.20 | 0.78 |
| 1" thick, R6.4 | " | 0.69 | 0.20 | 0.89 |
| 1.5" thick, R10.7 | " | 0.82 | 0.21 | 1.03 |
| 2" thick, R14.3 | " | 0.94 | 0.22 | 1.16 |
| Foamglass | | | | |
| 1" thick, R1.8 | S.F. | 1.90 | 0.20 | 2.10 |
| 2" thick, R5.26 | " | 2.40 | 0.22 | 2.62 |
| Wood fiber | | | | |
| 1" thick, R3.85 | S.F. | 2.10 | 0.20 | 2.30 |
| 2" thick, R7.7 | " | 2.55 | 0.22 | 2.77 |
| Particle board | | | | |
| 3/4" thick, R2.08 | S.F. | 0.56 | 0.20 | 0.76 |

| INSULATION | UNIT | MAT. | INST. | TOTAL |
|---|---|---|---|---|
| **07210.20 BOARD INSULATION** | | | | |
| 1" thick, R2.77 | S.F. | 0.62 | 0.20 | 0.82 |
| 2" thick, R5.50 | " | 0.64 | 0.22 | 0.86 |
| **07210.60 LOOSE FILL INSULATION** | | | | |
| Blown-in type | | | | |
| Fiberglass | | | | |
| 5" thick, R11 | S.F. | 0.21 | 0.20 | 0.41 |
| 6" thick, R13 | " | 0.26 | 0.24 | 0.50 |
| 9" thick, R19 | " | 0.30 | 0.35 | 0.65 |
| Rockwool, attic application | | | | |
| 6" thick, R13 | S.F. | 0.38 | 0.24 | 0.62 |
| 8" thick, R19 | " | 0.47 | 0.30 | 0.77 |
| 10" thick, R22 | " | 0.56 | 0.37 | 0.93 |
| 12" thick, R26 | " | 0.71 | 0.40 | 1.11 |
| 15" thick, R30 | " | 0.88 | 0.48 | 1.36 |
| Poured type | | | | |
| Fiberglass | | | | |
| 1" thick, R4 | S.F. | 0.19 | 0.15 | 0.34 |
| 2" thick, R8 | " | 0.36 | 0.17 | 0.53 |
| 3" thick, R12 | " | 0.53 | 0.20 | 0.73 |
| 4" thick, R16 | " | 0.70 | 0.24 | 0.94 |
| Mineral wool | | | | |
| 1" thick, R3 | S.F. | 0.21 | 0.15 | 0.36 |
| 2" thick, R6 | " | 0.40 | 0.17 | 0.57 |
| 3" thick, R9 | " | 0.61 | 0.20 | 0.81 |
| 4" thick, R12 | " | 0.70 | 0.24 | 0.94 |
| Vermiculite or perlite | | | | |
| 2" thick, R4.8 | S.F. | 0.48 | 0.17 | 0.65 |
| 3" thick, R7.2 | " | 0.68 | 0.20 | 0.88 |
| 4" thick, R9.6 | " | 0.89 | 0.24 | 1.13 |
| Masonry, poured vermiculite or perlite | | | | |
| 4" block | S.F. | 0.36 | 0.12 | 0.48 |
| 6" block | " | 0.46 | 0.15 | 0.61 |
| 8" block | " | 0.69 | 0.17 | 0.86 |
| 10" block | " | 0.82 | 0.19 | 1.01 |
| 12" block | " | 0.92 | 0.20 | 1.12 |
| **07210.70 SPRAYED INSULATION** | | | | |
| Foam, sprayed on | | | | |
| Polystyrene | | | | |
| 1" thick, R4 | S.F. | 0.34 | 0.24 | 0.58 |
| 2" thick, R8 | " | 0.68 | 0.32 | 1.00 |
| Urethane | | | | |
| 1" thick, R7.7 | S.F. | 0.40 | 0.24 | 0.64 |
| 2" thick, R15.4 | " | 0.68 | 0.32 | 1.00 |
| **07250.10 FIREPROOFING** | | | | |
| Sprayed on | | | | |
| 1" thick | | | | |
| On beams | S.F. | 0.38 | 0.54 | 0.92 |
| On columns | " | 0.40 | 0.48 | 0.88 |

# 07 THERMAL AND MOISTURE

## INSULATION

| 07250.10 FIREPROOFING | UNIT | MAT. | INST. | TOTAL |
|---|---|---|---|---|
| On decks | | | | |
| Flat surface | S.F. | 0.38 | 0.24 | 0.62 |
| Fluted surface | " | 0.41 | 0.30 | 0.71 |
| 1-1/2" thick | | | | |
| On beams | S.F. | 0.57 | 0.69 | 1.26 |
| On columns | " | 0.62 | 0.60 | 1.23 |
| On decks | | | | |
| Flat surface | S.F. | 0.57 | 0.30 | 0.87 |
| Fluted surface | " | 0.60 | 0.40 | 1.00 |

## SHINGLES AND TILES

| 07310.10 ASPHALT SHINGLES | UNIT | MAT. | INST. | TOTAL |
|---|---|---|---|---|
| Standard asphalt shingles, strip shingles | | | | |
| 210 lb/square | SQ. | 38.10 | 29.80 | 67.90 |
| 235 lb/square | " | 40.40 | 33.10 | 73.50 |
| 240 lb/square | " | 41.60 | 37.20 | 78.80 |
| 260 lb/square | " | 64.50 | 42.50 | 107.00 |
| 300 lb/square | " | 66.50 | 49.60 | 116.10 |
| 385 lb/square | " | 92.50 | 59.50 | 152.00 |
| Roll roofing, mineral surface | | | | |
| 90 lb | SQ. | 19.85 | 21.25 | 41.10 |
| 110 lb | " | 32.10 | 24.80 | 56.90 |
| 140 lb | " | 33.30 | 29.80 | 63.10 |

| 07310.30 METAL SHINGLES | UNIT | MAT. | INST. | TOTAL |
|---|---|---|---|---|
| Aluminum, .020" thick | | | | |
| Plain | SQ. | 180.00 | 59.50 | 239.50 |
| Colors | " | 230.00 | 59.50 | 289.50 |
| Steel, galvanized | | | | |
| 26 ga. | | | | |
| Plain | SQ. | 170.00 | 59.50 | 229.50 |
| Colors | " | 220.00 | 59.50 | 279.50 |
| 24 ga. | | | | |
| Plain | SQ. | 200.00 | 59.50 | 259.50 |
| Colors | " | 250.00 | 59.50 | 309.50 |
| Porcelain enamel, 22 ga. | | | | |
| Minimum | SQ. | 510.00 | 74.50 | 584.50 |
| Average | " | 580.00 | 74.50 | 654.50 |
| Maximum | " | 650.00 | 74.50 | 724.50 |

| SHINGLES AND TILES | UNIT | MAT. | INST. | TOTAL |
|---|---|---|---|---|
| **07310.60 SLATE SHINGLES** | | | | |
| Slate shingles | | | | |
| Pennsylvania | | | | |
| Ribbon | SQ. | 400.00 | 150.00 | 550.00 |
| Clear | " | 520.00 | 150.00 | 670.00 |
| Vermont | | | | |
| Black | SQ. | 440.00 | 150.00 | 590.00 |
| Grey | " | 490.00 | 150.00 | 640.00 |
| Green | " | 500.00 | 150.00 | 650.00 |
| Red | " | 510.00 | 150.00 | 660.00 |
| Replacement shingles | | | | |
| Small jobs | EA. | 6.95 | 9.90 | 16.85 |
| Large jobs | S.F. | 4.60 | 4.95 | 9.55 |
| **07310.70 WOOD SHINGLES** | | | | |
| Wood shingles, on roofs | | | | |
| White cedar, #1 shingles | | | | |
| 4" exposure | SQ. | 160.00 | 100.00 | 260.00 |
| 5" exposure | " | 140.00 | 74.50 | 214.50 |
| #2 shingles | | | | |
| 4" exposure | SQ. | 120.00 | 100.00 | 220.00 |
| 5" exposure | " | 98.00 | 74.50 | 172.50 |
| Resquared and rebutted | | | | |
| 4" exposure | SQ. | 140.00 | 100.00 | 240.00 |
| 5" exposure | " | 120.00 | 74.50 | 194.50 |
| On walls | | | | |
| White cedar, #1 shingles | | | | |
| 4" exposure | SQ. | 160.00 | 150.00 | 310.00 |
| 5" exposure | " | 140.00 | 120.00 | 260.00 |
| 6" exposure | " | 120.00 | 100.00 | 220.00 |
| #2 shingles | | | | |
| 4" exposure | SQ. | 120.00 | 150.00 | 270.00 |
| 5" exposure | " | 98.00 | 120.00 | 218.00 |
| 6" exposure | " | 81.00 | 100.00 | 181.00 |
| Add for fire retarding | " | | | 69.26 |
| **07310.80 WOOD SHAKES** | | | | |
| Shakes, hand split, 24" red cedar, on roofs | | | | |
| 5" exposure | SQ. | 150.00 | 150.00 | 300.00 |
| 7" exposure | " | 140.00 | 120.00 | 260.00 |
| 9" exposure | " | 130.00 | 100.00 | 230.00 |
| On walls | | | | |
| 6" exposure | SQ. | 150.00 | 150.00 | 300.00 |
| 8" exposure | " | 140.00 | 120.00 | 260.00 |
| 10" exposure | " | 130.00 | 100.00 | 230.00 |
| Add for fire retarding | " | | | 57.50 |

| ROOFING AND SIDING | UNIT | MAT. | INST. | TOTAL |
|---|---|---|---|---|
| **07410.10 MANUFACTURED ROOFS** | | | | |
| Aluminum roof panels, for structural steel framing | | | | |
| Corrugated | | | | |
| Unpainted finish | | | | |
| .024" | S.F. | 1.15 | 0.74 | 1.89 |
| .030" | " | 1.40 | 0.74 | 2.14 |
| Painted finish | | | | |
| .024" | S.F. | 1.50 | 0.74 | 2.24 |
| .030" | " | 1.85 | 0.74 | 2.59 |
| V-beam | | | | |
| Unpainted finish | | | | |
| .032" | S.F. | 1.55 | 0.74 | 2.29 |
| .040" | " | 1.90 | 0.74 | 2.64 |
| .050" | " | 2.35 | 0.74 | 3.09 |
| Painted finish | | | | |
| .032" | S.F. | 2.05 | 0.74 | 2.79 |
| .040" | " | 2.40 | 0.74 | 3.14 |
| .050" | " | 2.90 | 0.74 | 3.64 |
| Steel roof panels, for structural steel framing | | | | |
| Corrugated, painted | | | | |
| 18 ga. | S.F. | 3.50 | 0.74 | 4.24 |
| 20 ga. | " | 3.30 | 0.74 | 4.04 |
| 22 ga. | " | 2.90 | 0.74 | 3.64 |
| Box rib, painted | | | | |
| 18 ga. | S.F. | 4.10 | 0.78 | 4.88 |
| 20 ga. | " | 3.40 | 0.78 | 4.18 |
| 22 ga. | " | 3.05 | 0.78 | 3.83 |
| 4" rib, painted | | | | |
| 18 ga. | S.F. | 4.75 | 0.83 | 5.58 |
| 20 ga. | " | 4.00 | 0.83 | 4.83 |
| 22 ga. | " | 3.60 | 0.83 | 4.43 |
| Standing seam roof | | | | |
| 2" high seam, painted | | | | |
| 22 ga. | S.F. | 4.60 | 1.20 | 5.80 |
| 24 ga. | " | 4.50 | 1.20 | 5.70 |
| 26 ga. | " | 4.35 | 1.20 | 5.55 |
| **07410.30 MANUFACTURED WALLS** | | | | |
| Sandwich panels with 1-1/2" fiberglass insulation | | | | |
| Galvanized 18 ga. steel interior panels | | | | |
| Exterior panels | | | | |
| 16 ga. aluminum | S.F. | 4.35 | 4.50 | 8.85 |
| 18 ga. galvanized steel | " | 6.20 | 4.50 | 10.70 |
| 20 ga. painted steel | " | 6.30 | 4.50 | 10.80 |
| 20 ga. stainless steel | " | 6.45 | 4.50 | 10.95 |
| Metal liner panels, 1-3/8" thick, 24" wide | | | | |
| Galvanized | | | | |
| 22 ga. | S.F. | 2.30 | 1.15 | 3.45 |
| 20 ga. | " | 2.55 | 1.15 | 3.70 |
| 18 ga. | " | 3.10 | 1.15 | 4.25 |
| Primed | | | | |
| 22 ga. | S.F. | 1.80 | 1.15 | 2.95 |

# 07 THERMAL AND MOISTURE

| ROOFING AND SIDING | UNIT | MAT. | INST. | TOTAL |
|---|---|---|---|---|
| **07410.30 MANUFACTURED WALLS** | | | | |
| 20 ga. | S.F. | 2.15 | 1.15 | 3.30 |
| 18 ga. | " | 2.55 | 1.15 | 3.70 |
| **07440.10 AGGREGATE COATED PANELS** | | | | |
| Dryvit type system | | | | |
| 1" thick | S.F. | 1.65 | 1.15 | 2.80 |
| 1-1/2" thick | " | 1.70 | 1.20 | 2.90 |
| 2" thick | " | 1.80 | 1.40 | 3.20 |
| **07460.10 METAL SIDING PANELS** | | | | |
| Aluminum siding panels | | | | |
| Corrugated | | | | |
| Plain finish | | | | |
| .024" | S.F. | 1.05 | 1.35 | 2.40 |
| .032" | " | 1.25 | 1.35 | 2.60 |
| Painted finish | | | | |
| .024" | S.F. | 1.30 | 1.35 | 2.65 |
| .032" | " | 1.55 | 1.35 | 2.90 |
| V. beam | | | | |
| Plain finish | | | | |
| .032" | S.F. | 1.30 | 1.35 | 2.65 |
| .040" | " | 1.50 | 1.35 | 2.85 |
| .050" | " | 1.75 | 1.35 | 3.10 |
| Painted finish | | | | |
| .032" | S.F. | 1.60 | 1.35 | 2.95 |
| .040" | " | 1.90 | 1.35 | 3.25 |
| .050" | " | 2.75 | 1.35 | 4.10 |
| 4" rib | | | | |
| Plain finish | | | | |
| .032" | S.F. | 1.40 | 1.55 | 2.95 |
| .040" | " | 1.50 | 1.55 | 3.05 |
| .050" | " | 1.85 | 1.55 | 3.40 |
| Painted finish | | | | |
| .032" | S.F. | 1.70 | 1.55 | 3.25 |
| .040" | " | 1.85 | 1.55 | 3.40 |
| .050" | " | 2.10 | 1.55 | 3.65 |
| Steel siding panels | | | | |
| Corrugated | | | | |
| 22 ga. | S.F. | 1.65 | 2.25 | 3.90 |
| 24 ga. | " | 1.40 | 2.25 | 3.65 |
| 26 ga. | " | 1.15 | 2.25 | 3.40 |
| Box rib | | | | |
| 20 ga. | S.F. | 2.20 | 2.25 | 4.45 |
| 22 ga. | " | 1.85 | 2.25 | 4.10 |
| 24 ga. | " | 1.60 | 2.25 | 3.85 |
| 26 ga. | " | 1.15 | 2.25 | 3.40 |
| **07460.50 PLASTIC SIDING** | | | | |
| Horizontal vinyl siding, solid | | | | |
| 8" wide | | | | |
| Standard | S.F. | 0.73 | 1.20 | 1.93 |
| Insulated | " | 1.00 | 1.20 | 2.20 |

## ROOFING AND SIDING

| ROOFING AND SIDING | UNIT | MAT. | INST. | TOTAL |
|---|---|---|---|---|
| **07460.50 PLASTIC SIDING** | | | | |
| 10" wide | | | | |
| Standard | S.F. | 0.75 | 1.10 | 1.85 |
| Insulated | " | 1.10 | 1.10 | 2.20 |
| Vinyl moldings for doors and windows | L.F. | 0.47 | 1.25 | 1.72 |
| **07460.60 PLYWOOD SIDING** | | | | |
| Rough sawn cedar, 3/8" thick | S.F. | 0.97 | 1.05 | 2.02 |
| Fir, 3/8" thick | " | 0.57 | 1.05 | 1.62 |
| Texture 1-11, 5/8" thick | | | | |
| Cedar | S.F. | 1.40 | 1.10 | 2.50 |
| Fir | " | 0.91 | 1.10 | 2.01 |
| Redwood | " | 1.65 | 1.05 | 2.70 |
| Southern Yellow Pine | " | 0.81 | 1.10 | 1.91 |
| **07460.70 STEEL SIDING** | | | | |
| Ribbed, sheets, galvanized | | | | |
| 22 ga. | S.F. | 1.35 | 1.35 | 2.70 |
| 24 ga. | " | 1.15 | 1.35 | 2.50 |
| 26 ga. | " | 0.87 | 1.35 | 2.22 |
| 28 ga. | " | 0.73 | 1.35 | 2.08 |
| Primed | | | | |
| 24 ga. | S.F. | 1.60 | 1.35 | 2.95 |
| 26 ga. | " | 1.00 | 1.35 | 2.35 |
| 28 ga. | " | 0.87 | 1.35 | 2.22 |
| **07460.80 WOOD SIDING** | | | | |
| Beveled siding, cedar | | | | |
| A grade | | | | |
| 1/2 x 6 | S.F. | 2.10 | 1.55 | 3.65 |
| 1/2 x 8 | " | 2.15 | 1.25 | 3.40 |
| 3/4 x 10 | " | 2.10 | 1.05 | 3.15 |
| Clear | | | | |
| 1/2 x 6 | S.F. | 2.25 | 1.55 | 3.80 |
| 1/2 x 8 | " | 2.30 | 1.25 | 3.55 |
| 3/4 x 10 | " | 2.20 | 1.05 | 3.25 |
| B grade | | | | |
| 1/2 x 6 | S.F. | 1.80 | 1.55 | 3.35 |
| 1/2 x 8 | " | 2.05 | 12.35 | 14.40 |
| 3/4 x 10 | " | 1.90 | 1.05 | 2.95 |
| Board and batten | | | | |
| Cedar | | | | |
| 1x6 | S.F. | 2.60 | 1.55 | 4.15 |
| 1x8 | " | 2.40 | 1.25 | 3.65 |
| 1x10 | " | 2.15 | 1.10 | 3.25 |
| 1x12 | " | 1.90 | 1.00 | 2.90 |
| Pine | | | | |
| 1x6 | S.F. | 0.79 | 1.55 | 2.34 |
| 1x8 | " | 0.77 | 1.25 | 2.02 |
| 1x10 | " | 0.73 | 1.10 | 1.83 |
| 1x12 | " | 0.68 | 1.00 | 1.68 |

## ROOFING AND SIDING

| 07460.80 WOOD SIDING | UNIT | MAT. | INST. | TOTAL |
|---|---|---|---|---|
| Redwood | | | | |
| 1x6 | S.F. | 3.40 | 1.55 | 4.95 |
| 1x8 | " | 3.15 | 1.25 | 4.40 |
| 1x10 | " | 2.95 | 1.10 | 4.05 |
| 1x12 | " | 2.70 | 1.00 | 3.70 |
| Tongue and groove | | | | |
| Cedar | | | | |
| 1x4 | S.F. | 2.95 | 1.70 | 4.65 |
| 1x6 | " | 2.85 | 1.65 | 4.50 |
| 1x8 | " | 2.65 | 1.55 | 4.20 |
| 1x10 | " | 2.60 | 1.45 | 4.05 |
| Pine | | | | |
| 1x4 | S.F. | 0.89 | 1.70 | 2.59 |
| 1x6 | " | 0.84 | 1.65 | 2.49 |
| 1x8 | " | 0.78 | 1.55 | 2.33 |
| 1x10 | " | 0.73 | 1.45 | 2.18 |
| Redwood | | | | |
| 1x4 | S.F. | 3.10 | 1.70 | 4.80 |
| 1x6 | " | 3.00 | 1.65 | 4.65 |
| 1x8 | " | 2.90 | 1.55 | 4.45 |
| 1x10 | " | 2.80 | 1.45 | 4.25 |

## MEMBRANE ROOFING

| 07510.10 BUILT-UP ASPHALT ROOFING | UNIT | MAT. | INST. | TOTAL |
|---|---|---|---|---|
| Built-up roofing, asphalt felt, including gravel | | | | |
| 2 ply | SQ. | 26.90 | 74.50 | 101.40 |
| 3 ply | " | 36.80 | 100.00 | 136.80 |
| 4 ply | " | 52.50 | 120.00 | 172.50 |
| Walkway, for built-up roofs | | | | |
| 3' x 3' x | | | | |
| 1/2" thick | S.F. | 2.55 | 0.99 | 3.54 |
| 3/4" thick | " | 3.40 | 0.99 | 4.39 |
| 1" thick | " | 3.75 | 0.99 | 4.74 |
| Roof bonds | | | | |
| Asphalt felt | | | | |
| 10 yrs | SQ. | | | 20.41 |
| 20 yrs | " | | | 21.58 |
| Cant strip, 4" x 4" | | | | |
| Treated wood | L.F. | 1.15 | 0.85 | 2.00 |
| Foamglass | " | 0.61 | 0.74 | 1.35 |
| Mineral fiber | " | 0.26 | 0.74 | 1.00 |
| New gravel for built-up roofing, 400 lb/sq | SQ. | 18.95 | 59.50 | 78.45 |

## MEMBRANE ROOFING

| MEMBRANE ROOFING | UNIT | MAT. | INST. | TOTAL |
|---|---|---|---|---|
| **07510.10 BUILT-UP ASPHALT ROOFING** | | | | |
| Roof gravel (ballast) | C.Y. | 12.70 | 150.00 | 162.70 |
| Aluminum coating, top surfacing, for built-up roofing | SQ. | 20.05 | 49.60 | 69.65 |
| Remove & replace gravel, includes flood coat | " | 26.80 | 100.00 | 126.80 |
| **07530.10 SINGLE-PLY ROOFING** | | | | |
| Elastic sheet roofing | | | | |
| Neoprene, 1/16" thick | S.F. | 1.45 | 0.37 | 1.82 |
| EPDM rubber | | | | |
| 45 mil | S.F. | 0.80 | 0.37 | 1.17 |
| 60 mil | " | 1.10 | 0.37 | 1.47 |
| PVC | | | | |
| 45 mil | S.F. | 1.10 | 0.37 | 1.47 |
| 60 mil | " | 1.65 | 0.37 | 2.02 |
| Flashing | | | | |
| Pipe flashing, 90 mil thick | | | | |
| 1" pipe | EA. | 15.75 | 7.45 | 23.20 |
| 2" pipe | " | 16.90 | 7.45 | 24.35 |
| 3" pipe | " | 17.05 | 7.85 | 24.90 |
| 4" pipe | " | 18.45 | 7.85 | 26.30 |
| 5" pipe | " | 19.70 | 8.25 | 27.95 |
| 6" pipe | " | 21.50 | 8.25 | 29.75 |
| 8" pipe | " | 24.50 | 8.75 | 33.25 |
| 10" pipe | " | 28.00 | 9.90 | 37.90 |
| 12" pipe | " | 34.40 | 9.90 | 44.30 |
| Neoprene flashing, 60 mil thick strip | | | | |
| 6" wide | L.F. | 0.94 | 2.50 | 3.44 |
| 12" wide | " | 1.85 | 3.70 | 5.55 |
| 18" wide | " | 2.75 | 4.95 | 7.70 |
| 24" wide | " | 3.60 | 7.45 | 11.05 |
| Adhesives | | | | |
| Mastic sealer, applied at joints only | | | | |
| 1/4" bead | L.F. | 0.06 | 0.15 | 0.21 |
| Fluid applied roofing | | | | |
| Urethane, 2 components, elastomeric top membrane | | | | |
| 1" thick | S.F. | 1.65 | 0.50 | 2.15 |
| Vinyl liquid roofing, 2 coats, 2 mils per coat | " | 2.55 | 0.42 | 2.98 |
| Silicone roofing, 2 coats sprayed, 16 mil per coat | " | 1.90 | 0.50 | 2.40 |
| Inverted roof system | | | | |
| Insulated membrane with coarse gravel ballast | | | | |
| 3 ply with 2" polystyrene | S.F. | 3.45 | 0.50 | 3.95 |
| Ballast, 3/4" through 1-1/2" dia. river gravel, 100lb/sf | " | 0.23 | 29.80 | 30.03 |
| Walkway for membrane roofs, 1/2" thick | " | 1.30 | 0.99 | 2.29 |

# 07 THERMAL AND MOISTURE

## FLASHING AND SHEET METAL

| | UNIT | MAT. | INST. | TOTAL |
|---|---|---|---|---|
| **07610.10 METAL ROOFING** | | | | |
| Sheet metal roofing, copper, 16 oz, batten seam | SQ. | 460.00 | 200.00 | 660.00 |
| Standing seam | " | 450.00 | 190.00 | 640.00 |
| Aluminum roofing, natural finish | | | | |
| Corrugated, on steel frame | | | | |
| .0175" thick | SQ. | 69.50 | 85.00 | 154.50 |
| .0215" thick | " | 81.00 | 85.00 | 166.00 |
| .024" thick | " | 110.00 | 85.00 | 195.00 |
| .032" thick | " | 160.00 | 85.00 | 245.00 |
| V-beam, on steel frame | | | | |
| .032" thick | SQ. | 170.00 | 85.00 | 255.00 |
| .040" thick | " | 200.00 | 85.00 | 285.00 |
| .050" thick | " | 230.00 | 85.00 | 315.00 |
| Ridge cap | | | | |
| .019" thick | L.F. | 2.90 | 0.99 | 3.89 |
| Corrugated galvanized steel roofing, on steel frame | | | | |
| 28 ga. | SQ. | 81.00 | 85.00 | 166.00 |
| 26 ga. | " | 92.50 | 85.00 | 177.50 |
| 24 ga. | " | 100.00 | 85.00 | 185.00 |
| 22 ga. | " | 120.00 | 85.00 | 205.00 |
| 26 ga., factory insulated with 1" polystyrene | " | 230.00 | 120.00 | 350.00 |
| Ridge roll | | | | |
| 10" wide | L.F. | 1.25 | 0.99 | 2.24 |
| 20" wide | " | 2.55 | 1.20 | 3.75 |
| **07620.10 FLASHING AND TRIM** | | | | |
| Counter flashing | | | | |
| Aluminum, .032" | S.F. | 0.97 | 3.00 | 3.97 |
| Stainless steel, .015" | " | 2.80 | 3.00 | 5.80 |
| 16 oz. | " | 2.80 | 3.00 | 5.80 |
| 20 oz. | " | 3.70 | 3.00 | 6.70 |
| 24 oz. | " | 4.40 | 3.00 | 7.40 |
| 32 oz. | " | 5.55 | 3.00 | 8.55 |
| Valley flashing | | | | |
| Aluminum, .032" | S.F. | 0.97 | 1.85 | 2.82 |
| Stainless steel, .015 | " | 2.80 | 1.85 | 4.65 |
| Copper | | | | |
| 16 oz. | S.F. | 2.80 | 1.85 | 4.65 |
| 20 oz. | " | 3.70 | 2.50 | 6.20 |
| 24 oz. | " | 4.40 | 1.85 | 6.25 |
| 32 oz. | " | 5.55 | 1.85 | 7.40 |
| Base flashing | | | | |
| Aluminum, .040" | S.F. | 1.50 | 2.50 | 4.00 |
| Stainless steel, .018" | " | 3.35 | 2.50 | 5.85 |
| 16 oz. | " | 2.80 | 2.50 | 5.30 |
| 20 oz. | " | 3.70 | 1.85 | 5.55 |
| 24 oz. | " | 4.40 | 2.50 | 6.90 |
| 32 oz. | " | 5.55 | 2.50 | 8.05 |
| Waterstop, "T" section, 22 ga. | | | | |
| 1-1/2" x 3" | L.F. | 1.65 | 1.50 | 3.15 |
| 2" x 2" | " | 1.80 | 1.50 | 3.30 |
| 4" x 3" | " | 2.05 | 1.50 | 3.55 |

# 07 THERMAL AND MOISTURE

| FLASHING AND SHEET METAL | UNIT | MAT. | INST. | TOTAL |
|---|---|---|---|---|
| **07620.10 FLASHING AND TRIM** | | | | |
| 6" x 4" | L.F. | 2.15 | 1.50 | 3.65 |
| 8" x 4" | " | 2.70 | 1.50 | 4.20 |
| Scupper outlets | | | | |
| 10" x 10" x 4" | EA. | 17.35 | 7.45 | 24.80 |
| 22" x 4" x 4" | " | 21.35 | 7.45 | 28.80 |
| 8" x 8" x 5" | " | 17.35 | 7.45 | 24.80 |
| Flashing and trim, aluminum | | | | |
| .019" thick | S.F. | 0.79 | 2.15 | 2.94 |
| .032" thick | " | 0.97 | 2.15 | 3.12 |
| .040" thick | " | 1.50 | 2.30 | 3.80 |
| Neoprene sheet flashing, .060" thick | " | 1.25 | 1.85 | 3.10 |
| Copper, paper backed | | | | |
| 2 oz. | S.F. | 1.10 | 3.00 | 4.10 |
| Drainage boots, roof, cast iron | | | | |
| 2 x 3 | L.F. | 30.50 | 3.70 | 34.20 |
| 3 x 4 | " | 38.10 | 3.70 | 41.80 |
| 4 x 5 | " | 54.50 | 3.95 | 58.45 |
| 4 x 6 | " | 52.00 | 3.95 | 55.95 |
| 5 x 7 | " | 62.00 | 4.25 | 66.25 |
| Pitch pocket, copper, 16 oz. | | | | |
| 4 x 4 | EA. | 53.00 | 7.45 | 60.45 |
| 6 x 6 | " | 59.00 | 7.45 | 66.45 |
| 8 x 8 | " | 62.50 | 7.45 | 69.95 |
| 8 x 10 | " | 67.00 | 7.45 | 74.45 |
| 8 x 12 | " | 81.00 | 7.45 | 88.45 |
| Reglets, copper 10 oz. | L.F. | 2.10 | 2.00 | 4.10 |
| Stainless steel, .020" | " | 1.50 | 2.00 | 3.50 |
| Gravel stop | | | | |
| Aluminum, .032" | | | | |
| 4" | L.F. | 3.20 | 0.99 | 4.19 |
| 6" | " | 3.75 | 0.99 | 4.74 |
| 8" | " | 5.70 | 1.15 | 6.85 |
| 10" | " | 6.70 | 1.15 | 7.85 |
| Copper, 16 oz. | | | | |
| 4" | L.F. | 4.95 | 0.99 | 5.94 |
| 6" | " | 6.25 | 0.99 | 7.24 |
| 8" | " | 7.85 | 1.15 | 9.00 |
| 10" | " | 8.80 | 1.15 | 9.95 |
| **07620.20 GUTTERS AND DOWNSPOUTS** | | | | |
| Copper gutter and downspout | | | | |
| Downspouts, 16 oz. copper | | | | |
| Round | | | | |
| 3" dia. | L.F. | 4.70 | 2.00 | 6.70 |
| 4" dia. | " | 6.00 | 2.00 | 8.00 |
| Rectangular, corrugated | | | | |
| 2" x 3" | L.F. | 4.50 | 1.85 | 6.35 |
| 3" x 4" | " | 5.90 | 1.85 | 7.75 |
| Rectangular, flat surface | | | | |
| 2" x 3" | L.F. | 4.90 | 2.00 | 6.90 |
| 3" x 4" | " | 7.00 | 2.00 | 9.00 |
| Lead-coated copper downspouts | | | | |

## FLASHING AND SHEET METAL

| 07620.20 GUTTERS AND DOWNSPOUTS | UNIT | MAT. | INST. | TOTAL |
|---|---|---|---|---|
| Round | | | | |
| 3" dia. | L.F. | 5.75 | 1.85 | 7.60 |
| 4" dia. | " | 6.95 | 2.15 | 9.10 |
| Rectangular, corrugated | | | | |
| 2" x 3" | L.F. | 4.25 | 2.00 | 6.25 |
| 3" x 4" | " | 5.10 | 2.00 | 7.10 |
| Rectangular, plain | | | | |
| 2" x 3" | L.F. | 3.70 | 2.00 | 5.70 |
| 3" x 4" | " | 4.40 | 2.00 | 6.40 |
| Gutters, 16 oz. copper | | | | |
| Half round | | | | |
| 4" wide | L.F. | 4.35 | 3.00 | 7.35 |
| 5" wide | " | 5.20 | 3.30 | 8.50 |
| Type K | | | | |
| 4" wide | L.F. | 4.75 | 3.00 | 7.75 |
| 5" wide | " | 5.20 | 3.30 | 8.50 |
| Lead-coated copper gutters | | | | |
| Half round | | | | |
| 4" wide | L.F. | 4.05 | 3.00 | 7.05 |
| 6" wide | " | 6.35 | 3.30 | 9.65 |
| Type K | | | | |
| 4" wide | L.F. | 5.55 | 3.00 | 8.55 |
| 5" wide | " | 6.00 | 3.30 | 9.30 |
| Aluminum gutter and downspout | | | | |
| Downspouts | | | | |
| 2" x 3" | L.F. | 0.65 | 2.00 | 2.65 |
| 3" x 4" | " | 0.89 | 2.15 | 3.04 |
| 4" x 5" | " | 0.99 | 2.30 | 3.29 |
| Round | | | | |
| 3" dia. | L.F. | 1.05 | 2.00 | 3.05 |
| 4" dia. | " | 1.40 | 2.15 | 3.55 |
| Gutters, stock units | | | | |
| 4" wide | L.F. | 1.10 | 3.15 | 4.25 |
| 5" wide | " | 1.20 | 3.30 | 4.50 |
| Galvanized steel gutter and downspout | | | | |
| Downspouts, round corrugated | | | | |
| 3" dia. | L.F. | 0.93 | 2.00 | 2.93 |
| 4" dia. | " | 1.05 | 2.00 | 3.05 |
| 5" dia. | " | 1.25 | 2.15 | 3.40 |
| 6" dia. | " | 1.60 | 2.15 | 3.75 |
| Rectangular | | | | |
| 2" x 3" | L.F. | 0.93 | 2.00 | 2.93 |
| 3" x 4" | " | 1.50 | 1.85 | 3.35 |
| 4" x 4" | " | 1.85 | 1.85 | 3.70 |
| Gutters, stock units | | | | |
| 5" wide | | | | |
| Plain | L.F. | 1.05 | 3.30 | 4.35 |
| Painted | " | 1.15 | 3.30 | 4.45 |
| 6" wide | | | | |
| Plain | L.F. | 1.40 | 3.50 | 4.90 |
| Painted | " | 1.60 | 3.50 | 5.10 |

| ROOFING SPECIALTIES | UNIT | MAT. | INST. | TOTAL |
|---|---|---|---|---|
| **07700.10 MANUFACTURED SPECIALTIES** | | | | |
| Moisture relief vent | | | | |
| Aluminum | EA. | 12.70 | 4.25 | 16.95 |
| Copper | " | 22.50 | 4.25 | 26.75 |
| Expansion joint | | | | |
| Aluminum | | | | |
| Opening to 2.5" | L.F. | 6.65 | 2.15 | 8.80 |
| Opening to 3.5" | " | 7.30 | 2.30 | 9.60 |
| Copper, 16 oz. | | | | |
| Opening to 2.5" | L.F. | 11.45 | 2.15 | 13.60 |
| Opening to 3.5" | " | 14.65 | 2.30 | 16.95 |
| Butyl or neoprene | | | | |
| 4" wide | | | | |
| 16 oz. copper bellows | L.F. | 11.80 | 2.50 | 14.30 |
| 28 ga. stainless steel bellows | " | 10.75 | 2.50 | 13.25 |
| 6" wide | | | | |
| Copper bellows | L.F. | 11.45 | 2.70 | 14.15 |
| Stainless steel | | | | |
| Opening to 2.5" | L.F. | 11.30 | 2.15 | 13.45 |
| Opening to 3.5" | " | 13.85 | 2.30 | 16.15 |
| Smoke vent, 48" x 48" | | | | |
| Aluminum | EA. | 600.00 | 74.50 | 674.50 |
| Galvanized steel | " | 580.00 | 74.50 | 654.50 |
| Heat/smoke vent, 48" x 96" | | | | |
| Aluminum | EA. | 1,850 | 100.00 | 1,950 |
| Galvanized steel | " | 1,730 | 100.00 | 1,830 |
| Ridge vent strips | | | | |
| Mill finish | L.F. | 2.45 | 2.00 | 4.45 |
| Connectors | EA. | 11.50 | 7.45 | 18.95 |
| End cap | " | 23.70 | 8.50 | 32.20 |
| Soffit vents | | | | |
| Mill finish | | | | |
| 2-1/2" wide | L.F. | 0.30 | 1.20 | 1.50 |
| 3" wide | " | 0.36 | 1.20 | 1.56 |
| 6" wide | " | 0.63 | 1.20 | 1.83 |
| Roof hatches | | | | |
| Steel, plain, primed | | | | |
| 2'6" x 3'0" | EA. | 390.00 | 74.50 | 464.50 |
| 2'6" x 4'6" | " | 580.00 | 100.00 | 680.00 |
| 2'6" x 8'0" | " | 900.00 | 150.00 | 1,050 |
| Galvanized steel | | | | |
| 2'6" x 3'0" | EA. | 400.00 | 74.50 | 474.50 |
| 2'6" x 4'6" | " | 610.00 | 100.00 | 710.00 |
| 2'6" x 8'0" | " | 960.00 | 150.00 | 1,110 |
| Aluminum | | | | |
| 2'6" x 3'0" | EA. | 430.00 | 74.50 | 504.50 |
| 2'6" x 4'6" | " | 650.00 | 100.00 | 750.00 |
| 2'6" x 8'0" | " | 1,150 | 150.00 | 1,300 |
| Ceiling access doors | | | | |
| Swing up model, metal frame | | | | |
| Steel door | | | | |
| 2'6" x 2'6" | EA. | 290.00 | 29.80 | 319.80 |
| 2'6" x 3'0" | " | 320.00 | 29.80 | 349.80 |

## ROOFING SPECIALTIES

| 07700.10 MANUFACTURED SPECIALTIES | UNIT | MAT. | INST. | TOTAL |
|---|---|---|---|---|
| Aluminum door | | | | |
| 2'6" x 2'6" | EA. | 320.00 | 29.80 | 349.80 |
| 2'6" x 3'0" | " | 340.00 | 29.80 | 369.80 |
| Swing down model, metal frame | | | | |
| Steel door | | | | |
| 2'6" x 2'6" | EA. | 290.00 | 29.80 | 319.80 |
| 2'6" x 3'0" | " | 320.00 | 29.80 | 349.80 |
| Aluminum door | | | | |
| 2'6" x 2'6" | EA. | 310.00 | 29.80 | 339.80 |
| 2'6" x 3'0" | " | 330.00 | 29.80 | 359.80 |
| Gravity ventilators, with curb, base, damper and screen | | | | |
| Stationary siphon | | | | |
| 6" dia. | EA. | 21.95 | 19.85 | 41.80 |
| 12" dia. | " | 33.50 | 19.85 | 53.35 |
| 24" dia. | " | 150.00 | 29.80 | 179.80 |
| 36" dia. | " | 270.00 | 29.80 | 299.80 |
| Wind driven spinner | | | | |
| 6" dia. | EA. | 34.60 | 19.85 | 54.45 |
| 12" dia. | " | 46.20 | 19.85 | 66.05 |
| 24" dia. | " | 170.00 | 29.80 | 199.80 |
| 36" dia. | " | 360.00 | 29.80 | 389.80 |
| Stationary mushroom | | | | |
| 16" dia. | EA. | 220.00 | 29.80 | 249.80 |
| 30" dia. | " | 500.00 | 37.20 | 537.20 |
| 36" dia. | " | 640.00 | 49.60 | 689.60 |
| 42" dia. | " | 940.00 | 59.50 | 999.50 |

## SKYLIGHTS

| 07810.10 PLASTIC SKYLIGHTS | UNIT | MAT. | INST. | TOTAL |
|---|---|---|---|---|
| Single thickness, not including mounting curb | | | | |
| 2' x 4' | EA. | 210.00 | 37.20 | 247.20 |
| 4' x 4' | " | 280.00 | 49.60 | 329.60 |
| 5' x 5' | " | 380.00 | 74.50 | 454.50 |
| 6' x 8' | " | 810.00 | 100.00 | 910.00 |
| Double thickness, not including mounting curb | | | | |
| 2' x 4' | EA. | 280.00 | 37.20 | 317.20 |
| 4' x 4' | " | 350.00 | 49.60 | 399.60 |
| 5' x 5' | " | 510.00 | 74.50 | 584.50 |
| 6' x 8' | " | 910.00 | 100.00 | 1,010 |
| Metal framed skylights | | | | |
| Translucent panels, 2-1/2" thick | S.F. | 24.50 | 3.00 | 27.50 |
| Continuous vaults, 8' wide | | | | |
| Single glazed | S.F. | 33.80 | 3.70 | 37.50 |

## SKYLIGHTS

| 07810.10 PLASTIC SKYLIGHTS | UNIT | MAT. | INST. | TOTAL |
|---|---|---|---|---|
| Double glazed | S.F. | 53.50 | 4.25 | 57.75 |

## JOINT SEALERS

| 07920.10 CAULKING | UNIT | MAT. | INST. | TOTAL |
|---|---|---|---|---|
| Caulk exterior, two component | | | | |
| 1/4 x 1/2 | L.F. | 0.26 | 1.55 | 1.81 |
| 3/8 x 1/2 | " | 0.39 | 1.70 | 2.09 |
| 1/2 x 1/2 | " | 0.51 | 1.95 | 2.46 |
| Caulk interior, single component | | | | |
| 1/4 x 1/2 | L.F. | 0.17 | 1.45 | 1.62 |
| 3/8 x 1/2 | " | 0.24 | 1.65 | 1.89 |
| 1/2 x 1/2 | " | 0.32 | 1.80 | 2.12 |
| Butyl rubber fillers | | | | |
| 1/4" x 1/4" | L.F. | 0.56 | 0.62 | 1.18 |
| 1/2" x 1/2" | " | 0.74 | 1.05 | 1.79 |
| 1/2" x 3/4" | " | 1.20 | 1.25 | 2.45 |
| 3/4" x 3/4" | " | 1.50 | 1.25 | 2.75 |
| 1" x 1" | " | 1.80 | 1.35 | 3.15 |
| Seals, "O" ring type cord | | | | |
| 1/4" dia. | L.F. | 0.50 | 0.77 | 1.27 |
| 1/2" dia. | " | 1.50 | 0.81 | 2.31 |
| 1" dia. | " | 5.25 | 0.86 | 6.11 |
| 1-1/4" dia. | " | 6.65 | 0.91 | 7.56 |
| 1-1/2" dia. | " | 8.90 | 0.96 | 9.87 |
| 1-3/4" dia. | " | 13.10 | 1.00 | 14.10 |
| 2" dia. | " | 16.95 | 1.05 | 18.00 |
| Polyvinyl chloride, closed cell | | | | |
| 1/4" x 2" | L.F. | 0.34 | 1.10 | 1.44 |
| 1/4" x 6" | " | 1.05 | 1.40 | 2.45 |
| Silicon foam penetration seal | | | | |
| 1/4" x 1/2" | L.F. | 0.12 | 0.39 | 0.51 |
| 1/2" x 1/2" | " | 0.21 | 0.52 | 0.72 |
| 1/2" x 3/4" | " | 0.33 | 0.62 | 0.95 |
| 3/4" x 3/4" | " | 0.50 | 0.77 | 1.27 |
| 1/8" x 1" | " | 0.12 | 0.39 | 0.51 |
| 1/8" x 3" | " | 0.33 | 0.62 | 0.95 |
| 1/4" x 3" | " | 0.67 | 0.77 | 1.44 |
| 1/4" x 6" | " | 1.35 | 1.05 | 2.40 |
| 1/2" x 6" | " | 2.65 | 2.40 | 5.05 |
| 1/2" x 9" | " | 4.00 | 3.85 | 7.85 |
| 1/2" x 12" | " | 5.30 | 5.60 | 10.90 |

## JOINT SEALERS

### 07920.10 CAULKING

| | UNIT | MAT. | INST. | TOTAL |
|---|---|---|---|---|
| Oil base sealants and caulking | | | | |
| 1/4" x 1/4" | L.F. | 0.03 | 0.77 | **0.80** |
| 1/4" x 3/8" | " | 0.06 | 0.80 | **0.86** |
| 1/4" x 1/2" | " | 0.07 | 0.83 | **0.91** |
| 3/8" x 3/8" | " | 0.08 | 0.88 | **0.96** |
| 3/8" x 1/2" | " | 0.10 | 0.94 | **1.04** |
| 3/8" x 5/8" | " | 0.16 | 1.00 | **1.16** |
| 3/8" x 3/4" | " | 0.18 | 1.05 | **1.23** |
| 1/2" x 1/2" | " | 0.16 | 1.20 | **1.36** |
| 1/2" x 5/8" | " | 0.20 | 1.35 | **1.55** |
| 1/2" x 3/4" | " | 0.23 | 1.55 | **1.78** |
| 1/2" x 7/8" | " | 0.28 | 1.60 | **1.88** |
| 1/2" x 1" | " | 0.31 | 1.65 | **1.96** |
| 3/4" x 3/4" | " | 0.36 | 1.65 | **2.01** |
| 1" x 1" | " | 0.62 | 1.70 | **2.32** |
| Polyurethane compounds | | | | |
| 1/4" x 1/4" | L.F. | 0.11 | 0.77 | **0.88** |
| 1/4" x 3/8" | " | 0.21 | 0.80 | **1.01** |
| 1/4" x 1/2" | " | 0.28 | 0.83 | **1.12** |
| 3/8" x 3/8" | " | 0.31 | 0.88 | **1.19** |
| 3/8" x 1/2" | " | 0.38 | 0.94 | **1.32** |
| 3/8" x 5/8" | " | 0.48 | 1.00 | **1.48** |
| 3/8" x 3/4" | " | 0.59 | 1.05 | **1.64** |
| 1/2" x 1/2" | " | 0.57 | 1.20 | **1.77** |
| 1/2" x 5/8" | " | 0.61 | 1.35 | **1.96** |
| 1/2" x 3/4" | " | 0.70 | 1.55 | **2.25** |
| 1/2" x 7/8" | " | 0.89 | 1.60 | **2.49** |
| 1/2" x 1" | " | 1.10 | 1.70 | **2.80** |
| 3/4" x 3/4" | " | 1.20 | 1.65 | **2.85** |
| 3/4" x 1" | " | 1.25 | 1.70 | **2.95** |
| Backer rod, polyethylene | | | | |
| 1/4" | L.F. | 0.06 | 0.77 | **0.83** |
| 1/2" | " | 0.08 | 0.81 | **0.89** |
| 3/4" | " | 0.09 | 0.86 | **0.95** |
| 1" | " | 0.11 | 0.91 | **1.02** |

| METAL | UNIT | MAT. | INST. | TOTAL |
|---|---|---|---|---|
| **08110.10 METAL DOORS** | | | | |
| Flush hollow metal, standard duty, 20 ga., 1-3/8" | | | | |
| 2-6 x 6-8 | EA. | 150.00 | 34.30 | 184.30 |
| 2-8 x 6-8 | " | 170.00 | 34.30 | 204.30 |
| 3-0 x 6-8 | " | 190.00 | 34.30 | 224.30 |
| 1-3/4" | | | | |
| 2-6 x 6-8 | EA. | 170.00 | 34.30 | 204.30 |
| 2-8 x 6-8 | " | 180.00 | 34.30 | 214.30 |
| 3-0 x 6-8 | " | 200.00 | 34.30 | 234.30 |
| 2-6 x 7-0 | " | 190.00 | 34.30 | 224.30 |
| 2-8 x 7-0 | " | 190.00 | 34.30 | 224.30 |
| 3-0 x 7-0 | " | 210.00 | 34.30 | 244.30 |
| Heavy duty, 20 ga., unrated, 1-3/4" | | | | |
| 2-8 x 6-8 | EA. | 190.00 | 34.30 | 224.30 |
| 3-0 x 6-8 | " | 200.00 | 34.30 | 234.30 |
| 2-8 x 7-0 | " | 210.00 | 34.30 | 244.30 |
| 3-0 x 7-0 | " | 220.00 | 34.30 | 254.30 |
| 3-4 x 7-0 | " | 270.00 | 34.30 | 304.30 |
| 18 ga., 1-3/4", unrated door | | | | |
| 2-0 x 7-0 | EA. | 210.00 | 34.30 | 244.30 |
| 2-4 x 7-0 | " | 220.00 | 34.30 | 254.30 |
| 2-6 x 7-0 | " | 230.00 | 34.30 | 264.30 |
| 2-8 x 7-0 | " | 240.00 | 34.30 | 274.30 |
| 3-0 x 7-0 | " | 240.00 | 34.30 | 274.30 |
| 3-4 x 7-0 | " | 280.00 | 34.30 | 314.30 |
| 2", unrated door | | | | |
| 2-0 x 7-0 | EA. | 220.00 | 38.60 | 258.60 |
| 2-4 x 7-0 | " | 230.00 | 38.60 | 268.60 |
| 2-6 x 7-0 | " | 240.00 | 38.60 | 278.60 |
| 2-8 x 7-0 | " | 250.00 | 38.60 | 288.60 |
| 3-0 x 7-0 | " | 260.00 | 38.60 | 298.60 |
| 3-4 x 7-0 | " | 310.00 | 38.60 | 348.60 |
| Galvanized metal door | | | | |
| 3-0 x 7-0 | EA. | 310.00 | 38.60 | 348.60 |
| For lead lining in doors | " | | | 629.70 |
| For sound attenuation | " | | | 29.19 |
| Vision glass | | | | |
| 8" x 8" | EA. | 73.50 | 38.60 | 112.10 |
| 8" x 48" | " | 110.00 | 38.60 | 148.60 |
| Fixed metal louver | " | 100.00 | 30.90 | 130.90 |
| For fire rating, add | | | | |
| 3 hr door | EA. | | | 116.55 |
| 1-1/2 hr door | " | | | 48.98 |
| 3/4 hr door | " | | | 24.49 |
| 1' extra height, add to material, 20% | | | | |
| 1'6" extra height, add to material, 60% | | | | |
| For dutch doors with shelf, add to material, 100% | | | | |
| **08110.40 METAL DOOR FRAMES** | | | | |
| Hollow metal, stock, 18 ga., 4-3/4" x 1-3/4" | | | | |
| 2-0 x 7-0 | EA. | 73.50 | 38.60 | 112.10 |
| 2-4 x 7-0 | " | 74.50 | 38.60 | 113.10 |
| 2-6 x 7-0 | " | 76.00 | 38.60 | 114.60 |

| METAL | UNIT | MAT. | INST. | TOTAL |
|---|---|---|---|---|
| **08110.40 METAL DOOR FRAMES** | | | | |
| 2-8 x 7-0 | EA. | 77.00 | 38.60 | 115.60 |
| 3-0 x 7-0 | " | 78.00 | 38.60 | 116.60 |
| 4-0 x 7-0 | " | 83.00 | 51.50 | 134.50 |
| 5-0 x 7-0 | " | 84.00 | 51.50 | 135.50 |
| 6-0 x 7-0 | " | 90.00 | 51.50 | 141.50 |
| 16 ga., 6-3/4" x 1-3/4" | | | | |
| 2-0 x 7-0 | EA. | 78.00 | 42.30 | 120.30 |
| 2-4 x 7-0 | " | 80.50 | 42.30 | 122.80 |
| 2-6 x 7-0 | " | 81.50 | 42.30 | 123.80 |
| 2-8 x 7-0 | " | 84.00 | 42.30 | 126.30 |
| 3-0 x 7-0 | " | 90.00 | 42.30 | 132.30 |
| 4-0 x 7-0 | " | 93.50 | 56.50 | 150.00 |
| 6-0 x 7-0 | " | 95.50 | 56.50 | 152.00 |
| Transom frame | | | | |
| 3-8 x 1-6 | EA. | 60.50 | 42.30 | 102.80 |
| 6-4 x 1-6 | " | 84.00 | 42.30 | 126.30 |
| Transom sash | | | | |
| 3-0 x 1-4 | EA. | 410.00 | 42.30 | 452.30 |
| 3-4 x 1-4 | " | 410.00 | 42.30 | 452.30 |
| 6-0 x 1-4 | " | 460.00 | 42.30 | 502.30 |
| 1' extension of frame, add | " | | | 12.83 |
| 14 ga. frame, add | " | | | 12.83 |
| For fire rating, add | | | | |
| 3 hour | EA. | | | 34.99 |
| 1-1/2 hour | " | | | 27.98 |
| 3/4 hour | " | | | 24.49 |
| Lead lining in frame, add | " | | | 75.80 |
| Sidelights, complete | | | | |
| 1-0 x 7-2 | EA. | 240.00 | 42.30 | 282.30 |
| 1-4 x 7-2 | " | 270.00 | 42.30 | 312.30 |
| 1-0 x 8-8 | " | 280.00 | 42.30 | 322.30 |
| 1-6 x 8-8 | " | 290.00 | 42.30 | 332.30 |
| 16 ga., 4-3/4" x 1-3/4" | | | | |
| 2-0 x 7-0 | EA. | 74.50 | 42.30 | 116.80 |
| 2-4 x 7-0 | " | 76.00 | 42.30 | 118.30 |
| 2-6 x 7-0 | " | 77.00 | 42.30 | 119.30 |
| 2-8 x 7-0 | " | 78.00 | 42.30 | 120.30 |
| 3-0 x 7-0 | " | 79.50 | 42.30 | 121.80 |
| 4-0 x 7-0 | " | 81.50 | 56.50 | 138.00 |
| 6-0 x 7-0 | " | 87.50 | 56.50 | 144.00 |
| Transom frame | | | | |
| 3-4 x 1-6 | EA. | 60.50 | 42.30 | 102.80 |
| 3-8 x 1-6 | " | 60.50 | 42.30 | 102.80 |
| 6-4 x 1-6 | " | 84.00 | 42.30 | 126.30 |
| Transom sash | | | | |
| 3-0 x 1-4 | EA. | 400.00 | 42.30 | 442.30 |
| 3-4 x 1-4 | " | 400.00 | 42.30 | 442.30 |
| 6-0 x 1-4 | " | 450.00 | 42.30 | 492.30 |
| 1' extension of door frame, add | " | | | 12.83 |
| 14 ga., metal frame, add | " | | | 12.83 |
| For fire rating, add | | | | |
| 3 hour | EA. | | | 34.97 |

| METAL | UNIT | MAT. | INST. | TOTAL |
|---|---|---|---|---|
| **08110.40 METAL DOOR FRAMES** | | | | |
| 1-1/2 hour | EA. | | | 27.98 |
| 3/4 hour | " | | | 24.49 |
| Lead lining in frame, add | " | | | 73.47 |
| Sidelights, complete | | | | |
| 1-0 x 7-2 | EA. | 230.00 | 42.30 | 272.30 |
| 1-4 x 7-2 | " | 240.00 | 42.30 | 282.30 |
| 1-0 x 8-8 | " | 270.00 | 42.30 | 312.30 |
| 1-4 x 8-8 | " | 270.00 | 42.30 | 312.30 |
| 16 ga., 5-3/4" x 1-3/4" | | | | |
| 2-0 x 7-0 | EA. | 76.00 | 38.60 | 114.60 |
| 2-4 x 7-0 | " | 77.00 | 38.60 | 115.60 |
| 2-6 x 7-0 | " | 79.50 | 38.60 | 118.10 |
| 2-8 x 7-0 | " | 81.50 | 38.60 | 120.10 |
| 3-0 x 7-0 | " | 84.00 | 38.60 | 122.60 |
| 4-0 x 7-0 | " | 90.00 | 51.50 | 141.50 |
| 5-0 x 7-0 | " | 93.50 | 51.50 | 145.00 |
| 6-0 x 7-0 | " | 97.00 | 51.50 | 148.50 |
| Mullions, vertical | | | | |
| 5-1/4" x 1-3/4" | L.F. | 8.75 | 3.85 | 12.60 |
| 5-1/4" x 2" | " | 10.85 | 3.85 | 14.70 |
| Horizontal | | | | |
| 5-1/4" x 1-3/4" | L.F. | 8.75 | 3.85 | 12.60 |
| 5-1/4" x 2" | " | 10.85 | 3.85 | 14.70 |
| **08120.10 ALUMINUM DOORS** | | | | |
| Aluminum doors, commercial | | | | |
| Narrow stile | | | | |
| 2-6 x 7-0 | EA. | 480.00 | 170.00 | 650.00 |
| 3-0 x 7-0 | " | 510.00 | 170.00 | 680.00 |
| 3-6 x 7-0 | " | 540.00 | 170.00 | 710.00 |
| Pair | | | | |
| 5-0 x 7-0 | EA. | 820.00 | 340.00 | 1,160 |
| 6-0 x 7-0 | " | 830.00 | 340.00 | 1,170 |
| 7-0 x 7-0 | " | 870.00 | 340.00 | 1,210 |
| Wide stile | | | | |
| 2-6 x 7-0 | EA. | 750.00 | 170.00 | 920.00 |
| 3-0 x 7-0 | " | 770.00 | 170.00 | 940.00 |
| 3-6 x 7-0 | " | 790.00 | 170.00 | 960.00 |
| Pair | | | | |
| 5-0 x 7-0 | EA. | 1,400 | 340.00 | 1,740 |
| 6-0 x 7-0 | " | 1,460 | 340.00 | 1,800 |
| 7-0 x 7-0 | " | 1,520 | 340.00 | 1,860 |

| WOOD AND PLASTIC | UNIT | MAT. | INST. | TOTAL |
|---|---|---|---|---|
| **08210.10** WOOD DOORS | | | | |
| Solid core, 1-3/8" thick | | | | |
| Birch faced | | | | |
| 2-4 x 7-0 | EA. | 110.00 | 38.60 | 148.60 |
| 2-8 x 7-0 | " | 120.00 | 38.60 | 158.60 |
| 3-0 x 7-0 | " | 120.00 | 38.60 | 158.60 |
| 3-4 x 7-0 | " | 130.00 | 38.60 | 168.60 |
| 2-4 x 6-8 | " | 110.00 | 38.60 | 148.60 |
| 2-6 x 6-8 | " | 110.00 | 38.60 | 148.60 |
| 2-8 x 6-8 | " | 120.00 | 38.60 | 158.60 |
| 3-0 x 6-8 | " | 120.00 | 38.60 | 158.60 |
| Lauan faced | | | | |
| 2-4 x 6-8 | EA. | 94.50 | 38.60 | 133.10 |
| 2-8 x 6-8 | " | 100.00 | 38.60 | 138.60 |
| 3-0 x 6-8 | " | 110.00 | 38.60 | 148.60 |
| 3-4 x 6-8 | " | 100.00 | 38.60 | 138.60 |
| Tempered hardboard faced | | | | |
| 2-4 x 7-0 | EA. | 100.00 | 38.60 | 138.60 |
| 2-8 x 7-0 | " | 110.00 | 38.60 | 148.60 |
| 3-0 x 7-0 | " | 110.00 | 38.60 | 148.60 |
| 3-4 x 7-0 | " | 120.00 | 38.60 | 158.60 |
| Hollow core, 1-3/8" thick | | | | |
| Birch faced | | | | |
| 2-4 x 7-0 | EA. | 57.00 | 38.60 | 95.60 |
| 2-8 x 7-0 | " | 60.50 | 38.60 | 99.10 |
| 3-0 x 7-0 | " | 63.00 | 38.60 | 101.60 |
| 3-4 x 7-0 | " | 69.00 | 38.60 | 107.60 |
| Lauan faced | | | | |
| 2-4 x 6-8 | EA. | 36.20 | 38.60 | 74.80 |
| 2-6 x 6-8 | " | 38.50 | 38.60 | 77.10 |
| 2-8 x 6-8 | " | 40.80 | 38.60 | 79.40 |
| 3-0 x 6-8 | " | 44.30 | 38.60 | 82.90 |
| 3-4 x 6-8 | " | 46.60 | 38.60 | 85.20 |
| Tempered hardboard faced | | | | |
| 2-4 x 7-0 | EA. | 38.50 | 38.60 | 77.10 |
| 2-6 x 7-0 | " | 40.80 | 38.60 | 79.40 |
| 2-8 x 7-0 | " | 44.30 | 38.60 | 82.90 |
| 3-0 x 7-0 | " | 46.60 | 38.60 | 85.20 |
| 3-4 x 7-0 | " | 49.00 | 38.60 | 87.60 |
| Solid core, 1-3/4" thick | | | | |
| Birch faced | | | | |
| 2-4 x 7-0 | EA. | 130.00 | 38.60 | 168.60 |
| 2-6 x 7-0 | " | 140.00 | 38.60 | 178.60 |
| 2-8 x 7-0 | " | 140.00 | 38.60 | 178.60 |
| 3-0 x 7-0 | " | 150.00 | 38.60 | 188.60 |
| 3-4 x 7-0 | " | 150.00 | 38.60 | 188.60 |
| Lauan faced | | | | |
| 2-4 x 7-0 | EA. | 110.00 | 38.60 | 148.60 |
| 2-6 x 7-0 | " | 130.00 | 38.60 | 168.60 |
| 2-8 x 7-0 | " | 140.00 | 38.60 | 178.60 |
| 3-0 x 7-0 | " | 140.00 | 38.60 | 178.60 |
| 3-0 x 7-0 | " | 150.00 | 38.60 | 188.60 |
| Tempered hardboard faced | | | | |

# 08 DOORS AND WINDOWS

| WOOD AND PLASTIC | UNIT | MAT. | INST. | TOTAL |
|---|---|---|---|---|
| **08210.10 WOOD DOORS** | | | | |
| 2-4 x 7-0 | EA. | 110.00 | 38.60 | 148.60 |
| 2-6 x 7-0 | " | 130.00 | 38.60 | 168.60 |
| 2-8 x 7-0 | " | 140.00 | 38.60 | 178.60 |
| 3-0 x 7-0 | " | 140.00 | 38.60 | 178.60 |
| 3-4 x 7-0 | " | 150.00 | 38.60 | 188.60 |
| Hollow core, 1-3/4" thick | | | | |
| Birch faced | | | | |
| 2-4 x 7-0 | EA. | 58.50 | 38.60 | 97.10 |
| 2-6 x 7-0 | " | 60.50 | 38.60 | 99.10 |
| 2-8 x 7-0 | " | 64.00 | 38.60 | 102.60 |
| 3-0 x 7-0 | " | 66.50 | 38.60 | 105.10 |
| 3-4 x 7-0 | " | 70.00 | 38.60 | 108.60 |
| Lauan faced | | | | |
| 2-4 x 6-8 | EA. | 44.30 | 38.60 | 82.90 |
| 2-6 x 6-8 | " | 46.60 | 38.60 | 85.20 |
| 2-8 x 6-8 | " | 49.00 | 38.60 | 87.60 |
| 3-0 x 6-8 | " | 51.50 | 38.60 | 90.10 |
| 3-4 x 6-8 | " | 55.00 | 38.60 | 93.60 |
| Tempered hardboard | | | | |
| 2-4 x 7-0 | EA. | 44.30 | 38.60 | 82.90 |
| 2-6 x 7-0 | " | 46.60 | 38.60 | 85.20 |
| 2-8 x 7-0 | " | 49.00 | 38.60 | 87.60 |
| 3-0 x 7-0 | " | 51.50 | 38.60 | 90.10 |
| 3-4 x 7-0 | " | 55.00 | 38.60 | 93.60 |
| Add-on, louver | " | 14.00 | 30.90 | 44.90 |
| Glass | " | 51.50 | 30.90 | 82.40 |
| Exterior doors, 3-0 x 7-0 x 2-1/2", solid core | | | | |
| Carved | | | | |
| One face | EA. | 660.00 | 77.00 | 737.00 |
| Two faces | " | 990.00 | 77.00 | 1,067 |
| Closet doors, 1-3/4" thick | | | | |
| Bi-fold or bi-passing, includes frame and trim | | | | |
| Paneled | | | | |
| 4-0 x 6-8 | EA. | 240.00 | 51.50 | 291.50 |
| 6-0 x 6-8 | " | 290.00 | 51.50 | 341.50 |
| Louvered | | | | |
| 4-0 x 6-8 | EA. | 160.00 | 51.50 | 211.50 |
| 6-0 x 6-8 | " | 210.00 | 51.50 | 261.50 |
| Flush | | | | |
| 4-0 x 6-8 | EA. | 140.00 | 51.50 | 191.50 |
| 6-0 x 6-8 | " | 180.00 | 51.50 | 231.50 |
| Primed | | | | |
| 4-0 x 6-8 | EA. | 150.00 | 51.50 | 201.50 |
| 6-0 x 6-8 | " | 180.00 | 51.50 | 231.50 |
| **08210.90 WOOD FRAMES** | | | | |
| Frame, interior, pine | | | | |
| 2-6 x 6-8 | EA. | 26.00 | 44.10 | 70.10 |
| 2-8 x 6-8 | " | 27.20 | 44.10 | 71.30 |
| 3-0 x 6-8 | " | 28.30 | 44.10 | 72.40 |
| 5-0 x 6-8 | " | 29.40 | 44.10 | 73.50 |
| 6-0 x 6-8 | " | 30.60 | 44.10 | 74.70 |

# WOOD AND PLASTIC

| WOOD AND PLASTIC | UNIT | MAT. | INST. | TOTAL |
|---|---|---|---|---|
| **08210.90 WOOD FRAMES** | | | | |
| 2-6 x 7-0 | EA. | 28.30 | 44.10 | 72.40 |
| 2-8 x 7-0 | " | 30.60 | 44.10 | 74.70 |
| 3-0 x 7-0 | " | 31.70 | 44.10 | 75.80 |
| 5-0 x 7-0 | " | 34.40 | 62.00 | 96.40 |
| 6-0 x 7-0 | " | 35.70 | 62.00 | 97.70 |
| Exterior, custom, with threshold, including trim | | | | |
| Walnut | | | | |
| 3-0 x 7-0 | EA. | 180.00 | 77.00 | 257.00 |
| 6-0 x 7-0 | " | 220.00 | 77.00 | 297.00 |
| Oak | | | | |
| 3-0 x 7-0 | EA. | 160.00 | 77.00 | 237.00 |
| 6-0 x 7-0 | " | 200.00 | 77.00 | 277.00 |
| Pine | | | | |
| 2-4 x 7-0 | EA. | 57.50 | 62.00 | 119.50 |
| 2-6 x 7-0 | " | 60.00 | 62.00 | 122.00 |
| 2-8 x 7-0 | " | 61.00 | 62.00 | 123.00 |
| 3-0 x 7-0 | " | 68.00 | 62.00 | 130.00 |
| 3-4 x 7-0 | " | 74.50 | 62.00 | 136.50 |
| 6-0 x 7-0 | " | 140.00 | 100.00 | 240.00 |
| **08300.10 SPECIAL DOORS** | | | | |
| Vault door and frame, class 5, steel | EA. | 3,270 | 310.00 | 3,580 |
| Overhead door, coiling insulated | | | | |
| Chain gear, no frame, 12' x 12' | EA. | 1,630 | 390.00 | 2,020 |
| Aluminum, bronze glass panels, 12-9 x 13-0 | " | 1,870 | 310.00 | 2,180 |
| Garage door, flush insulated metal, primed, 9-0 x 7-0 | " | 630.00 | 100.00 | 730.00 |
| Sliding metal fire doors, motorized, fusible link, 3 hr. | | | | |
| 3-0 x 6-8 | EA. | 2,270 | 620.00 | 2,890 |
| 3-8 x 6-8 | " | 2,270 | 620.00 | 2,890 |
| 4-0 x 8-0 | " | 2,450 | 620.00 | 3,070 |
| 5-0 x 8-0 | " | 2,510 | 620.00 | 3,130 |
| Metal clad doors, including electric motor | | | | |
| Light duty | | | | |
| Minimum | S.F. | 24.50 | 5.15 | 29.65 |
| Maximum | " | 43.10 | 12.35 | 55.45 |
| Heavy duty | | | | |
| Minimum | S.F. | 49.00 | 15.45 | 64.45 |
| Maximum | " | 83.00 | 19.30 | 102.30 |
| Hangar doors, based on 150' openings | | | | |
| To 20' high | S.F. | 26.80 | 4.60 | 31.40 |
| 20' to 40' high | " | 24.50 | 2.90 | 27.40 |
| 40' to 60' high | " | 26.80 | 1.95 | 28.75 |
| 60' to 80' high | " | 31.50 | 1.15 | 32.65 |
| Over 80' high | " | 43.10 | 0.93 | 44.02 |
| Counter doors, (roll-up shutters), standard, manual | | | | |
| Opening, 4' high | | | | |
| 4' wide | EA. | 810.00 | 260.00 | 1,070 |
| 6' wide | " | 1,110 | 260.00 | 1,370 |
| 8' wide | " | 1,280 | 280.00 | 1,560 |
| 10' wide | " | 1,400 | 390.00 | 1,790 |
| 14' wide | " | 1,750 | 390.00 | 2,140 |

| WOOD AND PLASTIC | UNIT | MAT. | INST. | TOTAL |
|---|---|---|---|---|
| **08300.10 SPECIAL DOORS** | | | | |
| 6' high | | | | |
| 4' wide | EA. | 860.00 | 260.00 | 1,120 |
| 6' wide | " | 1,280 | 280.00 | 1,560 |
| 8' wide | " | 1,520 | 310.00 | 1,830 |
| 10' wide | " | 1,810 | 390.00 | 2,200 |
| 14' wide | " | 1,920 | 440.00 | 2,360 |
| For stainless steel, add to material, 80% | | | | |
| For motor operator, add | EA. | | | 886.34 |
| Service doors, (roll up shutters), standard, manual | | | | |
| Opening | | | | |
| 8' high x 8' wide | EA. | 1,040 | 170.00 | 1,210 |
| 10' high x 10' wide | " | 1,460 | 260.00 | 1,720 |
| 12' high x 12' wide | " | 1,870 | 390.00 | 2,260 |
| 14' high x 14' wide | " | 2,450 | 510.00 | 2,960 |
| 16' high x 14' wide | " | 2,800 | 510.00 | 3,310 |
| 20' high x 14' wide | " | 3,620 | 770.00 | 4,390 |
| 24' high x 16' wide | " | 4,900 | 690.00 | 5,590 |
| For motor operator | | | | |
| Up to 12-0 x 12-0, add | EA. | | | 1,003 |
| Over 12-0 x 12-0, add | " | | | 1,283 |
| Roll-up doors | | | | |
| 13-0 high x 14-0 wide | EA. | 760.00 | 440.00 | 1,200 |
| 12-0 high x 14-0 wide | " | 970.00 | 440.00 | 1,410 |
| Top coiling grilles, manually operated, steel or aluminum | | | | |
| Opening, 4' high x | | | | |
| 4' wide | EA. | 1,050 | 120.00 | 1,170 |
| 6' wide | " | 1,070 | 120.00 | 1,190 |
| 8' wide | " | 1,120 | 170.00 | 1,290 |
| 12' wide | " | 1,330 | 170.00 | 1,500 |
| 16' wide | " | 1,340 | 260.00 | 1,600 |
| 6' high x | | | | |
| 4' wide | EA. | 1,070 | 260.00 | 1,330 |
| 6' wide | " | 1,100 | 280.00 | 1,380 |
| 8' wide | " | 1,120 | 310.00 | 1,430 |
| 12' wide | " | 1,460 | 340.00 | 1,800 |
| 16' wide | " | 1,690 | 440.00 | 2,130 |
| Side coiling grilles, manually operated, aluminum | | | | |
| Opening, 8' high x | | | | |
| 18' wide | EA. | 7,000 | 2,890 | 9,890 |
| 24' wide | " | 8,160 | 3,300 | 11,460 |
| 12' high x | | | | |
| 12' wide | EA. | 5,830 | 2,890 | 8,720 |
| 18' wide | " | 8,860 | 3,300 | 12,160 |
| 24' wide | " | 19,130 | 3,850 | 22,980 |
| Accordion folding doors, tracks and fittings included | | | | |
| Vinyl covered, 2 layers | S.F. | 8.15 | 12.35 | 20.50 |
| Woven mahogany and vinyl | " | 10.50 | 12.35 | 22.85 |
| Economy vinyl | " | 7.00 | 12.35 | 19.35 |
| Rigid polyvinyl chloride | " | 11.65 | 12.35 | 24.00 |
| Sectional wood overhead doors, frames not included | | | | |
| Commercial grade, heavy duty, 1-3/4" thick, manual | | | | |
| 8' x 8' | EA. | 540.00 | 260.00 | 800.00 |

| WOOD AND PLASTIC | UNIT | MAT. | INST. | TOTAL |
|---|---|---|---|---|
| **08300.10 SPECIAL DOORS** | | | | |
| 10' x 10' | EA. | 830.00 | 280.00 | 1,110 |
| 12' x 12' | " | 1,120 | 310.00 | 1,430 |
| Chain hoist | | | | |
| 12' x 16' high | EA. | 1,630 | 510.00 | 2,140 |
| 14' x 14' high | " | 1,750 | 390.00 | 2,140 |
| 20' x 8' high | " | 1,520 | 620.00 | 2,140 |
| 16' high | " | 3,500 | 770.00 | 4,270 |
| Sectional metal overhead doors, complete | | | | |
| Residential grade, manual | | | | |
| 9' x 7' | EA. | 350.00 | 120.00 | 470.00 |
| 16' x 7' | " | 680.00 | 150.00 | 830.00 |
| Commercial grade | | | | |
| 8' x 8' | EA. | 500.00 | 260.00 | 760.00 |
| 10' x 10' | " | 760.00 | 280.00 | 1,040 |
| 12' x 12' | " | 1,000 | 310.00 | 1,310 |
| 20' x 14', with chain hoist | " | 2,570 | 620.00 | 3,190 |
| Sliding glass doors | | | | |
| Tempered plate glass, 1/4" thick | | | | |
| 6' wide | | | | |
| Economy grade | EA. | 630.00 | 100.00 | 730.00 |
| Premium grade | " | 720.00 | 100.00 | 820.00 |
| 12' wide | | | | |
| Economy grade | EA. | 1,000 | 150.00 | 1,150 |
| Premium grade | " | 1,460 | 150.00 | 1,610 |
| Insulating glass, 5/8" thick | | | | |
| 6' wide | | | | |
| Economy grade | EA. | 890.00 | 100.00 | 990.00 |
| Premium grade | " | 1,690 | 100.00 | 1,790 |
| 12' wide | | | | |
| Economy grade | EA. | 760.00 | 150.00 | 910.00 |
| Premium grade | " | 1,460 | 150.00 | 1,610 |
| 1" thick | | | | |
| 6' wide | | | | |
| Economy grade | EA. | 980.00 | 100.00 | 1,080 |
| Premium grade | " | 1,130 | 100.00 | 1,230 |
| 12' wide | | | | |
| Economy grade | EA. | 1,520 | 150.00 | 1,670 |
| Premium grade | " | 1,920 | 150.00 | 2,070 |
| Added costs | | | | |
| Custom quality, add to material, 30% | | | | |
| Vertical lift doors, channel frame construction | | | | |
| 20' high x | | | | |
| 10' wide | EA. | 12,360 | 520.00 | 12,880 |
| 15' wide | " | 13,600 | 520.00 | 14,120 |
| 20' wide | " | 14,830 | 930.00 | 15,760 |
| 25' wide | " | 16,070 | 930.00 | 17,000 |
| 25' high | | | | |
| 20' wide | EA. | 18,540 | 930.00 | 19,470 |
| 25' wide | " | 19,780 | 1,090 | 20,870 |
| 30' high | | | | |
| 25' wide | EA. | 21,010 | 1,090 | 22,100 |
| 30' wide | " | 23,490 | 1,090 | 24,580 |

# 08 DOORS AND WINDOWS

| WOOD AND PLASTIC | UNIT | MAT. | INST. | TOTAL |
|---|---|---|---|---|
| **08300.10 SPECIAL DOORS** | | | | |
| 35' wide | EA. | 25,960 | 1,090 | 27,050 |
| Residential storm door | | | | |
| Minimum | EA. | 99.00 | 51.50 | 150.50 |
| Average | " | 140.00 | 51.50 | 191.50 |
| Maximum | " | 300.00 | 77.00 | 377.00 |

| STOREFRONTS | UNIT | MAT. | INST. | TOTAL |
|---|---|---|---|---|
| **08410.10 STOREFRONTS** | | | | |
| Storefront, aluminum and glass | | | | |
| Minimum | S.F. | 12.35 | 4.20 | 16.55 |
| Average | " | 17.65 | 4.85 | 22.50 |
| Maximum | " | 29.40 | 5.65 | 35.05 |
| Entrance doors | | | | |
| 1/2" thick glass | | | | |
| 3' x 7' | EA. | 2,360 | 280.00 | 2,640 |
| 6' x 7' | " | 4,590 | 420.00 | 5,010 |
| 3/4" thick glass | | | | |
| 3' x 7' | EA. | 2,590 | 280.00 | 2,870 |
| 6' x 7' | " | 4,710 | 420.00 | 5,130 |
| 1" thick glass | | | | |
| 3' x 7' | EA. | 2,940 | 280.00 | 3,220 |
| 6' x 7' | " | 5,770 | 420.00 | 6,190 |
| For anodized color, add to material, 25% | | | | |
| Revolving doors | | | | |
| 7'diameter, 7' high | | | | |
| Minimum | EA. | 10,010 | 3,270 | 13,280 |
| Average | " | 11,770 | 5,230 | 17,000 |
| Maximum | " | 21,190 | 6,540 | 27,730 |

| METAL WINDOWS | UNIT | MAT. | INST. | TOTAL |
|---|---|---|---|---|
| **08510.10 STEEL WINDOWS** | | | | |
| Steel windows, primed | | | | |
| Casements | | | | |
| Operable | | | | |

| METAL WINDOWS | UNIT | MAT. | INST. | TOTAL |
|---|---|---|---|---|
| **08510.10 STEEL WINDOWS** | | | | |
| Minimum | S.F. | 17.50 | 2.00 | 19.50 |
| Maximum | " | 22.15 | 2.25 | 24.40 |
| Fixed sash | " | 12.85 | 1.70 | 14.55 |
| Double hung | " | 30.30 | 1.90 | 32.20 |
| Industrial windows | | | | |
| Horizontally pivoted sash | S.F. | 16.35 | 2.25 | 18.60 |
| Fixed sash | " | 15.15 | 1.90 | 17.05 |
| Security sash | | | | |
| Operable | S.F. | 25.70 | 2.25 | 27.95 |
| Fixed | " | 21.00 | 1.90 | 22.90 |
| Picture window | " | 15.15 | 1.90 | 17.05 |
| Projecting sash | | | | |
| Minimum | S.F. | 23.30 | 2.10 | 25.40 |
| Maximum | " | 26.80 | 2.10 | 28.90 |
| Mullions | L.F. | 7.60 | 1.70 | 9.30 |
| **08520.10 ALUMINUM WINDOWS** | | | | |
| Jalousie | | | | |
| 3-0 x 4-0 | EA. | 190.00 | 42.30 | 232.30 |
| 3-0 x 5-0 | " | 220.00 | 42.30 | 262.30 |
| Fixed window | | | | |
| 6 sf to 8 sf | S.F. | 8.65 | 4.85 | 13.50 |
| 12 sf to 16 sf | " | 8.05 | 3.75 | 11.80 |
| Projecting window | | | | |
| 6 sf to 8 sf | S.F. | 19.70 | 8.45 | 28.15 |
| 12 sf to 16 sf | " | 17.40 | 5.65 | 23.05 |
| Horizontal sliding | | | | |
| 6 sf to 8 sf | S.F. | 14.00 | 4.20 | 18.20 |
| 12 sf to 16 sf | " | 12.70 | 3.40 | 16.10 |
| Double hung | | | | |
| 6 sf to 8 sf | S.F. | 12.70 | 6.75 | 19.45 |
| 10 sf to 12 sf | " | 11.90 | 5.65 | 17.55 |
| Storm window, 0.5 cfm, up to | | | | |
| 60 u.i. (united inches) | EA. | 44.30 | 16.90 | 61.20 |
| 70 u.i. | " | 45.50 | 16.90 | 62.40 |
| 80 u.i. | " | 50.00 | 16.90 | 66.90 |
| 90 u.i. | " | 51.50 | 18.80 | 70.30 |
| 100 u.i. | " | 52.50 | 18.80 | 71.30 |
| 2.0 cfm, up to | | | | |
| 60 u.i. | EA. | 57.00 | 16.90 | 73.90 |
| 70 u.i. | " | 57.50 | 16.90 | 74.40 |
| 80 u.i. | " | 59.50 | 16.90 | 76.40 |
| 90 u.i. | " | 64.00 | 18.80 | 82.80 |
| 100 u.i. | " | 64.00 | 18.80 | 82.80 |

| WOOD AND PLASTIC | UNIT | MAT. | INST. | TOTAL |
|---|---|---|---|---|
| **08600.10 WOOD WINDOWS** | | | | |
| Double hung | | | | |
| 24" x 36" | | | | |
| Minimum | EA. | 120.00 | 30.90 | 150.90 |
| Average | " | 180.00 | 38.60 | 218.60 |
| Maximum | " | 240.00 | 51.50 | 291.50 |
| 24" x 48" | | | | |
| Minimum | EA. | 140.00 | 30.90 | 170.90 |
| Average | " | 210.00 | 38.60 | 248.60 |
| Maximum | " | 260.00 | 51.50 | 311.50 |
| 30" x 48" | | | | |
| Minimum | EA. | 150.00 | 34.30 | 184.30 |
| Average | " | 210.00 | 44.10 | 254.10 |
| Maximum | " | 280.00 | 62.00 | 342.00 |
| 30" x 60" | | | | |
| Minimum | EA. | 160.00 | 34.30 | 194.30 |
| Average | " | 270.00 | 44.10 | 314.10 |
| Maximum | " | 320.00 | 62.00 | 382.00 |
| Casement | | | | |
| 1 leaf, 22" x 38" high | | | | |
| Minimum | EA. | 170.00 | 30.90 | 200.90 |
| Average | " | 220.00 | 38.60 | 258.60 |
| Maximum | " | 250.00 | 51.50 | 301.50 |
| 2 leaf, 50" x 50" high | | | | |
| Minimum | EA. | 470.00 | 38.60 | 508.60 |
| Average | " | 560.00 | 51.50 | 611.50 |
| Maximum | " | 650.00 | 77.00 | 727.00 |
| 3 leaf, 71" x 62" high | | | | |
| Minimum | EA. | 690.00 | 38.60 | 728.60 |
| Average | " | 820.00 | 51.50 | 871.50 |
| Maximum | " | 960.00 | 77.00 | 1,037 |
| 4 leaf, 95" x 75" high | | | | |
| Minimum | EA. | 950.00 | 44.10 | 994.10 |
| Average | " | 1,150 | 62.00 | 1,212 |
| Maximum | " | 1,500 | 100.00 | 1,600 |
| 5 leaf, 119" x 75" high | | | | |
| Minimum | EA. | 1,100 | 44.10 | 1,144 |
| Average | " | 1,500 | 62.00 | 1,562 |
| Maximum | " | 1,910 | 100.00 | 2,010 |
| Picture window, fixed glass, 54" x 54" high | | | | |
| Minimum | EA. | 330.00 | 38.60 | 368.60 |
| Average | " | 420.00 | 44.10 | 464.10 |
| Maximum | " | 580.00 | 51.50 | 631.50 |
| 68" x 55" high | | | | |
| Minimum | EA. | 530.00 | 38.60 | 568.60 |
| Average | " | 610.00 | 44.10 | 654.10 |
| Maximum | " | 780.00 | 51.50 | 831.50 |
| Sliding, 40" x 31" high | | | | |
| Minimum | EA. | 150.00 | 30.90 | 180.90 |
| Average | " | 240.00 | 38.60 | 278.60 |
| Maximum | " | 300.00 | 51.50 | 351.50 |
| 52" x 39" high | | | | |
| Minimum | EA. | 200.00 | 38.60 | 238.60 |

| WOOD AND PLASTIC | UNIT | MAT. | INST. | TOTAL |
|---|---|---|---|---|
| **08600.10 WOOD WINDOWS** | | | | |
| Average | EA. | 310.00 | 44.10 | 354.10 |
| Maximum | " | 350.00 | 51.50 | 401.50 |
| 64" x 72" high | | | | |
| Minimum | EA. | 320.00 | 38.60 | 358.60 |
| Average | " | 530.00 | 51.50 | 581.50 |
| Maximum | " | 580.00 | 62.00 | 642.00 |
| Awning windows | | | | |
| 34" x 21" high | | | | |
| Minimum | EA. | 160.00 | 30.90 | 190.90 |
| Average | " | 190.00 | 38.60 | 228.60 |
| Maximum | " | 230.00 | 51.50 | 281.50 |
| 40" x 21" high | | | | |
| Minimum | EA. | 170.00 | 34.30 | 204.30 |
| Average | " | 210.00 | 44.10 | 254.10 |
| Maximum | " | 250.00 | 62.00 | 312.00 |
| 48" x 27" high | | | | |
| Minimum | EA. | 180.00 | 34.30 | 214.30 |
| Average | " | 250.00 | 44.10 | 294.10 |
| Maximum | " | 300.00 | 62.00 | 362.00 |
| 60" x 36" high | | | | |
| Minimum | EA. | 200.00 | 38.60 | 238.60 |
| Average | " | 410.00 | 51.50 | 461.50 |
| Maximum | " | 460.00 | 62.00 | 522.00 |
| Window frame, milled | | | | |
| Minimum | L.F. | 2.10 | 6.20 | 8.30 |
| Average | " | 2.65 | 7.70 | 10.35 |
| Maximum | " | 4.25 | 10.30 | 14.55 |

| HARDWARE | UNIT | MAT. | INST. | TOTAL |
|---|---|---|---|---|
| **08710.10 HINGES** | | | | |
| Hinges | | | | |
| 3 x 3 butts, steel, interior, plain bearing | PAIR | | | 11.78 |
| 4 x 4 butts, steel, standard | " | | | 17.66 |
| 5 x 4-1/2 butts, bronze/s. steel, heavy duty | " | | | 47.10 |
| Pivot hinges | | | | |
| Top pivot | EA. | | | 42.39 |
| Intermediate pivot | " | | | 45.92 |
| Bottom pivot | " | | | 94.19 |
| BHMA specifications | | | | |
| 3-1/2 x 3-1/2, full mortise butts | | | | |
| Plain bearing | PAIR | | | 14.13 |
| Ball bearing | " | | | 17.66 |

# 08 DOORS AND WINDOWS

| HARDWARE | UNIT | MAT. | INST. | TOTAL |
|---|---|---|---|---|
| **08710.10 HINGES** | | | | |
| Half surface butts | PAIR | | | 25.91 |
| 4 x 4 | | | | |
| Full mortise butts, plain bearing, standard duty | PAIR | | | 15.31 |
| Full mortise butts, ball bearing | " | | | 18.84 |
| Half surface butts | | | | |
| Standard duty | PAIR | | | 25.91 |
| Ball bearing | " | | | 25.91 |
| 4-1/2 x 4-1/2 | | | | |
| Full mortise butts, plain bearing | PAIR | | | 16.48 |
| Ball bearing, heavy duty | " | | | 41.21 |
| Half mortise and half surface butts | | | | |
| Plain bearing | PAIR | | | 25.91 |
| Full surface and half surface butts | | | | |
| Standard duty | PAIR | | | 40.04 |
| Heavy duty | " | | | 110.69 |
| Full mortise and full slide-in butts, ball bearing | " | | | 18.84 |
| Half mortise butts, ball bearing | | | | |
| Standard duty | PAIR | | | 88.31 |
| Heavy duty | " | | | 103.62 |
| 5 x 5, ball bearing | | | | |
| Full mortise butts | PAIR | | | 41.21 |
| Half mortise, full & half surface butts | " | | | 91.84 |
| Full mortise, full surface and half surface butts | " | | | 124.81 |
| 4 x 4 | | | | |
| Full mortise butts, plain bearing, standard duty | PAIR | | | 8.25 |
| 5 x 4-1/2 | | | | |
| Full mortise butts, ball bearing, heavy duty | PAIR | | | 47.10 |
| **08710.20 LOCKSETS** | | | | |
| Latchset, heavy duty | | | | |
| Cylindrical | EA. | 97.00 | 19.30 | 116.30 |
| Mortise | " | 87.50 | 30.90 | 118.40 |
| Lockset, heavy duty | | | | |
| Cylindrical | EA. | 120.00 | 19.30 | 139.30 |
| Mortise | " | 140.00 | 30.90 | 170.90 |
| Mortise locks and latchsets, chrome | | | | |
| Latchset passage or closet latch | EA. | 97.00 | 25.70 | 122.70 |
| Privacy (bath or bedroom) | " | 120.00 | 25.70 | 145.70 |
| Entry lockset | " | 150.00 | 25.70 | 175.70 |
| Classroom lockset (outside key operated) | " | 150.00 | 25.70 | 175.70 |
| Storeroom lock | " | 150.00 | 25.70 | 175.70 |
| Front door lock | " | 150.00 | 25.70 | 175.70 |
| Dormitory or exit lock | " | 150.00 | 25.70 | 175.70 |
| Preassembled locks and latches, brass | | | | |
| Latchset, passage or closet latch | EA. | 160.00 | 25.70 | 185.70 |
| Lockset | | | | |
| Privacy (bath or bathroom) | EA. | 190.00 | 25.70 | 215.70 |
| Entry lock | " | 230.00 | 25.70 | 255.70 |
| Classroom lock (outside key, operated) | " | 230.00 | 25.70 | 255.70 |
| Storeroom lock | " | 280.00 | 25.70 | 305.70 |
| Bored locks and latches, satin chrome plated | | | | |

| HARDWARE | UNIT | MAT. | INST. | TOTAL |
|---|---|---|---|---|
| **08710.20 LOCKSETS** | | | | |
| Latchset passage or closet latch | EA. | 85.00 | 25.70 | 110.70 |
| Lockset | | | | |
| Privacy (bath or bedroom) | EA. | 110.00 | 25.70 | 135.70 |
| Entry lock | " | 130.00 | 25.70 | 155.70 |
| Classroom lock | " | 130.00 | 25.70 | 155.70 |
| Corridor lock | " | 130.00 | 25.70 | 155.70 |
| Miscellaneous locks | | | | |
| Exit lock with alarm, single door | EA. | 400.00 | 120.00 | 520.00 |
| Electric strike | | | | |
| Rim mounted wrought steel | EA. | 330.00 | 77.00 | 407.00 |
| Mortised, wrought steel with bronze plating | " | 110.00 | 120.00 | 230.00 |
| Dead bolt | | | | |
| Bored, wrought brass, keyed both sides | EA. | 66.50 | 51.50 | 118.00 |
| Mortised, cast brass | " | 73.00 | 51.50 | 124.50 |
| Lockset, cipher, mechanical | " | 1,150 | 30.90 | 1,181 |
| **08710.30 CLOSERS** | | | | |
| Door closers | | | | |
| Surface mounted, traditional type, parallel arm | | | | |
| Standard | EA. | 110.00 | 38.60 | 148.60 |
| Heavy duty | " | 120.00 | 38.60 | 158.60 |
| Modern type, parallel arm, standard duty | " | 120.00 | 38.60 | 158.60 |
| Overhead, concealed, pivot hung, single acting | | | | |
| Interior | EA. | 200.00 | 38.60 | 238.60 |
| Exterior | " | 380.00 | 38.60 | 418.60 |
| Floor concealed, single acting, offset, pivoted | | | | |
| Interior | EA. | 360.00 | 100.00 | 460.00 |
| Exterior | " | 460.00 | 100.00 | 560.00 |
| **08710.40 DOOR TRIM** | | | | |
| Door bumper, bronze, wall type | EA. | 2.20 | 6.20 | 8.40 |
| Wall type, 4" dia. with convex rubber pad, aluminum | " | 20.00 | 6.20 | 26.20 |
| Floor type | | | | |
| Aluminum | EA. | 3.10 | 6.20 | 9.30 |
| Brass | " | 3.10 | 6.20 | 9.30 |
| Door holders | | | | |
| Wall type, bronze | EA. | 3.10 | 6.20 | 9.30 |
| Overhead | " | 31.20 | 15.45 | 46.65 |
| Floor type | " | 31.20 | 15.45 | 46.65 |
| Plunger type | " | 16.85 | 15.45 | 32.30 |
| Wall type, aluminum | " | 19.95 | 15.45 | 35.40 |
| Surface bolt | " | 14.95 | 6.20 | 21.15 |
| Panic device | | | | |
| Rim type with thumb piece | EA. | 400.00 | 77.00 | 477.00 |
| Mortise | " | 510.00 | 77.00 | 587.00 |
| Vertical rod | " | 590.00 | 77.00 | 667.00 |
| Labelled, rim type | " | 520.00 | 77.00 | 597.00 |
| Mortise | " | 630.00 | 77.00 | 707.00 |
| Vertical rod | " | 710.00 | 77.00 | 787.00 |
| Silencers, rubber type | " | 0.12 | 0.62 | 0.74 |

| HARDWARE | UNIT | MAT. | INST. | TOTAL |
|---|---|---|---|---|
| **08710.40 DOOR TRIM** | | | | |
| Dust proof strike with plate, brass | EA. | 11.25 | 10.30 | 21.55 |
| Flush bolt, lever extension, brass, rated | " | 19.95 | 6.20 | 26.15 |
| Surface bolt with strike, brass, 6" long | " | 14.95 | 6.20 | 21.15 |
| Door coordinator, labelled, brass, satin chrome | " | 72.50 | 22.05 | 94.55 |
| Door plates | | | | |
| Kick plate, aluminum, 3 beveled edges | | | | |
| 10" x 28" | EA. | 11.25 | 15.45 | 26.70 |
| 10" x 30" | " | 12.50 | 15.45 | 27.95 |
| 10" x 34" | " | 13.75 | 15.45 | 29.20 |
| 10" x 38" | " | 14.95 | 15.45 | 30.40 |
| Push plate, 4" x 16" | | | | |
| Aluminum | EA. | 17.65 | 6.20 | 23.85 |
| Bronze | " | 68.50 | 6.20 | 74.70 |
| Stainless steel | " | 55.50 | 6.20 | 61.70 |
| Armor plate, 40" x 34" | " | 55.00 | 12.35 | 67.35 |
| Pull handle, 4" x 16" | | | | |
| Aluminum | EA. | 31.80 | 6.20 | 38.00 |
| Bronze | " | 70.50 | 6.20 | 76.70 |
| Stainless steel | " | 64.50 | 6.20 | 70.70 |
| Hasp assembly | | | | |
| 3" | EA. | 0.88 | 5.15 | 6.03 |
| 4-1/2" | " | 1.20 | 6.85 | 8.05 |
| 6" | " | 1.85 | 8.80 | 10.65 |
| Electro-magnetic door holder | | | | |
| Wall mounted | EA. | 110.00 | 100.00 | 210.00 |
| Floor mounted | " | 170.00 | 100.00 | 270.00 |
| Smoke detector door holder | | | | |
| Photo electric type | EA. | 180.00 | 100.00 | 280.00 |
| Ionization type | " | 170.00 | 100.00 | 270.00 |
| Pneumatic operators, activated by rubber mats | | | | |
| Swing | | | | |
| Single | EA. | 840.00 | 260.00 | 1,100 |
| Double | " | 1,110 | 390.00 | 1,500 |
| Sliding | | | | |
| Single | EA. | 760.00 | 260.00 | 1,020 |
| Double | " | 1,370 | 390.00 | 1,760 |
| **08710.60 WEATHERSTRIPPING** | | | | |
| Weatherstrip, head and jamb, metal strip, neoprene bulb | | | | |
| Standard duty | L.F. | 2.60 | 1.70 | 4.30 |
| Heavy duty | " | 3.30 | 1.95 | 5.25 |
| Spring type | | | | |
| Metal doors | EA. | 34.00 | 77.00 | 111.00 |
| Wood doors | " | 34.00 | 100.00 | 134.00 |
| Sponge type with adhesive backing | " | 31.70 | 30.90 | 62.60 |
| Astragal | | | | |
| 1-3/4" x 13 ga., aluminum | L.F. | 3.35 | 2.55 | 5.90 |
| 1-3/8" x 5/8", oak | " | 3.15 | 2.05 | 5.20 |
| Thresholds | | | | |
| Bronze | L.F. | 32.70 | 7.70 | 40.40 |
| Aluminum | | | | |
| Plain | L.F. | 12.25 | 7.70 | 19.95 |

## HARDWARE

| 08710.60 WEATHERSTRIPPING | UNIT | MAT. | INST. | TOTAL |
|---|---|---|---|---|
| Vinyl insert | L.F. | 18.65 | 7.70 | 26.35 |
| Aluminum with grit | " | 17.50 | 7.70 | 25.20 |
| Steel | | | | |
| Plain | L.F. | 12.25 | 7.70 | 19.95 |
| Interlocking | " | 18.65 | 25.70 | 44.35 |

## GLAZING

| 08810.10 GLAZING | UNIT | MAT. | INST. | TOTAL |
|---|---|---|---|---|
| Sheet glass, 1/8" thick | S.F. | 4.55 | 1.90 | 6.45 |
| Plate glass, bronze or grey, 1/4" thick | " | 6.80 | 3.05 | 9.85 |
| Clear | " | 5.20 | 3.05 | 8.25 |
| Polished | " | 5.75 | 3.05 | 8.80 |
| Plexiglass | | | | |
| 1/8" thick | S.F. | 1.95 | 3.05 | 5.00 |
| 1/4" thick | " | 3.65 | 1.90 | 5.55 |
| Float glass, clear | | | | |
| 3/16" thick | S.F. | 3.00 | 2.80 | 5.80 |
| 1/4" thick | " | 3.55 | 3.05 | 6.60 |
| 5/16" thick | " | 6.90 | 3.40 | 10.30 |
| 3/8" thick | " | 8.15 | 4.20 | 12.35 |
| 1/2" thick | " | 13.10 | 5.65 | 18.75 |
| 5/8" thick | " | 15.05 | 6.75 | 21.80 |
| 3/4" thick | " | 16.40 | 8.45 | 24.85 |
| 1" thick | " | 28.90 | 11.25 | 40.15 |
| Tinted glass, polished plate, twin ground | | | | |
| 3/16" thick | S.F. | 5.20 | 2.80 | 8.00 |
| 1/4" thick | " | 5.20 | 3.05 | 8.25 |
| 3/8" thick | " | 8.40 | 4.20 | 12.60 |
| 1/2" thick | " | 13.60 | 5.65 | 19.25 |
| Total, full vision, all glass window system | | | | |
| To 10' high | | | | |
| Minimum | S.F. | 20.60 | 8.45 | 29.05 |
| Average | " | 21.85 | 8.45 | 30.30 |
| Maximum | " | 23.05 | 8.45 | 31.50 |
| 10' to 20' high | | | | |
| Minimum | S.F. | 25.50 | 8.45 | 33.95 |
| Average | " | 27.90 | 8.45 | 36.35 |
| Maximum | " | 30.30 | 8.45 | 38.75 |
| Insulated glass, bronze or gray | | | | |
| 1/2" thick | S.F. | 9.70 | 5.65 | 15.35 |
| 1" thick | " | 11.50 | 8.45 | 19.95 |
| Clear | | | | |

| GLAZING | UNIT | MAT. | INST. | TOTAL |
|---|---|---|---|---|
| **08810.10 GLAZING** | | | | |
| 1/2" thick | S.F. | 7.65 | 5.65 | 13.30 |
| 1" thick | " | 10.80 | 8.45 | 19.25 |
| Spandrel glass, polished bronze/grey, 1 side, 1/4" thick | " | 6.70 | 3.05 | 9.75 |
| Tempered glass (safety) | | | | |
| Clear sheet glass | | | | |
| 1/8" thick | S.F. | 5.95 | 1.90 | 7.85 |
| 3/16" thick | " | 6.70 | 2.60 | 9.30 |
| Clear float glass | | | | |
| 1/4" thick | S.F. | 6.20 | 2.80 | 9.00 |
| 5/16" thick | " | 11.35 | 3.40 | 14.75 |
| 3/8" thick | " | 16.35 | 4.20 | 20.55 |
| 1/2" thick | " | 25.40 | 5.65 | 31.05 |
| 5/8" thick | " | 27.70 | 6.75 | 34.45 |
| 3/4" thick | " | 43.90 | 11.25 | 55.15 |
| Tinted float glass | | | | |
| 3/16" thick | S.F. | 7.65 | 2.60 | 10.25 |
| 1/4" thick | " | 7.90 | 2.80 | 10.70 |
| 3/8" thick | " | 19.40 | 4.20 | 23.60 |
| 1/2" thick | " | 27.70 | 5.65 | 33.35 |
| Laminated glass | | | | |
| Float safety glass with polyvinyl plastic interlayer | | | | |
| 1/4", sheet or float | | | | |
| Two lites, 1/8" thick, clear glass | S.F. | 7.75 | 2.80 | 10.55 |
| 1/2" thick, float glass | | | | |
| Two lites, 1/4" thick, clear glass | S.F. | 11.55 | 5.65 | 17.20 |
| Tinted glass | " | 13.80 | 5.65 | 19.45 |
| Insulating glass, two lites, clear float glass | | | | |
| 1/2" thick | S.F. | 7.55 | 5.65 | 13.20 |
| 5/8" thick | " | 8.70 | 6.75 | 15.45 |
| 3/4" thick | " | 9.60 | 8.45 | 18.05 |
| 7/8" thick | " | 10.05 | 9.65 | 19.70 |
| 1" thick | " | 10.80 | 11.25 | 22.05 |
| Glass seal edge | | | | |
| 3/8" thick | S.F. | 6.35 | 5.65 | 12.00 |
| Tinted glass | | | | |
| 1/2" thick | S.F. | 13.00 | 5.65 | 18.65 |
| 1" thick | " | 13.95 | 11.25 | 25.20 |
| Tempered, clear | | | | |
| 1" thick | S.F. | 25.40 | 11.25 | 36.65 |
| Wire reinforced | " | 32.30 | 11.25 | 43.55 |
| Plate mirror glass | | | | |
| 1/4" thick | | | | |
| 15 sf | S.F. | 7.65 | 3.40 | 11.05 |
| Over 15 sf | " | 7.05 | 3.05 | 10.10 |
| Transparent, one way vision, 1/4" thick | " | 14.80 | 3.40 | 18.20 |
| Sheet mirror glass | | | | |
| 3/16" thick | S.F. | 6.90 | 3.40 | 10.30 |
| 1/4" thick | " | 7.40 | 2.80 | 10.20 |
| Wall tiles, 12" x 12" | | | | |
| Clear glass | S.F. | 2.20 | 1.90 | 4.10 |
| Veined glass | " | 2.55 | 1.90 | 4.45 |
| Wire glass, 1/4" thick | | | | |

## GLAZING

| GLAZING | UNIT | MAT. | INST. | TOTAL |
|---|---|---|---|---|
| **08810.10 GLAZING** | | | | |
| Clear | S.F. | 9.00 | 11.25 | 20.25 |
| Hammered | " | 9.10 | 11.25 | 20.35 |
| Obscure | " | 10.90 | 11.25 | 22.15 |
| Bullet resistant glass, plate glass with inter-leaved vinyl | | | | |
| 1-3/16" thick | | | | |
| To 15 sf | S.F. | 24.25 | 16.90 | 41.15 |
| Over 15 sf | " | 27.70 | 16.90 | 44.60 |
| 2" thick | | | | |
| To 15 sf | S.F. | 46.20 | 28.20 | 74.40 |
| Over 15 sf | " | 57.50 | 28.20 | 85.70 |
| Glazing accessories | | | | |
| Neoprene glazing gaskets | | | | |
| 1/4" glass | L.F. | 0.87 | 1.35 | 2.22 |
| 3/8" glass | " | 1.05 | 1.40 | 2.45 |
| 1/2" glass | " | 1.35 | 1.45 | 2.80 |
| 3/4" glass | " | 1.80 | 1.55 | 3.35 |
| 1" glass | " | 2.10 | 1.70 | 3.80 |
| Mullion section | | | | |
| 1/4" glass | L.F. | 0.38 | 0.68 | 1.06 |
| 3/8" glass | " | 0.49 | 0.84 | 1.34 |
| 1/2" glass | " | 0.72 | 0.97 | 1.69 |
| 3/4" glass | " | 1.10 | 1.15 | 2.25 |
| 1" glass | " | 1.45 | 1.35 | 2.80 |
| Molded corners | EA. | 1.60 | 22.55 | 24.15 |

## GLAZED CURTAIN WALLS

| GLAZED CURTAIN WALLS | UNIT | MAT. | INST. | TOTAL |
|---|---|---|---|---|
| **08910.10 GLAZED CURTAIN WALLS** | | | | |
| Curtain wall, aluminum system, framing sections | | | | |
| 2" x 3" | | | | |
| Jamb | L.F. | 6.55 | 2.80 | 9.35 |
| Horizontal | " | 6.65 | 2.80 | 9.45 |
| Mullion | " | 8.90 | 2.80 | 11.70 |
| 2" x 4" | | | | |
| Jamb | L.F. | 8.90 | 4.20 | 13.10 |
| Horizontal | " | 9.15 | 4.20 | 13.35 |
| Mullion | " | 8.90 | 4.20 | 13.10 |
| 3" x 5-1/2" | | | | |
| Jamb | L.F. | 11.75 | 4.20 | 15.95 |
| Horizontal | " | 13.10 | 4.20 | 17.30 |
| Mullion | " | 11.85 | 4.20 | 16.05 |
| 4" corner mullion | " | 15.70 | 5.65 | 21.35 |
| Coping sections | | | | |
| 1/8" x 8" | L.F. | 16.45 | 5.65 | 22.10 |

| GLAZED CURTAIN WALLS | UNIT | MAT. | INST. | TOTAL |
|---|---|---|---|---|
| **08910.10 GLAZED CURTAIN WALLS** | | | | |
| 1/8" x 9" | L.F. | 16.55 | 5.65 | 22.20 |
| 1/8" x 12-1/2" | " | 16.95 | 6.75 | 23.70 |
| Sill section | | | | |
| 1/8" x 6" | L.F. | 17.00 | 3.40 | 20.40 |
| 1/8" x 7" | " | 16.40 | 3.40 | 19.80 |
| 1/8" x 8-1/2" | " | 16.70 | 3.40 | 20.10 |
| Column covers, aluminum | | | | |
| 1/8" x 26" | L.F. | 16.40 | 8.45 | 24.85 |
| 1/8" x 34" | " | 16.40 | 8.90 | 25.30 |
| 1/8" x 38" | " | 16.55 | 8.90 | 25.45 |
| Doors | | | | |
| Aluminum framed, standard hardware | | | | |
| Narrow stile | | | | |
| 2-6 x 7-0 | EA. | 370.00 | 170.00 | 540.00 |
| 3-0 x 7-0 | " | 370.00 | 170.00 | 540.00 |
| 3-6 x 7-0 | " | 380.00 | 170.00 | 550.00 |
| Wide stile | | | | |
| 2-6 x 7-0 | EA. | 630.00 | 170.00 | 800.00 |
| 3-0 x 7-0 | " | 680.00 | 170.00 | 850.00 |
| 3-6 x 7-0 | " | 730.00 | 170.00 | 900.00 |
| Flush panel doors, to match adjacent wall panels | | | | |
| 2-6 x 7-0 | EA. | 530.00 | 210.00 | 740.00 |
| 3-0 x 7-0 | " | 560.00 | 210.00 | 770.00 |
| 3-6 x 7-0 | " | 580.00 | 210.00 | 790.00 |
| Wall panel, insulated | | | | |
| "U"=.08 | S.F. | 6.65 | 2.80 | 9.45 |
| "U"=.10 | " | 6.30 | 2.80 | 9.10 |
| "U"=.15 | " | 5.70 | 2.80 | 8.50 |
| Window wall system, complete | | | | |
| Minimum | S.F. | 11.80 | 3.40 | 15.20 |
| Average | " | 18.85 | 3.75 | 22.60 |
| Maximum | " | 43.60 | 4.85 | 48.45 |
| Added costs | | | | |
| For bronze, add 20% to material | | | | |
| For stainless steel, add 50% to material | | | | |

# 09 FINISHES

## SUPPORT SYSTEMS

| SUPPORT SYSTEMS | UNIT | MAT. | INST. | TOTAL |
|---|---|---|---|---|
| **09110.10 METAL STUDS** | | | | |
| Studs, non load bearing, galvanized | | | | |
| 2-1/2", 20 ga. | | | | |
| 12" o.c. | S.F. | 0.61 | 0.64 | 1.25 |
| 16" o.c. | " | 0.52 | 0.52 | 1.03 |
| 25 ga. | | | | |
| 12" o.c. | S.F. | 0.43 | 0.64 | 1.07 |
| 16" o.c. | " | 0.31 | 0.52 | 0.82 |
| 24" o.c. | " | 0.22 | 0.43 | 0.65 |
| 3-5/8", 20 ga. | | | | |
| 12" o.c. | S.F. | 0.74 | 0.77 | 1.51 |
| 16" o.c. | " | 0.62 | 0.62 | 1.24 |
| 24" o.c. | " | 0.51 | 0.52 | 1.02 |
| 25 ga. | | | | |
| 12" o.c. | S.F. | 0.42 | 0.77 | 1.19 |
| 16" o.c. | " | 0.36 | 0.62 | 0.98 |
| 24" o.c. | " | 0.30 | 0.52 | 0.81 |
| 4", 20 ga. | | | | |
| 12" o.c. | S.F. | 0.85 | 0.77 | 1.62 |
| 16" o.c. | " | 0.69 | 0.62 | 1.31 |
| 24" o.c. | " | 0.56 | 0.52 | 1.08 |
| 25 ga. | | | | |
| 12" o.c. | S.F. | 0.51 | 0.77 | 1.28 |
| 16" o.c. | " | 0.42 | 0.62 | 1.04 |
| 24" o.c. | " | 0.34 | 0.52 | 0.85 |
| 6", 20 ga. | | | | |
| 12" o.c. | S.F. | 1.05 | 0.96 | 2.02 |
| 16" o.c. | " | 0.85 | 0.77 | 1.62 |
| 24" o.c. | " | 0.68 | 0.64 | 1.32 |
| 25 ga. | | | | |
| 12" o.c. | S.F. | 0.63 | 0.96 | 1.60 |
| 16" o.c. | " | 0.52 | 0.77 | 1.29 |
| 24" o.c. | " | 0.43 | 0.64 | 1.07 |
| Load bearing studs, galvanized | | | | |
| 3-5/8", 16 ga. | | | | |
| 12" o.c. | S.F. | 1.60 | 0.77 | 2.37 |
| 16" o.c. | " | 1.35 | 0.62 | 1.97 |
| 18 ga. | | | | |
| 12" o.c. | S.F. | 1.45 | 0.52 | 1.97 |
| 16" o.c. | " | 1.20 | 0.62 | 1.82 |
| 4", 16 ga. | | | | |
| 12" o.c. | S.F. | 1.80 | 0.77 | 2.57 |
| 16" o.c. | " | 1.55 | 0.62 | 2.17 |
| 6", 16 ga. | | | | |
| 12" o.c. | S.F. | 2.20 | 0.96 | 3.17 |
| 16" o.c. | " | 1.90 | 0.77 | 2.67 |
| Furring | | | | |
| On beams and columns | | | | |
| 7/8" channel | L.F. | 0.32 | 2.05 | 2.37 |
| 1-1/2" channel | " | 0.38 | 2.40 | 2.78 |
| On ceilings | | | | |
| 3/4" furring channels | | | | |
| 12" o.c. | S.F. | 0.36 | 1.30 | 1.66 |

## SUPPORT SYSTEMS

| 09110.10 METAL STUDS | UNIT | MAT. | INST. | TOTAL |
|---|---|---|---|---|
| 16" o.c. | S.F. | 0.31 | 1.25 | 1.56 |
| 24" o.c. | " | 0.23 | 1.10 | 1.33 |
| 1-1/2" furring channels | | | | |
| 12" o.c. | S.F. | 0.44 | 1.40 | 1.84 |
| 16" o.c. | " | 0.38 | 1.30 | 1.68 |
| 24" o.c. | " | 0.32 | 1.20 | 1.52 |
| On walls | | | | |
| 3/4" furring channels | | | | |
| 12" o.c. | S.F. | 0.36 | 1.05 | 1.41 |
| 16" o.c. | " | 0.31 | 0.96 | 1.27 |
| 24" o.c. | " | 0.24 | 0.91 | 1.15 |
| 1-1/2" furring channels | | | | |
| 12" o.c. | S.F. | 0.44 | 1.10 | 1.54 |
| 16" o.c. | " | 0.38 | 1.05 | 1.43 |
| 24" o.c. | " | 0.32 | 0.96 | 1.28 |

## LATH AND PLASTER

| 09205.10 GYPSUM LATH | UNIT | MAT. | INST. | TOTAL |
|---|---|---|---|---|
| Gypsum lath, 1/2" thick | | | | |
| Clipped | S.Y. | 4.95 | 1.70 | 6.65 |
| Nailed | " | 4.75 | 1.95 | 6.70 |

| 09205.20 METAL LATH | UNIT | MAT. | INST. | TOTAL |
|---|---|---|---|---|
| Diamond expanded, galvanized | | | | |
| 2.5 lb., on walls | | | | |
| Nailed | S.Y. | 2.05 | 3.85 | 5.90 |
| Wired | " | 2.10 | 4.40 | 6.50 |
| On ceilings | | | | |
| Nailed | S.Y. | 2.05 | 4.40 | 6.45 |
| Wired | " | 2.10 | 5.15 | 7.25 |
| 3.4 lb., on walls | | | | |
| Nailed | S.Y. | 2.40 | 3.85 | 6.25 |
| Wired | " | 2.40 | 4.40 | 6.80 |
| On ceilings | | | | |
| Nailed | S.Y. | 2.40 | 4.40 | 6.80 |
| Wired | " | 2.40 | 5.15 | 7.55 |
| Flat rib | | | | |
| 2.75 lb., on walls | | | | |
| Nailed | S.Y. | 2.15 | 3.85 | 6.00 |
| Wired | " | 2.20 | 4.40 | 6.60 |
| On ceilings | | | | |

| LATH AND PLASTER | UNIT | MAT. | INST. | TOTAL |
|---|---|---|---|---|
| **09205.20 METAL LATH** | | | | |
| Nailed | S.Y. | 2.15 | 4.40 | 6.55 |
| Wired | " | 2.20 | 5.15 | 7.35 |
| 3.4 lb., on walls | | | | |
| Nailed | S.Y. | 2.50 | 3.85 | 6.35 |
| Wired | " | 2.55 | 4.40 | 6.95 |
| On ceilings | | | | |
| Nailed | S.Y. | 2.50 | 4.40 | 6.90 |
| Wired | " | 2.55 | 5.15 | 7.70 |
| Stucco lath | | | | |
| 1.8 lb. | S.Y. | 2.95 | 3.85 | 6.80 |
| 3.6 lb. | " | 3.40 | 3.85 | 7.25 |
| Paper backed | | | | |
| Minimum | S.Y. | 2.70 | 3.10 | 5.80 |
| Maximum | " | 4.20 | 4.40 | 8.60 |
| **09205.60 PLASTER ACCESSORIES** | | | | |
| Expansion joint, 3/4", 26 ga., galvanized, one piece | L.F. | 0.65 | 0.77 | 1.42 |
| Plaster corner beads, 3/4", galvanized | " | 0.38 | 0.88 | 1.26 |
| Casing bead, expanded flange, galvanized | " | 0.38 | 0.77 | 1.15 |
| Expanded wing, 1-1/4" wide, galvanized | " | 0.32 | 0.77 | 1.09 |
| Joint clips for lath | EA. | 0.17 | 0.15 | 0.32 |
| Metal base, galvanized, 2-1/2" high | L.F. | 0.51 | 1.05 | 1.56 |
| Stud clips for gypsum lath | EA. | 0.17 | 0.15 | 0.32 |
| Tie wire galvanized, 18 ga., 25 lb. hank | " | | | 32.33 |
| Sound deadening board, 1/4", nailed or clipped | S.F. | 0.32 | 0.52 | 0.83 |
| **09210.10 PLASTER** | | | | |
| Gypsum plaster, trowel finish, 2 coats | | | | |
| Ceilings | S.Y. | 3.75 | 9.50 | 13.25 |
| Walls | " | 3.75 | 8.95 | 12.70 |
| 3 coats | | | | |
| Ceilings | S.Y. | 5.20 | 13.20 | 18.40 |
| Walls | " | 5.20 | 11.70 | 16.90 |
| Vermiculite plaster | | | | |
| 2 coats | | | | |
| Ceilings | S.Y. | 4.25 | 14.50 | 18.75 |
| Walls | " | 4.25 | 13.20 | 17.45 |
| 3 coats | | | | |
| Ceilings | S.Y. | 6.70 | 17.90 | 24.60 |
| Walls | " | 6.70 | 16.00 | 22.70 |
| Keenes cement plaster | | | | |
| 2 coats | | | | |
| Ceilings | S.Y. | 5.65 | 11.70 | 17.35 |
| Walls | " | 5.65 | 10.15 | 15.80 |
| 3 coats | | | | |
| Ceilings | S.Y. | 6.90 | 13.20 | 20.10 |
| Walls | " | 6.90 | 11.70 | 18.60 |
| On columns, add to installation, 50% | " | | | |
| Chases, fascia, and soffits, add to installation, 50% | " | | | |
| Beams, add to installation, 50% | " | | | |

| LATH AND PLASTER | UNIT | MAT. | INST. | TOTAL |
|---|---|---|---|---|
| **09220.10 PORTLAND CEMENT PLASTER** | | | | |
| Stucco, portland, gray, 3 coat, 1" thick | | | | |
| Sand finish | S.Y. | 4.20 | 13.20 | 17.40 |
| Trowel finish | " | 4.20 | 13.80 | 18.00 |
| White cement | | | | |
| Sand finish | S.Y. | 4.80 | 13.80 | 18.60 |
| Trowel finish | " | 4.80 | 15.20 | 20.00 |
| Scratch coat | | | | |
| For ceramic tile | S.Y. | 1.50 | 3.05 | 4.55 |
| For quarry tile | " | 1.50 | 3.05 | 4.55 |
| Portland cement plaster | | | | |
| 2 coats, 1/2" | S.Y. | 3.05 | 6.10 | 9.15 |
| 3 coats, 7/8" | " | 3.60 | 7.60 | 11.20 |
| **09250.10 GYPSUM BOARD** | | | | |
| Drywall, plasterboard, 3/8" clipped to | | | | |
| Metal furred ceiling | S.F. | 0.27 | 0.34 | 0.61 |
| Columns and beams | " | 0.27 | 0.77 | 1.04 |
| Walls | " | 0.27 | 0.31 | 0.58 |
| Nailed or screwed to | | | | |
| Wood framed ceiling | S.F. | 0.27 | 0.31 | 0.58 |
| Columns and beams | " | 0.27 | 0.69 | 0.96 |
| Walls | " | 0.27 | 0.28 | 0.55 |
| 1/2", clipped to | | | | |
| Metal furred ceiling | S.F. | 0.28 | 0.34 | 0.62 |
| Columns and beams | " | 0.28 | 0.77 | 1.05 |
| Walls | " | 0.28 | 0.31 | 0.59 |
| Nailed or screwed to | | | | |
| Wood framed ceiling | S.F. | 0.28 | 0.31 | 0.59 |
| Columns and beams | " | 0.28 | 0.69 | 0.97 |
| Walls | " | 0.28 | 0.28 | 0.56 |
| 5/8", clipped to | | | | |
| Metal furred ceiling | S.F. | 0.31 | 0.39 | 0.70 |
| Columns and beams | " | 0.31 | 0.86 | 1.17 |
| Walls | " | 0.31 | 0.34 | 0.65 |
| Nailed or screwed to | | | | |
| Wood framed ceiling | S.F. | 0.31 | 0.39 | 0.70 |
| Columns and beams | " | 0.31 | 0.86 | 1.17 |
| Walls | " | 0.31 | 0.34 | 0.65 |
| Vinyl faced, clipped to metal studs | | | | |
| 1/2" | S.F. | 1.15 | 0.39 | 1.54 |
| 5/8" | " | 1.20 | 0.39 | 1.59 |
| Add for | | | | |
| Fire resistant | S.F. | | | 0.07 |
| Water resistant | " | | | 0.11 |
| Water and fire resistant | " | | | 0.14 |
| Taping and finishing joints | | | | |
| Minimum | S.F. | 0.03 | 0.21 | 0.24 |
| Average | " | 0.04 | 0.26 | 0.30 |
| Maximum | " | 0.06 | 0.31 | 0.37 |
| Casing bead | | | | |
| Minimum | L.F. | 0.10 | 0.88 | 0.98 |
| Average | " | 0.11 | 1.05 | 1.16 |

# 09 FINISHES

## LATH AND PLASTER

| 09250.10 GYPSUM BOARD | UNIT | MAT. | INST. | TOTAL |
|---|---|---|---|---|
| Maximum | L.F. | 0.14 | 1.55 | 1.69 |
| Corner bead | | | | |
| Minimum | L.F. | 0.11 | 0.88 | 0.99 |
| Average | " | 0.13 | 1.05 | 1.18 |
| Maximum | " | 0.17 | 1.55 | 1.72 |

## TILE

| 09310.10 CERAMIC TILE | UNIT | MAT. | INST. | TOTAL |
|---|---|---|---|---|
| Glazed wall tile, 4-1/4" x 4-1/4" | | | | |
| Minimum | S.F. | 1.60 | 2.15 | 3.75 |
| Average | " | 2.55 | 2.50 | 5.05 |
| Maximum | " | 6.95 | 3.00 | 9.95 |
| Base, 4-1/4" high | | | | |
| Minimum | L.F. | 2.50 | 3.75 | 6.25 |
| Average | " | 2.80 | 3.75 | 6.55 |
| Maximum | " | 4.40 | 3.75 | 8.15 |
| Unglazed floor tile | | | | |
| Portland cement bed, cushion edge, face mounted | | | | |
| 1" x 1" | S.F. | 4.85 | 2.75 | 7.60 |
| 1" x 2" | " | 7.75 | 2.60 | 10.35 |
| 2" x 2" | " | 5.10 | 2.50 | 7.60 |
| Adhesive bed, with white grout | | | | |
| 1" x 1" | S.F. | 4.25 | 2.75 | 7.00 |
| 1" x 2" | " | 4.45 | 2.60 | 7.05 |
| 2" x 2" | " | 4.50 | 2.50 | 7.00 |
| Organic adhesive bed, thin set, back mounted | | | | |
| 1" x 1" | S.F. | 4.75 | 2.75 | 7.50 |
| 1" x 2" | " | 4.85 | 2.60 | 7.45 |
| 2" x 2" | " | 5.10 | 2.50 | 7.60 |
| For group 2 colors, add to material, 10% | | | | |
| For group 3 colors, add to material, 20% | | | | |
| For abrasive surface, add to material, 25% | | | | |
| Unglazed wall tile | | | | |
| Organic adhesive, face mounted cushion edge | | | | |
| 1" x 1" | | | | |
| Minimum | S.F. | 2.25 | 2.50 | 4.75 |
| Average | " | 2.95 | 2.75 | 5.70 |
| Maximum | " | 4.40 | 3.00 | 7.40 |
| 1" x 2" | | | | |
| Minimum | S.F. | 2.50 | 2.40 | 4.90 |
| Average | " | 3.30 | 2.60 | 5.90 |
| Maximum | " | 4.85 | 2.85 | 7.70 |

# 09 FINISHES

| TILE | UNIT | MAT. | INST. | TOTAL |
|---|---|---|---|---|
| **09310.10 CERAMIC TILE** | | | | |
| 2" x 2" | | | | |
| Minimum | S.F. | 2.60 | 2.30 | 4.90 |
| Average | " | 2.95 | 2.50 | 5.45 |
| Maximum | " | 3.40 | 2.75 | 6.15 |
| Back mounted | | | | |
| 1" x 1" | | | | |
| Minimum | S.F. | 2.25 | 2.50 | 4.75 |
| Average | " | 2.95 | 2.75 | 5.70 |
| Maximum | " | 4.40 | 3.00 | 7.40 |
| 1" x 2" | | | | |
| Minimum | S.F. | 2.50 | 2.40 | 4.90 |
| Average | " | 3.30 | 2.60 | 5.90 |
| Maximum | " | 4.80 | 2.85 | 7.65 |
| 2" x 2" | | | | |
| Minimum | S.F. | 2.60 | 2.30 | 4.90 |
| Average | " | 2.95 | 2.50 | 5.45 |
| Maximum | " | 4.80 | 2.75 | 7.55 |
| For glazed finish, add to material, 25% | | | | |
| For glazed mosaic, add to material, 100% | | | | |
| For metallic colors, add to material, 125% | | | | |
| For exterior wall use, add to total, 25% | | | | |
| For exterior soffit, add to total, 25% | | | | |
| For portland cement bed, add to total, 25% | | | | |
| For dry set portland cement bed, add to total, 10% | | | | |
| Conductive floor tile, unglazed square edged | | | | |
| Portland cement bed | | | | |
| 1 x 1 | S.F. | 4.60 | 3.75 | 8.35 |
| 1-9/16 x 1-9/16 | " | 4.40 | 3.75 | 8.15 |
| Dry set | | | | |
| 1 x 1 | S.F. | 4.85 | 3.75 | 8.60 |
| 1-9/16 x 1-9/16 | " | 4.55 | 3.75 | 8.30 |
| Epoxy bed with epoxy joints | | | | |
| 1 x 1 | S.F. | 4.00 | 3.75 | 7.75 |
| 1-9/16 x 1-9/16 | " | 4.85 | 3.75 | 8.60 |
| For WWF in bed add to total, 15% | | | | |
| For abrasive surface, add to material, 40% | | | | |
| Ceramic accessories | | | | |
| Towel bar, 24" long | | | | |
| Minimum | EA. | 18.50 | 12.00 | 30.50 |
| Average | " | 24.25 | 15.00 | 39.25 |
| Maximum | " | 37.00 | 20.00 | 57.00 |
| Soap dish | | | | |
| Minimum | EA. | 16.15 | 20.00 | 36.15 |
| Average | " | 19.65 | 25.00 | 44.65 |
| Maximum | " | 27.70 | 30.00 | 57.70 |
| **09330.10 QUARRY TILE** | | | | |
| Floor | | | | |
| 4 x 4 x 1/2" | S.F. | 3.60 | 4.00 | 7.60 |
| 6 x 6 x 1/2" | " | 3.65 | 3.75 | 7.40 |
| 6 x 6 x 3/4" | " | 4.25 | 3.75 | 8.00 |

| TILE | UNIT | MAT. | INST. | TOTAL |
|---|---|---|---|---|
| **09330.10 QUARRY TILE** | | | | |
| Wall, applied to 3/4" portland cement bed | | | | |
| 4 x 4 x 1/2" | S.F. | 3.25 | 6.00 | 9.25 |
| 6 x 6 x 3/4" | " | 4.10 | 5.00 | 9.10 |
| Cove base | | | | |
| 5 x 6 x 1/2" straight top | L.F. | 3.20 | 5.00 | 8.20 |
| 6 x 6 x 3/4" round top | " | 3.35 | 5.00 | 8.35 |
| Stair treads 6 x 6 x 3/4" | " | 5.65 | 7.50 | 13.15 |
| Window sill 6 x 8 x 3/4" | " | 4.55 | 6.00 | 10.55 |
| For abrasive surface, add to material, 25% | | | | |
| **09410.10 TERRAZZO** | | | | |
| Floors on concrete, 1-3/4" thick, 5/8" topping | | | | |
| Gray cement | S.F. | 2.80 | 4.35 | 7.15 |
| White cement | " | 3.05 | 4.35 | 7.40 |
| Sand cushion, 3" thick, 5/8" top, 1/4" | | | | |
| Gray cement | S.F. | 3.30 | 5.05 | 8.35 |
| White cement | " | 3.70 | 5.05 | 8.75 |
| Monolithic terrazzo, 3-1/2" base slab, 5/8" topping | " | 2.40 | 3.80 | 6.20 |
| Terrazzo wainscot, cast-in-place, 1/2" thick | " | 4.95 | 7.60 | 12.55 |
| Base, cast in place, terrazzo cove type, 6" high | L.F. | 5.30 | 4.35 | 9.65 |
| Curb, cast in place, 6" wide x 6" high, polished top | " | 5.90 | 15.20 | 21.10 |
| For venetian type terrazzo, add to material, 10% | | | | |
| For abrasive heavy duty terrazzo, add to material, 15% | | | | |
| Divider strips | | | | |
| Zinc | L.F. | | | 1.02 |
| Brass | " | | | 1.91 |
| Stairs, cast-in-place, topping on concrete or metal | | | | |
| 1-1/2" thick treads, 12" wide | L.F. | 3.55 | 15.20 | 18.75 |
| Combined tread and riser | " | 5.30 | 38.00 | 43.30 |
| Precast terrazzo, thin set | | | | |
| Terrazzo tiles, non-slip surface | | | | |
| 9" x 9" x 1" thick | S.F. | 11.20 | 4.35 | 15.55 |
| 12" x 12" | | | | |
| 1" thick | S.F. | 12.10 | 4.05 | 16.15 |
| 1-1/2" thick | " | 12.60 | 4.35 | 16.95 |
| 18" x 18" x 1-1/2" thick | " | 16.50 | 4.35 | 20.85 |
| 24" x 24" x 1-1/2" thick | " | 21.20 | 3.60 | 24.80 |
| For white cement, add to material, 10% | | | | |
| For venetian type terrazzo, add to material, 25% | | | | |
| Terrazzo wainscot | | | | |
| 12" x 12" x 1" thick | S.F. | 12.35 | 7.60 | 19.95 |
| 18" x 18" x 1-1/2" thick | " | 19.45 | 8.70 | 28.15 |
| Base | | | | |
| 6" high | | | | |
| Straight | L.F. | 7.95 | 2.35 | 10.30 |
| Coved | " | 9.40 | 2.35 | 11.75 |
| 8" high | | | | |
| Straight | L.F. | 8.95 | 2.55 | 11.50 |
| Coved | " | 10.50 | 2.55 | 13.05 |
| Terrazzo curbs | | | | |
| 8" wide x 8" high | L.F. | 18.85 | 12.15 | 31.00 |

# 09 FINISHES

| TILE | UNIT | MAT. | INST. | TOTAL |
|---|---|---|---|---|
| **09410.10 TERRAZZO** | | | | |
| 6" wide x 6" high | L.F. | 17.05 | 10.15 | 27.20 |
| Precast terrazzo stair treads, 12" wide | | | | |
| 1-1/2" thick | | | | |
| Diamond pattern | L.F. | 20.00 | 5.55 | 25.55 |
| Non-slip surface | " | 22.40 | 5.55 | 27.95 |
| 2" thick | | | | |
| Diamond pattern | L.F. | 22.40 | 5.55 | 27.95 |
| Non-slip surface | " | 24.75 | 6.10 | 30.85 |
| Stair risers, 1" thick to 6" high | | | | |
| Straight sections | L.F. | 8.40 | 3.05 | 11.45 |
| Cove sections | " | 9.90 | 3.05 | 12.95 |
| Combined tread and riser | | | | |
| Straight sections | | | | |
| 1-1/2" tread, 3/4" riser | L.F. | 41.00 | 8.70 | 49.70 |
| 3" tread, 1" riser | " | 43.20 | 8.70 | 51.90 |
| Curved sections | | | | |
| 2" tread, 1" riser | L.F. | 46.20 | 10.15 | 56.35 |
| 3" tread, 1" riser | " | 47.70 | 10.15 | 57.85 |
| Stair stringers, notched for treads and risers | | | | |
| 1" thick | L.F. | 21.80 | 7.60 | 29.40 |
| 2" thick | " | 22.60 | 10.15 | 32.75 |
| Landings, structural, nonslip | | | | |
| 1-1/2" thick | S.F. | 17.05 | 5.05 | 22.10 |
| 3" thick | " | 18.25 | 6.10 | 24.35 |
| Conductive terrazzo, spark proof industrial floor | | | | |
| Epoxy terrazzo | | | | |
| Floor | S.F. | 7.90 | 1.90 | 9.80 |
| Base | " | 7.90 | 2.55 | 10.45 |
| Polyacrylate | | | | |
| Floor | S.F. | 6.35 | 1.90 | 8.25 |
| Base | " | 6.35 | 2.55 | 8.90 |
| Polyester | | | | |
| Floor | S.F. | 2.25 | 1.20 | 3.45 |
| Base | " | 2.25 | 1.50 | 3.75 |
| Synthetic latex mastic | | | | |
| Floor | S.F. | 3.65 | 1.90 | 5.55 |
| Base | " | 3.65 | 2.55 | 6.20 |

| ACOUSTICAL TREATMENT | UNIT | MAT. | INST. | TOTAL |
|---|---|---|---|---|
| **09510.10 CEILINGS AND WALLS** | | | | |
| Acoustical panels, suspension system not included | | | | |
| Fiberglass panels | | | | |

| ACOUSTICAL TREATMENT | UNIT | MAT. | INST. | TOTAL |
|---|---|---|---|---|
| 09510.10 CEILINGS AND WALLS | | | | |
| 5/8" thick | | | | |
| 2' x 2' | S.F. | 0.58 | 0.44 | 1.02 |
| 2' x 4' | " | 0.58 | 0.34 | 0.92 |
| 3/4" thick | | | | |
| 2' x 2' | S.F. | 0.70 | 0.44 | 1.14 |
| 2' x 4' | " | 0.70 | 0.34 | 1.04 |
| Glass cloth faced fiberglass panels | | | | |
| 3/4" thick | S.F. | 1.60 | 0.52 | 2.12 |
| 1" thick | " | 1.75 | 0.52 | 2.27 |
| Mineral fiber panels | | | | |
| 5/8" thick | | | | |
| 2' x 2' | S.F. | 0.70 | 0.44 | 1.14 |
| 2' x 4' | " | 0.70 | 0.34 | 1.04 |
| 3/4" thick | | | | |
| 2' x 2' | S.F. | 0.93 | 0.44 | 1.37 |
| 2' x 4' | " | 0.93 | 0.34 | 1.27 |
| For aluminum faced panels, add to material, 80% | | | | |
| For vinyl faced panels, add to total, 125% | | | | |
| For fire rated panels, add to material, 75% | | | | |
| Wood fiber panels | | | | |
| 2" thick | | | | |
| 2' x 2' | S.F. | 1.65 | 0.52 | 2.17 |
| 2' x 4' | " | 1.65 | 0.39 | 2.04 |
| For flameproofing, add to material, 10% | | | | |
| For sculptured finish, add to material, 15% | | | | |
| Air distributing panels | | | | |
| 3/4" thick | S.F. | 1.65 | 0.77 | 2.42 |
| 5/8" thick | " | 1.40 | 0.62 | 2.02 |
| Acoustical tiles, suspension system not included | | | | |
| Fiberglass tile, 12" x 12" | | | | |
| 5/8" thick | S.F. | 0.41 | 0.56 | 0.97 |
| 3/4" thick | " | 0.44 | 0.69 | 1.13 |
| Glass cloth faced fiberglass tile | | | | |
| 3/4" thick | S.F. | 1.60 | 0.69 | 2.29 |
| 3" thick | " | 1.60 | 0.77 | 2.37 |
| Mineral fiber tile, 12" x 12" | | | | |
| 5/8" thick | | | | |
| Standard | S.F. | 0.97 | 0.62 | 1.59 |
| Vinyl faced | " | 1.10 | 0.62 | 1.72 |
| 3/4" thick | | | | |
| Standard | S.F. | 1.20 | 0.62 | 1.82 |
| Vinyl faced | " | 1.40 | 0.62 | 2.02 |
| Fire rated | " | 2.45 | 0.62 | 3.07 |
| Aluminum or mylar faced | " | 2.80 | 0.62 | 3.42 |
| Wood fiber tile, 12" x 12" | | | | |
| 1/2" thick | S.F. | 0.79 | 0.62 | 1.41 |
| 3/4" thick | " | 1.10 | 0.62 | 1.72 |
| For flameproofing, add to material, 10% | | | | |
| For sculptured 3 dimensional, add to material, 50% | | | | |
| Metal pan units, 24 ga. steel | | | | |
| 12" x 12" | S.F. | 5.45 | 1.25 | 6.70 |
| 12" x 24" | " | 5.25 | 1.05 | 6.30 |

| ACOUSTICAL TREATMENT | UNIT | MAT. | INST. | TOTAL |
|---|---|---|---|---|
| **09510.10 CEILINGS AND WALLS** | | | | |
| Aluminum, .025" thick | | | | |
| 12" x 12" | S.F. | 3.55 | 1.25 | 4.80 |
| 12" x 24" | " | 3.65 | 1.05 | 4.70 |
| Anodized aluminum, 0.25" thick | | | | |
| 12" x 12" | S.F. | 4.10 | 1.25 | 5.35 |
| 12" x 24" | " | 4.85 | 1.05 | 5.90 |
| Stainless steel, 24 ga. | | | | |
| 12" x 12" | S.F. | 7.80 | 1.25 | 9.05 |
| 12" x 24" | " | 8.05 | 1.05 | 9.10 |
| For flameproof sound absorbing pads, add to material | " | | | 1.21 |
| Metal ceiling systems | | | | |
| .020" thick panels | | | | |
| 10', 12', and 16' lengths | S.F. | 3.00 | 0.88 | 3.88 |
| Custom lengths, 3' to 20' | " | 3.05 | 0.88 | 3.93 |
| .025" thick panels | | | | |
| 32 sf, 38 sf, and 52 sf pieces | S.F. | 3.45 | 1.05 | 4.50 |
| Custom lengths, 10 sf to 65 sf | " | 3.55 | 1.05 | 4.60 |
| Carriers, black, add | " | | | 1.82 |
| Recess filler strip, add | " | | | 0.61 |
| Custom lengths, add | " | | | 0.91 |
| Sound absorption walls, with fabric cover | | | | |
| 2-6" x 9' x 3/4" | S.F. | 4.10 | 1.05 | 5.15 |
| 2' x 9' x 1" | " | 4.45 | 1.05 | 5.50 |
| Starter spline | L.F. | 0.98 | 0.77 | 1.75 |
| Internal spline | " | 0.85 | 0.77 | 1.62 |
| Acoustical treatment | | | | |
| Barriers for plenums | | | | |
| Leaded vinyl | | | | |
| 0.48 lb per sf | S.F. | 1.75 | 1.45 | 3.20 |
| 0.87 lb per sf | " | 1.75 | 1.55 | 3.30 |
| Aluminum foil, fiberglass reinforcement | | | | |
| Minimum | S.F. | 0.69 | 1.05 | 1.74 |
| Maximum | " | 0.69 | 1.55 | 2.24 |
| Aluminum mesh, paper backed | " | 0.62 | 1.05 | 1.67 |
| Fibered cement sheet, 3/16" thick | " | 1.30 | 1.10 | 2.40 |
| Sheet lead, 1/64" thick | " | 1.65 | 0.77 | 2.42 |
| Sound attenuation blanket | | | | |
| 1" thick | S.F. | 0.38 | 3.10 | 3.48 |
| 1-1/2" thick | " | 0.49 | 3.10 | 3.59 |
| 2" thick | " | 0.67 | 3.10 | 3.77 |
| 3" thick | " | 0.79 | 3.45 | 4.24 |
| Ceiling suspension systems | | | | |
| T bar system | | | | |
| 2' x 4' | S.F. | 0.70 | 0.31 | 1.01 |
| 2' x 2' | " | 0.75 | 0.34 | 1.09 |
| Concealed Z bar suspension system, 12" module | " | 0.65 | 0.52 | 1.17 |
| For 1-1/2" carrier channels, 4' o.c., add | " | | | 0.23 |
| Carrier channel for recessed light fixtures | " | | | 0.41 |

| FLOORING | UNIT | MAT. | INST. | TOTAL |
|---|---|---|---|---|
| **09550.10 WOOD FLOORING** | | | | |
| Wood strip flooring, unfinished | | | | |
| Fir floor | | | | |
| C and better | | | | |
| Vertical grain | S.F. | 2.25 | 1.05 | 3.30 |
| Flat grain | " | 2.15 | 1.05 | 3.20 |
| Oak floor | | | | |
| Minimum | S.F. | 2.55 | 1.45 | 4.00 |
| Average | " | 3.05 | 1.45 | 4.50 |
| Maximum | " | 3.80 | 1.45 | 5.25 |
| Maple floor | | | | |
| 25/32" x 2-1/4" | | | | |
| Minimum | S.F. | 2.80 | 1.45 | 4.25 |
| Maximum | " | 3.25 | 1.45 | 4.70 |
| 33/32" x 3-1/4" | | | | |
| Minimum | S.F. | 3.55 | 1.45 | 5.00 |
| Maximum | " | 4.05 | 1.45 | 5.50 |
| Added costs | | | | |
| For factory finish, add to material, 10% | | | | |
| For random width floor, add to total, 20% | | | | |
| For simulated pegs, add to total, 10% | | | | |
| Wood block industrial flooring | | | | |
| Creosoted | | | | |
| 2" thick | S.F. | 2.65 | 0.81 | 3.46 |
| 2-1/2" thick | " | 2.80 | 0.96 | 3.77 |
| 3" thick | " | 2.90 | 1.05 | 3.95 |
| Parquet, 5/16", white oak | | | | |
| Finished | S.F. | 5.85 | 1.55 | 7.40 |
| Unfinished | " | 2.10 | 1.55 | 3.65 |
| Gym floor, 2 ply felt, 25/32" maple, finished, in mastic | " | 4.95 | 1.70 | 6.65 |
| Over wood sleepers | " | 5.55 | 1.95 | 7.50 |
| Finishing, sand, fill, finish, and wax | " | 0.47 | 0.77 | 1.24 |
| Refinish sand, seal, and 2 coats of polyurethane | " | 0.78 | 1.05 | 1.83 |
| Clean and wax floors | " | 0.13 | 0.15 | 0.28 |
| **09630.10 UNIT MASONRY FLOORING** | | | | |
| Clay brick | | | | |
| 9 x 4-1/2 x 3" thick | | | | |
| Glazed | S.F. | 5.55 | 2.55 | 8.10 |
| Unglazed | " | 5.30 | 2.55 | 7.85 |
| 8 x 4 x 3/4" thick | | | | |
| Glazed | S.F. | 5.00 | 2.70 | 7.70 |
| Unglazed | " | 4.90 | 2.70 | 7.60 |
| For herringbone pattern, add to labor, 15% | | | | |
| **09660.10 RESILIENT TILE FLOORING** | | | | |
| Solid vinyl tile, 1/8" thick, 12" x 12" | | | | |
| Marble patterns | S.F. | 1.65 | 0.77 | 2.42 |
| Solid colors | " | 2.15 | 0.77 | 2.92 |
| Travertine patterns | " | 2.40 | 0.77 | 3.17 |
| Conductive resilient flooring, vinyl tile | | | | |
| 1/8" thick, 12" x 12" | S.F. | 3.75 | 0.88 | 4.63 |

| FLOORING | UNIT | MAT. | INST. | TOTAL |
|---|---|---|---|---|
| **09665.10 RESILIENT SHEET FLOORING** | | | | |
| Vinyl sheet flooring | | | | |
| Minimum | S.F. | 1.45 | 0.31 | 1.76 |
| Average | " | 2.10 | 0.38 | 2.48 |
| Maximum | " | 4.40 | 0.52 | 4.92 |
| Cove, to 6" | L.F. | 0.61 | 0.62 | 1.23 |
| Fluid applied resilient flooring | | | | |
| Polyurethane, poured in place, 3/8" thick | S.F. | 5.95 | 2.55 | 8.50 |
| Vinyl sheet goods, backed | | | | |
| 0.070" thick | S.F. | 2.20 | 0.39 | 2.59 |
| 0.093" thick | " | 3.40 | 0.39 | 3.79 |
| 0.125" thick | " | 3.95 | 0.39 | 4.34 |
| 0.250" thick | " | 4.50 | 0.39 | 4.89 |
| **09678.10 RESILIENT BASE AND ACCESSORIES** | | | | |
| Wall base, vinyl | | | | |
| Group 1 | | | | |
| 4" high | L.F. | 0.58 | 1.05 | 1.63 |
| 6" high | " | 0.80 | 1.05 | 1.85 |
| Group 2 | | | | |
| 4" high | L.F. | 0.51 | 1.05 | 1.56 |
| 6" high | " | 0.60 | 1.05 | 1.65 |
| Group 3 | | | | |
| 4" high | L.F. | 0.56 | 1.05 | 1.61 |
| 6" high | " | 0.67 | 1.05 | 1.72 |
| Stair accessories | | | | |
| Treads, 1/4" x 12", rubber diamond surface | | | | |
| Marbled | L.F. | 4.60 | 2.55 | 7.15 |
| Plain | " | 4.50 | 2.55 | 7.05 |
| Grit strip safety tread, 12" wide, colors | | | | |
| 3/16" thick | L.F. | 6.35 | 2.55 | 8.90 |
| 5/16" thick | " | 7.95 | 2.55 | 10.50 |
| Risers, 7" high, 1/8" thick, colors | | | | |
| Flat | L.F. | 1.60 | 1.55 | 3.15 |
| Coved | " | 1.85 | 1.55 | 3.40 |
| Nosing, rubber | | | | |
| 3/16" thick, 3" wide | | | | |
| Black | L.F. | 1.85 | 1.55 | 3.40 |
| Colors | " | 1.95 | 1.55 | 3.50 |
| 6" wide | | | | |
| Black | L.F. | 2.95 | 2.55 | 5.50 |
| Colors | " | 3.25 | 2.55 | 5.80 |

| CARPET | UNIT | MAT. | INST. | TOTAL |
|---|---|---|---|---|
| **09680.10 FLOOR LEVELING** | | | | |
| Repair and level floors to receive new flooring | | | | |
| Minimum | S.Y. | 0.52 | 1.05 | 1.57 |
| Average | " | 2.05 | 2.55 | 4.60 |
| Maximum | " | 2.65 | 3.10 | 5.75 |
| **09682.10 CARPET PADDING** | | | | |
| Carpet padding | | | | |
| Foam rubber, waffle type, 0.3" thick | S.Y. | 3.90 | 1.55 | 5.45 |
| Jute padding | | | | |
| Minimum | S.Y. | 2.85 | 1.40 | 4.25 |
| Average | " | 3.40 | 1.55 | 4.95 |
| Maximum | " | 4.50 | 1.70 | 6.20 |
| Sponge rubber cushion | | | | |
| Minimum | S.Y. | 3.55 | 1.40 | 4.95 |
| Average | " | 3.75 | 1.55 | 5.30 |
| Maximum | " | 4.35 | 1.70 | 6.05 |
| Urethane cushion, 3/8" thick | | | | |
| Minimum | S.Y. | 2.25 | 1.40 | 3.65 |
| Average | " | 2.60 | 1.55 | 4.15 |
| Maximum | " | 3.05 | 1.70 | 4.75 |
| **09685.10 CARPET** | | | | |
| Carpet, acrylic | | | | |
| 24 oz., light traffic | S.Y. | 17.85 | 1.70 | 19.55 |
| 28 oz., medium traffic | " | 22.60 | 1.70 | 24.30 |
| Residential | | | | |
| Nylon | | | | |
| 15 oz., light traffic | S.Y. | 12.85 | 1.70 | 14.55 |
| 28 oz., medium traffic | " | 15.40 | 1.70 | 17.10 |
| Commercial | | | | |
| Nylon | | | | |
| 28 oz., medium traffic | S.Y. | 16.40 | 1.70 | 18.10 |
| 35 oz., heavy traffic | " | 19.60 | 1.70 | 21.30 |
| Wool | | | | |
| 30 oz., medium traffic | S.Y. | 26.80 | 1.70 | 28.50 |
| 36 oz., medium traffic | " | 28.20 | 1.70 | 29.90 |
| 42 oz., heavy traffic | " | 37.40 | 1.70 | 39.10 |
| Carpet tile | | | | |
| Foam backed | | | | |
| Minimum | S.F. | 2.05 | 0.31 | 2.36 |
| Average | " | 2.40 | 0.34 | 2.74 |
| Maximum | " | 3.75 | 0.39 | 4.14 |
| Tufted loop or shag | | | | |
| Minimum | S.F. | 2.20 | 0.31 | 2.51 |
| Average | " | 2.65 | 0.34 | 2.99 |
| Maximum | " | 4.30 | 0.39 | 4.69 |
| Clean and vacuum carpet | | | | |
| Minimum | S.Y. | 0.18 | 0.12 | 0.30 |
| Average | " | 0.30 | 0.21 | 0.51 |
| Maximum | " | 0.40 | 0.31 | 0.71 |

# 09 FINISHES

| CARPET | UNIT | MAT. | INST. | TOTAL |
|---|---|---|---|---|
| **09700.10 SPECIAL FLOORING** | | | | |
| Epoxy flooring, marble chips | | | | |
| Epoxy with colored quartz chips in 1/4" base | S.F. | 7.00 | 1.70 | 8.70 |
| Heavy duty epoxy topping, 3/16" thick | " | 6.15 | 1.70 | 7.85 |
| Epoxy terrazzo | | | | |
| 1/4" thick chemical resistant | S.F. | 7.15 | 1.95 | 9.10 |

| PAINTING | UNIT | MAT. | INST. | TOTAL |
|---|---|---|---|---|
| **09910.10 EXTERIOR PAINTING** | | | | |
| Exterior painting | | | | |
| Wood surfaces, 1 coat primer, two coats paint | | | | |
| Door and frame | EA. | 3.55 | 52.00 | 55.55 |
| Windows | S.F. | 0.08 | 0.78 | 0.86 |
| Wood trim | " | 0.18 | 0.78 | 0.96 |
| Wood siding | " | 0.18 | 0.39 | 0.57 |
| Hardboard surfaces | | | | |
| One coat primer, two coats paint | S.F. | 0.18 | 0.39 | 0.57 |
| Asbestos cement surfaces | | | | |
| One coat primer, two coats paint | S.F. | 0.20 | 0.39 | 0.59 |
| Galvanized surfaces, galvanized primer | | | | |
| One coat primer, two coats paint | S.F. | 0.19 | 0.37 | 0.56 |
| Stucco surfaces, acrylic primer, acrylic latex paint | | | | |
| One coat primer, two coats paint | S.F. | 0.19 | 0.52 | 0.71 |
| Concrete masonry unit surfaces, brush work | | | | |
| One coat filler, one coat paint | S.F. | 0.21 | 0.39 | 0.60 |
| Two coats epoxy | " | 0.31 | 0.52 | 0.83 |
| Texture coating | " | 0.29 | 0.31 | 0.60 |
| Concrete surfaces | | | | |
| One coat filler, one coat paint | S.F. | 0.18 | 0.39 | 0.57 |
| Two coats paint | " | 0.24 | 0.52 | 0.76 |
| Structural steel | | | | |
| One field coat paint, brush work | | | | |
| Light framing | S.F. | 0.10 | 0.26 | 0.36 |
| Heavy framing | " | 0.10 | 0.16 | 0.26 |
| One field coat paint, spray work | | | | |
| Light framing | S.F. | 0.08 | 0.08 | 0.16 |
| Heavy framing | " | 0.08 | 0.06 | 0.14 |
| Miscellaneous steel items, spray work, one coat | | | | |
| Exposed decking | S.F. | 0.10 | 0.26 | 0.36 |
| Joist | " | 0.10 | 0.39 | 0.49 |
| Columns | " | 0.10 | 0.39 | 0.49 |
| Pipes, one coat primer, one coat paint | | | | |
| 4" dia. | L.F. | 0.08 | 0.39 | 0.47 |

## PAINTING

| PAINTING | UNIT | MAT. | INST. | TOTAL |
|---|---|---|---|---|
| **09910.10 EXTERIOR PAINTING** | | | | |
| 8" dia. | L.F. | 0.18 | 0.52 | 0.70 |
| 12" dia. | " | 0.32 | 0.78 | 1.10 |
| Paint letters on pipe with brush | " | 0.36 | 0.78 | 1.14 |
| Paint pipe insulation cloth cover | S.F. | 0.21 | 0.39 | 0.60 |
| Sprinkler system piping | L.F. | 0.22 | 0.26 | 0.48 |
| Miscellaneous surfaces | | | | |
| Stair pipe rails | | | | |
| Two rails | L.F. | 0.16 | 1.05 | 1.21 |
| One rail | " | 0.09 | 0.62 | 0.71 |
| Stair to 4' wide, including rails, per riser | EA. | 0.59 | 4.45 | 5.04 |
| Gratings and frames | S.F. | 0.16 | 1.05 | 1.21 |
| Ladders | L.F. | 0.18 | 0.89 | 1.07 |
| Miscellaneous exposed metal | S.F. | 0.16 | 0.31 | 0.47 |
| **09920.10 INTERIOR PAINTING** | | | | |
| Walls, concrete and masonry, brush, primer, acrylic | | | | |
| One coat primer, one coat paint | S.F. | 0.10 | 0.39 | 0.49 |
| Two coats paint | " | 0.16 | 0.52 | 0.68 |
| Plywood, paint | " | 0.06 | 0.17 | 0.23 |
| Natural finish | " | 0.07 | 0.20 | 0.27 |
| Wood, paint | " | 0.07 | 0.20 | 0.27 |
| Natural finish | " | 0.09 | 0.22 | 0.31 |
| Metal | | | | |
| One coat filler | S.F. | 0.16 | 0.20 | 0.35 |
| One coat primer, one coat paint | " | 0.18 | 0.39 | 0.57 |
| Two coats paint | " | 0.22 | 0.52 | 0.74 |
| Plaster or gypsum board, paint | " | 0.09 | 0.17 | 0.26 |
| Epoxy | " | 0.11 | 0.20 | 0.30 |
| Ceilings, one coat paint, wood | " | 0.08 | 0.22 | 0.30 |
| Concrete | " | 0.11 | 0.20 | 0.30 |
| Plaster | " | 0.08 | 0.17 | 0.25 |
| Miscellaneous metal, brushwork | " | 0.16 | 0.39 | 0.55 |
| **09920.30 DOORS AND MILLWORK** | | | | |
| Painting, doors | | | | |
| Minimum | S.F. | 0.12 | 1.05 | 1.17 |
| Average | " | 0.22 | 1.55 | 1.77 |
| Maximum | " | 0.37 | 2.05 | 2.42 |
| Cabinets, shelves, and millwork | | | | |
| Minimum | S.F. | 0.09 | 0.52 | 0.61 |
| Average | " | 0.18 | 0.89 | 1.07 |
| Maximum | " | 0.28 | 1.55 | 1.83 |
| **09920.60 WINDOWS** | | | | |
| Painting, windows | | | | |
| Minimum | S.F. | 0.06 | 0.62 | 0.68 |
| Average | " | 0.09 | 0.78 | 0.87 |
| Maximum | " | 0.13 | 1.25 | 1.38 |

| PAINTING | UNIT | MAT. | INST. | TOTAL |
|---|---|---|---|---|
| **09955.10 WALL COVERING** | | | | |
| Vinyl wall covering | | | | |
| Medium duty | S.F. | 0.91 | 0.45 | 1.36 |
| Heavy duty | " | 1.20 | 0.52 | 1.72 |
| Over pipes and irregular shapes | | | | |
| Lightweight, 13 oz. | S.F. | 0.91 | 0.62 | 1.53 |
| Medium weight, 25 oz. | " | 1.05 | 0.69 | 1.74 |
| Heavy weight, 34 oz. | " | 1.20 | 0.78 | 1.98 |
| Cork wall covering | | | | |
| 1' x 1' squares | | | | |
| 1/4" thick | S.F. | 1.30 | 0.78 | 2.08 |
| 1/2" thick | " | 1.65 | 0.78 | 2.43 |
| 3/4" thick | " | 1.90 | 0.78 | 2.68 |
| Wall fabrics | | | | |
| Natural fabrics, grass cloths | | | | |
| Minimum | S.F. | 0.75 | 0.48 | 1.23 |
| Average | " | 0.94 | 0.52 | 1.46 |
| Maximum | " | 2.95 | 0.62 | 3.57 |
| Flexible gypsum coated wall fabric, fire resistant | " | 0.94 | 0.31 | 1.25 |
| Vinyl corner guards | | | | |
| 3/4" x 3/4" x 8' | EA. | 4.45 | 3.90 | 8.35 |
| 2-3/4" x 2-3/4" x 4' | " | 2.65 | 3.90 | 6.55 |
| **09980.10 PAINTING PREPARATION** | | | | |
| Cleaning, light | | | | |
| Wood | S.F. | 0.03 | 0.08 | 0.11 |
| Plaster or gypsum wallboard | " | 0.03 | 0.07 | 0.10 |
| Prepare sprinkler piping for painting | L.F. | 0.03 | 0.06 | 0.09 |
| Prepare insulated pipe for painting | S.F. | 0.03 | 0.08 | 0.11 |
| Normal painting prep, masonry and concrete | | | | |
| Unpainted | S.F. | 0.03 | 0.05 | 0.08 |
| Painted | " | 0.03 | 0.08 | 0.11 |
| Plaster or gypsum | | | | |
| Unpainted | S.F. | 0.03 | 0.05 | 0.08 |
| Painted | " | 0.03 | 0.08 | 0.11 |
| Wood | | | | |
| Unpainted | S.F. | 0.03 | 0.05 | 0.08 |
| Painted | " | 0.03 | 0.08 | 0.11 |
| Painted steel, light rusting | " | 0.04 | 0.10 | 0.14 |
| Sandblasting | | | | |
| Brush off blast | S.F. | 0.12 | 0.21 | 0.33 |
| Commercial blast | " | 0.24 | 0.52 | 0.76 |
| Near white metal blast | " | 0.36 | 0.89 | 1.25 |
| White metal blast | " | 0.48 | 1.05 | 1.53 |
| **09980.15 PAINT** | | | | |
| Paint, enamel | | | | |
| 600 sf per gal. | GAL | | | 27.08 |
| 550 sf per gal. | " | | | 21.19 |
| 500 sf per gal. | " | | | 15.90 |
| 450 sf per gal. | " | | | 14.72 |
| 350 sf per gal. | " | | | 14.13 |

## 09 FINISHES

| PAINTING | UNIT | MAT. | INST. | TOTAL |
|---|---|---|---|---|
| **09980.15 PAINT** | | | | |
| Filler, 60 sf per gal. | GAL | | | 17.07 |
| Latex, 400 sf per gal. | " | | | 13.54 |
| Aluminum | | | | |
| 400 sf per gal. | GAL | | | 20.01 |
| 500 sf per gal. | " | | | 38.86 |
| Red lead, 350 sf per gal. | " | | | 32.97 |
| Primer | | | | |
| 400 sf per gal. | GAL | | | 20.72 |
| 300 sf per gal. | " | | | 20.59 |
| Latex base, interior, white | " | | | 17.23 |
| Sealer and varnish | | | | |
| 400 sf per gal. | GAL | | | 15.92 |
| 425 sf per gal. | " | | | 23.29 |
| 600 sf per gal. | " | | | 31.20 |

# 10 SPECIALTIES

| SPECIALTIES | UNIT | MAT. | INST. | TOTAL |
|---|---|---|---|---|
| **10110.10 CHALKBOARDS** | | | | |
| Chalkboard, metal frame, 1/4" thick | | | | |
| 48"x60" | EA. | 190.00 | 30.90 | 220.90 |
| 48"x96" | " | 310.00 | 34.30 | 344.30 |
| 48"x144" | " | 470.00 | 38.60 | 508.60 |
| 48"x192" | " | 610.00 | 44.10 | 654.10 |
| Liquid chalkboard | | | | |
| 48"x60" | EA. | 220.00 | 30.90 | 250.90 |
| 48"x96" | " | 340.00 | 34.30 | 374.30 |
| 48"x144" | " | 530.00 | 38.60 | 568.60 |
| 48"x192" | " | 670.00 | 44.10 | 714.10 |
| Map rail, deluxe | L.F. | 5.85 | 1.55 | 7.40 |
| Average | PCT. | | | 2.95 |
| **10165.10 TOILET PARTITIONS** | | | | |
| Toilet partition, plastic laminate | | | | |
| Ceiling mounted | EA. | 640.00 | 100.00 | 740.00 |
| Floor mounted | " | 580.00 | 77.00 | 657.00 |
| Metal | | | | |
| Ceiling mounted | EA. | 470.00 | 100.00 | 570.00 |
| Floor mounted | " | 440.00 | 77.00 | 517.00 |
| Wheel chair partition, plastic laminate | | | | |
| Ceiling mounted | EA. | 810.00 | 100.00 | 910.00 |
| Floor mounted | " | 760.00 | 77.00 | 837.00 |
| Painted metal | | | | |
| Ceiling mounted | EA. | 640.00 | 100.00 | 740.00 |
| Floor mounted | " | 580.00 | 77.00 | 657.00 |
| Urinal screen, plastic laminate | | | | |
| Wall hung | EA. | 300.00 | 38.60 | 338.60 |
| Floor mounted | " | 260.00 | 38.60 | 298.60 |
| Porcelain enameled steel, floor mounted | " | 340.00 | 38.60 | 378.60 |
| Painted metal, floor mounted | " | 230.00 | 38.60 | 268.60 |
| Stainless steel, floor mounted | " | 430.00 | 38.60 | 468.60 |
| Metal toilet partitions | | | | |
| Toilet partitions, front door and side divider, floor mounted | | | | |
| Porcelain enameled steel | EA. | 720.00 | 77.00 | 797.00 |
| Painted steel | " | 430.00 | 77.00 | 507.00 |
| Stainless steel | " | 820.00 | 77.00 | 897.00 |
| **10185.10 SHOWER STALLS** | | | | |
| Shower receptors | | | | |
| Precast, terrazzo | | | | |
| 32" x 32" | EA. | 240.00 | 28.50 | 268.50 |
| 32" x 48" | " | 310.00 | 34.20 | 344.20 |
| Concrete | | | | |
| 32" x 32" | EA. | 140.00 | 28.50 | 168.50 |
| 48" x 48" | " | 180.00 | 38.00 | 218.00 |
| Shower door, trim and hardware | | | | |
| Economy, 24" wide, chrome frame, tempered glass | EA. | 170.00 | 34.20 | 204.20 |
| Porcelain enameled steel, flush | " | 310.00 | 34.20 | 344.20 |
| Baked enameled steel, flush | " | 180.00 | 34.20 | 214.20 |
| Aluminum frame, tempered glass, 48" wide, sliding | " | 400.00 | 42.70 | 442.70 |

| SPECIALTIES | UNIT | MAT. | INST. | TOTAL |
|---|---|---|---|---|
| **10185.10 SHOWER STALLS** | | | | |
| Folding | EA. | 370.00 | 42.70 | 412.70 |
| Aluminum frame and tempered glass, molded plastic | | | | |
| Complete with receptor and door | | | | |
| 32" x 32" | EA. | 500.00 | 85.50 | 585.50 |
| 36" x 36" | " | 560.00 | 85.50 | 645.50 |
| 40" x 40" | " | 630.00 | 97.50 | 727.50 |
| Shower compartment, precast concrete receptor | | | | |
| Single entry type | | | | |
| Porcelain enameled steel | EA. | 1,520 | 340.00 | 1,860 |
| Baked enameled steel | " | 870.00 | 340.00 | 1,210 |
| Stainless steel | " | 1,750 | 340.00 | 2,090 |
| Double entry type | | | | |
| Porcelain enameled steel | EA. | 4,310 | 430.00 | 4,740 |
| Baked enameled steel | " | 2,800 | 430.00 | 3,230 |
| Stainless steel | " | 4,370 | 430.00 | 4,800 |
| **10190.10 CUBICLES** | | | | |
| Hospital track | | | | |
| Ceiling hung | L.F. | 3.75 | 3.45 | 7.20 |
| Suspended | " | 4.20 | 4.40 | 8.60 |
| Hospital metal dividers, galvanized steel | | | | |
| Baked enamel finish | | | | |
| 54" high | | | | |
| 10" glass light | L.F. | 81.50 | 15.45 | 96.95 |
| 14" glass light | " | 81.50 | 15.45 | 96.95 |
| 24" glass light | " | 77.50 | 15.45 | 92.95 |
| 60" high | | | | |
| 10" glass light | L.F. | 86.50 | 17.15 | 103.65 |
| 14" glass light | " | 86.50 | 17.15 | 103.65 |
| 24" glass light | " | 81.00 | 17.15 | 98.15 |
| Stainless steel | | | | |
| 54" high | | | | |
| 10" glass light | L.F. | 140.00 | 17.15 | 157.15 |
| 14" glass light | " | 150.00 | 17.15 | 167.15 |
| 24" glass light | " | 180.00 | 17.15 | 197.15 |
| 60" high | | | | |
| 14" glass light | L.F. | 160.00 | 19.30 | 179.30 |
| 14" glass light | " | 160.00 | 19.30 | 179.30 |
| 24" glass light | " | 200.00 | 19.30 | 219.30 |
| **10210.10 VENTS AND WALL LOUVERS** | | | | |
| Block vent, 8"x16"x4" aluminum, w/screen, mill finish | EA. | 92.00 | 11.25 | 103.25 |
| Standard | " | 50.00 | 10.55 | 60.55 |
| Vents w/screen, 4" deep, 8" wide, 5" high | | | | |
| Modular | EA. | 59.00 | 10.55 | 69.55 |
| Aluminum gable louvers | S.F. | 10.50 | 5.65 | 16.15 |
| Vent screen aluminum, 4" wide, continuous | L.F. | 2.90 | 1.15 | 4.05 |
| Louvers, aluminum, anodized, fixed blade | | | | |
| Horizontal line | S.F. | 33.80 | 8.45 | 42.25 |
| Vertical line | " | 33.80 | 8.45 | 42.25 |
| Wall louver, aluminum mill finish | | | | |

| SPECIALTIES | UNIT | MAT. | INST. | TOTAL |
|---|---|---|---|---|
| **10210.10 VENTS AND WALL LOUVERS** | | | | |
| Under, 2 sf | S.F. | 22.15 | 4.20 | 26.35 |
| 2 to 4 sf | " | 18.65 | 3.75 | 22.40 |
| 5 to 10 sf | " | 18.05 | 3.75 | 21.80 |
| Galvanized steel | | | | |
| Under 2 sf | S.F. | 21.00 | 4.20 | 25.20 |
| 2 to 4 sf | " | 15.15 | 3.75 | 18.90 |
| 5 to 10 sf | " | 14.00 | 3.75 | 17.75 |
| **10225.10 DOOR LOUVERS** | | | | |
| Fixed, 1" thick, enameled steel | | | | |
| 8"x8" | EA. | 33.80 | 3.85 | 37.65 |
| 12"x8" | " | 38.50 | 3.85 | 42.35 |
| 12"x12" | " | 43.10 | 4.40 | 47.50 |
| 16"x12" | " | 62.00 | 4.75 | 66.75 |
| 18"x12" | " | 62.00 | 7.70 | 69.70 |
| 20"x8" | " | 69.00 | 4.40 | 73.40 |
| 20"x12" | " | 78.00 | 8.80 | 86.80 |
| 20"x16" | " | 80.50 | 10.30 | 90.80 |
| 20"x20" | " | 83.00 | 12.35 | 95.35 |
| 24"x12" | " | 69.50 | 10.30 | 79.80 |
| 24"x16" | " | 73.00 | 11.05 | 84.05 |
| 24"x18" | " | 85.00 | 11.90 | 96.90 |
| 24"x20" | " | 88.50 | 12.85 | 101.35 |
| 24"x24" | " | 88.50 | 14.05 | 102.55 |
| 26"x26" | " | 110.00 | 19.30 | 129.30 |
| **10270.40 ACCESS & PEDESTAL FLOOR** | | | | |
| Panels, no covering, 2'x2' | | | | |
| Plain | S.F. | 6.40 | 0.39 | 6.79 |
| Perforated | " | 8.75 | 15.45 | 24.20 |
| Pedestals | | | | |
| For 6" to 12" clearance | EA. | 4.90 | 3.10 | 8.00 |
| Stringers | | | | |
| 2' | L.F. | 0.83 | 1.45 | 2.28 |
| 6' | " | 0.83 | 1.05 | 1.88 |
| Accessories | | | | |
| Ramp assembly | S.F. | 30.30 | 1.25 | 31.55 |
| Elevated floor assembly | " | 47.20 | 1.15 | 48.35 |
| Handrail | L.F. | 36.70 | 15.45 | 52.15 |
| Fascia plate | " | 17.50 | 7.70 | 25.20 |
| For carpet tiles, add | S.F. | | | 4.09 |
| For vinyl flooring, add | " | | | 5.25 |
| RF shielding components, floor liner | | | | |
| Hot rolled steel sheet | | | | |
| 14 ga. | S.F. | 2.00 | 0.77 | 2.77 |
| 11 ga. | " | 2.10 | 2.40 | 4.50 |

# 10 SPECIALTIES

| SPECIALTIES | UNIT | MAT. | INST. | TOTAL |
|---|---|---|---|---|
| **10290.10 PEST CONTROL** | | | | |
| Termite control | | | | |
| Under slab spraying | | | | |
| Minimum | S.F. | 0.12 | 0.06 | 0.18 |
| Average | " | 0.16 | 0.12 | 0.28 |
| Maximum | " | 0.21 | 0.24 | 0.45 |
| **10350.10 FLAGPOLES** | | | | |
| Installed in concrete base | | | | |
| Fiberglass | | | | |
| 25' high | EA. | 820.00 | 210.00 | 1,030 |
| 50' high | " | 3,930 | 510.00 | 4,440 |
| Aluminum | | | | |
| 25' high | EA. | 990.00 | 210.00 | 1,200 |
| 50' high | " | 3,620 | 510.00 | 4,130 |
| Bonderized steel | | | | |
| 25' high | EA. | 1,520 | 240.00 | 1,760 |
| 50' high | " | 3,850 | 620.00 | 4,470 |
| Freestanding tapered, fiberglass | | | | |
| 30' high | EA. | 930.00 | 220.00 | 1,150 |
| 40' high | " | 1,720 | 280.00 | 2,000 |
| 50' high | " | 3,380 | 310.00 | 3,690 |
| 60' high | " | 4,550 | 360.00 | 4,910 |
| Wall mounted, with collar, brushed aluminum finish | | | | |
| 15' long | EA. | 580.00 | 150.00 | 730.00 |
| 18' long | " | 700.00 | 150.00 | 850.00 |
| 20' long | " | 760.00 | 160.00 | 920.00 |
| 24' long | " | 820.00 | 180.00 | 1,000 |
| Outrigger, wall, including base | | | | |
| 10' long | EA. | 170.00 | 210.00 | 380.00 |
| 20' long | " | 220.00 | 260.00 | 480.00 |
| **10400.10 IDENTIFYING DEVICES** | | | | |
| Directory and bulletin boards | | | | |
| Open face boards | | | | |
| Chrome plated steel frame | S.F. | 19.25 | 15.45 | 34.70 |
| Aluminum framed | " | 16.90 | 15.45 | 32.35 |
| Bronze framed | " | 18.65 | 15.45 | 34.10 |
| Stainless steel framed | " | 36.20 | 15.45 | 51.65 |
| Tack board, aluminum framed | " | 10.85 | 15.45 | 26.30 |
| Visual aid board, aluminum framed | " | 10.85 | 15.45 | 26.30 |
| Glass encased boards, hinged and keyed | | | | |
| Aluminum framed | S.F. | 42.00 | 38.60 | 80.60 |
| Bronze framed | " | 45.50 | 38.60 | 84.10 |
| Stainless steel framed | " | 110.00 | 38.60 | 148.60 |
| Chrome plated steel framed | " | 100.00 | 38.60 | 138.60 |
| Metal plaque | | | | |
| Cast bronze | S.F. | 270.00 | 25.70 | 295.70 |
| Aluminum | " | 260.00 | 25.70 | 285.70 |
| Metal engraved plaque | | | | |
| Porcelain steel | S.F. | 340.00 | 25.70 | 365.70 |

| SPECIALTIES | UNIT | MAT. | INST. | TOTAL |
|---|---|---|---|---|
| **10400.10 IDENTIFYING DEVICES** | | | | |
| Stainless steel | S.F. | 310.00 | 25.70 | 335.70 |
| Brass | " | 340.00 | 25.70 | 365.70 |
| Aluminum | " | 160.00 | 25.70 | 185.70 |
| Metal built-up plaque | | | | |
| Bronze | S.F. | 270.00 | 30.90 | 300.90 |
| Copper and bronze | " | 270.00 | 30.90 | 300.90 |
| Copper and aluminum | " | 300.00 | 30.90 | 330.90 |
| Metal nameplate plaques | | | | |
| Cast bronze | S.F. | 270.00 | 19.30 | 289.30 |
| Cast aluminum | " | 240.00 | 19.30 | 259.30 |
| Engraved, 1-1/2" x 6" | | | | |
| Bronze | EA. | 84.00 | 19.30 | 103.30 |
| Aluminum | " | 84.00 | 19.30 | 103.30 |
| Letters, on masonry or concrete, aluminum, satin finish | | | | |
| 1/2" thick | | | | |
| 2" high | EA. | 9.90 | 12.35 | 22.25 |
| 4" high | " | 11.10 | 15.45 | 26.55 |
| 6" high | " | 14.55 | 17.15 | 31.70 |
| 3/4" thick | | | | |
| 8" high | EA. | 19.25 | 19.30 | 38.55 |
| 10" high | " | 24.50 | 22.05 | 46.55 |
| 1" thick | | | | |
| 12" high | EA. | 32.10 | 25.70 | 57.80 |
| 14" high | " | 36.20 | 30.90 | 67.10 |
| 16" high | " | 50.00 | 38.60 | 88.60 |
| For polished aluminum add, 15% | | | | |
| For clear anodized aluminum add, 15% | | | | |
| For colored anodic aluminum add, 30% | | | | |
| For profiled and color enameled letters add, 50% | | | | |
| Cast bronze, satin finish letters | | | | |
| 3/8" thick | | | | |
| 2" high | EA. | 18.05 | 12.35 | 30.40 |
| 4" high | " | 19.60 | 15.45 | 35.05 |
| 1/2" thick, 6" high | " | 25.40 | 17.15 | 42.55 |
| 5/8" thick, 8" high | " | 32.90 | 19.30 | 52.20 |
| 1" thick | | | | |
| 10" high | EA. | 49.70 | 22.05 | 71.75 |
| 12" high | " | 63.50 | 25.70 | 89.20 |
| 14" high | " | 71.00 | 30.90 | 101.90 |
| 16" high | " | 90.00 | 38.60 | 128.60 |
| Interior door signs, adhesive, flexible | | | | |
| 2" x 8" | EA. | 12.25 | 6.05 | 18.30 |
| 4" x 4" | " | 12.25 | 6.05 | 18.30 |
| 6" x 7" | " | 15.95 | 6.05 | 22.00 |
| 6" x 9" | " | 21.00 | 6.05 | 27.05 |
| 10" x 9" | " | 28.00 | 6.05 | 34.05 |
| 10" x 12" | " | 36.20 | 6.05 | 42.25 |
| Hard plastic type, no frame | | | | |
| 3" x 8" | EA. | 25.70 | 6.05 | 31.75 |
| 4" x 4" | " | 24.50 | 6.05 | 30.55 |
| 4" x 12" | " | 30.30 | 6.05 | 36.35 |
| Hard plastic type, with frame | | | | |

| SPECIALTIES | UNIT | MAT. | INST. | TOTAL |
|---|---|---|---|---|
| **10400.10 IDENTIFYING DEVICES** | | | | |
| 3" x 8" | EA. | 81.50 | 6.05 | 87.55 |
| 4" x 4" | " | 66.50 | 6.05 | 72.55 |
| 4" x 12" | " | 110.00 | 6.05 | 116.05 |
| **10450.10 CONTROL** | | | | |
| Access control, 7' high, indoor or outdoor impenetrability | | | | |
| Remote or card control, type B | EA. | 2,800 | 420.00 | 3,220 |
| Free passage, type B | " | 2,570 | 420.00 | 2,990 |
| Remote or card control, type AA | " | 3,730 | 420.00 | 4,150 |
| Free passage, type AA | " | 3,380 | 420.00 | 3,800 |
| **10500.10 LOCKERS** | | | | |
| Locker bench, floor mounted, laminated maple | | | | |
| 4' | EA. | 70.00 | 25.70 | 95.70 |
| 6' | " | 100.00 | 25.70 | 125.70 |
| Wardrobe locker, 12" x 60" x 15", baked on enamel | | | | |
| 1-tier | EA. | 110.00 | 15.45 | 125.45 |
| 2-tier | " | 120.00 | 15.45 | 135.45 |
| 3-tier | " | 130.00 | 16.25 | 146.25 |
| 4-tier | " | 150.00 | 16.25 | 166.25 |
| 12" x 72" x 15", baked on enamel | | | | |
| 1-tier | EA. | 110.00 | 15.45 | 125.45 |
| 2-tier | " | 140.00 | 15.45 | 155.45 |
| 4-tier | " | 170.00 | 16.25 | 186.25 |
| 5-tier | " | 170.00 | 16.25 | 186.25 |
| 15" x 60" x 15", baked on enamel | | | | |
| 1-tier | EA. | 150.00 | 15.45 | 165.45 |
| 4-tier | " | 160.00 | 16.25 | 176.25 |
| Wardrobe locker, single tier type | | | | |
| 12" x 15" x 72" | EA. | 110.00 | 30.90 | 140.90 |
| 18" x 15" x 72" | " | 140.00 | 32.50 | 172.50 |
| 12" x 18" x 72" | " | 130.00 | 34.30 | 164.30 |
| 18" x 18" x 72" | " | 160.00 | 36.30 | 196.30 |
| Double tier type | | | | |
| 12" x 15" x 36" | EA. | 80.50 | 15.45 | 95.95 |
| 18" x 15" x 36" | " | 83.00 | 15.45 | 98.45 |
| 12" x 18" x 36" | " | 74.50 | 15.45 | 89.95 |
| 18" x 18" x 36" | " | 85.00 | 15.45 | 100.45 |
| Two person unit | | | | |
| 18" x 15" x 72" | EA. | 220.00 | 51.50 | 271.50 |
| 18" x 18" x 72" | " | 130.00 | 62.00 | 192.00 |
| Duplex unit | | | | |
| 15" x 15" x 72" | EA. | 220.00 | 30.90 | 250.90 |
| 15" x 21" x 72" | " | 230.00 | 30.90 | 260.90 |
| Basket lockers, basket sets with baskets | | | | |
| 24 basket set | SET | 470.00 | 150.00 | 620.00 |
| 30 basket set | " | 560.00 | 190.00 | 750.00 |
| 36 basket set | " | 620.00 | 260.00 | 880.00 |
| 42 basket set | " | 700.00 | 310.00 | 1,010 |

| SPECIALTIES | UNIT | MAT. | INST. | TOTAL |
|---|---|---|---|---|
| **10520.10 FIRE PROTECTION** | | | | |
| Portable fire extinguishers | | | | |
| Water pump tank type | | | | |
| 2.5 gal. | | | | |
| Red enameled galvanized | EA. | 77.00 | 16.15 | 93.15 |
| Red enameled copper | " | 88.50 | 16.15 | 104.65 |
| Polished copper | " | 130.00 | 16.15 | 146.15 |
| Carbon dioxide type, red enamel steel | | | | |
| Squeeze grip with hose and horn | | | | |
| 2.5 lb | EA. | 42.00 | 16.15 | 58.15 |
| 5 lb | " | 85.00 | 18.60 | 103.60 |
| 10 lb | " | 140.00 | 24.20 | 164.20 |
| 15 lb | " | 170.00 | 30.30 | 200.30 |
| 20 lb | " | 190.00 | 30.30 | 220.30 |
| Wheeled type | | | | |
| 125 lb | EA. | 1,120 | 48.40 | 1,168 |
| 250 lb | " | 1,150 | 48.40 | 1,198 |
| 500 lb | " | 2,160 | 48.40 | 2,208 |
| Dry chemical, pressurized type | | | | |
| Red enameled steel | | | | |
| 2.5 lb | EA. | 23.30 | 16.15 | 39.45 |
| 5 lb | " | 35.00 | 18.60 | 53.60 |
| 10 lb | " | 49.00 | 24.20 | 73.20 |
| 20 lb | " | 85.00 | 30.30 | 115.30 |
| 30 lb | " | 230.00 | 30.30 | 260.30 |
| Chrome plated steel, 2.5 lb | " | 50.00 | 16.15 | 66.15 |
| Other type extinguishers | | | | |
| 2.5 gal, stainless steel, pressurized water tanks | EA. | 49.00 | 16.15 | 65.15 |
| Soda and acid type | " | 95.50 | 16.15 | 111.65 |
| Cartridge operated, water type | " | 160.00 | 16.15 | 176.15 |
| Loaded stream, water type | " | 76.00 | 16.15 | 92.15 |
| Foam type | " | 63.00 | 16.15 | 79.15 |
| 40 gal, wheeled foam type | " | 3,380 | 48.40 | 3,428 |
| Fire extinguisher cabinets | | | | |
| Enameled steel | | | | |
| 8" x 12" x 27" | EA. | 120.00 | 48.40 | 168.40 |
| 8" x 16" x 38" | " | 160.00 | 48.40 | 208.40 |
| Aluminum | | | | |
| 8" x 12" x 27" | EA. | 160.00 | 48.40 | 208.40 |
| 8" x 16" x 38" | " | 190.00 | 48.40 | 238.40 |
| 8" x 12" x 27" | " | 270.00 | 48.40 | 318.40 |
| Stainless steel | | | | |
| 8" x 16" x 38" | EA. | 350.00 | 48.40 | 398.40 |
| **10550.10 POSTAL SPECIALTIES** | | | | |
| Mail chutes | | | | |
| Single mail chute | | | | |
| Finished aluminum | L.F. | 440.00 | 77.00 | 517.00 |
| Bronze | " | 640.00 | 77.00 | 717.00 |
| Single mail chute receiving box | | | | |
| Finished aluminum | EA. | 690.00 | 150.00 | 840.00 |
| Bronze | " | 1,280 | 150.00 | 1,430 |
| Twin mail chute, double parallel | | | | |

| SPECIALTIES | UNIT | MAT. | INST. | TOTAL |
|---|---|---|---|---|
| **10550.10 POSTAL SPECIALTIES** | | | | |
| Finished aluminum | FLR | 820.00 | 150.00 | 970.00 |
| Bronze | " | 1,280 | 150.00 | 1,430 |
| Receiving box, 36" x 20" x 12" | | | | |
| Finished aluminum | EA. | 1,430 | 260.00 | 1,690 |
| Bronze | " | 2,040 | 260.00 | 2,300 |
| Locked receiving mail box | | | | |
| Finished aluminum | EA. | 690.00 | 150.00 | 840.00 |
| Bronze | " | 1,170 | 150.00 | 1,320 |
| Commercial postal accessories for mail chutes | | | | |
| Letter slot, brass | EA. | 230.00 | 51.50 | 281.50 |
| Bulk mail slot, brass | " | 650.00 | 51.50 | 701.50 |
| Mail boxes | | | | |
| Residential postal accessories | | | | |
| Letter slot | EA. | 67.50 | 15.45 | 82.95 |
| Rural letter box | " | 47.80 | 38.60 | 86.40 |
| Apartment house, keyed, 3.5" x 4.5" x 16" | " | 70.00 | 10.30 | 80.30 |
| Ranch style | " | 90.00 | 15.45 | 105.45 |
| Commercial postal accessories | | | | |
| Letter box, with combination lock | EA. | 57.00 | 11.05 | 68.05 |
| Key lock | " | 52.00 | 11.05 | 63.05 |
| Mail box, aluminum w/glass front, 4x5 | | | | |
| Horizontal rear load | EA. | 71.00 | 8.80 | 79.80 |
| Vertical front load | " | 55.00 | 8.80 | 63.80 |
| **10600.10 MOVABLE PARTITIONS** | | | | |
| Partition, movable office type, 2-1/2" thick, vinyl-gypsum | S.F. | 9.30 | 1.55 | 10.85 |
| Enameled steel frame, with 1/4" thick clear glass | " | 12.85 | 1.55 | 14.40 |
| Door frame and hardware for movable partitions | EA. | 370.00 | 100.00 | 470.00 |
| Add for acoustic movable partition | S.F. | | | 0.80 |
| Accordion partition, 12' high | | | | |
| Vinyl | S.F. | 5.70 | 5.15 | 10.85 |
| Acoustical | " | 6.75 | 5.15 | 11.90 |
| Standard office cubicles, 8' high, steel framed | | | | |
| Baked enamel finish | | | | |
| 100% flush | L.F. | 88.50 | 7.70 | 96.20 |
| 75% flush and 25% glass | " | 110.00 | 8.60 | 118.60 |
| 50% flush and 50% glass | " | 150.00 | 8.60 | 158.60 |
| 100% glass | " | 190.00 | 10.30 | 200.30 |
| Natural hardwood panels | | | | |
| 100% flush | L.F. | 94.50 | 10.30 | 104.80 |
| 50% flush and 50% glass | " | 130.00 | 11.05 | 141.05 |
| Plastic laminated panels | | | | |
| 100% flush | L.F. | 110.00 | 10.30 | 120.30 |
| 75% flush and 25% glass | " | 140.00 | 11.05 | 151.05 |
| 50% flush and 50% glass | " | 170.00 | 11.05 | 181.05 |
| Vinyl covered panels | | | | |
| 100% flush | L.F. | 100.00 | 10.65 | 110.65 |
| 75% flush and 25% glass | " | 130.00 | 11.45 | 141.45 |
| 50% and 50% glass | " | 150.00 | 11.45 | 161.45 |
| Aluminum framed | | | | |
| Enameled or anodized aluminum panels | | | | |

| SPECIALTIES | UNIT | MAT. | INST. | TOTAL |
|---|---|---|---|---|
| **10600.10 MOVABLE PARTITIONS** | | | | |
| 100% flush | L.F. | 77.00 | 7.70 | 84.70 |
| 75% flush and 25% glass | " | 73.50 | 8.60 | 82.10 |
| 50% flush and 50% glass | " | 68.00 | 8.60 | 76.60 |
| Vinyl covered panels | | | | |
| 100% flush | L.F. | 83.00 | 10.65 | 93.65 |
| 75% flush and 25% glass | " | 100.00 | 11.45 | 111.45 |
| 50% flush and 50% glass | " | 120.00 | 11.45 | 131.45 |
| 60" high partitions, steel framed | | | | |
| Enameled panels | L.F. | 40.80 | 6.85 | 47.65 |
| Natural hardwood panels, two sides | " | 180.00 | 7.20 | 187.20 |
| Plastic laminated panels | " | 90.00 | 7.20 | 97.20 |
| Vinyl covered panels | " | 120.00 | 6.85 | 126.85 |
| Aluminum framed | | | | |
| Anodized or baked enamel panels | L.F. | 59.00 | 6.85 | 65.85 |
| Natural hardwood panels | " | 180.00 | 7.20 | 187.20 |
| Plastic laminated panels | " | 90.00 | 7.20 | 97.20 |
| Vinyl covered panels | " | 120.00 | 6.85 | 126.85 |
| Wire mesh partitions | | | | |
| Wall panels | | | | |
| 4' x 7' | EA. | 89.00 | 19.30 | 108.30 |
| 4' x 8' | " | 98.00 | 20.60 | 118.60 |
| 4' x 10' | " | 120.00 | 23.75 | 143.75 |
| Wall filler panels | | | | |
| 1' x 7' | EA. | 51.50 | 19.30 | 70.80 |
| 1' x 8' | " | 52.50 | 20.60 | 73.10 |
| 1' x 10' | " | 70.00 | 22.05 | 92.05 |
| 2' x 7' | " | 50.00 | 19.30 | 69.30 |
| 2' x 8' | " | 53.50 | 20.60 | 74.10 |
| 2' x 10' | " | 63.00 | 22.05 | 85.05 |
| 3' x 7' | " | 66.50 | 19.30 | 85.80 |
| 3' x 8' | " | 72.50 | 20.60 | 93.10 |
| 3' x 10' | " | 88.50 | 22.05 | 110.55 |
| Ceiling panels | | | | |
| 10' x 2' | EA. | 60.50 | 44.10 | 104.60 |
| 10' x 4' | " | 120.00 | 62.00 | 182.00 |
| Wall panel with service window | | | | |
| 5' wide | | | | |
| 7' high | EA. | 290.00 | 19.30 | 309.30 |
| 8' high | " | 300.00 | 20.60 | 320.60 |
| 10' high | " | 310.00 | 22.05 | 332.05 |
| Doors | | | | |
| Sliding | | | | |
| 3' x 7' | EA. | 350.00 | 77.00 | 427.00 |
| 3' x 8' | " | 320.00 | 88.00 | 408.00 |
| 3' x 10' | " | 330.00 | 120.00 | 450.00 |
| 4' x 7' | " | 330.00 | 88.00 | 418.00 |
| 4' x 8' | " | 340.00 | 120.00 | 460.00 |
| 4' x 10' | " | 370.00 | 150.00 | 520.00 |
| 5' x 7' | " | 360.00 | 120.00 | 480.00 |
| 5' x 8' | " | 430.00 | 150.00 | 580.00 |
| 5' x 10' | " | 500.00 | 150.00 | 650.00 |
| Swing door | | | | |

# 10 SPECIALTIES

| SPECIALTIES | UNIT | MAT. | INST. | TOTAL |
|---|---|---|---|---|
| **10600.10 MOVABLE PARTITIONS** | | | | |
| 3' x 7' | EA. | 260.00 | 77.00 | 337.00 |
| 4' x 7' | " | 290.00 | 88.00 | 378.00 |
| Swing door, with 1' transom | | | | |
| 3' x 7' | EA. | 280.00 | 88.00 | 368.00 |
| 4' x 7' | " | 310.00 | 120.00 | 430.00 |
| Swing door, with 3' transom | | | | |
| 3' x 7' | EA. | 240.00 | 120.00 | 360.00 |
| 4' x 7' | " | 270.00 | 150.00 | 420.00 |
| **10670.10 SHELVING** | | | | |
| Shelving, enamel, closed side and back, 12" x 36" | | | | |
| 5 shelves | EA. | 110.00 | 51.50 | 161.50 |
| 8 shelves | " | 140.00 | 68.50 | 208.50 |
| Open | | | | |
| 5 shelves | EA. | 77.00 | 51.50 | 128.50 |
| 8 shelves | " | 77.00 | 68.50 | 145.50 |
| Metal storage shelving, baked enamel | | | | |
| 7 shelf unit, 72" or 84" high | | | | |
| 10" shelf | L.F. | 31.50 | 30.90 | 62.40 |
| 12" shelf | " | 32.70 | 32.50 | 65.20 |
| 15" shelf | " | 37.30 | 34.30 | 71.60 |
| 18" shelf | " | 37.30 | 36.30 | 73.60 |
| 24" shelf | " | 43.10 | 38.60 | 81.70 |
| 30" shelf | " | 46.60 | 41.20 | 87.80 |
| 36" shelf | " | 56.00 | 44.10 | 100.10 |
| 4 shelf unit, 40" high | | | | |
| 10" shelf | L.F. | 26.80 | 25.70 | 52.50 |
| 12" shelf | " | 28.00 | 28.10 | 56.10 |
| 15" shelf | " | 35.00 | 30.90 | 65.90 |
| 18" shelf | " | 35.00 | 32.50 | 67.50 |
| 24" shelf | " | 39.60 | 34.30 | 73.90 |
| 3 shelf unit, 32" high | | | | |
| 10" shelf | L.F. | 22.15 | 15.45 | 37.60 |
| 12" shelf | " | 22.15 | 16.25 | 38.40 |
| 15" shelf | " | 24.50 | 17.15 | 41.65 |
| 18" shelf | " | 25.70 | 18.15 | 43.85 |
| 24" shelf | " | 29.20 | 19.30 | 48.50 |
| Single shelf unit, attached to masonry | | | | |
| 10" shelf | L.F. | 7.95 | 5.15 | 13.10 |
| 12" shelf | " | 8.75 | 5.60 | 14.35 |
| 15" shelf | " | 9.00 | 5.95 | 14.95 |
| 18" shelf | " | 10.15 | 6.30 | 16.45 |
| 24" shelf | " | 12.25 | 6.70 | 18.95 |
| For stainless steel, add to material, 150% | | | | |
| For attachment to gypsum board, add to labor, 50% | | | | |
| Built-in wood shelves | | | | |
| Posts and trimmed plywood | L.F. | 1.80 | 4.40 | 6.20 |
| Solid clear pine | " | 3.05 | 4.75 | 7.80 |
| Closet shelf, pine with rod | " | 2.05 | 4.75 | 6.80 |
| For lumber edge band, add to material | " | | | 1.05 |
| For prefinished shelves, add to material, 225% | | | | |

| SPECIALTIES | UNIT | MAT. | INST. | TOTAL |
|---|---|---|---|---|
| **10750.10 TELEPHONE ENCLOSURES** | | | | |
| Telephone enclosure, wall mounted, shelf, 28" x 30" x 15" | EA. | 700.00 | 77.00 | 777.00 |
| Directory shelf, stainless steel, 3 binders | " | 580.00 | 51.50 | 631.50 |
| **10800.10 BATH ACCESSORIES** | | | | |
| Ash receiver, wall mounted, aluminum | EA. | 120.00 | 15.45 | 135.45 |
| Grab bar, 1-1/2" dia., stainless steel, wall mounted | | | | |
| 24" long | EA. | 33.80 | 15.45 | 49.25 |
| 36" long | " | 49.00 | 16.25 | 65.25 |
| 42" long | " | 56.00 | 17.15 | 73.15 |
| 48" long | " | 67.50 | 18.15 | 85.65 |
| 52" long | " | 73.50 | 19.30 | 92.80 |
| 1" dia., stainless steel | | | | |
| 12" long | EA. | 26.80 | 13.45 | 40.25 |
| 18" long | " | 29.20 | 14.05 | 43.25 |
| 24" long | " | 31.50 | 15.45 | 46.95 |
| 30" long | " | 35.00 | 16.25 | 51.25 |
| 36" long | " | 39.60 | 17.15 | 56.75 |
| 48" long | " | 40.80 | 18.15 | 58.95 |
| Hand dryer, surface mounted, 110 volt | " | 380.00 | 38.60 | 418.60 |
| Medicine cabinet, 16 x 22, baked enamel, steel, lighted | " | 60.50 | 12.35 | 72.85 |
| With mirror, lighted | " | 110.00 | 20.60 | 130.60 |
| Mirror, 1/4" plate glass, up to 10 sf | S.F. | 6.05 | 3.10 | 9.15 |
| Mirror, stainless steel frame | | | | |
| 18"x24" | EA. | 95.50 | 10.30 | 105.80 |
| 18"x32" | " | 100.00 | 12.35 | 112.35 |
| 18"x36" | " | 110.00 | 15.45 | 125.45 |
| 24"x30" | " | 120.00 | 15.45 | 135.45 |
| 24"x36" | " | 140.00 | 17.15 | 157.15 |
| 24"x48" | " | 240.00 | 25.70 | 265.70 |
| 24"x60" | " | 250.00 | 30.90 | 280.90 |
| 30"x30" | " | 140.00 | 30.90 | 170.90 |
| 30"x72" | " | 380.00 | 38.60 | 418.60 |
| 48"x72" | " | 400.00 | 51.50 | 451.50 |
| With shelf, 18"x24" | " | 140.00 | 12.35 | 152.35 |
| Sanitary napkin dispenser, stainless steel, wall mounted | " | 330.00 | 20.60 | 350.60 |
| Shower rod, 1" diameter | | | | |
| Chrome finish over brass | EA. | 55.00 | 15.45 | 70.45 |
| Stainless steel | " | 52.50 | 15.45 | 67.95 |
| Soap dish, stainless steel, wall mounted | " | 66.50 | 20.60 | 87.10 |
| Toilet tissue dispenser, stainless, wall mounted | | | | |
| Single roll | EA. | 32.70 | 7.70 | 40.40 |
| Double roll | " | 51.50 | 8.80 | 60.30 |
| Towel dispenser, stainless steel | | | | |
| Flush mounted | EA. | 94.50 | 17.15 | 111.65 |
| Surface mounted | " | 88.50 | 15.45 | 103.95 |
| Combination towel dispenser and waste receptacle | " | 290.00 | 20.60 | 310.60 |
| Towel bar, stainless steel | | | | |
| 18" long | EA. | 28.00 | 12.35 | 40.35 |
| 24" long | " | 32.70 | 14.05 | 46.75 |
| 30" long | " | 35.00 | 15.45 | 50.45 |
| 36" long | " | 38.50 | 17.15 | 55.65 |
| Toothbrush and tumbler holder | " | 24.50 | 10.30 | 34.80 |

# 10 SPECIALTIES

| SPECIALTIES | UNIT | MAT. | INST. | TOTAL |
|---|---|---|---|---|
| **10800.10 BATH ACCESSORIES** | | | | |
| Waste receptacle, stainless steel, wall mounted | EA. | 200.00 | 25.70 | 225.70 |
| **10900.10 WARDROBE SPECIALTIES** | | | | |
| Hospital wardrobe units, 24" x 24" x 76", with door | | | | |
| Baked enameled steel | EA. | 1,460 | 170.00 | 1,630 |
| Hardwood | " | 780.00 | 170.00 | 950.00 |
| Stainless steel | " | 2,450 | 170.00 | 2,620 |
| Plastic laminated | " | 870.00 | 170.00 | 1,040 |
| Dormitory wardrobe units, 24" x 76", with door | | | | |
| Hardwood | EA. | 780.00 | 170.00 | 950.00 |
| Plastic laminated | " | 690.00 | 170.00 | 860.00 |
| Hat and coat rack | | | | |
| Single tier | | | | |
| Baked enameled steel | L.F. | 93.00 | 7.70 | 100.70 |
| Stainless steel | " | 110.00 | 7.70 | 117.70 |
| Aluminum | " | 100.00 | 7.70 | 107.70 |
| Double tier | | | | |
| Baked enameled steel | L.F. | 160.00 | 8.80 | 168.80 |
| Stainless steel | " | 220.00 | 8.80 | 228.80 |
| Aluminum | " | 190.00 | 8.80 | 198.80 |

| ARCHITECTURAL EQUIPMENT | UNIT | MAT. | INST. | TOTAL |
|---|---|---|---|---|
| **11010.10 MAINTENANCE EQUIPMENT** | | | | |
| Vacuum cleaning system | | | | |
| 3 valves | | | | |
| 1.5 hp | EA. | 870.00 | 340.00 | 1,210 |
| 2.5 hp | " | 1,280 | 440.00 | 1,720 |
| 5 valves | " | 1,690 | 620.00 | 2,310 |
| 7 valves | " | 1,920 | 770.00 | 2,690 |
| **11020.10 SECURITY EQUIPMENT** | | | | |
| Bulletproof teller window | | | | |
| 4' x 4' | EA. | 8,400 | 510.00 | 8,910 |
| 5' x 4' | " | 9,210 | 620.00 | 9,830 |
| Bulletproof partitions | | | | |
| Up to 12' high, 2.5" thick | S.F. | 42.00 | 2.05 | 44.05 |
| Counter for banks | | | | |
| Minimum | L.F. | 480.00 | 62.00 | 542.00 |
| Maximum | " | 2,160 | 100.00 | 2,260 |
| Drive-up window | | | | |
| Minimum | EA. | 5,710 | 440.00 | 6,150 |
| Maximum | " | 34,980 | 1,030 | 36,010 |
| Night depository | | | | |
| Minimum | EA. | 7,930 | 440.00 | 8,370 |
| Maximum | " | 16,910 | 1,030 | 17,940 |
| Office safes, 30" x 20" x 20", 1 hr rating | " | 1,520 | 77.00 | 1,597 |
| 30" x 16" x 15", 2 hr rating | " | 1,460 | 62.00 | 1,522 |
| 30" x 28" x 20", H&G rating | " | 7,000 | 38.60 | 7,039 |
| Service windows, pass through painted steel | | | | |
| 24" x 36" | EA. | 3,620 | 310.00 | 3,930 |
| 48" x 40" | " | 5,660 | 390.00 | 6,050 |
| 72" x 40" | " | 6,530 | 620.00 | 7,150 |
| Special doors and windows | | | | |
| 3' x 7' bulletproof door with frame | EA. | 4,260 | 440.00 | 4,700 |
| 12" x 12" vision panel | " | 2,450 | 220.00 | 2,670 |
| Surveillance system | | | | |
| Minimum | EA. | 4,140 | 620.00 | 4,760 |
| Maximum | " | 32,070 | 3,090 | 35,160 |
| Vault door, 3' wide, 6'6" high | | | | |
| 3-1/2" thick | EA. | 15,160 | 3,860 | 19,020 |
| 7" thick | " | 18,080 | 5,150 | 23,230 |
| 10" thick | " | 21,570 | 6,180 | 27,750 |
| Insulated vault door | | | | |
| 2 hr rating | | | | |
| 32" wide | EA. | 2,450 | 310.00 | 2,760 |
| 40" wide | " | 2,680 | 330.00 | 3,010 |
| 4 hr rating | | | | |
| 32" wide | EA. | 2,620 | 340.00 | 2,960 |
| 40" wide | " | 2,740 | 390.00 | 3,130 |
| 6 hr rating | | | | |
| 32" wide | EA. | 2,800 | 340.00 | 3,140 |
| 40" wide | " | 3,090 | 390.00 | 3,480 |
| Insulated file room door | | | | |
| 1 hr rating | | | | |
| 32" wide | EA. | 2,800 | 310.00 | 3,110 |

# 11 EQUIPMENT

| ARCHITECTURAL EQUIPMENT | UNIT | MAT. | INST. | TOTAL |
|---|---|---|---|---|
| **11020.10 SECURITY EQUIPMENT** | | | | |
| 40" wide | EA. | 2,970 | 340.00 | 3,310 |
| **11060.10 THEATER EQUIPMENT** | | | | |
| Roll out stage, steel frame, wood floor | | | | |
| Manual | S.F. | 24.50 | 1.95 | 26.45 |
| Electric | " | 26.80 | 3.10 | 29.90 |
| Portable stages | | | | |
| 8" high | S.F. | 7.80 | 1.55 | 9.35 |
| 18" high | " | 9.00 | 1.70 | 10.70 |
| 36" high | " | 9.90 | 1.80 | 11.70 |
| 48" high | " | 10.50 | 1.95 | 12.45 |
| Band risers | | | | |
| Minimum | S.F. | 21.60 | 1.55 | 23.15 |
| Maximum | " | 35.00 | 1.55 | 36.55 |
| Chairs for risers | | | | |
| Minimum | EA. | 59.50 | 1.10 | 60.60 |
| Maximum | " | 110.00 | 1.10 | 111.10 |
| **11080.10 POLICE EQUIPMENT** | | | | |
| Firing range equipment, rifle | | | | |
| 3 position | EA. | 9,330 | 1,030 | 10,360 |
| 4 position | " | 12,240 | 1,540 | 13,780 |
| 5 position | " | 15,160 | 1,720 | 16,880 |
| 6 position | " | 18,660 | 1,820 | 20,480 |
| **11090.10 CHECKROOM EQUIPMENT** | | | | |
| Motorized checkroom equipment | | | | |
| No shelf system, 6'4" height | | | | |
| 7'6" length | EA. | 1,810 | 310.00 | 2,120 |
| 14'6" length | " | 2,040 | 310.00 | 2,350 |
| 28' length | " | 2,800 | 310.00 | 3,110 |
| One shelf, 6'8" height | | | | |
| 7'6" length | EA. | 2,220 | 310.00 | 2,530 |
| 14'6" length | " | 2,680 | 310.00 | 2,990 |
| 28' length | " | 3,960 | 310.00 | 4,270 |
| Two shelves, 7'5" height | | | | |
| 7'6" length | EA. | 2,680 | 310.00 | 2,990 |
| 14'6" length | " | 3,270 | 310.00 | 3,580 |
| 28' length | " | 4,660 | 310.00 | 4,970 |
| Three shelves, 8' height | | | | |
| 7'6" length | EA. | 2,800 | 620.00 | 3,420 |
| 14'6" length | " | 3,380 | 620.00 | 4,000 |
| 28' length | " | 4,780 | 620.00 | 5,400 |
| Four shelves, 8'7" height | | | | |
| 7'6" length | EA. | 2,800 | 620.00 | 3,420 |
| 14'6" length | " | 3,500 | 620.00 | 4,120 |
| 28' length | " | 4,780 | 620.00 | 5,400 |
| 200 lb | | | | |

# 11 EQUIPMENT

## ARCHITECTURAL EQUIPMENT

| ARCHITECTURAL EQUIPMENT | UNIT | MAT. | INST. | TOTAL |
|---|---|---|---|---|
| **11110.10 LAUNDRY EQUIPMENT** | | | | |
| High capacity, heavy duty | | | | |
| Washer extractors | | | | |
| 135 lb | | | | |
| Standard | EA. | 41,400 | 260.00 | 41,660 |
| Pass through | " | 47,810 | 260.00 | 48,070 |
| 200 lb | | | | |
| Standard | EA. | 46,060 | 260.00 | 46,320 |
| Pass through | " | 51,890 | 260.00 | 52,150 |
| 110 lb dryer | " | 8,160 | 260.00 | 8,420 |
| Hand operated presser | " | 5,600 | 340.00 | 5,940 |
| Mushroom press | " | 3,620 | 340.00 | 3,960 |
| Spreader feeders | | | | |
| 2 station | EA. | 53,640 | 340.00 | 53,980 |
| 4 station | " | 58,890 | 620.00 | 59,510 |
| Delivery carts | | | | |
| 12 bushel | EA. | 360.00 | 3.85 | 363.85 |
| 16 bushel | " | 420.00 | 4.10 | 424.10 |
| 18 bushel | " | 430.00 | 4.40 | 434.40 |
| 30 bushel | " | 620.00 | 5.15 | 625.15 |
| 40 bushel | " | 760.00 | 6.20 | 766.20 |
| Low capacity | | | | |
| Pressers | | | | |
| Air operated | EA. | 4,430 | 120.00 | 4,550 |
| Hand operated | " | 3,620 | 120.00 | 3,740 |
| Extractor, low capacity | " | 5,190 | 120.00 | 5,310 |
| Ironer, 48" | " | 2,450 | 62.00 | 2,512 |
| Coin washers | | | | |
| 10 lb capacity | EA. | 1,100 | 62.00 | 1,162 |
| 20 lb capacity | " | 2,800 | 62.00 | 2,862 |
| Coin dryer | " | 450.00 | 38.60 | 488.60 |
| Coin dry cleaner, 20 lb | " | 22,160 | 120.00 | 22,280 |
| **11161.10 LOADING DOCK EQUIPMENT** | | | | |
| Dock leveler, 10 ton capacity | | | | |
| 6' x 8' | EA. | 3,620 | 310.00 | 3,930 |
| 7' x 8' | " | 4,080 | 310.00 | 4,390 |
| Bumpers, laminated rubber | | | | |
| 4-1/2" thick | | | | |
| 6" x 14" | EA. | 56.50 | 6.20 | 62.70 |
| 6" x 36" | " | 57.50 | 6.85 | 64.35 |
| 10" x 14" | " | 60.50 | 7.70 | 68.20 |
| 10" x 24" | " | 69.00 | 8.80 | 77.80 |
| 10" x 36" | " | 78.00 | 10.30 | 88.30 |
| 12" x 14" | " | 66.50 | 8.15 | 74.65 |
| 12" x 24" | " | 71.00 | 9.65 | 80.65 |
| 12" x 36" | " | 88.50 | 11.45 | 99.95 |
| 6" thick | | | | |
| 10" x 14" | EA. | 71.00 | 8.80 | 79.80 |
| 10" x 24" | " | 93.50 | 10.65 | 104.15 |
| 10" x 36" | " | 97.00 | 15.45 | 112.45 |
| Extruded rubber bumpers | | | | |

# 11 EQUIPMENT

| ARCHITECTURAL EQUIPMENT | UNIT | MAT. | INST. | TOTAL |
|---|---|---|---|---|
| **11161.10 LOADING DOCK EQUIPMENT** | | | | |
| T-section, 22" x 22" x 3" | EA. | 90.00 | 6.20 | 96.20 |
| Molded rubber bumpers | | | | |
| 24" x 12" x 3" thick | EA. | 55.00 | 15.45 | 70.45 |
| Door seal, 12" x 12", vinyl covered | L.F. | 31.50 | 7.70 | 39.20 |
| Dock boards, heavy duty, 5' x 5' | | | | |
| 5000 lb | | | | |
| Minimum | EA. | 890.00 | 260.00 | 1,150 |
| Maximum | " | 1,520 | 260.00 | 1,780 |
| 9000 lb | | | | |
| Minimum | EA. | 1,010 | 260.00 | 1,270 |
| Maximum | " | 1,690 | 280.00 | 1,970 |
| 15,000 lb | " | 1,120 | 280.00 | 1,400 |
| Truck shelters | | | | |
| Minimum | EA. | 1,060 | 240.00 | 1,300 |
| Maximum | " | 1,750 | 440.00 | 2,190 |
| **11170.10 WASTE HANDLING** | | | | |
| Incinerator, electric | | | | |
| 100 lb/hr | | | | |
| Minimum | EA. | 16,090 | 320.00 | 16,410 |
| Maximum | " | 23,320 | 320.00 | 23,640 |
| 400 lb/hr | | | | |
| Minimum | EA. | 28,570 | 630.00 | 29,200 |
| Maximum | " | 45,480 | 630.00 | 46,110 |
| 1000 lb/hr | | | | |
| Minimum | EA. | 47,810 | 960.00 | 48,770 |
| Maximum | " | 65,310 | 960.00 | 66,270 |
| Incinerator, medical-waste | | | | |
| 25 lb/hr, 2-7 x 4-0 | EA. | 14,580 | 630.00 | 15,210 |
| 50 lb/hr, 2-11 x 4-11 | " | 16,090 | 630.00 | 16,720 |
| 75 lb/hr, 3-8 x 5-0 | " | 23,320 | 1,270 | 24,590 |
| 100 lb/hr, 3-8 x 6-0 | " | 25,660 | 1,270 | 26,930 |
| Industrial compactor | | | | |
| 1 cy | EA. | 8,110 | 350.00 | 8,460 |
| 3 cy | " | 16,330 | 450.00 | 16,780 |
| 5 cy | " | 18,660 | 630.00 | 19,290 |
| Trash chutes steel, including sprinklers | | | | |
| 18" dia. | L.F. | 55.00 | 150.00 | 205.00 |
| 24" dia. | " | 71.00 | 160.00 | 231.00 |
| 30" dia. | " | 90.00 | 170.00 | 260.00 |
| 36" dia. | " | 120.00 | 180.00 | 300.00 |
| Refuse bottom hopper | EA. | 1,750 | 170.00 | 1,920 |
| **11400.10 FOOD SERVICE EQUIPMENT** | | | | |
| Unit kitchens | | | | |
| 30" compact kitchen | | | | |
| Refrigerator, with range, sink | EA. | 2,680 | 160.00 | 2,840 |
| Sink only | " | 2,450 | 110.00 | 2,560 |
| Range only | " | 2,270 | 79.00 | 2,349 |
| Cabinet for upper wall section | " | 310.00 | 45.30 | 355.30 |

| ARCHITECTURAL EQUIPMENT | UNIT | MAT. | INST. | TOTAL |
|---|---|---|---|---|
| **11400.10 FOOD SERVICE EQUIPMENT** | | | | |
| Stainless shield, for rear wall | EA. | 100.00 | 12.65 | 112.65 |
| Side wall | " | 53.50 | 12.65 | 66.15 |
| 42" compact kitchen | | | | |
| Refrigerator with range, sink | EA. | 3,320 | 180.00 | 3,500 |
| Sink only | " | 2,860 | 160.00 | 3,020 |
| Cabinet for upper wall section | " | 440.00 | 53.00 | 493.00 |
| Stainless shield, for rear wall | " | 140.00 | 13.20 | 153.20 |
| Side wall | " | 53.50 | 13.20 | 66.70 |
| 54" compact kitchen | | | | |
| Refrigerator, oven, range, sink | EA. | 3,960 | 230.00 | 4,190 |
| Cabinet for upper wall section | " | 570.00 | 63.50 | 633.50 |
| Stainless shield, for | | | | |
| Rear wall | EA. | 200.00 | 14.40 | 214.40 |
| Side wall | " | 53.50 | 14.40 | 67.90 |
| 60" compact kitchen | | | | |
| Refrigerator, oven, range, sink | EA. | 4,610 | 230.00 | 4,840 |
| Cabinet for upper wall section | " | 650.00 | 63.50 | 713.50 |
| Stainless shield, for | | | | |
| Rear wall | EA. | 200.00 | 14.40 | 214.40 |
| Side wall | " | 53.50 | 14.40 | 67.90 |
| 72" compact kitchen | | | | |
| Refrigerator, oven, range, sink | EA. | 4,490 | 260.00 | 4,750 |
| Cabinet for upper wall section | " | 1,280 | 63.50 | 1,344 |
| Stainless shield for | | | | |
| Rear wall | EA. | 230.00 | 15.85 | 245.85 |
| Side wall | " | 53.50 | 15.85 | 69.35 |
| Bake oven | | | | |
| Single deck | | | | |
| Minimum | EA. | 1,520 | 39.60 | 1,560 |
| Maximum | " | 3,500 | 79.00 | 3,579 |
| Double deck | | | | |
| Minimum | EA. | 1,750 | 53.00 | 1,803 |
| Maximum | " | 6,060 | 79.00 | 6,139 |
| Triple deck | | | | |
| Minimum | EA. | 2,450 | 53.00 | 2,503 |
| Maximum | " | 7,350 | 110.00 | 7,460 |
| Convection type oven, electric, 40" x 45" x 57" | | | | |
| Minimum | EA. | 2,450 | 39.60 | 2,490 |
| Maximum | " | 4,200 | 79.00 | 4,279 |
| Broiler, without oven, 69" x 26" x 39" | | | | |
| Minimum | EA. | 2,220 | 39.60 | 2,260 |
| Maximum | " | 4,200 | 53.00 | 4,253 |
| Coffee urns, 10 gallons | | | | |
| Minimum | EA. | 4,780 | 110.00 | 4,890 |
| Maximum | " | 7,700 | 160.00 | 7,860 |
| Fryer, with submerger | | | | |
| Single | | | | |
| Minimum | EA. | 720.00 | 63.50 | 783.50 |
| Maximum | " | 2,220 | 110.00 | 2,330 |
| Double | | | | |
| Minimum | EA. | 1,280 | 79.00 | 1,359 |
| Maximum | " | 4,080 | 110.00 | 4,190 |

## ARCHITECTURAL EQUIPMENT

### 11400.10 FOOD SERVICE EQUIPMENT

| ARCHITECTURAL EQUIPMENT | UNIT | MAT. | INST. | TOTAL |
|---|---|---|---|---|
| Griddle, counter | | | | |
| 3' long | | | | |
| Minimum | EA. | 1,220 | 53.00 | 1,273 |
| Maximum | " | 3,620 | 63.50 | 3,684 |
| 5' long | | | | |
| Minimum | EA. | 1,870 | 79.00 | 1,949 |
| Maximum | " | 4,660 | 110.00 | 4,770 |
| Kettles, steam, jacketed | | | | |
| 20 gallons | | | | |
| Minimum | EA. | 1,980 | 79.00 | 2,059 |
| Maximum | " | 4,550 | 160.00 | 4,710 |
| 40 gallons | | | | |
| Minimum | EA. | 2,330 | 79.00 | 2,409 |
| Maximum | " | 5,010 | 160.00 | 5,170 |
| 60 gallons | | | | |
| Minimum | EA. | 3,030 | 79.00 | 3,109 |
| Maximum | " | 6,710 | 160.00 | 6,870 |
| Range | | | | |
| Heavy duty, single oven, open top | | | | |
| Minimum | EA. | 1,690 | 39.60 | 1,730 |
| Maximum | " | 4,310 | 110.00 | 4,420 |
| Fry top | | | | |
| Minimum | EA. | 2,220 | 39.60 | 2,260 |
| Maximum | " | 4,660 | 110.00 | 4,770 |
| Steamers, electric | | | | |
| 27 kw | | | | |
| Minimum | EA. | 6,410 | 79.00 | 6,489 |
| Maximum | " | 12,830 | 110.00 | 12,940 |
| 18 kw | | | | |
| Minimum | EA. | 4,720 | 79.00 | 4,799 |
| Maximum | " | 8,110 | 110.00 | 8,220 |
| Dishwasher, rack type | | | | |
| Single tank, 190 racks/hr | EA. | 10,030 | 160.00 | 10,190 |
| Double tank | | | | |
| 234 racks/hr | EA. | 14,580 | 180.00 | 14,760 |
| 265 racks/hr | " | 16,090 | 210.00 | 16,300 |
| Dishwasher, automatic 100 meals/hr | " | 47,110 | 110.00 | 47,220 |
| Disposals | | | | |
| 100 gal/hr | EA. | 1,110 | 110.00 | 1,220 |
| 120 gal/hr | " | 1,570 | 110.00 | 1,680 |
| 250 gal/hr | " | 4,020 | 110.00 | 4,130 |
| Exhaust hood for dishwasher, gutter 4 sides, s-steel | | | | |
| 4'x4'x2' | EA. | 1,460 | 120.00 | 1,580 |
| 4'x7'x2' | " | 2,450 | 130.00 | 2,580 |
| Food preparation machines | | | | |
| Vertical cutter mixers | | | | |
| 25 quart | EA. | 6,240 | 110.00 | 6,350 |
| 40 quart | " | 7,520 | 110.00 | 7,630 |
| 80 quart | " | 8,570 | 160.00 | 8,730 |
| 130 quart | " | 9,450 | 260.00 | 9,710 |
| Choppers | | | | |
| 5 lb | EA. | 2,200 | 79.00 | 2,279 |

## ARCHITECTURAL EQUIPMENT

### 11400.10 FOOD SERVICE EQUIPMENT

| ARCHITECTURAL EQUIPMENT | UNIT | MAT. | INST. | TOTAL |
|---|---|---|---|---|
| 16 lb | EA. | 2,970 | 110.00 | 3,080 |
| 40 lb | " | 4,140 | 160.00 | 4,300 |
| Mixers, floor models | | | | |
| 20 quart | EA. | 2,430 | 39.60 | 2,470 |
| 60 quart | " | 7,110 | 39.60 | 7,150 |
| 80 quart | " | 9,450 | 45.30 | 9,495 |
| 140 quart | " | 10,730 | 63.50 | 10,794 |
| Ice cube maker | | | | |
| 50 lb per day | | | | |
| Minimum | EA. | 970.00 | 320.00 | 1,290 |
| Maximum | " | 1,980 | 320.00 | 2,300 |
| 500 lb per day | | | | |
| Minimum | EA. | 3,790 | 530.00 | 4,320 |
| Maximum | " | 6,880 | 530.00 | 7,410 |
| Ice flakers | | | | |
| 300 lb per day | EA. | 2,620 | 320.00 | 2,940 |
| 600 lb per day | " | 3,090 | 530.00 | 3,620 |
| 1000 lb per day | " | 3,620 | 700.00 | 4,320 |
| 2000 lb per day | " | 8,110 | 790.00 | 8,900 |
| Refrigerated cases | | | | |
| Dairy products | | | | |
| Multi deck type | L.F. | 530.00 | 21.10 | 551.10 |
| For rear sliding doors, add | " | | | 60.64 |
| Delicatessen case, service deli | | | | |
| Single deck | L.F. | 500.00 | 160.00 | 660.00 |
| Multi deck | " | 550.00 | 200.00 | 750.00 |
| Meat case | | | | |
| Single deck | L.F. | 360.00 | 190.00 | 550.00 |
| Multi deck | " | 400.00 | 200.00 | 600.00 |
| Produce case | | | | |
| Single deck | L.F. | 380.00 | 190.00 | 570.00 |
| Multi deck | " | 430.00 | 200.00 | 630.00 |
| Bottle coolers | | | | |
| 6' long | | | | |
| Minimum | EA. | 1,460 | 630.00 | 2,090 |
| Maximum | " | 2,970 | 630.00 | 3,600 |
| 10' long | | | | |
| Minimum | EA. | 1,980 | 1,060 | 3,040 |
| Maximum | " | 3,850 | 1,060 | 4,910 |
| Frozen food cases | | | | |
| Chest type | L.F. | 390.00 | 190.00 | 580.00 |
| Reach-in, glass door | " | 800.00 | 200.00 | 1,000 |
| Island case, single | " | 400.00 | 190.00 | 590.00 |
| Multi deck | " | 870.00 | 200.00 | 1,070 |
| Ice storage bins | | | | |
| 500 lb capacity | EA. | 1,110 | 450.00 | 1,560 |
| 1000 lb capacity | " | 2,160 | 910.00 | 3,070 |

### 11450.10 RESIDENTIAL EQUIPMENT

| ARCHITECTURAL EQUIPMENT | UNIT | MAT. | INST. | TOTAL |
|---|---|---|---|---|
| Compactor, 4 to 1 compaction | EA. | 1,060 | 79.00 | 1,139 |
| Dishwasher, built-in | | | | |

# 11 EQUIPMENT

| ARCHITECTURAL EQUIPMENT | UNIT | MAT. | INST. | TOTAL |
|---|---|---|---|---|
| **11450.10 RESIDENTIAL EQUIPMENT** | | | | |
| 2 cycles | EA. | 890.00 | 160.00 | 1,050 |
| 4 or more cycles | " | 1,400 | 160.00 | 1,560 |
| Disposal | | | | |
| Garbage disposer | EA. | 360.00 | 110.00 | 470.00 |
| Heaters, electric, built-in | | | | |
| Ceiling type | EA. | 190.00 | 110.00 | 300.00 |
| Wall type | | | | |
| Minimum | EA. | 130.00 | 79.00 | 209.00 |
| Maximum | " | 240.00 | 110.00 | 350.00 |
| Hood for range, 2-speed, vented | | | | |
| 30" wide | EA. | 130.00 | 110.00 | 240.00 |
| 42" wide | " | 500.00 | 110.00 | 610.00 |
| Ice maker, automatic | | | | |
| 30 lb per day | EA. | 620.00 | 45.30 | 665.30 |
| 50 lb per day | " | 1,400 | 160.00 | 1,560 |
| Folding access stairs, disappearing metal stair | | | | |
| 8' long | EA. | 1,460 | 45.30 | 1,505 |
| 11' long | " | 1,520 | 45.30 | 1,565 |
| 12' long | " | 1,630 | 45.30 | 1,675 |
| Wood frame, wood stair | | | | |
| 22" x 54" x 8'9" long | EA. | 78.00 | 31.70 | 109.70 |
| 25" x 54" x 10' long | " | 100.00 | 31.70 | 131.70 |
| Ranges electric | | | | |
| Built-in, 30", 1 oven | EA. | 870.00 | 110.00 | 980.00 |
| 2 oven | " | 1,340 | 110.00 | 1,450 |
| Counter top, 4 burner, standard | " | 490.00 | 79.00 | 569.00 |
| With grill | " | 1,460 | 79.00 | 1,539 |
| Free standing, 21", 1 oven | " | 560.00 | 110.00 | 670.00 |
| 30", 1 oven | " | 720.00 | 63.50 | 783.50 |
| 2 oven | " | 2,160 | 63.50 | 2,224 |
| Water softener | | | | |
| 30 grains per gallon | EA. | 730.00 | 110.00 | 840.00 |
| 70 grains per gallon | " | 1,400 | 160.00 | 1,560 |
| **11470.10 DARKROOM EQUIPMENT** | | | | |
| Dryers | | | | |
| 36" x 25" x 68" | EA. | 7,110 | 170.00 | 7,280 |
| 48" x 25" x 68" | " | 7,350 | 170.00 | 7,520 |
| Processors, film | | | | |
| Black and white | EA. | 10,550 | 170.00 | 10,720 |
| Color negatives | " | 11,660 | 170.00 | 11,830 |
| Prints | " | 12,590 | 170.00 | 12,760 |
| Transparencies | " | 13,990 | 170.00 | 14,160 |
| Sinks with cabinet and/or stand | | | | |
| 5" sink with stand | | | | |
| 24" x 48" | EA. | 770.00 | 85.50 | 855.50 |
| 32" x 64" | " | 960.00 | 110.00 | 1,070 |
| 38" x 52" | " | 960.00 | 110.00 | 1,070 |
| 42" x 132" | " | 1,260 | 170.00 | 1,430 |
| 48" x 52" | " | 1,160 | 170.00 | 1,330 |
| 5" sink with cabinet | | | | |

# 11 EQUIPMENT

| ARCHITECTURAL EQUIPMENT | UNIT | MAT. | INST. | TOTAL |
|---|---|---|---|---|
| **11470.10 DARKROOM EQUIPMENT** | | | | |
| 24" x 48" | EA. | 1,570 | 85.50 | 1,656 |
| 32" x 64" | " | 1,690 | 110.00 | 1,800 |
| 38" x 52" | " | 1,660 | 110.00 | 1,770 |
| 42" x 132" | " | 3,440 | 170.00 | 3,610 |
| 48" x 52" | " | 2,160 | 170.00 | 2,330 |
| 10" sink with stand | | | | |
| 24" x 48" | EA. | 960.00 | 85.50 | 1,046 |
| 32" x 64" | " | 1,130 | 110.00 | 1,240 |
| 38" x 52" | " | 1,120 | 110.00 | 1,230 |
| 10" sink with cabinet | | | | |
| 24" x 48" | EA. | 1,620 | 85.50 | 1,706 |
| 38" x 52" | " | 1,870 | 110.00 | 1,980 |
| **11480.10 ATHLETIC EQUIPMENT** | | | | |
| Basketball backboard | | | | |
| Fixed | EA. | 610.00 | 390.00 | 1,000 |
| Swing-up | " | 1,400 | 620.00 | 2,020 |
| Portable, hydraulic | " | 7,750 | 150.00 | 7,900 |
| Suspended type, standard | " | 2,740 | 620.00 | 3,360 |
| For glass backboard, add | " | | | 839.60 |
| For electrically operated, add | " | | | 1,096 |
| Bleacher, telescoping, manual | | | | |
| 15 tier, minimum | SEAT | 77.00 | 6.20 | 83.20 |
| Maximum | " | 88.50 | 6.20 | 94.70 |
| 20 tier, minimum | " | 78.00 | 6.85 | 84.85 |
| Maximum | " | 90.00 | 6.85 | 96.85 |
| 30 tier, minimum | " | 79.50 | 10.30 | 89.80 |
| Maximum | " | 91.00 | 10.30 | 101.30 |
| Boxing ring elevated, complete, 22' x 22' | EA. | 100.00 | 3,090 | 3,190 |
| Gym divider curtain | | | | |
| Minimum | S.F. | 3.15 | 0.41 | 3.56 |
| Maximum | " | 4.30 | 0.41 | 4.71 |
| Scoreboards, single face | | | | |
| Minimum | EA. | 2,860 | 310.00 | 3,170 |
| Maximum | " | 18,420 | 1,540 | 19,960 |
| Parallel bars, wall mounted | | | | |
| Minimum | EA. | 960.00 | 310.00 | 1,270 |
| Maximum | " | 2,680 | 510.00 | 3,190 |
| **11500.10 INDUSTRIAL EQUIPMENT** | | | | |
| Vehicular paint spray booth, solid back, 14'4" x 9'6" | | | | |
| 24' deep | EA. | 9,560 | 310.00 | 9,870 |
| 26'6" deep | " | 9,680 | 310.00 | 9,990 |
| 28'6" deep | " | 9,850 | 310.00 | 10,160 |
| Drive through, 14'9" x 9'6" | | | | |
| 24' deep | EA. | 9,850 | 310.00 | 10,160 |
| 26'6" deep | " | 10,090 | 310.00 | 10,400 |
| 28'6" deep | " | 10,200 | 310.00 | 10,510 |
| Water wash, paint spray booth | | | | |
| 5' x 11'2" x 10'8" | EA. | 4,960 | 310.00 | 5,270 |
| 6' x 11'2" x 10'8" | " | 5,070 | 310.00 | 5,380 |
| 8' x 11'2" x 10'8" | " | 6,410 | 310.00 | 6,720 |

# 11 EQUIPMENT

| ARCHITECTURAL EQUIPMENT | UNIT | MAT. | INST. | TOTAL |
|---|---|---|---|---|
| **11500.10 INDUSTRIAL EQUIPMENT** | | | | |
| 10' x 11'2" x 11'2" | EA. | 7,110 | 310.00 | 7,420 |
| 12' x 12'2" x 11'2" | " | 8,690 | 310.00 | 9,000 |
| 14' x 12'2" x 11'2" | " | 10,200 | 310.00 | 10,510 |
| 16' x 12'2" x 11'2" | " | 11,490 | 310.00 | 11,800 |
| 20' x 12'2" x 11'2" | " | 13,760 | 310.00 | 14,070 |
| Dry type spray booth, with paint arrestors | | | | |
| 5'4" x 7'2" x 6'8" | EA. | 2,160 | 310.00 | 2,470 |
| 6'4" x 7'2" x 6'8" | " | 2,620 | 310.00 | 2,930 |
| 8'4" x 7'2" x 9'2" | " | 2,970 | 310.00 | 3,280 |
| 10'4" x 7'2" x 9'2" | " | 3,440 | 310.00 | 3,750 |
| 12'4" x 7'6" x 9'2" | " | 4,020 | 310.00 | 4,330 |
| 14'4" x 7'6" x 9'8" | " | 4,610 | 310.00 | 4,920 |
| 16'4" x 7'7" x 9'8" | " | 5,190 | 310.00 | 5,500 |
| 20'4" x 7'7" x 10'8" | " | 5,830 | 310.00 | 6,140 |
| Air compressor, electric | | | | |
| 1 hp | | | | |
| 115 volt | EA. | 2,620 | 210.00 | 2,830 |
| 5 hp | | | | |
| 115 volt | EA. | 5,420 | 310.00 | 5,730 |
| 230 volt | " | 4,610 | 310.00 | 4,920 |
| Hydraulic lifts | | | | |
| 8,000 lb capacity | EA. | 410.00 | 770.00 | 1,180 |
| 11,000 lb capacity | " | 5,660 | 1,240 | 6,900 |
| 24,000 lb capacity | " | 7,990 | 2,060 | 10,050 |
| Power tools | | | | |
| Band saws | | | | |
| 10" | EA. | 550.00 | 25.70 | 575.70 |
| 14" | " | 960.00 | 30.90 | 990.90 |
| Motorized shaper | " | 510.00 | 23.75 | 533.75 |
| Motorized lathe | " | 610.00 | 25.70 | 635.70 |
| Bench saws | | | | |
| 9" saw | EA. | 1,100 | 20.60 | 1,121 |
| 10" saw | " | 1,690 | 22.05 | 1,712 |
| 12" saw | " | 2,160 | 25.70 | 2,186 |
| Electric grinders | | | | |
| 1/3 hp | EA. | 210.00 | 12.35 | 222.35 |
| 1/2 hp | " | 300.00 | 13.45 | 313.45 |
| 3/4 hp | " | 360.00 | 13.45 | 373.45 |
| **11600.10 LABORATORY EQUIPMENT** | | | | |
| Cabinets, base | | | | |
| Minimum | L.F. | 95.50 | 25.70 | 121.20 |
| Maximum | " | 170.00 | 25.70 | 195.70 |
| Full storage, 7' high | | | | |
| Minimum | L.F. | 140.00 | 25.70 | 165.70 |
| Maximum | " | 220.00 | 25.70 | 245.70 |
| Wall | | | | |
| Minimum | L.F. | 63.00 | 30.90 | 93.90 |
| Maximum | " | 110.00 | 30.90 | 140.90 |
| Counter tops | | | | |
| Minimum | S.F. | 13.40 | 3.85 | 17.25 |
| Average | " | 31.50 | 4.40 | 35.90 |

# 11 EQUIPMENT

| ARCHITECTURAL EQUIPMENT | UNIT | MAT. | INST. | TOTAL |
|---|---|---|---|---|
| **11600.10 LABORATORY EQUIPMENT** | | | | |
| Maximum | S.F. | 51.50 | 5.15 | 56.65 |
| Tables | | | | |
| Open underneath | S.F. | 72.50 | 15.45 | 87.95 |
| Doors underneath | " | 130.00 | 19.30 | 149.30 |
| Medical laboratory equipment | | | | |
| Analyzer | | | | |
| Chloride | EA. | 420.00 | 15.85 | 435.85 |
| Blood | " | 18,660 | 26.40 | 18,686 |
| Bath, water, utility, countertop unit | " | 300.00 | 31.70 | 331.70 |
| Hot plate, lab, countertop | " | 100.00 | 28.80 | 128.80 |
| Stirrer | " | 170.00 | 28.80 | 198.80 |
| Incubator, anaerobic, 23x23x36" | " | 4,900 | 160.00 | 5,060 |
| Dry heat bath | " | 360.00 | 53.00 | 413.00 |
| Incinerator, for sterilizing | " | 150.00 | 3.15 | 153.15 |
| Meter, serum protein | " | 650.00 | 3.95 | 653.95 |
| Ph analog, general purpose | " | 490.00 | 4.55 | 494.55 |
| Refrigerator, blood bank, undercounter type 153 litres | " | 3,620 | 53.00 | 3,673 |
| 5.4 cf, undercounter type | " | 3,030 | 53.00 | 3,083 |
| Refrigerator/freezer, 4.4 cf, undercounter type | " | 340.00 | 53.00 | 393.00 |
| Sealer, impulse, free standing, 20x12x4" | " | 190.00 | 10.55 | 200.55 |
| Timer, electric, 1-60 minutes, bench or wall mounted | " | 49.00 | 17.60 | 66.60 |
| Glassware washer - dryer, undercounter | " | 5,540 | 400.00 | 5,940 |
| Balance, torsion suspension, tabletop, 4.5 lb capacity | " | 400.00 | 17.60 | 417.60 |
| Binocular microscope, with in base illuminator | " | 1,570 | 12.20 | 1,582 |
| Centrifuge, table model, 19x16x13" | " | 910.00 | 12.65 | 922.65 |
| Clinical model, with four place head | " | 480.00 | 7.05 | 487.05 |
| **11700.10 MEDICAL EQUIPMENT** | | | | |
| Hospital equipment, lights | | | | |
| Examination, portable | EA. | 760.00 | 26.40 | 786.40 |
| Meters | | | | |
| Air flow meter | EA. | 44.30 | 17.60 | 61.90 |
| Oxygen flow meters | " | 30.30 | 13.20 | 43.50 |
| Racks | | | | |
| 40 chart, revolving open frame; mobile caddy | EA. | 870.00 | 26.40 | 896.40 |
| Scales. | | | | |
| Clinical, metric with measure rod, 350 lb | EA. | 270.00 | 28.80 | 298.80 |
| Physical therapy | | | | |
| Chair, hydrotherapy | EA. | 220.00 | 5.15 | 225.15 |
| Diathermy, shortwave, portable, on casters | " | 1,980 | 12.35 | 1,992 |
| Exercise bicycle, floor standing, 35" x 15" | " | 550.00 | 10.30 | 560.30 |
| Hydrocollator, 4 pack, portable, 129 x 90 x 160" | " | 220.00 | 4.40 | 224.40 |
| Lamp, infrared, mobile with variable heat control | " | 290.00 | 23.75 | 313.75 |
| Ultra violet, base mounted | " | 270.00 | 23.75 | 293.75 |
| Mirror, posture training, 27" wide and 72" high | " | 300.00 | 7.70 | 307.70 |
| Parallel bars, adjustable | " | 1,340 | 38.60 | 1,379 |
| Platform mat 10'x6', 1" thick | " | 360.00 | 7.70 | 367.70 |
| Pulley, duplex, wall mounted | " | 260.00 | 100.00 | 360.00 |
| Rack, crutch, wall mounted, 66 x 16 x 13" | " | 560.00 | 30.90 | 590.90 |
| Stimulator, galvanic-faradic, hand held | " | 63.00 | 2.05 | 65.05 |
| Ultrasound, muscle stimulator, portable, 13x13x8" | " | 1,460 | 2.55 | 1,463 |
| Sandbag set, velcro straps, saddle bag type | " | 79.50 | 4.40 | 83.90 |

## ARCHITECTURAL EQUIPMENT

### 11700.10 MEDICAL EQUIPMENT

| ARCHITECTURAL EQUIPMENT | UNIT | MAT. | INST. | TOTAL |
|---|---|---|---|---|
| Whirlpool, 85 gallon | EA. | 1,980 | 150.00 | 2,130 |
| 65 gallon capacity | " | 1,690 | 150.00 | 1,840 |
| Radiology | | | | |
| Radiographic table, motor driven tilting table | EA. | 21,570 | 3,090 | 24,660 |
| Fluoroscope image/tv system | " | 60,640 | 6,180 | 66,820 |
| Processor for washing and drying radiographs | | | | |
| Water filter unit, 30" x 48-1/2" x 37-1/2" | EA. | 19.45 | 530.00 | 549.45 |
| Cassette transfer cabinet | " | 610.00 | 26.40 | 636.40 |
| Base storage cabinets, sectional design | | | | |
| With back splash, 24" deep and 35" high | L.F. | 1,490 | 26.40 | 1,516 |
| Wall storage cabinets | " | 110.00 | 39.60 | 149.60 |
| Steam sterilizers | | | | |
| For heat and moisture stable materials | EA. | 2,220 | 31.70 | 2,252 |
| For fast drying after sterilization | " | 1,520 | 39.60 | 1,560 |
| Compact unit | " | 1,070 | 39.60 | 1,110 |
| Semi-automatic | " | 4,430 | 160.00 | 4,590 |
| Floor loading | | | | |
| Single door | EA. | 80,470 | 260.00 | 80,730 |
| Double door | " | 101,460 | 320.00 | |
| Utensil washer, sanitizer | " | 4,550 | 240.00 | 4,790 |
| Automatic washer/sterilizer | " | 17,490 | 630.00 | 18,120 |
| 16 x 16 x 26", including generator & accessories | " | 16,910 | 1,060 | 17,970 |
| Steam generator, elec., 10 kw to 180 kw | " | 15,740 | 630.00 | 16,370 |
| Surgical scrub | | | | |
| Minimum | EA. | 1,730 | 110.00 | 1,840 |
| Maximum | " | 3,620 | 110.00 | 3,730 |
| Gas sterilizers | | | | |
| Automatic, free standing, 21x19x29" | EA. | 3,090 | 320.00 | 3,410 |
| Surgical tables | | | | |
| Minimum | EA. | 8,050 | 450.00 | 8,500 |
| Maximum | " | 36,150 | 630.00 | 36,780 |
| Surgical lights, ceiling mounted | | | | |
| Minimum | EA. | 3,030 | 530.00 | 3,560 |
| Maximum | " | 4,900 | 630.00 | 5,530 |
| Water stills | | | | |
| 4 liters/hr | EA. | 1,570 | 110.00 | 1,680 |
| 8 liters/hr | " | 2,270 | 110.00 | 2,380 |
| 19 liters/hr | " | 6,300 | 260.00 | 6,560 |
| X-ray equipment | | | | |
| Mobile unit | | | | |
| Minimum | EA. | 5,360 | 160.00 | 5,520 |
| Maximum | " | 26,820 | 320.00 | 27,140 |
| Film viewers | | | | |
| Minimum | EA. | 110.00 | 53.00 | 163.00 |
| Maximum | " | 480.00 | 110.00 | 590.00 |
| Autopsy table | | | | |
| Minimum | EA. | 3,030 | 320.00 | 3,350 |
| Maximum | " | 6,010 | 320.00 | 6,330 |
| Incubators | | | | |
| 15 cf | EA. | 2,800 | 160.00 | 2,960 |
| 29 cf | " | 3,380 | 260.00 | 3,640 |
| Infant transport, portable | " | 7,110 | 170.00 | 7,280 |

# 11 EQUIPMENT

| ARCHITECTURAL EQUIPMENT | UNIT | MAT. | INST. | TOTAL |
|---|---|---|---|---|
| **11700.10 MEDICAL EQUIPMENT** | | | | |
| Beds | | | | |
| Stretcher, with pad, 30" x 78" | EA. | 1,080 | 79.00 | 1,159 |
| Transfer, for patient transport | " | 1,870 | 79.00 | 1,949 |
| Headwall | | | | |
| Aluminum, with back frame and console | EA. | 2,510 | 160.00 | 2,670 |
| **11700.20 DENTAL EQUIPMENT** | | | | |
| Dental care equipment | | | | |
| Drill console with accessories | EA. | 8,860 | 530.00 | 9,390 |
| Amalgamator | " | 220.00 | 15.85 | 235.85 |
| Lathe | " | 230.00 | 10.55 | 240.55 |
| Finish polisher | " | 170.00 | 21.10 | 191.10 |
| Model trimmer | " | 280.00 | 14.40 | 294.40 |
| Motor, wall mounted | " | 310.00 | 14.40 | 324.40 |
| Cleaner, ultrasonic | " | 1,150 | 31.70 | 1,182 |
| Curing unit, bench mounted | " | 1,570 | 53.00 | 1,623 |
| Oral evacuation system, dual pump | " | 8,570 | 39.60 | 8,610 |
| Sterilizer, table top, self contained | " | 600.00 | 17.60 | 617.60 |
| Dental lights | | | | |
| Light, floor or ceiling mounted | EA. | 1,460 | 160.00 | 1,620 |
| X-ray unit | | | | |
| Portable | EA. | 2,200 | 79.00 | 2,279 |
| Wall mounted with remote control | " | 4,550 | 260.00 | 4,810 |
| Illuminator, single panel | " | 6,470 | 450.00 | 6,920 |
| X-ray film processor | " | 1,060 | 260.00 | 1,320 |
| Shield, portable x-ray, lead lined | " | 230.00 | 21.10 | 251.10 |

# 12 FURNISHINGS

| INTERIOR | UNIT | MAT. | INST. | TOTAL |
|---|---|---|---|---|
| **12302.10 CASEWORK** | | | | |
| Kitchen base cabinet, prefinished, 24" deep, 35" high | | | | |
| 12" wide | EA. | 130.00 | 30.90 | 160.90 |
| 18" wide | " | 150.00 | 30.90 | 180.90 |
| 24" wide | " | 170.00 | 34.30 | 204.30 |
| 27" wide | " | 200.00 | 34.30 | 234.30 |
| 36" wide | " | 220.00 | 38.60 | 258.60 |
| 48" wide | " | 250.00 | 38.60 | 288.60 |
| Corner cabinet, 36" wide | " | 170.00 | 38.60 | 208.60 |
| Wall cabinet, 12" deep, 12" high | | | | |
| 30" wide | EA. | 95.50 | 30.90 | 126.40 |
| 36" wide | " | 100.00 | 30.90 | 130.90 |
| 15" high | | | | |
| 30" wide | EA. | 100.00 | 34.30 | 134.30 |
| 36" wide | " | 110.00 | 34.30 | 144.30 |
| 24" high | | | | |
| 30" wide | EA. | 120.00 | 34.30 | 154.30 |
| 36" wide | " | 150.00 | 34.30 | 184.30 |
| 30" high | | | | |
| 12" wide | EA. | 91.00 | 38.60 | 129.60 |
| 18" wide | " | 100.00 | 38.60 | 138.60 |
| 24" wide | " | 110.00 | 38.60 | 148.60 |
| 27" wide | " | 130.00 | 38.60 | 168.60 |
| 30" wide | " | 150.00 | 44.10 | 194.10 |
| 36" wide | " | 160.00 | 44.10 | 204.10 |
| Corner cabinet, 30" high | | | | |
| 24" wide | EA. | 140.00 | 51.50 | 191.50 |
| 30" wide | " | 180.00 | 51.50 | 231.50 |
| 36" wide | " | 200.00 | 51.50 | 251.50 |
| Wardrobe | " | 460.00 | 77.00 | 537.00 |
| Vanity with top, laminated plastic | | | | |
| 24" wide | EA. | 290.00 | 77.00 | 367.00 |
| 30" wide | " | 350.00 | 77.00 | 427.00 |
| 36" wide | " | 370.00 | 100.00 | 470.00 |
| 48" wide | " | 410.00 | 120.00 | 530.00 |
| **12390.10 COUNTER TOPS** | | | | |
| Stainless steel, counter top, with backsplash | S.F. | 73.50 | 7.70 | 81.20 |
| Acid-proof, kemrock surface | " | 19.80 | 5.15 | 24.95 |
| **12500.10 WINDOW TREATMENT** | | | | |
| Drapery tracks, wall or ceiling mounted | | | | |
| Basic traverse rod | | | | |
| 50 to 90" | EA. | 29.20 | 15.45 | 44.65 |
| 84 to 156" | " | 43.70 | 17.15 | 60.85 |
| 136 to 250" | " | 56.00 | 17.15 | 73.15 |
| 165 to 312" | " | 81.50 | 19.30 | 100.80 |
| Traverse rod with stationary curtain rod | | | | |
| 30 to 50" | EA. | 35.00 | 15.45 | 50.45 |
| 50 to 90" | " | 43.10 | 15.45 | 58.55 |
| 84 to 156" | " | 63.00 | 17.15 | 80.15 |
| 136 to 250" | " | 72.50 | 19.30 | 91.80 |
| Double traverse rod | | | | |

# 12 FURNISHINGS

| INTERIOR | UNIT | MAT. | INST. | TOTAL |
|---|---|---|---|---|
| **12500.10 WINDOW TREATMENT** | | | | |
| 30 to 50" | EA. | 42.00 | 15.45 | 57.45 |
| 50 to 84" | " | 56.00 | 15.45 | 71.45 |
| 84 to 156" | " | 69.00 | 17.15 | 86.15 |
| 136 to 250" | " | 84.00 | 19.30 | 103.30 |
| **12510.10 BLINDS** | | | | |
| Venetian blinds | | | | |
| 2" slats | S.F. | 3.25 | 0.77 | 4.02 |
| 1" slats | " | 5.60 | 0.77 | 6.37 |
| **12690.40 FLOOR MATS** | | | | |
| Recessed entrance mat, 3/8" thick, aluminum link | S.F. | 14.55 | 15.45 | 30.00 |
| Steel, flexible | " | 7.35 | 15.45 | 22.80 |

# 13 SPECIAL

| CONSTRUCTION | UNIT | MAT. | INST. | TOTAL |
|---|---|---|---|---|
| **13056.10 VAULTS** | | | | |
| Floor safes | | | | |
| Class C | | | | |
| 1.0 cf | EA. | 470.00 | 25.70 | 495.70 |
| 1.3 cf | " | 530.00 | 38.60 | 568.60 |
| 1.9 cf | " | 580.00 | 51.50 | 631.50 |
| 5.2 cf | " | 1,400 | 51.50 | 1,452 |
| **13121.10 PRE-ENGINEERED BUILDINGS** | | | | |
| Pre-engineered metal building, 40'x100' | | | | |
| 14' eave height | S.F. | 7.45 | 2.55 | 10.00 |
| 16' eave height | " | 8.30 | 2.95 | 11.25 |
| 20' eave height | " | 9.45 | 3.80 | 13.25 |
| 60'x100' | | | | |
| 14' eave height | S.F. | 6.75 | 2.55 | 9.30 |
| 16' eave height | " | 7.10 | 2.95 | 10.05 |
| 20' eave height | " | 7.95 | 3.80 | 11.75 |
| 80'x100' | | | | |
| 14' eave height | S.F. | 6.75 | 2.55 | 9.30 |
| 16' eave height | " | 7.00 | 2.95 | 9.95 |
| 20' eave height | " | 8.05 | 3.80 | 11.85 |
| 100'x100' | | | | |
| 14' eave height | S.F. | 6.75 | 2.55 | 9.30 |
| 16' eave height | " | 7.10 | 2.95 | 10.05 |
| 20' eave height | " | 7.80 | 3.80 | 11.60 |
| 100'x150' | | | | |
| 14' eave height | S.F. | 6.20 | 2.55 | 8.75 |
| 16' eave height | " | 6.65 | 2.95 | 9.60 |
| 20' eave height | " | 7.35 | 3.80 | 11.15 |
| 120'x150' | | | | |
| 14' eave height | S.F. | 6.20 | 2.55 | 8.75 |
| 16' eave height | " | 6.55 | 2.95 | 9.50 |
| 20' eave height | " | 7.10 | 3.80 | 10.90 |
| 140'x150' | | | | |
| 14' eave height | S.F. | 5.85 | 2.55 | 8.40 |
| 16' eave height | " | 6.30 | 2.95 | 9.25 |
| 20' eave height | " | 7.00 | 3.80 | 10.80 |
| 160'x200' | | | | |
| 14' eave height | S.F. | 5.70 | 2.55 | 8.25 |
| 16' eave height | " | 5.75 | 2.95 | 8.70 |
| 20' eave height | " | 6.55 | 3.80 | 10.35 |
| 200'x200' | | | | |
| 14' eave height | S.F. | 5.70 | 2.55 | 8.25 |
| 16' eave height | " | 5.95 | 2.95 | 8.90 |
| 20' eave height | " | 6.40 | 3.80 | 10.20 |
| Hollow metal door and frame, 6' x 7' | EA. | | | 582.75 |
| Sectional steel overhead door, manually operated | | | | |
| 8' x 8' | EA. | | | 629.70 |
| 12' x 12' | " | | | 944.61 |
| Roll-up steel door, manually operated | | | | |
| 10' x 10' | EA. | | | 1,167 |
| 12' x 12' | " | | | 1,400 |
| For gravity ridge ventilator with birdscreen | " | | | 431.46 |

| CONSTRUCTION | UNIT | MAT. | INST. | TOTAL |
|---|---|---|---|---|
| **13121.10 PRE-ENGINEERED BUILDINGS** | | | | |
| 9" throat x 10' | EA. | | | 466.42 |
| 12" throat x 10' | " | | | 489.84 |
| For 20" rotary vent with damper | " | | | 198.25 |
| For 4' x 3' fixed louver | " | | | 209.90 |
| For 4' x 3' aluminum sliding window | " | | | 215.78 |
| For 3' x 9' fiberglass panels | " | | | 86.29 |
| Liner panel, 26 ga, painted steel | S.F. | 1.35 | 0.84 | 2.19 |
| Wall panel insulated, 26 ga. steel, foam core | " | 4.30 | 0.84 | 5.14 |
| Roof panel, 26 ga. painted steel | " | 1.10 | 0.48 | 1.58 |
| Plastic (sky light) | " | 3.15 | 0.48 | 3.63 |
| Insulation, 3-1/2" thick blanket, R11 | " | 0.83 | 0.23 | 1.05 |
| **13152.10 SWIMMING POOL EQUIPMENT** | | | | |
| Diving boards | | | | |
| 14' long | | | | |
| Aluminum | EA. | 940.00 | 130.00 | 1,070 |
| Fiberglass | " | 580.00 | 130.00 | 710.00 |
| Ladders, heavy duty | | | | |
| 2 steps | | | | |
| Minimum | EA. | 380.00 | 48.40 | 428.40 |
| Maximum | " | 600.00 | 48.40 | 648.40 |
| 4 steps | | | | |
| Minimum | EA. | 420.00 | 60.50 | 480.50 |
| Maximum | " | 790.00 | 60.50 | 850.50 |
| Lifeguard chair | | | | |
| Minimum | EA. | 870.00 | 240.00 | 1,110 |
| Maximum | " | 1,400 | 240.00 | 1,640 |
| Lights, underwater | | | | |
| 12 volt, with transformer | EA. | 280.00 | 60.50 | 340.50 |
| 110 volt | | | | |
| Minimum | EA. | 260.00 | 60.50 | 320.50 |
| Maximum | " | 820.00 | 60.50 | 880.50 |
| Ground fault interrupter for 110 volt, each light | " | 100.00 | 20.15 | 120.15 |
| Pool covers | | | | |
| Reinforced polyethylene | S.F. | 0.47 | 1.85 | 2.32 |
| Vinyl water tube | | | | |
| Minimum | S.F. | 0.52 | 1.85 | 2.37 |
| Maximum | " | 0.81 | 1.85 | 2.66 |
| Slides with water tube | | | | |
| Minimum | EA. | 280.00 | 200.00 | 480.00 |
| Maximum | " | 790.00 | 200.00 | 990.00 |

| ELEVATORS | UNIT | MAT. | INST. | TOTAL |
|---|---|---|---|---|
| **14210.10 ELEVATORS** | | | | |
| Passenger elevators, electric, geared | | | | |
| Based on a shaft of 6 stops and 6 openings | | | | |
| 50 fpm, 2000 lb | EA. | 62,970 | 1,310 | 64,280 |
| 100 fpm, 2000 lb | " | 65,310 | 1,450 | 66,760 |
| 150 fpm | | | | |
| 2000 lb | EA. | 72,300 | 1,630 | 73,930 |
| 3000 lb | " | 90,960 | 1,870 | 92,830 |
| 4000 lb | " | 94,460 | 2,180 | 96,640 |
| 200 fpm | | | | |
| 2500 lb | EA. | 88,050 | 1,870 | 89,920 |
| 3000 lb | " | 90,380 | 2,010 | 92,390 |
| 4000 lb | " | 95,630 | 2,180 | 97,810 |
| 250 fpm | | | | |
| 2500 lb | EA. | 90,960 | 1,870 | 92,830 |
| 3000 lb | " | 95,630 | 2,010 | 97,640 |
| 4000 lb | " | 97,960 | 2,180 | |
| 300 fpm | | | | |
| 2500 lb | EA. | 90,960 | 1,870 | 92,830 |
| 3000 lb | " | 94,460 | 2,010 | 96,470 |
| 4000 lb | " | 97,960 | 1,310 | 99,270 |
| For each additional; 50 fpm, add per stop, $2000 | | | | |
| 500 lb, add per stop, $2000 | | | | |
| Opening, add per stop, $4500 | | | | |
| Stop, add per stop, $4000 | | | | |
| Bonderized steel door, add per opening, $150 | | | | |
| Colored aluminum door, add per opening, $850 | | | | |
| Stainless steel door, add per opening, $600 | | | | |
| Cast bronze door, add per opening, $1100 | | | | |
| Two speed door, add per opening, $360 | | | | |
| Bi-parting door, add per opening, $850 | | | | |
| Custom cab interior add, $4800 | | | | |
| Based on a shaft of 8 stops and 8 openings | | | | |
| 300 fpm | | | | |
| 3000 lb | EA. | 108,450 | 2,620 | |
| 3500 lb | " | 110,790 | 2,620 | |
| 4000 lb | " | 116,620 | 2,910 | |
| 5000 lb | " | 130,610 | 3,110 | |
| 400 fpm | | | | |
| 3000 lb | EA. | 114,280 | 2,620 | |
| 3500 lb | " | 116,030 | 2,620 | |
| 4000 lb | " | 12,830 | 2,910 | 15,740 |
| 5000 lb | " | 139,940 | 3,110 | |
| 600 fpm | | | | |
| 3000 lb | EA. | 163,260 | 2,910 | |
| 3500 lb | " | 165,600 | 3,110 | |
| 4000 lb | " | 166,760 | 3,190 | |
| 5000 lb | " | 174,340 | 3,270 | |
| 800 fpm | | | | |
| 3000 lb | EA. | 192,420 | 2,910 | |
| 3500 lb | " | 194,750 | 3,110 | |
| 4000 lb | " | 197,080 | 3,190 | |
| 5000 lb | " | 204,080 | 3,270 | |

# 14 CONVEYING

## ELEVATORS

### 14210.10 ELEVATORS

| ELEVATORS | UNIT | MAT. | INST. | TOTAL |
|---|---|---|---|---|
| For each additional; 100 fpm add per stop, $11,000 | | | | |
| 500 lb, add per stop, $5500 | | | | |
| Opening add per stop, $11,000 | | | | |
| Stop add per stop, $4800 | | | | |
| Bypass floor, add per each, $2000 | | | | |
| Bonderized steel door, add per opening, $150 | | | | |
| Colored aluminum door, add per opening, $900 | | | | |
| Stainless steel door, add per opening, $600 | | | | |
| Cast bronze door, add per opening, $600 | | | | |
| Two speed bi-parting door, add per opening, $1000 | | | | |
| Custom cab interior, add $5000 | | | | |
| Hydraulic, based on a shaft of 3 stops, 3 openings | | | | |
| 50 fpm | | | | |
| 2000 lb | EA. | 43,180 | 1,090 | 44,270 |
| 2500 lb | " | 46,210 | 1,090 | 47,300 |
| 3000 lb | " | 48,760 | 1,140 | 49,900 |
| 100 fpm | | | | |
| 2000 lb | EA. | 47,420 | 1,090 | 48,510 |
| 2500 lb | " | 49,970 | 1,140 | 51,110 |
| 3000 lb | " | 53,120 | 1,190 | 54,310 |
| 150 fpm | | | | |
| 2000 lb | EA. | 51,180 | 1,090 | 52,270 |
| 2500 lb | " | 56,030 | 1,140 | 57,170 |
| 3000 lb | " | 59,910 | 1,250 | 61,160 |
| For each additional; 50 fpm add per stop, $3500 | | | | |
| 500 lb, add per stop, $3500 | | | | |
| Opening, add, $4200 | | | | |
| Stop, add per stop, $5300 | | | | |
| Bonderized steel door, add per opening, $400 | | | | |
| Colored aluminum door, add per opening, $1500 | | | | |
| Stainless steel door, add per opening, $650 | | | | |
| Cast bronze door, add per opening, $1200 | | | | |
| Two speed door, add per opening, $400 | | | | |
| Bi-parting door, add per opening, $900 | | | | |
| Custom cab interior, add per cab, $5000 | | | | |
| Small elevators, 4 to 6 passenger capacity | | | | |
| Electric, push | | | | |
| 2 stops | EA. | 15,770 | 1,090 | 16,860 |
| 3 stops | " | 19,410 | 1,190 | 20,600 |
| 4 stops | " | 22,440 | 1,310 | 23,750 |
| Freight elevators, electric | | | | |
| Based on a shaft of 6 stops and 6 openings | | | | |
| 50 fpm | | | | |
| 3500 lb | EA. | 124,900 | 1,450 | |
| 4000 lb | " | 125,370 | 1,450 | |
| 5000 lb | " | 126,560 | 1,630 | |
| 100 fpm | | | | |
| 3500 lb | EA. | 131,440 | 1,630 | |
| 4000 lb | " | 132,030 | 1,630 | |
| 5000 lb | " | 128,820 | 1,870 | |
| 200 fpm | | | | |
| 3500 lb | EA. | 129,420 | 1,870 | |

# 14 CONVEYING

| ELEVATORS | UNIT | MAT. | INST. | TOTAL |
|---|---|---|---|---|
| **14210.10 ELEVATORS** | | | | |
| 4000 lb | EA. | 130,010 | 1,870 | |
| 5000 lb | " | 131,200 | 2,180 | |
| For elevator with manual door, deduct 15% | | | | |
| For variable voltage control, add 20% | | | | |
| Based on shaft of 8 stops and 8 openings | | | | |
| 100 fpm | | | | |
| 4000 lb | EA. | 131,800 | 1,630 | |
| 6000 lb | " | 127,280 | 1,680 | |
| 8000 lb | " | 129,650 | 1,770 | |
| 150 fpm | | | | |
| 4000 lb | EA. | 127,280 | 1,870 | |
| 6000 lb | " | 129,650 | 1,920 | |
| 8000 lb | " | 130,840 | 2,040 | |
| 200 fpm | | | | |
| 4000 lb | EA. | 128,460 | 2,180 | |
| 6000 lb | " | 130,840 | 2,250 | |
| 8000 lb | " | 136,790 | 2,380 | |
| For each additional; 50 fpm, add per stop, $2000 | | | | |
| 500 lb, add per stop, $600 | | | | |
| Opening, add per stop, $7000 | | | | |
| Stop, add per stop, $5500 | | | | |
| For variable voltage, add 20% | | | | |
| Hydraulic, based on 3 stops and 3 openings | | | | |
| 50 fpm | | | | |
| 3000 lb | EA. | 53,360 | 930.00 | 54,290 |
| 4000 lb | " | 58,220 | 970.00 | 59,190 |
| 6000 lb | " | 67,920 | 1,010 | 68,930 |
| 100 fpm | | | | |
| 3000 lb | EA. | 60,160 | 930.00 | 61,090 |
| 4000 lb | " | 64,280 | 970.00 | 65,250 |
| 6000 lb | " | 75,070 | 1,010 | 76,080 |
| 150 fpm | | | | |
| 3000 lb | EA. | 66,100 | 930.00 | 67,030 |
| 4000 lb | " | 70,950 | 970.00 | 71,920 |
| 6000 lb | " | 81,380 | 1,010 | 82,390 |
| For each additional; 50 fpm, add per stop, $2000 | | | | |
| 500 lb, add per stop, $600 | | | | |
| Opening, add per stop, $5500 | | | | |
| Stop, add per stop, $5500 | | | | |
| For elevator with manual door deduct from total, 15% | | | | |
| **14300.10 ESCALATORS** | | | | |
| Escalators | | | | |
| 32" wide, floor to floor | | | | |
| 12' high | EA. | 82,800 | 2,180 | 84,980 |
| 15' high | " | 89,790 | 2,620 | 92,410 |
| 18' high | " | 96,790 | 3,270 | |
| 22' high | " | 104,950 | 4,360 | |
| 25' high | " | 116,620 | 5,230 | |
| 48" wide | | | | |
| 12' high | EA. | 92,130 | 2,250 | 94,380 |
| 15' high | " | 100,290 | 2,720 | |

# 14 CONVEYING

## ELEVATORS

| 14300.10 ESCALATORS | UNIT | MAT. | INST. | TOTAL |
|---|---|---|---|---|
| 18' high | EA. | 107,290 | 3,440 | |
| 22' high | " | 120,120 | 4,670 | |
| 25' high | " | 128,280 | 5,230 | |

## LIFTS

| 14410.10 PERSONNEL LIFTS | UNIT | MAT. | INST. | TOTAL |
|---|---|---|---|---|
| Electrically operated, 1 or 2 person lift | | | | |
| With attached foot platforms | | | | |
| 3 stops | EA. | | | 14,578 |
| 5 stops | " | | | 18,659 |
| 7 stops | " | | | 22,157 |
| For each additional stop, add $1250 | | | | |
| Residential stair climber, per story | EA. | 4,310 | 260.00 | 4,570 |

| 14450.10 VEHICLE LIFTS | UNIT | MAT. | INST. | TOTAL |
|---|---|---|---|---|
| Automotive hoist, one post, semi-hydraulic, 8,000 lb | EA. | 3,380 | 1,310 | 4,690 |
| Full hydraulic, 8,000 lb | " | 3,210 | 1,310 | 4,520 |
| 2 post, semi-hydraulic, 10,000 lb | " | 3,500 | 1,870 | 5,370 |
| Full hydraulic | | | | |
| 10,000 lb | EA. | 3,730 | 1,870 | 5,600 |
| 13,000 lb | " | 4,610 | 3,270 | 7,880 |
| 18,500 lb | " | 6,010 | 3,270 | 9,280 |
| 24,000 lb | " | 8,110 | 3,270 | 11,380 |
| 26,000 lb | " | 9,390 | 3,270 | 12,660 |
| Pneumatic hoist, fully hydraulic | | | | |
| 11,000 lb | EA. | 4,490 | 4,360 | 8,850 |
| 24,000 lb | " | 7,870 | 4,360 | 12,230 |

## MATERIAL HANDLING

| 14560.10 CHUTES | UNIT | MAT. | INST. | TOTAL |
|---|---|---|---|---|
| Linen chutes, stainless steel, with supports | | | | |
| 18" dia. | L.F. | 84.00 | 2.40 | 86.40 |

# 14 CONVEYING

## MATERIAL HANDLING

| 14560.10 CHUTES | UNIT | MAT. | INST. | TOTAL |
|---|---|---|---|---|
| 24" dia. | L.F. | 110.00 | 2.60 | 112.60 |
| 30" dia. | " | 110.00 | 2.80 | 112.80 |
| Hopper | EA. | 1,340 | 22.55 | 1,363 |
| Skylight | " | 810.00 | 33.80 | 843.80 |
| Sprinkler unit at top | " | 300.00 | 37.60 | 337.60 |
| For galvanized metal, deduct from material cost, 35% | | | | |
| For aluminum, deduct from material cost, 25% | | | | |

| 14580.10 PNEUMATIC SYSTEMS | UNIT | MAT. | INST. | TOTAL |
|---|---|---|---|---|
| Pneumatic message tube system | | | | |
| Average, 20 station job | | | | |
| 3" round system | E.A. | 216,210 | 2,880 | |
| 4" round system | " | 270,260 | 3,170 | |
| 6" round system | " | 408,390 | 3,520 | |
| 4" x 7" oval system | " | 426,410 | 6,340 | |
| Trash and linen tube system | | | | |
| 10 stations | EA. | 138,130 | 6,540 | |
| 15 stations | " | 174,170 | 8,720 | |
| 20 stations | " | 233,640 | 10,060 | |
| 30 stations | " | 306,290 | 11,890 | |

## HOISTS AND CRANES

| 14600.10 INDUSTRIAL HOISTS | UNIT | MAT. | INST. | TOTAL |
|---|---|---|---|---|
| Industrial hoists, electric, light to medium duty | | | | |
| 500 lb | EA. | 4,730 | 160.00 | 4,890 |
| 1000 lb | " | 4,970 | 170.00 | 5,140 |
| 2000 lb | " | 5,090 | 180.00 | 5,270 |
| 3000 lb | " | 5,190 | 190.00 | 5,380 |
| 4000 lb | " | 5,860 | 200.00 | 6,060 |
| 5000 lb | " | 8,290 | 210.00 | 8,500 |
| 6000 lb | " | 8,560 | 220.00 | 8,780 |
| 7500 lb | " | 9,160 | 230.00 | 9,390 |
| 10,000 lb | " | 15,460 | 230.00 | 15,690 |
| 15,000 lb | " | 22,960 | 240.00 | 23,200 |
| 20,000 lb | " | 27,120 | 260.00 | 27,380 |
| 25,000 lb | " | 29,740 | 290.00 | 30,030 |
| 30,000 lb | " | 29,150 | 320.00 | 29,470 |
| Heavy duty | | | | |
| 500 lb | EA. | 6,980 | 160.00 | 7,140 |
| 1000 lb | " | 7,200 | 170.00 | 7,370 |
| 2000 lb | " | 8,010 | 180.00 | 8,190 |

| HOISTS AND CRANES | UNIT | MAT. | INST. | TOTAL |
|---|---|---|---|---|
| **14600.10 INDUSTRIAL HOISTS** | | | | |
| 3000 lb | EA. | 8,450 | 190.00 | 8,640 |
| 4000 lb | " | 8,920 | 200.00 | 9,120 |
| 5000 lb | " | 9,280 | 210.00 | 9,490 |
| 6000 lb | " | 9,520 | 220.00 | 9,740 |
| 7500 lb | " | 13,080 | 230.00 | 13,310 |
| 10,000 lb | " | 13,680 | 230.00 | 13,910 |
| 15,000 lb | " | 15,110 | 240.00 | 15,350 |
| 20,000 lb | " | 18,670 | 260.00 | 18,930 |
| 25,000 lb | " | 27,120 | 290.00 | 27,410 |
| 30,000 lb | " | 28,310 | 320.00 | 28,630 |
| Air powered hoists | | | | |
| 500 lb | EA. | 2,620 | 160.00 | 2,780 |
| 1000 lb | " | 2,910 | 160.00 | 3,070 |
| 2000 lb | " | 3,180 | 170.00 | 3,350 |
| 4000 lb | " | 3,720 | 190.00 | 3,910 |
| 6000 lb | " | 4,430 | 240.00 | 4,670 |
| Overhead traveling bridge crane | | | | |
| Single girder, 20' span | | | | |
| 3 ton | EA. | 14,870 | 650.00 | 15,520 |
| 5 ton | " | 16,360 | 650.00 | 17,010 |
| 7.5 ton | " | 20,700 | 650.00 | 21,350 |
| 10 ton | " | 20,700 | 820.00 | 21,520 |
| 15 ton | " | 25,570 | 820.00 | 26,390 |
| 30' span | | | | |
| 3 ton | EA. | 15,820 | 650.00 | 16,470 |
| 5 ton | " | 17,490 | 650.00 | 18,140 |
| 10 ton | " | 22,360 | 820.00 | 23,180 |
| 15 ton | " | 26,760 | 820.00 | 27,580 |
| Double girder, 40' span | | | | |
| 3 ton | EA. | 30,690 | 1,450 | 32,140 |
| 5 ton | " | 31,520 | 1,450 | 32,970 |
| 7.5 ton | " | 32,590 | 1,450 | 34,040 |
| 10 ton | " | 34,260 | 1,870 | 36,130 |
| 15 ton | " | 38,540 | 1,870 | 40,410 |
| 25 ton | " | 57,100 | 1,870 | 58,970 |
| 50' span | | | | |
| 3 ton | EA. | 33,480 | 1,450 | 34,930 |
| 5 ton | " | 34,670 | 1,450 | 36,120 |
| 7.5 ton | " | 35,920 | 1,450 | 37,370 |
| 10 ton | " | 39,020 | 1,870 | 40,890 |
| 15 ton | " | 45,200 | 1,870 | 47,070 |
| 25 ton | " | 61,850 | 1,870 | 63,720 |
| **14650.10 JIB CRANES** | | | | |
| Self supporting, swinging 8' boom, 200 deg rotation | | | | |
| 1000 lb | EA. | 1,370 | 280.00 | 1,650 |
| 2000 lb | " | 1,430 | 280.00 | 1,710 |
| 3000 lb | " | 1,670 | 560.00 | 2,230 |
| 4000 lb | " | 1,900 | 560.00 | 2,460 |
| 6000 lb | " | 2,140 | 560.00 | 2,700 |
| 10,000 lb | " | 2,820 | 560.00 | 3,380 |
| Wall mounted, 180 deg rotation | | | | |

# 14 CONVEYING

| HOISTS AND CRANES | UNIT | MAT. | INST. | TOTAL |
|---|---|---|---|---|
| **14650.10 JIB CRANES** | | | | |
| 2000 lb | EA. | 760.00 | 280.00 | 1,040 |
| 3000 lb | " | 900.00 | 280.00 | 1,180 |
| 4000 lb | " | 1,010 | 560.00 | 1,570 |
| 6000 lb | " | 1,140 | 560.00 | 1,700 |
| 10,000 lb | " | 2,260 | 560.00 | 2,820 |

| BASIC MATERIALS | UNIT | MAT. | INST. | TOTAL |
|---|---|---|---|---|
| **15100.10 SPECIALTIES** | | | | |
| Wall penetration | | | | |
| Concrete wall, 6" thick | | | | |
| 2" dia. | EA. | 0.00 | 8.05 | 8.05 |
| 4" dia. | " | 0.00 | 12.10 | 12.10 |
| 8" dia. | " | 0.00 | 17.30 | 17.30 |
| 12" thick | | | | |
| 2" dia. | EA. | 0.00 | 11.00 | 11.00 |
| 4" dia. | " | 0.00 | 17.30 | 17.30 |
| 8" dia. | " | 0.00 | 26.90 | 26.90 |
| Non-destructive testing, piping systems | | | | |
| X-ray of welds | | | | |
| 3" dia. pipe | EA. | 44.70 | 34.20 | 78.90 |
| 4" dia. pipe | " | 38.50 | 34.20 | 72.70 |
| 6" dia. pipe | " | 32.80 | 34.20 | 67.00 |
| 8" dia. pipe | " | 35.10 | 42.70 | 77.80 |
| 10" dia. pipe | " | 37.40 | 42.70 | 80.10 |
| Liquid penetration of welds | | | | |
| 2" dia. pipe | EA. | 22.65 | 21.35 | 44.00 |
| 3" dia. pipe | " | 23.80 | 21.35 | 45.15 |
| 4" dia. pipe | " | 26.00 | 21.35 | 47.35 |
| 6" dia. pipe | " | 26.00 | 21.35 | 47.35 |
| 8" dia. pipe | " | 31.70 | 21.35 | 53.05 |
| 10" dia. pipe | " | 31.70 | 21.35 | 53.05 |
| **15120.10 BACKFLOW PREVENTERS** | | | | |
| Backflow preventer, flanged, cast iron, with valves | | | | |
| 3" pipe | EA. | 2,040 | 170.00 | 2,210 |
| 4" pipe | " | 2,600 | 190.00 | 2,790 |
| 6" pipe | " | 3,850 | 280.00 | 4,130 |
| 8" pipe | " | 7,130 | 340.00 | 7,470 |
| Threaded | | | | |
| 3/4" pipe | EA. | 330.00 | 21.35 | 351.35 |
| 2" pipe | " | 700.00 | 34.20 | 734.20 |
| Reduced pressure assembly, bronze, threaded | | | | |
| 3/4" | EA. | 430.00 | 21.35 | 451.35 |
| 1" | " | 550.00 | 24.40 | 574.40 |
| 1-1/4" | " | 790.00 | 28.50 | 818.50 |
| 1-1/2" | " | 860.00 | 34.20 | 894.20 |
| **15140.10 PIPE HANGERS, HEAVY** | | | | |
| Hangers | | | | |
| 1/2" pipe, clevis pipe hanger | | | | |
| Black steel | EA. | 1.80 | 11.40 | 13.20 |
| Galvanized | " | 2.05 | 11.40 | 13.45 |
| 3/4" pipe, clevis pipe hanger | | | | |
| Black steel | EA. | 1.80 | 11.40 | 13.20 |
| Galvanized | " | 2.10 | 11.40 | 13.50 |
| 1" pipe, clevis pipe hanger | | | | |
| Black steel | EA. | 1.90 | 11.40 | 13.30 |
| Galvanized | " | 2.20 | 11.40 | 13.60 |

# 15 MECHANICAL

## BASIC MATERIALS

| BASIC MATERIALS | UNIT | MAT. | INST. | TOTAL |
|---|---|---|---|---|
| **15140.10 PIPE HANGERS, HEAVY** | | | | |
| U bolt | EA. | 1.20 | 3.40 | 4.60 |
| 2" pipe, clevis pipe hanger | | | | |
| Black steel | EA. | 2.85 | 11.40 | 14.25 |
| Galvanized | " | 3.15 | 11.40 | 14.55 |
| 3" pipe, clevis pipe hanger | | | | |
| Black steel | EA. | 4.75 | 11.40 | 16.15 |
| Galvanized | " | 8.50 | 11.40 | 19.90 |
| 4" pipe, clevis pipe hanger | | | | |
| Black steel | EA. | 6.65 | 11.40 | 18.05 |
| Galvanized | " | 8.40 | 11.40 | 19.80 |
| 6" pipe, clevis pipe hanger | | | | |
| Black steel | EA. | 10.35 | 13.65 | 24.00 |
| Galvanized | " | 15.40 | 13.65 | 29.05 |
| 12" pipe, clevis pipe hanger | | | | |
| Black steel | EA. | 16.55 | 13.65 | 30.20 |
| Galvanized | " | 23.45 | 13.65 | 37.10 |
| Threaded rod, galvanized | | | | |
| 3/8" | L.F. | | | 0.32 |
| 1/2" | " | | | 0.52 |
| Hex nuts, galvanized | | | | |
| 3/8" | EA. | | | 0.03 |
| 1/2" | " | | | 0.07 |
| 1" | " | | | 0.56 |
| C-clamp, steel, with lock nut | | | | |
| 3/8" | EA. | 1.20 | 4.25 | 5.45 |
| 1/2" | " | 1.50 | 4.25 | 5.75 |
| **15140.11 PIPE HANGERS, LIGHT** | | | | |
| A band, black iron | | | | |
| 1/2" | EA. | 0.43 | 2.45 | 2.88 |
| 1" | " | 0.54 | 2.55 | 3.09 |
| 1-1/4" | " | 0.60 | 2.65 | 3.25 |
| 1-1/2" | " | 0.69 | 2.85 | 3.54 |
| 2" | " | 0.77 | 3.10 | 3.87 |
| 2-1/2" | " | 1.05 | 3.40 | 4.45 |
| 3" | " | 1.10 | 3.80 | 4.90 |
| 4" | " | 1.60 | 4.25 | 5.85 |
| 5" | " | 2.05 | 4.55 | 6.60 |
| 6" | " | 2.85 | 4.90 | 7.75 |
| 8" | " | 4.90 | 5.70 | 10.60 |
| Copper | | | | |
| 1/2" | EA. | 0.60 | 2.45 | 3.05 |
| 3/4" | " | 0.60 | 2.55 | 3.15 |
| 1" | " | 0.63 | 2.55 | 3.18 |
| 1-1/4" | " | 0.65 | 2.65 | 3.30 |
| 1-1/2" | " | 0.74 | 2.85 | 3.59 |
| 2" | " | 0.87 | 3.10 | 3.97 |
| 2-1/2" | " | 1.15 | 3.40 | 4.55 |
| 3" | " | 1.30 | 3.80 | 5.10 |
| 4" | " | 1.85 | 4.25 | 6.10 |
| 2 hole clips, galvanized | | | | |
| 3/4" | EA. | 0.14 | 2.30 | 2.44 |

| BASIC MATERIALS | UNIT | MAT. | INST. | TOTAL |
|---|---|---|---|---|
| **15140.11 PIPE HANGERS, LIGHT** | | | | |
| 1" | EA. | 0.17 | 2.35 | 2.52 |
| 1-1/4" | " | 0.21 | 2.45 | 2.66 |
| 1-1/2" | " | 0.27 | 2.55 | 2.82 |
| 2" | " | 0.34 | 2.65 | 2.99 |
| 2-1/2" | " | 0.63 | 2.75 | 3.38 |
| 3" | " | 0.92 | 2.85 | 3.77 |
| 4" | " | 1.95 | 3.10 | 5.05 |
| Perforated strap | | | | |
| 3/4" | | | | |
| Galvanized, 20 ga. | L.F. | 0.21 | 1.70 | 1.91 |
| Copper, 22 ga. | " | 0.27 | 1.70 | 1.97 |
| J-Hooks | | | | |
| 1/2" | EA. | 0.24 | 1.55 | 1.79 |
| 3/4" | " | 0.27 | 1.55 | 1.82 |
| 1" | " | 0.28 | 1.65 | 1.93 |
| 1-1/4" | " | 0.29 | 1.65 | 1.94 |
| 1-1/2" | " | 0.30 | 1.70 | 2.00 |
| 2" | " | 0.31 | 1.70 | 2.01 |
| 3" | " | 0.37 | 1.80 | 2.17 |
| 4" | " | 0.37 | 1.80 | 2.17 |
| PVC coated hangers, galvanized, 28 ga. | | | | |
| 1-1/2" x 12" | EA. | 0.72 | 2.30 | 3.02 |
| 2" x 12" | " | 0.80 | 2.45 | 3.25 |
| 3" x 12" | " | 0.89 | 2.65 | 3.54 |
| 4" x 12" | " | 0.97 | 2.85 | 3.82 |
| Copper, 30 ga. | | | | |
| 1-1/2" x 12" | EA. | 0.94 | 2.30 | 3.24 |
| 2" x 12" | " | 1.10 | 2.45 | 3.55 |
| 3" x 12" | " | 1.25 | 2.65 | 3.90 |
| 4" x 12" | " | 1.35 | 2.85 | 4.20 |
| Wire hook hangers | | | | |
| Black wire, 1/2" x | | | | |
| 4" | EA. | 0.21 | 1.70 | 1.91 |
| 6" | " | 0.24 | 1.80 | 2.04 |
| Copper wire hooks | | | | |
| 1/2" x | | | | |
| 4" | EA. | 0.32 | 1.70 | 2.02 |
| 6" | " | 0.38 | 1.80 | 2.18 |

| INSULATION | UNIT | MAT. | INST. | TOTAL |
|---|---|---|---|---|
| **15260.10 FIBERGLASS PIPE INSULATION** | | | | |
| Fiberglass insulation on 1/2" pipe | | | | |
| 1" thick | L.F. | 1.05 | 1.15 | 2.20 |

| INSULATION | UNIT | MAT. | INST. | TOTAL |
|---|---|---|---|---|
| **15260.10 FIBERGLASS PIPE INSULATION** | | | | |
| 1-1/2" thick | L.F. | 2.55 | 1.40 | 3.95 |
| 3/4" pipe | | | | |
| 1" thick | L.F. | 2.25 | 1.15 | 3.40 |
| 1-1/2" thick | " | 2.70 | 1.40 | 4.10 |
| 1" pipe | | | | |
| 1" thick | L.F. | 2.30 | 1.15 | 3.45 |
| 1-1/2" thick | " | 2.95 | 1.40 | 4.35 |
| 2" pipe | | | | |
| 1" thick | L.F. | 3.30 | 1.40 | 4.70 |
| 1-1/2" thick | " | 3.85 | 1.55 | 5.40 |
| 3" pipe | | | | |
| 1" thick | L.F. | 3.75 | 1.65 | 5.40 |
| 1-1/2" thick | " | 4.55 | 1.70 | 6.25 |
| 6" pipe | | | | |
| 1" thick | L.F. | 5.75 | 1.80 | 7.55 |
| 2" thick | " | 8.60 | 1.90 | 10.50 |
| 10" pipe | | | | |
| 2" thick | L.F. | 12.80 | 1.80 | 14.60 |
| 3" thick | " | 18.70 | 1.90 | 20.60 |
| **15260.60 EXTERIOR PIPE INSULATION** | | | | |
| Fiberglass insulation, aluminum jacket | | | | |
| 1/2" pipe | | | | |
| 1" thick | L.F. | 2.85 | 2.65 | 5.50 |
| 1-1/2" thick | " | 3.35 | 2.85 | 6.20 |
| 3/4" pipe | | | | |
| 1" thick | L.F. | 2.95 | 2.65 | 5.60 |
| 1-1/2" thick | " | 3.45 | 2.85 | 6.30 |
| 1" pipe | | | | |
| 1" thick | L.F. | 3.05 | 2.65 | 5.70 |
| 1-1/2" thick | " | 3.75 | 2.85 | 6.60 |
| 2" pipe | | | | |
| 1" thick | L.F. | 3.85 | 3.10 | 6.95 |
| 1-1/2" thick | " | 4.65 | 3.25 | 7.90 |
| 3" pipe | | | | |
| 1" thick | L.F. | 4.30 | 3.40 | 7.70 |
| 1-1/2" thick | " | 5.10 | 3.60 | 8.70 |
| 6" pipe | | | | |
| 1" thick | L.F. | 5.60 | 3.80 | 9.40 |
| 2" thick | " | 9.30 | 4.00 | 13.30 |
| 10" pipe | | | | |
| 2" thick | L.F. | 14.15 | 3.80 | 17.95 |
| 3" thick | " | 20.40 | 4.00 | 24.40 |
| **15260.90 PIPE INSULATION FITTINGS** | | | | |
| Insulation protection saddle | | | | |
| 1" thick covering | | | | |
| 1/2" pipe | EA. | 3.50 | 13.65 | 17.15 |
| 3/4" pipe | " | 3.50 | 13.65 | 17.15 |
| 1" pipe | " | 3.50 | 13.65 | 17.15 |
| 2" pipe | " | 4.85 | 13.65 | 18.50 |

| INSULATION | UNIT | MAT. | INST. | TOTAL |
|---|---|---|---|---|
| **15260.90 PIPE INSULATION FITTINGS** | | | | |
| 3" pipe | EA. | 5.45 | 15.55 | 21.00 |
| 6" pipe | " | 6.90 | 21.35 | 28.25 |
| 1-1/2" thick covering | | | | |
| 3/4" pipe | EA. | 5.65 | 13.65 | 19.30 |
| 1" pipe | " | 5.65 | 13.65 | 19.30 |
| 2" pipe | " | 6.35 | 13.65 | 20.00 |
| 3" pipe | " | 7.15 | 13.65 | 20.80 |
| 6" pipe | " | 10.55 | 21.35 | 31.90 |
| 10" pipe | " | 13.25 | 28.50 | 41.75 |
| **15280.10 EQUIPMENT INSULATION** | | | | |
| Equipment insulation, 2" thick, cellular glass | S.F. | 1.70 | 2.15 | 3.85 |
| Urethane, rigid, field applied jacket, plastered finish | " | 1.85 | 4.25 | 6.10 |
| Fiberglass, rigid, with vapor barrier | " | 1.70 | 1.90 | 3.60 |
| **15290.10 DUCTWORK INSULATION** | | | | |
| Fiberglass duct insulation, plain blanket | | | | |
| 1-1/2" thick | S.F. | 0.27 | 0.43 | 0.70 |
| 2" thick | " | 0.31 | 0.57 | 0.88 |
| With vapor barrier | | | | |
| 1-1/2" thick | S.F. | 0.33 | 0.43 | 0.76 |
| 2" thick | " | 0.39 | 0.57 | 0.96 |
| Rigid with vapor barrier | | | | |
| 2" thick | S.F. | 1.15 | 1.15 | 2.30 |
| 3" thick | " | 1.85 | 1.35 | 3.20 |
| 4" thick | " | 2.45 | 1.70 | 4.15 |
| 6" thick | " | 3.35 | 2.30 | 5.65 |
| Weatherproof, polystyrene, 3" thick, w/vapor barrier | " | 2.50 | 3.40 | 5.90 |
| Urethane board with vapor barrier | " | 3.50 | 4.25 | 7.75 |

| FIRE PROTECTION | UNIT | MAT. | INST. | TOTAL |
|---|---|---|---|---|
| **15330.10 WET SPRINKLER SYSTEM** | | | | |
| Sprinkler head, 212 deg, brass, exposed piping | EA. | 5.30 | 13.65 | 18.95 |
| Chrome, concealed piping | " | 5.10 | 19.00 | 24.10 |
| Water motor alarm | " | 67.00 | 57.00 | 124.00 |
| Fire department inlet connection | " | 100.00 | 68.50 | 168.50 |
| Wall plate for fire dept connection | " | 51.00 | 28.50 | 79.50 |
| Swing check valve flanged iron body, 4" | " | 110.00 | 110.00 | 220.00 |
| Check valve, 6" | " | 490.00 | 170.00 | 660.00 |
| Wet pipe valve, flange to groove, 4" | " | 430.00 | 38.00 | 468.00 |

## FIRE PROTECTION

| | UNIT | MAT. | INST. | TOTAL |
|---|---|---|---|---|
| **15330.10 WET SPRINKLER SYSTEM** | | | | |
| Flange to flange | | | | |
| 6" | EA. | 540.00 | 57.00 | 597.00 |
| 8" | " | 710.00 | 110.00 | 820.00 |
| Alarm valve, flange to flange, (wet valve) | | | | |
| 4" | EA. | 510.00 | 38.00 | 548.00 |
| 8" | " | 780.00 | 280.00 | 1,060 |
| Inspector's test connection | " | 31.70 | 28.50 | 60.20 |
| Wall hydrant, polished brass, 2-1/2" x 2-1/2", single | " | 250.00 | 24.40 | 274.40 |
| 2-way | " | 570.00 | 24.40 | 594.40 |
| 3-way | " | 1,120 | 24.40 | 1,144 |
| Wet valve trim, includes retard chamber & gauges, 4"-6" | " | 380.00 | 28.50 | 408.50 |
| Retard pressure switch for wet systems | " | 710.00 | 68.50 | 778.50 |
| Air maintenance device | " | 200.00 | 28.50 | 228.50 |
| Wall hydrant non-freeze, 8" thick wall, vacuum breaker | " | 20.40 | 17.10 | 37.50 |
| 12" thick wall | " | 22.65 | 17.10 | 39.75 |
| **15330.50 DRY SPRINKLER SYSTEM** | | | | |
| Dry pipe valve, flange to flange | | | | |
| 4" | EA. | 810.00 | 68.50 | 878.50 |
| 6" | " | 1,050 | 85.50 | 1,136 |
| Trim, 4" and 6", includes gauges | " | 380.00 | 28.50 | 408.50 |
| Field testing and flushing | " | 0.00 | 280.00 | 280.00 |
| Disinfection | " | 0.00 | 280.00 | 280.00 |
| Pressure switch double circuit, open/close contacts | " | 190.00 | 85.50 | 275.50 |
| Low air | | | | |
| Supervisory unit | EA. | 630.00 | 57.00 | 687.00 |
| Pressure switch | " | 200.00 | 28.50 | 228.50 |
| **15330.90 HALON TYPE SYSTEM** | | | | |
| Halon type fire protection system, per computer room | | | | |
| Minimum | EA. | | | 2,604 |
| Average | " | | | 3,510 |
| Maximum | " | | | 6,454 |

## PLUMBING

| | UNIT | MAT. | INST. | TOTAL |
|---|---|---|---|---|
| **15410.05 C.I. PIPE, ABOVE GROUND** | | | | |
| No hub pipe | | | | |
| 1-1/2" pipe | L.F. | 4.30 | 2.45 | 6.75 |
| 2" pipe | " | 4.70 | 2.85 | 7.55 |
| 3" pipe | " | 6.65 | 3.40 | 10.05 |
| 4" pipe | " | 9.05 | 5.70 | 14.75 |
| 6" pipe | " | 13.60 | 6.85 | 20.45 |

| PLUMBING | UNIT | MAT. | INST. | TOTAL |
|---|---|---|---|---|
| **15410.05 C.I. PIPE, ABOVE GROUND** | | | | |
| 8" pipe | L.F. | 20.40 | 11.40 | 31.80 |
| 10" pipe | " | 31.70 | 13.65 | 45.35 |
| No hub fittings, 1-1/2" pipe | | | | |
| 1/4 bend | EA. | 5.10 | 11.40 | 16.50 |
| 1/8 bend | " | 3.60 | 11.40 | 15.00 |
| Sanitary tee | " | 6.25 | 17.10 | 23.35 |
| Wye | " | 6.80 | 17.10 | 23.90 |
| 2" pipe | | | | |
| 1/4 bend | EA. | 5.20 | 13.65 | 18.85 |
| 1/8 bend | " | 3.75 | 13.65 | 17.40 |
| Sanitary tee | " | 6.45 | 22.75 | 29.20 |
| Coupling | " | | | 6.79 |
| Wye | " | 7.25 | 28.50 | 35.75 |
| 3" pipe | | | | |
| 1/4 bend | EA. | 6.55 | 17.10 | 23.65 |
| 1/8 bend | " | 5.90 | 17.10 | 23.00 |
| Sanitary tee | " | 7.70 | 21.35 | 29.05 |
| Coupling | " | | | 7.92 |
| Wye | " | 7.95 | 28.50 | 36.45 |
| 4" pipe | | | | |
| 1/4 bend | EA. | 9.05 | 17.10 | 26.15 |
| 1/8 bend | " | 7.70 | 17.10 | 24.80 |
| Sanitary tee | " | 11.30 | 28.50 | 39.80 |
| Coupling | " | | | 8.84 |
| Wye | " | 11.30 | 28.50 | 39.80 |
| 8" deep | " | 15.85 | 17.10 | 32.95 |
| 6" pipe | | | | |
| 1/4 bend | EA. | 21.50 | 28.50 | 50.00 |
| 1/8 bend | " | 15.85 | 28.50 | 44.35 |
| Sanitary tee | " | 34.00 | 34.20 | 68.20 |
| Coupling | " | | | 18.11 |
| Wye | " | 27.20 | 34.20 | 61.40 |
| 8" pipe | | | | |
| 1/4 bend | EA. | 44.20 | 28.50 | 72.70 |
| 1/8 bend | " | 39.60 | 28.50 | 68.10 |
| Sanitary tee | " | 79.50 | 42.70 | 122.20 |
| Coupling | " | | | 20.38 |
| Wye | " | 54.50 | 34.20 | 88.70 |
| 10" pipe | | | | |
| 1/4 bend | EA. | 84.00 | 28.50 | 112.50 |
| 1/8 bend | " | 77.00 | 28.50 | 105.50 |
| Coupling | " | | | 28.31 |
| Wye | " | 110.00 | 57.00 | 167.00 |
| 10x6" wye | " | 88.50 | 57.00 | 145.50 |
| **15410.06 C.I. PIPE, BELOW GROUND** | | | | |
| No hub pipe | | | | |
| 1-1/2" pipe | L.F. | 4.30 | 1.70 | 6.00 |
| 2" pipe | " | 4.70 | 1.90 | 6.60 |
| 3" pipe | " | 6.65 | 2.15 | 8.80 |
| 4" pipe | " | 9.05 | 2.85 | 11.90 |
| 6" pipe | " | 13.60 | 3.10 | 16.70 |

# 15 MECHANICAL

| PLUMBING | UNIT | MAT. | INST. | TOTAL |
|---|---|---|---|---|
| **15410.06 C.I. PIPE, BELOW GROUND** | | | | |
| 8" pipe | L.F. | 20.40 | 3.80 | 24.20 |
| 10" pipe | " | 31.70 | 4.25 | 35.95 |
| Fittings, 1-1/2" | | | | |
| 1/4 bend | EA. | 5.10 | 9.75 | 14.85 |
| 1/8 bend | " | 3.95 | 9.75 | 13.70 |
| Wye | " | 6.80 | 13.65 | 20.45 |
| 2" | | | | |
| 1/4 bend | EA. | 5.20 | 11.40 | 16.60 |
| 1/8 bend | " | 3.75 | 11.40 | 15.15 |
| 3" | | | | |
| 1/4 bend | EA. | 6.55 | 13.65 | 20.20 |
| 1/8 bend | " | 5.90 | 13.65 | 19.55 |
| Wye | " | 7.95 | 21.35 | 29.30 |
| 4" | | | | |
| 1/4 bend | EA. | 9.05 | 13.65 | 22.70 |
| 1/8 bend | " | 7.70 | 13.65 | 21.35 |
| Wye | " | 11.30 | 21.35 | 32.65 |
| 6" | | | | |
| 1/4 bend | EA. | 21.50 | 21.35 | 42.85 |
| 1/8 bend | " | 15.85 | 21.35 | 37.20 |
| 8" | | | | |
| 1/4 bend | EA. | 44.20 | 21.35 | 65.55 |
| 1/8 bend | " | 39.60 | 21.35 | 60.95 |
| Wye | " | 54.50 | 28.50 | 83.00 |
| 10" | | | | |
| 1/4 bend | EA. | 84.00 | 21.35 | 105.35 |
| 1/8 bend | " | 77.00 | 21.35 | 98.35 |
| Plug | " | | | 20.38 |
| Wye | " | 110.00 | 42.70 | 152.70 |
| **15410.09 SERVICE WEIGHT PIPE** | | | | |
| Service weight pipe, single hub | | | | |
| 3" x 5' | EA. | 27.20 | 7.25 | 34.45 |
| 4" x 5' | " | 32.80 | 7.60 | 40.40 |
| 6" x 5' | " | 52.00 | 8.55 | 60.55 |
| 1/8 bend | | | | |
| 3" | EA. | 5.90 | 13.65 | 19.55 |
| 4" | " | 8.15 | 15.55 | 23.70 |
| 6" | " | 13.00 | 17.10 | 30.10 |
| 1/4 bend | | | | |
| 3" | EA. | 7.35 | 13.65 | 21.00 |
| 4" | " | 10.30 | 15.55 | 25.85 |
| 6" | " | 17.55 | 17.10 | 34.65 |
| Sweep | | | | |
| 3" | EA. | 11.30 | 13.65 | 24.95 |
| 4" | " | 15.85 | 15.55 | 31.40 |
| 6" | " | 28.30 | 17.10 | 45.40 |
| Sanitary T | | | | |
| 3" | EA. | 11.90 | 24.40 | 36.30 |
| 4" | " | 14.70 | 28.50 | 43.20 |
| 6" | " | 29.40 | 31.10 | 60.50 |

| PLUMBING | UNIT | MAT. | INST. | TOTAL |
|---|---|---|---|---|
| **15410.09 SERVICE WEIGHT PIPE** | | | | |
| Wye | | | | |
| 3" | EA. | 13.00 | 19.00 | 32.00 |
| 4" | " | 16.40 | 20.10 | 36.50 |
| 6" | " | 34.00 | 24.40 | 58.40 |
| **15410.10 COPPER PIPE** | | | | |
| Type "K" copper | | | | |
| 1/2" | L.F. | 1.90 | 1.05 | 2.95 |
| 3/4" | " | 3.50 | 1.15 | 4.65 |
| 1" | " | 4.40 | 1.20 | 5.60 |
| 1-1/4" | " | 5.55 | 1.30 | 6.85 |
| 1-1/2" | " | 7.00 | 1.40 | 8.40 |
| 2" | " | 9.60 | 1.55 | 11.15 |
| 2-1/2" | " | 12.45 | 1.70 | 14.15 |
| 3" | " | 16.85 | 1.80 | 18.65 |
| 4" | " | 27.20 | 1.90 | 29.10 |
| DWV, copper | | | | |
| 1-1/4" | L.F. | 3.50 | 1.40 | 4.90 |
| 1-1/2" | " | 4.55 | 1.55 | 6.10 |
| 2" | " | 5.10 | 1.70 | 6.80 |
| 3" | " | 9.60 | 1.90 | 11.50 |
| 4" | " | 15.85 | 2.15 | 18.00 |
| 6" | " | 51.00 | 2.45 | 53.45 |
| Refrigeration tubing, copper, sealed | | | | |
| 1/8" | L.F. | 0.51 | 1.35 | 1.86 |
| 3/16" | " | 0.54 | 1.40 | 1.94 |
| 1/4" | " | 0.63 | 1.50 | 2.13 |
| 5/16" | " | 0.81 | 1.55 | 2.36 |
| 3/8" | " | 0.91 | 1.65 | 2.56 |
| 1/2" | " | 1.20 | 1.70 | 2.90 |
| 7/8" | " | 2.05 | 1.95 | 4.00 |
| 1-1/8" | " | 3.30 | 2.30 | 5.60 |
| 1-3/8" | " | 3.95 | 2.65 | 6.60 |
| Type "L" copper | | | | |
| 1/4" | L.F. | 1.15 | 1.00 | 2.15 |
| 3/8" | " | 1.25 | 1.00 | 2.25 |
| 1/2" | " | 1.90 | 1.05 | 2.95 |
| 3/4" | " | 2.15 | 1.15 | 3.30 |
| 1" | " | 2.95 | 1.20 | 4.15 |
| 1-1/4" | " | 3.95 | 1.30 | 5.25 |
| 1-1/2" | " | 4.75 | 1.40 | 6.15 |
| 2" | " | 7.10 | 1.55 | 8.65 |
| Type "M" copper | | | | |
| 1/2" | L.F. | 1.35 | 1.05 | 2.40 |
| 3/4" | " | 1.90 | 1.15 | 3.05 |
| 1" | " | 2.40 | 1.20 | 3.60 |
| 1-1/4" | " | 3.40 | 1.30 | 4.70 |
| 2" | " | 6.55 | 1.55 | 8.10 |
| **15410.11 COPPER FITTINGS** | | | | |
| Coupling, with stop | | | | |
| 1/4" | EA. | 0.40 | 11.40 | 11.80 |

| PLUMBING | UNIT | MAT. | INST. | TOTAL |
|---|---|---|---|---|
| **15410.11 COPPER FITTINGS** | | | | |
| 3/8" | EA. | 0.65 | 13.65 | 14.30 |
| 1/2" | " | 0.79 | 14.85 | 15.64 |
| 5/8" | " | 1.00 | 17.10 | 18.10 |
| 3/4" | " | 1.15 | 19.00 | 20.15 |
| 1" | " | 2.25 | 20.10 | 22.35 |
| 3" | " | 20.40 | 34.20 | 54.60 |
| 4" | " | 40.80 | 42.70 | 83.50 |
| Reducing coupling | | | | |
| 1/4" x 1/8" | EA. | 1.30 | 13.65 | 14.95 |
| 3/8" x 1/4" | " | 1.40 | 14.85 | 16.25 |
| 1/2" x | | | | |
| 3/8" | EA. | 1.35 | 17.10 | 18.45 |
| 1/4" | " | 1.35 | 17.10 | 18.45 |
| 1/8" | " | 1.35 | 17.10 | 18.45 |
| 3/4" x | | | | |
| 3/8" | EA. | 1.85 | 19.00 | 20.85 |
| 1/2" | " | 1.90 | 19.00 | 20.90 |
| 1" x | | | | |
| 3/8" | EA. | 3.85 | 21.35 | 25.20 |
| 1" x 1/2" | " | 3.95 | 21.35 | 25.30 |
| 1" x 3/4" | " | 4.05 | 21.35 | 25.40 |
| 1-1/4" x | | | | |
| 1/2" | EA. | 4.65 | 22.75 | 27.40 |
| 3/4" | " | 4.75 | 22.75 | 27.50 |
| 1" | " | 5.00 | 22.75 | 27.75 |
| 1-1/2" x | | | | |
| 1/2" | EA. | 7.95 | 24.40 | 32.35 |
| 3/4" | " | 8.05 | 24.40 | 32.45 |
| 1" | " | 8.15 | 24.40 | 32.55 |
| 1-1/4" | " | 8.25 | 24.40 | 32.65 |
| 2" x | | | | |
| 1/2" | EA. | 11.90 | 28.50 | 40.40 |
| 3/4" | " | 12.10 | 28.50 | 40.60 |
| 1" | " | 12.45 | 28.50 | 40.95 |
| 1-1/4" | " | 12.70 | 28.50 | 41.20 |
| 1-1/2" | " | 12.80 | 28.50 | 41.30 |
| 2-1/2" x | | | | |
| 1" | EA. | 26.00 | 34.20 | 60.20 |
| 1-1/4" | " | 26.60 | 34.20 | 60.80 |
| 1-1/2" | " | 26.60 | 34.20 | 60.80 |
| 2" | " | 27.20 | 34.20 | 61.40 |
| 3" x | | | | |
| 1-1/2" | EA. | 32.80 | 42.70 | 75.50 |
| 2" | " | 34.00 | 42.70 | 76.70 |
| 2-1/2" | " | 35.10 | 42.70 | 77.80 |
| 4" x | | | | |
| 2" | EA. | 69.00 | 48.80 | 117.80 |
| 2-1/2" | " | 70.00 | 48.80 | 118.80 |
| 3" | " | 71.50 | 48.80 | 120.30 |
| Slip coupling | | | | |
| 1/4" | EA. | 0.34 | 11.40 | 11.74 |
| 1/2" | " | 0.52 | 13.65 | 14.17 |

| PLUMBING | UNIT | MAT. | INST. | TOTAL |
|---|---|---|---|---|
| **15410.11 COPPER FITTINGS** | | | | |
| 3/4" | EA. | 1.00 | 17.10 | 18.10 |
| 1" | " | 2.15 | 19.00 | 21.15 |
| 1-1/4" | " | 3.95 | 21.35 | 25.30 |
| 1-1/2" | " | 5.30 | 22.75 | 28.05 |
| 2" | " | 7.70 | 28.50 | 36.20 |
| 2-1/2" | " | 10.55 | 28.50 | 39.05 |
| 3" | " | 20.40 | 34.20 | 54.60 |
| 4" | " | 39.60 | 42.70 | 82.30 |
| Coupling with drain | | | | |
| 1/2" | EA. | 4.65 | 17.10 | 21.75 |
| 3/4" | " | 6.80 | 19.00 | 25.80 |
| 1" | " | 8.50 | 21.35 | 29.85 |
| Reducer | | | | |
| 3/8" x 1/4" | EA. | 1.50 | 13.65 | 15.15 |
| 1/2" x 3/8" | " | 1.60 | 13.65 | 15.25 |
| 3/4" x | | | | |
| 1/4" | EA. | 2.40 | 15.55 | 17.95 |
| 3/8" | " | 2.50 | 15.55 | 18.05 |
| 1/2" | " | 2.60 | 15.55 | 18.15 |
| 1" x | | | | |
| 1/2" | EA. | 3.95 | 17.10 | 21.05 |
| 3/4" | " | 4.05 | 17.10 | 21.15 |
| 1-1/4" x | | | | |
| 1/2" | EA. | 5.30 | 19.00 | 24.30 |
| 3/4" | " | 6.70 | 19.00 | 25.70 |
| 1" | " | 6.80 | 19.00 | 25.80 |
| 1-1/2" x | | | | |
| 1/2" | EA. | 7.95 | 21.35 | 29.30 |
| 3/4" | " | 8.00 | 21.35 | 29.35 |
| 1" | " | 8.15 | 21.35 | 29.50 |
| 1-1/4" | " | 8.25 | 21.35 | 29.60 |
| 2" x | | | | |
| 1/2" | EA. | 12.45 | 24.40 | 36.85 |
| 3/4" | " | 12.70 | 24.40 | 37.10 |
| 1" | " | 12.70 | 24.40 | 37.10 |
| 1-1/4" | " | 13.00 | 24.40 | 37.40 |
| 1-1/2" | " | 13.00 | 24.40 | 37.40 |
| 2-1/2" x | | | | |
| 1" | EA. | 27.20 | 28.50 | 55.70 |
| 1-1/4" | " | 28.30 | 28.50 | 56.80 |
| 1-1/2" | " | 28.30 | 28.50 | 56.80 |
| 2" | " | 29.40 | 28.50 | 57.90 |
| 3" x | | | | |
| 1-1/4" | EA. | 36.20 | 34.20 | 70.40 |
| 1-1/2" | " | 36.20 | 34.20 | 70.40 |
| 2" | " | 37.40 | 34.20 | 71.60 |
| 2-1/2" | " | 37.40 | 34.20 | 71.60 |
| 4" x | | | | |
| 2" | EA. | 61.00 | 42.70 | 103.70 |
| 3" | " | 63.50 | 42.70 | 106.20 |
| Female adapters | | | | |
| 1/4" | EA. | 5.20 | 13.65 | 18.85 |

| PLUMBING | UNIT | MAT. | INST. | TOTAL |
|---|---|---|---|---|
| **15410.11 COPPER FITTINGS** | | | | |
| 3/8" | EA. | 5.45 | 15.55 | 21.00 |
| 1/2" | " | 2.70 | 17.10 | 19.80 |
| 3/4" | " | 3.95 | 19.00 | 22.95 |
| 1" | " | 6.70 | 19.00 | 25.70 |
| 1-1/4" | " | 10.75 | 21.35 | 32.10 |
| 1-1/2" | " | 15.85 | 21.35 | 37.20 |
| 2" | " | 20.40 | 22.75 | 43.15 |
| 2-1/2" | " | 65.50 | 24.40 | 89.90 |
| 3" | " | 90.50 | 28.50 | 119.00 |
| 4" | " | 150.00 | 34.20 | 184.20 |
| Increasing female adapters | | | | |
| 1/8" x | | | | |
| 3/8" | EA. | 3.95 | 13.65 | 17.60 |
| 1/2" | " | 3.95 | 13.65 | 17.60 |
| 1/4" x 1/2" | " | 3.75 | 14.85 | 18.60 |
| 3/8" x 1/2" | " | 3.75 | 15.55 | 19.30 |
| 1/2" X | | | | |
| 3/4" | EA. | 5.30 | 17.10 | 22.40 |
| 1" | " | 9.85 | 17.10 | 26.95 |
| 3/4" X | | | | |
| 1" | EA. | 9.05 | 19.00 | 28.05 |
| 1-1/4" | " | 16.40 | 19.00 | 35.40 |
| 1" x | | | | |
| 1-1/4" | EA. | 16.40 | 19.00 | 35.40 |
| 1-1/2" | " | 19.05 | 19.00 | 38.05 |
| 1-1/4" x | | | | |
| 1-1/2" | EA. | 19.15 | 21.35 | 40.50 |
| 2" | " | 27.20 | 21.35 | 48.55 |
| 1-1/2" x 2" | " | 28.30 | 22.75 | 51.05 |
| Reducing female adapters | | | | |
| 3/8" x 1/4" | EA. | 4.30 | 15.55 | 19.85 |
| 1/2" x | | | | |
| 1/4" | EA. | 4.30 | 17.10 | 21.40 |
| 3/8" | " | 4.55 | 17.10 | 21.65 |
| 3/4" x 1/2" | " | 5.55 | 19.00 | 24.55 |
| 1" x | | | | |
| 1/2" | EA. | 8.95 | 19.00 | 27.95 |
| 3/4" | " | 8.95 | 19.00 | 27.95 |
| 1-1/4" x | | | | |
| 1/2" | EA. | 17.00 | 21.35 | 38.35 |
| 3/4" | " | 17.00 | 21.35 | 38.35 |
| 1" | " | 17.00 | 21.35 | 38.35 |
| 1-1/2" x | | | | |
| 1" | EA. | 18.10 | 22.75 | 40.85 |
| 1-1/4" | " | 18.10 | 22.75 | 40.85 |
| 2" x | | | | |
| 1" | EA. | 26.00 | 24.40 | 50.40 |
| 1-1/4" | " | 27.20 | 24.40 | 51.60 |
| 1-1/2" | " | 28.30 | 24.40 | 52.70 |
| Female fitting adapters | | | | |
| 1/2" | EA. | 4.20 | 17.10 | 21.30 |
| 3/4" | " | 5.45 | 17.10 | 22.55 |

| PLUMBING | UNIT | MAT. | INST. | TOTAL |
|---|---|---|---|---|
| **15410.11 COPPER FITTINGS** | | | | |
| 3/4" x 1/2" | EA. | 9.05 | 18.00 | 27.05 |
| 1" | " | 8.50 | 19.00 | 27.50 |
| 1-1/4" | " | 13.00 | 20.10 | 33.10 |
| 1-1/2" | " | 17.00 | 21.35 | 38.35 |
| 2" | " | 18.10 | 22.75 | 40.85 |
| Male adapters | | | | |
| 1/4" | EA. | 5.00 | 15.55 | 20.55 |
| 3/8" | " | 2.85 | 15.55 | 18.40 |
| 3" | " | 59.00 | 28.50 | 87.50 |
| 4" | " | 87.00 | 34.20 | 121.20 |
| Increasing male adapters | | | | |
| 3/8" x 1/2" | EA. | 2.95 | 15.55 | 18.50 |
| 1/2" x | | | | |
| 3/4" | EA. | 4.30 | 17.10 | 21.40 |
| 1" | " | 7.95 | 17.10 | 25.05 |
| 3/4" x | | | | |
| 1" | EA. | 7.95 | 18.00 | 25.95 |
| 1-1/4" | " | 11.30 | 18.00 | 29.30 |
| 1" x 1-1/4" | " | 11.30 | 19.00 | 30.30 |
| 1-1/2" x | | | | |
| 3/4" | EA. | 14.70 | 20.10 | 34.80 |
| 1" | " | 14.95 | 20.10 | 35.05 |
| 1-1/4" | " | 15.15 | 20.10 | 35.25 |
| 2" x | | | | |
| 1" | EA. | 24.90 | 21.35 | 46.25 |
| 1-1/4" | " | 24.90 | 21.35 | 46.25 |
| 1-1/2" | " | 26.00 | 21.35 | 47.35 |
| 2" x 2-1/2" | " | 84.00 | 22.75 | 106.75 |
| Copper pipe fittings | | | | |
| 1/2" | | | | |
| 90 deg ell | EA. | 0.81 | 7.60 | 8.41 |
| 45 deg ell | " | 1.00 | 7.60 | 8.60 |
| Tee | " | 1.40 | 9.75 | 11.15 |
| Cap | " | 0.54 | 3.80 | 4.34 |
| Coupling | " | 0.59 | 7.60 | 8.19 |
| Union | " | 4.05 | 8.55 | 12.60 |
| 3/4" | | | | |
| 90 deg ell | EA. | 1.75 | 8.55 | 10.30 |
| 45 deg ell | " | 2.05 | 8.55 | 10.60 |
| Tee | " | 2.95 | 11.40 | 14.35 |
| Cap | " | 1.10 | 4.00 | 5.10 |
| Coupling | " | 1.20 | 8.55 | 9.75 |
| Union | " | 6.00 | 9.75 | 15.75 |
| 1" | | | | |
| 90 deg ell | EA. | 4.05 | 11.40 | 15.45 |
| 45 deg ell | " | 5.30 | 11.40 | 16.70 |
| Tee | " | 6.70 | 13.65 | 20.35 |
| Cap | " | 2.00 | 5.70 | 7.70 |
| Coupling | " | 2.95 | 11.40 | 14.35 |
| Union | " | 7.85 | 11.40 | 19.25 |
| 1-1/4" | | | | |
| 90 deg ell | EA. | 5.55 | 9.75 | 15.30 |

| PLUMBING | UNIT | MAT. | INST. | TOTAL |
|---|---|---|---|---|
| **15410.11 COPPER FITTINGS** | | | | |
| 45 deg ell | EA. | 6.90 | 9.75 | 16.65 |
| Tee | " | 9.05 | 17.10 | 26.15 |
| Cap | " | 1.60 | 5.70 | 7.30 |
| Union | " | 13.00 | 12.20 | 25.20 |
| 1-1/2" | | | | |
| 90 deg ell | EA. | 7.25 | 12.20 | 19.45 |
| 45 deg ell | " | 8.60 | 12.20 | 20.80 |
| Tee | " | 11.90 | 19.00 | 30.90 |
| Cap | " | 1.60 | 5.70 | 7.30 |
| Coupling | " | 5.30 | 11.40 | 16.70 |
| Union | " | 19.80 | 15.55 | 35.35 |
| 2" | | | | |
| 90 deg ell | EA. | 14.15 | 13.65 | 27.80 |
| 45 deg ell | " | 13.00 | 21.35 | 34.35 |
| Tee | " | 20.40 | 21.35 | 41.75 |
| Cap | " | 3.30 | 6.85 | 10.15 |
| Coupling | " | 8.60 | 13.65 | 22.25 |
| Union | " | 21.50 | 17.10 | 38.60 |
| 2-1/2" | | | | |
| 90 deg ell | EA. | 27.20 | 17.10 | 44.30 |
| 45 deg ell | " | 23.80 | 17.10 | 40.90 |
| Tee | " | 27.20 | 24.40 | 51.60 |
| Cap | " | 6.70 | 8.55 | 15.25 |
| Coupling | " | 13.00 | 17.10 | 30.10 |
| Union | " | 39.60 | 19.00 | 58.60 |
| **15410.18 GLASS PIPE** | | | | |
| Glass pipe | | | | |
| 1-1/2" dia. | L.F. | 10.75 | 6.85 | 17.60 |
| 2" dia. | " | 12.45 | 7.60 | 20.05 |
| 3" dia. | " | 22.10 | 8.55 | 30.65 |
| 4" dia. | " | 31.70 | 9.75 | 41.45 |
| 6" dia. | " | 55.50 | 11.40 | 66.90 |
| **15410.30 PVC/CPVC PIPE** | | | | |
| PVC schedule 40 | | | | |
| 1/2" pipe | L.F. | 1.00 | 1.40 | 2.40 |
| 3/4" pipe | " | 1.15 | 1.55 | 2.70 |
| 1" pipe | " | 1.35 | 1.70 | 3.05 |
| 1-1/4" pipe | " | 1.50 | 1.90 | 3.40 |
| 1-1/2" pipe | " | 1.60 | 2.15 | 3.75 |
| 2" pipe | " | 1.90 | 2.45 | 4.35 |
| 2-1/2" pipe | " | 2.85 | 2.85 | 5.70 |
| 3" pipe | " | 3.30 | 3.40 | 6.70 |
| 4" pipe | " | 5.20 | 4.25 | 9.45 |
| 6" pipe | " | 9.60 | 8.55 | 18.15 |
| 8" pipe | " | 14.70 | 11.40 | 26.10 |
| PVC schedule 80 pipe | | | | |
| 1-1/2" pipe | L.F. | 1.70 | 2.15 | 3.85 |
| 2" pipe | " | 2.15 | 2.45 | 4.60 |
| 3" pipe | " | 3.30 | 3.40 | 6.70 |

| PLUMBING | UNIT | MAT. | INST. | TOTAL |
|---|---|---|---|---|
| **15410.30 PVC/CPVC PIPE** | | | | |
| 4" pipe | L.F. | 4.55 | 4.25 | 8.80 |
| Polypropylene, acid resistant, DWV pipe | | | | |
| Schedule 40 | | | | |
| 1-1/2" pipe | L.F. | 3.40 | 2.45 | 5.85 |
| 2" pipe | " | 4.60 | 2.85 | 7.45 |
| 3" pipe | " | 8.50 | 3.40 | 11.90 |
| 4" pipe | " | 12.00 | 4.25 | 16.25 |
| 6" pipe | " | 21.55 | 8.55 | 30.10 |
| Polyethylene pipe and fittings | | | | |
| SDR-21 | | | | |
| 3" pipe | L.F. | 2.25 | 4.25 | 6.50 |
| 4" pipe | " | 3.40 | 5.70 | 9.10 |
| 6" pipe | " | 5.65 | 8.55 | 14.20 |
| 8" pipe | " | 9.05 | 9.75 | 18.80 |
| 10" pipe | " | 10.20 | 11.40 | 21.60 |
| **15410.33 ABS DWV PIPE** | | | | |
| Schedule 40 ABS | | | | |
| 1-1/2" pipe | L.F. | 1.25 | 1.70 | 2.95 |
| 2" pipe | " | 1.60 | 1.90 | 3.50 |
| 3" pipe | " | 2.95 | 2.45 | 5.40 |
| 4" pipe | " | 4.40 | 3.40 | 7.80 |
| 6" pipe | " | 6.70 | 4.25 | 10.95 |
| **15410.35 PLASTIC PIPE** | | | | |
| Fiberglass reinforced pipe | | | | |
| 2" pipe | L.F. | 9.05 | 2.65 | 11.70 |
| 3" pipe | " | 10.30 | 2.85 | 13.15 |
| 4" pipe | " | 13.60 | 3.10 | 16.70 |
| 6" pipe | " | 19.80 | 3.40 | 23.20 |
| 8" pipe | " | 29.40 | 5.70 | 35.10 |
| 10" pipe | " | 63.50 | 6.85 | 70.35 |
| 12" pipe | " | 54.50 | 8.55 | 63.05 |
| **15410.70 STAINLESS STEEL PIPE** | | | | |
| Stainless steel, schedule 40, threaded | | | | |
| 1/2" pipe | L.F. | 4.05 | 4.90 | 8.95 |
| 3/4" pipe | " | 5.30 | 5.00 | 10.30 |
| 1" pipe | " | 6.70 | 5.25 | 11.95 |
| 1-1/2" pipe | " | 9.60 | 5.70 | 15.30 |
| 2" pipe | " | 15.85 | 6.20 | 22.05 |
| 2-1/2" pipe | " | 20.40 | 6.85 | 27.25 |
| 3" pipe | " | 29.40 | 7.60 | 37.00 |
| 4" pipe | " | 37.40 | 8.55 | 45.95 |
| **15410.80 STEEL PIPE** | | | | |
| Black steel, extra heavy pipe, threaded | | | | |
| 1/2" pipe | L.F. | 1.10 | 1.35 | 2.45 |
| 3/4" pipe | " | 1.35 | 1.35 | 2.70 |

| PLUMBING | UNIT | MAT. | INST. | TOTAL |
|---|---|---|---|---|
| **15410.80 STEEL PIPE** | | | | |
| 1" pipe | L.F. | 1.80 | 1.70 | 3.50 |
| 1-1/2" pipe | " | 2.95 | 1.90 | 4.85 |
| 2-1/2" pipe | " | 6.25 | 4.25 | 10.50 |
| 3" pipe | " | 8.50 | 5.70 | 14.20 |
| 4" pipe | " | 13.50 | 6.85 | 20.35 |
| 5" pipe | " | 18.10 | 8.55 | 26.65 |
| 6" pipe | " | 22.65 | 8.55 | 31.20 |
| 8" pipe | " | 33.40 | 11.40 | 44.80 |
| 10" pipe | " | 46.40 | 13.65 | 60.05 |
| 12" pipe | " | 59.00 | 17.10 | 76.10 |
| Fittings, malleable iron, threaded, 1/2" pipe | | | | |
| 90 deg ell | EA. | 1.35 | 11.40 | 12.75 |
| 45 deg ell | " | 2.15 | 11.40 | 13.55 |
| Tee | " | 1.80 | 17.10 | 18.90 |
| 3/4" pipe | | | | |
| 90 deg ell | EA. | 1.80 | 11.40 | 13.20 |
| 45 deg ell | " | 2.95 | 17.10 | 20.05 |
| Tee | " | 3.00 | 17.10 | 20.10 |
| 1" pipe | | | | |
| 90 deg ell | EA. | 2.95 | 13.65 | 16.60 |
| 45 deg ell | " | 3.30 | 13.65 | 16.95 |
| Tee | " | 4.40 | 19.00 | 23.40 |
| 1-1/2" pipe | | | | |
| 90 deg ell | EA. | 5.60 | 17.10 | 22.70 |
| 45 deg ell | " | 6.55 | 17.10 | 23.65 |
| Tee | " | 8.10 | 24.40 | 32.50 |
| 2-1/2" pipe | | | | |
| 90 deg ell | EA. | 22.65 | 42.70 | 65.35 |
| 45 deg ell | " | 28.30 | 42.70 | 71.00 |
| Tee | " | 30.60 | 57.00 | 87.60 |
| 3" pipe | | | | |
| 90 deg ell | EA. | 32.80 | 57.00 | 89.80 |
| 45 deg ell | " | 37.40 | 57.00 | 94.40 |
| Tee | " | 41.90 | 85.50 | 127.40 |
| 4" pipe | | | | |
| 90 deg ell | EA. | 65.50 | 68.50 | 134.00 |
| 45 deg ell | " | 67.00 | 68.50 | 135.50 |
| Tee | " | 96.00 | 110.00 | 206.00 |
| 6" pipe | | | | |
| 90 deg ell | EA. | 190.00 | 68.50 | 258.50 |
| 45 deg ell | " | 230.00 | 68.50 | 298.50 |
| Tee | " | 260.00 | 110.00 | 370.00 |
| 8" pipe | | | | |
| 90 deg ell | EA. | 190.00 | 140.00 | 330.00 |
| 45 deg ell | " | 240.00 | 140.00 | 380.00 |
| Tee | " | 280.00 | 210.00 | 490.00 |
| 10" pipe | | | | |
| 90 deg ell | EA. | 220.00 | 170.00 | 390.00 |
| 45 deg ell | " | 250.00 | 170.00 | 420.00 |
| Tee | " | 320.00 | 210.00 | 530.00 |
| 12" pipe | | | | |
| 90 deg ell | EA. | 250.00 | 210.00 | 460.00 |

| PLUMBING | UNIT | MAT. | INST. | TOTAL |
|---|---|---|---|---|
| **15410.80 STEEL PIPE** | | | | |
| 45 deg ell | EA. | 290.00 | 210.00 | 500.00 |
| Tee | " | 370.00 | 280.00 | 650.00 |
| **15410.82 GALVANIZED STEEL PIPE** | | | | |
| Galvanized pipe | | | | |
| 1/2" pipe | L.F. | 1.25 | 3.40 | 4.65 |
| 3/4" pipe | " | 1.60 | 4.25 | 5.85 |
| 1" pipe | " | 2.25 | 4.90 | 7.15 |
| 1-1/4" pipe | " | 2.70 | 5.70 | 8.40 |
| 1-1/2" pipe | " | 3.30 | 6.85 | 10.15 |
| 2" pipe | " | 4.75 | 8.55 | 13.30 |
| 2-1/2" pipe | " | 5.90 | 11.40 | 17.30 |
| 3" pipe | " | 10.20 | 12.20 | 22.40 |
| 4" pipe | " | 15.85 | 14.25 | 30.10 |
| 6" pipe | " | 22.10 | 28.50 | 50.60 |
| **15430.23 CLEANOUTS** | | | | |
| Cleanout, wall | | | | |
| 2" | EA. | 52.50 | 22.75 | 75.25 |
| 3" | " | 59.00 | 22.75 | 81.75 |
| 4" | " | 78.00 | 28.50 | 106.50 |
| 6" | " | 140.00 | 34.20 | 174.20 |
| 8" | " | 160.00 | 42.70 | 202.70 |
| Floor | | | | |
| 2" | EA. | 57.50 | 28.50 | 86.00 |
| 3" | " | 72.50 | 28.50 | 101.00 |
| 4" | " | 80.00 | 34.20 | 114.20 |
| 6" | " | 110.00 | 42.70 | 152.70 |
| 8" | " | 200.00 | 48.80 | 248.80 |
| **15430.24 GREASE TRAPS** | | | | |
| Grease traps, cast iron, 3" pipe | | | | |
| 35 gpm, 70 lb capacity | EA. | 1,870 | 340.00 | 2,210 |
| 50 gpm, 100 lb capacity | " | 2,380 | 430.00 | 2,810 |
| **15430.25 HOSE BIBBS** | | | | |
| Hose bibb | | | | |
| 1/2" | EA. | 5.00 | 11.40 | 16.40 |
| 3/4" | " | 5.40 | 11.40 | 16.80 |
| **15430.60 VALVES** | | | | |
| Gate valve, 125 lb, bronze, soldered | | | | |
| 1/2" | EA. | 12.45 | 8.55 | 21.00 |
| 3/4" | " | 14.70 | 8.55 | 23.25 |
| 1" | " | 18.10 | 11.40 | 29.50 |
| 1-1/2" | " | 31.70 | 13.65 | 45.35 |
| 2" | " | 44.20 | 17.10 | 61.30 |
| 2-1/2" | " | 90.50 | 21.35 | 111.85 |

| PLUMBING | UNIT | MAT. | INST. | TOTAL |
|---|---|---|---|---|
| **15430.60 VALVES** | | | | |
| Threaded | | | | |
| 1/4", 125 lb | EA. | 10.20 | 13.65 | 23.85 |
| 1/2" | | | | |
| 125 lb | EA. | 13.60 | 13.65 | 27.25 |
| 300 lb | " | 31.70 | 13.65 | 45.35 |
| 3/4" | | | | |
| 125 lb | EA. | 15.85 | 13.65 | 29.50 |
| 300 lb | " | 40.80 | 13.65 | 54.45 |
| 1" | | | | |
| 125 lb | EA. | 20.40 | 13.65 | 34.05 |
| 300 lb | " | 56.50 | 17.10 | 73.60 |
| 1-1/2" | | | | |
| 125 lb | EA. | 30.60 | 17.10 | 47.70 |
| 300 lb | " | 110.00 | 19.00 | 129.00 |
| 2" | | | | |
| 125 lb | EA. | 43.00 | 24.40 | 67.40 |
| 300 lb | " | 160.00 | 28.50 | 188.50 |
| Cast iron, flanged | | | | |
| 2", 150 lb | EA. | 190.00 | 28.50 | 218.50 |
| 2-1/2" | | | | |
| 125 lb | EA. | 180.00 | 28.50 | 208.50 |
| 250 lb | " | 360.00 | 28.50 | 388.50 |
| 3" | | | | |
| 125 lb | EA. | 220.00 | 34.20 | 254.20 |
| 250 lb | " | 450.00 | 34.20 | 484.20 |
| 4" | | | | |
| 125 lb | EA. | 260.00 | 48.80 | 308.80 |
| 250 lb | " | 610.00 | 48.80 | 658.80 |
| 6" | | | | |
| 125 lb | EA. | 400.00 | 68.50 | 468.50 |
| 250 lb | " | 1,020 | 68.50 | 1,089 |
| 8" | | | | |
| 125 lb | EA. | 790.00 | 85.50 | 875.50 |
| 250 lb | " | 1,920 | 85.50 | 2,006 |
| OS&Y, flanged | | | | |
| 2" | | | | |
| 125 lb | EA. | 140.00 | 28.50 | 168.50 |
| 250 lb | " | 320.00 | 28.50 | 348.50 |
| 2-1/2" | | | | |
| 125 lb | EA. | 150.00 | 28.50 | 178.50 |
| 250 lb | " | 360.00 | 34.20 | 394.20 |
| 3" | | | | |
| 125 lb | EA. | 150.00 | 34.20 | 184.20 |
| 250 lb | " | 400.00 | 34.20 | 434.20 |
| 4" | | | | |
| 125 lb | EA. | 210.00 | 57.00 | 267.00 |
| 250 lb | " | 610.00 | 57.00 | 667.00 |
| 6" | | | | |
| 125 lb | EA. | 330.00 | 68.50 | 398.50 |
| 250 lb | " | 990.00 | 68.50 | 1,059 |
| Ball valve, bronze, 250 lb, threaded | | | | |
| 1/2" | EA. | 7.95 | 13.65 | 21.60 |

# 15 MECHANICAL

| PLUMBING | UNIT | MAT. | INST. | TOTAL |
|---|---|---|---|---|
| **15430.60 VALVES** | | | | |
| 3/4" | EA. | 11.90 | 13.65 | 25.55 |
| 1" | " | 13.60 | 17.10 | 30.70 |
| 1-1/4" | " | 22.10 | 19.00 | 41.10 |
| 1-1/2" | " | 35.10 | 21.35 | 56.45 |
| 2" | " | 39.60 | 24.40 | 64.00 |
| Angle valve, bronze, 150 lb, threaded | | | | |
| 1/2" | EA. | 43.00 | 12.20 | 55.20 |
| 3/4" | " | 56.50 | 13.65 | 70.15 |
| 1" | " | 81.50 | 13.65 | 95.15 |
| 1-1/4" | " | 97.50 | 17.10 | 114.60 |
| 1-1/2" | " | 120.00 | 19.00 | 139.00 |
| Balancing valve, with meter connections, circuit setter | | | | |
| 1/2" | EA. | 39.60 | 13.65 | 53.25 |
| 3/4" | " | 41.90 | 15.55 | 57.45 |
| 1" | " | 54.50 | 17.10 | 71.60 |
| 1-1/4" | " | 74.50 | 19.00 | 93.50 |
| 1-1/2" | " | 88.50 | 22.75 | 111.25 |
| 2" | " | 130.00 | 28.50 | 158.50 |
| 2-1/2" | " | 250.00 | 34.20 | 284.20 |
| 3" | " | 370.00 | 42.70 | 412.70 |
| 4" | " | 520.00 | 57.00 | 577.00 |
| Pressure reducing valve, bronze, threaded, 250 lb | | | | |
| 1/2" | EA. | 250.00 | 21.35 | 271.35 |
| 3/4" | " | 250.00 | 21.35 | 271.35 |
| 1" | " | 270.00 | 21.35 | 291.35 |
| 1-1/4" | " | 320.00 | 24.40 | 344.40 |
| 1-1/2" | " | 360.00 | 28.50 | 388.50 |
| Pressure regulating valve, bronze, class 300 | | | | |
| 1" | EA. | 520.00 | 21.35 | 541.35 |
| 1-1/2" | " | 700.00 | 26.30 | 726.30 |
| 2" | " | 790.00 | 34.20 | 824.20 |
| 3" | " | 890.00 | 48.80 | 938.80 |
| 4" | " | 1,120 | 68.50 | 1,189 |
| 5" | " | 1,700 | 85.50 | 1,786 |
| 6" | " | 1,720 | 110.00 | 1,830 |
| Solar water temperature regulating valve | | | | |
| 3/4" | EA. | 320.00 | 28.50 | 348.50 |
| 1" | " | 320.00 | 34.20 | 354.20 |
| 1-1/4" | " | 350.00 | 38.00 | 388.00 |
| 1-1/2" | " | 380.00 | 42.70 | 422.70 |
| 2" | " | 480.00 | 48.80 | 528.80 |
| 2-1/2" | " | 910.00 | 85.50 | 995.50 |
| Tempering valve, threaded | | | | |
| 3/4" | EA. | 48.70 | 11.40 | 60.10 |
| 1" | " | 120.00 | 13.65 | 133.65 |
| 1-1/4" | " | 150.00 | 17.10 | 167.10 |
| 1-1/2" | " | 200.00 | 17.10 | 217.10 |
| 2" | " | 250.00 | 21.35 | 271.35 |
| 2-1/2" | " | 310.00 | 28.50 | 338.50 |
| 3" | " | 380.00 | 34.20 | 414.20 |
| 4" | " | 850.00 | 48.80 | 898.80 |
| Thermostatic mixing valve, threaded | | | | |

| PLUMBING | UNIT | MAT. | INST. | TOTAL |
|---|---|---|---|---|
| **15430.60 VALVES** | | | | |
| 1/2" | EA. | 57.50 | 12.20 | 69.70 |
| 3/4" | " | 61.00 | 13.65 | 74.65 |
| 1" | " | 220.00 | 14.85 | 234.85 |
| 1-1/2" | " | 240.00 | 17.10 | 257.10 |
| 2" | " | 310.00 | 21.35 | 331.35 |
| Sweat connection | | | | |
| 1/2" | EA. | 68.00 | 12.20 | 80.20 |
| 3/4" | " | 79.50 | 13.65 | 93.15 |
| Mixing valve, sweat connection | | | | |
| 1/2" | EA. | 27.20 | 12.20 | 39.40 |
| 3/4" | " | 30.60 | 13.65 | 44.25 |
| Liquid level gauge, aluminum body | | | | |
| 3/4" | EA. | 220.00 | 13.65 | 233.65 |
| 4125 psi, pvc body | | | | |
| 3/4" | EA. | 220.00 | 13.65 | 233.65 |
| 150 psi, crs body | | | | |
| 3/4" | EA. | 170.00 | 13.65 | 183.65 |
| 1" | " | 190.00 | 13.65 | 203.65 |
| 175 psi, bronze body, 1/2" | " | 350.00 | 12.20 | 362.20 |
| **15430.65 VACUUM BREAKERS** | | | | |
| Vacuum breaker, atmospheric, threaded connection | | | | |
| 3/4" | EA. | 22.65 | 13.65 | 36.30 |
| 1" | " | 34.00 | 13.65 | 47.65 |
| Anti-siphon, brass | | | | |
| 3/4" | EA. | 26.00 | 13.65 | 39.65 |
| 1" | " | 35.10 | 13.65 | 48.75 |
| 1-1/4" | " | 55.50 | 17.10 | 72.60 |
| 1-1/2" | " | 67.00 | 19.00 | 86.00 |
| 2" | " | 95.00 | 21.35 | 116.35 |
| **15430.68 STRAINERS** | | | | |
| Strainer, Y pattern, 125 psi, cast iron body, threaded | | | | |
| 3/4" | EA. | 11.30 | 12.20 | 23.50 |
| 1" | " | 13.60 | 13.65 | 27.25 |
| 1-1/4" | " | 18.10 | 17.10 | 35.20 |
| 1-1/2" | " | 22.65 | 17.10 | 39.75 |
| 2" | " | 34.00 | 21.35 | 55.35 |
| 250 psi, brass body, threaded | | | | |
| 3/4" | EA. | 26.00 | 13.65 | 39.65 |
| 1" | " | 36.20 | 13.65 | 49.85 |
| 1-1/4" | " | 45.30 | 17.10 | 62.40 |
| 1-1/2" | " | 65.50 | 17.10 | 82.60 |
| 2" | " | 110.00 | 21.35 | 131.35 |
| Cast iron body, threaded | | | | |
| 3/4" | EA. | 16.55 | 13.65 | 30.20 |
| 1" | " | 21.05 | 13.65 | 34.70 |
| 1-1/4" | " | 28.10 | 17.10 | 45.20 |
| 1-1/2" | " | 37.20 | 17.10 | 54.30 |
| 2" | " | 47.20 | 21.35 | 68.55 |

| PLUMBING | UNIT | MAT. | INST. | TOTAL |
|---|---|---|---|---|
| **15430.70 DRAINS, ROOF & FLOOR** | | | | |
| Floor drain, cast iron, with cast iron top | | | | |
| 2" | EA. | 45.30 | 28.50 | 73.80 |
| 3" | " | 69.00 | 28.50 | 97.50 |
| 4" | " | 90.50 | 28.50 | 119.00 |
| 6" | " | 170.00 | 34.20 | 204.20 |
| Roof drain, cast iron | | | | |
| 2" | EA. | 120.00 | 28.50 | 148.50 |
| 3" | " | 140.00 | 28.50 | 168.50 |
| 4" | " | 150.00 | 28.50 | 178.50 |
| 5" | " | 210.00 | 34.20 | 244.20 |
| 6" | " | 220.00 | 34.20 | 254.20 |
| **15430.80 TRAPS** | | | | |
| Bucket trap, threaded | | | | |
| 3/4" | EA. | 90.50 | 21.35 | 111.85 |
| 1" | " | 280.00 | 22.75 | 302.75 |
| 1-1/4" | " | 340.00 | 26.30 | 366.30 |
| 1-1/2" | " | 500.00 | 31.10 | 531.10 |
| Inverted bucket steam trap, threaded | | | | |
| 3/4" | EA. | 96.00 | 21.35 | 117.35 |
| 1" | " | 240.00 | 21.35 | 261.35 |
| 1-1/4" | " | 330.00 | 19.00 | 349.00 |
| 1-1/2" | " | 360.00 | 28.50 | 388.50 |
| Float trap, 15 psi | | | | |
| 3/4" | EA. | 79.50 | 21.35 | 100.85 |
| 1" | " | 96.00 | 22.75 | 118.75 |
| 1-1/4" | " | 110.00 | 24.40 | 134.40 |
| 1-1/2" | " | 150.00 | 28.50 | 178.50 |
| 2" | " | 260.00 | 34.20 | 294.20 |
| Float and thermostatic trap, 15 psi | | | | |
| 3/4" | EA. | 90.50 | 21.35 | 111.85 |
| 1" | " | 96.00 | 22.75 | 118.75 |
| 1-1/4" | " | 160.00 | 24.40 | 184.40 |
| 1-1/2" | " | 190.00 | 28.50 | 218.50 |
| 2" | " | 340.00 | 34.20 | 374.20 |
| Steam trap, cast iron body, threaded, 125 psi | | | | |
| 3/4" | EA. | 110.00 | 21.35 | 131.35 |
| 1" | " | 130.00 | 22.75 | 152.75 |
| 1-1/4" | " | 190.00 | 24.40 | 214.40 |
| 1-1/2" | " | 310.00 | 28.50 | 338.50 |

# 15 MECHANICAL

| PLUMBING FIXTURES | UNIT | MAT. | INST. | TOTAL |
|---|---|---|---|---|
| **15440.10 BATHS** | | | | |
| Bath tub, 5' long | | | | |
| Minimum | EA. | 340.00 | 110.00 | 450.00 |
| Average | " | 750.00 | 170.00 | 920.00 |
| Maximum | " | 1,280 | 340.00 | 1,620 |
| 6' long | | | | |
| Minimum | EA. | 390.00 | 110.00 | 500.00 |
| Average | " | 800.00 | 170.00 | 970.00 |
| Maximum | " | 1,400 | 340.00 | 1,740 |
| Square tub, whirlpool, 4'x4' | | | | |
| Minimum | EA. | 1,190 | 170.00 | 1,360 |
| Average | " | 1,630 | 340.00 | 1,970 |
| Maximum | " | 3,380 | 430.00 | 3,810 |
| 5'x5' | | | | |
| Minimum | EA. | 1,210 | 170.00 | 1,380 |
| Average | " | 1,870 | 340.00 | 2,210 |
| Maximum | " | 3,500 | 430.00 | 3,930 |
| 6'x6' | | | | |
| Minimum | EA. | 1,410 | 170.00 | 1,580 |
| Average | " | 2,040 | 340.00 | 2,380 |
| Maximum | " | 4,200 | 430.00 | 4,630 |
| For trim and rough-in | | | | |
| Minimum | EA. | 110.00 | 110.00 | 220.00 |
| Average | " | 160.00 | 170.00 | 330.00 |
| Maximum | " | 230.00 | 340.00 | 570.00 |
| **15440.15 FAUCETS** | | | | |
| Kitchen | | | | |
| Minimum | EA. | 92.00 | 57.00 | 149.00 |
| Average | " | 150.00 | 68.50 | 218.50 |
| Maximum | " | 190.00 | 85.50 | 275.50 |
| Bath | | | | |
| Minimum | EA. | 150.00 | 57.00 | 207.00 |
| Average | " | 210.00 | 68.50 | 278.50 |
| Maximum | " | 240.00 | 85.50 | 325.50 |
| Lavatory, domestic | | | | |
| Minimum | EA. | 94.50 | 57.00 | 151.50 |
| Average | " | 200.00 | 68.50 | 268.50 |
| Maximum | " | 300.00 | 85.50 | 385.50 |
| Hospital, patient rooms | | | | |
| Minimum | EA. | 230.00 | 85.50 | 315.50 |
| Average | " | 330.00 | 110.00 | 440.00 |
| Maximum | " | 430.00 | 170.00 | 600.00 |
| Operating room | | | | |
| Minimum | EA. | 270.00 | 85.50 | 355.50 |
| Average | " | 360.00 | 110.00 | 470.00 |
| Maximum | " | 520.00 | 170.00 | 690.00 |
| Washroom | | | | |
| Minimum | EA. | 130.00 | 57.00 | 187.00 |
| Average | " | 220.00 | 68.50 | 288.50 |
| Maximum | " | 330.00 | 85.50 | 415.50 |
| Handicapped | | | | |
| Minimum | EA. | 170.00 | 68.50 | 238.50 |

| PLUMBING FIXTURES | UNIT | MAT. | INST. | TOTAL |
|---|---|---|---|---|
| **15440.15 FAUCETS** | | | | |
| Average | EA. | 240.00 | 85.50 | 325.50 |
| Maximum | " | 360.00 | 110.00 | 470.00 |
| Shower | | | | |
| Minimum | EA. | 110.00 | 57.00 | 167.00 |
| Average | " | 210.00 | 68.50 | 278.50 |
| Maximum | " | 330.00 | 85.50 | 415.50 |
| For trim and rough-in | | | | |
| Minimum | EA. | 49.00 | 68.50 | 117.50 |
| Average | " | 73.50 | 85.50 | 159.00 |
| Maximum | " | 110.00 | 170.00 | 280.00 |
| **15440.18 HYDRANTS** | | | | |
| Wall hydrant | | | | |
| 8" thick | EA. | 68.00 | 57.00 | 125.00 |
| 12" thick | " | 79.50 | 68.50 | 148.00 |
| 18" thick | " | 90.50 | 76.00 | 166.50 |
| 24" thick | " | 100.00 | 85.50 | 185.50 |
| Ground hydrant | | | | |
| 2' deep | EA. | 170.00 | 42.70 | 212.70 |
| 4' deep | " | 180.00 | 48.80 | 228.80 |
| 6' deep | " | 220.00 | 57.00 | 277.00 |
| 8' deep | " | 250.00 | 85.50 | 335.50 |
| **15440.20 LAVATORIES** | | | | |
| Lavatory, counter top, porcelain enamel on cast iron | | | | |
| Minimum | EA. | 120.00 | 68.50 | 188.50 |
| Average | " | 190.00 | 85.50 | 275.50 |
| Maximum | " | 280.00 | 110.00 | 390.00 |
| Wall hung, china | | | | |
| Minimum | EA. | 150.00 | 68.50 | 218.50 |
| Average | " | 200.00 | 85.50 | 285.50 |
| Maximum | " | 490.00 | 110.00 | 600.00 |
| Handicapped | | | | |
| Minimum | EA. | 290.00 | 85.50 | 375.50 |
| Average | " | 340.00 | 110.00 | 450.00 |
| Maximum | " | 510.00 | 170.00 | 680.00 |
| For trim and rough-in | | | | |
| Minimum | EA. | 120.00 | 85.50 | 205.50 |
| Average | " | 140.00 | 110.00 | 250.00 |
| Maximum | " | 240.00 | 170.00 | 410.00 |
| **15440.30 SHOWERS** | | | | |
| Shower, fiberglass, 36"x34"x84" | | | | |
| Minimum | EA. | 480.00 | 240.00 | 720.00 |
| Average | " | 660.00 | 340.00 | 1,000 |
| Maximum | " | 960.00 | 340.00 | 1,300 |
| Steel, 1 piece, 36"x36" | | | | |
| Minimum | EA. | 440.00 | 240.00 | 680.00 |
| Average | " | 660.00 | 340.00 | 1,000 |
| Maximum | " | 780.00 | 340.00 | 1,120 |
| Receptor, molded stone, 36"x36" | | | | |
| Minimum | EA. | 150.00 | 110.00 | 260.00 |

| PLUMBING FIXTURES | UNIT | MAT. | INST. | TOTAL |
|---|---|---|---|---|
| **15440.30 SHOWERS** | | | | |
| Average | EA. | 250.00 | 170.00 | 420.00 |
| Maximum | " | 380.00 | 280.00 | 660.00 |
| For trim and rough-in | | | | |
| Minimum | EA. | 110.00 | 160.00 | 270.00 |
| Average | " | 150.00 | 190.00 | 340.00 |
| Maximum | " | 190.00 | 340.00 | 530.00 |
| **15440.40 SINKS** | | | | |
| Service sink, 24"x29" | | | | |
| Minimum | EA. | 440.00 | 85.50 | 525.50 |
| Average | " | 550.00 | 110.00 | 660.00 |
| Maximum | " | 730.00 | 170.00 | 900.00 |
| Kitchen sink, single, stainless steel, single bowl | | | | |
| Minimum | EA. | 130.00 | 68.50 | 198.50 |
| Average | " | 170.00 | 85.50 | 255.50 |
| Maximum | " | 240.00 | 110.00 | 350.00 |
| Double bowl | | | | |
| Minimum | EA. | 170.00 | 85.50 | 255.50 |
| Average | " | 200.00 | 110.00 | 310.00 |
| Maximum | " | 260.00 | 170.00 | 430.00 |
| Porcelain enamel, cast iron, single bowl | | | | |
| Minimum | EA. | 130.00 | 68.50 | 198.50 |
| Average | " | 150.00 | 85.50 | 235.50 |
| Maximum | " | 210.00 | 110.00 | 320.00 |
| Double bowl | | | | |
| Minimum | EA. | 180.00 | 85.50 | 265.50 |
| Average | " | 200.00 | 110.00 | 310.00 |
| Maximum | " | 270.00 | 170.00 | 440.00 |
| Mop sink, 24"x36"x10" | | | | |
| Minimum | EA. | 320.00 | 68.50 | 388.50 |
| Average | " | 420.00 | 85.50 | 505.50 |
| Maximum | " | 510.00 | 110.00 | 620.00 |
| Washing machine box | | | | |
| Minimum | EA. | 110.00 | 85.50 | 195.50 |
| Average | " | 150.00 | 110.00 | 260.00 |
| Maximum | " | 200.00 | 170.00 | 370.00 |
| For trim and rough-in | | | | |
| Minimum | EA. | 170.00 | 110.00 | 280.00 |
| Average | " | 270.00 | 170.00 | 440.00 |
| Maximum | " | 340.00 | 230.00 | 570.00 |
| **15440.50 URINALS** | | | | |
| Urinal, flush valve, floor mounted | | | | |
| Minimum | EA. | 460.00 | 85.50 | 545.50 |
| Average | " | 560.00 | 110.00 | 670.00 |
| Maximum | " | 730.00 | 170.00 | 900.00 |
| Wall mounted | | | | |
| Minimum | EA. | 130.00 | 85.50 | 215.50 |
| Average | " | 340.00 | 110.00 | 450.00 |
| Maximum | " | 440.00 | 170.00 | 610.00 |
| For trim and rough-in | | | | |
| Minimum | EA. | 93.00 | 85.50 | 178.50 |

| PLUMBING FIXTURES | UNIT | MAT. | INST. | TOTAL |
|---|---|---|---|---|
| **15440.50 URINALS** | | | | |
| Average | EA. | 140.00 | 170.00 | 310.00 |
| Maximum | " | 170.00 | 230.00 | 400.00 |
| **15440.60 WATER CLOSETS** | | | | |
| Water closet flush tank, floor mounted | | | | |
| Minimum | EA. | 170.00 | 85.50 | 255.50 |
| Average | " | 450.00 | 110.00 | 560.00 |
| Maximum | " | 870.00 | 170.00 | 1,040 |
| Handicapped | | | | |
| Minimum | EA. | 260.00 | 110.00 | 370.00 |
| Average | " | 460.00 | 170.00 | 630.00 |
| Maximum | " | 700.00 | 340.00 | 1,040 |
| Bowl, with flush valve, floor mounted | | | | |
| Minimum | EA. | 310.00 | 85.50 | 395.50 |
| Average | " | 350.00 | 110.00 | 460.00 |
| Maximum | " | 680.00 | 170.00 | 850.00 |
| Wall mounted | | | | |
| Minimum | EA. | 310.00 | 85.50 | 395.50 |
| Average | " | 370.00 | 110.00 | 480.00 |
| Maximum | " | 710.00 | 170.00 | 880.00 |
| For trim and rough-in | | | | |
| Minimum | EA. | 110.00 | 85.50 | 195.50 |
| Average | " | 130.00 | 110.00 | 240.00 |
| Maximum | " | 170.00 | 170.00 | 340.00 |
| **15440.70 WATER HEATERS** | | | | |
| Water heater, electric | | | | |
| 6 gal | EA. | 200.00 | 57.00 | 257.00 |
| 10 gal | " | 210.00 | 57.00 | 267.00 |
| 15 gal | " | 220.00 | 57.00 | 277.00 |
| 20 gal | " | 240.00 | 68.50 | 308.50 |
| 30 gal | " | 260.00 | 68.50 | 328.50 |
| 40 gal | " | 270.00 | 68.50 | 338.50 |
| 52 gal | " | 300.00 | 85.50 | 385.50 |
| 66 gal | " | 470.00 | 85.50 | 555.50 |
| 80 gal | " | 510.00 | 85.50 | 595.50 |
| 100 gal | " | 630.00 | 110.00 | 740.00 |
| 120 gal | " | 800.00 | 110.00 | 910.00 |
| Oil fired | | | | |
| 20 gal | EA. | 830.00 | 170.00 | 1,000 |
| 50 gal | " | 1,290 | 240.00 | 1,530 |
| **15440.90 MISCELLANEOUS FIXTURES** | | | | |
| Electric water cooler | | | | |
| Floor mounted | EA. | 480.00 | 110.00 | 590.00 |
| Wall mounted | " | 580.00 | 110.00 | 690.00 |
| Wash fountain | | | | |
| Wall mounted | EA. | 800.00 | 170.00 | 970.00 |
| Circular, floor supported | " | 1,520 | 340.00 | 1,860 |
| Deluge shower and eye wash | " | 570.00 | 170.00 | 740.00 |

| PLUMBING FIXTURES | UNIT | MAT. | INST. | TOTAL |
|---|---|---|---|---|
| **15440.95 FIXTURE CARRIERS** | | | | |
| Water fountain, wall carrier | | | | |
| Minimum | EA. | 90.50 | 34.20 | 124.70 |
| Average | " | 110.00 | 42.70 | 152.70 |
| Maximum | " | 140.00 | 57.00 | 197.00 |
| Lavatory, wall carrier | | | | |
| Minimum | EA. | 90.50 | 34.20 | 124.70 |
| Average | " | 110.00 | 42.70 | 152.70 |
| Maximum | " | 140.00 | 57.00 | 197.00 |
| Sink, industrial, wall carrier | | | | |
| Minimum | EA. | 96.00 | 34.20 | 130.20 |
| Average | " | 120.00 | 42.70 | 162.70 |
| Maximum | " | 140.00 | 57.00 | 197.00 |
| Toilets, water closets, wall carrier | | | | |
| Minimum | EA. | 120.00 | 34.20 | 154.20 |
| Average | " | 140.00 | 42.70 | 182.70 |
| Maximum | " | 180.00 | 57.00 | 237.00 |
| Floor support | | | | |
| Minimum | EA. | 56.50 | 28.50 | 85.00 |
| Average | " | 68.00 | 34.20 | 102.20 |
| Maximum | " | 79.50 | 42.70 | 122.20 |
| Urinals, wall carrier | | | | |
| Minimum | EA. | 73.50 | 34.20 | 107.70 |
| Average | " | 96.00 | 42.70 | 138.70 |
| Maximum | " | 120.00 | 57.00 | 177.00 |
| Floor support | | | | |
| Minimum | EA. | 62.50 | 28.50 | 91.00 |
| Average | " | 90.50 | 34.20 | 124.70 |
| Maximum | " | 110.00 | 42.70 | 152.70 |
| **15450.30 PUMPS** | | | | |
| In-line pump, bronze, centrifugal | | | | |
| 5 gpm, 20' head | EA. | 470.00 | 21.35 | 491.35 |
| 20 gpm, 40' head | " | 870.00 | 21.35 | 891.35 |
| 50 gpm | | | | |
| 50' head | EA. | 1,140 | 42.70 | 1,183 |
| Cast iron, centrifugal | | | | |
| 50 gpm, 200' head | EA. | 2,410 | 42.70 | 2,453 |
| 100 gpm | | | | |
| 100' head | EA. | 2,230 | 57.00 | 2,287 |
| Centrifugal, close coupled, c.i., single stage | | | | |
| 50 gpm, 100' head | EA. | 980.00 | 42.70 | 1,023 |
| 100 gpm, 100' head | " | 1,190 | 57.00 | 1,247 |
| Base mounted | | | | |
| 50 gpm, 100' head | EA. | 1,590 | 42.70 | 1,633 |
| 100 gpm, 50' head | " | 1,760 | 57.00 | 1,817 |
| 200 gpm, 100' head | " | 2,250 | 85.50 | 2,336 |
| 300 gpm, 175' head | " | 2,920 | 85.50 | 3,006 |
| Condensate pump, simplex | | | | |
| 1000 sf EDR, 2 gpm | EA. | 820.00 | 280.00 | 1,100 |
| 2000 sf EDR, 3 gpm | " | 830.00 | 280.00 | 1,110 |
| 4000 sf EDR, 6 gpm | " | 840.00 | 310.00 | 1,150 |
| 6000 sf EDR, 9 gpm | " | 850.00 | 310.00 | 1,160 |

# 15 MECHANICAL

| PLUMBING FIXTURES | UNIT | MAT. | INST. | TOTAL |
|---|---|---|---|---|
| **15450.30 PUMPS** | | | | |
| Duplex, bronze | | | | |
| 8000 sf EDR, 12 gpm | EA. | 1,170 | 310.00 | 1,480 |
| 10,000 sf EDR, 15 gpm | " | 1,210 | 430.00 | 1,640 |
| 15,000 sf EDR, 23 gpm | " | 1,460 | 490.00 | 1,950 |
| 20,000 sf EDR, 30 gpm | " | 1,690 | 680.00 | 2,370 |
| 25,000 sf EDR, 38 gpm | " | 1,750 | 680.00 | 2,430 |
| **15450.40 STORAGE TANKS** | | | | |
| Hot water storage tank, cement lined | | | | |
| 10 gallon | EA. | 2,720 | 110.00 | 2,830 |
| 70 gallon | " | 2,720 | 170.00 | 2,890 |
| 200 gallon | " | 2,830 | 240.00 | 3,070 |
| 900 gallon | " | 3,280 | 430.00 | 3,710 |
| 1100 gallon | " | 3,740 | 430.00 | 4,170 |
| 2000 gallon | " | 7,470 | 430.00 | 7,900 |
| **15480.10 SPECIAL SYSTEMS** | | | | |
| Air compressor, air cooled, two stage | | | | |
| 5.0 cfm, 175 psi | EA. | 2,390 | 680.00 | 3,070 |
| 10 cfm, 175 psi | " | 2,920 | 760.00 | 3,680 |
| 20 cfm, 175 psi | " | 4,020 | 810.00 | 4,830 |
| 50 cfm, 125 psi | " | 5,860 | 900.00 | 6,760 |
| 80 cfm, 125 psi | " | 8,400 | 980.00 | 9,380 |
| Single stage, 125 psi | | | | |
| 1.0 cfm | EA. | 2,200 | 490.00 | 2,690 |
| 1.5 cfm | " | 2,240 | 490.00 | 2,730 |
| 2.0 cfm | " | 2,290 | 490.00 | 2,780 |
| Automotive compressor, hose reel, air and water, 50' hose | " | 1,030 | 280.00 | 1,310 |
| Lube equipment, 3 reel, with pumps | " | 5,360 | 1,370 | 6,730 |
| Tire changer | | | | |
| Truck | EA. | 8,400 | 490.00 | 8,890 |
| Passenger car | " | 1,980 | 260.00 | 2,240 |
| Air hose reel, includes, 50' hose | " | 530.00 | 260.00 | 790.00 |
| Hose reel, 5 reel, motor oil, gear oil, lube, air & water | " | 4,900 | 1,370 | 6,270 |
| Water hose reel, 50' hose | " | 530.00 | 260.00 | 790.00 |
| Pump, air operated, for motor or gear oil, fits 55 gal drum | " | 690.00 | 34.20 | 724.20 |
| For chassis lube | " | 1,120 | 34.20 | 1,154 |
| Fuel dispensing pump, lighted dial, one product | | | | |
| One hose | EA. | 2,450 | 280.00 | 2,730 |
| Two hose | " | 4,310 | 280.00 | 4,590 |
| Two products, two hose | " | 4,550 | 280.00 | 4,830 |

| HEATING & VENTILATING | UNIT | MAT. | INST. | TOTAL |
|---|---|---|---|---|
| **15555.10 BOILERS** | | | | |
| Cast iron, gas fired, hot water | | | | |
| 115 mbh | EA. | 1,390 | 1,090 | 2,480 |
| 175 mbh | " | 1,620 | 1,190 | 2,810 |
| 235 mbh | " | 2,430 | 1,310 | 3,740 |
| 940 mbh | " | 7,160 | 2,620 | 9,780 |
| 1600 mbh | " | 9,820 | 3,270 | 13,090 |
| 3000 mbh | " | 20,790 | 4,360 | 25,150 |
| 6000 mbh | " | 48,500 | 6,540 | 55,040 |
| Steam | | | | |
| 115 mbh | EA. | 2,000 | 1,090 | 3,090 |
| 175 mbh | " | 2,550 | 1,190 | 3,740 |
| 235 mbh | " | 2,790 | 1,310 | 4,100 |
| 940 mbh | " | 7,040 | 2,620 | 9,660 |
| 1600 mbh | " | 12,700 | 3,270 | 15,970 |
| 3000 mbh | " | 24,250 | 4,360 | 28,610 |
| 6000 mbh | " | 51,970 | 6,540 | 58,510 |
| Electric, hot water | | | | |
| 115 mbh | EA. | 4,040 | 650.00 | 4,690 |
| 175 mbh | " | 4,850 | 650.00 | 5,500 |
| 235 mbh | " | 6,580 | 650.00 | 7,230 |
| 940 mbh | " | 11,550 | 1,310 | 12,860 |
| 1600 mbh | " | 17,320 | 2,620 | 19,940 |
| 3000 mbh | " | 30,030 | 3,270 | 33,300 |
| 6000 mbh | " | 45,040 | 4,360 | 49,400 |
| Steam | | | | |
| 115 mbh | EA. | 6,350 | 650.00 | 7,000 |
| 175 mbh | " | 7,280 | 650.00 | 7,930 |
| 235 mbh | " | 8,550 | 650.00 | 9,200 |
| 940 mbh | " | 17,320 | 1,310 | 18,630 |
| 1600 mbh | " | 28,870 | 2,620 | 31,490 |
| 3000 mbh | " | 33,490 | 3,270 | 36,760 |
| 6000 mbh | " | 51,970 | 4,360 | 56,330 |
| Oil fired, hot water | | | | |
| 115 mbh | EA. | 1,960 | 870.00 | 2,830 |
| 175 mbh | " | 2,430 | 1,010 | 3,440 |
| 235 mbh | " | 3,180 | 1,190 | 4,370 |
| 940 mbh | " | 9,350 | 2,180 | 11,530 |
| 1600 mbh | " | 16,230 | 2,620 | 18,850 |
| 3000 mbh | " | 27,720 | 3,270 | 30,990 |
| 6000 mbh | " | 51,970 | 6,540 | 58,510 |
| Steam | | | | |
| 115 mbh | EA. | 2,060 | 870.00 | 2,930 |
| 175 mbh | " | 2,490 | 1,010 | 3,500 |
| 235 mbh | " | 3,270 | 1,190 | 4,460 |
| 940 mbh | " | 10,550 | 2,180 | 12,730 |
| 1600 mbh | " | 17,220 | 2,620 | 19,840 |
| 3000 mbh | " | 29,830 | 3,270 | 33,100 |
| 6000 mbh | " | 53,120 | 6,540 | 59,660 |
| **15610.10 FURNACES** | | | | |
| Electric, hot air | | | | |
| 40 mbh | EA. | 550.00 | 170.00 | 720.00 |

# 15 MECHANICAL

## HEATING & VENTILATING

| 15610.10 FURNACES | UNIT | MAT. | INST. | TOTAL |
|---|---|---|---|---|
| 60 mbh | EA. | 640.00 | 180.00 | 820.00 |
| 80 mbh | " | 590.00 | 190.00 | 780.00 |
| 100 mbh | " | 950.00 | 200.00 | 1,150 |
| 125 mbh | " | 1,040 | 210.00 | 1,250 |
| 160 mbh | " | 1,520 | 210.00 | 1,730 |
| 200 mbh | " | 2,790 | 220.00 | 3,010 |
| 400 mbh | " | 4,970 | 230.00 | 5,200 |
| Gas fired hot air | | | | |
| 40 mbh | EA. | 550.00 | 170.00 | 720.00 |
| 60 mbh | " | 580.00 | 180.00 | 760.00 |
| 80 mbh | " | 670.00 | 190.00 | 860.00 |
| 100 mbh | " | 700.00 | 200.00 | 900.00 |
| 125 mbh | " | 760.00 | 210.00 | 970.00 |
| 160 mbh | " | 910.00 | 210.00 | 1,120 |
| 200 mbh | " | 1,620 | 220.00 | 1,840 |
| 400 mbh | " | 5,770 | 230.00 | 6,000 |
| Oil fired hot air | | | | |
| 40 mbh | EA. | 690.00 | 170.00 | 860.00 |
| 60 mbh | " | 830.00 | 180.00 | 1,010 |
| 80 mbh | " | 910.00 | 190.00 | 1,100 |
| 100 mbh | " | 1,060 | 200.00 | 1,260 |
| 125 mbh | " | 1,180 | 210.00 | 1,390 |
| 160 mbh | " | 1,270 | 210.00 | 1,480 |
| 200 mbh | " | 2,910 | 220.00 | 3,130 |
| 400 mbh | " | 5,770 | 230.00 | 6,000 |

## REFRIGERATION

| 15670.10 CONDENSING UNITS | UNIT | MAT. | INST. | TOTAL |
|---|---|---|---|---|
| Air cooled condenser, single circuit | | | | |
| 3 ton | EA. | 1,360 | 57.00 | 1,417 |
| 5 ton | " | 2,260 | 57.00 | 2,317 |
| 7.5 ton | " | 3,620 | 160.00 | 3,780 |
| 20 ton | " | 7,640 | 170.00 | 7,810 |
| 25 ton | " | 8,920 | 170.00 | 9,090 |
| 30 ton | " | 10,270 | 170.00 | 10,440 |
| With low ambient dampers | | | | |
| 3 ton | EA. | 1,590 | 85.50 | 1,676 |
| 5 ton | " | 2,490 | 85.50 | 2,576 |
| 7.5 ton | " | 3,850 | 170.00 | 4,020 |
| 20 ton | " | 7,930 | 230.00 | 8,160 |
| 25 ton | " | 9,170 | 230.00 | 9,400 |
| 30 ton | " | 10,530 | 230.00 | 10,760 |

# 15 MECHANICAL

## REFRIGERATION

| 15670.10 CONDENSING UNITS | UNIT | MAT. | INST. | TOTAL |
|---|---|---|---|---|
| Dual circuit | | | | |
| 10 ton | EA. | 3,400 | 170.00 | 3,570 |
| 15 ton | " | 4,980 | 240.00 | 5,220 |
| 20 ton | " | 8,490 | 240.00 | 8,730 |
| 25 ton | " | 9,740 | 240.00 | 9,980 |
| 30 ton | " | 11,320 | 240.00 | 11,560 |
| With low ambient dampers | | | | |
| 15 ton | EA. | 5,550 | 240.00 | 5,790 |
| 20 ton | " | 9,060 | 240.00 | 9,300 |
| 25 ton | " | 10,300 | 240.00 | 10,540 |
| 30 ton | " | 11,100 | 240.00 | 11,340 |

| 15680.10 CHILLERS | UNIT | MAT. | INST. | TOTAL |
|---|---|---|---|---|
| Chiller, reciprocal | | | | |
| Air cooled, remote condenser, starter | | | | |
| 20 ton | EA. | 15,850 | 440.00 | 16,290 |
| 25 ton | " | 17,890 | 440.00 | 18,330 |
| 30 ton | " | 18,910 | 440.00 | 19,350 |
| 40 ton | " | 27,970 | 650.00 | 28,620 |
| Water cooled, with starter | | | | |
| 20 ton | EA. | 13,590 | 440.00 | 14,030 |
| 25 ton | " | 15,060 | 440.00 | 15,500 |
| 30 ton | " | 18,120 | 650.00 | 18,770 |
| 40 ton | " | 24,910 | 650.00 | 25,560 |
| Packaged, air cooled, with starter | | | | |
| 20 ton | EA. | 15,170 | 330.00 | 15,500 |
| 25 ton | " | 16,420 | 330.00 | 16,750 |
| 30 ton | " | 19,130 | 330.00 | 19,460 |
| 40 ton | " | 21,960 | 330.00 | 22,290 |

## HEAT TRANSFER

| 15780.10 COMPUTER ROOM A/C | UNIT | MAT. | INST. | TOTAL |
|---|---|---|---|---|
| Air cooled, alarm, high efficiency filter, elec. heat | | | | |
| 3 ton | EA. | 8,200 | 260.00 | 8,460 |
| 5 ton | " | 8,800 | 280.00 | 9,080 |
| 7.5 ton | " | 14,980 | 340.00 | 15,320 |
| 10 ton | " | 15,450 | 430.00 | 15,880 |
| 15 ton | " | 17,120 | 490.00 | 17,610 |
| Steam heat | | | | |
| 3 ton | EA. | 9,510 | 260.00 | 9,770 |
| 5 ton | " | 10,100 | 280.00 | 10,380 |
| 7.5 ton | " | 16,050 | 340.00 | 16,390 |

## HEAT TRANSFER

| HEAT TRANSFER | UNIT | MAT. | INST. | TOTAL |
|---|---|---|---|---|
| **15780.10 COMPUTER ROOM A/C** | | | | |
| 10 ton | EA. | 16,520 | 430.00 | 16,950 |
| 15 ton | " | 18,310 | 490.00 | 18,800 |
| Hot water heat | | | | |
| 3 ton | EA. | 9,510 | 260.00 | 9,770 |
| 5 ton | " | 10,100 | 280.00 | 10,380 |
| 7.5 ton | " | 16,050 | 340.00 | 16,390 |
| 10 ton | " | 16,520 | 430.00 | 16,950 |
| 15 ton | " | 18,310 | 490.00 | 18,800 |
| Air cooled condenser, low ambient damper | | | | |
| 3 ton | EA. | 2,140 | 68.50 | 2,209 |
| 5 ton | " | 2,380 | 85.50 | 2,466 |
| 7.5 ton | " | 3,450 | 170.00 | 3,620 |
| 10 ton | " | 3,450 | 240.00 | 3,690 |
| 15 ton | " | 3,800 | 200.00 | 4,000 |
| Water cooled, high efficiency filter, alarm, elec. heat | | | | |
| 3 ton | EA. | 9,390 | 240.00 | 9,630 |
| 5 ton | " | 10,100 | 280.00 | 10,380 |
| 7.5 ton | " | 16,050 | 430.00 | 16,480 |
| 10 ton | " | 16,640 | 490.00 | 17,130 |
| 15 ton | " | 19,500 | 570.00 | 20,070 |
| Steam heat | | | | |
| 3 ton | EA. | 10,700 | 240.00 | 10,940 |
| 5 ton | " | 12,230 | 280.00 | 12,510 |
| 7.5 ton | " | 17,240 | 430.00 | 17,670 |
| 10 ton | " | 17,830 | 490.00 | 18,320 |
| 15 ton | " | 20,690 | 570.00 | 21,260 |
| Hot water heat | | | | |
| 3 ton | EA. | 10,700 | 240.00 | 10,940 |
| 5 ton | " | 11,410 | 280.00 | 11,690 |
| 7.5 ton | " | 17,240 | 430.00 | 17,670 |
| 10 ton | " | 17,830 | 490.00 | 18,320 |
| 15 ton | " | 20,690 | 570.00 | 21,260 |
| **15780.20 ROOFTOP UNITS** | | | | |
| Packaged, single zone rooftop unit, with roof curb | | | | |
| 2 ton | EA. | 2,380 | 340.00 | 2,720 |
| 3 ton | " | 2,600 | 340.00 | 2,940 |
| 4 ton | " | 3,510 | 430.00 | 3,940 |
| 5 ton | " | 3,850 | 570.00 | 4,420 |
| 7.5 ton | " | 5,380 | 680.00 | 6,060 |
| **15820.10 DEHUMIDIFIERS** | | | | |
| Dessicant dehumidifier, 1125 cfm | EA. | | | 12,454 |
| **15830.20 FAN COIL UNITS** | | | | |
| Fan coil unit, 2 pipe, complete | | | | |
| 200 cfm ceiling hung | EA. | 570.00 | 110.00 | 680.00 |
| Floor mounted | " | 530.00 | 85.50 | 615.50 |
| 300 cfm, ceiling hung | " | 610.00 | 140.00 | 750.00 |
| Floor mounted | " | 580.00 | 110.00 | 690.00 |

## HEAT TRANSFER

| HEAT TRANSFER | UNIT | MAT. | INST. | TOTAL |
|---|---|---|---|---|
| **15830.20 FAN COIL UNITS** | | | | |
| 400 cfm, ceiling hung | EA. | 640.00 | 160.00 | 800.00 |
| Floor mounted | " | 610.00 | 110.00 | 720.00 |
| 500 cfm, ceiling hung | " | 740.00 | 170.00 | 910.00 |
| Floor mounted | " | 710.00 | 130.00 | 840.00 |
| 600 cfm, ceiling hung | " | 940.00 | 190.00 | 1,130 |
| Floor mounted | " | 870.00 | 160.00 | 1,030 |
| **15830.70 UNIT HEATERS** | | | | |
| Steam unit heater, horizontal | | | | |
| 12,500 btuh, 200 cfm | EA. | 200.00 | 57.00 | 257.00 |
| 17,000 btuh, 300 cfm | " | 220.00 | 57.00 | 277.00 |
| 40,000 btuh, 500 cfm | " | 270.00 | 57.00 | 327.00 |
| 60,000 btuh, 700 cfm | " | 320.00 | 57.00 | 377.00 |
| 70,000 btuh, 1000 cfm | " | 430.00 | 85.50 | 515.50 |
| Vertical | | | | |
| 12,500 btuh, 200 cfm | EA. | 220.00 | 57.00 | 277.00 |
| 17,000 btuh, 300 cfm | " | 350.00 | 57.00 | 407.00 |
| 40,000 btuh, 500 cfm | " | 460.00 | 57.00 | 517.00 |
| 60,000 btuh, 700 cfm | " | 520.00 | 57.00 | 577.00 |
| 70,000 btuh, 1000 cfm | " | 570.00 | 57.00 | 627.00 |
| Gas unit heater, horizontal | | | | |
| 27,400 btuh | EA. | 430.00 | 140.00 | 570.00 |
| 38,000 btuh | " | 490.00 | 140.00 | 630.00 |
| 56,000 btuh | " | 540.00 | 140.00 | 680.00 |
| 82,200 btuh | " | 600.00 | 140.00 | 740.00 |
| 103,900 btuh | " | 630.00 | 210.00 | 840.00 |
| 125,700 btuh | " | 660.00 | 210.00 | 870.00 |
| 133,200 btuh | " | 600.00 | 210.00 | 810.00 |
| 149,000 btuh | " | 780.00 | 210.00 | 990.00 |
| 172,000 btuh | " | 830.00 | 210.00 | 1,040 |
| 190,000 btuh | " | 900.00 | 210.00 | 1,110 |
| 225,000 btuh | " | 990.00 | 210.00 | 1,200 |
| Hot water unit heater, horizontal | | | | |
| 12,500 btuh, 200 cfm | EA. | 220.00 | 57.00 | 277.00 |
| 17,000 btuh, 300 cfm | " | 280.00 | 57.00 | 337.00 |
| 25,000 btuh, 500 cfm | " | 310.00 | 57.00 | 367.00 |
| 30,000 btuh, 700 cfm | " | 350.00 | 57.00 | 407.00 |
| 50,000 btuh, 1000 cfm | " | 400.00 | 85.50 | 485.50 |
| 60,000 btuh, 1300 cfm | " | 500.00 | 85.50 | 585.50 |
| Vertical | | | | |
| 12,500 btuh, 200 cfm | EA. | 250.00 | 57.00 | 307.00 |
| 17,000 btuh, 300 cfm | " | 330.00 | 57.00 | 387.00 |
| 25,000 btuh, 500 cfm | " | 360.00 | 57.00 | 417.00 |
| 30,000 btuh, 700 cfm | " | 400.00 | 57.00 | 457.00 |
| 50,000 btuh, 1000 cfm | " | 450.00 | 57.00 | 507.00 |
| 60,000 btuh, 1300 cfm | " | 500.00 | 57.00 | 557.00 |
| Cabinet unit heaters, ceiling, exposed, hot water | | | | |
| 200 cfm | EA. | 670.00 | 110.00 | 780.00 |
| 300 cfm | " | 710.00 | 140.00 | 850.00 |
| 400 cfm | " | 770.00 | 160.00 | 930.00 |
| 600 cfm | " | 790.00 | 180.00 | 970.00 |
| 800 cfm | " | 1,100 | 210.00 | 1,310 |

## HEAT TRANSFER

### 15830.70 UNIT HEATERS

| HEAT TRANSFER | UNIT | MAT. | INST. | TOTAL |
|---|---|---|---|---|
| 1000 cfm | EA. | 1,250 | 240.00 | 1,490 |
| 1200 cfm | " | 1,360 | 280.00 | 1,640 |
| 2000 cfm | " | 2,380 | 380.00 | 2,760 |

## AIR HANDLING

### 15855.10 AIR HANDLING UNITS

| AIR HANDLING | UNIT | MAT. | INST. | TOTAL |
|---|---|---|---|---|
| Air handling unit, medium pressure, single zone | | | | |
| 1500 cfm | EA. | 2,260 | 210.00 | 2,470 |
| 3000 cfm | " | 2,830 | 380.00 | 3,210 |
| 4000 cfm | " | 3,170 | 430.00 | 3,600 |
| 5000 cfm | " | 3,620 | 460.00 | 4,080 |
| 6000 cfm | " | 4,080 | 490.00 | 4,570 |
| 7000 cfm | " | 4,530 | 530.00 | 5,060 |
| 8500 cfm | " | 5,320 | 570.00 | 5,890 |
| 10,500 cfm | " | 6,000 | 680.00 | 6,680 |
| 12,500 cfm | " | 7,700 | 760.00 | 8,460 |
| Rooftop air handling units | | | | |
| 4950 cfm | EA. | 7,130 | 380.00 | 7,510 |
| 7370 cfm | " | 8,150 | 490.00 | 8,640 |
| 9790 cfm | " | 9,620 | 570.00 | 10,190 |
| 14,300 cfm | " | 13,590 | 490.00 | 14,080 |
| 21,725 cfm | " | 19,250 | 490.00 | 19,740 |
| 33,000 cfm | " | 27,170 | 570.00 | 27,740 |

### 15870.20 EXHAUST FANS

| AIR HANDLING | UNIT | MAT. | INST. | TOTAL |
|---|---|---|---|---|
| Belt drive roof exhaust fans | | | | |
| 640 cfm, 2618 fpm | EA. | 680.00 | 42.70 | 722.70 |
| 940 cfm, 2604 fpm | " | 880.00 | 42.70 | 922.70 |
| 1050 cfm, 3325 fpm | " | 780.00 | 42.70 | 822.70 |
| 1170 cfm, 2373 fpm | " | 1,130 | 42.70 | 1,173 |
| 2440 cfm, 4501 fpm | " | 890.00 | 42.70 | 932.70 |
| 2760 cfm, 4950 fpm | " | 990.00 | 42.70 | 1,033 |
| 3890 cfm, 6769 fpm | " | 1,120 | 42.70 | 1,163 |
| 2380 cfm, 3382 fpm | " | 1,250 | 42.70 | 1,293 |
| 2880 cfm, 3859 fpm | " | 1,300 | 42.70 | 1,343 |
| 3200 cfm, 4173 fpm | " | 1,310 | 57.00 | 1,367 |
| 3660 cfm, 3437 fpm | " | 1,340 | 57.00 | 1,397 |
| Direct drive fans | | | | |
| 60 to 390 cfm | EA. | 550.00 | 42.70 | 592.70 |
| 145 to 590 cfm | " | 670.00 | 42.70 | 712.70 |
| 295 to 860 cfm | " | 810.00 | 42.70 | 852.70 |

## AIR HANDLING

| AIR HANDLING | UNIT | MAT. | INST. | TOTAL |
|---|---|---|---|---|
| **15870.20 EXHAUST FANS** | | | | |
| 235 to 1300 cfm | EA. | 870.00 | 42.70 | 912.70 |
| 415 to 1630 cfm | " | 990.00 | 42.70 | 1,033 |
| 590 to 2045 cfm | " | 1,130 | 42.70 | 1,173 |

## AIR DISTRIBUTION

| AIR DISTRIBUTION | UNIT | MAT. | INST. | TOTAL |
|---|---|---|---|---|
| **15890.10 METAL DUCTWORK** | | | | |
| Rectangular duct | | | | |
| Galvanized steel | | | | |
| Minimum | Lb. | 1.80 | 3.10 | 4.90 |
| Average | " | 2.15 | 3.80 | 5.95 |
| Maximum | " | 2.55 | 5.70 | 8.25 |
| Aluminum | | | | |
| Minimum | Lb. | 4.75 | 6.85 | 11.60 |
| Average | " | 5.45 | 8.55 | 14.00 |
| Maximum | " | 6.80 | 11.40 | 18.20 |
| Fittings | | | | |
| Minimum | EA. | 2.70 | 11.40 | 14.10 |
| Average | " | 5.90 | 17.10 | 23.00 |
| Maximum | " | 14.15 | 34.20 | 48.35 |
| For work | | | | |
| 10-20' high, add per pound, $.30 | | | | |
| 30-50', add per pound, $.50 | | | | |
| **15890.30 FLEXIBLE DUCTWORK** | | | | |
| Flexible duct, 1.25" fiberglass | | | | |
| 5" dia. | L.F. | 1.70 | 1.70 | 3.40 |
| 6" dia. | " | 2.05 | 1.90 | 3.95 |
| 7" dia. | " | 2.25 | 2.00 | 4.25 |
| 8" dia. | " | 2.60 | 2.15 | 4.75 |
| 10" dia. | " | 2.95 | 2.45 | 5.40 |
| 12" dia. | " | 3.60 | 2.65 | 6.25 |
| 14" dia. | " | 4.40 | 2.85 | 7.25 |
| 16" dia. | " | 5.45 | 3.10 | 8.55 |
| Flexible duct connector, 3" wide fabric | " | 1.55 | 5.70 | 7.25 |
| **15895.10 ROOF CURBS** | | | | |
| 8" high, insulated, with liner and raised can | | | | |
| 15" x 15" | EA. | 59.00 | 17.10 | 76.10 |
| 21" x 21" | " | 73.50 | 17.10 | 90.60 |
| 36" x 36" | " | 110.00 | 24.40 | 134.40 |
| 48" x 48" | " | 300.00 | 26.30 | 326.30 |
| 60" x 60" | " | 550.00 | 34.20 | 584.20 |

# 15 MECHANICAL

| AIR DISTRIBUTION | UNIT | MAT. | INST. | TOTAL |
|---|---|---|---|---|
| **15895.10 ROOF CURBS** | | | | |
| 72" x 72" | EA. | 880.00 | 42.70 | 922.70 |
| **15910.10 DAMPERS** | | | | |
| Horizontal parallel aluminum backdraft damper | | | | |
| 12" x 12" | EA. | 29.70 | 8.55 | 38.25 |
| 16" x 16" | " | 41.60 | 9.75 | 51.35 |
| 24" x 24" | " | 65.50 | 17.10 | 82.60 |
| 36" x 36" | " | 150.00 | 24.40 | 174.40 |
| 48" x 48" | " | 270.00 | 34.20 | 304.20 |
| "Up", parallel dampers | | | | |
| 12" x 12" | EA. | 52.50 | 8.55 | 61.05 |
| 16" x 16" | " | 73.50 | 9.75 | 83.25 |
| 24" x 24" | " | 90.50 | 17.10 | 107.60 |
| 36" x 36" | " | 150.00 | 24.40 | 174.40 |
| 48" x 48" | " | 300.00 | 34.20 | 334.20 |
| "Down", parallel dampers | | | | |
| 12" x 12" | EA. | 52.50 | 8.55 | 61.05 |
| 16" x 16" | " | 73.50 | 9.75 | 83.25 |
| 24" x 24" | " | 90.50 | 17.10 | 107.60 |
| 36" x 36" | " | 150.00 | 24.40 | 174.40 |
| 48" x 48" | " | 300.00 | 34.20 | 334.20 |
| Fire damper, 1.5 hr rating | | | | |
| 12" x 12" | EA. | 17.85 | 17.10 | 34.95 |
| 16" x 16" | " | 28.50 | 17.10 | 45.60 |
| 24" x 24" | " | 35.70 | 17.10 | 52.80 |
| 36" x 36" | " | 59.50 | 34.20 | 93.70 |
| 48" x 48" | " | 110.00 | 48.80 | 158.80 |
| **15940.10 DIFFUSERS** | | | | |
| Ceiling diffusers, round, baked enamel finish | | | | |
| 6" dia. | EA. | 29.70 | 11.40 | 41.10 |
| 12" dia. | " | 53.50 | 14.25 | 67.75 |
| 16" dia. | " | 77.50 | 15.55 | 93.05 |
| 20" dia. | " | 110.00 | 17.10 | 127.10 |
| Rectangular | | | | |
| 6x6" | EA. | 32.80 | 11.40 | 44.20 |
| 12x12" | " | 57.50 | 17.10 | 74.60 |
| 18x18" | " | 89.50 | 17.10 | 106.60 |
| 24x24" | " | 160.00 | 21.35 | 181.35 |
| Lay in, flush mounted, perforated face, with grid | | | | |
| 6x6/24x24 | EA. | 54.50 | 13.65 | 68.15 |
| 12x12/24x24 | " | 61.00 | 13.65 | 74.65 |
| 18x18/24x24 | " | 63.50 | 13.65 | 77.15 |
| Two-way slot diffuser with balancing damper, 4' | " | 68.00 | 34.20 | 102.20 |
| **15940.20 RELIEF VENTILATORS** | | | | |
| Intake ventilator, aluminum, with screen, no curbs | | | | |
| 12" x 12" | EA. | 110.00 | 28.50 | 138.50 |
| 16" x 16" | " | 160.00 | 34.20 | 194.20 |

| AIR DISTRIBUTION | UNIT | MAT. | INST. | TOTAL |
|---|---|---|---|---|
| **15940.20 RELIEF VENTILATORS** | | | | |
| 36" x 36" | EA. | 620.00 | 57.00 | 677.00 |
| 48" x 48" | " | 1,020 | 68.50 | 1,089 |
| **15940.40 REGISTERS AND GRILLES** | | | | |
| Lay in flush mounted, perforated face, return | | | | |
| 6x6/24x24 | EA. | 27.20 | 13.65 | 40.85 |
| 8x8/24x24 | " | 27.20 | 13.65 | 40.85 |
| 9x9/24x24 | " | 29.40 | 13.65 | 43.05 |
| 10x10/24x24 | " | 31.70 | 13.65 | 45.35 |
| 12x12/24x24 | " | 31.70 | 13.65 | 45.35 |
| Rectangular, ceiling return, single deflection | | | | |
| 10x10 | EA. | 13.60 | 17.10 | 30.70 |
| 12x12 | " | 15.85 | 17.10 | 32.95 |
| 14x14 | " | 19.25 | 17.10 | 36.35 |
| 16x8 | " | 15.85 | 17.10 | 32.95 |
| 16x16 | " | 15.85 | 17.10 | 32.95 |
| 24x12 | " | 43.00 | 17.10 | 60.10 |
| 24x18 | " | 55.50 | 17.10 | 72.60 |
| 36x24 | " | 110.00 | 19.00 | 129.00 |
| 36x30 | " | 160.00 | 19.00 | 179.00 |
| Wall, return air register | | | | |
| 12x12 | EA. | 22.60 | 8.55 | 31.15 |
| 16x16 | " | 33.30 | 8.55 | 41.85 |
| 18x18 | " | 39.20 | 8.55 | 47.75 |
| 20x20 | " | 46.40 | 8.55 | 54.95 |
| 24x24 | " | 64.00 | 8.55 | 72.55 |
| Ceiling, return air grille | | | | |
| 6x6 | EA. | 8.35 | 11.40 | 19.75 |
| 8x8 | " | 9.50 | 13.65 | 23.15 |
| 10x10 | " | 10.70 | 13.65 | 24.35 |
| Ceiling, exhaust grille, aluminum egg crate | | | | |
| 6x6 | EA. | 10.70 | 11.40 | 22.10 |
| 8x8 | " | 10.70 | 13.65 | 24.35 |
| 10x10 | " | 11.90 | 13.65 | 25.55 |
| 12x12 | " | 15.45 | 17.10 | 32.55 |
| 14x14 | " | 20.20 | 17.10 | 37.30 |
| 16x16 | " | 23.80 | 17.10 | 40.90 |
| 18x18 | " | 28.50 | 17.10 | 45.60 |
| **15940.80 PENTHOUSE LOUVERS** | | | | |
| Penthouse louvers | | | | |
| 12" high, extruded aluminum, 4" louver | | | | |
| 6' perimeter | EA. | 270.00 | 85.50 | 355.50 |
| 12' perimeter | " | 660.00 | 85.50 | 745.50 |
| 20' perimeter | " | 1,360 | 230.00 | 1,590 |
| 16" high x 4' perimeter | " | 230.00 | 85.50 | 315.50 |
| 6' perimeter | " | 320.00 | 85.50 | 405.50 |
| 12' perimeter | " | 740.00 | 85.50 | 825.50 |
| 20' perimeter | " | 1,590 | 230.00 | 1,820 |
| 20" high x 4' perimeter | " | 290.00 | 85.50 | 375.50 |
| 6' perimeter | " | 340.00 | 85.50 | 425.50 |
| 12' perimeter | " | 830.00 | 85.50 | 915.50 |

| AIR DISTRIBUTION | UNIT | MAT. | INST. | TOTAL |
|---|---|---|---|---|
| **15940.80 PENTHOUSE LOUVERS** | | | | |
| 20' perimeter | EA. | 1,700 | 230.00 | 1,930 |
| 24" high x 4' perimeter | " | 320.00 | 85.50 | 405.50 |
| 6' perimeter | " | 380.00 | 85.50 | 465.50 |
| 12' perimeter | " | 920.00 | 85.50 | 1,006 |
| 20' perimeter | " | 1,910 | 230.00 | 2,140 |

| CONTROLS | UNIT | MAT. | INST. | TOTAL |
|---|---|---|---|---|
| **15950.10 HVAC CONTROLS** | | | | |
| Pressure gauge, direct reading gage cock and siphon | EA. | 51.00 | 21.35 | 72.35 |
| Control valve, 1", modulating | | | | |
| 2-way | EA. | 510.00 | 28.50 | 538.50 |
| 3-way | " | 580.00 | 42.70 | 622.70 |
| Self contained control valve with sensing element, 3/4" | " | 93.00 | 21.35 | 114.35 |
| Instrument air system 2-2 hp compressors, rcvr refrigerant dryer | " | | | 4,302 |
| Thermostat primary control device | " | | | 92.84 |
| Humidistat primary control device | " | | | 70.20 |
| Timers primary control device, indoor/outdoor, 24 hour | " | | | 144.92 |
| Thermometer, dir. reading, 3 dial | " | | | 70.20 |
| Control dampers, round | | | | |
| 6" dia. | EA. | 52.00 | 13.65 | 65.65 |
| 8" dia | " | 76.00 | 13.65 | 89.65 |
| 10" dia | " | 100.00 | 13.65 | 113.65 |
| 12" dia | " | 130.00 | 13.65 | 143.65 |
| 12" dia | " | 190.00 | 17.10 | 207.10 |
| 18" dia | " | 200.00 | 17.10 | 217.10 |
| 20" dia | " | 270.00 | 17.10 | 287.10 |
| Rectangular, parallel blade standard leakage | | | | |
| 12" x 12" | EA. | 38.50 | 17.10 | 55.60 |
| 16" x 16" | " | 56.50 | 17.10 | 73.60 |
| 20" x 20" | " | 68.00 | 17.10 | 85.10 |
| 48" x 48" | " | 200.00 | 48.80 | 248.80 |
| 48" x 60" | " | 260.00 | 57.00 | 317.00 |
| 48" x 72" | " | 320.00 | 57.00 | 377.00 |
| Low leakage | | | | |
| 12" x 12" | EA. | 74.50 | 17.10 | 91.60 |
| 16" x 16" | " | 95.00 | 17.10 | 112.10 |
| 36" x 36" | " | 240.00 | 28.50 | 268.50 |
| 48" x 48" | " | 430.00 | 48.80 | 478.80 |
| 48" x 72" | " | 670.00 | 57.00 | 727.00 |
| Rectangular, opposed horizontal blade | | | | |
| 12" x 12" | EA. | 49.80 | 17.10 | 66.90 |
| 16" x 16" | " | 70.00 | 17.10 | 87.10 |

| CONTROLS | UNIT | MAT. | INST. | TOTAL |
|---|---|---|---|---|
| **15950.10 HVAC CONTROLS** | | | | |
| 24" x 24" | EA. | 93.00 | 17.10 | 110.10 |
| 36" x 36" | " | 150.00 | 28.50 | 178.50 |
| 48" x 72" | " | 430.00 | 57.00 | 487.00 |

| BASIC MATERIALS | UNIT | MAT. | INST. | TOTAL |
|---|---|---|---|---|
| **16110.12 CABLE TRAY** | | | | |
| Cable tray, 6" | | | | |
| Ventilated cover | L.F. | 11.70 | 2.35 | 14.05 |
| Solid cover | " | 4.75 | 1.20 | 5.95 |
| | " | 3.70 | 1.20 | 4.90 |
| **16110.20 CONDUIT SPECIALTIES** | | | | |
| Rod beam clamp, 1/2" | EA. | 3.60 | 2.00 | 5.60 |
| Hanger rod | | | | |
| 3/8" | L.F. | 0.75 | 1.60 | 2.35 |
| 1/2" | " | 1.20 | 2.00 | 3.20 |
| All thread rod | | | | |
| 1/4" | L.F. | 0.44 | 1.20 | 1.64 |
| 3/8" | " | 0.74 | 1.60 | 2.34 |
| 1/2" | " | 1.20 | 2.00 | 3.20 |
| 5/8" | " | 1.95 | 3.15 | 5.10 |
| Hanger channel, 1-1/2" | | | | |
| No holes | EA. | 2.95 | 1.20 | 4.15 |
| Holes | " | 3.25 | 1.20 | 4.45 |
| Channel strap | | | | |
| 1/2" | EA. | 0.78 | 2.00 | 2.78 |
| 3/4" | " | 0.92 | 2.00 | 2.92 |
| 1" | " | 1.00 | 2.00 | 3.00 |
| 1-1/4" | " | 1.10 | 3.15 | 4.25 |
| 1-1/2" | " | 1.25 | 3.15 | 4.40 |
| 2" | " | 1.45 | 3.15 | 4.60 |
| 2-1/2" | " | 1.60 | 4.85 | 6.45 |
| 3" | " | 1.75 | 4.85 | 6.60 |
| 3-1/2" | " | 2.15 | 4.85 | 7.00 |
| 4" | " | 2.45 | 5.75 | 8.20 |
| 5" | " | 3.40 | 5.75 | 9.15 |
| 6" | " | 4.05 | 5.75 | 9.80 |
| Conduit penetrations, roof and wall, 8" thick | | | | |
| 1/2" | EA. | 0.00 | 24.35 | 24.35 |
| 3/4" | " | 0.00 | 24.35 | 24.35 |
| 1" | " | 0.00 | 31.70 | 31.70 |
| 1-1/4" | " | 0.00 | 31.70 | 31.70 |
| 1-1/2" | " | 0.00 | 31.70 | 31.70 |
| 2" | " | 0.00 | 63.50 | 63.50 |
| 2-1/2" | " | 0.00 | 63.50 | 63.50 |
| 3" | " | 0.00 | 63.50 | 63.50 |
| 3-1/2" | " | 0.00 | 79.00 | 79.00 |
| 4" | " | 0.00 | 79.00 | 79.00 |
| Fireproofing, for conduit penetrations | | | | |
| 1/2" | EA. | 1.70 | 19.80 | 21.50 |
| 3/4" | " | 1.75 | 19.80 | 21.55 |
| 1" | " | 1.80 | 19.80 | 21.60 |
| 1-1/4" | " | 2.45 | 31.10 | 33.55 |
| 1-1/2" | " | 2.50 | 28.80 | 31.30 |
| 2" | " | 2.55 | 28.80 | 31.35 |
| 2-1/2" | " | 4.85 | 35.60 | 40.45 |
| 3" | " | 5.05 | 38.40 | 43.45 |
| 3-1/2" | " | 5.90 | 49.50 | 55.40 |

# 16 ELECTRICAL

## BASIC MATERIALS

| BASIC MATERIALS | UNIT | MAT. | INST. | TOTAL |
|---|---|---|---|---|
| **16110.20 CONDUIT SPECIALTIES** | | | | |
| 4“ | EA. | 7.10 | 60.00 | 67.10 |
| **16110.21 ALUMINUM CONDUIT** | | | | |
| Aluminum conduit | | | | |
| 1/2” | L.F. | 0.91 | 1.20 | 2.11 |
| 3/4“ | ” | 1.15 | 1.60 | 2.75 |
| 1“ | ” | 1.20 | 2.00 | 3.20 |
| 1-1/4“ | ” | 2.25 | 2.35 | 4.60 |
| 1-1/2“ | ” | 2.80 | 3.15 | 5.95 |
| 2“ | ” | 3.70 | 3.50 | 7.20 |
| 2-1/2“ | ” | 5.85 | 3.95 | 9.80 |
| 3“ | ” | 7.70 | 4.20 | 11.90 |
| 3-1/2“ | ” | 9.25 | 4.85 | 14.10 |
| 4“ | ” | 10.95 | 5.75 | 16.70 |
| 5“ | ” | 15.70 | 7.20 | 22.90 |
| 6“ | ” | 20.70 | 7.90 | 28.60 |
| **16110.22 EMT CONDUIT** | | | | |
| EMT conduit | | | | |
| 1/2“ | L.F. | 0.19 | 1.20 | 1.39 |
| 3/4” | “ | 0.28 | 1.60 | 1.88 |
| 1” | “ | 0.42 | 2.00 | 2.42 |
| 1-1/4” | “ | 0.60 | 2.35 | 2.95 |
| 1-1/2” | “ | 0.70 | 3.15 | 3.85 |
| 2” | “ | 0.89 | 3.50 | 4.39 |
| 2-1/2” | “ | 2.00 | 3.95 | 5.95 |
| 3” | “ | 2.45 | 4.85 | 7.30 |
| 3-1/2” | “ | 3.50 | 5.75 | 9.25 |
| 4” | “ | 4.10 | 7.20 | 11.30 |
| **16110.23 FLEXIBLE CONDUIT** | | | | |
| Flexible conduit, steel | | | | |
| 3/8” | L.F. | 0.21 | 1.20 | 1.41 |
| 1/2 | “ | 0.22 | 1.20 | 1.42 |
| 3/4” | “ | 0.32 | 1.60 | 1.92 |
| 1” | “ | 0.62 | 1.60 | 2.22 |
| 1-1/4” | “ | 0.78 | 2.00 | 2.78 |
| 1-1/2” | “ | 1.05 | 2.35 | 3.40 |
| 2” | “ | 1.35 | 3.15 | 4.50 |
| 2-1/2” | “ | 1.60 | 3.50 | 5.10 |
| 3” | “ | 1.95 | 4.20 | 6.15 |
| **16110.24 GALVANIZED CONDUIT** | | | | |
| Galvanized rigid steel conduit | | | | |
| 1/2” | L.F. | 0.80 | 1.60 | 2.40 |
| 3/4“ | ” | 1.00 | 2.00 | 3.00 |
| 1“ | ” | 1.35 | 2.35 | 3.70 |

| BASIC MATERIALS | UNIT | MAT. | INST. | TOTAL |
|---|---|---|---|---|
| **16110.24 GALVANIZED CONDUIT** | | | | |
| 1-1/4" | L.F. | 1.80 | 3.15 | 4.95 |
| 1-1/2" | " | 2.20 | 3.50 | 5.70 |
| 2" | " | 2.80 | 3.95 | 6.75 |
| 2-1/2" | " | 4.50 | 5.75 | 10.25 |
| 3" | " | 6.00 | 7.20 | 13.20 |
| 3-1/2" | " | 6.05 | 7.55 | 13.60 |
| 4" | " | 8.90 | 8.35 | 17.25 |
| 5" | " | 18.90 | 11.30 | 30.20 |
| 6" | " | 27.20 | 15.10 | 42.30 |
| **16110.27 PLASTIC COATED CONDUIT** | | | | |
| Rigid steel conduit, plastic coated | | | | |
| 1/2" | L.F. | 2.40 | 2.00 | 4.40 |
| 3/4" | " | 2.80 | 2.35 | 5.15 |
| 1" | " | 3.60 | 3.15 | 6.75 |
| 1-1/4" | " | 4.55 | 3.95 | 8.50 |
| 1-1/2" | " | 5.50 | 4.85 | 10.35 |
| 2" | " | 7.25 | 5.75 | 13.00 |
| 2-1/2" | " | 11.05 | 7.55 | 18.60 |
| 3" | " | 13.90 | 8.80 | 22.70 |
| 3-1/2" | " | 16.85 | 9.90 | 26.75 |
| 4" | " | 20.60 | 12.20 | 32.80 |
| 5" | " | 35.30 | 15.10 | 50.40 |
| **16110.28 STEEL CONDUIT** | | | | |
| Intermediate metal conduit (IMC) | | | | |
| 1/2" | L.F. | 0.74 | 1.20 | 1.94 |
| 3/4" | " | 0.89 | 1.60 | 2.49 |
| 1" | " | 1.20 | 2.00 | 3.20 |
| 1-1/4" | " | 1.60 | 2.35 | 3.95 |
| 1-1/2" | " | 1.85 | 3.15 | 5.00 |
| 2" | " | 2.60 | 3.50 | 6.10 |
| 2-1/2" | " | 4.05 | 4.75 | 8.80 |
| 3" | " | 5.45 | 5.75 | 11.20 |
| 3-1/2" | " | 7.20 | 7.20 | 14.40 |
| 4" | " | 8.55 | 7.55 | 16.10 |
| **16110.35 WIREMOLD** | | | | |
| Wiremold raceway with fittings, surface mounted | | | | |
| #200 | L.F. | 0.50 | 1.20 | 1.70 |
| #500 | " | 0.60 | 1.20 | 1.80 |
| #700 | " | 0.65 | 1.60 | 2.25 |
| #800 | " | 1.10 | 1.60 | 2.70 |
| Fittings, #200, 90 degree flat elbow | EA. | 0.97 | 2.00 | 2.97 |
| Internal elbow | " | 1.90 | 2.00 | 3.90 |
| Extension adapter | " | 3.40 | 2.35 | 5.75 |
| #200, #500, #700 | | | | |
| Single pole switch and box | EA. | 6.40 | 15.85 | 22.25 |
| Duplex receptacle with box | " | 7.45 | 12.20 | 19.65 |
| #500, #700 | | | | |

# 16 ELECTRICAL

| BASIC MATERIALS | UNIT | MAT. | INST. | TOTAL |
|---|---|---|---|---|
| **16110.35 WIREMOLD** | | | | |
| 90 deg. flat elbow | EA. | 0.83 | 3.15 | 3.98 |
| Internal elbow | " | 1.10 | 3.15 | 4.25 |
| Junction box | " | 5.45 | 5.30 | 10.75 |
| Fixture box | " | 5.50 | 5.30 | 10.80 |
| Shallow switch and receptacle box | " | 4.50 | 3.15 | 7.65 |
| #800 | | | | |
| 90 deg. flat elbow | EA. | 1.00 | 3.15 | 4.15 |
| Internal elbow | " | 1.00 | 3.15 | 4.15 |
| Junction box | " | 3.20 | 4.85 | 8.05 |
| **16110.80 WIREWAYS** | | | | |
| Wireway, hinge cover type | | | | |
| 2-1/2" x 2-1/2" | | | | |
| 1' section | EA. | 8.50 | 6.10 | 14.60 |
| 2' | " | 12.05 | 7.55 | 19.60 |
| 3' | " | 16.40 | 9.90 | 26.30 |
| 5' | " | 27.10 | 15.10 | 42.20 |
| 10' | " | 56.50 | 26.40 | 82.90 |
| 4" x 4" | | | | |
| 1' | EA. | 8.25 | 9.90 | 18.15 |
| 2' | " | 13.65 | 9.90 | 23.55 |
| 3' | " | 20.20 | 12.20 | 32.40 |
| 4' | " | 27.50 | 12.20 | 39.70 |
| 10' | " | 61.50 | 31.70 | 93.20 |
| **16120.43 COPPER CONDUCTORS** | | | | |
| Copper conductors, type THW, solid | | | | |
| #14 | L.F. | 0.07 | 0.16 | 0.23 |
| #12 | " | 0.08 | 0.20 | 0.28 |
| #10 | " | 0.09 | 0.24 | 0.33 |
| Stranded | | | | |
| #14 | L.F. | 0.09 | 0.16 | 0.25 |
| #12 | " | 0.10 | 0.20 | 0.30 |
| #10 | " | 0.13 | 0.24 | 0.37 |
| #8 | " | 0.20 | 0.32 | 0.52 |
| #6 | " | 0.27 | 0.36 | 0.63 |
| #4 | " | 0.39 | 0.40 | 0.79 |
| #3 | " | 0.53 | 0.40 | 0.93 |
| #2 | " | 0.67 | 0.48 | 1.15 |
| #1 | " | 0.87 | 0.56 | 1.43 |
| 1/0 | " | 1.05 | 0.63 | 1.68 |
| 2/0 | " | 1.20 | 0.79 | 1.99 |
| 3/0 | " | 1.50 | 0.99 | 2.49 |
| 4/0 | " | 1.90 | 1.10 | 3.00 |
| 250 MCM | " | 2.30 | 1.20 | 3.50 |
| 300 MCM | " | 2.75 | 1.30 | 4.05 |
| 350 MCM | " | 3.35 | 1.60 | 4.95 |
| 400 MCM | " | 3.80 | 1.75 | 5.55 |
| 500 MCM | " | 4.55 | 2.05 | 6.60 |
| 600 MCM | " | 6.50 | 2.35 | 8.85 |

| BASIC MATERIALS | UNIT | MAT. | INST. | TOTAL |
|---|---|---|---|---|
| **16120.43 COPPER CONDUCTORS** | | | | |
| 750 MCM | L.F. | 8.25 | 2.65 | 10.90 |
| 1000 MCM | " | 9.15 | 3.00 | 12.15 |
| THHN-THWN, solid | | | | |
| #14 | L.F. | 0.07 | 0.16 | 0.23 |
| #12 | " | 0.09 | 0.20 | 0.29 |
| #10 | " | 0.11 | 0.24 | 0.35 |
| Stranded | | | | |
| #14 | L.F. | 0.09 | 0.16 | 0.25 |
| #12 | " | 0.11 | 0.20 | 0.31 |
| #10 | " | 0.13 | 0.24 | 0.37 |
| #8 | " | 0.20 | 0.32 | 0.52 |
| #6 | " | 0.27 | 0.36 | 0.63 |
| #4 | " | 0.39 | 0.40 | 0.79 |
| #2 | " | 0.67 | 0.48 | 1.15 |
| #1 | " | 0.85 | 0.56 | 1.41 |
| 1/0 | " | 1.05 | 0.63 | 1.68 |
| 2/0 | " | 1.20 | 0.79 | 1.99 |
| 3/0 | " | 1.50 | 0.99 | 2.49 |
| 4/0 | " | 1.90 | 1.10 | 3.00 |
| 250 MCM | " | 2.30 | 1.20 | 3.50 |
| 350 MCM | " | 3.35 | 1.60 | 4.95 |
| **16120.47 SHEATHED CABLE** | | | | |
| Non-metallic sheathed cable | | | | |
| Type NM cable with ground | | | | |
| #14/2 | L.F. | 0.14 | 0.59 | 0.73 |
| #12/2 | " | 0.20 | 0.63 | 0.83 |
| #10/2 | " | 0.32 | 0.70 | 1.02 |
| #8/2 | " | 0.64 | 0.79 | 1.43 |
| #6/2 | " | 0.98 | 0.99 | 1.97 |
| #14/3 | " | 0.22 | 1.00 | 1.22 |
| #12/3 | " | 0.32 | 1.05 | 1.37 |
| #10/3 | " | 0.47 | 1.05 | 1.52 |
| #8/3 | " | 1.00 | 1.10 | 2.10 |
| #6/3 | " | 1.40 | 1.10 | 2.50 |
| #4/3 | " | 2.10 | 1.25 | 3.35 |
| #2/3 | " | 3.05 | 1.40 | 4.45 |
| Type U.F. cable with ground | | | | |
| #14/2 | L.F. | 0.18 | 0.63 | 0.81 |
| #12/2 | " | 0.23 | 0.75 | 0.98 |
| #10/2 | " | 0.32 | 0.79 | 1.11 |
| #8/2 | " | 1.00 | 0.91 | 1.91 |
| #6/2 | " | 1.20 | 1.05 | 2.25 |
| #14/3 | " | 0.24 | 0.79 | 1.03 |
| #12/3 | " | 0.34 | 0.87 | 1.21 |
| #10/3 | " | 0.49 | 0.99 | 1.48 |
| #8/3 | " | 1.30 | 1.10 | 2.40 |
| #6/3 | " | 1.90 | 1.25 | 3.15 |

# 16 ELECTRICAL

| BASIC MATERIALS | UNIT | MAT. | INST. | TOTAL |
|---|---|---|---|---|
| **16130.60 PULL AND JUNCTION BOXES** | | | | |
| 4" | | | | |
| Octagon box | EA. | 1.25 | 4.55 | 5.80 |
| Box extension | " | 1.80 | 2.35 | 4.15 |
| Plaster ring | " | 0.97 | 2.35 | 3.32 |
| Cover blank | " | 0.40 | 2.35 | 2.75 |
| Square box | " | 1.50 | 4.55 | 6.05 |
| Box extension | " | 1.65 | 2.35 | 4.00 |
| Plaster ring | " | 0.74 | 2.35 | 3.09 |
| Cover blank | " | 0.46 | 2.35 | 2.81 |
| 4-11/16" | | | | |
| Square box | EA. | 2.70 | 4.55 | 7.25 |
| Box extension | " | 3.60 | 2.35 | 5.95 |
| Plaster ring | " | 2.10 | 2.35 | 4.45 |
| Cover blank | " | 0.81 | 2.35 | 3.16 |
| Switch and device boxes | | | | |
| 2 gang | EA. | 7.85 | 4.55 | 12.40 |
| 3 gang | " | 9.65 | 4.55 | 14.20 |
| 4 gang | " | 13.65 | 6.35 | 20.00 |
| Device covers | | | | |
| 2 gang | EA. | 3.85 | 2.35 | 6.20 |
| 3 gang | " | 4.55 | 2.35 | 6.90 |
| 4 gang | " | 6.55 | 2.35 | 8.90 |
| Handy box | " | 1.15 | 4.55 | 5.70 |
| Extension | " | 1.60 | 2.35 | 3.95 |
| Switch cover | " | 0.46 | 2.35 | 2.81 |
| Switch box with knockout | " | 1.30 | 5.75 | 7.05 |
| Weatherproof cover, spring type | " | 4.55 | 3.15 | 7.70 |
| Cover plate, dryer receptacle 1 gang plastic | " | 2.10 | 3.95 | 6.05 |
| For 4" receptacle, 2 gang | " | 3.60 | 3.95 | 7.55 |
| Duplex receptacle cover plate, plastic | " | 0.40 | 2.35 | 2.75 |
| 4", vertical bracket box, 1-1/2" with | | | | |
| RMX clamps | EA. | 1.90 | 5.75 | 7.65 |
| BX clamps | " | 2.85 | 5.75 | 8.60 |
| 4", octagon device cover | | | | |
| 1 switch | EA. | 1.30 | 2.35 | 3.65 |
| 1 duplex recept | " | 1.20 | 2.35 | 3.55 |
| 4", square face bracket boxes, 1-1/2" | | | | |
| RMX | EA. | 3.15 | 5.75 | 8.90 |
| BX | " | 3.35 | 5.75 | 9.10 |
| **16130.65 PULL BOXES AND CABINETS** | | | | |
| Galvanized pull boxes, screw cover | | | | |
| 4x4x4 | EA. | 4.15 | 7.55 | 11.70 |
| 4x6x4 | " | 5.15 | 7.55 | 12.70 |
| 6x6x4 | " | 6.25 | 7.55 | 13.80 |
| 6x8x4 | " | 9.00 | 7.55 | 16.55 |
| 8x8x4 | " | 10.55 | 9.90 | 20.45 |
| 8x10x4 | " | 12.20 | 9.60 | 21.80 |
| 8x12x4 | " | 13.65 | 9.90 | 23.55 |
| Screw cover | | | | |
| 10x10x4 | EA. | 13.65 | 12.20 | 25.85 |

| BASIC MATERIALS | UNIT | MAT. | INST. | TOTAL |
|---|---|---|---|---|
| **16130.65 PULL BOXES AND CABINETS** | | | | |
| 12x12x6 | EA. | 20.85 | 17.60 | 38.45 |
| 12x15x6 | " | 23.60 | 17.60 | 41.20 |
| 12x18x6 | " | 27.60 | 19.80 | 47.40 |
| 15x18x6 | " | 31.20 | 22.65 | 53.85 |
| 18x24x6 | " | 56.50 | 24.35 | 80.85 |
| 18x30x6 | " | 62.50 | 28.80 | 91.30 |
| 24x36x6 | " | 97.00 | 28.80 | 125.80 |
| **16130.80 RECEPTACLES** | | | | |
| 125 volt, 20a, duplex, grounding type, standard grade | EA. | 5.35 | 7.90 | 13.25 |
| Ground fault interrupter type | " | 40.90 | 11.75 | 52.65 |
| 250 volt, 20a, 2 pole, single receptacle, ground type | " | 6.20 | 7.90 | 14.10 |
| 120/208v, 4 pole, single receptacle, twist lock | | | | |
| 20a | EA. | 13.40 | 13.75 | 27.15 |
| 50a | " | 25.60 | 13.75 | 39.35 |
| 125/250v, 3 pole, flush receptacle | | | | |
| 30a | EA. | 8.50 | 11.75 | 20.25 |
| 50a | " | 9.80 | 11.75 | 21.55 |
| 60a | " | 39.60 | 13.75 | 53.35 |
| 277 v, 20a, 2 pole, grounding type, twist lock | " | 9.80 | 7.90 | 17.70 |
| Dryer receptacle, 250v, 30a/50a, 3 wire | " | 16.45 | 11.75 | 28.20 |
| Clock receptacle, 2 pole, grounding type | " | 6.85 | 7.90 | 14.75 |
| 125v, 20a single recept. grounding type | | | | |
| Standard grade | EA. | 5.00 | 7.90 | 12.90 |
| Specification | " | 7.75 | 7.90 | 15.65 |
| Hospital | " | 8.00 | 7.90 | 15.90 |
| Isolated ground orange | " | 13.90 | 9.90 | 23.80 |
| Duplex | | | | |
| Specification grade | EA. | 8.05 | 7.90 | 15.95 |
| Hospital | " | 10.40 | 7.90 | 18.30 |
| Isolated ground orange | " | 15.65 | 9.90 | 25.55 |
| GFI hospital grade recepts, 20a, 125v, duplex | " | 47.30 | 11.75 | 59.05 |
| **16198.10 ELECTRIC MANHOLES** | | | | |
| Precast, handhole, 4' deep | | | | |
| 2'x2' | EA. | 310.00 | 140.00 | 450.00 |
| 3'x3' | " | 430.00 | 220.00 | 650.00 |
| 4'x4' | " | 920.00 | 410.00 | 1,330 |
| Power manhole, complete, precast, 8' deep | | | | |
| 4'x4' | EA. | 1,390 | 560.00 | 1,950 |
| 6'x6' | " | 1,830 | 790.00 | 2,620 |
| 8'x8' | " | 2,120 | 830.00 | 2,950 |
| 6' deep, 9' x 12' | " | 2,460 | 990.00 | 3,450 |
| Cast in place, power manhole, 8' deep | | | | |
| 4'x4' | EA. | 1,640 | 560.00 | 2,200 |
| 6'x6' | " | 2,020 | 790.00 | 2,810 |
| 8'x8' | " | 2,260 | 830.00 | 3,090 |

# 16 ELECTRICAL

## BASIC MATERIALS

| BASIC MATERIALS | UNIT | MAT. | INST. | TOTAL |
|---|---|---|---|---|
| **16199.10 UTILITY POLES & FITTINGS** | | | | |
| Wood pole, creosoted | | | | |
| 25' | EA. | 160.00 | 93.00 | 253.00 |
| 40' | " | 740.00 | 150.00 | 890.00 |
| 55' | " | 1,020 | 300.00 | 1,320 |
| Treated, wood preservative, 6"x6" | | | | |
| 8' | EA. | 14.85 | 19.80 | 34.65 |
| 16' | " | 74.50 | 63.50 | 138.00 |
| 20' | " | 150.00 | 79.00 | 229.00 |
| Aluminum, brushed, no base | | | | |
| 8' | EA. | 280.00 | 79.00 | 359.00 |
| 20' | " | 760.00 | 130.00 | 890.00 |
| 40' | " | 2,490 | 250.00 | 2,740 |
| Steel, no base | | | | |
| 10' | EA. | 340.00 | 99.00 | 439.00 |
| 20' | " | 500.00 | 150.00 | 650.00 |
| 35' | " | 1,440 | 250.00 | 1,690 |
| Concrete, no base | | | | |
| 13' | EA. | 520.00 | 220.00 | 740.00 |
| 30' | " | 1,490 | 480.00 | 1,970 |
| 50' | " | 3,400 | 720.00 | 4,120 |
| 60' | " | 4,320 | 790.00 | 5,110 |

## POWER GENERATION

| POWER GENERATION | UNIT | MAT. | INST. | TOTAL |
|---|---|---|---|---|
| **16210.10 GENERATORS** | | | | |
| Diesel generator, with auto transfer switch | | | | |
| 50kw | EA. | 21,170 | 1,220 | 22,390 |
| 125kw | " | 31,880 | 1,980 | 33,860 |
| 300kw | " | 52,220 | 3,960 | 56,180 |
| 750kw | " | 152,250 | 7,920 | |
| **16320.10 TRANSFORMERS** | | | | |
| Floor mounted, single phase, int. dry, 480v-120/240v | | | | |
| 3 kva | EA. | 240.00 | 72.00 | 312.00 |
| 5 kva | " | 330.00 | 120.00 | 450.00 |
| 7.5 kva | " | 490.00 | 140.00 | 630.00 |
| 10 kva | " | 560.00 | 150.00 | 710.00 |
| 15 kva | " | 720.00 | 170.00 | 890.00 |
| 100 kva | " | 2,620 | 460.00 | 3,080 |
| Three phase, 480v-120/280v | | | | |
| 15 kva | EA. | 920.00 | 240.00 | 1,160 |
| 30 kva | " | 1,180 | 370.00 | 1,550 |

## POWER GENERATION

| | UNIT | MAT. | INST. | TOTAL |
|---|---|---|---|---|
| **16320.10 TRANSFORMERS** | | | | |
| 45 kva | EA. | 1,570 | 430.00 | 2,000 |
| 225 kva | " | 5,180 | 610.00 | 5,790 |
| **16350.10 CIRCUIT BREAKERS** | | | | |
| Molded case, 240v, 15-60a, bolt-on | | | | |
| 1 pole | EA. | 8.35 | 9.90 | 18.25 |
| 2 pole | " | 18.55 | 13.75 | 32.30 |
| 70-100a, 2 pole | " | 49.00 | 21.10 | 70.10 |
| 15-60a, 3 pole | " | 60.00 | 15.85 | 75.85 |
| 70-100a, 3 pole | " | 87.50 | 24.35 | 111.85 |
| 480v, 2 pole | | | | |
| 15-60a | EA. | 130.00 | 11.75 | 141.75 |
| 70-100a | " | 170.00 | 15.85 | 185.85 |
| 3 pole | | | | |
| 15-60a | EA. | 170.00 | 15.85 | 185.85 |
| 70-100a | " | 200.00 | 17.60 | 217.60 |
| 70-225a | " | 440.00 | 24.35 | 464.35 |
| Load center circuit breakers, 240v | | | | |
| 1 pole, 10-60a | EA. | 8.10 | 9.90 | 18.00 |
| 2 pole | | | | |
| 10-60a | EA. | 18.55 | 15.85 | 34.40 |
| 70-100a | " | 17.80 | 26.40 | 44.20 |
| 110-150a | " | 110.00 | 28.80 | 138.80 |
| 3 pole | | | | |
| 10-60a | EA. | 63.50 | 19.80 | 83.30 |
| 70-100a | " | 95.50 | 28.80 | 124.30 |
| Load center, G.F.I. breakers, 240v | | | | |
| 1 pole, 15-30a | EA. | 63.50 | 11.75 | 75.25 |
| 2 pole, 15-30a | " | 110.00 | 15.85 | 125.85 |
| Key operated breakers, 240v, 1 pole, 10-30a | " | 10.60 | 11.75 | 22.35 |
| Tandem breakers, 240v | | | | |
| 1 pole, 15-30a | EA. | 17.20 | 15.85 | 33.05 |
| 2 pole, 15-30a | " | 33.10 | 21.10 | 54.20 |
| Bolt-on, G.F.I. breakers, 240v, 1 pole, 15-30a | " | 79.00 | 13.75 | 92.75 |
| **16360.10 SAFETY SWITCHES** | | | | |
| Fused, 3 phase, 30 amp, 600v, heavy duty | | | | |
| NEMA 1 | EA. | 120.00 | 45.30 | 165.30 |
| NEMA 3r | " | 210.00 | 45.30 | 255.30 |
| NEMA 4 | " | 530.00 | 63.50 | 593.50 |
| NEMA 12 | " | 200.00 | 69.00 | 269.00 |
| 60a | | | | |
| NEMA 1 | EA. | 150.00 | 45.30 | 195.30 |
| NEMA 3r | " | 240.00 | 45.30 | 285.30 |
| NEMA 4 | " | 600.00 | 63.50 | 663.50 |
| NEMA 12 | " | 210.00 | 69.00 | 279.00 |
| 100a | | | | |
| NEMA 1 | EA. | 280.00 | 69.00 | 349.00 |
| NEMA 3r | " | 380.00 | 69.00 | 449.00 |
| NEMA 4 | " | 1,170 | 79.00 | 1,249 |
| NEMA 12 | " | 320.00 | 99.00 | 419.00 |
| 200a | | | | |

# 16 ELECTRICAL

| POWER GENERATION | UNIT | MAT. | INST. | TOTAL |
|---|---|---|---|---|
| **16360.10 SAFETY SWITCHES** | | | | |
| NEMA 1 | EA. | 370.00 | 99.00 | 469.00 |
| NEMA 3r | " | 530.00 | 99.00 | 629.00 |
| NEMA 4 | " | 1,630 | 110.00 | 1,740 |
| NEMA 12 | " | 520.00 | 140.00 | 660.00 |
| Non-fused, 240-600v, heavy duty, 3 phase, 30 amp | | | | |
| NEMA 1 | EA. | 64.50 | 45.30 | 109.80 |
| NEMA 3r | " | 120.00 | 45.30 | 165.30 |
| NEMA 4 | " | 470.00 | 69.00 | 539.00 |
| NEMA 12 | " | 140.00 | 69.00 | 209.00 |
| 60a | | | | |
| NEMA1 | EA. | 110.00 | 45.30 | 155.30 |
| NEMA 3r | " | 200.00 | 45.30 | 245.30 |
| NEMA 4 | " | 560.00 | 69.00 | 629.00 |
| NEMA 12 | " | 170.00 | 69.00 | 239.00 |
| 100a | | | | |
| NEMA 1 | EA. | 180.00 | 69.00 | 249.00 |
| NEMA 3r | " | 290.00 | 69.00 | 359.00 |
| NEMA 4 | " | 1,140 | 99.00 | 1,239 |
| NEMA 12 | " | 250.00 | 99.00 | 349.00 |
| 200a, NEMA 1 | " | 310.00 | 99.00 | 409.00 |
| 600a, NEMA 12 | " | 1,500 | 490.00 | 1,990 |
| **16365.10 FUSES** | | | | |
| Fuse, one-time, 250v | | | | |
| 30a | EA. | 0.97 | 2.00 | 2.97 |
| 60a | " | 1.20 | 2.00 | 3.20 |
| 100a | " | 5.25 | 2.00 | 7.25 |
| 200a | " | 13.10 | 2.00 | 15.10 |
| 400a | " | 23.60 | 2.00 | 25.60 |
| 600a | " | 34.00 | 2.00 | 36.00 |
| 600v | | | | |
| 30a | EA. | 3.90 | 2.00 | 5.90 |
| 60a | " | 5.25 | 2.00 | 7.25 |
| 100a | " | 11.80 | 2.00 | 13.80 |
| 200a | " | 22.25 | 2.00 | 24.25 |
| 400a | " | 43.20 | 2.00 | 45.20 |
| **16395.10 GROUNDING** | | | | |
| Ground rods, copper clad, 1/2" x | | | | |
| 6' | EA. | 7.35 | 26.40 | 33.75 |
| 8' | " | 9.50 | 28.80 | 38.30 |
| 10' | " | 11.75 | 39.60 | 51.35 |
| 5/8" x | | | | |
| 5' | EA. | 7.50 | 24.35 | 31.85 |
| 6' | " | 8.85 | 28.80 | 37.65 |
| 8' | " | 11.00 | 39.60 | 50.60 |
| 10' | " | 14.30 | 49.50 | 63.80 |
| 3/4" x | | | | |
| 8' | EA. | 17.85 | 28.80 | 46.65 |
| 10' | " | 21.10 | 31.70 | 52.80 |

# 16 ELECTRICAL

| POWER GENERATION | UNIT | MAT. | INST. | TOTAL |
|---|---|---|---|---|
| **16395.10 GROUNDING** | | | | |
| Ground rod clamp | | | | |
| 5/8" | EA. | 2.35 | 4.85 | 7.20 |
| 3/4" | " | 3.05 | 4.85 | 7.90 |

| SERVICE AND DISTRIBUTION | UNIT | MAT. | INST. | TOTAL |
|---|---|---|---|---|
| **16425.10 SWITCHBOARDS** | | | | |
| Switchboard, 90" high, no main disconnect, 208/120v | | | | |
| 400a | EA. | 1,510 | 310.00 | 1,820 |
| 600a | " | 2,350 | 320.00 | 2,670 |
| 1000a | " | 2,960 | 320.00 | 3,280 |
| 1200a | " | 3,160 | 400.00 | 3,560 |
| 1600a | " | 3,470 | 470.00 | 3,940 |
| 2000a | " | 3,730 | 560.00 | 4,290 |
| 2500a | " | 4,110 | 630.00 | 4,740 |
| 277/480v | | | | |
| 600a | EA. | 2,700 | 320.00 | 3,020 |
| 800a | " | 2,960 | 320.00 | 3,280 |
| 1600a | " | 3,730 | 470.00 | 4,200 |
| 2000a | " | 3,980 | 560.00 | 4,540 |
| 2500a | " | 4,240 | 630.00 | 4,870 |
| 3000a | " | 4,880 | 1,090 | 5,970 |
| 4000a | " | 5,910 | 1,170 | 7,080 |
| **16470.10 PANELBOARDS** | | | | |
| 3 phase, 480/277v, main lugs only, 120a, 30 circuits | EA. | 610.00 | 140.00 | 750.00 |
| 277/480v, 4 wire, flush surface | | | | |
| 225a, 30 circuits | EA. | 650.00 | 160.00 | 810.00 |
| 400a, 30 circuits | " | 740.00 | 200.00 | 940.00 |
| 600a, 42 circuits | " | 920.00 | 240.00 | 1,160 |
| 208/120v, main circuit breaker, 3 phase, 4 wire | | | | |
| 100a | | | | |
| 12 circuits | EA. | 530.00 | 200.00 | 730.00 |
| 20 circuits | " | 650.00 | 250.00 | 900.00 |
| 30 circuits | " | 960.00 | 280.00 | 1,240 |
| 400a | | | | |
| 30 circuits | EA. | 2,020 | 590.00 | 2,610 |
| 42 circuits | " | 2,430 | 630.00 | 3,060 |
| 600a, 42 circuits | " | 2,700 | 720.00 | 3,420 |
| 120/208v, flush, 3 ph., 4 wire, main only | | | | |
| 100a | | | | |
| 12 circuits | EA. | 370.00 | 200.00 | 570.00 |
| 20 circuits | " | 510.00 | 250.00 | 760.00 |

| SERVICE AND DISTRIBUTION | UNIT | MAT. | INST. | TOTAL |
|---|---|---|---|---|
| **16470.10 PANELBOARDS** | | | | |
| 30 circuits | EA. | 740.00 | 280.00 | 1,020 |
| 400a | | | | |
| 30 circuits | EA. | 1,480 | 590.00 | 2,070 |
| 42 circuits | " | 2,170 | 630.00 | 2,800 |
| 600a, 42 circuits | " | 3,370 | 720.00 | 4,090 |
| **16480.10 MOTOR CONTROLS** | | | | |
| Motor generator set, 3 phase, 480/277v, w/controls | | | | |
| 10kw | EA. | 7,330 | 1,090 | 8,420 |
| 15kw | " | 9,560 | 1,220 | 10,780 |
| 20kw | " | 10,610 | 1,270 | 11,880 |
| 40kw | " | 14,970 | 1,510 | 16,480 |
| 100kw | " | 24,490 | 2,440 | 26,930 |
| 200kw | " | 55,980 | 2,880 | 58,860 |
| 300kw | " | 69,970 | 3,170 | 73,140 |
| 2 pole, 230 volt starter, w/NEMA-1 | | | | |
| 1 hp, 9 amp, size 00 | EA. | 110.00 | 39.60 | 149.60 |
| 2 hp, 18amp, size 0 | " | 140.00 | 39.60 | 179.60 |
| 3 hp, 27amp, size 1 | " | 150.00 | 39.60 | 189.60 |
| 5 hp, 45amp, size 1p | " | 190.00 | 39.60 | 229.60 |
| 7-1/2 hp, 45a, size 2 | " | 330.00 | 39.60 | 369.60 |
| 15 hp, 90a, size 3 | " | 500.00 | 39.60 | 539.60 |
| **16490.10 SWITCHES** | | | | |
| Fused interrupter load, 35kv | | | | |
| 20A | | | | |
| 1 pole | EA. | 14,130 | 630.00 | 14,760 |
| 2 pole | " | 15,310 | 670.00 | 15,980 |
| 3 way | " | 16,480 | 670.00 | 17,150 |
| 4 way | " | 17,660 | 720.00 | 18,380 |
| 30a, 1 pole | " | 14,130 | 630.00 | 14,760 |
| 3 way | " | 16,480 | 670.00 | 17,150 |
| 4 way | " | 17,660 | 720.00 | 18,380 |
| Weatherproof switch, including box & cover, 20a | | | | |
| 1 pole | EA. | 15,310 | 630.00 | 15,940 |
| 2 pole | " | 16,480 | 670.00 | 17,150 |
| 3 way | " | 17,660 | 720.00 | 18,380 |
| 4 way | " | 18,840 | 720.00 | 19,560 |
| Specification grade toggle switches, 20a, 120-277v | | | | |
| Single pole | EA. | 10.10 | 7.90 | 18.00 |
| Double pole | " | 11.60 | 11.75 | 23.35 |
| 3 way | " | 10.70 | 9.90 | 20.60 |
| 4 way | " | 27.30 | 11.75 | 39.05 |
| Switch plates, plastic ivory | | | | |
| 1 gang | EA. | 0.38 | 3.15 | 3.53 |
| 2 gang | " | 0.77 | 3.95 | 4.72 |
| 3 gang | " | 1.15 | 4.75 | 5.90 |
| 4 gang | " | 1.95 | 5.75 | 7.70 |
| 5 gang | " | 4.40 | 6.35 | 10.75 |
| 6 gang | " | 5.20 | 7.20 | 12.40 |

## SERVICE AND DISTRIBUTION

| | UNIT | MAT. | INST. | TOTAL |
|---|---|---|---|---|
| **16490.10 SWITCHES** | | | | |
| Stainless steel | | | | |
| 1 gang | EA. | 1.30 | 3.15 | 4.45 |
| 2 gang | " | 2.50 | 3.95 | 6.45 |
| 3 gang | " | 4.40 | 4.85 | 9.25 |
| 4 gang | " | 6.10 | 5.75 | 11.85 |
| 5 gang | " | 8.35 | 6.35 | 14.70 |
| 6 gang | " | 10.40 | 7.20 | 17.60 |
| Brass | | | | |
| 1 gang | EA. | 2.80 | 3.15 | 5.95 |
| 2 gang | " | 5.55 | 3.95 | 9.50 |
| 3 gang | " | 10.10 | 4.85 | 14.95 |
| 4 gang | " | 13.60 | 5.75 | 19.35 |
| 5 gang | " | 16.85 | 6.35 | 23.20 |
| 6 gang | " | 20.10 | 7.20 | 27.30 |
| **16490.20 TRANSFER SWITCHES** | | | | |
| Automatic transfer switch 600v, 3 pole | | | | |
| 30a | EA. | 1,440 | 140.00 | 1,580 |
| 100a | " | 3,000 | 190.00 | 3,190 |
| 400a | " | 6,610 | 400.00 | 7,010 |
| 800a | " | 11,050 | 720.00 | 11,770 |
| 1200a | " | 18,020 | 910.00 | 18,930 |
| 2600a | " | 42,040 | 1,670 | 43,710 |
| **16490.80 SAFETY SWITCHES** | | | | |
| Safety switch, 600v, 3 pole, heavy duty, NEMA-1 | | | | |
| 30a | EA. | 120.00 | 39.60 | 159.60 |
| 60a | " | 150.00 | 45.30 | 195.30 |
| 100a | " | 290.00 | 63.50 | 353.50 |
| 200a | " | 370.00 | 99.00 | 469.00 |
| 400a | " | 1,050 | 220.00 | 1,270 |
| 600a | " | 1,720 | 320.00 | 2,040 |
| 800a | " | 3,070 | 420.00 | 3,490 |
| 1200a | " | 3,810 | 570.00 | 4,380 |

## LIGHTING

| | UNIT | MAT. | INST. | TOTAL |
|---|---|---|---|---|
| **16510.05 INTERIOR LIGHTING** | | | | |
| Recessed fluorescent fixtures, 2'x2' | | | | |
| 2 lamp | EA. | 52.50 | 28.80 | 81.30 |
| 4 lamp | " | 70.50 | 28.80 | 99.30 |
| 2 lamp w/flange | " | 66.00 | 39.60 | 105.60 |

| LIGHTING | UNIT | MAT. | INST. | TOTAL |
|---|---|---|---|---|
| **16510.05 INTERIOR LIGHTING** | | | | |
| 4 lamp w/flange | EA. | 83.00 | 39.60 | 122.60 |
| 1'x4' | | | | |
| 2 lamp | EA. | 53.50 | 26.40 | 79.90 |
| 3 lamp | " | 73.50 | 26.40 | 99.90 |
| 2 lamp w/flange | " | 66.00 | 28.80 | 94.80 |
| 3 lamp w/flange | " | 86.00 | 28.80 | 114.80 |
| 2'x4' | | | | |
| 2 lamp | EA. | 66.00 | 28.80 | 94.80 |
| 3 lamp | " | 77.00 | 28.80 | 105.80 |
| 4 lamp | " | 75.00 | 28.80 | 103.80 |
| 2 lamp w/flange | " | 78.50 | 39.60 | 118.10 |
| 3 lamp w/flange | " | 89.50 | 39.60 | 129.10 |
| 4 lamp w/flange | " | 88.00 | 39.60 | 127.60 |
| 4'x4' | | | | |
| 4 lamp | EA. | 160.00 | 39.60 | 199.60 |
| 6 lamp | " | 180.00 | 39.60 | 219.60 |
| 8 lamp | " | 190.00 | 39.60 | 229.60 |
| 4 lamp w/flange | " | 190.00 | 60.00 | 250.00 |
| 6 lamp w/flange | " | 240.00 | 60.00 | 300.00 |
| 8 lamp, w/flange | " | 220.00 | 60.00 | 280.00 |
| Surface mounted incandescent fixtures | | | | |
| 40w | EA. | 47.70 | 26.40 | 74.10 |
| 75w | " | 52.00 | 26.40 | 78.40 |
| 100w | " | 61.50 | 26.40 | 87.90 |
| 150w | " | 72.00 | 26.40 | 98.40 |
| Pendant | | | | |
| 40w | EA. | 52.00 | 31.70 | 83.70 |
| 75w | " | 57.00 | 31.70 | 88.70 |
| 100w | " | 65.50 | 31.70 | 97.20 |
| 150w | " | 74.50 | 31.70 | 106.20 |
| Recessed incandescent fixtures | | | | |
| 40w | EA. | 88.00 | 60.00 | 148.00 |
| 75w | " | 92.00 | 60.00 | 152.00 |
| 100w | " | 94.50 | 60.00 | 154.50 |
| 150w | " | 110.00 | 60.00 | 170.00 |
| Exit lights, 120v | | | | |
| Recessed | EA. | 68.00 | 49.50 | 117.50 |
| Back mount | " | 47.70 | 28.80 | 76.50 |
| Universal mount | " | 54.50 | 28.80 | 83.30 |
| Emergency battery units, 6v-120v, 50 unit | " | 100.00 | 60.00 | 160.00 |
| With 1 head | " | 120.00 | 60.00 | 180.00 |
| With 2 heads | " | 140.00 | 60.00 | 200.00 |
| Light track single circuit | | | | |
| 2' | EA. | 24.20 | 19.80 | 44.00 |
| 4' | " | 45.20 | 19.80 | 65.00 |
| 8' | " | 80.50 | 39.60 | 120.10 |
| 12' | " | 130.00 | 60.00 | 190.00 |
| Fixtures, square | | | | |
| R-20 | EA. | 64.50 | 5.75 | 70.25 |
| R-30 | " | 67.50 | 5.75 | 73.25 |
| Mini spot | " | 89.00 | 5.75 | 94.75 |
| Surface mounted fluorescent, wrap around lens | | | | |

## LIGHTING

### 16510.05 INTERIOR LIGHTING

| LIGHTING | UNIT | MAT. | INST. | TOTAL |
|---|---|---|---|---|
| 1 lamp | EA. | 48.70 | 31.70 | 80.40 |
| 2 lamps | " | 53.00 | 35.20 | 88.20 |
| 4 lamps | " | 83.50 | 39.60 | 123.10 |
| Wall mounted fluorescent | | | | |
| 2-20w lamps | EA. | 38.90 | 19.80 | 58.70 |
| 2-30w lamps | " | 41.70 | 19.80 | 61.50 |
| 2-40w lamps | " | 48.70 | 26.40 | 75.10 |
| Indirect, with wood shielding, 2049w lamps | | | | |
| 4' | EA. | 83.50 | 39.60 | 123.10 |
| 8' | " | 120.00 | 63.50 | 183.50 |
| Industrial fluorescent, 2 lamp | | | | |
| 4' | EA. | 83.50 | 28.80 | 112.30 |
| 8' | " | 130.00 | 53.00 | 183.00 |
| Strip fluorescent | | | | |
| 4' | | | | |
| 1 lamp | EA. | 27.80 | 26.40 | 54.20 |
| 2 lamps | " | 34.70 | 26.40 | 61.10 |
| 8' | | | | |
| 1 lamp | EA. | 55.50 | 28.80 | 84.30 |
| 2 lamps | " | 62.50 | 35.20 | 97.70 |
| Wire guard for strip fixture, 4' long | " | 13.90 | 13.75 | 27.65 |
| Strip fluorescent, 8' long, two 4' lamps | " | 83.50 | 53.00 | 136.50 |
| With four 4' lamps | " | 110.00 | 63.50 | 173.50 |
| Wet location fluorescent, plastic housing | | | | |
| 4' long | | | | |
| 1 lamp | EA. | 97.50 | 39.60 | 137.10 |
| 2 lamps | " | 100.00 | 53.00 | 153.00 |
| 8' long | | | | |
| 2 lamps | EA. | 170.00 | 63.50 | 233.50 |
| 4 lamps | " | 230.00 | 69.00 | 299.00 |
| Parabolic troffer, 2'x2' | | | | |
| With 2 "U" lamps | EA. | 120.00 | 39.60 | 159.60 |
| With 3 "U" lamps | " | 140.00 | 45.30 | 185.30 |
| 2'x4' | | | | |
| With 2 40w lamps | EA. | 160.00 | 45.30 | 205.30 |
| With 3 40w lamps | " | 180.00 | 53.00 | 233.00 |
| With 4 40w lamps | " | 190.00 | 53.00 | 243.00 |
| 1'x4' | | | | |
| With 1 T-12 lamp, 9 cell | EA. | 120.00 | 28.80 | 148.80 |
| With 2 T-12 lamps | " | 130.00 | 35.20 | 165.20 |
| With 1 T-12 lamp, 20 cell | " | 130.00 | 28.80 | 158.80 |
| With 2 T-12 lamps | " | 140.00 | 35.20 | 175.20 |
| Steel sided surface fluorescent, 2'x4' | | | | |
| 3 lamps | EA. | 100.00 | 53.00 | 153.00 |
| 4 lamps | " | 110.00 | 53.00 | 163.00 |
| Outdoor sign fluor., 1 lamp, remote ballast | | | | |
| 4' long | EA. | 1,990 | 240.00 | 2,230 |
| 6' long | " | 2,380 | 320.00 | 2,700 |
| Recess mounted, commercial, 2'x2', 13" high | | | | |
| 100w | EA. | 660.00 | 160.00 | 820.00 |
| 250w | " | 730.00 | 180.00 | 910.00 |
| High pressure sodium, hi-bay open | | | | |

| LIGHTING | UNIT | MAT. | INST. | TOTAL |
|---|---|---|---|---|
| **16510.05 INTERIOR LIGHTING** | | | | |
| 400w | EA. | 350.00 | 69.00 | 419.00 |
| 1000w | " | 610.00 | 96.00 | 706.00 |
| Enclosed | | | | |
| 400w | EA. | 550.00 | 96.00 | 646.00 |
| 1000w | " | 810.00 | 120.00 | 930.00 |
| Metal halide hi-bay, open | | | | |
| 400w | EA. | 320.00 | 69.00 | 389.00 |
| 1000w | " | 540.00 | 96.00 | 636.00 |
| Enclosed | | | | |
| 400w | EA. | 510.00 | 96.00 | 606.00 |
| 1000w | " | 770.00 | 120.00 | 890.00 |
| High pressure sodium, low bay, surface mounted | | | | |
| 100w | EA. | 220.00 | 39.60 | 259.60 |
| 150w | " | 240.00 | 45.30 | 285.30 |
| 250w | " | 280.00 | 53.00 | 333.00 |
| 400w | " | 350.00 | 63.50 | 413.50 |
| Metal halide, low bay, pendant mounted | | | | |
| 175w | EA. | 240.00 | 53.00 | 293.00 |
| 250w | " | 320.00 | 63.50 | 383.50 |
| 400w | " | 400.00 | 88.00 | 488.00 |
| Indirect luminare, square, metal halide, freestanding | | | | |
| 175w | EA. | 640.00 | 39.60 | 679.60 |
| 250w | " | 650.00 | 39.60 | 689.60 |
| 400w | " | 710.00 | 39.60 | 749.60 |
| High pressure sodium | | | | |
| 150w | EA. | 650.00 | 39.60 | 689.60 |
| 250w | " | 690.00 | 39.60 | 729.60 |
| 400w | " | 740.00 | 39.60 | 779.60 |
| Round, metal halide | | | | |
| 175w | EA. | 610.00 | 39.60 | 649.60 |
| 250w | " | 640.00 | 39.60 | 679.60 |
| 400w | " | 680.00 | 39.60 | 719.60 |
| High pressure sodium | | | | |
| 150w | EA. | 610.00 | 39.60 | 649.60 |
| 250w | " | 670.00 | 39.60 | 709.60 |
| 400w | " | 690.00 | 39.60 | 729.60 |
| Wall mounted, metal halide | | | | |
| 175w | EA. | 500.00 | 99.00 | 599.00 |
| 250w | " | 540.00 | 99.00 | 639.00 |
| 400w | " | 890.00 | 130.00 | 1,020 |
| High pressure sodium | | | | |
| 150w | EA. | 580.00 | 99.00 | 679.00 |
| 250w | " | 710.00 | 99.00 | 809.00 |
| 400w | " | 920.00 | 130.00 | 1,050 |
| Wall pack lithonia, high pressure sodium | | | | |
| 35w | EA. | 150.00 | 35.20 | 185.20 |
| 55w | " | 170.00 | 39.60 | 209.60 |
| 150w | " | 190.00 | 63.50 | 253.50 |
| 250w | " | 240.00 | 69.00 | 309.00 |
| Low pressure sodium | | | | |
| 35w | EA. | 220.00 | 69.00 | 289.00 |
| 55w | " | 330.00 | 79.00 | 409.00 |

| LIGHTING | UNIT | MAT. | INST. | TOTAL |
|---|---|---|---|---|
| **16510.05 INTERIOR LIGHTING** | | | | |
| Wall pack hubbell, high pressure sodium | | | | |
| 35w | EA. | 170.00 | 35.20 | 205.20 |
| 150w | " | 210.00 | 63.50 | 273.50 |
| 250w | " | 270.00 | 69.00 | 339.00 |
| Compact fluorescent | | | | |
| 2-7w | EA. | 100.00 | 39.60 | 139.60 |
| 2-13w | " | 110.00 | 53.00 | 163.00 |
| 1-18w | " | 140.00 | 53.00 | 193.00 |
| Handball & racquet ball court, 2'x2', metal halide | | | | |
| 250w | EA. | 360.00 | 99.00 | 459.00 |
| 400w | " | 450.00 | 110.00 | 560.00 |
| High pressure sodium | | | | |
| 250w | EA. | 400.00 | 99.00 | 499.00 |
| 400w | " | 430.00 | 110.00 | 540.00 |
| Bollard light, 42" w/found., high pressure sodium | | | | |
| 70w | EA. | 610.00 | 100.00 | 710.00 |
| 100w | " | 620.00 | 100.00 | 720.00 |
| 150w | " | 630.00 | 100.00 | 730.00 |
| Light fixture lamps | | | | |
| Lamp | | | | |
| 20w med. bipin base, cool white, 24" | EA. | 4.50 | 5.75 | 10.25 |
| 30w cool white, rapid start, 36" | " | 5.85 | 5.75 | 11.60 |
| 40w cool white "U", 3" | " | 11.15 | 5.75 | 16.90 |
| 40w cool white, rapid start, 48" | " | 2.70 | 5.75 | 8.45 |
| 70w high pressure sodium, mogul base | " | 58.00 | 7.90 | 65.90 |
| 75w slimline, 96" | " | 7.45 | 7.90 | 15.35 |
| 100w | | | | |
| Incandescent, 100a, inside frost | EA. | 2.30 | 3.95 | 6.25 |
| Mercury vapor, clear, mogul base | " | 28.30 | 7.90 | 36.20 |
| High pressure sodium, mogul base | " | 61.50 | 7.90 | 69.40 |
| 150w | | | | |
| Par 38 flood or spot, incandescent | EA. | 6.95 | 3.95 | 10.90 |
| High pressure sodium, 1/2 mogul base | " | 64.50 | 7.90 | 72.40 |
| 175w | | | | |
| Mercury vapor, clear, mogul base | EA. | 48.80 | 7.90 | 56.70 |
| Metal halide, clear, mogul base | " | 21.85 | 7.90 | 29.75 |
| 250w | | | | |
| High pressure sodium, mogul base | EA. | 68.00 | 7.90 | 75.90 |
| Mercury vapor, clear, mogul base | " | 38.60 | 7.90 | 46.50 |
| Metal halide, clear, mogul base | " | 61.50 | 7.90 | 69.40 |
| High pressure sodium, mogul base | " | 68.00 | 7.90 | 75.90 |
| 400w | | | | |
| Mercury vapor, clear, mogul base | EA. | 30.80 | 7.90 | 38.70 |
| Metal halide, clear, mogul base | " | 58.00 | 7.90 | 65.90 |
| High pressure sodium, mogul base | " | 73.00 | 7.90 | 80.90 |
| 1000w | | | | |
| Mercury vapor, clear, mogul base | EA. | 67.00 | 9.90 | 76.90 |
| High pressure sodium, mogul base | " | 170.00 | 9.90 | 179.90 |
| **16510.30 EXTERIOR LIGHTING** | | | | |
| Exterior light fixtures | | | | |
| Rectangle, high pressure sodium | | | | |

# 16 ELECTRICAL

| LIGHTING | UNIT | MAT. | INST. | TOTAL |
|---|---|---|---|---|
| **16510.30 EXTERIOR LIGHTING** | | | | |
| 70w | EA. | 390.00 | 99.00 | 489.00 |
| 100w | " | 400.00 | 100.00 | 500.00 |
| 150w | " | 440.00 | 100.00 | 540.00 |
| 250w | " | 590.00 | 110.00 | 700.00 |
| 400w | " | 660.00 | 140.00 | 800.00 |
| Flood, rectangular, high pressure sodium | | | | |
| 70w | EA. | 400.00 | 99.00 | 499.00 |
| 100w | " | 420.00 | 100.00 | 520.00 |
| 150w | " | 470.00 | 100.00 | 570.00 |
| 400w | " | 700.00 | 140.00 | 840.00 |
| 1000w | " | 1,060 | 180.00 | 1,240 |
| Round | | | | |
| 400w | EA. | 720.00 | 140.00 | 860.00 |
| 1000w | " | 1,130 | 180.00 | 1,310 |
| Round, metal halide | | | | |
| 400w | EA. | 800.00 | 140.00 | 940.00 |
| 1000w | " | 1,190 | 180.00 | 1,370 |
| Light fixture arms, cobra head, 6', high press. sodium | | | | |
| 100w | EA. | 440.00 | 79.00 | 519.00 |
| 150w | " | 690.00 | 99.00 | 789.00 |
| 250w | " | 720.00 | 99.00 | 819.00 |
| 400w | " | 740.00 | 120.00 | 860.00 |
| Flood, metal halide | | | | |
| 400w | EA. | 760.00 | 140.00 | 900.00 |
| 1000w | " | 1,150 | 180.00 | 1,330 |
| 1500w | " | 1,560 | 240.00 | 1,800 |
| Mercury vapor | | | | |
| 250w | EA. | 450.00 | 110.00 | 560.00 |
| 400w | " | 780.00 | 140.00 | 920.00 |
| Incandescent | | | | |
| 300w | EA. | 81.00 | 69.00 | 150.00 |
| 500w | " | 140.00 | 79.00 | 219.00 |
| 1000w | " | 150.00 | 130.00 | 280.00 |
| **16610.30 UNINTERRUPTIBLE POWER** | | | | |
| Uninterruptible power systems, (U.P.S.) | EA. | 11,270 | 320.00 | 11,590 |
| 5 kva | " | 13,930 | 440.00 | 14,370 |
| 7.5 kva | " | 16,980 | 630.00 | 17,610 |
| 10 kva | " | 23,780 | 870.00 | 24,650 |
| 15 kva | " | 27,100 | 910.00 | 28,010 |
| 20 kva | " | 36,230 | 950.00 | 37,180 |
| 25 kva | " | 45,170 | 990.00 | 46,160 |
| 30 kva | " | 48,010 | 1,030 | 49,040 |
| 35 kva | " | 50,880 | 1,070 | 51,950 |
| 40 kva | " | 53,780 | 1,110 | 54,890 |
| 45 kva | " | 56,590 | 1,150 | 57,740 |
| 50 kva | " | 59,440 | 1,190 | 60,630 |
| 62.5 kva | " | 68,380 | 1,270 | 69,650 |
| 75 kva | " | 77,330 | 1,380 | 78,710 |
| 100 kva | " | 99,630 | 1,430 | |
| 150 kva | " | 149,450 | 1,980 | |

# 16 ELECTRICAL

| LIGHTING | UNIT | MAT. | INST. | TOTAL |
|---|---|---|---|---|
| **16610.30 UNINTERRUPTIBLE POWER** | | | | |
| 200 kva | EA. | 199,270 | 2,180 | |
| 300 kva | " | 298,900 | 2,960 | |
| 400 kva | " | 398,530 | 3,560 | |
| 500 kva | " | 498,170 | 4,340 | |
| **16670.10 LIGHTNING PROTECTION** | | | | |
| Lightning protection | | | | |
| Copper point, nickel plated, 12' | | | | |
| 1/2" dia. | EA. | 32.10 | 39.60 | 71.70 |
| 5/8" dia. | " | 37.90 | 39.60 | 77.50 |

| COMMUNICATIONS | UNIT | MAT. | INST. | TOTAL |
|---|---|---|---|---|
| **16720.10 FIRE ALARM SYSTEMS** | | | | |
| Master fire alarm box, pedestal mounted | EA. | 4,410 | 630.00 | 5,040 |
| Master fire alarm box | " | 2,270 | 240.00 | 2,510 |
| Box light | " | 130.00 | 19.80 | 149.80 |
| Ground assembly for box | " | 63.00 | 26.40 | 89.40 |
| Bracket for pole type box | " | 82.00 | 28.80 | 110.80 |
| Pull station | | | | |
| Waterproof | EA. | 63.00 | 19.80 | 82.80 |
| Manual | " | 37.80 | 15.85 | 53.65 |
| Horn, waterproof | " | 56.50 | 39.60 | 96.10 |
| Interior alarm | " | 37.80 | 28.80 | 66.60 |
| Coded transmitter, automatic | " | 670.00 | 79.00 | 749.00 |
| Control panel, 8 zone | " | 1,130 | 320.00 | 1,450 |
| Battery charger and cabinet | " | 400.00 | 79.00 | 479.00 |
| Batteries, nickel cadmium or lead calcium | " | 160.00 | 200.00 | 360.00 |
| CO2 pressure switch connection | " | 63.00 | 28.80 | 91.80 |
| Annunciator panels | | | | |
| Fire detection annunciator, remote type, 8 zone | EA. | 230.00 | 72.00 | 302.00 |
| 12 zone | " | 290.00 | 79.00 | 369.00 |
| 16 zone | " | 370.00 | 99.00 | 469.00 |
| Fire alarm systems | | | | |
| Bell | EA. | 69.50 | 24.35 | 93.85 |
| Weatherproof bell | " | 85.50 | 26.40 | 111.90 |
| Horn | " | 40.30 | 28.80 | 69.10 |
| Siren | " | 410.00 | 79.00 | 489.00 |
| Chime | " | 50.50 | 24.35 | 74.85 |
| Audio/visual | " | 73.00 | 28.80 | 101.80 |
| Strobe light | " | 67.50 | 28.80 | 96.30 |
| Smoke detector | " | 110.00 | 26.40 | 136.40 |
| Heat detection | " | 18.90 | 19.80 | 38.70 |
| Thermal detector | " | 17.65 | 19.80 | 37.45 |

# 16 ELECTRICAL

| COMMUNICATIONS | UNIT | MAT. | INST. | TOTAL |
|---|---|---|---|---|
| **16720.10 FIRE ALARM SYSTEMS** | | | | |
| Ionization detector | EA. | 88.50 | 21.10 | 109.60 |
| Duct detector | " | 300.00 | 110.00 | 410.00 |
| Test switch | " | 50.50 | 19.80 | 70.30 |
| Remote indicator | " | 25.50 | 22.65 | 48.15 |
| Door holder | " | 110.00 | 28.80 | 138.80 |
| Telephone jack | " | 18.90 | 11.75 | 30.65 |
| Fireman phone | " | 270.00 | 39.60 | 309.60 |
| Speaker | " | 54.00 | 31.70 | 85.70 |
| Remote fire alarm annunciator panel | | | | |
| 24 zone | EA. | 1,390 | 260.00 | 1,650 |
| 48 zone | " | 2,770 | 520.00 | 3,290 |
| Control panel | | | | |
| 12 zone | EA. | 950.00 | 120.00 | 1,070 |
| 16 zone | " | 1,230 | 180.00 | 1,410 |
| 24 zone | " | 1,890 | 260.00 | 2,150 |
| 48 zone | " | 3,530 | 630.00 | 4,160 |
| Power supply | " | 220.00 | 60.00 | 280.00 |
| Status command | " | 5,990 | 200.00 | 6,190 |
| Printer | " | 2,330 | 60.00 | 2,390 |
| Transponder | " | 190.00 | 35.60 | 225.60 |
| Transformer | " | 140.00 | 26.40 | 166.40 |
| Transceiver | " | 190.00 | 28.80 | 218.80 |
| Relays | " | 75.50 | 19.80 | 95.30 |
| Flow switch | " | 240.00 | 79.00 | 319.00 |
| Tamper switch | " | 150.00 | 120.00 | 270.00 |
| End of line resistor | " | 10.70 | 13.75 | 24.45 |
| Printed ckt. card | " | 94.50 | 19.80 | 114.30 |
| Central processing unit | " | 12,350 | 240.00 | 12,590 |
| UPS backup to c.p.u. | " | 11,980 | 360.00 | 12,340 |
| Smoke detector, fixed temp. & rate of rise comb. | " | 190.00 | 63.50 | 253.50 |
| **16720.50 SECURITY SYSTEMS** | | | | |
| Sensors | EA. | | | |
| Balanced magnetic door switch, surface mounted | " | 91.00 | 19.80 | 110.80 |
| With remote test | " | 120.00 | 39.60 | 159.60 |
| Flush mounted | " | 87.00 | 73.50 | 160.50 |
| Mounted bracket | " | 6.65 | 13.75 | 20.40 |
| Mounted bracket spacer | " | 6.00 | 13.75 | 19.75 |
| Photoelectric sensor, for fence | " | | | |
| 6 beam | " | 9,790 | 110.00 | 9,900 |
| 9 beam | " | 11,960 | 170.00 | 12,130 |
| Photoelectric sensor, 12 volt dc | " | | | |
| 500' range | " | 290.00 | 63.50 | 353.50 |
| 800' range | " | 320.00 | 79.00 | 399.00 |
| Capacitance wire grid kit | " | | | |
| Surface | " | 80.00 | 39.60 | 119.60 |
| Duct | " | 60.00 | 63.50 | 123.50 |
| Tube grid kit | " | 100.00 | 19.80 | 119.80 |
| Vibration sensor, 30 max per zone | " | 120.00 | 19.80 | 139.80 |
| Audio sensor, 30 max per zone | " | 130.00 | 19.80 | 149.80 |
| Inertia sensor | " | | | |
| Outdoor | " | 93.50 | 28.80 | 122.30 |

## COMMUNICATIONS

### 16720.50 SECURITY SYSTEMS

| COMMUNICATIONS | UNIT | MAT. | INST. | TOTAL |
|---|---|---|---|---|
| Indoor | EA. | 60.00 | 19.80 | 79.80 |
| Ultrasonic transmitter, 20 max per zone | | | | |
| Omni-directional | EA. | 67.00 | 63.50 | 130.50 |
| Directional | " | 73.50 | 53.00 | 126.50 |
| Transceiver | | | | |
| Omni-directional | EA. | 73.50 | 39.60 | 113.10 |
| Directional | " | 80.00 | 39.60 | 119.60 |
| Passive infra-red sensor, 20 max per zone | " | 520.00 | 63.50 | 583.50 |
| Access/secure control unit, balanced magnetic switch | " | 320.00 | 63.50 | 383.50 |
| Photoelectric sensor | " | 530.00 | 63.50 | 593.50 |
| Photoelectric fence sensor | " | 550.00 | 63.50 | 613.50 |
| Capacitance sensor | " | 630.00 | 69.00 | 699.00 |
| Audio and vibration sensor | " | 560.00 | 63.50 | 623.50 |
| Inertia sensor | " | 750.00 | 63.50 | 813.50 |
| Ultrasonic sensor | " | 850.00 | 69.00 | 919.00 |
| Infra-red sensor | " | 530.00 | 79.00 | 609.00 |
| Monitor panel, with access/secure tone, standard | " | 360.00 | 69.00 | 429.00 |
| High security | " | 530.00 | 79.00 | 609.00 |
| Emergency power indicator | " | 220.00 | 19.80 | 239.80 |
| Monitor rack with 115v power supply | | | | |
| 1 zone | EA. | 300.00 | 39.60 | 339.60 |
| 10 zone | " | 1,540 | 99.00 | 1,639 |
| Monitor cabinet, wall mounted | | | | |
| 1 zone | EA. | 470.00 | 39.60 | 509.60 |
| 5 zone | " | 1,670 | 63.50 | 1,734 |
| 10 zone | " | 730.00 | 69.00 | 799.00 |
| 20 zone | " | 2,320 | 79.00 | 2,399 |
| Floor mounted, 50 zone | " | 2,420 | 160.00 | 2,580 |
| Security system accessories | | | | |
| Tamper assembly for monitor cabinet | EA. | 60.00 | 17.60 | 77.60 |
| Monitor panel blank | " | 8.00 | 13.75 | 21.75 |
| Audible alarm | " | 68.00 | 19.80 | 87.80 |
| Audible alarm control | " | 280.00 | 13.75 | 293.75 |
| Termination screw, terminal cabinet | | | | |
| 25 pair | EA. | 200.00 | 63.50 | 263.50 |
| 50 pair | " | 320.00 | 99.00 | 419.00 |
| 150 pair | " | 530.00 | 200.00 | 730.00 |
| Universal termination, cabinets & panel | | | | |
| Remote test | EA. | 46.80 | 69.00 | 115.80 |
| No remote test | " | 33.40 | 28.80 | 62.20 |
| High security line supervision termination | " | 240.00 | 39.60 | 279.60 |
| Door cord for capacitance sensor, 12" | " | 8.00 | 19.80 | 27.80 |
| Insulation block kit for capacitance sensor | " | 37.40 | 13.75 | 51.15 |
| Termination block for capacitance sensor | " | 8.15 | 13.75 | 21.90 |
| Guard alert display | " | 870.00 | 24.35 | 894.35 |
| Uninterrupted power supply | | | | |
| Plug-in 40kva transformer | | | | |
| 12 volt | EA. | 33.40 | 13.75 | 47.15 |
| 18 volt | " | 22.70 | 13.75 | 36.45 |
| 24 volt | " | 16.05 | 13.75 | 29.80 |
| Test relay | " | 51.50 | 13.75 | 65.25 |
| Coaxial cable, 50 ohm | L.F. | 0.21 | 0.24 | 0.45 |

# 16 ELECTRICAL

| COMMUNICATIONS | UNIT | MAT. | INST. | TOTAL |
|---|---|---|---|---|
| **16720.50 SECURITY SYSTEMS** | | | | |
| Door openers | EA. | 60.00 | 19.80 | 79.80 |
| Push buttons | | | | |
| Standard | EA. | 13.35 | 13.75 | 27.10 |
| Weatherproof | " | 20.05 | 17.60 | 37.65 |
| Bells | " | 51.00 | 28.80 | 79.80 |
| Horns | | | | |
| Standard | EA. | 53.50 | 39.60 | 93.10 |
| Weatherproof | " | 100.00 | 49.50 | 149.50 |
| Chimes | " | 77.50 | 26.40 | 103.90 |
| Flasher | " | 56.00 | 24.35 | 80.35 |
| Motion detectors | " | 230.00 | 60.00 | 290.00 |
| Intercom units | " | 53.50 | 28.80 | 82.30 |
| Remote annunciator | " | 2,400 | 200.00 | 2,600 |
| **16740.10 TELEPHONE SYSTEMS** | | | | |
| Communication cable | | | | |
| 25 pair | L.F. | 0.40 | 1.00 | 1.40 |
| 100 pair | " | 2.05 | 1.15 | 3.20 |
| 400 pair | " | 6.85 | 1.75 | 8.60 |
| Cable tap in manhole or junction box | | | | |
| 25 pair cable | EA. | 3.45 | 150.00 | 153.45 |
| 100 pair cable | " | 13.85 | 600.00 | 613.85 |
| 400 pair cable | " | 55.50 | 2,440 | 2,496 |
| Cable terminations, manhole or junction box | | | | |
| 25 pair cable | EA. | 3.45 | 150.00 | 153.45 |
| 100 pair cable | " | 13.85 | 600.00 | 613.85 |
| 400 pair cable | " | 43.20 | 2,440 | 2,483 |
| **16780.50 TELEVISION SYSTEMS** | | | | |
| TV outlet, self terminating, w/cover plate | EA. | 7.25 | 12.20 | 19.45 |
| Thru splitter | " | 15.85 | 63.50 | 79.35 |
| End of line | " | 13.20 | 53.00 | 66.20 |
| In line splitter multitap | | | | |
| 4 way | EA. | 26.50 | 72.00 | 98.50 |
| 2 way | " | 19.80 | 67.50 | 87.30 |
| Equipment cabinet | " | 66.00 | 63.50 | 129.50 |
| Antenna | | | | |
| Broad band uhf | EA. | 130.00 | 140.00 | 270.00 |
| Lightning arrester | " | 35.70 | 28.80 | 64.50 |
| TV cable | L.F. | 0.30 | 0.20 | 0.50 |
| **16850.10 ELECTRIC HEATING** | | | | |
| Baseboard heater | | | | |
| 2', 375w | EA. | 40.10 | 39.60 | 79.70 |
| 3', 500w | " | 49.50 | 39.60 | 89.10 |
| 4', 750w | " | 61.50 | 45.30 | 106.80 |
| 5', 935w | " | 81.50 | 53.00 | 134.50 |
| 6', 1125w | " | 91.00 | 63.50 | 154.50 |
| 7', 1310w | " | 110.00 | 72.00 | 182.00 |
| 8', 1500w | " | 130.00 | 79.00 | 209.00 |

| COMMUNICATIONS | UNIT | MAT. | INST. | TOTAL |
|---|---|---|---|---|
| **16850.10 ELECTRIC HEATING** | | | | |
| 9', 1680w | EA. | 140.00 | 88.00 | 228.00 |
| 10', 1875w | " | 150.00 | 90.50 | 240.50 |
| Unit heater wall mounted | | | | |
| 1500w | EA. | 180.00 | 63.50 | 243.50 |
| 2500w | " | 200.00 | 72.00 | 272.00 |
| 4000w | " | 260.00 | 90.50 | 350.50 |
| Thermostat | | | | |
| Integral | EA. | 30.70 | 19.80 | 50.50 |
| Line voltage | " | 31.10 | 19.80 | 50.90 |
| Electric heater connection | " | 1.30 | 9.90 | 11.20 |

BNi® Building News

# Man-Hour Tables

The man-hour productivities used to develop the labor costs are listed in the following section of this book. These productivities represent typical installation labor for thousands of construction items. The data takes into account all activities involved in normal construction under commonly experienced working conditions. As with the Costbook pages, these items are listed according to the CSI MASTERFORMAT. In order to best use the information in this book, please review this sample page and read the "Features in this Book" section.

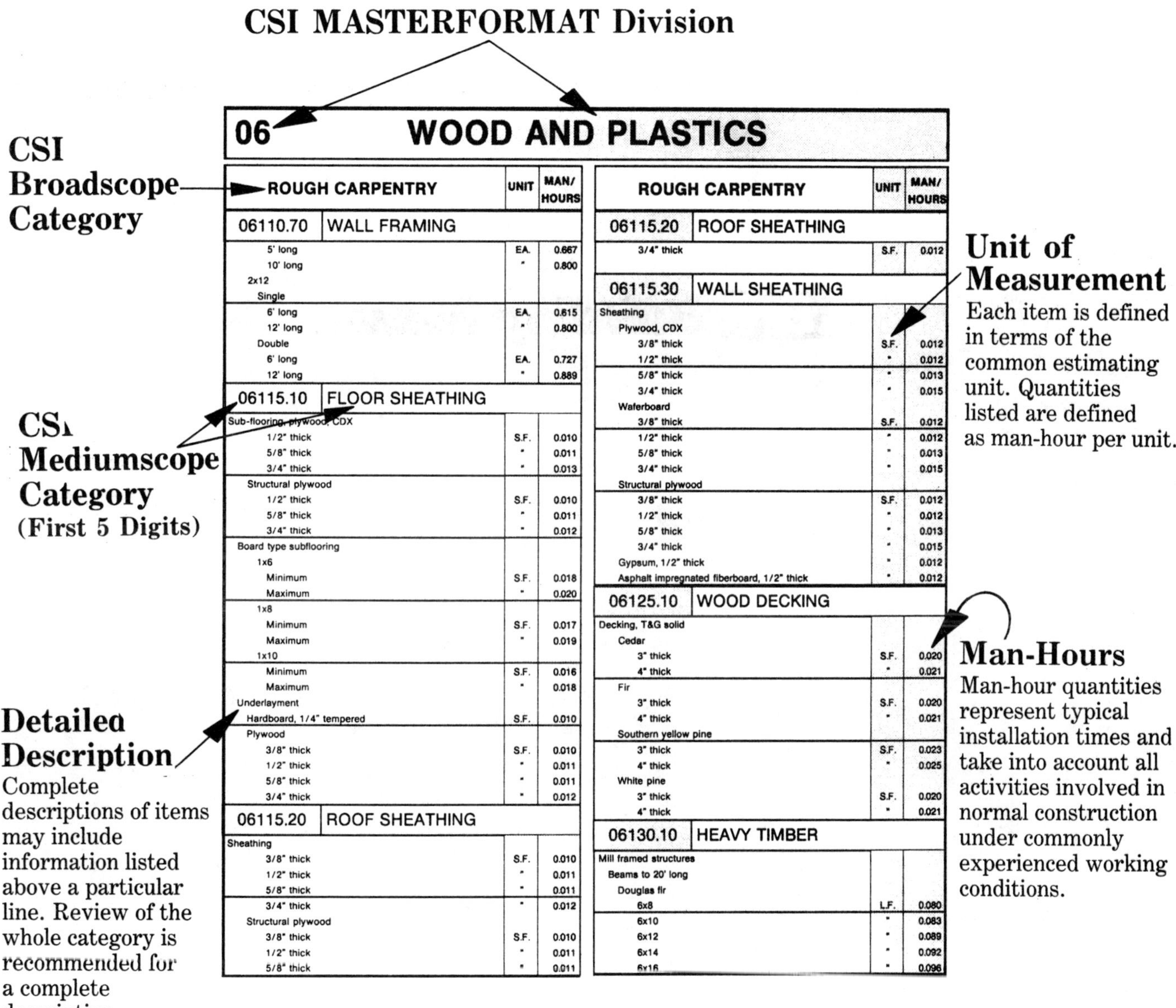


**06 WOOD AND PLASTICS**

| ROUGH CARPENTRY | UNIT | MAN/ HOURS |
|---|---|---|
| **06110.70 WALL FRAMING** | | |
| 5' long | EA. | 0.667 |
| 10' long | " | 0.800 |
| 2x12 | | |
| Single | | |
| 6' long | EA. | 0.615 |
| 12' long | " | 0.800 |
| Double | | |
| 6' long | EA. | 0.727 |
| 12' long | " | 0.889 |
| **06115.10 FLOOR SHEATHING** | | |
| Sub-flooring, plywood, CDX | | |
| 1/2" thick | S.F. | 0.010 |
| 5/8" thick | " | 0.011 |
| 3/4" thick | " | 0.013 |
| Structural plywood | | |
| 1/2" thick | S.F. | 0.010 |
| 5/8" thick | " | 0.011 |
| 3/4" thick | " | 0.012 |
| Board type subflooring | | |
| 1x6 | | |
| Minimum | S.F. | 0.018 |
| Maximum | " | 0.020 |
| 1x8 | | |
| Minimum | S.F. | 0.017 |
| Maximum | " | 0.019 |
| 1x10 | | |
| Minimum | S.F. | 0.016 |
| Maximum | " | 0.018 |
| Underlayment | | |
| Hardboard, 1/4" tempered | S.F. | 0.010 |
| Plywood | | |
| 3/8" thick | S.F. | 0.010 |
| 1/2" thick | " | 0.011 |
| 5/8" thick | " | 0.011 |
| 3/4" thick | " | 0.012 |
| **06115.20 ROOF SHEATHING** | | |
| Sheathing | | |
| 3/8" thick | S.F. | 0.010 |
| 1/2" thick | " | 0.011 |
| 5/8" thick | " | 0.011 |
| 3/4" thick | " | 0.012 |
| Structural plywood | | |
| 3/8" thick | S.F. | 0.010 |
| 1/2" thick | " | 0.011 |
| 5/8" thick | " | 0.011 |

| ROUGH CARPENTRY | UNIT | MAN/ HOURS |
|---|---|---|
| **06115.20 ROOF SHEATHING** | | |
| 3/4" thick | S.F. | 0.012 |
| **06115.30 WALL SHEATHING** | | |
| Sheathing | | |
| Plywood, CDX | | |
| 3/8" thick | S.F. | 0.012 |
| 1/2" thick | " | 0.012 |
| 5/8" thick | " | 0.013 |
| 3/4" thick | " | 0.015 |
| Waferboard | | |
| 3/8" thick | S.F. | 0.012 |
| 1/2" thick | " | 0.012 |
| 5/8" thick | " | 0.013 |
| 3/4" thick | " | 0.015 |
| Structural plywood | | |
| 3/8" thick | S.F. | 0.012 |
| 1/2" thick | " | 0.012 |
| 5/8" thick | " | 0.013 |
| 3/4" thick | " | 0.015 |
| Gypsum, 1/2" thick | " | 0.012 |
| Asphalt impregnated fiberboard, 1/2" thick | " | 0.012 |
| **06125.10 WOOD DECKING** | | |
| Decking, T&G solid | | |
| Cedar | | |
| 3" thick | S.F. | 0.020 |
| 4" thick | " | 0.021 |
| Fir | | |
| 3" thick | S.F. | 0.020 |
| 4" thick | " | 0.021 |
| Southern yellow pine | | |
| 3" thick | S.F. | 0.023 |
| 4" thick | " | 0.025 |
| White pine | | |
| 3" thick | S.F. | 0.020 |
| 4" thick | " | 0.021 |
| **06130.10 HEAVY TIMBER** | | |
| Mill framed structures | | |
| Beams to 20' long | | |
| Douglas fir | | |
| 6x8 | L.F. | 0.080 |
| 6x10 | " | 0.083 |
| 6x12 | " | 0.089 |
| 6x14 | " | 0.092 |
| 6x16 | " | 0.096 |

BNi® Building News

# 02 SITEWORK

## SOIL TESTS

| 02010.10 SOIL BORING | UNIT | MAN/HOURS |
|---|---|---|
| Borings, uncased, stable earth | | |
| 2-1/2" dia. | | |
| Minimum | L.F. | 0.200 |
| Average | " | 0.300 |
| Maximum | " | 0.480 |
| 4" dia. | | |
| Minimum | L.F. | 0.218 |
| Average | " | 0.343 |
| Maximum | " | 0.600 |
| Cased, including samples | | |
| 2-1/2" dia. | | |
| Minimum | L.F. | 0.240 |
| Average | " | 0.400 |
| Maximum | " | 0.800 |
| 4" dia. | | |
| Minumum | L.F. | 0.480 |
| Average | " | 0.686 |
| Maximum | " | 0.960 |
| Drilling in rock | | |
| No sampling | | |
| Minimum | L.F. | 0.436 |
| Average | " | 0.632 |
| Maximum | " | 0.857 |
| With casing and sampling | | |
| Minimum | L.F. | 0.600 |
| Average | " | 0.800 |
| Maximum | " | 1.200 |
| Test pits | | |
| Light soil | | |
| Minimum | EA. | 3.000 |
| Average | " | 4.000 |
| Maximum | " | 8.000 |
| Heavy soil | | |
| Minimum | EA. | 4.800 |
| Average | " | 6.000 |
| Maximum | " | 12.000 |

## DEMOLITION

| 02060.10 BUILDING DEMOLITION | UNIT | MAN/HOURS |
|---|---|---|
| Building, complete with disposal | | |
| Wood frame | C.F. | 0.003 |
| Concrete | " | 0.004 |
| Steel frame | " | 0.005 |
| Partition removal | | |
| Concrete block partitions | | |
| 4" thick | S.F. | 0.040 |
| 8" thick | S.F. | 0.053 |
| 12" thick | " | 0.073 |
| Brick masonry partitions | | |
| 4" thick | S.F. | 0.040 |
| 8" thick | " | 0.050 |
| 12" thick | " | 0.067 |
| 16" thick | " | 0.100 |
| Cast in place concrete partitions | | |
| Unreinforced | | |
| 6" thick | S.F. | 0.160 |
| 8" thick | " | 0.171 |
| 10" thick | " | 0.200 |
| 12" thick | " | 0.240 |
| Reinforced | | |
| 6" thick | S.F. | 0.185 |
| 8" thick | " | 0.240 |
| 10" thick | " | 0.267 |
| 12" thick | " | 0.320 |
| Terra cotta | | |
| To 6" thick | S.F. | 0.040 |
| Stud partitions | | |
| Metal or wood, with drywall both sides | S.F. | 0.040 |
| Metal studs, both sides, lath and plaster | " | 0.053 |
| Concrete, elevated slabs, mesh reinforcing | | |
| Under 5 cf | C.F. | 0.800 |
| Over 5 cf | " | 0.667 |
| Bar reinforcing | | |
| Under 5 cf | C.F. | 1.333 |
| Over 5 cf | " | 1.000 |
| Walls, concrete, bar reinforcing | | |
| Small jobs | C.F. | 0.533 |
| Large jobs | " | 0.444 |
| Brick walls, not including toothing | | |
| 4" thick | S.F. | 0.040 |
| 8" thick | " | 0.050 |
| 12" thick | " | 0.067 |
| 16" thick | " | 0.100 |
| Concrete block walls, not including toothing | | |
| 4" thick | S.F. | 0.044 |
| 6" thick | " | 0.047 |
| 8" thick | " | 0.050 |
| 10" thick | " | 0.057 |
| 12" thick | " | 0.067 |
| Rubbish handling | | |
| Load in dumpster or truck | | |
| Minimum | C.F. | 0.018 |
| Maximum | " | 0.027 |
| For use of elevators, add | | |
| Minimum | C.F. | 0.004 |
| Maximum | " | 0.008 |
| Rubbish hauling | | |
| Hand loaded on trucks, 2 mile trip | C.Y. | 0.320 |
| Machine loaded on trucks, 2 mile trip | " | 0.240 |

# 02 SITEWORK

## HIGHWAY DEMOLITION

| 02065.10 PAVEMENT DEMOLITION | UNIT | MAN/HOURS |
|---|---|---|
| Bituminous pavement, up to 3" thick | | |
| On streets | S.Y. | 0.096 |
| On pipe trench | " | 0.120 |
| Concrete pavement, 6" thick | | |
| No reinforcement | S.Y. | 0.160 |
| With wire mesh | " | 0.240 |
| With rebars | " | 0.300 |
| 9" thick | | |
| No reinforcement | S.Y. | 0.200 |
| With wire mesh | " | 0.300 |
| With rebars | " | 0.400 |
| 12" thick | | |
| No reinforcement | S.Y. | 0.240 |
| With wire mesh | " | 0.343 |
| With rebars | " | 0.480 |
| Sidewalk, 4" thick, with disposal | " | 0.080 |
| Removal of pavement markings by waterblasting | S.F. | 0.004 |

| 02065.15 SAW CUTTING PAVEMENT | UNIT | MAN/HOURS |
|---|---|---|
| Pavement, bituminous | | |
| 2" thick | L.F. | 0.016 |
| 3" thick | " | 0.020 |
| 4" thick | " | 0.025 |
| 5" thick | " | 0.027 |
| 6" thick | " | 0.029 |
| Concrete pavement, with wire mesh | | |
| 4" thick | L.F. | 0.031 |
| 5" thick | " | 0.033 |
| 6" thick | " | 0.036 |
| 8" thick | " | 0.040 |
| 10" thick | " | 0.044 |
| Plain concrete, unreinforced | | |
| 4" thick | L.F. | 0.027 |
| 5" thick | " | 0.031 |
| 6" thick | " | 0.033 |
| 8" thick | " | 0.036 |
| 10" thick | " | 0.040 |

| 02065.80 CURB & GUTTER | UNIT | MAN/HOURS |
|---|---|---|
| Removal, plain concrete curb | L.F. | 0.060 |
| Plain concrete curb and 2' gutter | " | 0.083 |

| 02065.85 GUARDRAILS | UNIT | MAN/HOURS |
|---|---|---|
| Remove standard guardrail | | |
| Steel | L.F. | 0.080 |
| Wood | " | 0.062 |

| 02075.80 CORE DRILLING | UNIT | MAN/HOURS |
|---|---|---|
| Concrete | | |
| 6" thick | | |
| 3" dia. | EA. | 0.571 |
| 4" dia. | " | 0.667 |
| 6" dia. | EA. | 0.800 |
| 8" dia. | " | 1.333 |
| 8" thick | | |
| 3" dia. | EA. | 0.800 |
| 4" dia. | " | 1.000 |
| 6" dia. | " | 1.143 |
| 8" dia. | " | 1.600 |
| 10" thick | | |
| 3" dia. | EA. | 1.000 |
| 4" dia. | " | 1.143 |
| 6" dia. | " | 1.333 |
| 8" dia. | " | 2.000 |
| 12" thick | | |
| 3" dia. | EA. | 1.333 |
| 4" dia. | " | 1.600 |
| 6" dia. | " | 2.000 |
| 8" dia. | " | 2.667 |

## HAZARDOUS WASTE

| 02080.10 ASBESTOS REMOVAL | UNIT | MAN/HOURS |
|---|---|---|
| Enclosure using wood studs & poly, install & remove | S.F. | 0.020 |

## SITE DEMOLITION

| 02105.10 CATCH BASINS/MANHOLES | UNIT | MAN/HOURS |
|---|---|---|
| Abandon catch basin or manhole (fill with sand) | EA. | 4.800 |
| Remove and reset frame and cover | " | 2.400 |
| Remove catch basin, to 10' deep | | |
| Masonry | EA. | 6.000 |
| Concrete | " | 8.000 |

| 02105.20 FENCES | UNIT | MAN/HOURS |
|---|---|---|
| Remove fencing | | |
| Chain link, 8' high | | |

## SITE DEMOLITION

| SITE DEMOLITION | UNIT | MAN/ HOURS |
|---|---|---|
| **02105.20 FENCES** | | |
| For disposal | L.F. | 0.040 |
| For reuse | " | 0.100 |
| Wood | | |
| 4' high | S.F. | 0.027 |
| 6' high | " | 0.032 |
| 8' high | " | 0.040 |
| Masonry | | |
| 8" thick | | |
| 4' high | S.F. | 0.080 |
| 6' high | " | 0.100 |
| 8' high | " | 0.114 |
| 12" thick | | |
| 4' high | S.F. | 0.133 |
| 6' high | " | 0.160 |
| 8' high | " | 0.200 |
| 12' high | " | 0.267 |
| **02105.30 HYDRANTS** | | |
| Remove fire hydrant | EA. | 4.000 |
| Remove and reset fire hydrant | " | 12.000 |
| **02105.42 DRAINAGE PIPING** | | |
| Remove drainage pipe, not including excavation | | |
| 12" dia. | L.F. | 0.100 |
| 18" dia. | " | 0.126 |
| 24" dia. | " | 0.160 |
| 36" dia. | " | 0.200 |
| **02105.43 GAS PIPING** | | |
| Remove welded steel pipe, not including excavation | | |
| 4" dia. | L.F. | 0.150 |
| 5" dia. | " | 0.240 |
| 6" dia. | " | 0.300 |
| 8" dia. | " | 0.480 |
| 10" dia. | " | 0.600 |
| **02105.45 SANITARY PIPING** | | |
| Remove sewer pipe, not including excavation | | |
| 4" dia. | L.F. | 0.096 |
| 6" dia. | " | 0.109 |
| 8" dia. | " | 0.120 |
| 10" dia. | " | 0.126 |
| 12" dia. | " | 0.133 |
| 15" dia. | " | 0.141 |
| 18" dia. | " | 0.160 |
| 24" dia. | " | 0.200 |
| 30" dia. | " | 0.240 |
| 36" dia. | " | 0.300 |
| **02105.48 WATER PIPING** | | |
| Remove water pipe, not including excavation | | |
| 4" dia. | L.F. | 0.109 |

| SITE DEMOLITION | UNIT | MAN/ HOURS |
|---|---|---|
| **02105.48 WATER PIPING** | | |
| 6" dia. | L.F. | 0.114 |
| 8" dia. | " | 0.126 |
| 10" dia. | " | 0.133 |
| 12" dia. | " | 0.141 |
| 14" dia. | " | 0.150 |
| 16" dia. | " | 0.160 |
| 18" dia. | " | 0.171 |
| 20" dia. | " | 0.185 |
| Remove valves | | |
| 6" | EA. | 1.200 |
| 10" | " | 1.333 |
| 14" | " | 1.500 |
| 18" | " | 2.000 |
| **02105.60 UNDERGROUND TANKS** | | |
| Remove underground storage tank, and backfill | | |
| 50 to 250 gals | EA. | 8.000 |
| 600 gals | " | 8.000 |
| 1000 gals | " | 12.000 |
| 4000 gals | " | 19.200 |
| 5000 gals | " | 19.200 |
| 10,000 gals | " | 32.000 |
| 12,000 gals | " | 40.000 |
| 15,000 gals | " | 48.000 |
| 20,000 gals | " | 60.000 |
| **02105.66 SEPTIC TANKS** | | |
| Remove septic tank | | |
| 1000 gals | EA. | 2.000 |
| 2000 gals | " | 2.400 |
| 5000 gals | " | 3.000 |
| 15,000 gals | " | 24.000 |
| 25,000 gals | " | 32.000 |
| 40,000 gals | " | 48.000 |
| **02105.80 WALLS, EXTERIOR** | | |
| Concrete wall | | |
| Light reinforcing | | |
| 6" thick | S.F. | 0.120 |
| 8" thick | " | 0.126 |
| 10" thick | " | 0.133 |
| 12" thick | " | 0.150 |
| Medium reinforcing | | |
| 6" thick | S.F. | 0.126 |
| 8" thick | " | 0.133 |
| 10" thick | " | 0.150 |
| 12" thick | " | 0.171 |
| Heavy reinforcing | | |
| 6" thick | S.F. | 0.141 |
| 8" thick | " | 0.150 |
| 10" thick | " | 0.171 |
| 12" thick | " | 0.200 |

# 02 SITEWORK

## SITE DEMOLITION

| 02105.80 WALLS, EXTERIOR | UNIT | MAN/HOURS |
|---|---|---|
| Masonry | | |
| No reinforcing | | |
| 8" thick | S.F. | 0.053 |
| 12" thick | " | 0.060 |
| 16" thick | " | 0.069 |
| Horizontal reinforcing | | |
| 8" thick | S.F. | 0.060 |
| 12" thick | " | 0.065 |
| 16" thick | " | 0.077 |
| Vertical reinforcing | | |
| 8" thick | S.F. | 0.077 |
| 12" thick | " | 0.089 |
| 16" thick | " | 0.109 |
| Remove concrete headwall | | |
| 15" pipe | EA. | 1.714 |
| 18" pipe | " | 2.000 |
| 24" pipe | " | 2.182 |
| 30" pipe | " | 2.400 |
| 36" pipe | " | 2.667 |
| 48" pipe | " | 3.429 |
| 60" pipe | " | 4.800 |

| 02110.10 CLEARING AND GRUBBING | UNIT | MAN/HOURS |
|---|---|---|
| Clear wooded area | | |
| Light density | ACRE | 60.000 |
| Medium density | " | 80.000 |
| Heavy density | " | 96.000 |

| 02110.50 TREE CUTTING & CLEARING | UNIT | MAN/HOURS |
|---|---|---|
| Cut trees and clear out stumps | | |
| 9" to 12" dia. | EA. | 4.800 |
| To 24" dia. | " | 6.000 |
| 24" dia. and up | " | 8.000 |
| Loading and trucking | | |
| For machine load, per load, round trip | | |
| 1 mile | EA. | 0.960 |
| 3 mile | " | 1.091 |
| 5 mile | " | 1.200 |
| 10 mile | " | 1.600 |
| 20 mile | " | 2.400 |
| Hand loaded, round trip | | |
| 1 mile | EA. | 2.000 |
| 3 mile | " | 2.286 |
| 5 mile | " | 2.667 |
| 10 mile | " | 3.200 |
| 20 mile | " | 4.000 |
| Tree trimming for pole line construction | | |
| Light cutting | L.F. | 0.012 |
| Medium cutting | " | 0.016 |
| Heavy cutting | " | 0.024 |

## DEWATERING

| 02144.10 WELLPOINT SYSTEMS | UNIT | MAN/HOURS |
|---|---|---|
| Pumping, gas driven, 50' hose | | |
| 3" header pipe | DAY | 8.000 |
| 6" header pipe | " | 10.000 |
| Wellpoint system per job | | |
| 6" header pipe | L.F. | 0.032 |
| 8" header pipe | " | 0.040 |
| 10" header pipe | " | 0.053 |
| Jetting wellpoint system | | |
| 14' long | EA. | 0.533 |
| 18' long | " | 0.667 |
| Sand filter for wellpoints | L.F. | 0.013 |
| Replacement of wellpoint components | EA. | 0.160 |

## SHORING AND UNDERPINNING

| 02162.10 TRENCH SHEETING | UNIT | MAN/HOURS |
|---|---|---|
| Closed timber, including pull and salvage, excavation | | |
| 8' deep | S.F. | 0.064 |
| 10' deep | " | 0.067 |
| 12' deep | " | 0.071 |
| 14' deep | " | 0.075 |
| 16' deep | " | 0.080 |
| 18' deep | " | 0.098 |
| 20' deep | " | 0.091 |

| 02170.10 COFFERDAMS | UNIT | MAN/HOURS |
|---|---|---|
| Cofferdam, steel, driven from shore | | |
| 15' deep | S.F. | 0.137 |
| 20' deep | " | 0.128 |
| 25' deep | " | 0.120 |
| 30' deep | " | 0.113 |
| 40' deep | " | 0.107 |
| Driven from barge | | |
| 20' deep | S.F. | 0.148 |
| 30' deep | " | 0.137 |
| 40' deep | " | 0.128 |
| 50' deep | " | 0.120 |

# 02 SITEWORK

| EARTHWORK | UNIT | MAN/HOURS |
|---|---|---|
| **02210.10 HAULING MATERIAL** | | |
| Haul material by 10 cy dump truck, round trip distance | | |
| 1 mile | C.Y. | 0.044 |
| 2 mile | " | 0.053 |
| 5 mile | " | 0.073 |
| 10 mile | " | 0.080 |
| 20 mile | " | 0.089 |
| 30 mile | " | 0.107 |
| Site grading, cut & fill, sandy clay, 200' haul, 75 hp dozer | " | 0.032 |
| Spread topsoil by equipment on site | " | 0.036 |
| Site grading (cut and fill to 6") less than 1 acre | | |
| 75 hp dozer | C.Y. | 0.053 |
| 1.5 cy backhoe/loader | " | 0.080 |

| EARTHWORK | UNIT | MAN/HOURS |
|---|---|---|
| **02210.30 BULK EXCAVATION** | | |
| Excavation, by small dozer | | |
| Large areas | C.Y. | 0.016 |
| Small areas | " | 0.027 |
| Trim banks | " | 0.040 |
| Drag line | | |
| 1-1/2 cy bucket | | |
| Sand or gravel | C.Y. | 0.040 |
| Light clay | " | 0.053 |
| Heavy clay | " | 0.060 |
| Unclassified | " | 0.064 |
| 2 cy bucket | | |
| Sand or gravel | C.Y. | 0.037 |
| Light clay | " | 0.048 |
| Heavy clay | " | 0.053 |
| Unclassified | " | 0.056 |
| 2-1/2 cy bucket | | |
| Sand or gravel | C.Y. | 0.034 |
| Light clay | " | 0.044 |
| Heavy clay | " | 0.048 |
| Unclassified | " | 0.051 |
| 3 cy bucket | | |
| Sand or gravel | C.Y. | 0.030 |
| Light clay | " | 0.040 |
| Heavy clay | " | 0.044 |
| Unclassified | " | 0.046 |
| Hydraulic excavator | | |
| 1 cy capacity | | |
| Light material | C.Y. | 0.040 |
| Medium material | " | 0.048 |
| Wet material | " | 0.060 |
| Blasted rock | " | 0.069 |
| 1-1/2 cy capacity | | |
| Light material | C.Y. | 0.010 |
| Medium material | " | 0.013 |
| Wet material | " | 0.016 |
| Blasted rock | " | 0.020 |
| 2 cy capacity | | |
| Light material | C.Y. | 0.009 |
| Medium material | " | 0.011 |
| Wet material | " | 0.013 |

| EARTHWORK | UNIT | MAN/HOURS |
|---|---|---|
| **02210.30 BULK EXCAVATION** | | |
| Blasted rock | C.Y. | 0.016 |
| Wheel mounted front-end loader | | |
| 7/8 cy capacity | | |
| Light material | C.Y. | 0.020 |
| Medium material | " | 0.023 |
| Wet material | " | 0.027 |
| Blasted rock | " | 0.032 |
| 1-1/2 cy capacity | | |
| Light material | C.Y. | 0.011 |
| Medium material | " | 0.012 |
| Wet material | " | 0.013 |
| Blasted rock | " | 0.015 |
| 2-1/2 cy capacity | | |
| Light material | C.Y. | 0.009 |
| Medium material | " | 0.010 |
| Wet material | " | 0.011 |
| Blasted rock | " | 0.011 |
| 3-1/2 cy capacity | | |
| Light material | C.Y. | 0.009 |
| Medium material | " | 0.009 |
| Wet material | " | 0.010 |
| Blasted rock | " | 0.011 |
| 6 cy capacity | | |
| Light material | C.Y. | 0.005 |
| Medium material | " | 0.006 |
| Wet material | " | 0.006 |
| Blasted rock | " | 0.007 |
| Track mounted front-end loader | | |
| 1-1/2 cy capacity | | |
| Light material | C.Y. | 0.013 |
| Medium material | " | 0.015 |
| Wet material | " | 0.016 |
| Blasted rock | " | 0.018 |
| 2-3/4 cy capacity | | |
| Light material | C.Y. | 0.008 |
| Medium material | " | 0.009 |
| Wet material | " | 0.010 |
| Blasted rock | " | 0.011 |
| **02220.10 BORROW** | | |
| Borrow fill, F.O.B. at pit | | |
| Sand, haul to site, round trip | | |
| 10 mile | C.Y. | 0.080 |
| 20 mile | " | 0.133 |
| 30 mile | " | 0.200 |
| Place borrow fill and compact | | |
| Less than 1 in 4 slope | C.Y. | 0.040 |
| Greater than 1 in 4 slope | " | 0.053 |

| EARTHWORK | UNIT | MAN/HOURS |
|---|---|---|
| **02220.40 BUILDING EXCAVATION** | | |
| Structural excavation, unclassified earth | | |
| 3/8 cy backhoe | C.Y. | 0.107 |
| 3/4 cy backhoe | " | 0.080 |
| 1 cy backhoe | " | 0.067 |

# 02 SITEWORK

| EARTHWORK | UNIT | MAN/ HOURS |
|---|---|---|
| **02220.40 BUILDING EXCAVATION** | | |
| Foundation backfill and compaction by machine | C.Y. | 0.160 |
| **02220.50 UTILITY EXCAVATION** | | |
| Trencher, sandy clay, 8" wide trench | | |
| 18" deep | L.F. | 0.018 |
| 24" deep | " | 0.020 |
| 36" deep | " | 0.023 |
| Trench backfill, 95% compaction | | |
| Tamp by hand | C.Y. | 0.500 |
| Vibratory compaction | " | 0.400 |
| Trench backfilling, with borrow sand, place & compact | " | 0.400 |
| **02220.60 TRENCHING** | | |
| Trenching and continuous footing excavation | | |
| By gradall | | |
| 1 cy capacity | | |
| Light soil | C.Y. | 0.023 |
| Medium soil | " | 0.025 |
| Heavy/wet soil | " | 0.027 |
| Loose rock | " | 0.029 |
| Blasted rock | " | 0.031 |
| By hydraulic excavator | | |
| 1/2 cy capacity | | |
| Light soil | C.Y. | 0.027 |
| Medium soil | " | 0.029 |
| Heavy/wet soil | " | 0.032 |
| Loose rock | " | 0.036 |
| Blasted rock | " | 0.040 |
| 1 cy capacity | | |
| Light soil | C.Y. | 0.019 |
| Medium soil | " | 0.020 |
| Heavy/wet soil | " | 0.021 |
| Loose rock | " | 0.023 |
| Blasted rock | " | 0.025 |
| 1-1/2 cy capacity | | |
| Light soil | C.Y. | 0.017 |
| Medium soil | " | 0.018 |
| Heavy/wet soil | " | 0.019 |
| Loose rock | " | 0.020 |
| Blasted rock | " | 0.021 |
| 2 cy capacity | | |
| Light soil | C.Y. | 0.016 |
| Medium soil | " | 0.017 |
| Heavy/wet soil | " | 0.018 |
| Loose rock | " | 0.019 |
| Blasted rock | " | 0.020 |
| 2-1/2 cy capacity | | |
| Light soil | C.Y. | 0.015 |
| Medium soil | " | 0.015 |
| Heavy/wet soil | " | 0.016 |
| Loose rock | " | 0.017 |
| Blasted rock | " | 0.018 |

| EARTHWORK | UNIT | MAN/ HOURS |
|---|---|---|
| **02220.60 TRENCHING** | | |
| Trencher, chain, 1' wide to 4' deep | | |
| Light soil | C.Y. | 0.020 |
| Medium soil | " | 0.023 |
| Heavy soil | " | 0.027 |
| Hand excavation | | |
| Bulk, wheeled 100' | | |
| Normal soil | C.Y. | 0.889 |
| Sand or gravel | " | 0.800 |
| Medium clay | " | 1.143 |
| Heavy clay | " | 1.600 |
| Loose rock | " | 2.000 |
| Trenches, up to 2' deep | | |
| Normal soil | C.Y. | 1.000 |
| Sand or gravel | " | 0.889 |
| Medium clay | " | 1.333 |
| Heavy clay | " | 2.000 |
| Loose rock | " | 2.667 |
| Trenches, to 6' deep | | |
| Normal soil | C.Y. | 1.143 |
| Sand or gravel | " | 1.000 |
| Medium clay | " | 1.600 |
| Heavy clay | " | 2.667 |
| Loose rock | " | 4.000 |
| Backfill trenches | | |
| With compaction | | |
| By hand | C.Y. | 0.667 |
| By 60 hp tracked dozer | " | 0.020 |
| By 200 hp tracked dozer | " | 0.009 |
| By small front-end loader | " | 0.023 |
| Spread dumped fill or gravel, no compaction | | |
| 6" layers | S.Y. | 0.013 |
| 12" layers | " | 0.016 |
| Compaction in 6" layers | | |
| By hand with air tamper | S.Y. | 0.016 |
| Backfill trenches, sand bedding, no compaction | | |
| By hand | C.Y. | 0.667 |
| By small front-end loader | " | 0.023 |
| **02220.70 ROADWAY EXCAVATION** | | |
| Roadway excavation | | |
| 1/4 mile haul | C.Y. | 0.016 |
| 2 mile haul | " | 0.027 |
| 5 mile haul | " | 0.040 |
| Excavation of open ditches | " | 0.011 |
| Trim banks, swales or ditches | S.Y. | 0.013 |
| Bulk swale excavation by dragline | | |
| Small jobs | C.Y. | 0.060 |
| Large jobs | " | 0.034 |
| Spread base course | " | 0.020 |
| Roll and compact | " | 0.027 |

# 02 SITEWORK

## EARTHWORK

| EARTHWORK | UNIT | MAN/HOURS |
|---|---|---|
| **02220.71 BASE COURSE** | | |
| Base course, crushed stone | | |
| 3" thick | S.Y. | 0.004 |
| 4" thick | " | 0.004 |
| 6" thick | " | 0.005 |
| 8" thick | " | 0.005 |
| 10" thick | " | 0.006 |
| 12" thick | " | 0.007 |
| Base course, bank run gravel | | |
| 4" deep | S.Y. | 0.004 |
| 6" deep | " | 0.005 |
| 8" deep | " | 0.005 |
| 10" deep | " | 0.005 |
| 12" deep | " | 0.006 |
| Prepare and roll sub base | | |
| Minimum | S.Y. | 0.004 |
| Average | " | 0.005 |
| Maximum | " | 0.007 |
| **02220.90 HAND EXCAVATION** | | |
| Excavation | | |
| To 2' deep | | |
| Normal soil | C.Y. | 0.889 |
| Sand and gravel | " | 0.800 |
| Medium clay | " | 1.000 |
| Heavy clay | " | 1.143 |
| Loose rock | " | 1.333 |
| To 6' deep | | |
| Normal soil | C.Y. | 1.143 |
| Sand and gravel | " | 1.000 |
| Medium clay | " | 1.333 |
| Heavy clay | " | 1.600 |
| Loose rock | " | 2.000 |
| Backfilling foundation without compaction, 6" lifts | " | 0.500 |
| Compaction of backfill around structures or in trench | | |
| By hand with air tamper | C.Y. | 0.571 |
| By hand with vibrating plate tamper | " | 0.533 |
| 1 ton roller | " | 0.400 |
| Miscellaneous hand labor | | |
| Trim slopes, sides of excavation | S.F. | 0.001 |
| Trim bottom of excavation | " | 0.002 |
| Excavation around obstructions and services | C.Y. | 2.667 |
| **02240.05 SOIL STABILIZATION** | | |
| Straw bale secured with rebar | L.F. | 0.027 |
| Filter barrier, 18" high filter fabric | " | 0.080 |
| Sediment fence, 36" fabric with 6" mesh | " | 0.100 |
| Soil stabilization with tar paper, burlap, straw and stakes | S.F. | 0.001 |
| **02240.30 GEOTEXTILE** | | |
| Filter cloth, light reinforcement | | |
| Woven | | |
| 12'-6" wide x 50' long | S.F. | 0.001 |
| Various lengths | " | 0.001 |

| EARTHWORK | UNIT | MAN/HOURS |
|---|---|---|
| **02240.30 GEOTEXTILE** | | |
| Non-woven | | |
| 14'-8" wide x 430' long | S.F. | 0.001 |
| Various lengths | " | 0.001 |
| **02270.10 SLOPE PROTECTION** | | |
| Gabions, stone filled | | |
| 6" deep | S.Y. | 0.200 |
| 9" deep | " | 0.229 |
| 12" deep | " | 0.267 |
| 18" deep | " | 0.320 |
| 36" deep | " | 0.533 |
| **02270.40 RIPRAP** | | |
| Riprap | | |
| Crushed stone blanket, max size 2-1/2" | TON | 0.533 |
| Stone, quarry run, 300 lb. stones | " | 0.492 |
| 400 lb. stones | " | 0.457 |
| 500 lb. stones | " | 0.427 |
| 750 lb. stones | " | 0.400 |
| Dry concrete riprap in bags 3" thick, 80 lb. per bag | BAG | 0.027 |
| **02280.20 SOIL TREATMENT** | | |
| Soil treatment, termite control pretreatment | | |
| Under slabs | S.F. | 0.004 |
| By walls | " | 0.005 |
| **02290.30 WEED CONTROL** | | |
| Weed control, bromicil, 15 lb./acre, wettable powder | ACRE | 4.000 |
| Vegetation control, by application of plant killer | S.Y. | 0.003 |
| Weed killer, lawns and fields | " | 0.002 |

## TUNNELING

| TUNNELING | UNIT | MAN/HOURS |
|---|---|---|
| **02300.10 PIPE JACKING** | | |
| Pipe casing, horizontal jacking | | |
| 18" dia. | L.F. | 0.711 |
| 21" dia. | " | 0.762 |
| 24" dia. | " | 0.800 |
| 27" dia. | " | 0.800 |
| 30" dia. | " | 0.842 |
| 36" dia. | " | 0.914 |
| 42" dia. | " | 1.000 |
| 48" dia. | " | 1.067 |

| PILES AND CAISSONS | UNIT | MAN/ HOURS |
|---|---|---|
| **02360.50 PRESTRESSED PILING** | | |
| Prestressed concrete piling, less than 60' long | | |
| 10" sq. | L.F. | 0.040 |
| 12" sq. | " | 0.042 |
| 14" sq. | " | 0.043 |
| 16" sq. | " | 0.044 |
| 18" sq. | " | 0.047 |
| 20" sq. | " | 0.048 |
| 24" sq. | " | 0.049 |
| More than 60' long | | |
| 12" sq. | L.F. | 0.034 |
| 14" sq. | " | 0.035 |
| 16" sq. | " | 0.036 |
| 18" sq. | " | 0.036 |
| 20" sq. | " | 0.037 |
| 24" sq. | " | 0.038 |
| Straight cylinder, less than 60' long | | |
| 12" dia. | L.F. | 0.044 |
| 14" dia. | " | 0.045 |
| 16" dia. | " | 0.046 |
| 18" dia. | " | 0.047 |
| 20" dia. | " | 0.048 |
| 24" dia. | " | 0.049 |
| More than 60' long | | |
| 12" dia. | L.F. | 0.035 |
| 14" dia. | " | 0.036 |
| 16" dia. | " | 0.036 |
| 18" dia. | " | 0.037 |
| 20" dia. | " | 0.038 |
| 24" dia. | " | 0.038 |
| Concrete sheet piling | | |
| 12" thick x 20' long | S.F. | 0.096 |
| 25' long | " | 0.087 |
| 30' long | " | 0.080 |
| 35' long | " | 0.074 |
| 40' long | " | 0.069 |
| 16" thick x 40' long | " | 0.053 |
| 45' long | " | 0.051 |
| 50' long | " | 0.048 |
| 55' long | " | 0.046 |
| 60' long | " | 0.044 |
| **02360.60 STEEL PILES** | | |
| H-section piles | | |
| 8x8 | | |
| 36 lb/ft | | |
| 30' long | L.F. | 0.080 |
| 40' long | " | 0.064 |
| 50' long | " | 0.053 |
| 10x10 | | |
| 42 lb/ft | | |
| 30' long | L.F. | 0.080 |
| 40' long | " | 0.064 |
| 50' long | " | 0.053 |
| 57 lb/ft | | |

| PILES AND CAISSONS | UNIT | MAN/ HOURS |
|---|---|---|
| **02360.60 STEEL PILES** | | |
| 30' long | L.F. | 0.080 |
| 40' long | " | 0.064 |
| 50' long | " | 0.053 |
| 12x12 | | |
| 53 lb/ft | | |
| 30' long | L.F. | 0.087 |
| 40' long | " | 0.069 |
| 50' long | " | 0.053 |
| 74 lb/ft | | |
| 30' long | L.F. | 0.087 |
| 40' long | " | 0.069 |
| 50' long | " | 0.053 |
| 14x14 | | |
| 73 lb/ft | | |
| 40' long | L.F. | 0.087 |
| 50' long | " | 0.069 |
| 60' long | " | 0.053 |
| 89 lb/ft | | |
| 40' long | L.F. | 0.087 |
| 50' long | " | 0.069 |
| 60' long | " | 0.053 |
| 102 lb/ft | | |
| 40' long | L.F. | 0.087 |
| 50' long | " | 0.069 |
| 60' long | " | 0.053 |
| 117 lb/ft | | |
| 40' long | L.F. | 0.091 |
| 50' long | " | 0.071 |
| 60' long | " | 0.055 |
| Splice | | |
| 8" | EA. | 1.333 |
| 10" | " | 1.600 |
| 12" | " | 1.600 |
| 14" | " | 2.000 |
| Driving cap | | |
| 8" | EA. | 0.800 |
| 10" | " | 1.000 |
| 12" | " | 1.000 |
| 14" | " | 1.143 |
| Standard point | | |
| 8" | EA. | 0.800 |
| 10" | " | 1.000 |
| 12" | " | 1.143 |
| 14" | " | 1.333 |
| Heavy duty point | | |
| 8" | EA. | 0.889 |
| 10" | " | 1.143 |
| 12" | " | 1.333 |
| 14" | " | 1.600 |
| Tapered friction piles, with fluted steel casing, up to 50' | | |
| With 4000 psi concrete no reinforcing | | |
| 12" dia. | L.F. | 0.048 |
| 14" dia. | " | 0.049 |
| 16" dia. | " | 0.051 |

| PILES AND CAISSONS | UNIT | MAN/HOURS |
|---|---|---|
| **02360.60 STEEL PILES** | | |
| 18" dia. | L.F. | 0.056 |
| **02360.65 STEEL PIPE PILES** | | |
| Concrete filled, 3000# concrete, up to 40' | | |
| 8" dia. | L.F. | 0.069 |
| 10" dia. | " | 0.071 |
| 12" dia. | " | 0.074 |
| 14" dia. | " | 0.077 |
| 16" dia. | " | 0.080 |
| 18" dia. | " | 0.083 |
| Pipe piles, non-filled | | |
| 8" dia. | L.F. | 0.053 |
| 10" dia. | " | 0.055 |
| 12" dia. | " | 0.056 |
| 14" dia. | " | 0.060 |
| 16" dia. | " | 0.062 |
| 18" dia. | " | 0.064 |
| Splice | | |
| 8" dia. | EA. | 1.600 |
| 10" dia. | " | 1.600 |
| 12" dia. | " | 2.000 |
| 14" dia. | " | 2.000 |
| 16" dia. | " | 2.667 |
| 18" dia. | " | 2.667 |
| Standard point | | |
| 8" dia. | EA. | 1.600 |
| 10" dia. | " | 1.600 |
| 12" dia. | " | 2.000 |
| 14" dia. | " | 2.000 |
| 16" dia. | " | 2.667 |
| 18" dia. | " | 2.667 |
| Heavy duty point | | |
| 8" dia. | EA. | 2.000 |
| 10" dia. | " | 2.000 |
| 12" dia. | " | 2.667 |
| 14" dia. | " | 2.667 |
| 16" dia. | " | 3.200 |
| 18" dia. | " | 3.200 |
| **02360.70 STEEL SHEET PILING** | | |
| Steel sheet piling,12" wide | | |
| 20' long | S.F. | 0.096 |
| 35' long | " | 0.069 |
| 50' long | " | 0.048 |
| Over 50' long | " | 0.044 |
| **02360.80 WOOD AND TIMBER PILES** | | |
| Treated wood piles, 12" butt, 8" tip | | |
| 25' long | L.F. | 0.096 |
| 30' long | " | 0.080 |

| PILES AND CAISSONS | UNIT | MAN/HOURS |
|---|---|---|
| **02360.80 WOOD AND TIMBER PILES** | | |
| 35' long | L.F. | 0.069 |
| 40' long | " | 0.060 |
| 12" butt, 7" tip | | |
| 40' long | L.F. | 0.060 |
| 45' long | " | 0.053 |
| 50' long | " | 0.048 |
| 55' long | " | 0.044 |
| 60' long | " | 0.040 |
| **02380.10 CAISSONS** | | |
| Caisson, including 3000# concrete, in stable ground | | |
| 18" dia. | L.F. | 0.192 |
| 24" dia. | " | 0.200 |
| 30" dia. | " | 0.240 |
| 36" dia. | " | 0.274 |
| 48" dia. | " | 0.320 |
| 60" dia. | " | 0.436 |
| 72" dia. | " | 0.533 |
| 84" dia. | " | 0.686 |
| Wet ground, casing required but pulled | | |
| 18" dia. | L.F. | 0.240 |
| 24" dia. | " | 0.267 |
| 30" dia. | " | 0.300 |
| 36" dia. | " | 0.320 |
| 48" dia. | " | 0.400 |
| 60" dia. | " | 0.533 |
| 72" dia. | " | 0.800 |
| 84" dia. | " | 1.200 |
| Soft rock | | |
| 18" dia. | L.F. | 0.686 |
| 24" dia. | " | 1.200 |
| 30" dia. | " | 1.600 |
| 36" dia. | " | 2.400 |
| 48" dia. | " | 3.200 |
| 60" dia. | " | 4.800 |
| 72" dia. | " | 5.333 |
| 84" dia. | " | 6.000 |

| RAILROAD WORK | UNIT | MAN/HOURS |
|---|---|---|
| **02450.10 RAILROAD WORK** | | |
| Rail | | |
| 90 lb | L.F. | 0.010 |
| 100 lb | " | 0.010 |
| 115 lb | " | 0.010 |
| 132 lb | " | 0.010 |

| RAILROAD WORK | UNIT | MAN/ HOURS |
|---|---|---|
| **02450.10 RAILROAD WORK** | | |
| Rail relay | | |
| 90 lb | L.F. | 0.010 |
| 100 lb | " | 0.010 |
| 115 lb | " | 0.010 |
| 132 lb | " | 0.010 |
| New angle bars, per pair | | |
| 90 lb | EA. | 0.012 |
| 100 lb | " | 0.012 |
| 115 lb | " | 0.012 |
| 132 lb | " | 0.012 |
| Angle bar relay | | |
| 90 lb | EA. | 0.012 |
| 100 lb | " | 0.012 |
| 115 lb | " | 0.012 |
| 132 lb | " | 0.012 |
| New tie plates | | |
| 90 lb | EA. | 0.009 |
| 100 lb | " | 0.009 |
| 115 lb | " | 0.009 |
| 132 lb | " | 0.009 |
| Tie plate relay | | |
| 90 lb | EA. | 0.009 |
| 100 lb | " | 0.009 |
| 115 lb | " | 0.009 |
| 132 lb | " | 0.009 |
| Track accessories | | |
| Wooden cross ties, 8' | EA. | 0.060 |
| Concrete cross ties, 8' | " | 0.120 |
| Tie plugs, 5" | " | 0.006 |
| Track bolts and nuts, 1" | " | 0.006 |
| Lockwashers, 1" | " | 0.004 |
| Track spikes, 6" | " | 0.024 |
| Wooden switch ties | B.F. | 0.006 |
| Rail anchors | EA. | 0.022 |
| Ballast | TON | 0.120 |
| Gauge rods | EA. | 0.096 |
| Compromise splice bars | " | 0.160 |
| Turnout | | |
| 90 lb | EA. | 24.000 |
| 100 lb | " | 24.000 |
| 110 lb | " | 24.000 |
| 115 lb | " | 24.000 |
| 132 lb | " | 24.000 |
| Turnout relay | | |
| 90 lb | EA. | 24.000 |
| 100 lb | " | 24.000 |
| 110 lb | " | 24.000 |
| 115 lb | " | 24.000 |
| 132 lb | " | 24.000 |
| Railroad track in place, complete | | |
| New rail | | |
| 90 lb | L.F. | 0.240 |
| 100 lb | " | 0.240 |
| 110 lb | " | 0.240 |

| RAILROAD WORK | UNIT | MAN/ HOURS |
|---|---|---|
| **02450.10 RAILROAD WORK** | | |
| 115 lb | L.F. | 0.240 |
| 132 lb | " | 0.240 |
| Rail relay | | |
| 90 lb | L.F. | 0.240 |
| 100 lb | " | 0.240 |
| 110 lb | " | 0.240 |
| 115 lb | " | 0.240 |
| 132 lb | " | 0.240 |
| No. 8 turnout | | |
| 90 lb | EA. | 32.000 |
| 100 lb | " | 32.000 |
| 110 lb | " | 32.000 |
| 115 lb | " | 32.000 |
| 132 lb | " | 32.000 |
| No. 8 turnout relay | | |
| 90 lb | EA. | 32.000 |
| 100 lb | " | 32.000 |
| 110 lb | " | 32.000 |
| 115 lb | " | 32.000 |
| 132 lb | " | 32.000 |
| Railroad crossings, asphalt, based on 8" thick x 20' | | |
| Including track and approach | | |
| 12' roadway | EA. | 6.000 |
| 15' roadway | " | 6.857 |
| 18' roadway | " | 8.000 |
| 21' roadway | " | 9.600 |
| 24' roadway | " | 12.000 |
| Precast concrete inserts | | |
| 12' roadway | EA. | 2.400 |
| 15' roadway | " | 3.000 |
| 18' roadway | " | 4.000 |
| 21' roadway | " | 4.800 |
| 24' roadway | " | 5.333 |
| Molded rubber, with headers | | |
| 12' roadway | EA. | 2.400 |
| 15' roadway | " | 3.000 |
| 18' roadway | " | 4.000 |
| 21' roadway | " | 4.800 |
| 24' roadway | " | 5.333 |

| PAVING AND SURFACING | UNIT | MAN/ HOURS |
|---|---|---|
| **02510.20 ASPHALT SURFACES** | | |
| Asphalt wearing surface, for flexible pavement | | |
| 1" thick | S.Y. | 0.016 |
| 1-1/2" thick | " | 0.019 |

## PAVING AND SURFACING

| PAVING AND SURFACING | UNIT | MAN/HOURS |
|---|---|---|
| **02510.20 ASPHALT SURFACES** | | |
| 2" thick | S.Y. | 0.024 |
| 3" thick | " | 0.032 |
| Bituminous sidewalk, no base | | |
| 2" thick | S.Y. | 0.028 |
| 3" thick | " | 0.030 |
| **02520.10 CONCRETE PAVING** | | |
| Concrete paving, reinforced, 5000 psi concrete | | |
| 6" thick | S.Y. | 0.150 |
| 7" thick | " | 0.160 |
| 8" thick | " | 0.171 |
| 9" thick | " | 0.185 |
| 10" thick | " | 0.200 |
| 11" thick | " | 0.218 |
| 12" thick | " | 0.240 |
| 15" thick | " | 0.300 |
| Concrete paving, for pipe trench, reinforced | | |
| 7" thick | S.Y. | 0.240 |
| 8" thick | " | 0.267 |
| 9" thick | " | 0.300 |
| 10" thick | " | 0.343 |
| Fibrous concrete | | |
| 5" thick | S.Y. | 0.185 |
| 8" thick | " | 0.200 |
| Roller compacted concrete, (RCC), place and compact | | |
| 8" thick | S.Y. | 0.240 |
| 12" thick | " | 0.300 |
| Steel edge forms up to | | |
| 12" deep | L.F. | 0.027 |
| 15" deep | " | 0.032 |
| Paving finishes | | |
| Belt dragged | S.Y. | 0.040 |
| Curing | " | 0.008 |
| **02545.10 ASPHALT REPAIR** | | |
| Coal tar emulsion seal coat, rubber additive, fuel resistant | S.Y. | 0.011 |
| Bituminous surface treatment, single | " | 0.008 |
| Double | " | 0.001 |
| Bituminous prime coat | " | 0.001 |
| Tack coat | " | 0.001 |
| Crack sealing, concrete paving | L.F. | 0.005 |
| Bituminous paving for pipe trench, 4" thick | S.Y. | 0.160 |
| Polypropylene, nonwoven paving fabric | " | 0.004 |
| Rubberized asphalt | " | 0.073 |
| Asphalt slurry seal | " | 0.047 |
| **02580.10 PAVEMENT MARKINGS** | | |
| Pavement line marking, paint | | |
| 4" wide | L.F. | 0.002 |
| 6" wide | " | 0.004 |
| 8" wide | " | 0.007 |
| Reflective paint, 4" wide | " | 0.007 |
| Airfield markings, retro-reflective | | |

| PAVING AND SURFACING | UNIT | MAN/HOURS |
|---|---|---|
| **02580.10 PAVEMENT MARKINGS** | | |
| White | L.F. | 0.007 |
| Yellow | " | 0.007 |
| Preformed tape, 4" wide | | |
| Inlaid reflective | L.F. | 0.001 |
| Reflective paint | " | 0.002 |
| Thermoplastic | | |
| White | L.F. | 0.004 |
| Yellow | " | 0.004 |
| 12" wide, thermoplastic, white | " | 0.011 |
| Directional arrows, reflective preformed tape | EA. | 0.800 |
| Messages, reflective preformed tape (per letter) | " | 0.400 |
| Handicap symbol, preformed tape | " | 0.800 |
| Parking stall painting | " | 0.160 |

## UTILITIES

| UTILITIES | UNIT | MAN/HOURS |
|---|---|---|
| **02605.30 MANHOLES** | | |
| Precast sections, 48" dia. | | |
| Base section | EA. | 2.000 |
| 1'0" riser | " | 1.600 |
| 1'4" riser | " | 1.714 |
| 2'8" riser | " | 1.846 |
| 4'0" riser | " | 2.000 |
| 2'8" cone top | " | 2.400 |
| Precast manholes, 48" dia. | | |
| 4' deep | EA. | 4.800 |
| 6' deep | " | 6.000 |
| 7' deep | " | 6.857 |
| 8' deep | " | 8.000 |
| 10' deep | " | 9.600 |
| Cast-in-place, 48" dia., with frame and cover | | |
| 5' deep | EA. | 12.000 |
| 6' deep | " | 13.714 |
| 8' deep | " | 16.000 |
| 10' deep | " | 19.200 |
| Brick manholes, 48" dia. with cover, 8" thick | | |
| 4' deep | EA. | 8.000 |
| 6' deep | " | 8.889 |
| 8' deep | " | 10.000 |
| 10' deep | " | 11.429 |
| 12' deep | " | 13.333 |
| 14' deep | " | 16.000 |
| Inverts for manholes | | |
| Single channel | EA. | 3.200 |
| Triple channel | " | 4.000 |
| Frames and covers, 24" diameter | | |

# 02 SITEWORK

| UTILITIES | UNIT | MAN/ HOURS |
|---|---|---|
| **02605.30 MANHOLES** | | |
| 300 lb | EA. | 0.800 |
| 400 lb | " | 0.889 |
| 500 lb | " | 1.143 |
| Watertight, 350 lb | " | 2.667 |
| For heavy equipment, 1200 lb | " | 4.000 |
| Steps for manholes | | |
| 7" x 9" | EA. | 0.160 |
| 8" x 9" | " | 0.178 |
| Curb inlet, 4' throat, cast in place | | |
| 12"-30" pipe | EA. | 12.000 |
| 36"-48" pipe | " | 13.714 |
| Raise exist frame and cover, when repaving | " | 4.800 |
| **02610.10 CAST IRON FLANGED PIPE** | | |
| Cast iron flanged sections | | |
| 4" pipe, with one bolt set | | |
| 3' section | EA. | 0.218 |
| 4' section | " | 0.240 |
| 5' section | " | 0.267 |
| 6' section | " | 0.300 |
| 8' section | " | 0.343 |
| 10' section | " | 0.480 |
| 12' section | " | 0.800 |
| 15' section | " | 1.200 |
| 18' section | " | 1.600 |
| 6" pipe, with one bolt set | | |
| 3' section | EA. | 0.240 |
| 4' section | " | 0.282 |
| 5' section | " | 0.320 |
| 6' section | " | 0.369 |
| 8' section | " | 0.533 |
| 10' section | " | 0.600 |
| 12' section | " | 0.800 |
| 15' section | " | 1.200 |
| 18' section | " | 1.714 |
| 8" pipe, with one bolt set | | |
| 3' section | EA. | 0.300 |
| 4' section | " | 0.343 |
| 5' section | " | 0.400 |
| 6' section | " | 0.480 |
| 8' section | " | 0.686 |
| 10' section | " | 0.800 |
| 12' section | " | 1.200 |
| 15' section | " | 1.600 |
| 18' section | " | 2.000 |
| 10" pipe, with one bolt set | | |
| 3' section | EA. | 0.308 |
| 4' section | " | 0.353 |
| 5' section | " | 0.414 |
| 6' section | " | 0.500 |
| 8' section | " | 0.727 |
| 10' section | " | 0.857 |
| 12' section | " | 1.333 |
| 15' section | " | 1.714 |

| UTILITIES | UNIT | MAN/ HOURS |
|---|---|---|
| **02610.10 CAST IRON FLANGED PIPE** | | |
| 18' section | EA. | 2.400 |
| 12" pipe, with one bolt set | | |
| 3' section | EA. | 0.333 |
| 4' section | " | 0.387 |
| 5' section | " | 0.462 |
| 6' section | " | 0.545 |
| 8' section | " | 0.800 |
| 10' section | " | 0.923 |
| 12' section | " | 1.500 |
| 15' section | " | 2.000 |
| 18' section | " | 2.667 |
| **02610.11 CAST IRON FITTINGS** | | |
| Mechanical joint, with 2 bolt kits | | |
| 90 deg bend | | |
| 4" | EA. | 0.533 |
| 6" | " | 0.615 |
| 8" | " | 0.800 |
| 10" | " | 1.143 |
| 12" | " | 1.600 |
| 14" | " | 2.000 |
| 16" | " | 2.667 |
| 45 deg bend | | |
| 4" | EA. | 0.533 |
| 6" | " | 0.615 |
| 8" | " | 0.800 |
| 10" | " | 1.143 |
| 12" | " | 1.600 |
| 14" | " | 2.000 |
| 16" | " | 2.667 |
| Tee, with 3 bolt kits | | |
| 4" x 4" | EA. | 0.800 |
| 6" x 6" | " | 1.000 |
| 8" x 8" | " | 1.333 |
| 10" x 10" | " | 2.000 |
| 12" x 12" | " | 2.667 |
| Wye, with 3 bolt kits | | |
| 6" x 6" | EA. | 1.000 |
| 8" x 8" | " | 1.333 |
| 10" x 10" | " | 2.000 |
| 12" x 12" | " | 2.667 |
| Reducer, with 2 bolt kits | | |
| 6" x 4" | EA. | 1.000 |
| 8" x 6" | " | 1.333 |
| 10" x 8" | " | 2.000 |
| 12" x 10" | " | 2.667 |
| Flanged, 90 deg bend, 125 lb. | | |
| 4" | EA. | 0.667 |
| 6" | " | 0.800 |
| 8" | " | 1.000 |
| 10" | " | 1.333 |
| 12" | " | 2.000 |
| 14" | " | 2.667 |

| UTILITIES | UNIT | MAN/ HOURS |
|---|---|---|
| **02610.11 CAST IRON FITTINGS** | | |
| 16" | EA. | 2.667 |
| Tee | | |
| 4" | EA. | 1.000 |
| 6" | " | 1.143 |
| 8" | " | 1.333 |
| 10" | " | 1.600 |
| 12" | " | 2.000 |
| 14" | " | 2.667 |
| 16" | " | 4.000 |
| **02610.13 GATE VALVES** | | |
| Gate valve, (AWWA) mechanical joint, with adjustable box | | |
| 4" valve | EA. | 0.800 |
| 6" valve | " | 0.960 |
| 8" valve | " | 1.200 |
| 10" valve | " | 1.412 |
| 12" valve | " | 1.714 |
| 14" valve | " | 2.000 |
| 16" valve | " | 2.182 |
| 18" valve | " | 2.400 |
| Flanged, with box, post indicator (AWWA) | | |
| 4" valve | EA. | 0.960 |
| 6" valve | " | 1.091 |
| 8" valve | " | 1.333 |
| 10" valve | " | 1.600 |
| 12" valve | " | 2.000 |
| 14" valve | " | 2.400 |
| 16" valve | " | 3.000 |
| **02610.15 WATER METERS** | | |
| Water meter, displacement type | | |
| 1" | EA. | 0.800 |
| 1-1/2" | " | 0.889 |
| 2" | " | 1.000 |
| **02610.17 CORPORATION STOPS** | | |
| Stop for flared copper service pipe | | |
| 3/4" | EA. | 0.400 |
| 1" | " | 0.444 |
| 1-1/4" | " | 0.533 |
| 1-1/2" | " | 0.667 |
| 2" | " | 0.800 |
| **02610.40 DUCTILE IRON PIPE** | | |
| Ductile iron pipe, cement lined, slip-on joints | | |
| 4" | L.F. | 0.067 |
| 6" | " | 0.071 |
| 8" | " | 0.075 |
| 10" | " | 0.080 |
| 12" | " | 0.096 |
| 14" | " | 0.120 |
| 16" | " | 0.133 |

| UTILITIES | UNIT | MAN/ HOURS |
|---|---|---|
| **02610.40 DUCTILE IRON PIPE** | | |
| 18" | L.F. | 0.150 |
| 20" | " | 0.171 |
| Mechanical joint pipe | | |
| 4" | L.F. | 0.092 |
| 6" | " | 0.100 |
| 8" | " | 0.109 |
| 10" | " | 0.120 |
| 12" | " | 0.160 |
| 14" | " | 0.185 |
| 16" | " | 0.218 |
| 18" | " | 0.240 |
| 20" | " | 0.267 |
| Fittings, mechanical joint | | |
| 90 degree elbow | | |
| 4" | EA. | 0.533 |
| 6" | " | 0.615 |
| 8" | " | 0.800 |
| 10" | " | 1.143 |
| 12" | " | 1.600 |
| 14" | " | 2.000 |
| 16" | " | 2.667 |
| 18" | " | 3.200 |
| 20" | " | 4.000 |
| 45 degree elbow | | |
| 4" | EA. | 0.533 |
| 6" | " | 0.615 |
| 8" | " | 0.800 |
| 10" | " | 1.143 |
| 12" | " | 1.600 |
| 14" | " | 2.000 |
| 16" | " | 2.667 |
| 18" | " | 4.000 |
| 20" | " | 4.000 |
| Tee | | |
| 4"x3" | EA. | 1.000 |
| 4"x4" | " | 1.000 |
| 6"x3" | " | 1.143 |
| 6"x4" | " | 1.143 |
| 6"x6" | " | 1.143 |
| 8"x4" | " | 1.333 |
| 8"x6" | " | 1.333 |
| 8"x8" | " | 1.333 |
| 10"x4" | " | 1.600 |
| 10"x6" | " | 1.600 |
| 10"x8" | " | 1.600 |
| 10"x10" | " | 1.600 |
| 12"x4" | " | 2.000 |
| 12"x6" | " | 2.000 |
| 12"x8" | " | 2.000 |
| 12"x10" | " | 2.000 |
| 12"x12" | " | 2.133 |
| 14"x4" | " | 2.286 |
| 14"x6" | " | 2.286 |
| 14"x8" | " | 2.286 |

| UTILITIES | UNIT | MAN/ HOURS |
|---|---|---|
| **02610.40 DUCTILE IRON PIPE** | | |
| 14"x10" | EA. | 2.286 |
| 14"x12" | " | 2.462 |
| 14"x14" | " | 2.462 |
| 16"x4" | " | 2.667 |
| 16"x6" | " | 2.667 |
| 16"x8" | " | 2.667 |
| 16"x10" | " | 2.667 |
| 16"x12" | " | 2.667 |
| 16"x14" | " | 2.667 |
| 16"x16" | " | 2.667 |
| 18"x6" | " | 2.909 |
| 18"x8" | " | 2.909 |
| 18"x10" | " | 2.909 |
| 18"x12" | " | 2.909 |
| 18"x14" | " | 2.909 |
| 18"x16" | " | 2.909 |
| 18"x18" | " | 2.909 |
| 20"x6" | " | 3.200 |
| 20"x8" | " | 3.200 |
| 20"x10" | " | 3.200 |
| 20"x12" | " | 3.200 |
| 20"x14" | " | 3.200 |
| 20"x16" | " | 3.200 |
| 20"x18" | " | 3.200 |
| 20"x20" | " | 3.200 |
| Cross | | |
| 4"x3" | EA. | 1.333 |
| 4"x4" | " | 1.333 |
| 6"x3" | " | 1.600 |
| 6"x4" | " | 1.600 |
| 6"x6" | " | 1.600 |
| 8"x4" | " | 1.778 |
| 8"x6" | " | 1.778 |
| 8"x8" | " | 1.778 |
| 10"x4" | " | 2.000 |
| 10"x6" | " | 2.000 |
| 10"x8" | " | 2.000 |
| 10"x10" | " | 2.000 |
| 12"x4" | " | 2.286 |
| 12"x6" | " | 2.286 |
| 12"x8" | " | 2.286 |
| 12"x10" | " | 2.462 |
| 12"x12" | " | 2.462 |
| 14"x4" | " | 2.667 |
| 14"x6" | " | 2.667 |
| 14"x8" | " | 2.667 |
| 14"x10" | " | 2.667 |
| 14"x12" | " | 2.909 |
| 14"x14" | " | 2.909 |
| 16"x4" | " | 3.200 |
| 16"x6" | " | 3.200 |
| 16"x8" | " | 3.200 |
| 16"x10" | " | 3.200 |
| 16"x12" | " | 3.200 |

| UTILITIES | UNIT | MAN/ HOURS |
|---|---|---|
| **02610.40 DUCTILE IRON PIPE** | | |
| 16"x14" | EA. | 3.200 |
| 16"x16" | " | 3.200 |
| 18"x6" | " | 3.556 |
| 18"x8" | " | 3.556 |
| 18"x10" | " | 3.556 |
| 18"x12" | " | 3.556 |
| 18"x14" | " | 3.556 |
| 18"x16" | " | 3.556 |
| 18"x18" | " | 3.556 |
| 20"x6" | " | 3.810 |
| 20"x8" | " | 3.810 |
| 20"x10" | " | 3.810 |
| 20"x12" | " | 3.810 |
| 20"x14" | " | 3.810 |
| 20"x16" | " | 3.810 |
| 20"x18" | " | 4.000 |
| 20"x20" | " | 4.000 |
| **02610.60 PLASTIC PIPE** | | |
| PVC, class 150 pipe | | |
| 4" dia. | L.F. | 0.060 |
| 6" dia. | " | 0.065 |
| 8" dia. | " | 0.069 |
| 10" dia. | " | 0.075 |
| 12" dia. | " | 0.080 |
| Schedule 40 pipe | | |
| 1-1/2" dia. | L.F. | 0.047 |
| 2" dia. | " | 0.050 |
| 2-1/2" dia. | " | 0.053 |
| 3" dia. | " | 0.057 |
| 4" dia. | " | 0.067 |
| 6" dia. | " | 0.080 |
| 90 degree elbows | | |
| 1" | EA. | 0.133 |
| 1-1/2" | " | 0.133 |
| 2" | " | 0.145 |
| 2-1/2" | " | 0.160 |
| 3" | " | 0.178 |
| 4" | " | 0.200 |
| 6" | " | 0.267 |
| 45 degree elbows | | |
| 1" | EA. | 0.133 |
| 1-1/2" | " | 0.133 |
| 2" | " | 0.145 |
| 2-1/2" | " | 0.160 |
| 3" | " | 0.178 |
| 4" | " | 0.200 |
| 6" | " | 0.267 |
| Tees | | |
| 1" | EA. | 0.160 |
| 1-1/2" | " | 0.160 |
| 2" | " | 0.178 |
| 2-1/2" | " | 0.200 |

| UTILITIES | UNIT | MAN/ HOURS |
|---|---|---|
| **02610.60 PLASTIC PIPE** | | |
| 3" | EA. | 0.229 |
| 4" | " | 0.267 |
| 6" | " | 0.320 |
| Couplings | | |
| 1" | EA. | 0.133 |
| 1-1/2" | " | 0.133 |
| 2" | " | 0.145 |
| 2-1/2" | " | 0.160 |
| 3" | " | 0.178 |
| 4" | " | 0.200 |
| 6" | " | 0.267 |
| Drainage pipe | | |
| PVC schedule 80 | | |
| 1" dia. | L.F. | 0.047 |
| 1-1/2" dia. | " | 0.047 |
| ABS, 2" dia. | " | 0.050 |
| 2-1/2" dia. | " | 0.053 |
| 3" dia. | " | 0.057 |
| 4" dia. | " | 0.067 |
| 6" dia. | " | 0.080 |
| 8" dia. | " | 0.063 |
| 10" dia. | " | 0.075 |
| 12" dia. | " | 0.080 |
| 90 degree elbows | | |
| 1" | EA. | 0.133 |
| 1-1/2" | " | 0.133 |
| 2" | " | 0.145 |
| 2-1/2" | " | 0.160 |
| 3" | " | 0.178 |
| 4" | " | 0.200 |
| 6" | " | 0.267 |
| 45 degree elbows | | |
| 1" | EA. | 0.133 |
| 1-1/2" | " | 0.133 |
| 2" | " | 0.145 |
| 2-1/2" | " | 0.160 |
| 3" | " | 0.178 |
| 4" | " | 0.200 |
| 6" | " | 0.267 |
| Tees | | |
| 1" | EA. | 0.160 |
| 1-1/2" | " | 0.160 |
| 2" | " | 0.178 |
| 2-1/2" | " | 0.200 |
| 3" | " | 0.229 |
| 4" | " | 0.267 |
| 6" | " | 0.320 |
| Couplings | | |
| 1" | EA. | 0.133 |
| 1-1/2" | " | 0.133 |
| 2" | " | 0.145 |
| 2-1/2" | " | 0.160 |
| 3" | " | 0.178 |
| 4" | " | 0.200 |

| UTILITIES | UNIT | MAN/ HOURS |
|---|---|---|
| **02610.60 PLASTIC PIPE** | | |
| 6" | EA. | 0.267 |
| Pressure pipe | | |
| PVC, class 200 pipe | | |
| 3/4" | L.F. | 0.040 |
| 1" | " | 0.042 |
| 1-1/4" | " | 0.044 |
| 1-1/2" | " | 0.047 |
| 2" | " | 0.050 |
| 2-1/2" | " | 0.053 |
| 3" | " | 0.057 |
| 4" | " | 0.067 |
| 6" | " | 0.080 |
| 8" | " | 0.069 |
| 90 degree elbows | | |
| 3/4" | EA. | 0.133 |
| 1" | " | 0.133 |
| 1-1/4" | " | 0.133 |
| 1-1/2" | " | 0.133 |
| 2" | " | 0.145 |
| 2-1/2" | " | 0.160 |
| 3" | " | 0.178 |
| 4" | " | 0.200 |
| 6" | " | 0.267 |
| 8" | " | 0.400 |
| 45 degree elbows | | |
| 3/4" | EA. | 0.133 |
| 1" | " | 0.133 |
| 1-1/4" | " | 0.133 |
| 1-1/2" | " | 0.133 |
| 2" | " | 0.145 |
| 2-1/2" | " | 0.160 |
| 3" | " | 0.178 |
| 4" | " | 0.200 |
| 6" | " | 0.267 |
| 8" | " | 0.400 |
| Tees | | |
| 3/4" | EA. | 0.160 |
| 1" | " | 0.160 |
| 1-1/4" | " | 0.160 |
| 1-1/2" | " | 0.160 |
| 2" | " | 0.178 |
| 2-1/2" | " | 0.200 |
| 3" | " | 0.229 |
| 4" | " | 0.267 |
| 6" | " | 0.320 |
| 8" | " | 0.444 |
| Couplings | | |
| 3/4" | EA. | 0.133 |
| 1" | " | 0.133 |
| 1-1/4" | " | 0.133 |
| 1-1/2" | " | 0.133 |
| 2" | " | 0.145 |
| 2-1/2" | " | 0.160 |
| 3" | " | 0.178 |

# 02 SITEWORK

| UTILITIES | UNIT | MAN/HOURS |
|---|---|---|
| **02610.60 PLASTIC PIPE** | | |
| 4" | EA. | 0.178 |
| 6" | " | 0.200 |
| 8" | " | 0.267 |
| **02610.90 VITRIFIED CLAY PIPE** | | |
| Vitrified clay pipe, extra strength | | |
| 6" dia. | L.F. | 0.109 |
| 8" dia. | " | 0.114 |
| 10" dia. | " | 0.120 |
| 12" dia. | " | 0.160 |
| 15" dia. | " | 0.240 |
| 18" dia. | " | 0.267 |
| 24" dia. | " | 0.343 |
| 30" dia. | " | 0.480 |
| 36" dia. | " | 0.686 |
| **02630.10 TAPPING SADDLES & SLEEVES** | | |
| Tapping saddle, tap size to 2" | | |
| 4" saddle | EA. | 0.400 |
| 6" saddle | " | 0.500 |
| 8" saddle | " | 0.667 |
| 10" saddle | " | 0.800 |
| 12" saddle | " | 1.143 |
| 14" saddle | " | 1.600 |
| Tapping sleeve | | |
| 4x4 | EA. | 0.533 |
| 6x4 | " | 0.615 |
| 6x6 | " | 0.615 |
| 8x4 | " | 0.800 |
| 8x6 | " | 0.800 |
| 10x4 | " | 0.960 |
| 10x6 | " | 0.960 |
| 10x8 | " | 0.960 |
| 10x10 | " | 1.000 |
| 12x4 | " | 1.000 |
| 12x6 | " | 1.091 |
| 12x8 | " | 1.200 |
| 12x10 | " | 1.333 |
| 12x12 | " | 1.500 |
| Tapping valve, mechanical joint | | |
| 4" valve | EA. | 3.000 |
| 6" valve | " | 4.000 |
| 8" valve | " | 6.000 |
| 10" valve | " | 8.000 |
| 12" valve | " | 12.000 |
| Tap hole in pipe | | |
| 4" hole | EA. | 1.000 |
| 6" hole | " | 1.600 |
| 8" hole | " | 2.667 |
| 10" hole | " | 3.200 |
| 12" hole | " | 4.000 |

| UTILITIES | UNIT | MAN/HOURS |
|---|---|---|
| **02640.15 VALVE BOXES** | | |
| Valve box, adjustable, for valves up to 20" | | |
| 3' deep | EA. | 0.267 |
| 4' deep | " | 0.320 |
| 5' deep | " | 0.400 |
| **02640.19 THRUST BLOCKS** | | |
| Thrust block, 3000# concrete | | |
| 1/4 c.y. | EA. | 1.333 |
| 1/2 c.y. | " | 1.600 |
| 3/4 c.y. | " | 2.667 |
| 1 c.y. | " | 5.333 |
| **02645.10 FIRE HYDRANTS** | | |
| Standard, 3 way post, 6" mechanical joint | | |
| 2' deep | EA. | 8.000 |
| 4' deep | " | 9.600 |
| 6' deep | " | 12.000 |
| 8' deep | " | 13.714 |
| **02665.10 CHILLED WATER SYSTEMS** | | |
| Chilled water pipe, 2" thick insulation, w/casing | | |
| Align and tack weld on sleepers | | |
| 1-1/2" dia. | L.F. | 0.022 |
| 3" dia. | " | 0.034 |
| 4" dia. | " | 0.048 |
| 6" dia. | " | 0.060 |
| 8" dia. | " | 0.069 |
| 10" dia. | " | 0.080 |
| 12" dia. | " | 0.096 |
| 14" dia. | " | 0.104 |
| 16" dia. | " | 0.120 |
| Align and tack weld on trench bottom | | |
| 18" dia. | L.F. | 0.133 |
| 20" dia. | " | 0.150 |
| Preinsulated fittings | | |
| Align and tack weld on sleepers | | |
| Elbows | | |
| 1-1/2" | EA. | 0.500 |
| 3" | " | 0.800 |
| 4" | " | 1.000 |
| 6" | " | 1.333 |
| 8" | " | 1.600 |
| Tees | | |
| 1-1/2" | EA. | 0.533 |
| 3" | " | 0.889 |
| 4" | " | 1.143 |
| 6" | " | 1.600 |
| 8" | " | 2.000 |
| Reducers | | |
| 3" | EA. | 0.667 |
| 4" | " | 0.800 |
| 6" | " | 1.000 |

| UTILITIES | UNIT | MAN/HOURS |
|---|---|---|
| **02665.10 CHILLED WATER SYSTEMS** | | |
| 8" | EA. | 1.333 |
| Anchors, not including concrete | | |
| 4" | EA. | 1.000 |
| 6" | " | 1.000 |
| Align and tack weld on trench bottom | | |
| Elbows | | |
| 10" | EA. | 1.500 |
| 12" | " | 1.714 |
| 14" | " | 1.846 |
| 16" | " | 2.000 |
| 18" | " | 2.182 |
| 20" | " | 2.400 |
| Tees | | |
| 10" | EA. | 1.500 |
| 12" | " | 1.714 |
| 14" | " | 1.846 |
| 16" | " | 2.000 |
| 18" | " | 2.182 |
| 20" | " | 2.400 |
| Reducers | | |
| 10" | EA. | 1.000 |
| 12" | " | 1.091 |
| 14" | " | 1.200 |
| 16" | " | 1.333 |
| 18" | " | 1.500 |
| 20" | " | 1.714 |
| Anchors, not including concrete | | |
| 10" | EA. | 1.000 |
| 12" | " | 1.091 |
| 14" | " | 1.200 |
| 16" | " | 1.333 |
| 18" | " | 1.500 |
| 20" | " | 1.714 |
| **02670.10 WELLS** | | |
| Domestic water, drilled and cased | | |
| 4" dia. | L.F. | 1.600 |
| 6" dia. | " | 1.846 |
| 8" dia. | " | 2.400 |
| **02685.10 GAS DISTRIBUTION** | | |
| Gas distribution lines | | |
| Polyethylene, 60 psi coils | | |
| 1-1/4" dia. | L.F. | 0.053 |
| 1-1/2" dia. | " | 0.057 |
| 2" dia. | " | 0.067 |
| 3" dia. | " | 0.080 |
| 30' pipe lengths | | |
| 3" dia. | L.F. | 0.089 |
| 4" dia. | " | 0.100 |
| 6" dia. | " | 0.133 |
| 8" dia. | " | 0.160 |
| Steel, schedule 40, plain end | | |

| UTILITIES | UNIT | MAN/HOURS |
|---|---|---|
| **02685.10 GAS DISTRIBUTION** | | |
| 1" dia. | L.F. | 0.067 |
| 2" dia. | " | 0.073 |
| 3" dia. | " | 0.080 |
| 4" dia. | " | 0.160 |
| 5" dia. | " | 0.171 |
| 6" dia. | " | 0.200 |
| 8" dia. | " | 0.218 |
| Natural gas meters, direct digital reading, threaded | | |
| 250 cfh @ 5 lbs | EA. | 1.600 |
| 425 cfh @ 10 lbs | " | 1.600 |
| 800 cfh @ 20 lbs | " | 2.000 |
| 1000 cfh @ 25 lbs | " | 2.000 |
| 1,400 cfh @ 100 lbs | " | 2.667 |
| 2,300 cfh @ 100 lbs | " | 4.000 |
| 5,000 cfh @ 100 lbs | " | 8.000 |
| Gas pressure regulators | | |
| Threaded | | |
| 3/4" | EA. | 1.000 |
| 1" | " | 1.333 |
| 1-1/4" | " | 1.333 |
| 1-1/2" | " | 1.333 |
| 2" | " | 1.600 |
| Flanged | | |
| 3" | EA. | 2.000 |
| 4" | " | 2.667 |
| **02690.10 STORAGE TANKS** | | |
| Oil storage tank, underground | | |
| Steel | | |
| 500 gals | EA. | 3.000 |
| 1,000 gals | " | 4.000 |
| 4,000 gals | " | 8.000 |
| 5,000 gals | " | 12.000 |
| 10,000 gals | " | 24.000 |
| Fiberglass, double wall | | |
| 550 gals | EA. | 4.000 |
| 1,000 gals | " | 4.000 |
| 2,000 gals | " | 6.000 |
| 4,000 gals | " | 12.000 |
| 6,000 gals | " | 16.000 |
| 8,000 gals | " | 24.000 |
| 10,000 gals | " | 30.000 |
| 12,000 gals | " | 40.000 |
| 15,000 gals | " | 53.333 |
| 20,000 gals | " | 60.000 |
| Above ground | | |
| Steel | | |
| 275 gals | EA. | 2.400 |
| 500 gals | " | 4.000 |
| 1,000 gals | " | 4.800 |
| 1,500 gals | " | 6.000 |
| 2,000 gals | " | 8.000 |
| 5,000 gals | " | 12.000 |

# 02 SITEWORK

## UTILITIES

| | UNIT | MAN/HOURS |
|---|---|---|
| **02690.10 STORAGE TANKS** | | |
| Fill cap | EA. | 0.800 |
| Vent cap | " | 0.800 |
| Level indicator | " | 0.800 |
| **02695.80 STEAM METERS** | | |
| In-line turbine, direct reading, 300 lb, flanged | | |
| 2" | EA. | 1.000 |
| 3" | " | 1.333 |
| 4" | " | 1.600 |
| Threaded, 2" | | |
| 5" line | EA. | 8.000 |
| 6" line | " | 8.000 |
| 8" line | " | 8.000 |
| 10" line | " | 8.000 |
| 12" line | " | 8.000 |
| 14" line | " | 8.000 |
| 16" line | " | 8.000 |

## SEWERAGE AND DRAINAGE

| | UNIT | MAN/HOURS |
|---|---|---|
| **02720.10 CATCH BASINS** | | |
| Standard concrete catch basin | | |
| Cast in place, 3'8" x 3'8", 6" thick wall | | |
| 2' deep | EA. | 6.000 |
| 3' deep | " | 6.000 |
| 4' deep | " | 8.000 |
| 5' deep | " | 8.000 |
| 6' deep | " | 9.600 |
| 4'x4', 8" thick wall, cast in place | | |
| 2' deep | EA. | 6.000 |
| 3' deep | " | 6.000 |
| 4' deep | " | 8.000 |
| 5' deep | " | 8.000 |
| 6' deep | " | 9.600 |
| Frames and covers, cast iron | | |
| Round | | |
| 24" dia. | EA. | 2.000 |
| 26" dia. | " | 2.000 |
| 28" dia. | " | 2.000 |
| Rectangular | | |
| 23"x23" | EA. | 2.000 |
| 27"x20" | " | 2.000 |
| 24"x24" | " | 2.000 |
| 26"x26" | " | 2.000 |
| Curb inlet frames and covers | | |
| 27"x27" | EA. | 2.000 |

## SEWERAGE AND DRAINAGE

| | UNIT | MAN/HOURS |
|---|---|---|
| **02720.10 CATCH BASINS** | | |
| 24"x36" | EA. | 2.000 |
| 24"x25" | " | 2.000 |
| 24"x22" | " | 2.000 |
| 20"x22" | " | 2.000 |
| Airfield catch basin frame and grating, galvanized | | |
| 2'x4' | EA. | 2.000 |
| 2'x2' | " | 2.000 |
| **02720.40 STORM DRAINAGE** | | |
| Concrete pipe | | |
| Plain, bell and spigot joint, class II | | |
| 6" pipe | L.F. | 0.109 |
| 8" pipe | " | 0.120 |
| 10" pipe | " | 0.126 |
| 12" pipe | " | 0.133 |
| 15" pipe | " | 0.141 |
| 18" pipe | " | 0.150 |
| 21" pipe | " | 0.160 |
| 24" pipe | " | 0.171 |
| Reinforced, class III, tongue and groove joint | | |
| 12" pipe | L.F. | 0.133 |
| 15" pipe | " | 0.141 |
| 18" pipe | " | 0.150 |
| 21" pipe | " | 0.160 |
| 24" pipe | " | 0.171 |
| 27" pipe | " | 0.185 |
| 30" pipe | " | 0.200 |
| 36" pipe | " | 0.218 |
| 42" pipe | " | 0.240 |
| 48" pipe | " | 0.267 |
| 54" pipe | " | 0.300 |
| 60" pipe | " | 0.343 |
| 66" pipe | " | 0.400 |
| 72" pipe | " | 0.480 |
| Flared end-section, concrete | | |
| 12" pipe | L.F. | 0.133 |
| 15" pipe | " | 0.141 |
| 18" pipe | " | 0.150 |
| 24" pipe | " | 0.171 |
| 30" pipe | " | 0.200 |
| 36" pipe | " | 0.218 |
| 42" pipe | " | 0.240 |
| 48" pipe | " | 0.267 |
| 54" pipe | " | 0.300 |
| Corrugated metal pipe, coated, paved invert | | |
| 16 ga. | | |
| 8" pipe | L.F. | 0.080 |
| 10" pipe | " | 0.083 |
| 12" pipe | " | 0.086 |
| 15" pipe | " | 0.092 |
| 18" pipe | " | 0.100 |
| 21" pipe | " | 0.109 |
| 24" pipe | " | 0.120 |

# 02 SITEWORK

| SEWERAGE AND DRAINAGE | UNIT | MAN/ HOURS |
|---|---|---|
| **02720.40 STORM DRAINAGE** | | |
| 30" pipe | L.F. | 0.133 |
| 36" pipe | " | 0.150 |
| 12 ga., 48" pipe | " | 0.171 |
| 10 ga. | | |
| 60" pipe | L.F. | 0.200 |
| 72" pipe | " | 0.240 |
| Galvanized or aluminum, plain | | |
| 16 ga. | | |
| 8" pipe | L.F. | 0.080 |
| 10" pipe | " | 0.083 |
| 12" pipe | " | 0.086 |
| 15" pipe | " | 0.092 |
| 18" pipe | " | 0.100 |
| 24" pipe | " | 0.120 |
| 30" pipe | " | 0.133 |
| 36" pipe | " | 0.150 |
| 12 ga., 48" pipe | " | 0.171 |
| 10 ga., 60" pipe | " | 0.200 |
| Galvanized or aluminum, coated oval arch | | |
| 16 ga. | | |
| 17" x 13" | L.F. | 0.109 |
| 21" x 15" | " | 0.120 |
| 14 ga. | | |
| 28" x 20" | L.F. | 0.133 |
| 35" x 24" | " | 0.171 |
| 12 ga. | | |
| 42" x 29" | L.F. | 0.200 |
| 57" x 38" | " | 0.240 |
| 64" x 43" | " | 0.253 |
| Oval arch culverts, plain | | |
| 16 ga. | | |
| 17" x 13" | L.F. | 0.109 |
| 21" x 15" | " | 0.120 |
| 14 ga. | | |
| 28" x 20" | L.F. | 0.133 |
| 35" x 24" | " | 0.171 |
| 12 ga. | | |
| 57" x 38" | L.F. | 0.200 |
| 64" x 43" | " | 0.240 |
| 71" x 47" | " | 0.253 |
| Nestable corrugated metal pipe | | |
| 16 ga. | | |
| 10" pipe | L.F. | 0.083 |
| 12" pipe | " | 0.086 |
| 15" pipe | " | 0.092 |
| 18" pipe | " | 0.100 |
| 24" pipe | " | 0.120 |
| 30" pipe | " | 0.133 |
| 14 ga., 36" pipe | " | 0.150 |
| Headwalls, cast in place, 30 deg wingwall | | |
| 12" pipe | EA. | 2.000 |
| 15" pipe | " | 2.000 |
| 18" pipe | " | 2.286 |
| 24" pipe | " | 2.286 |

| SEWERAGE AND DRAINAGE | UNIT | MAN/ HOURS |
|---|---|---|
| **02720.40 STORM DRAINAGE** | | |
| 30" pipe | EA. | 2.667 |
| 36" pipe | " | 4.000 |
| 42" pipe | " | 4.000 |
| 48" pipe | " | 5.333 |
| 54" pipe | " | 6.667 |
| 60" pipe | " | 8.000 |
| 4" cleanout for storm drain | | |
| 4" pipe | EA. | 1.000 |
| 6" pipe | " | 1.000 |
| 8" pipe | " | 1.000 |
| Connect new drain line | | |
| To existing manhole | EA. | 2.667 |
| To new manhole | " | 1.600 |
| **02720.70 UNDERDRAIN** | | |
| Drain tile, clay | | |
| 6" pipe | L.F. | 0.053 |
| 8" pipe | " | 0.056 |
| 12" pipe | " | 0.060 |
| Porous concrete, standard strength | | |
| 6" pipe | L.F. | 0.053 |
| 8" pipe | " | 0.056 |
| 12" pipe | " | 0.060 |
| 15" pipe | " | 0.067 |
| 18" pipe | " | 0.080 |
| Corrugated metal pipe, perforated type | | |
| 6" pipe | L.F. | 0.060 |
| 8" pipe | " | 0.063 |
| 10" pipe | " | 0.067 |
| 12" pipe | " | 0.071 |
| 18" pipe | " | 0.075 |
| Perforated clay pipe | | |
| 6" pipe | L.F. | 0.069 |
| 8" pipe | " | 0.071 |
| 12" pipe | " | 0.073 |
| Drain tile, concrete | | |
| 6" pipe | L.F. | 0.053 |
| 8" pipe | " | 0.056 |
| 12" pipe | " | 0.060 |
| Perforated rigid PVC underdrain pipe | | |
| 4" pipe | L.F. | 0.040 |
| 6" pipe | " | 0.048 |
| 8" pipe | " | 0.053 |
| 10" pipe | " | 0.060 |
| 12" pipe | " | 0.069 |
| Underslab drainage, crushed stone | | |
| 3" thick | S.F. | 0.008 |
| 4" thick | " | 0.009 |
| 6" thick | " | 0.010 |
| 8" thick | " | 0.010 |
| Plastic filter fabric for drain lines | " | 0.008 |
| Gravel fill in trench, crushed or bank run, 1/2" to 3/4" | C.Y. | 0.600 |

# 02 SITEWORK

## SEWERAGE AND DRAINAGE

| 02730.10 SANITARY SEWERS | UNIT | MAN/HOURS |
|---|---|---|
| Clay | | |
| 6" pipe | L.F. | 0.080 |
| 8" pipe | " | 0.086 |
| 10" pipe | " | 0.092 |
| 12" pipe | " | 0.100 |
| PVC | | |
| 4" pipe | L.F. | 0.060 |
| 6" pipe | " | 0.063 |
| 8" pipe | " | 0.067 |
| 10" pipe | " | 0.071 |
| 12" pipe | " | 0.075 |
| Cleanout | | |
| 4" pipe | EA. | 1.000 |
| 6" pipe | " | 1.000 |
| 8" pipe | " | 1.000 |
| Connect new sewer line | | |
| To existing manhole | EA. | 2.667 |
| To new manhole | " | 1.600 |

| 02740.10 DRAINAGE FIELDS | UNIT | MAN/HOURS |
|---|---|---|
| Perforated PVC pipe, for drain field | | |
| 4" pipe | L.F. | 0.053 |
| 6" pipe | " | 0.057 |

| 02740.50 SEPTIC TANKS | UNIT | MAN/HOURS |
|---|---|---|
| Septic tank, precast concrete | | |
| 1000 gals | EA. | 4.000 |
| 2000 gals | " | 6.000 |
| 5000 gals | " | 12.000 |
| 25,000 gals | " | 48.000 |
| 40,000 gals | " | 80.000 |
| Leaching pit, precast concrete, 72" diameter | | |
| 3' deep | EA. | 3.000 |
| 6' deep | " | 3.429 |
| 8' deep | " | 4.000 |

| 02760.10 PIPELINE RESTORATION | UNIT | MAN/HOURS |
|---|---|---|
| Relining existing water main | | |
| 6" dia. | L.F. | 0.240 |
| 8" dia. | " | 0.253 |
| 10" dia. | " | 0.267 |
| 12" dia. | " | 0.282 |
| 14" dia. | " | 0.300 |
| 16" dia. | " | 0.320 |
| 18" dia. | " | 0.343 |
| 20" dia. | " | 0.369 |
| 24" dia. | " | 0.400 |
| 36" dia. | " | 0.480 |
| 48" dia. | " | 0.533 |
| 72" dia. | " | 0.600 |
| Replacing in line gate valves | | |
| 6" valve | EA. | 3.200 |
| 8" valve | " | 4.000 |

## SEWERAGE AND DRAINAGE

| 02760.10 PIPELINE RESTORATION | UNIT | MAN/HOURS |
|---|---|---|
| 10" valve | EA. | 4.800 |
| 12" valve | " | 6.000 |
| 16" valve | " | 6.857 |
| 18" valve | " | 8.000 |
| 20" valve | " | 9.600 |
| 24" valve | " | 12.000 |
| 36" valve | " | 16.000 |

## SITE IMPROVEMENTS

| 02830.10 CHAIN LINK FENCE | UNIT | MAN/HOURS |
|---|---|---|
| Chain link fence, 9 ga., galvanized, with posts 10' o.c. | | |
| 4' high | L.F. | 0.057 |
| 5' high | " | 0.073 |
| 6' high | " | 0.100 |
| 7' high | " | 0.123 |
| 8' high | " | 0.160 |
| For barbed wire with hangers, add | | |
| 3 strand | L.F. | 0.040 |
| 6 strand | " | 0.067 |
| Corner or gate post, 3" post | | |
| 4' high | EA. | 0.267 |
| 5' high | " | 0.296 |
| 6' high | " | 0.348 |
| 7' high | " | 0.400 |
| 8' high | " | 0.444 |
| 4" post | | |
| 4' high | EA. | 0.296 |
| 5' high | " | 0.348 |
| 6' high | " | 0.400 |
| 7' high | " | 0.444 |
| 8' high | " | 0.500 |
| Gate with gate posts, galvanized, 3' wide | | |
| 4' high | EA. | 2.000 |
| 5' high | " | 2.667 |
| 6' high | " | 2.667 |
| 7' high | " | 4.000 |
| 8' high | " | 4.000 |
| Fabric, galvanized chain link, 2" mesh, 9 ga. | | |
| 4' high | L.F. | 0.027 |
| 5' high | " | 0.032 |
| 6' high | " | 0.040 |
| 8' high | " | 0.053 |
| Line post, no rail fitting, galvanized, 2-1/2" dia. | | |
| 4' high | EA. | 0.229 |
| 5' high | " | 0.250 |

| SITE IMPROVEMENTS | UNIT | MAN/ HOURS |
|---|---|---|
| **02830.10 CHAIN LINK FENCE** | | |
| 6' high | EA. | 0.267 |
| 7' high | " | 0.320 |
| 8' high | " | 0.400 |
| 1-7/8" H beam | | |
| 4' high | EA. | 0.229 |
| 5' high | " | 0.250 |
| 6' high | " | 0.267 |
| 7' high | " | 0.320 |
| 8' high | " | 0.400 |
| 2-1/4" H beam | | |
| 4' high | EA. | 0.229 |
| 5' high | " | 0.250 |
| 6' high | " | 0.267 |
| 7' high | " | 0.320 |
| 8' high | " | 0.400 |
| Vinyl coated, 9 ga., with posts 10' o.c. | | |
| 4' high | L.F. | 0.057 |
| 5' high | " | 0.073 |
| 6' high | " | 0.100 |
| 7' high | " | 0.123 |
| 8' high | " | 0.160 |
| For barbed wire w/hangers, add | | |
| 3 strand | L.F. | 0.040 |
| 6 Strand | " | 0.067 |
| Corner, or gate post, 4' high | | |
| 3" dia. | EA. | 0.267 |
| 4" dia. | " | 0.267 |
| 6" dia. | " | 0.320 |
| Gate, with posts, 3' wide | | |
| 4' high | EA. | 2.000 |
| 5' high | " | 2.667 |
| 6' high | " | 2.667 |
| 7' high | " | 4.000 |
| 8' high | " | 4.000 |
| Line post, no rail fitting, 2-1/2" dia. | | |
| 4' high | EA. | 0.229 |
| 5' high | " | 0.250 |
| 6' high | " | 0.267 |
| 7' high | " | 0.320 |
| 8' high | " | 0.400 |
| Corner post, no top rail fitting, 4" dia. | | |
| 4' high | EA. | 0.267 |
| 5' high | " | 0.296 |
| 6' high | " | 0.348 |
| 7' high | " | 0.400 |
| 8' high | " | 0.444 |
| Fabric, vinyl, chain link, 2" mesh, 9 ga. | | |
| 4' high | L.F. | 0.027 |
| 5' high | " | 0.032 |
| 6' high | " | 0.040 |
| 8' high | " | 0.053 |
| Swing gates, galvanized, 4' high | | |
| Single gate | | |
| 3' wide | EA. | 2.000 |

| SITE IMPROVEMENTS | UNIT | MAN/ HOURS |
|---|---|---|
| **02830.10 CHAIN LINK FENCE** | | |
| 4' wide | EA. | 2.000 |
| Double gate | | |
| 10' wide | EA. | 3.200 |
| 12' wide | " | 3.200 |
| 14' wide | " | 3.200 |
| 16' wide | " | 3.200 |
| 18' wide | " | 4.571 |
| 20' wide | " | 4.571 |
| 22' wide | " | 4.571 |
| 24' wide | " | 5.333 |
| 26' wide | " | 5.333 |
| 28' wide | " | 6.400 |
| 30' wide | " | 6.400 |
| 5' high | | |
| Single gate | | |
| 3' wide | EA. | 2.667 |
| 4' wide | " | 2.667 |
| Double gate | | |
| 10' wide | EA. | 4.000 |
| 12' wide | " | 4.000 |
| 14' wide | " | 4.000 |
| 16' wide | " | 4.000 |
| 18' wide | " | 4.571 |
| 20' wide | " | 4.571 |
| 22' wide | " | 4.571 |
| 24' wide | " | 5.333 |
| 26' wide | " | 5.333 |
| 28' wide | " | 6.400 |
| 30' wide | " | 6.400 |
| 6' high | | |
| Single gate | | |
| 3' wide | EA. | 2.667 |
| 4' wide | " | 2.667 |
| Double gate | | |
| 10' wide | EA. | 4.000 |
| 12' wide | " | 4.000 |
| 14' wide | " | 4.000 |
| 16' wide | " | 4.000 |
| 18' wide | " | 4.571 |
| 20' wide | " | 4.571 |
| 22' wide | " | 4.571 |
| 24' wide | " | 5.333 |
| 26' wide | " | 5.333 |
| 28' wide | " | 6.400 |
| 30' wide | " | 6.400 |
| 7' high | | |
| Single gate | | |
| 3' wide | EA. | 4.000 |
| 4' wide | " | 4.000 |
| Double gate | | |
| 10' wide | EA. | 5.333 |
| 12' wide | " | 5.333 |
| 14' wide | " | 5.333 |
| 16' wide | " | 5.333 |

| SITE IMPROVEMENTS | UNIT | MAN/ HOURS |
|---|---|---|
| **02830.10 CHAIN LINK FENCE** | | |
| 18' wide | EA. | 6.400 |
| 20' wide | " | 6.400 |
| 22' wide | " | 6.400 |
| 24' wide | " | 8.000 |
| 26' wide | " | 8.000 |
| 28' wide | " | 10.000 |
| 30' wide | " | 10.000 |
| 8' high | | |
| Single gate | | |
| 3' wide | EA. | 4.000 |
| 4' wide | " | 4.000 |
| Double gate | | |
| 10' wide | EA. | 5.333 |
| 12' wide | " | 5.333 |
| 14' wide | " | 5.333 |
| 16' wide | " | 5.333 |
| 18' wide | " | 6.400 |
| 20' wide | " | 6.400 |
| 22' wide | " | 6.400 |
| 24' wide | " | 8.000 |
| 26' wide | " | 8.000 |
| 28' wide | " | 10.000 |
| 30' wide | " | 10.000 |
| Vinyl coated swing gates, 4' high | | |
| Single gate | | |
| 3' wide | EA. | 2.000 |
| 4' wide | " | 2.000 |
| Double gate | | |
| 10' wide | EA. | 3.200 |
| 12' wide | " | 3.200 |
| 14' wide | " | 3.200 |
| 16' wide | " | 3.200 |
| 18' wide | " | 4.571 |
| 20' wide | " | 4.571 |
| 22' wide | " | 4.571 |
| 24' wide | " | 5.333 |
| 26' wide | " | 5.333 |
| 28' wide | " | 6.400 |
| 30' wide | " | 6.400 |
| 5' high | | |
| Single gate | | |
| 3' wide | EA. | 2.667 |
| 4' wide | " | 2.667 |
| Double gate | | |
| 10' wide | EA. | 4.000 |
| 12' wide | " | 4.000 |
| 14' wide | " | 4.000 |
| 16' wide | " | 4.000 |
| 18' wide | " | 4.571 |
| 20' wide | " | 4.571 |
| 22' wide | " | 4.571 |
| 24' wide | " | 5.333 |
| 26' wide | " | 5.333 |
| 28' wide | " | 6.400 |

| SITE IMPROVEMENTS | UNIT | MAN/ HOURS |
|---|---|---|
| **02830.10 CHAIN LINK FENCE** | | |
| 30' wide | EA. | 6.400 |
| 6' high | | |
| Single gate | | |
| 3' wide | EA. | 2.667 |
| 4' wide | " | 2.667 |
| Double gate | | |
| 10' wide | EA. | 4.000 |
| 12' wide | " | 4.000 |
| 14' wide | " | 4.000 |
| 16' wide | " | 4.000 |
| 18' wide | " | 4.571 |
| 20' wide | " | 4.571 |
| 22' wide | " | 4.571 |
| 24' wide | " | 5.333 |
| 26' wide | " | 5.333 |
| 28' wide | " | 6.400 |
| 30' wide | " | 6.400 |
| 7' high | | |
| Single gate | | |
| 3' wide | EA. | 4.000 |
| 4' wide | " | 4.000 |
| Double gate | | |
| 10' wide | EA. | 5.333 |
| 12' wide | " | 5.333 |
| 14' wide | " | 5.333 |
| 16' wide | " | 5.333 |
| 18' wide | " | 6.400 |
| 20' wide | " | 6.400 |
| 22' wide | " | 6.400 |
| 24' wide | " | 8.000 |
| 26' wide | " | 8.000 |
| 28' wide | " | 10.000 |
| 30' wide | " | 10.000 |
| 8' high | | |
| Single gate | | |
| 3' wide | EA. | 4.000 |
| 4' wide | " | 4.000 |
| Double gate | | |
| 10' wide | EA. | 5.333 |
| 12' wide | " | 5.333 |
| 14' wide | " | 5.333 |
| 16' wide | " | 5.333 |
| 18' wide | " | 6.400 |
| 20' wide | " | 6.400 |
| 22' wide | " | 6.400 |
| 24' wide | " | 8.000 |
| 28' wide | " | 8.000 |
| 30' wide | " | 10.000 |
| Motor operator for gates, no wiring | " | |
| Drilling fence post holes | | |
| In soil | | |
| By hand | EA. | 0.400 |
| By machine auger | " | 0.200 |
| In rock | | |

# 02 SITEWORK

| SITE IMPROVEMENTS | UNIT | MAN/HOURS |
|---|---|---|
| **02830.10 CHAIN LINK FENCE** | | |
| By jackhammer | EA. | 2.667 |
| By rock drill | " | 0.800 |
| Aluminum privacy slats, installed vertically | S.F. | 0.020 |
| Post hole, dig by hand | EA. | 0.533 |
| Set fence post in concrete | " | 0.400 |
| **02830.70 RECREATIONAL COURTS** | | |
| Walls, galvanized steel | | |
| 8' high | L.F. | 0.160 |
| 10' high | " | 0.178 |
| 12' high | " | 0.211 |
| Vinyl coated | | |
| 8' high | L.F. | 0.160 |
| 10' high | " | 0.178 |
| 12' high | " | 0.211 |
| Gates, galvanized steel | | |
| Single, 3' transom | | |
| 3'x7' | EA. | 4.000 |
| 4'x7' | " | 4.571 |
| 5'x7' | " | 5.333 |
| 6'x7' | " | 6.400 |
| Double, 3' transom | | |
| 10'x7' | EA. | 16.000 |
| 12'x7' | " | 17.778 |
| 14'x7' | " | 20.000 |
| Double, no transom | | |
| 10'x10' | EA. | 13.333 |
| 12'x10' | " | 16.000 |
| 14'x10' | " | 17.778 |
| Vinyl coated | | |
| Single, 3' transom | | |
| 3'x7' | EA. | 4.000 |
| 4'x7' | " | 4.571 |
| 5'x7' | " | 5.333 |
| 6'x7' | " | 6.400 |
| Double, 3' | | |
| 10'x7' | EA. | 16.000 |
| 12'x7' | " | 17.778 |
| 14'x7' | " | 20.000 |
| Double, no transom | | |
| 10'x10' | EA. | 13.333 |
| 12'x10' | " | 16.000 |
| 14'x10' | " | 17.778 |
| Wire and miscellaneous metal fences | | |
| Chicken wire, post 4' o.c. | | |
| 2" mesh | | |
| 4' high | L.F. | 0.040 |
| 6' high | " | 0.053 |
| Galvanized steel | | |
| 12 gauge, 2" by 4" mesh, posts 5' o.c. | | |
| 3' high | L.F. | 0.040 |
| 5' high | " | 0.050 |
| 14 gauge, 1" by 2" mesh, posts 5' o.c. | | |

| SITE IMPROVEMENTS | UNIT | MAN/HOURS |
|---|---|---|
| **02830.70 RECREATIONAL COURTS** | | |
| 3' high | L.F. | 0.040 |
| 5' high | " | 0.050 |
| **02840.30 GUARDRAILS** | | |
| Pipe bollard, steel pipe, concrete filled, painted | | |
| 6" dia. | EA. | 0.667 |
| 8" dia. | " | 1.000 |
| 12" dia. | " | 2.667 |
| Corrugated steel, guardrail, galvanized | L.F. | 0.040 |
| End section, wrap around or flared | EA. | 0.800 |
| Timber guardrail, 4" x 8" | L.F. | 0.030 |
| Guard rail, 3 cables, 3/4" dia. | | |
| Steel posts | L.F. | 0.120 |
| Wood posts | " | 0.096 |
| Steel box beam | | |
| 6" x 6" | L.F. | 0.133 |
| 6" x 8" | " | |
| Concrete posts | EA. | 0.400 |
| Barrel type impact barrier | " | 0.800 |
| Light shield, 6' high | L.F. | 0.160 |
| **02840.40 PARKING BARRIERS** | | |
| Timber, treated, 8' long | | |
| 4" x 4" | EA. | 0.667 |
| 6" x 6" | " | 0.800 |
| Precast concrete, 4' long, with dowels | | |
| 12" x 6" | EA. | 0.400 |
| 12" x 8" | " | 0.444 |
| **02840.60 SIGNAGE** | | |
| Traffic signs | | |
| Reflectorized signs per OSHA standards, including post | | |
| Stop, 24"x24" | EA. | 0.533 |
| Yield, 30" triangle | " | 0.533 |
| Speed limit, 12"x18" | " | 0.533 |
| Directional, 12"x18" | " | 0.533 |
| Exit, 12"x18" | " | 0.533 |
| Entry, 12"x18" | " | 0.533 |
| Warning, 24"x24" | " | 0.533 |
| Informational, 12"x18" | " | 0.533 |
| Handicap parking, 12"x18" | " | 0.533 |
| **02860.40 RECREATIONAL FACILITIES** | | |
| Bleachers, outdoor, portable, per seat | | |
| 10 tiers | | |
| Minimum | EA. | 0.150 |
| Maximum | " | 0.200 |
| 20 tiers | | |
| Minimum | EA. | 0.141 |
| Maximum | " | 0.185 |
| Grandstands, fixed, wood seat, steel frame, per seat | | |
| 15 tiers | | |
| Minimum | EA. | 0.240 |

# 02 SITEWORK

| SITE IMPROVEMENTS | UNIT | MAN/ HOURS |
|---|---|---|
| **02860.40 RECREATIONAL FACILITIES** | | |
| Maximum | EA. | 0.400 |
| 30 tiers | | |
| Minimum | EA. | 0.218 |
| Maximum | " | 0.343 |
| Seats | | |
| Seat backs only | | |
| Fiberglass | EA. | 0.080 |
| Steel and wood seat | " | 0.080 |
| Seat restoration, fiberglass on wood | | |
| Seats | EA. | 0.160 |
| Plain bench, no backs | " | 0.067 |
| Benches | | |
| Park, precast concrete with backs | | |
| 4' long | EA. | 2.667 |
| 8' long | " | 4.000 |
| Fiberglass, with backs | | |
| 4' long | EA. | 2.000 |
| 8' long | " | 2.667 |
| Wood, with backs and fiberglass supports | | |
| 4' long | EA. | 2.000 |
| 8' long | " | 2.667 |
| Steel frame, 6' long | | |
| All steel | EA. | 2.000 |
| Hardwood boards | " | 2.000 |
| Players bench (no back), steel frame, fir seat, 10' long | " | 2.667 |
| Soccer goal posts | PAIR | |
| Running track | | |
| Gravel and cinders over stone base | S.Y. | 0.060 |
| Rubber-cork base resilient pavement | " | 0.480 |
| For colored surfaces, add | " | 0.048 |
| Colored rubberized asphalt | " | 0.600 |
| Artificial resilient mat over asphalt | " | 1.200 |
| Tennis courts | | |
| Bituminous pavement, 2-1/2" thick | S.Y. | 0.150 |
| Colored sealer, acrylic emulsion | | |
| 3 coats | S.Y. | 0.053 |
| For 2 color seal coating, add | " | 0.008 |
| For preparing old courts, add | " | 0.005 |
| Net, nylon, 42' long | EA. | 1.000 |
| Paint markings on asphalt, 2 coats | " | 8.000 |
| Playground equipment | | |
| Basketball backboard | | |
| Minimum | EA. | 2.000 |
| Maximum | " | 2.286 |
| Bike rack, 10' long | " | 1.600 |
| Golf shelter, fiberglass | " | 2.000 |
| Ground socket for movable posts | | |
| Minimum | EA. | 0.500 |
| Maximum | " | 0.500 |
| Horizontal monkey ladder, 14' long | " | 1.333 |
| Posts, tether ball | " | 0.400 |
| Multiple purpose, 10' long | " | 0.800 |
| See-saw, steel | | |
| Minimum | EA. | 3.200 |

| SITE IMPROVEMENTS | UNIT | MAN/ HOURS |
|---|---|---|
| **02860.40 RECREATIONAL FACILITIES** | | |
| Average | EA. | 4.000 |
| Maximum | " | 5.333 |
| Slide | | |
| Minimum | EA. | 6.400 |
| Maximum | " | 7.273 |
| Swings, plain seats | | |
| 8' high | | |
| Minimum | EA. | 5.333 |
| Maximum | " | 6.154 |
| 12' high | | |
| Minimum | EA. | 6.154 |
| Maximum | " | 8.889 |
| **02870.10 PREFABRICATED PLANTERS** | | |
| Concrete precast, circular | | |
| 24" dia., 18" high | EA. | 0.800 |
| 42" dia., 30" high | " | 1.000 |
| Fiberglass, circular | | |
| 36" dia., 27" high | EA. | 0.400 |
| 60" dia., 39" high | " | 0.444 |
| Tapered, circular | | |
| 24" dia., 36" high | EA. | 0.364 |
| 40" dia., 36" high | " | 0.400 |
| Square | | |
| 2' by 2', 17" high | EA. | 0.364 |
| 4' by 4', 39" high | " | 0.444 |
| Rectangular | | |
| 4' by 1', 18" high | EA. | 0.400 |

| LANDSCAPING | UNIT | MAN/ HOURS |
|---|---|---|
| **02910.10 SHRUB & TREE MAINTENANCE** | | |
| Moving shrubs on site | | |
| 12" ball | EA. | 1.000 |
| 24" ball | " | 1.333 |
| 3' high | " | 0.800 |
| 4' high | " | 0.889 |
| 5' high | " | 1.000 |
| 18" spread | " | 1.143 |
| 30" spread | " | 1.333 |
| Moving trees on site | | |
| 24" ball | EA. | 1.200 |
| 48" ball | " | 1.600 |
| Trees | | |
| 3' high | EA. | 0.480 |
| 6' high | " | 0.533 |
| 8' high | " | 0.600 |

| LANDSCAPING | UNIT | MAN/ HOURS |
|---|---|---|
| **02910.10 SHRUB & TREE MAINTENANCE** | | |
| 10' high | EA. | 0.800 |
| Palm trees | | |
| 7' high | EA. | 0.600 |
| 10' high | " | 0.800 |
| 20' high | " | 2.400 |
| 40' high | " | 4.800 |
| Guying trees | | |
| 4" dia. | EA. | 0.400 |
| 8" dia. | " | 0.500 |
| **02920.10 TOPSOIL** | | |
| Spread topsoil, with equipment | | |
| Minimum | C.Y. | 0.080 |
| Maximum | " | 0.100 |
| By hand | | |
| Minimum | C.Y. | 0.800 |
| Maximum | " | 1.000 |
| Area preparation for seeding (grade, rake and clean) | | |
| Square yard | S.Y. | 0.006 |
| By acre | ACRE | 32.000 |
| Remove topsoil and stockpile on site | | |
| 4" deep | C.Y. | 0.067 |
| 6" deep | " | 0.062 |
| Spreading topsoil from stock pile | | |
| By loader | C.Y. | 0.073 |
| By hand | " | 0.800 |
| Top dress by hand | S.Y. | 0.008 |
| Place imported top soil | | |
| By loader | | |
| 4" deep | S.Y. | 0.008 |
| 6" deep | " | 0.009 |
| By hand | | |
| 4" deep | S.Y. | 0.089 |
| 6" deep | " | 0.100 |
| Plant bed preparation, 18" deep | | |
| With backhoe/loader | S.Y. | 0.020 |
| By hand | " | 0.133 |
| **02930.30 SEEDING** | | |
| Mechanical seeding, 175 lb/acre | | |
| By square yard | S.Y. | 0.002 |
| By acre | ACRE | 8.000 |
| 450 lb/acre | | |
| By square yard | S.Y. | 0.002 |
| By acre | ACRE | 10.000 |
| Seeding by hand, 10 lb per 100 s.y. | | |
| By square yard | S.Y. | 0.003 |
| By acre | ACRE | 13.333 |
| Reseed disturbed areas | S.F. | 0.004 |
| **02950.10 PLANTS** | | |
| Euonymus coloratus, 18" (purple wintercreeper) | EA. | 0.133 |
| Hedera Helix, 2-1/4" pot (English ivy) | " | 0.133 |

| LANDSCAPING | UNIT | MAN/ HOURS |
|---|---|---|
| **02950.10 PLANTS** | | |
| Liriope muscari, 2" clumps | EA. | 0.080 |
| Santolina, 12" | " | 0.080 |
| Vinca major or minor, 3" pot | " | 0.080 |
| Cortaderia argentia, 2 gallon (pampas grass) | " | 0.080 |
| Ophiopogan japonicus, 1 quart (4" pot) | " | 0.080 |
| Ajuga reptans, 2-3/4" pot (carpet bugle) | " | 0.080 |
| Pachysandra terminalis, 2-3/4" pot (Japanese spurge) | " | 0.080 |
| **02950.30 SHRUBS** | | |
| Juniperus conferia litoralis, 18"-24" (Shore Juniper) | EA. | 0.320 |
| Horizontalis plumosa, 18"-24" (Andorra Juniper) | " | 0.320 |
| Sabina tamar-iscfolia-tamarix juniper, 18"-24" | " | 0.320 |
| Chin San Hose, 18"-24" (San Hose Juniper) | " | 0.320 |
| Sargenti, 18"-24" (Sargent's Juniper) | " | 0.320 |
| Nandina domestica, 18"-24" (Heavenly Bamboo) | " | 0.320 |
| Raphiolepis Indica Springtime, 18"-24" Indian Hawthorn | " | 0.320 |
| Osmanthus Heterophyllus Gulftide, 18"-24" (Osmanthus) | " | 0.320 |
| Ilex Cornuta Burfordi Nana, 18"-24" (Dwarf Burford Holly) | " | 0.320 |
| Glabra, 18"-24" (Inkberry Holly) | " | 0.320 |
| Azalea, Indica types, 18"-24" | " | 0.320 |
| Kurume types, 18"-24" | " | 0.320 |
| Berberis Julianae, 18"-24" (Wintergreen Barberry) | " | 0.320 |
| Pieris Japonica Japanese, 18"-24" (Japanese Pieris) | " | 0.320 |
| Ilex Cornuta Rotunda, 18"-24" (Dwarf Chinese Holly) | " | 0.320 |
| Juniperus Horizontalis Plumosa, 24"-30" (Andorra Juniper) | " | 0.400 |
| Rhodopendrow Hybrids, 24"-30" | " | 0.400 |
| Aucuba Japonica Varigata, 24"-30" (Gold Dust Aucuba) | " | 0.400 |
| Ilex Crenata Willow Leaf, 24"-30" (Japanese Holly) | " | 0.400 |
| Cleyera Japonica, 30"-36" (Japanese Cleyera) | " | 0.500 |
| Pittosporum Tobira, 30"-36" | " | 0.500 |
| Prumus Laurocerasus, 30"-36" | " | 0.500 |
| Ilex Cornuta Burfordi, 30"-36" (Burford Holly) | " | 0.500 |
| Abelia Grandiflora, 24"-36" (Yew Podocarpus) | " | 0.400 |
| Podocarpos Macrophylla, 24"-36" (Yew Podocarpus) | " | 0.400 |
| Pyracantha Coccinea Lalandi, 3'-4' (Firethorn) | " | 0.500 |
| Photinia Frazieri, 3'-4' (Red Photinia) | " | 0.500 |
| Forsythia Suspensa, 3'-4' (Weeping Forsythia) | " | 0.500 |
| Camellia Japonica, 3'-4' (Common Camellia) | " | 0.500 |
| Juniperus Chin Torulosa, 3'-4' (Hollywood Juniper) | " | 0.500 |
| Cupressocyparis Leylandi, 3'-4' | " | 0.500 |
| Ilex Opaca Fosteri, 5'-6' (Foster's Holly) | " | 0.667 |
| Opaca, 5'-6' (American Holly) | " | 0.667 |
| Nyrica Cerifera, 4'-5' (Southern Wax Myrtles) | " | 0.571 |
| Ligustrum Japonicum, 4'-5' (Japanese Privet) | " | 0.571 |
| **02950.60 TREES** | | |
| Cornus Florida, 5'-6' (White flowering Dogwood) | EA. | 0.667 |
| Prunus Serrulata Kwanzan, 6'-8' (Kwanzan Cherry) | " | 0.800 |
| Caroliniana, 6'-8' (Carolina Cherry Laurel) | " | 0.800 |
| Cercis Canadensis, 6'-8' (Eastern Redbud) | " | 0.800 |
| Koelreuteria Paniculata, 8'-10' (Goldenrain tree) | " | 1.000 |
| Acer Platanoides, 1-3/4"-2" (11'-13') (Norway Maple) | " | 1.333 |

| LANDSCAPING | UNIT | MAN/ HOURS |
|---|---|---|
| **02950.60 TREES** | | |
| Rubrum, 1-3/4"-2" (11'-13') (Red Maple) | EA. | 1.333 |
| Saccharum, 1-3/4"-2" (Sugar Maple) | " | 1.333 |
| Fraxinus Pennsylvanica, 1-3/4"-2" Laneolata-Green Ash | " | 1.333 |
| Celtis Occidentalis, 1-3/4"-2" (American Hackberry) | " | 1.333 |
| Glenditsia Triacantos Inermis, 2" | " | 1.333 |
| Prunus Cerasifera 'Thundercloud', 6'-8' | " | 0.800 |
| Yeodensis, 6'-8' (Yoshino Cherry) | " | 0.800 |
| Lagerstroemia Indica, 8'-10' (Crapemyrtle) | " | 1.000 |
| Crataegus Phaenopyrum, 8'-10' Washington Hawthorn | " | 1.000 |
| Quercus Borealis, 1-3/4"-2" (Northern Red Oak) | " | 1.333 |
| Quercus Acutissima, 1-3/4"-2" (8'-10') (Sawtooth Oak) | " | 1.333 |
| Saliz Babylonica, 1-3/4"-2" (Weeping Willow) | " | 1.333 |
| Tilia Cordata Greenspire, 1-3/4"-2" (10'-12') | " | 1.333 |
| Malus, 2"-2-1/2" (8'-10') (Flowering Crabapple) | " | 1.333 |
| Platanus Occidentalis, (12'-14') | " | 1.600 |
| Pyrus Calleryana Bradford, 2"-2-1/2" (Bradford Pear) | " | 1.333 |
| Quercus Palustris, 2"-2-1/2" (12'-14') (Pin Oak) | " | 1.333 |
| Phellos, 2-1/2"-3" (Willow Oak) | " | 1.600 |
| Nigra, 2"-2-1/2" (Water Oak) | " | 1.333 |
| Magnolia Soulangeana, 4'-5' (Saucer Magnolia) | " | 0.667 |
| Grandiflora, 6'-8' (Southern Magnolia) | " | 0.800 |
| Cedrus Deodara, 10'-12' (Deodare Cedar) | " | 1.333 |
| Ginkgo Biloba, 10'-12' (2"-2-1/2") (Maidenhair Tree) | " | 1.333 |
| Pinus Thunbergi, 5'-6' (Japanese Black Pine) | " | 0.667 |
| Strobus, 6'-8' (White Pine) | " | 0.800 |
| Taeda, 6'-8' (Loblolly Pine) | " | 0.800 |
| Quercus Virginiana, 2"-2-1/2" (live oak) | " | 1.600 |
| **02970.10 FERTILIZING** | | |
| Fertilizing (23#/1000 sf) | | |
| By square yard | S.Y. | 0.002 |
| By acre | ACRE | 10.000 |
| Liming (70#/1000 sf) | | |
| By square yard | S.Y. | 0.003 |
| By acre | ACRE | 13.333 |
| **02980.10 LANDSCAPE ACCESSORIES** | | |
| Steel edging, 3/16" x 4" | L.F. | 0.010 |
| Landscaping stepping stones, 15"x15", white | EA. | 0.040 |
| Wood chip mulch | C.Y. | 0.533 |
| 2" thick | S.Y. | 0.016 |
| 4" thick | " | 0.023 |
| 6" thick | " | 0.029 |
| Gravel mulch, 3/4" stone | C.Y. | 0.800 |
| White marble chips, 1" deep | S.F. | 0.008 |
| Peat moss | | |
| 2" thick | S.Y. | 0.018 |
| 4" thick | " | 0.027 |
| 6" thick | " | 0.033 |
| Landscaping timbers, treated lumber | | |
| 4" x 4" | L.F. | 0.027 |
| 6" x 6" | " | 0.029 |
| 8" x 8" | " | 0.033 |

| FORMWORK | UNIT | MAN/HOURS |
|---|---|---|
| **03110.05 BEAM FORMWORK** | | |
| Beam forms, job built | | |
| Beam bottoms | | |
| 1 use | S.F. | 0.133 |
| 2 uses | " | 0.127 |
| 3 uses | " | 0.123 |
| 4 uses | " | 0.118 |
| 5 uses | " | 0.114 |
| Beam sides | | |
| 1 use | S.F. | 0.089 |
| 2 uses | " | 0.084 |
| 3 uses | " | 0.080 |
| 4 uses | " | 0.076 |
| 5 uses | " | 0.073 |
| **03110.10 BOX CULVERT FORMWORK** | | |
| Box culverts, job built | | |
| 6' x 6' | | |
| 1 use | S.F. | 0.080 |
| 2 uses | " | 0.076 |
| 3 uses | " | 0.073 |
| 4 uses | " | 0.070 |
| 5 uses | " | 0.067 |
| 8' x 12' | | |
| 1 use | S.F. | 0.067 |
| 2 uses | " | 0.064 |
| 3 uses | " | 0.062 |
| 4 uses | " | 0.059 |
| 5 uses | " | 0.057 |
| **03110.15 COLUMN FORMWORK** | | |
| Column, square forms, job built | | |
| 8" x 8" columns | | |
| 1 use | S.F. | 0.160 |
| 2 uses | " | 0.154 |
| 3 uses | " | 0.148 |
| 4 uses | " | 0.143 |
| 5 uses | " | 0.138 |
| 12" x 12" columns | | |
| 1 use | S.F. | 0.145 |
| 2 uses | " | 0.140 |
| 3 uses | " | 0.136 |
| 4 uses | " | 0.131 |
| 5 uses | " | 0.127 |
| 16" x 16" columns | | |
| 1 use | S.F. | 0.133 |
| 2 uses | " | 0.129 |
| 3 uses | " | 0.125 |
| 4 uses | " | 0.121 |
| 5 uses | " | 0.118 |
| 24" x 24" columns | | |
| 1 use | S.F. | 0.123 |
| 2 uses | " | 0.119 |

| FORMWORK | UNIT | MAN/HOURS |
|---|---|---|
| **03110.15 COLUMN FORMWORK** | | |
| 3 uses | S.F. | 0.116 |
| 4 uses | " | 0.113 |
| 5 uses | " | 0.110 |
| 36" x 36" columns | | |
| 1 use | S.F. | 0.114 |
| 2 uses | " | 0.111 |
| 3 uses | " | 0.108 |
| 4 uses | " | 0.105 |
| 5 uses | " | 0.103 |
| Round fiber forms, 1 use | | |
| 10" dia. | L.F. | 0.160 |
| 12" dia. | " | 0.163 |
| 14" dia. | " | 0.170 |
| 16" dia. | " | 0.178 |
| 18" dia. | " | 0.190 |
| 24" dia. | " | 0.205 |
| 30" dia. | " | 0.222 |
| 36" dia. | " | 0.242 |
| 42" dia. | " | 0.267 |
| **03110.18 CURB FORMWORK** | | |
| Curb forms | | |
| Straight, 6" high | | |
| 1 use | L.F. | 0.080 |
| 2 uses | " | 0.076 |
| 3 uses | " | 0.073 |
| 4 uses | " | 0.070 |
| 5 uses | " | 0.067 |
| Curved, 6" high | | |
| 1 use | L.F. | 0.100 |
| 2 uses | " | 0.094 |
| 3 uses | " | 0.089 |
| 4 uses | " | 0.085 |
| 5 uses | " | 0.082 |
| **03110.20 ELEVATED SLAB FORMWORK** | | |
| Elevated slab formwork | | |
| Slab, with drop panels | | |
| 1 use | S.F. | 0.064 |
| 2 uses | " | 0.062 |
| 3 uses | " | 0.059 |
| 4 uses | " | 0.057 |
| 5 uses | " | 0.055 |
| Floor slab, hung from steel beams | | |
| 1 use | S.F. | 0.062 |
| 2 uses | " | 0.059 |
| 3 uses | " | 0.057 |
| 4 uses | " | 0.055 |
| 5 uses | " | 0.053 |
| Floor slab, with pans or domes | | |
| 1 use | S.F. | 0.073 |
| 2 uses | " | 0.070 |

# 03 CONCRETE

| FORMWORK | UNIT | MAN/ HOURS |
|---|---|---|
| **03110.20 ELEVATED SLAB FORMWORK** | | |
| 3 uses | S.F. | 0.067 |
| 4 uses | " | 0.064 |
| 5 uses | " | 0.062 |
| Equipment curbs, 12" high | | |
| 1 use | L.F. | 0.080 |
| 2 uses | " | 0.076 |
| 3 uses | " | 0.073 |
| 4 uses | " | 0.070 |
| 5 uses | " | 0.067 |
| **03110.25 EQUIPMENT PAD FORMWORK** | | |
| Equipment pad, job built | | |
| 1 use | S.F. | 0.100 |
| 2 uses | " | 0.094 |
| 3 uses | " | 0.089 |
| 4 uses | " | 0.084 |
| 5 uses | " | 0.080 |
| **03110.35 FOOTING FORMWORK** | | |
| Wall footings, job built, continuous | | |
| 1 use | S.F. | 0.080 |
| 2 uses | " | 0.076 |
| 3 uses | " | 0.073 |
| 4 uses | " | 0.070 |
| 5 uses | " | 0.067 |
| Column footings, spread | | |
| 1 use | S.F. | 0.100 |
| 2 uses | " | 0.094 |
| 3 uses | " | 0.089 |
| 4 uses | " | 0.084 |
| 5 uses | " | 0.080 |
| **03110.50 GRADE BEAM FORMWORK** | | |
| Grade beams, job built | | |
| 1 use | S.F. | 0.080 |
| 2 uses | " | 0.076 |
| 3 uses | " | 0.073 |
| 4 uses | " | 0.070 |
| 5 uses | " | 0.067 |
| **03110.53 PILE CAP FORMWORK** | | |
| Pile cap forms, job built | | |
| Square | | |
| 1 use | S.F. | 0.100 |
| 2 uses | " | 0.094 |
| 3 uses | " | 0.089 |
| 4 uses | " | 0.084 |
| 5 uses | " | 0.080 |
| Triangular | | |
| 1 use | S.F. | 0.114 |
| 2 uses | " | 0.107 |
| 3 uses | " | 0.100 |
| 4 uses | " | 0.094 |

| FORMWORK | UNIT | MAN/ HOURS |
|---|---|---|
| **03110.53 PILE CAP FORMWORK** | | |
| 5 uses | S.F. | 0.089 |
| **03110.55 SLAB/MAT FORMWORK** | | |
| Mat foundations, job built | | |
| 1 use | S.F. | 0.100 |
| 2 uses | " | 0.094 |
| 3 uses | " | 0.089 |
| 4 uses | " | 0.084 |
| 5 uses | " | 0.080 |
| Edge forms | | |
| 6" high | | |
| 1 use | L.F. | 0.073 |
| 2 uses | " | 0.070 |
| 3 uses | " | 0.067 |
| 4 uses | " | 0.064 |
| 5 uses | " | 0.062 |
| 12" high | | |
| 1 use | L.F. | 0.080 |
| 2 uses | " | 0.076 |
| 3 uses | " | 0.073 |
| 4 uses | " | 0.070 |
| 5 uses | " | 0.067 |
| Formwork for openings | | |
| 1 use | S.F. | 0.160 |
| 2 uses | " | 0.145 |
| 3 uses | " | 0.133 |
| 4 uses | " | 0.123 |
| 5 uses | " | 0.114 |
| **03110.60 STAIR FORMWORK** | | |
| Stairway forms, job built | | |
| 1 use | S.F. | 0.160 |
| 2 uses | " | 0.145 |
| 3 uses | " | 0.133 |
| 4 uses | " | 0.123 |
| 5 uses | " | 0.114 |
| Stairs, elevated | | |
| 1 use | S.F. | 0.160 |
| 2 uses | " | 0.133 |
| 3 uses | " | 0.114 |
| 4 uses | " | 0.107 |
| 5 uses | " | 0.100 |
| **03110.65 WALL FORMWORK** | | |
| Wall forms, exterior, job built | | |
| Up to 8' high wall | | |
| 1 use | S.F. | 0.080 |
| 2 uses | " | 0.076 |
| 3 uses | " | 0.073 |
| 4 uses | " | 0.070 |
| 5 uses | " | 0.067 |

# 03 CONCRETE

| FORMWORK | UNIT | MAN/ HOURS |
|---|---|---|
| **03110.65 WALL FORMWORK** | | |
| Over 8' high wall | | |
| 1 use | S.F. | 0.100 |
| 2 uses | " | 0.094 |
| 3 uses | " | 0.089 |
| 4 uses | " | 0.084 |
| 5 uses | " | 0.080 |
| Over 16' high wall | | |
| 1 use | S.F. | 0.114 |
| 2 uses | " | 0.107 |
| 3 uses | " | 0.100 |
| 4 uses | " | 0.094 |
| 5 uses | " | 0.089 |
| Radial wall forms | | |
| 1 use | S.F. | 0.123 |
| 2 uses | " | 0.114 |
| 3 uses | " | 0.107 |
| 4 uses | " | 0.100 |
| 5 uses | " | 0.094 |
| Retaining wall forms | | |
| 1 use | S.F. | 0.089 |
| 2 uses | " | 0.084 |
| 3 uses | " | 0.080 |
| 4 uses | " | 0.076 |
| 5 uses | " | 0.073 |
| Radial retaining wall forms | | |
| 1 use | S.F. | 0.133 |
| 2 uses | " | 0.123 |
| 3 uses | " | 0.114 |
| 4 uses | " | 0.107 |
| 5 uses | " | 0.100 |
| Column pier and pilaster | | |
| 1 use | S.F. | 0.160 |
| 2 uses | " | 0.145 |
| 3 uses | " | 0.133 |
| 4 uses | " | 0.123 |
| 5 uses | " | 0.114 |
| Interior wall forms | | |
| Up to 8' high | | |
| 1 use | S.F. | 0.073 |
| 2 uses | " | 0.070 |
| 3 uses | " | 0.067 |
| 4 uses | " | 0.064 |
| 5 uses | " | 0.062 |
| Over 8' high | | |
| 1 use | S.F. | 0.089 |
| 2 uses | " | 0.084 |
| 3 uses | " | 0.080 |
| 4 uses | " | 0.076 |
| 5 uses | " | 0.073 |
| Over 16' high | | |
| 1 use | S.F. | 0.100 |
| 2 uses | " | 0.094 |
| 3 uses | " | 0.089 |
| 4 uses | " | 0.084 |

| FORMWORK | UNIT | MAN/ HOURS |
|---|---|---|
| **03110.65 WALL FORMWORK** | | |
| 5 uses | S.F. | 0.080 |
| Radial wall forms | | |
| 1 use | S.F. | 0.107 |
| 2 uses | " | 0.100 |
| 3 uses | " | 0.094 |
| 4 uses | " | 0.089 |
| 5 uses | " | 0.084 |
| Curved wall forms, 24" sections | | |
| 1 use | S.F. | 0.160 |
| 2 uses | " | 0.145 |
| 3 uses | " | 0.133 |
| 4 uses | " | 0.123 |
| 5 uses | " | 0.114 |
| PVC form liner, per side, smooth finish | | |
| 1 use | S.F. | 0.067 |
| 2 uses | " | 0.064 |
| 3 uses | " | 0.062 |
| 4 uses | " | 0.057 |
| 5 uses | " | 0.053 |
| **03110.90 MISCELLANEOUS FORMWORK** | | |
| Keyway forms (5 uses) | | |
| 2 x 4 | L.F. | 0.040 |
| 2 x 6 | " | 0.044 |
| Bulkheads | | |
| Walls, with keyways | | |
| 2 piece | L.F. | 0.073 |
| 3 piece | " | 0.080 |
| Elevated slab, with keyway | | |
| 2 piece | L.F. | 0.067 |
| 3 piece | " | 0.073 |
| Ground slab, with keyway | | |
| 2 piece | L.F. | 0.057 |
| 3 piece | " | 0.062 |
| Chamfer strips | | |
| Wood | | |
| 1/2" wide | L.F. | 0.018 |
| 3/4" wide | " | 0.018 |
| 1" wide | " | 0.018 |
| PVC | | |
| 1/2" wide | L.F. | 0.018 |
| 3/4" wide | " | 0.018 |
| 1" wide | " | 0.018 |
| Radius | | |
| 1" | L.F. | 0.019 |
| 1-1/2" | " | 0.019 |
| Reglets | | |
| Galvanized steel, 24 ga. | L.F. | 0.032 |
| Metal formwork | | |
| Straight edge forms | | |
| 4" high | L.F. | 0.050 |
| 6" high | " | 0.053 |
| 8" high | " | 0.057 |

# 03 CONCRETE

| FORMWORK | UNIT | MAN/ HOURS |
|---|---|---|
| **03110.90 MISCELLANEOUS FORMWORK** | | |
| 12" high | L.F. | 0.062 |
| 16" high | " | 0.067 |
| Curb form, S-shape | | |
| 12" x | | |
| 1'-6" | L.F. | 0.114 |
| 2' | " | 0.107 |
| 2'-6" | " | 0.100 |
| 3' | " | 0.089 |

| REINFORCEMENT | UNIT | MAN/ HOURS |
|---|---|---|
| **03210.05 BEAM REINFORCING** | | |
| Beam-girders | | |
| #3 - #4 | TON | 20.000 |
| #5 - #6 | " | 16.000 |
| #7 - #8 | " | 13.333 |
| #9 - #10 | " | 11.429 |
| #11 - #12 | " | 10.667 |
| #13 - #14 | " | 10.000 |
| Galvanized | | |
| #3 - #4 | TON | 20.000 |
| #5 - #6 | " | 16.000 |
| #7 - #8 | " | 13.333 |
| #9 - #10 | " | 11.429 |
| #11 - #12 | " | 10.667 |
| #13 - #14 | " | 10.000 |
| Bond beams | | |
| #3 - #4 | TON | 26.667 |
| #5 - #6 | " | 20.000 |
| #7 - #8 | " | 17.778 |
| Galvanized | | |
| #3 - #4 | TON | 26.667 |
| #5 - #6 | " | 20.000 |
| #7 - #8 | " | 17.778 |
| **03210.10 BOX CULVERT REINFORCING** | | |
| Box culverts | | |
| #3 - #4 | TON | 10.000 |
| #5 - #6 | " | 8.889 |
| #7 - #8 | " | 8.000 |
| #9 - #10 | " | 7.273 |
| #11 - #12 | " | 6.667 |
| Galvanized | | |
| #3 - #4 | TON | 10.000 |
| #5 - #6 | " | 8.889 |
| #7 - #8 | " | 8.000 |
| #9 - #10 | " | 7.273 |

| REINFORCEMENT | UNIT | MAN/ HOURS |
|---|---|---|
| **03210.10 BOX CULVERT REINFORCING** | | |
| #11 - #12 | TON | 6.667 |
| **03210.15 COLUMN REINFORCING** | | |
| Columns | | |
| #3 - #4 | TON | 22.857 |
| #5 - #6 | " | 17.778 |
| #7 - #8 | " | 16.000 |
| #9 - #10 | " | 14.545 |
| #11 - #12 | " | 13.333 |
| #13 - #14 | " | 12.308 |
| #15 - #16 | " | 11.429 |
| Galvanized | | |
| #3 - #4 | TON | 22.857 |
| #5 - #6 | " | 17.778 |
| #7 - #8 | " | 16.000 |
| #9 - #10 | " | 14.545 |
| #11 - #12 | " | 13.333 |
| #13 - #14 | " | 12.308 |
| #15 - #16 | " | 11.429 |
| Spirals | | |
| 8" to 24" dia. | TON | 20.000 |
| 24" to 48" dia. | " | 17.778 |
| 48" to 84" dia. | " | 16.000 |
| **03210.20 ELEVATED SLAB REINFORCING** | | |
| Elevated slab | | |
| #3 - #4 | TON | 10.000 |
| #5 - #6 | " | 8.889 |
| #7 - #8 | " | 8.000 |
| #9 - #10 | " | 7.273 |
| #11 - #12 | " | 6.667 |
| Galvanized | | |
| #3 - #4 | TON | 10.000 |
| #5 - #6 | " | 8.889 |
| #7 - #8 | " | 8.000 |
| #9 - #10 | " | 7.273 |
| #11 - #12 | " | 6.667 |
| **03210.25 EQUIP. PAD REINFORCING** | | |
| Equipment pad | | |
| #3 - #4 | TON | 16.000 |
| #5 - #6 | " | 14.545 |
| #7 - #8 | " | 13.333 |
| #9 - #10 | " | 12.308 |
| #11 - #12 | " | 11.429 |
| **03210.35 FOOTING REINFORCING** | | |
| Footings | | |
| Grade 50 | | |
| #3 - #4 | TON | 13.333 |
| #5 - #6 | " | 11.429 |
| #7 - #8 | " | 10.000 |

# 03 CONCRETE

| REINFORCEMENT | UNIT | MAN/HOURS |
|---|---|---|
| **03210.35 FOOTING REINFORCING** | | |
| #9 - #10 | TON | 8.889 |
| Grade 60 | | |
| #3 - #4 | TON | 13.333 |
| #5 - #6 | " | 11.429 |
| #7 - #8 | " | 10.000 |
| #9 - #10 | " | 8.889 |
| Grade 70 | | |
| #3 - #4 | TON | 13.333 |
| #5 - #6 | " | 11.429 |
| #7 - #8 | " | 10.000 |
| #9 - #10 | " | 8.889 |
| #11- #12 | " | 8.000 |
| Straight dowels, 24" long | | |
| 1" dia. (#8) | EA. | 0.080 |
| 3/4" dia. (#6) | " | 0.080 |
| 5/8" dia. (#5) | " | 0.067 |
| 1/2" dia. (#4) | " | 0.057 |
| **03210.45 FOUNDATION REINFORCING** | | |
| Foundations | | |
| #3 - #4 | TON | 13.333 |
| #5 - #6 | " | 11.429 |
| #7 - #8 | " | 10.000 |
| #9 - #10 | " | 8.889 |
| #11 - #12 | " | 8.000 |
| Galvanized | | |
| #3 - #4 | TON | 13.333 |
| #5 - #6 | " | 11.429 |
| #7 - #8 | " | 10.000 |
| #9 - #10 | " | 8.889 |
| #11 - #12 | " | 8.000 |
| **03210.50 GRADE BEAM REINFORCING** | | |
| Grade beams | | |
| #3 - #4 | TON | 12.308 |
| #5 - #6 | " | 10.667 |
| #7 - #8 | " | 9.412 |
| #9 - #10 | " | 8.421 |
| #11 - #12 | " | 7.619 |
| Galvanized | | |
| #3 - #4 | TON | 12.308 |
| #5 - #6 | " | 10.667 |
| #7 - #8 | " | 9.412 |
| #9 - #10 | " | 8.421 |
| #11 - #12 | " | 7.619 |
| **03210.53 PILE CAP REINFORCING** | | |
| Pile caps | | |
| #3 - #4 | TON | 20.000 |
| #5 - #6 | " | 17.778 |
| #7 - #8 | " | 16.000 |
| #9 - #10 | " | 14.545 |
| #11 - #12 | " | 13.333 |

| REINFORCEMENT | UNIT | MAN/HOURS |
|---|---|---|
| **03210.53 PILE CAP REINFORCING** | | |
| Galvanized | | |
| #3 - #4 | TON | 20.000 |
| #5 - #6 | " | 17.778 |
| #7 - #8 | " | 16.000 |
| #9 - #10 | " | 14.545 |
| #11 - #12 | " | 13.333 |
| **03210.55 SLAB/MAT REINFORCING** | | |
| Bars, slabs | | |
| #3 - #4 | TON | 13.333 |
| #5 - #6 | " | 11.429 |
| #7 - #8 | " | 10.000 |
| #9 - #10 | " | 8.889 |
| #11 - #12 | " | 8.000 |
| Galvanized | | |
| #3 - #4 | TON | 13.333 |
| #5 - #6 | " | 11.429 |
| #7 - #8 | " | 10.000 |
| #9 - #10 | " | 8.889 |
| #11 - #12 | " | 8.000 |
| Wire mesh, slabs | | |
| Galvanized | | |
| 4x4 | | |
| W1.4xW1.4 | S.F. | 0.005 |
| W2.0xW2.0 | " | 0.006 |
| W2.9xW2.9 | " | 0.006 |
| W4.0xW4.0 | " | 0.007 |
| 6x6 | | |
| W1.4xW1.4 | S.F. | 0.004 |
| W2.0xW2.0 | " | 0.004 |
| W2.9xW2.9 | " | 0.005 |
| W4.0xW4.0 | " | 0.005 |
| Standard | | |
| 2x2 | | |
| W.9xW.9 | S.F. | 0.005 |
| 4x4 | | |
| W1.4xW1.4 | S.F. | 0.005 |
| W2.0xW2.0 | " | 0.006 |
| W2.9xW2.9 | " | 0.006 |
| W4.0xW4.0 | " | 0.007 |
| 6x6 | | |
| W1.4xW1.4 | S.F. | 0.004 |
| W2.0xW2.0 | " | 0.004 |
| W2.9xW2.9 | " | 0.005 |
| W4.0xW4.0 | " | 0.005 |
| **03210.60 STAIR REINFORCING** | | |
| Stairs | | |
| #3 - #4 | TON | 16.000 |
| #5 - #6 | " | 13.333 |
| #7 - #8 | " | 11.429 |
| #9 - #10 | " | 10.000 |
| Galvanized | | |

# 03 CONCRETE

| REINFORCEMENT | UNIT | MAN/ HOURS |
|---|---|---|
| **03210.60 STAIR REINFORCING** | | |
| #3 - #4 | TON | 16.000 |
| #5 - #6 | " | 13.333 |
| #7 - #8 | " | 11.429 |
| #9 - #10 | " | 10.000 |
| **03210.65 WALL REINFORCING** | | |
| Walls | | |
| #3 - #4 | TON | 11.429 |
| #5 - #6 | " | 10.000 |
| #7 - #8 | " | 8.889 |
| #9 - #10 | " | 8.000 |
| Galvanized | | |
| #3 - #4 | TON | 11.429 |
| #5 - #6 | " | 10.000 |
| #7 - #8 | " | 8.889 |
| #9 - #10 | " | 8.000 |
| Masonry wall (horizontal) | | |
| #3 - #4 | TON | 32.000 |
| #5 - #6 | " | 26.667 |
| Galvanized | | |
| #3 - #4 | TON | 32.000 |
| #5 - #6 | " | 26.667 |
| Masonry wall (vertical) | | |
| #3 - #4 | TON | 40.000 |
| #5 - #6 | " | 32.000 |
| Galvanized | | |
| #3 - #4 | TON | 40.000 |
| #5 - #6 | " | 32.000 |

| ACCESSORIES | UNIT | MAN/ HOURS |
|---|---|---|
| **03250.40 CONCRETE ACCESSORIES** | | |
| Expansion joint, poured | | |
| Asphalt | | |
| 1/2" x 1" | L.F. | 0.016 |
| 1" x 2" | " | 0.017 |
| Liquid neoprene, cold applied | | |
| 1/2" x 1" | L.F. | 0.016 |
| 1" x 2" | " | 0.018 |
| Polyurethane, 2 parts | | |
| 1/2" x 1" | L.F. | 0.027 |
| 1" x 2" | " | 0.029 |
| Rubberized asphalt, cold | | |
| 1/2" x 1" | L.F. | 0.016 |
| 1" x 2" | " | 0.017 |
| Hot, fuel resistant | | |
| 1/2" x 1" | L.F. | 0.016 |

| ACCESSORIES | UNIT | MAN/ HOURS |
|---|---|---|
| **03250.40 CONCRETE ACCESSORIES** | | |
| 1" x 2" | L.F. | 0.017 |
| Expansion joint, premolded, in slabs | | |
| Asphalt | | |
| 1/2" x 6" | L.F. | 0.020 |
| 1" x 12" | " | 0.027 |
| Cork | | |
| 1/2" x 6" | L.F. | 0.020 |
| 1" x 12" | " | 0.027 |
| Neoprene sponge | | |
| 1/2" x 6" | L.F. | 0.020 |
| 1" x 12" | " | 0.027 |
| Polyethylene foam | | |
| 1/2" x 6" | L.F. | 0.020 |
| 1" x 12" | " | 0.027 |
| Polyurethane foam | | |
| 1/2" x 6" | L.F. | 0.020 |
| 1" x 12" | " | 0.027 |
| Polyvinyl chloride foam | | |
| 1/2" x 6" | L.F. | 0.020 |
| 1" x 12" | " | 0.027 |
| Rubber, gray sponge | | |
| 1/2" x 6" | L.F. | 0.020 |
| 1" x 12" | " | 0.027 |
| Asphalt felt control joints or bond breaker, screed joints | | |
| 4" slab | L.F. | 0.016 |
| 6" slab | " | 0.018 |
| 8" slab | " | 0.020 |
| 10" slab | " | 0.023 |
| Keyed cold expansion and control joints, 24 ga. | | |
| 4" slab | L.F. | 0.050 |
| 5" slab | " | 0.050 |
| 6" slab | " | 0.053 |
| 8" slab | " | 0.057 |
| 10" slab | " | 0.062 |
| Waterstops | | |
| Polyvinyl chloride | | |
| Ribbed | | |
| 3/16" thick x | | |
| 4" wide | L.F. | 0.040 |
| 6" wide | " | 0.044 |
| 1/2" thick x | | |
| 9" wide | L.F. | 0.050 |
| Ribbed with center bulb | | |
| 3/16" thick x 9" wide | L.F. | 0.050 |
| 3/8" thick x 9" wide | " | 0.050 |
| Dumbbell type, 3/8" thick x 6" wide | " | 0.044 |
| Plain, 3/8" thick x 9" wide | " | 0.050 |
| Center bulb, 3/8" thick x 9" wide | " | 0.050 |
| Rubber | | |
| Flat dumbbell | | |
| 3/8" thick x | | |
| 6" wide | L.F. | 0.044 |
| 9" wide | " | 0.050 |
| Center bulb | | |

| ACCESSORIES | UNIT | MAN/HOURS |
|---|---|---|
| **03250.40 CONCRETE ACCESSORIES** | | |
| 3/8" thick x | | |
| 6" wide | L.F. | 0.044 |
| 9" wide | " | 0.050 |
| Vapor barrier | | |
| 4 mil polyethylene | S.F. | 0.003 |
| 6 mil polyethylene | " | 0.003 |
| Gravel porous fill, under floor slabs, 3/4" stone | C.Y. | 1.333 |
| Reinforcing accessories | | |
| Beam bolsters | | |
| 1-1/2" high, plain | L.F. | 0.008 |
| Galvanized | " | 0.008 |
| 3" high | | |
| Plain | L.F. | 0.010 |
| Galvanized | " | 0.010 |
| Slab bolsters | | |
| 1" high | | |
| Plain | L.F. | 0.004 |
| Galvanized | " | 0.004 |
| 2" high | | |
| Plain | L.F. | 0.004 |
| Galvanized | " | 0.004 |
| Chairs, high chairs | | |
| 3" high | | |
| Plain | EA. | 0.020 |
| Galvanized | " | 0.020 |
| 5" high | | |
| Plain | EA. | 0.021 |
| Galvanized | " | 0.021 |
| 8" high | | |
| Plain | EA. | 0.023 |
| Galvanized | " | 0.023 |
| 12" high | | |
| Plain | EA. | 0.027 |
| Galvanized | " | 0.027 |
| Continuous, high chair | | |
| 3" high | | |
| Plain | L.F. | 0.005 |
| Galvanized | " | 0.005 |
| 5" high | | |
| Plain | L.F. | 0.006 |
| Galvanized | " | 0.006 |
| 8" high | | |
| Plain | L.F. | 0.006 |
| Galvanized | " | 0.006 |
| 12" high | | |
| Plain | L.F. | 0.007 |
| Galvanized | " | 0.007 |

| CAST-IN-PLACE CONCRETE | UNIT | MAN/HOURS |
|---|---|---|
| **03350.10 CONCRETE FINISHES** | | |
| Floor finishes | | |
| Broom | S.F. | 0.011 |
| Screed | " | 0.010 |
| Darby | " | 0.010 |
| Steel float | " | 0.013 |
| Granolithic topping | | |
| 1/2" thick | S.F. | 0.036 |
| 1" thick | " | 0.040 |
| 2" thick | " | 0.044 |
| Wall finishes | | |
| Burlap rub, with cement paste | S.F. | 0.013 |
| Float finish | " | 0.020 |
| Etch with acid | " | 0.013 |
| Sandblast | | |
| Minimum | S.F. | 0.016 |
| Maximum | " | 0.016 |
| Bush hammer | | |
| Green concrete | S.F. | 0.040 |
| Cured concrete | " | 0.062 |
| Break ties and patch holes | " | 0.016 |
| Carborundum | | |
| Dry rub | S.F. | 0.027 |
| Wet rub | " | 0.040 |
| Floor hardeners | | |
| Metallic | | |
| Light service | S.F. | 0.010 |
| Heavy service | " | 0.013 |
| Non-metallic | | |
| Light service | S.F. | 0.010 |
| Heavy service | " | 0.013 |
| Rusticated concrete finish | | |
| Beveled edge | L.F. | 0.044 |
| Square edge | " | 0.057 |
| Solid board concrete finish | | |
| Standard | S.F. | 0.067 |
| Rustic | " | 0.080 |
| **03360.10 PNEUMATIC CONCRETE** | | |
| Pneumatic applied concrete (gunite) | | |
| 2" thick | S.F. | 0.030 |
| 3" thick | " | 0.040 |
| 4" thick | " | 0.048 |
| Finish surface | | |
| Minimum | S.F. | 0.040 |
| Maximum | " | 0.080 |
| **03370.10 CURING CONCRETE** | | |
| Sprayed membrane | | |
| Slabs | S.F. | 0.002 |
| Walls | " | 0.002 |
| Curing paper | | |
| Slabs | S.F. | 0.002 |
| Walls | " | 0.002 |

# 03 CONCRETE

| CAST-IN-PLACE CONCRETE | UNIT | MAN/ HOURS |
|---|---|---|
| **03370.10 CURING CONCRETE** | | |
| Burlap | | |
| 7.5 oz. | S.F. | 0.003 |
| 12 oz. | " | 0.003 |

| PLACING CONCRETE | UNIT | MAN/ HOURS |
|---|---|---|
| **03380.05 BEAM CONCRETE** | | |
| Beams and girders | | |
| 2500# or 3000# concrete | | |
| By crane | C.Y. | 0.960 |
| By pump | " | 0.873 |
| By hand buggy | " | 0.800 |
| 3500# or 4000# concrete | | |
| By crane | C.Y. | 0.960 |
| By pump | " | 0.873 |
| By hand buggy | " | 0.800 |
| 5000# concrete | | |
| By crane | C.Y. | 0.960 |
| By pump | " | 0.873 |
| By hand buggy | " | 0.800 |
| Bond beam, 3000# concrete | | |
| By pump | | |
| 8" high | | |
| 4" wide | L.F. | 0.019 |
| 6" wide | " | 0.022 |
| 8" wide | " | 0.024 |
| 10" wide | " | 0.027 |
| 12" wide | " | 0.030 |
| 16" high | | |
| 8" wide | L.F. | 0.030 |
| 10" wide | " | 0.034 |
| 12" wide | " | 0.040 |
| By crane | | |
| 8" high | | |
| 4" wide | L.F. | 0.021 |
| 6" wide | " | 0.023 |
| 8" wide | " | 0.024 |
| 10" wide | " | 0.027 |
| 12" wide | " | 0.030 |
| 16" high | | |
| 8" wide | L.F. | 0.030 |
| 10" wide | " | 0.032 |
| 12" wide | " | 0.037 |

| PLACING CONCRETE | UNIT | MAN/ HOURS |
|---|---|---|
| **03380.15 COLUMN CONCRETE** | | |
| Columns | | |
| 2500# or 3000# concrete | | |
| By crane | C.Y. | 0.873 |
| By pump | " | 0.800 |
| 3500# or 4000# concrete | | |
| By crane | C.Y. | 0.873 |
| By pump | " | 0.800 |
| 5000# concrete | | |
| By crane | C.Y. | 0.873 |
| By pump | " | 0.800 |
| **03380.20 ELEVATED SLAB CONCRETE** | | |
| Elevated slab | | |
| 2500# or 3000# concrete | | |
| By crane | C.Y. | 0.480 |
| By pump | " | 0.369 |
| By hand buggy | " | 0.800 |
| 3500# or 4000# concrete | | |
| By crane | C.Y. | 0.480 |
| By pump | " | 0.369 |
| By hand buggy | " | 0.800 |
| 5000# concrete | | |
| By crane | C.Y. | 0.480 |
| By pump | " | 0.369 |
| By hand buggy | " | 0.800 |
| Topping | | |
| 2500# or 3000# concrete | | |
| By crane | C.Y. | 0.480 |
| By pump | " | 0.369 |
| By hand buggy | " | 0.800 |
| 3500# or 4000# concrete | | |
| By crane | C.Y. | 0.480 |
| By pump | " | 0.369 |
| By hand buggy | " | 0.800 |
| 5000# concrete | | |
| By crane | C.Y. | 0.480 |
| By pump | " | 0.369 |
| By hand buggy | " | 0.800 |
| **03380.25 EQUIPMENT PAD CONCRETE** | | |
| Equipment pad | | |
| 2500# or 3000# concrete | | |
| By chute | C.Y. | 0.267 |
| By pump | " | 0.686 |
| By crane | " | 0.800 |
| 3500# or 4000# concrete | | |
| By chute | C.Y. | 0.267 |
| By pump | " | 0.686 |
| By crane | " | 0.800 |
| 5000# concrete | | |
| By chute | C.Y. | 0.267 |
| By pump | " | 0.686 |
| By crane | " | 0.800 |

# 03 CONCRETE

| PLACING CONCRETE | UNIT | MAN/HOURS |
|---|---|---|
| **03380.35 FOOTING CONCRETE** | | |
| Continuous footing | | |
| 2500# or 3000# concrete | | |
| By chute | C.Y. | 0.267 |
| By pump | " | 0.600 |
| By crane | " | 0.686 |
| 3500# or 4000# concrete | | |
| By chute | C.Y. | 0.267 |
| By pump | " | 0.600 |
| By crane | " | 0.686 |
| 5000# concrete | | |
| By chute | C.Y. | 0.267 |
| By pump | " | 0.600 |
| By crane | " | 0.686 |
| Spread footing | | |
| 2500# or 3000# concrete | | |
| Under 5 cy | | |
| By chute | C.Y. | 0.267 |
| By pump | " | 0.640 |
| By crane | " | 0.738 |
| Over 5 cy | | |
| By chute | C.Y. | 0.200 |
| By pump | " | 0.565 |
| By crane | " | 0.640 |
| 3500# or 4000# concrete | | |
| Under 5 c.y. | | |
| By chute | C.Y. | 0.267 |
| By pump | " | 0.640 |
| By crane | " | 0.738 |
| Over 5 c.y. | | |
| By chute | C.Y. | 0.200 |
| By pump | " | 0.565 |
| By crane | " | 0.640 |
| 5000# concrete | | |
| Under 5 c.y. | | |
| By chute | C.Y. | 0.267 |
| By pump | " | 0.640 |
| By crane | " | 0.738 |
| Over 5 c.y. | | |
| By chute | C.Y. | 0.200 |
| By pump | " | 0.565 |
| By crane | " | 0.640 |
| **03380.50 GRADE BEAM CONCRETE** | | |
| Grade beam | | |
| 2500# or 3000# concrete | | |
| By chute | C.Y. | 0.267 |
| By crane | " | 0.686 |
| By pump | " | 0.600 |
| By hand buggy | " | 0.800 |
| 3500# or 4000# concrete | | |
| By chute | C.Y. | 0.267 |
| By crane | " | 0.686 |
| By pump | " | 0.600 |

| PLACING CONCRETE | UNIT | MAN/HOURS |
|---|---|---|
| **03380.50 GRADE BEAM CONCRETE** | | |
| By hand buggy | C.Y. | 0.800 |
| 5000# concrete | | |
| By chute | C.Y. | 0.267 |
| By crane | " | 0.686 |
| By pump | " | 0.600 |
| By hand buggy | " | 0.800 |
| **03380.53 PILE CAP CONCRETE** | | |
| Pile cap | | |
| 2500# or 3000 concrete | | |
| By chute | C.Y. | 0.267 |
| By crane | " | 0.800 |
| By pump | " | 0.686 |
| By hand buggy | " | 0.800 |
| 3500# or 4000# concrete | | |
| By chute | C.Y. | 0.267 |
| By crane | " | 0.800 |
| By pump | " | 0.686 |
| By hand buggy | " | 0.800 |
| 5000# concrete | | |
| By chute | C.Y. | 0.267 |
| By crane | " | 0.800 |
| By pump | " | 0.686 |
| By hand buggy | " | 0.800 |
| **03380.55 SLAB/MAT CONCRETE** | | |
| Slab on grade | | |
| 2500# or 3000# concrete | | |
| By chute | C.Y. | 0.200 |
| By crane | " | 0.400 |
| By pump | " | 0.343 |
| By hand buggy | " | 0.533 |
| 3500# or 4000# concrete | | |
| By chute | C.Y. | 0.200 |
| By crane | " | 0.400 |
| By pump | " | 0.343 |
| By hand buggy | " | 0.533 |
| 5000# concrete | | |
| By chute | C.Y. | 0.200 |
| By crane | " | 0.400 |
| By pump | " | 0.343 |
| By hand buggy | " | 0.533 |
| Foundation mat | | |
| 2500# or 3000# concrete, over 20 cy | | |
| By chute | C.Y. | 0.160 |
| By crane | " | 0.343 |
| By pump | " | 0.300 |
| By hand buggy | " | 0.400 |
| **03380.58 SIDEWALKS** | | |
| Walks, cast in place with wire mesh, base not incl. | | |
| 4" thick | S.F. | 0.027 |
| 5" thick | " | 0.032 |
| 6" thick | " | 0.040 |

# 03 CONCRETE

| PLACING CONCRETE | UNIT | MAN/ HOURS |
|---|---|---|
| **03380.60 STAIR CONCRETE** | | |
| Stairs | | |
| 2500# or 3000# concrete | | |
| By chute | C.Y. | 0.267 |
| By crane | " | 0.800 |
| By pump | " | 0.686 |
| By hand buggy | " | 0.800 |
| 3500# or 4000# concrete | | |
| By chute | C.Y. | 0.267 |
| By crane | " | 0.800 |
| By pump | " | 0.686 |
| By hand buggy | " | 0.800 |
| 5000# concrete | | |
| By chute | C.Y. | 0.267 |
| By crane | " | 0.800 |
| By pump | " | 0.686 |
| By hand buggy | " | 0.800 |
| **03380.65 WALL CONCRETE** | | |
| Walls | | |
| 2500# or 3000# concrete | | |
| To 4' | | |
| By chute | C.Y. | 0.229 |
| By crane | " | 0.800 |
| By pump | " | 0.738 |
| To 8' | | |
| By crane | C.Y. | 0.873 |
| By pump | " | 0.800 |
| To 16' | | |
| By crane | C.Y. | 0.960 |
| By pump | " | 0.873 |
| Over 16' | | |
| By crane | C.Y. | 1.067 |
| By pump | " | 0.960 |
| 3500# or 4000# concrete | | |
| To 4' | | |
| By chute | C.Y. | 0.229 |
| By crane | " | 0.800 |
| By pump | " | 0.738 |
| To 8' | | |
| By crane | C.Y. | 0.873 |
| By pump | " | 0.800 |
| To 16' | | |
| By crane | C.Y. | 0.960 |
| By pump | " | 0.873 |
| Over 16' | | |
| By crane | C.Y. | 1.067 |
| By pump | " | 0.960 |
| 5000# concrete | | |
| To 4' | | |
| By chute | C.Y. | 0.229 |
| By crane | " | 0.800 |
| By pump | " | 0.738 |
| To 8' | | |
| By crane | C.Y. | 0.873 |

| PLACING CONCRETE | UNIT | MAN/ HOURS |
|---|---|---|
| **03380.65 WALL CONCRETE** | | |
| By pump | C.Y. | 0.800 |
| To 16' | | |
| By crane | C.Y. | 0.960 |
| By pump | " | 0.873 |
| Filled block (CMU) | | |
| 3000# concrete, by pump | | |
| 4" wide | S.F. | 0.034 |
| 6" wide | " | 0.040 |
| 8" wide | " | 0.048 |
| 10" wide | " | 0.056 |
| 12" wide | " | 0.069 |
| Pilasters, 3000# concrete | C.F. | 0.960 |
| Wall cavity, 2" thick, 3000# concrete | S.F. | 0.032 |

| PRECAST CONCRETE | UNIT | MAN/ HOURS |
|---|---|---|
| **03400.10 PRECAST BEAMS** | | |
| Prestressed, double tee, 24" deep, 8' wide | | |
| 35' span | | |
| 115 psf | S.F. | 0.008 |
| 140 psf | " | 0.008 |
| 40' span | | |
| 80 psf | S.F. | 0.009 |
| 143 psf | " | 0.009 |
| 45' span | | |
| 50 psf | S.F. | 0.007 |
| 70 psf | " | 0.007 |
| 100 psf | " | 0.007 |
| 130 psf | " | 0.007 |
| 50' span | | |
| 75 psf | S.F. | 0.007 |
| 100 psf | " | 0.007 |
| Precast beams, girders and joists | | |
| 1000 lb/lf live load | | |
| 10' span | L.F. | 0.160 |
| 20' span | " | 0.096 |
| 30' span | " | 0.080 |
| 3000 lb/lf live load | | |
| 10' span | L.F. | 0.160 |
| 20' span | " | 0.096 |
| 30' span | " | 0.080 |
| 5000 lb/lf live load | | |
| 10' span | L.F. | 0.160 |
| 20' span | " | 0.096 |
| 30' span | " | 0.080 |

# 03 CONCRETE

| PRECAST CONCRETE | UNIT | MAN/ HOURS |
|---|---|---|
| **03400.20 PRECAST COLUMNS** | | |
| Prestressed concrete columns | | |
| 10" x 10" | | |
| 10' long | EA. | 0.960 |
| 15' long | " | 1.000 |
| 20' long | " | 1.067 |
| 25' long | " | 1.143 |
| 30' long | " | 1.200 |
| 12" x 12" | | |
| 20' long | EA. | 1.200 |
| 25' long | " | 1.297 |
| 30' long | " | 1.371 |
| 16" x 16" | | |
| 20' long | EA. | 1.200 |
| 25' long | " | 1.297 |
| 30' long | " | 1.371 |
| 20" x 20" | | |
| 20' long | EA. | 1.263 |
| 25' long | " | 1.333 |
| 30' long | " | 1.412 |
| 24" x 24" | | |
| 20' long | EA. | 1.333 |
| 25' long | " | 1.412 |
| 30' long | " | 1.500 |
| 28" x 28" | | |
| 20' long | EA. | 1.500 |
| 25' long | " | 1.600 |
| 30' long | " | 1.714 |
| 32" x 32" | | |
| 20' long | EA. | 1.600 |
| 25' long | " | 1.714 |
| 30' long | " | 1.846 |
| 36" x 36" | | |
| 20' long | EA. | 1.714 |
| 25' long | " | 1.846 |
| 30' long | " | 2.000 |
| **03400.30 PRECAST SLABS** | | |
| Prestressed flat slab | | |
| 6" thick, 4' wide | | |
| 20' span | | |
| 80 psf | S.F. | 0.020 |
| 110 psf | " | 0.020 |
| 25' span | | |
| 80 psf | S.F. | 0.019 |
| Cored slab | | |
| 6" thick, 4' wide | | |
| 20' span | | |
| 80 psf | S.F. | 0.020 |
| 100 psf | " | 0.020 |
| 130 psf | " | 0.020 |
| 8" thick, 4' wide | | |
| 25' span | | |
| 70 psf | S.F. | 0.019 |

| PRECAST CONCRETE | UNIT | MAN/ HOURS |
|---|---|---|
| **03400.30 PRECAST SLABS** | | |
| 125 psf | S.F. | 0.019 |
| 170 psf | " | 0.019 |
| 30' span | | |
| 70 psf | S.F. | 0.016 |
| 90 psf | " | 0.016 |
| 35' span | | |
| 70 psf | S.F. | 0.015 |
| 10" thick, 4' wide | | |
| 30' span | | |
| 75 psf | S.F. | 0.016 |
| 100 psf | " | 0.016 |
| 130 psf | " | 0.016 |
| 35' span | | |
| 60 psf | S.F. | 0.015 |
| 80 psf | " | 0.015 |
| 120 psf | " | 0.015 |
| 40' span | | |
| 65 psf | S.F. | 0.012 |
| Slabs, roof and floor members, 4' wide | | |
| 6" thick, 25' span | S.F. | 0.019 |
| 8" thick, 30' span | " | 0.015 |
| 10" thick, 40' span | " | 0.013 |
| Tee members | | |
| Multiple tee, roof and floor | | |
| Minimum | S.F. | 0.012 |
| Maximum | " | 0.024 |
| Double tee wall member | | |
| Minimum | S.F. | 0.014 |
| Maximum | " | 0.027 |
| Single tee | | |
| Short span, roof members | | |
| Minimum | S.F. | 0.015 |
| Maximum | " | 0.030 |
| Long span, roof members | | |
| Minimum | S.F. | 0.012 |
| Maximum | " | 0.024 |
| **03400.40 PRECAST WALLS** | | |
| Wall panel, 8' x 20' | | |
| Gray cement | | |
| Liner finish | | |
| 4" wall | S.F. | 0.014 |
| 5" wall | " | 0.014 |
| 6" wall | " | 0.015 |
| 8" wall | " | 0.015 |
| Sandblast finish | | |
| 4" wall | S.F. | 0.014 |
| 5" wall | " | 0.014 |
| 6" wall | " | 0.015 |
| 8" wall | " | 0.015 |
| White cement | | |
| Liner finish | | |
| 4" wall | S.F. | 0.014 |
| 5" wall | " | 0.014 |

# 03 CONCRETE

| PRECAST CONCRETE | UNIT | MAN/ HOURS |
|---|---|---|
| **03400.40 PRECAST WALLS** | | |
| 6" wall | S.F. | 0.015 |
| 8" wall | " | 0.015 |
| Sandblast finish | | |
| 4" wall | S.F. | 0.014 |
| 5" wall | " | 0.014 |
| 6" wall | " | 0.015 |
| 8" wall | " | 0.015 |
| Double tee wall panel, 24" deep | | |
| Gray cement | | |
| Liner finish | S.F. | 0.016 |
| Sandblast finish | " | 0.016 |
| White cement | | |
| Form liner finish | S.F. | 0.016 |
| Sandblast finish | " | 0.016 |
| Partition panels | | |
| 4" wall | S.F. | 0.016 |
| 5" wall | " | 0.016 |
| 6" wall | " | 0.016 |
| 8" wall | " | 0.016 |
| Cladding panels | | |
| 4" wall | S.F. | 0.017 |
| 5" wall | " | 0.017 |
| 6" wall | " | 0.017 |
| 8" wall | " | 0.017 |
| Sandwich panel, 2.5" cladding panel, 2" insulation | | |
| 5" wall | S.F. | 0.017 |
| 6" wall | " | 0.017 |
| 8" wall | " | 0.017 |
| **03400.90 PRECAST SPECIALTIES** | | |
| Precast concrete, coping, 4' to 8' long | | |
| 12" wide | L.F. | 0.060 |
| 10" wide | " | 0.069 |
| Splash block, 30"x12"x4" | EA. | 0.400 |
| Stair unit, per riser | " | 0.400 |
| Sun screen and trellis, 8' long, 12" high | | |
| 4" thick blades | EA. | 0.300 |
| 5" thick blades | " | 0.300 |
| 6" thick blades | " | 0.320 |
| 8" thick blades | " | 0.320 |
| Bearing pads for precast members, 2" wide strips | | |
| 1/8" thick | L.F. | 0.003 |
| 1/4" thick | " | 0.003 |
| 1/2" thick | " | 0.003 |
| 3/4" thick | " | 0.004 |
| 1" thick | " | 0.004 |
| 1-1/2" thick | " | 0.004 |

| CEMENTITOUS TOPPINGS | UNIT | MAN/ HOURS |
|---|---|---|
| **03550.10 CONCRETE TOPPINGS** | | |
| Gypsum fill | | |
| 2" thick | S.F. | 0.005 |
| 2-1/2" thick | " | 0.005 |
| 3" thick | " | 0.005 |
| 3-1/2" thick | " | 0.005 |
| 4" thick | " | 0.006 |
| Formboard | | |
| Mineral fiber board | | |
| 1" thick | S.F. | 0.020 |
| 1-1/2" thick | " | 0.023 |
| Cement fiber board | | |
| 1" thick | S.F. | 0.027 |
| 1-1/2" thick | " | 0.031 |
| Glass fiber board | | |
| 1" thick | S.F. | 0.020 |
| 1-1/2" thick | " | 0.023 |
| Poured deck | | |
| Vermiculite or perlite | | |
| 1 to 4 mix | C.Y. | 0.800 |
| 1 to 6 mix | " | 0.738 |
| Vermiculite or perlite | | |
| 2" thick | | |
| 1 to 4 mix | S.F. | 0.005 |
| 1 to 6 mix | " | 0.005 |
| 3" thick | | |
| 1 to 4 mix | S.F. | 0.007 |
| 1 to 6 mix | " | 0.007 |
| Concrete plank, lightweight | | |
| 2" thick | S.F. | 0.024 |
| 2-1/2" thick | " | 0.024 |
| 3-1/2" thick | " | 0.027 |
| 4" thick | " | 0.027 |
| Channel slab, lightweight, straight | | |
| 2-3/4" thick | S.F. | 0.024 |
| 3-1/2" thick | " | 0.024 |
| 3-3/4" thick | " | 0.024 |
| 4-3/4" thick | " | 0.027 |
| Gypsum plank | | |
| 2" thick | S.F. | 0.024 |
| 3" thick | " | 0.024 |
| Cement fiber, T and G planks | | |
| 1" thick | S.F. | 0.022 |
| 1-1/2" thick | " | 0.022 |
| 2" thick | " | 0.024 |
| 2-1/2" thick | " | 0.024 |
| 3" thick | " | 0.024 |
| 3-1/2" thick | " | 0.027 |
| 4" thick | " | 0.027 |

| GROUT | UNIT | MAN/HOURS |
|---|---|---|
| **03600.10 GROUTING** | | |
| Grouting for bases | | |
| Nonshrink | | |
| Metallic grout | | |
| 1" deep | S.F. | 0.160 |
| 2" deep | " | 0.178 |
| Non-metallic grout | | |
| 1" deep | S.F. | 0.160 |
| 2" deep | " | 0.178 |
| Fluid type | | |
| Non-metallic | | |
| 1" deep | S.F. | 0.160 |
| 2" deep | " | 0.178 |
| Grouting for joints | | |
| Portland cement grout (1 cement to 3 sand, by volume) | | |
| 1/2" joint thickness | | |
| 6" wide joints | L.F. | 0.027 |
| 8" wide joints | " | 0.032 |
| 1" joint thickness | | |
| 4" wide joints | L.F. | 0.025 |
| 6" wide joints | " | 0.028 |
| 8" wide joints | " | 0.033 |
| Nonshrink, nonmetallic grout | | |
| 1/2" joint thickness | | |
| 4" wide joint | L.F. | 0.023 |
| 6" wide joint | " | 0.027 |
| 8" wide joint | " | 0.032 |
| 1" joint thickness | | |
| 4" wide joint | L.F. | 0.025 |
| 6" wide joint | " | 0.028 |
| 8" wide joint | " | 0.033 |

| CONCRETE RESTORATION | UNIT | MAN/HOURS |
|---|---|---|
| **03730.10 CONCRETE REPAIR** | | |
| Epoxy grout floor patch, 1/4" thick | S.F. | 0.080 |
| Epoxy gel grout | " | 0.800 |
| Injection valve, 1 way, threaded plastic | EA. | 0.160 |
| Grout crack seal, 2 component | C.F. | 0.800 |
| Grout, non shrink | " | 0.800 |
| Concrete, epoxy modified | | |
| Sand mix | C.F. | 0.320 |
| Gravel mix | " | 0.296 |
| Concrete repair | | |
| Soffit repair | | |
| 16" wide | L.F. | 0.160 |
| 18" wide | " | 0.167 |
| 24" wide | " | 0.178 |

| CONCRETE RESTORATION | UNIT | MAN/HOURS |
|---|---|---|
| **03730.10 CONCRETE REPAIR** | | |
| 30" wide | L.F. | 0.190 |
| 32" wide | " | 0.200 |
| Edge repair | | |
| 2" spall | L.F. | 0.200 |
| 3" spall | " | 0.211 |
| 4" spall | " | 0.216 |
| 6" spall | " | 0.222 |
| 8" spall | " | 0.235 |
| 9" spall | " | 0.267 |
| Crack repair, 1/8" crack | " | 0.080 |
| Reinforcing steel repair | | |
| 1 bar, 4 ft | | |
| #4 bar | L.F. | 0.100 |
| #5 bar | " | 0.100 |
| #6 bar | " | 0.107 |
| #8 bar | " | 0.107 |
| #9 bar | " | 0.114 |
| #11 bar | " | 0.114 |
| Pile repairs | | |
| Polyethylene wrap | | |
| 30 mil thick | | |
| 60" wide | S.F. | 0.267 |
| 72" wide | " | 0.320 |
| 60 mil thick | | |
| 60" wide | S.F. | 0.267 |
| 80" wide | " | 0.364 |
| Pile spall, average repair 3' | | |
| 18" x 18" | EA. | 0.667 |
| 20" x 20" | " | 0.800 |

| MORTAR AND GROUT | UNIT | MAN/ HOURS |
|---|---|---|
| **04100.10 MASONRY GROUT** | | |
| Grout, non shrink, non-metallic, trowelable | C.F. | 0.016 |
| Grout door frame, hollow metal | | |
| Single | EA. | 0.600 |
| Double | " | 0.632 |
| Grout-filled concrete block (CMU) | | |
| 4" wide | S.F. | 0.020 |
| 6" wide | " | 0.022 |
| 8" wide | " | 0.024 |
| 12" wide | " | 0.025 |
| Grout-filled individual CMU cells | | |
| 4" wide | L.F. | 0.012 |
| 6" wide | " | 0.012 |
| 8" wide | " | 0.012 |
| 10" wide | " | 0.014 |
| 12" wide | " | 0.014 |
| Bond beams or lintels, 8" deep | | |
| 6" thick | L.F. | 0.022 |
| 8" thick | " | 0.024 |
| 10" thick | " | 0.027 |
| 12" thick | " | 0.030 |
| Cavity walls | | |
| 2" thick | S.F. | 0.032 |
| 3" thick | " | 0.032 |
| 4" thick | " | 0.034 |
| 6" thick | " | 0.040 |
| **04150.10 MASONRY ACCESSORIES** | | |
| Foundation vents | EA. | 0.320 |
| Bar reinforcing | | |
| Horizontal | | |
| #3 - #4 | Lb. | 0.032 |
| #5 - #6 | " | 0.027 |
| Vertical | | |
| #3 - #4 | Lb. | 0.040 |
| #5 - #6 | " | 0.032 |
| Horizontal joint reinforcing | | |
| Truss type | | |
| 4" wide, 6" wall | L.F. | 0.003 |
| 6" wide, 8" wall | " | 0.003 |
| 8" wide, 10" wall | " | 0.003 |
| 10" wide, 12" wall | " | 0.004 |
| 12" wide, 14" wall | " | 0.004 |
| Ladder type | | |
| 4" wide, 6" wall | L.F. | 0.003 |
| 6" wide, 8" wall | " | 0.003 |
| 8" wide, 10" wall | " | 0.003 |
| 10" wide, 12" wall | " | 0.003 |
| Rectangular wall ties | | |
| 3/16" dia., galvanized | | |
| 2" x 6" | EA. | 0.013 |
| 2" x 8" | " | 0.013 |
| 2" x 10" | " | 0.013 |
| 2" x 12" | " | 0.013 |

| MORTAR AND GROUT | UNIT | MAN/ HOURS |
|---|---|---|
| **04150.10 MASONRY ACCESSORIES** | | |
| 4" x 6" | EA. | 0.016 |
| 4" x 8" | " | 0.016 |
| 4" x 10" | " | 0.016 |
| 4" x 12" | " | 0.016 |
| 1/4" dia., galvanized | | |
| 2" x 6" | EA. | 0.013 |
| 2" x 8" | " | 0.013 |
| 2" x 10" | " | 0.013 |
| 2" x 12" | " | 0.013 |
| 4" x 6" | " | 0.016 |
| 4" x 8" | " | 0.016 |
| 4" x 10" | " | 0.016 |
| 4" x 12" | " | 0.016 |
| "Z" type wall ties, galvanized | | |
| 6" long | | |
| 1/8" dia. | EA. | 0.013 |
| 3/16" dia. | " | 0.013 |
| 1/4" dia. | " | 0.013 |
| 8" long | | |
| 1/8" dia. | EA. | 0.013 |
| 3/16" dia. | " | 0.013 |
| 1/4" dia. | " | 0.013 |
| 10" long | | |
| 1/8" dia. | EA. | 0.013 |
| 3/16" dia. | " | 0.013 |
| 1/4" dia. | " | 0.013 |
| Dovetail anchor slots | | |
| Galvanized steel, filled | | |
| 24 ga. | L.F. | 0.020 |
| 20 ga. | " | 0.020 |
| 16 oz. copper, foam filled | " | 0.020 |
| Dovetail anchors | | |
| 16 ga. | | |
| 3-1/2" long | EA. | 0.013 |
| 5-1/2" long | " | 0.013 |
| 12 ga. | | |
| 3-1/2" long | EA. | 0.013 |
| 5-1/2" long | " | 0.013 |
| Dovetail, triangular galvanized ties, 12 ga. | | |
| 3" x 3" | EA. | 0.013 |
| 5" x 5" | " | 0.013 |
| 7" x 7" | " | 0.013 |
| 7" x 9" | " | 0.013 |
| Brick anchors | | |
| Corrugated, 3-1/2" long | | |
| 16 ga. | EA. | 0.013 |
| 12 ga. | " | 0.013 |
| Non-corrugated, 3-1/2" long | | |
| 16 ga. | EA. | 0.013 |
| 12 ga. | " | 0.013 |
| Cavity wall anchors, corrugated, galvanized | | |
| 5" long | | |
| 16 ga. | EA. | 0.013 |
| 12 ga. | " | 0.013 |

## MORTAR AND GROUT

### 04150.10 MASONRY ACCESSORIES

| | UNIT | MAN/HOURS |
|---|---|---|
| 7" long | | |
| 28 ga. | EA. | 0.013 |
| 24 ga. | " | 0.013 |
| 22 ga. | " | 0.013 |
| 16 ga. | " | 0.013 |
| Mesh ties, 16 ga., 3" wide | | |
| 8" long | EA. | 0.013 |
| 12" long | " | 0.013 |
| 20" long | " | 0.013 |
| 24" long | " | 0.013 |

### 04150.20 MASONRY CONTROL JOINTS

| | UNIT | MAN/HOURS |
|---|---|---|
| Control joint, cross shaped PVC | L.F. | 0.020 |
| Closed cell joint filler | | |
| 1/2" | L.F. | 0.020 |
| 3/4" | " | 0.020 |
| Rubber, for | | |
| 4" wall | L.F. | 0.020 |
| 6" wall | " | 0.021 |
| 8" wall | " | 0.022 |
| PVC, for | | |
| 4" wall | L.F. | 0.020 |
| 6" wall | " | 0.021 |
| 8" wall | " | 0.022 |

### 04150.50 MASONRY FLASHING

| | UNIT | MAN/HOURS |
|---|---|---|
| Through-wall flashing | | |
| 5 oz. coated copper | S.F. | 0.067 |
| 0.030" elastomeric | " | 0.053 |

## UNIT MASONRY

### 04210.10 BRICK MASONRY

| | UNIT | MAN/HOURS |
|---|---|---|
| Standard size brick, running bond | | |
| Face brick, red (6.4/sf) | | |
| Veneer | S.F. | 0.133 |
| Cavity wall | " | 0.114 |
| 9" solid wall | " | 0.229 |
| Common brick (6.4/sf) | | |
| Select common for veneers | S.F. | 0.133 |
| Back-up | | |
| 4" thick | S.F. | 0.100 |
| 8" thick | " | 0.160 |
| Firewall | | |
| 12" thick | S.F. | 0.267 |
| 16" thick | " | 0.364 |

## UNIT MASONRY

### 04210.10 BRICK MASONRY

| | UNIT | MAN/HOURS |
|---|---|---|
| Glazed brick (7.4/sf) | | |
| Veneer | S.F. | 0.145 |
| Buff or gray face brick (6.4/sf) | | |
| Veneer | S.F. | 0.133 |
| Cavity wall | " | 0.114 |
| Jumbo or oversize brick (3/sf) | | |
| 4" veneer | S.F. | 0.080 |
| 4" back-up | " | 0.067 |
| 8" back-up | " | 0.114 |
| 12" firewall | " | 0.200 |
| 16" firewall | " | 0.267 |
| Norman brick, red face, (4.5/sf) | | |
| 4" veneer | S.F. | 0.100 |
| Cavity wall | " | 0.089 |
| Chimney, standard brick, including flue | | |
| 16" x 16" | L.F. | 0.800 |
| 16" x 20" | " | 0.800 |
| 16" x 24" | " | 0.800 |
| 20" x 20" | " | 1.000 |
| 20" x 24" | " | 1.000 |
| 20" x 32" | " | 1.143 |
| Window sill, face brick on edge | " | 0.200 |

### 04210.20 STRUCTURAL TILE

| | UNIT | MAN/HOURS |
|---|---|---|
| Structural glazed tile | | |
| 6T series, 5-1/2" x 12" | | |
| Glazed on one side | | |
| 2" thick | S.F. | 0.080 |
| 4" thick | " | 0.080 |
| 6" thick | " | 0.089 |
| 8" thick | " | 0.100 |
| Glazed on two sides | | |
| 4" thick | S.F. | 0.100 |
| 6" thick | " | 0.114 |
| Special shapes | | |
| Group 1 | S.F. | 0.160 |
| Group 2 | " | 0.160 |
| Group 3 | " | 0.160 |
| Group 4 | " | 0.160 |
| Group 5 | " | 0.160 |
| Fire rated | | |
| 4" thick, 1 hr rating | S.F. | 0.080 |
| 6" thick, 2 hr rating | " | 0.089 |
| 8W series, 8" x 16" | | |
| Glazed on one side | | |
| 2" thick | S.F. | 0.053 |
| 4" thick | " | 0.053 |
| 6" thick | " | 0.062 |
| 8" thick | " | 0.062 |
| Glazed on two sides | | |
| 4" thick | S.F. | 0.067 |
| 6" thick | " | 0.080 |
| 8" thick | " | 0.080 |
| Special shapes | | |

# 04 MASONRY

| UNIT MASONRY | UNIT | MAN/ HOURS |
|---|---|---|
| **04210.20 STRUCTURAL TILE** | | |
| Group 1 | S.F. | 0.114 |
| Group 2 | " | 0.114 |
| Group 3 | " | 0.114 |
| Group 4 | " | 0.114 |
| Group 5 | " | 0.114 |
| Fire rated | | |
| 4" thick, 1 hr rating | S.F. | 0.114 |
| 6" thick, 2 hr rating | " | 0.114 |
| **04210.60 PAVERS, MASONRY** | | |
| Brick walk laid on sand, sand joints | | |
| Laid flat, (4.5 per sf) | S.F. | 0.089 |
| Laid on edge, (7.2 per sf) | " | 0.133 |
| Precast concrete patio blocks | | |
| 2" thick | | |
| Natural | S.F. | 0.027 |
| Colors | " | 0.027 |
| Exposed aggregates, local aggregate | | |
| Natural | S.F. | 0.027 |
| Colors | " | 0.027 |
| Granite or limestone aggregate | " | 0.027 |
| White tumblestone aggregate | " | 0.027 |
| Stone pavers, set in mortar | | |
| Bluestone | | |
| 1" thick | | |
| Irregular | S.F. | 0.200 |
| Snapped rectangular | " | 0.160 |
| 1-1/2" thick, random rectangular | " | 0.200 |
| 2" thick, random rectangular | " | 0.229 |
| Slate | | |
| Natural cleft | | |
| Irregular, 3/4" thick | S.F. | 0.229 |
| Random rectangular | | |
| 1-1/4" thick | S.F. | 0.200 |
| 1-1/2" thick | " | 0.222 |
| Granite blocks | | |
| 3" thick, 3" to 6" wide | | |
| 4" to 12" long | S.F. | 0.267 |
| 6" to 15" long | " | 0.229 |
| Crushed stone, white marble, 3" thick | " | 0.016 |
| **04220.10 CONCRETE MASONRY UNITS** | | |
| Hollow, load bearing | | |
| 4" | S.F. | 0.059 |
| 6" | " | 0.062 |
| 8" | " | 0.067 |
| 10" | " | 0.073 |
| 12" | " | 0.080 |
| Solid, load bearing | | |
| 4" | S.F. | 0.059 |
| 6" | " | 0.062 |
| 8" | " | 0.067 |
| 10" | " | 0.073 |
| 12" | " | 0.080 |

| UNIT MASONRY | UNIT | MAN/ HOURS |
|---|---|---|
| **04220.10 CONCRETE MASONRY UNITS** | | |
| Back-up block, 8" x 16" | | |
| 2" | S.F. | 0.046 |
| 4" | " | 0.047 |
| 6" | " | 0.050 |
| 8" | " | 0.053 |
| 10" | " | 0.057 |
| 12" | " | 0.062 |
| Foundation wall, 8" x 16" | | |
| 6" | S.F. | 0.057 |
| 8" | " | 0.062 |
| 10" | " | 0.067 |
| 12" | " | 0.073 |
| Solid | | |
| 6" | S.F. | 0.062 |
| 8" | " | 0.067 |
| 10" | " | 0.073 |
| 12" | " | 0.080 |
| Exterior, styrofoam inserts, standard weight, 8" x 16" | | |
| 6" | S.F. | 0.062 |
| 8" | " | 0.067 |
| 10" | " | 0.073 |
| 12" | " | 0.080 |
| Lightweight | | |
| 6" | S.F. | 0.062 |
| 8" | " | 0.067 |
| 10" | " | 0.073 |
| 12" | " | 0.080 |
| Acoustical slotted block | | |
| 4" | S.F. | 0.073 |
| 6" | " | 0.073 |
| 8" | " | 0.080 |
| Filled cavities | | |
| 4" | S.F. | 0.089 |
| 6" | " | 0.094 |
| 8" | " | 0.100 |
| Hollow, split face | | |
| 4" | S.F. | 0.059 |
| 6" | " | 0.062 |
| 8" | " | 0.067 |
| 10" | " | 0.073 |
| 12" | " | 0.080 |
| Split rib profile | | |
| 4" | S.F. | 0.073 |
| 6" | " | 0.073 |
| 8" | " | 0.080 |
| 10" | " | 0.080 |
| 12" | " | 0.080 |
| High strength block, 3500 psi | | |
| 2" | S.F. | 0.059 |
| 4" | " | 0.062 |
| 6" | " | 0.062 |
| 8" | " | 0.067 |
| 10" | " | 0.073 |
| 12" | " | 0.080 |

| UNIT MASONRY | UNIT | MAN/HOURS |
|---|---|---|
| **04220.10 CONCRETE MASONRY UNITS** | | |
| Solar screen concrete block | | |
| 4" thick | | |
| 6" x 6" | S.F. | 0.178 |
| 8" x 8" | " | 0.160 |
| 12" x 12" | " | 0.123 |
| 8" thick | | |
| 8" x 16" | S.F. | 0.114 |
| Glazed block | | |
| Cove base, glazed 1 side, 2" | L.F. | 0.089 |
| 4" | " | 0.089 |
| 6" | " | 0.100 |
| 8" | " | 0.100 |
| Single face | | |
| 2" | S.F. | 0.067 |
| 4" | " | 0.067 |
| 6" | " | 0.073 |
| 8" | " | 0.080 |
| 10" | " | 0.089 |
| 12" | " | 0.094 |
| Double face | | |
| 4" | S.F. | 0.084 |
| 6" | " | 0.089 |
| 8" | " | 0.100 |
| Corner or bullnose | | |
| 2" | EA. | 0.100 |
| 4" | " | 0.114 |
| 6" | " | 0.114 |
| 8" | " | 0.133 |
| 10" | " | 0.145 |
| 12" | " | 0.160 |
| Gypsum unit masonry | | |
| Partition blocks (12"x30") | | |
| Solid | | |
| 2" | S.F. | 0.032 |
| Hollow | | |
| 3" | S.F. | 0.032 |
| 4" | " | 0.033 |
| 6" | " | 0.036 |
| Vertical reinforcing | | |
| 4' o.c., add 5% to labor | | |
| 2'8" o.c., add 15% to labor | | |
| Interior partitions, add 10% to labor | | |
| **04220.90 BOND BEAMS & LINTELS** | | |
| Bond beam, no grout or reinforcement | | |
| 8" x 16" x | | |
| 4" thick | L.F. | 0.062 |
| 6" thick | " | 0.064 |
| 8" thick | " | 0.067 |
| 10" thick | " | 0.070 |
| 12" thick | " | 0.073 |
| Beam lintel, no grout or reinforcement | | |
| 8" x 16" x | | |
| 10" thick | L.F. | 0.080 |

| UNIT MASONRY | UNIT | MAN/HOURS |
|---|---|---|
| **04220.90 BOND BEAMS & LINTELS** | | |
| 12" thick | L.F. | 0.089 |
| Precast masonry lintel | | |
| 6 lf, 8" high x | | |
| 4" thick | L.F. | 0.133 |
| 6" thick | " | 0.133 |
| 8" thick | " | 0.145 |
| 10" thick | " | 0.145 |
| 10 lf, 8" high x | | |
| 4" thick | L.F. | 0.080 |
| 6" thick | " | 0.080 |
| 8" thick | " | 0.089 |
| 10" thick | " | 0.089 |
| Steel angles and plates | | |
| Minimum | Lb. | 0.011 |
| Maximum | " | 0.020 |
| Various size angle lintels | | |
| 1/4" stock | | |
| 3" x 3" | L.F. | 0.050 |
| 3" x 3-1/2" | " | 0.050 |
| 3/8" stock | | |
| 3" x 4" | L.F. | 0.050 |
| 3-1/2" x 4" | " | 0.050 |
| 4" x 4" | " | 0.050 |
| 5" x 3-1/2" | " | 0.050 |
| 6" x 3-1/2" | " | 0.050 |
| 1/2" stock | | |
| 6" x 4" | L.F. | 0.050 |
| **04240.10 CLAY TILE** | | |
| Hollow clay tile, for back-up, 12" x 12" | | |
| Scored face | | |
| Load bearing | | |
| 4" thick | S.F. | 0.057 |
| 6" thick | " | 0.059 |
| 8" thick | " | 0.062 |
| 10" thick | " | 0.064 |
| 12" thick | " | 0.067 |
| Non-load bearing | | |
| 3" thick | S.F. | 0.055 |
| 4" thick | " | 0.057 |
| 6" thick | " | 0.059 |
| 8" thick | " | 0.062 |
| 12" thick | " | 0.067 |
| Partition, 12" x 12" | | |
| In walls | | |
| 3" thick | S.F. | 0.067 |
| 4" thick | " | 0.067 |
| 6" thick | " | 0.070 |
| 8" thick | " | 0.073 |
| 10" thick | " | 0.076 |
| 12" thick | " | 0.080 |
| Clay tile floors | | |
| 4" thick | S.F. | 0.044 |

# 04 MASONRY

## UNIT MASONRY

| UNIT MASONRY | UNIT | MAN/HOURS |
|---|---|---|
| **04240.10 CLAY TILE** | | |
| 6" thick | S.F. | 0.047 |
| 8" thick | " | 0.050 |
| 10" thick | " | 0.053 |
| 12" thick | " | 0.057 |
| Terra cotta | | |
| Coping, 10" or 12" wide, 3" thick | L.F. | 0.160 |
| **04270.10 GLASS BLOCK** | | |
| Glass block, 4" thick | | |
| 6" x 6" | S.F. | 0.267 |
| 8" x 8" | " | 0.200 |
| 12" x 12" | " | 0.160 |
| **04295.10 PARGING/MASONRY PLASTER** | | |
| Parging | | |
| 1/2" thick | S.F. | 0.053 |
| 3/4" thick | " | 0.067 |
| 1" thick | " | 0.080 |

## STONE

| STONE | UNIT | MAN/HOURS |
|---|---|---|
| **04400.10 STONE** | | |
| Rubble stone | | |
| Walls set in mortar | | |
| 8" thick | S.F. | 0.200 |
| 12" thick | " | 0.320 |
| 18" thick | " | 0.400 |
| 24" thick | " | 0.533 |
| Dry set wall | | |
| 8" thick | S.F. | 0.133 |
| 12" thick | " | 0.200 |
| 18" thick | " | 0.267 |
| 24" thick | " | 0.320 |
| Cut stone | | |
| Imported marble | | |
| Facing panels | | |
| 3/4" thick | S.F. | 0.320 |
| 1-1/2" thick | " | 0.364 |
| 2-1/4" thick | " | 0.444 |
| Base | | |
| 1" thick | | |
| 4" high | L.F. | 0.400 |
| 6" high | " | 0.400 |
| Columns, solid | | |
| Plain faced | C.F. | 5.333 |
| Fluted | " | 5.333 |
| Flooring, travertine, minimum | S.F. | 0.123 |

## STONE

| STONE | UNIT | MAN/HOURS |
|---|---|---|
| **04400.10 STONE** | | |
| Average | S.F. | 0.160 |
| Maximum | " | 0.178 |
| Domestic marble | | |
| Facing panels | | |
| 7/8" thick | S.F. | 0.320 |
| 1-1/2" thick | " | 0.364 |
| 2-1/4" thick | " | 0.444 |
| Stairs | | |
| 12" treads | L.F. | 0.400 |
| 6" risers | " | 0.267 |
| Thresholds, 7/8" thick, 3' long, 4" to 6" wide | | |
| Plain | EA. | 0.667 |
| Beveled | " | 0.667 |
| Window sill | | |
| 6" wide, 2" thick | L.F. | 0.320 |
| Stools | | |
| 5" wide, 7/8" thick | L.F. | 0.320 |
| Limestone panels up to 12' x 5', smooth finish | | |
| 2" thick | S.F. | 0.096 |
| 3" thick | " | 0.096 |
| 4" thick | " | 0.096 |
| Miscellaneous limestone items | | |
| Steps, 14" wide, 6" deep | L.F. | 0.533 |
| Coping, smooth finish | C.F. | 0.267 |
| Sills, lintels, jambs, smooth finish | " | 0.320 |
| Granite veneer facing panels, polished | | |
| 7/8" thick | | |
| Black | S.F. | 0.320 |
| Gray | " | 0.320 |
| Base | | |
| 4" high | L.F. | 0.160 |
| 6" high | " | 0.178 |
| Curbing, straight, 6" x 16" | " | 0.400 |
| Radius curbs, radius over 5' | " | 0.533 |
| Ashlar veneer | | |
| 4" thick, random | S.F. | 0.320 |
| Pavers, 4" x 4" split | | |
| Gray | S.F. | 0.160 |
| Pink | " | 0.160 |
| Black | " | 0.160 |
| Slate, panels | | |
| 1" thick | S.F. | 0.320 |
| 2" thick | " | 0.364 |
| Sills or stools | | |
| 1" thick | | |
| 6" wide | L.F. | 0.320 |
| 10" wide | " | 0.348 |
| 2" thick | | |
| 6" wide | L.F. | 0.364 |
| 10" wide | " | 0.400 |

# 04 MASONRY

| MASONRY RESTORATION | UNIT | MAN/ HOURS |
|---|---|---|
| **04520.10 RESTORATION AND CLEANING** | | |
| Masonry cleaning | | |
| Washing brick | | |
| Smooth surface | S.F. | 0.013 |
| Rough surface | " | 0.018 |
| Steam clean masonry | | |
| Smooth face | | |
| Minimum | S.F. | 0.010 |
| Maximum | " | 0.015 |
| Rough face | | |
| Minimum | S.F. | 0.013 |
| Maximum | " | 0.020 |
| Sandblast masonry | | |
| Minimum | S.F. | 0.016 |
| Maximum | " | 0.027 |
| Pointing masonry | | |
| Brick | S.F. | 0.032 |
| Concrete block | " | 0.023 |
| Cut and repoint | | |
| Brick | | |
| Minimum | S.F. | 0.040 |
| Maximum | " | 0.080 |
| Stone work | L.F. | 0.062 |
| Cut and recaulk | | |
| Oil base caulks | L.F. | 0.053 |
| Butyl caulks | " | 0.053 |
| Polysulfides and acrylics | " | 0.053 |
| Silicones | " | 0.053 |
| Cement and sand grout on walls, to 1/8" thick | | |
| Minimum | S.F. | 0.032 |
| Maximum | " | 0.040 |
| Brick removal and replacement | | |
| Minimum | EA. | 0.100 |
| Average | " | 0.133 |
| Maximum | " | 0.400 |
| **04550.10 REFRACTORIES** | | |
| Flue liners | | |
| Rectangular | | |
| 8" x 12" | L.F. | 0.133 |
| 12" x 12" | " | 0.145 |
| 12" x 18" | " | 0.160 |
| 16" x 16" | " | 0.178 |
| 18" x 18" | " | 0.190 |
| 20" x 20" | " | 0.200 |
| 24" x 24" | " | 0.229 |
| Round | | |
| 18" dia. | L.F. | 0.190 |
| 24" dia. | " | 0.229 |

# 05 METALS

## METAL FASTENING

| 05050.10 STRUCTURAL WELDING | UNIT | MAN/HOURS |
|---|---|---|
| Welding | | |
| Single pass | | |
| 1/8" | L.F. | 0.040 |
| 3/16" | " | 0.053 |
| 1/4" | " | 0.067 |
| Miscellaneous steel shapes | | |
| Plain | Lb. | 0.002 |
| Galvanized | " | 0.003 |
| Plates | | |
| Plain | Lb. | 0.002 |
| Galvanized | " | 0.003 |

| 05050.95 METAL LINTELS | UNIT | MAN/HOURS |
|---|---|---|
| Lintels, steel | | |
| Plain | Lb. | 0.020 |
| Galvanized | " | 0.020 |

| 05120.10 STRUCTURAL STEEL | UNIT | MAN/HOURS |
|---|---|---|
| Beams and girders, A-36 | | |
| Welded | TON | 4.800 |
| Bolted | " | 4.364 |
| Columns | | |
| Pipe | | |
| 6" dia. | Lb. | 0.005 |
| 12" dia. | " | 0.004 |
| Purlins and girts | | |
| Welded | TON | 8.000 |
| Bolted | " | 6.857 |
| Column base plates | | |
| Up to 150 lb each | Lb. | 0.005 |
| Over 150 lb each | " | 0.007 |
| Structural pipe | | |
| 3" to 5" o.d. | TON | 9.600 |
| 6" to 12" o.d. | " | 6.857 |
| Structural tube | | |
| 6" square | | |
| Light sections | TON | 9.600 |
| Heavy sections | " | 6.857 |
| 6" wide rectangular | | |
| Light sections | TON | 8.000 |
| Heavy sections | " | 6.000 |
| Greater than 6" wide rectangular | | |
| Light sections | TON | 8.000 |
| Heavy sections | " | 6.000 |
| Miscellaneous structural shapes | | |
| Steel angle | TON | 12.000 |
| Steel plate | " | 8.000 |
| Trusses, field welded | | |
| 60 lb/lf | TON | 6.000 |
| 100 lb/lf | " | 4.800 |
| 150 lb/lf | " | 4.000 |
| Bolted | | |

## METAL FASTENING

| 05120.10 STRUCTURAL STEEL | UNIT | MAN/HOURS |
|---|---|---|
| 60 lb/lf | TON | 5.333 |
| 100 lb/lf | " | 4.364 |
| 150 lb/lf | " | 3.692 |

| 05200.10 METAL JOISTS | UNIT | MAN/HOURS |
|---|---|---|
| Joist | | |
| DLH series | TON | 3.200 |
| K series | " | 3.200 |
| LH series | " | 3.200 |

| 05300.10 METAL DECKING | UNIT | MAN/HOURS |
|---|---|---|
| Roof, 1-1/2" deep, non-composite | | |
| 16 ga. | | |
| Primed | S.F. | 0.008 |
| Galvanized | " | 0.008 |
| 18 ga. | | |
| Primed | S.F. | 0.008 |
| Galvanized | " | 0.008 |
| 20 ga. | | |
| Primed | S.F. | 0.008 |
| Galvanized | " | 0.008 |
| 22 ga. | | |
| Primed | S.F. | 0.008 |
| Galvanized | " | 0.008 |
| Open type decking, galvanized | | |
| 1-1/2" deep | | |
| 18 ga. | S.F. | 0.008 |
| 20 ga. | " | 0.008 |
| 22 ga. | " | 0.008 |
| 3" deep | | |
| 16 ga. | S.F. | 0.009 |
| 18 ga. | " | 0.009 |
| 20 ga. | " | 0.009 |
| 22 ga. | " | 0.009 |
| 4-1/2" deep | | |
| 16 ga. | S.F. | 0.010 |
| 18 ga. | " | 0.010 |
| 6" deep | | |
| 16 ga. | S.F. | 0.011 |
| 18 ga. | " | 0.011 |
| 7-1/2" deep | | |
| 16 ga. | S.F. | 0.011 |
| 18 ga. | " | 0.011 |
| Cellular type | | |
| 1-1/2" deep, galvanized | | |
| 18-18 ga. | S.F. | 0.010 |
| 22-18 ga. | " | 0.010 |
| 3" deep, galvanized | | |
| 16-16 ga. | S.F. | 0.011 |
| 18-16 ga. | " | 0.011 |
| 18-18 ga. | " | 0.011 |
| 20-18 ga. | " | 0.011 |
| 4-1/2" deep, galvanized | | |

## METAL FASTENING

### 05300.10 METAL DECKING

| METAL FASTENING | UNIT | MAN/HOURS |
|---|---|---|
| 16-16 ga. | S.F. | 0.011 |
| 18-16 ga. | " | 0.011 |
| 18-18 ga. | " | 0.011 |
| 20-18 ga. | " | 0.011 |
| Composite deck, non-cellular, galvanized | | |
| 1-1/2" deep | | |
| 18 ga. | S.F. | 0.009 |
| 20 ga. | " | 0.009 |
| 22 ga. | " | 0.009 |
| 3" deep | | |
| 18 ga. | S.F. | 0.009 |
| 20 ga. | " | 0.009 |
| 22 ga. | " | 0.009 |
| Slab form floor deck | | |
| 9/16" deep | | |
| 28 ga. | S.F. | 0.009 |
| 1-5/16" deep | | |
| 24 ga. | S.F. | 0.009 |
| 22 ga. | " | 0.009 |

## COLD FORMED FRAMING

### 05410.10 METAL FRAMING

| COLD FORMED FRAMING | UNIT | MAN/HOURS |
|---|---|---|
| Furring channel, galvanized | | |
| Beams and columns, 3/4" | | |
| 12" o.c. | S.F. | 0.080 |
| 16" o.c. | " | 0.073 |
| Walls, 3/4" | | |
| 12" o.c. | S.F. | 0.040 |
| 16" o.c. | " | 0.033 |
| 24" o.c. | " | 0.027 |
| 1-1/2" | | |
| 12" o.c. | S.F. | 0.040 |
| 16" o.c. | " | 0.033 |
| 24" o.c. | " | 0.027 |
| Stud, load bearing | | |
| 16" o.c. | | |
| 16 ga. | | |
| 2-1/2" | S.F. | 0.036 |
| 3-5/8" | " | 0.036 |
| 4" | " | 0.036 |
| 6" | " | 0.040 |
| 18 ga. | | |
| 2-1/2" | S.F. | 0.036 |
| 3-5/8" | " | 0.036 |
| 4" | " | 0.036 |

## COLD FORMED FRAMING

### 05410.10 METAL FRAMING

| COLD FORMED FRAMING | UNIT | MAN/HOURS |
|---|---|---|
| 6" | S.F. | 0.040 |
| 8" | " | 0.040 |
| 20 ga. | | |
| 2-1/2" | S.F. | 0.036 |
| 3-5/8" | " | 0.036 |
| 4" | " | 0.036 |
| 6" | " | 0.040 |
| 8" | " | 0.040 |
| 24" o.c. | | |
| 16 ga. | | |
| 2-1/2" | S.F. | 0.031 |
| 3-5/8" | " | 0.031 |
| 4" | " | 0.031 |
| 6" | " | 0.033 |
| 8" | " | 0.033 |
| 18 ga. | | |
| 2-1/2" | S.F. | 0.031 |
| 3-5/8" | " | 0.031 |
| 4" | " | 0.031 |
| 6" | " | 0.033 |
| 8" | " | 0.033 |
| 20 ga. | | |
| 2-1/2" | S.F. | 0.031 |
| 3-5/8" | " | 0.031 |
| 4" | " | 0.031 |
| 6" | " | 0.033 |
| 8" | " | 0.033 |

## METAL FABRICATIONS

### 05510.10 STAIRS

| METAL FABRICATIONS | UNIT | MAN/HOURS |
|---|---|---|
| Stock unit, steel, complete, per riser | | |
| Tread | | |
| 3'-6" wide | EA. | 1.000 |
| 4' wide | " | 1.143 |
| 5' wide | " | 1.333 |
| Metal pan stair, cement filled, per riser | | |
| 3'-6" wide | EA. | 0.800 |
| 4' wide | " | 0.889 |
| 5' wide | " | 1.000 |
| Landing, steel pan | S.F. | 0.200 |
| Cast iron tread, steel stringers, stock units, per riser | | |
| Tread | | |
| 3'-6" wide | EA. | 1.000 |
| 4' wide | " | 1.143 |
| 5' wide | " | 1.333 |

# 05 METALS

## METAL FABRICATIONS

| METAL FABRICATIONS | UNIT | MAN/ HOURS |
|---|---|---|
| **05510.10 STAIRS** | | |
| Stair treads, abrasive, 12" x 3'-6" | | |
| Cast iron | | |
| 3/8" | EA. | 0.400 |
| 1/2" | " | 0.400 |
| Cast aluminum | | |
| 5/16" | EA. | 0.400 |
| 3/8" | " | 0.400 |
| 1/2" | " | 0.400 |
| **05515.10 LADDERS** | | |
| Ladder, 18" wide | | |
| With cage | L.F. | 0.533 |
| Without cage | " | 0.400 |
| **05520.10 RAILINGS** | | |
| Railing, pipe | | |
| 1-1/4" diameter, welded steel | | |
| 2-rail | | |
| Primed | L.F. | 0.160 |
| Galvanized | " | 0.160 |
| 3-rail | | |
| Primed | L.F. | 0.200 |
| Galvanized | " | 0.200 |
| Wall mounted, single rail, welded steel | | |
| Primed | L.F. | 0.123 |
| Galvanized | " | 0.123 |
| 1-1/2" diameter, welded steel | | |
| 2-rail | | |
| Primed | L.F. | 0.160 |
| Galvanized | " | 0.160 |
| 3-rail | | |
| Primed | L.F. | 0.200 |
| Galvanized | " | 0.200 |
| Wall mounted, single rail, welded steel | | |
| Primed | L.F. | 0.123 |
| Galvanized | " | 0.123 |
| 2" diameter, welded steel | | |
| 2-rail | | |
| Primed | L.F. | 0.178 |
| Galvanized | " | 0.178 |
| 3-rail | | |
| Primed | L.F. | 0.229 |
| Galvanized | " | 0.229 |
| Wall mounted, single rail, welded steel | | |
| Primed | L.F. | 0.133 |
| Galvanized | " | 0.133 |
| **05530.10 METAL GRATING** | | |
| Floor plate, checkered, steel | | |
| 1/4" | | |
| Primed | S.F. | 0.011 |
| Galvanized | " | 0.011 |

| METAL FABRICATIONS | UNIT | MAN/ HOURS |
|---|---|---|
| **05530.10 METAL GRATING** | | |
| 3/8" | | |
| Primed | S.F. | 0.012 |
| Galvanized | " | 0.012 |
| Aluminum grating, pressure-locked bearing bars | | |
| 3/4" x 1/8" | S.F. | 0.020 |
| 1" x 1/8" | " | 0.020 |
| 1-1/4" x 1/8" | " | 0.020 |
| 1-1/4" x 3/16" | " | 0.020 |
| 1-1/2" x 1/8" | " | 0.020 |
| 1-3/4" x 3/16" | " | 0.020 |
| Miscellaneous expenses | | |
| Cutting | | |
| Minimum | L.F. | 0.053 |
| Maximum | " | 0.080 |
| Banding | | |
| Minimum | L.F. | 0.133 |
| Maximum | " | 0.160 |
| Toe plates | | |
| Minimum | L.F. | 0.160 |
| Maximum | " | 0.200 |
| Steel grating, primed | | |
| 3/4" x 1/8" | S.F. | 0.027 |
| 1" x 1/8" | " | 0.027 |
| 1-1/4" x 1/8" | " | 0.027 |
| 1-1/4" x 3/16" | " | 0.027 |
| 1-1/2" x 1/8" | " | 0.027 |
| 1-3/4" x 3/16" | " | 0.027 |
| Galvanized | | |
| 3/4" x 1/8" | S.F. | 0.027 |
| 1" x 1/8" | " | 0.027 |
| 1-1/4" x 1/8" | " | 0.027 |
| 1-1/4" x 3/16" | " | 0.027 |
| 1-1/2" x 1/8" | " | 0.027 |
| 1-3/4" x 3/16" | " | 0.027 |
| Miscellaneous expenses | | |
| Cutting | | |
| Minimum | L.F. | 0.057 |
| Maximum | " | 0.089 |
| Banding | | |
| Minimum | L.F. | 0.145 |
| Maximum | " | 0.178 |
| Toe plates | | |
| Minimum | L.F. | 0.178 |
| Maximum | " | 0.229 |
| **05540.10 CASTINGS** | | |
| Miscellaneous castings | | |
| Light sections | Lb. | 0.016 |
| Heavy sections | " | 0.011 |
| Manhole covers and frames | | |
| Regular, city type | | |
| 18" dia. | | |
| 100 lb | EA. | 1.600 |
| 24" dia. | | |

# 05 METALS

## METAL FABRICATIONS

| 05540.10 CASTINGS | UNIT | MAN/ HOURS |
|---|---|---|
| 200 lb | EA. | 1.600 |
| 300 lb | " | 1.778 |
| 400 lb | " | 1.778 |
| 26" dia., 475 lb | " | 2.000 |
| 30" dia., 600 lb | " | 2.286 |
| 8" square, 75 lb | " | 0.320 |
| 24" square | | |
| 126 lb | EA. | 1.600 |
| 500 lb | " | 2.000 |
| Watertight type | | |
| 20" dia., 200 lb | EA. | 2.000 |
| 24" dia., 350 lb | " | 2.667 |
| Steps, cast iron | | |
| 7" x 9" | EA. | 0.160 |
| 8" x 9" | " | 0.178 |
| Manhole covers and frames, aluminum | | |
| 12" x 12" | EA. | 0.320 |
| 18" x 18" | " | 0.320 |
| 24" x 24" | " | 0.400 |
| Corner protection | | |
| Steel angle guard with anchors | | |
| 2" x 2" x 3/16" | L.F. | 0.114 |
| 2" x 3" x 1/4" | " | 0.114 |
| 3" x 3" x 5/16" | " | 0.114 |
| 3" x 4" x 5/16" | " | 0.123 |
| 4" x 4" x 5/16" | " | 0.123 |

## MISC. FABRICATIONS

| 05580.10 METAL SPECIALTIES | UNIT | MAN/ HOURS |
|---|---|---|
| Kick plate | | |
| 4" high x 1/4" thick | | |
| Primed | L.F. | 0.160 |
| Galvanized | " | 0.160 |
| 6" high x 1/4" thick | | |
| Primed | L.F. | 0.178 |
| Galvanized | " | 0.178 |

| 05700.10 ORNAMENTAL METAL | UNIT | MAN/ HOURS |
|---|---|---|
| Railings, vertical square bars, 6" o.c., with shaped top rails | | |
| Steel | L.F. | 0.400 |
| Aluminum | " | 0.400 |
| Bronze | " | 0.533 |
| Stainless steel | " | 0.533 |
| Laminated metal or wood handrails with metal supports | | |
| 2-1/2" round or oval shape | L.F. | 0.400 |

## MISC. FABRICATIONS

| 05700.10 ORNAMENTAL METAL | UNIT | MAN/ HOURS |
|---|---|---|
| Grilles and louvers | | |
| Fixed type louvers | | |
| 4 through 10 sf | S.F. | 0.133 |
| Over 10 sf | " | 0.100 |
| Movable type louvers | | |
| 4 through 10 sf | S.F. | 0.133 |
| Over 10 sf | " | 0.100 |
| Aluminum louvers | | |
| Residential use, fixed type, with screen | | |
| 8" x 8" | EA. | 0.400 |
| 12" x 12" | " | 0.400 |
| 12" x 18" | " | 0.400 |
| 14" x 24" | " | 0.400 |
| 18" x 24" | " | 0.400 |
| 30" x 24" | " | 0.444 |

| 05800.10 EXPANSION CONTROL | UNIT | MAN/ HOURS |
|---|---|---|
| Expansion joints with covers, floor assembly type | | |
| With 1" space | | |
| Aluminum | L.F. | 0.133 |
| Bronze | " | 0.133 |
| Stainless steel | " | 0.133 |
| With 2" space | | |
| Aluminum | L.F. | 0.133 |
| Bronze | " | 0.133 |
| Stainless steel | " | 0.133 |
| Ceiling and wall assembly type | | |
| With 1" space | | |
| Aluminum | L.F. | 0.160 |
| Bronze | " | 0.160 |
| Stainless steel | " | 0.160 |
| With 2" space | | |
| Aluminum | L.F. | 0.160 |
| Bronze | " | 0.160 |
| Stainless steel | " | 0.160 |
| Exterior roof and wall, aluminum | | |
| Roof to roof | | |
| With 1" space | L.F. | 0.133 |
| With 2" space | " | 0.133 |
| Roof to wall | | |
| With 1" space | L.F. | 0.145 |
| With 2" space | " | 0.145 |
| Flat wall to wall | | |
| With 1" space | L.F. | 0.133 |
| With 2" space | " | 0.133 |
| Corner to flat wall | | |
| With 1" space | L.F. | 0.160 |
| With 2" in space | " | 0.160 |

# 06 WOOD AND PLASTICS

## FASTENERS AND ADHESIVES

### 06050.10 ACCESSORIES

| FASTENERS AND ADHESIVES | UNIT | MAN/ HOURS |
|---|---|---|
| Column/post base, cast aluminum | | |
| 4" x 4" | EA. | 0.200 |
| 6" x 6" | " | 0.200 |
| Bridging, metal, per pair | | |
| 12" o.c. | EA. | 0.080 |
| 16" o.c. | " | 0.073 |
| Anchors | | |
| Bolts, threaded two ends, with nuts and washers | | |
| 1/2" dia. | | |
| 4" long | EA. | 0.050 |
| 7-1/2" long | " | 0.050 |
| 3/4" dia. | | |
| 7-1/2" long | EA. | 0.050 |
| 15" long | " | 0.050 |
| Framing anchors | | |
| 10 gauge | EA. | 0.067 |
| Bolts, carriage | | |
| 1/4 x 4 | EA. | 0.080 |
| 5/16 x 6 | " | 0.084 |
| 3/8 x 6 | " | 0.084 |
| 1/2 x 6 | " | 0.084 |
| Joist and beam hangers | | |
| 18 ga. | | |
| 2 x 4 | EA. | 0.080 |
| 2 x 6 | " | 0.080 |
| 2 x 8 | " | 0.080 |
| 2 x 10 | " | 0.089 |
| 2 x 12 | " | 0.100 |
| 16 ga. | | |
| 3 x 6 | EA. | 0.089 |
| 3 x 8 | " | 0.089 |
| 3 x 10 | " | 0.094 |
| 3 x 12 | " | 0.107 |
| 3 x 14 | " | 0.114 |
| 4 x 6 | " | 0.089 |
| 4 x 8 | " | 0.089 |
| 4 x 10 | " | 0.094 |
| 4 x 12 | " | 0.107 |
| 4 x 14 | " | 0.114 |
| Rafter anchors, 18 ga., 1-1/2" wide | | |
| 5-1/4" long | EA. | 0.067 |
| 10-3/4" long | " | 0.067 |
| Shear plates | | |
| 2-5/8" dia. | EA. | 0.062 |
| 4" dia. | " | 0.067 |
| Sill anchors | | |
| Embedded in concrete | EA. | 0.080 |
| Split rings | | |
| 2-1/2" dia. | EA. | 0.089 |
| 4" dia. | " | 0.100 |
| Strap ties, 14 ga., 1-3/8" wide | | |
| 12" long | EA. | 0.067 |
| 18" long | " | 0.073 |
| 24" long | " | 0.080 |
| 36" long | EA. | 0.089 |
| Toothed rings | | |
| 2-5/8" dia. | EA. | 0.133 |
| 4" dia. | " | 0.160 |

## ROUGH CARPENTRY

### 06110.10 BLOCKING

| ROUGH CARPENTRY | UNIT | MAN/ HOURS |
|---|---|---|
| Steel construction | | |
| Walls | | |
| 2x4 | L.F. | 0.053 |
| 2x6 | " | 0.062 |
| 2x8 | " | 0.067 |
| 2x10 | " | 0.073 |
| 2x12 | " | 0.080 |
| Ceilings | | |
| 2x4 | L.F. | 0.062 |
| 2x6 | " | 0.073 |
| 2x8 | " | 0.080 |
| 2x10 | " | 0.089 |
| 2x12 | " | 0.100 |
| Wood construction | | |
| Walls | | |
| 2x4 | L.F. | 0.044 |
| 2x6 | " | 0.050 |
| 2x8 | " | 0.053 |
| 2x10 | " | 0.057 |
| 2x12 | " | 0.062 |
| Ceilings | | |
| 2x4 | L.F. | 0.050 |
| 2x6 | " | 0.057 |
| 2x8 | " | 0.062 |
| 2x10 | " | 0.067 |
| 2x12 | " | 0.073 |

### 06110.20 CEILING FRAMING

| ROUGH CARPENTRY | UNIT | MAN/ HOURS |
|---|---|---|
| Ceiling joists | | |
| 12" o.c. | | |
| 2x4 | S.F. | 0.019 |
| 2x6 | " | 0.020 |
| 2x8 | " | 0.021 |
| 2x10 | " | 0.022 |
| 2x12 | " | 0.024 |
| 16" o.c. | | |
| 2x4 | S.F. | 0.015 |
| 2x6 | " | 0.016 |
| 2x8 | " | 0.017 |

# 06 WOOD AND PLASTICS

| ROUGH CARPENTRY | UNIT | MAN/HOURS |
|---|---|---|
| **06110.20 CEILING FRAMING** | | |
| 2x10 | S.F. | 0.017 |
| 2x12 | " | 0.018 |
| 24" o.c. | | |
| 2x4 | S.F. | 0.013 |
| 2x6 | " | 0.013 |
| 2x8 | " | 0.014 |
| 2x10 | " | 0.015 |
| 2x12 | " | 0.016 |
| Headers and nailers | | |
| 2x4 | L.F. | 0.026 |
| 2x6 | " | 0.027 |
| 2x8 | " | 0.029 |
| 2x10 | " | 0.031 |
| 2x12 | " | 0.033 |
| Sister joists for ceilings | | |
| 2x4 | L.F. | 0.057 |
| 2x6 | " | 0.067 |
| 2x8 | " | 0.080 |
| 2x10 | " | 0.100 |
| 2x12 | " | 0.133 |
| **06110.30 FLOOR FRAMING** | | |
| Floor joists | | |
| 12" o.c. | | |
| 2x6 | S.F. | 0.016 |
| 2x8 | " | 0.016 |
| 2x10 | " | 0.017 |
| 2x12 | " | 0.017 |
| 2x14 | " | 0.017 |
| 3x6 | " | 0.017 |
| 3x8 | " | 0.017 |
| 3x10 | " | 0.018 |
| 3x12 | " | 0.019 |
| 3x14 | " | 0.020 |
| 4x6 | " | 0.017 |
| 4x8 | " | 0.017 |
| 4x10 | " | 0.018 |
| 4x12 | " | 0.019 |
| 4x14 | " | 0.020 |
| 16" o.c. | | |
| 2x6 | S.F. | 0.013 |
| 2x8 | " | 0.014 |
| 2x10 | " | 0.014 |
| 2x12 | " | 0.014 |
| 2x14 | " | 0.015 |
| 3x6 | " | 0.014 |
| 3x8 | " | 0.014 |
| 3x10 | " | 0.015 |
| 3x12 | " | 0.015 |
| 3x14 | " | 0.016 |
| 4x6 | " | 0.014 |
| 4x8 | " | 0.014 |
| 4x10 | " | 0.015 |
| 4x12 | " | 0.015 |

| ROUGH CARPENTRY | UNIT | MAN/HOURS |
|---|---|---|
| **06110.30 FLOOR FRAMING** | | |
| 4x14 | S.F. | 0.016 |
| Sister joists for floors | | |
| 2x4 | L.F. | 0.050 |
| 2x6 | " | 0.057 |
| 2x8 | " | 0.067 |
| 2x10 | " | 0.080 |
| 2x12 | " | 0.100 |
| 3x6 | " | 0.080 |
| 3x8 | " | 0.089 |
| 3x10 | " | 0.100 |
| 3x12 | " | 0.114 |
| 4x6 | " | 0.080 |
| 4x8 | " | 0.089 |
| 4x10 | " | 0.100 |
| 4x12 | " | 0.114 |
| **06110.40 FURRING** | | |
| Furring, wood strips | | |
| Walls | | |
| On masonry or concrete walls | | |
| 1x2 furring | | |
| 12" o.c. | S.F. | 0.025 |
| 16" o.c. | " | 0.023 |
| 24" o.c. | " | 0.021 |
| 1x3 furring | | |
| 12" o.c. | S.F. | 0.025 |
| 16" o.c. | " | 0.023 |
| 24" o.c. | " | 0.021 |
| On wood walls | | |
| 1x2 furring | | |
| 12" o.c. | S.F. | 0.018 |
| 16" o.c. | " | 0.016 |
| 24" o.c. | " | 0.015 |
| 1x3 furring | | |
| 12" o.c. | S.F. | 0.018 |
| 16" o.c. | " | 0.016 |
| 24" o.c. | " | 0.015 |
| Ceilings | | |
| On masonry or concrete ceilings | | |
| 1x2 furring | | |
| 12" o.c. | S.F. | 0.044 |
| 16" o.c. | " | 0.040 |
| 24" o.c. | " | 0.036 |
| 1x3 furring | | |
| 12" o.c. | S.F. | 0.044 |
| 16" o.c. | " | 0.040 |
| 24" o.c. | " | 0.036 |
| On wood ceilings | | |
| 1x2 furring | | |
| 12" o.c. | S.F. | 0.030 |
| 16" o.c. | " | 0.027 |
| 24" o.c. | " | 0.024 |
| 1x3 | | |

# 06 WOOD AND PLASTICS

| ROUGH CARPENTRY | UNIT | MAN/ HOURS |
|---|---|---|
| **06110.40 FURRING** | | |
| 12" o.c. | S.F. | 0.030 |
| 16" o.c. | " | 0.027 |
| 24" o.c. | " | 0.024 |
| **06110.50 ROOF FRAMING** | | |
| Roof framing | | |
| Rafters, gable end | | |
| 0-2 pitch (flat to 2-in-12) | | |
| 12" o.c. | | |
| 2x4 | S.F. | 0.017 |
| 2x6 | " | 0.017 |
| 2x8 | " | 0.018 |
| 2x10 | " | 0.019 |
| 2x12 | " | 0.020 |
| 16" o.c. | | |
| 2x6 | S.F. | 0.014 |
| 2x8 | " | 0.015 |
| 2x10 | " | 0.015 |
| 2x12 | " | 0.016 |
| 24" o.c. | | |
| 2x6 | S.F. | 0.012 |
| 2x8 | " | 0.013 |
| 2x10 | " | 0.013 |
| 2x12 | " | 0.013 |
| 4-6 pitch (4-in-12 to 6-in-12) | | |
| 12" o.c. | | |
| 2x4 | S.F. | 0.017 |
| 2x6 | " | 0.018 |
| 2x8 | " | 0.019 |
| 2x10 | " | 0.020 |
| 2x12 | " | 0.021 |
| 16" o.c. | | |
| 2x6 | S.F. | 0.015 |
| 2x8 | " | 0.015 |
| 2x10 | " | 0.016 |
| 2x12 | " | 0.017 |
| 24" o.c. | | |
| 2x6 | S.F. | 0.013 |
| 2x8 | " | 0.013 |
| 2x10 | " | 0.014 |
| 2x12 | " | 0.015 |
| 8-12 pitch (8-in-12 to 12-in-12) | | |
| 12" o.c. | | |
| 2x4 | S.F. | 0.018 |
| 2x6 | " | 0.019 |
| 2x8 | " | 0.020 |
| 2x10 | " | 0.021 |
| 2x12 | " | 0.022 |
| 16" o.c. | | |
| 2x6 | S.F. | 0.015 |
| 2x8 | " | 0.016 |
| 2x10 | " | 0.017 |
| 2x12 | " | 0.017 |

| ROUGH CARPENTRY | UNIT | MAN/ HOURS |
|---|---|---|
| **06110.50 ROOF FRAMING** | | |
| 24" o.c. | | |
| 2x6 | S.F. | 0.013 |
| 2x8 | " | 0.013 |
| 2x10 | " | 0.014 |
| 2x12 | " | 0.014 |
| Ridge boards | | |
| 2x6 | L.F. | 0.040 |
| 2x8 | " | 0.044 |
| 2x10 | " | 0.050 |
| 2x12 | " | 0.057 |
| Hip rafters | | |
| 2x6 | L.F. | 0.029 |
| 2x8 | " | 0.030 |
| 2x10 | " | 0.031 |
| 2x12 | " | 0.032 |
| Jack rafters | | |
| 4-6 pitch (4-in-12 to 6-in-12) | | |
| 16" o.c. | | |
| 2x6 | S.F. | 0.024 |
| 2x8 | " | 0.024 |
| 2x10 | " | 0.026 |
| 2x12 | " | 0.027 |
| 24" o.c. | | |
| 2x6 | S.F. | 0.018 |
| 2x8 | " | 0.019 |
| 2x10 | " | 0.020 |
| 2x12 | " | 0.020 |
| 8-12 pitch (8-in-12 to 12-in-12) | | |
| 16" o.c. | | |
| 2x6 | S.F. | 0.025 |
| 2x8 | " | 0.026 |
| 2x10 | " | 0.027 |
| 2x12 | " | 0.028 |
| 24" o.c. | | |
| 2x6 | S.F. | 0.019 |
| 2x8 | " | 0.020 |
| 2x10 | " | 0.020 |
| 2x12 | " | 0.021 |
| Sister rafters | | |
| 2x4 | L.F. | 0.057 |
| 2x6 | " | 0.067 |
| 2x8 | " | 0.080 |
| 2x10 | " | 0.100 |
| 2x12 | " | 0.133 |
| Fascia boards | | |
| 2x4 | L.F. | 0.040 |
| 2x6 | " | 0.040 |
| 2x8 | " | 0.044 |
| 2x10 | " | 0.044 |
| 2x12 | " | 0.050 |
| Cant strips | | |
| Fiber | | |
| 3x3 | L.F. | 0.023 |
| 4x4 | " | 0.024 |

# 06 WOOD AND PLASTICS

| ROUGH CARPENTRY | UNIT | MAN/HOURS |
|---|---|---|
| **06110.50 ROOF FRAMING** | | |
| Wood | | |
| 3x3 | L.F. | 0.024 |
| **06110.60 SLEEPERS** | | |
| Sleepers, over concrete | | |
| 12" o.c. | | |
| 1x2 | S.F. | 0.018 |
| 1x3 | " | 0.019 |
| 2x4 | " | 0.022 |
| 2x6 | " | 0.024 |
| 16" o.c. | | |
| 1x2 | S.F. | 0.016 |
| 1x3 | " | 0.016 |
| 2x4 | " | 0.019 |
| 2x6 | " | 0.020 |
| **06110.65 SOFFITS** | | |
| Soffit framing | | |
| 2x3 | L.F. | 0.057 |
| 2x4 | " | 0.062 |
| 2x6 | " | 0.067 |
| 2x8 | " | 0.073 |
| **06110.70 WALL FRAMING** | | |
| Framing wall, studs | | |
| 12" o.c. | | |
| 2x3 | S.F. | 0.015 |
| 2x4 | " | 0.015 |
| 2x6 | " | 0.016 |
| 2x8 | " | 0.017 |
| 16" o.c. | | |
| 2x3 | S.F. | 0.013 |
| 2x4 | " | 0.013 |
| 2x6 | " | 0.013 |
| 2x8 | " | 0.014 |
| 24" o.c. | | |
| 2x3 | S.F. | 0.011 |
| 2x4 | " | 0.011 |
| 2x6 | " | 0.011 |
| 2x8 | " | 0.012 |
| Plates, top or bottom | | |
| 2x3 | L.F. | 0.024 |
| 2x4 | " | 0.025 |
| 2x6 | " | 0.027 |
| 2x8 | " | 0.029 |
| Headers, door or window | | |
| 2x6 | | |
| Single | | |
| 3' long | EA. | 0.400 |
| 6' long | " | 0.500 |
| Double | | |
| 3' long | EA. | 0.444 |

| ROUGH CARPENTRY | UNIT | MAN/HOURS |
|---|---|---|
| **06110.70 WALL FRAMING** | | |
| 6' long | EA. | 0.571 |
| 2x8 | | |
| Single | | |
| 4' long | EA. | 0.500 |
| 8' long | " | 0.615 |
| Double | | |
| 4' long | EA. | 0.571 |
| 8' long | " | 0.727 |
| 2x10 | | |
| Single | | |
| 5' long | EA. | 0.615 |
| 10' long | " | 0.800 |
| Double | | |
| 5' long | EA. | 0.667 |
| 10' long | " | 0.800 |
| 2x12 | | |
| Single | | |
| 6' long | EA. | 0.615 |
| 12' long | " | 0.800 |
| Double | | |
| 6' long | EA. | 0.727 |
| 12' long | " | 0.889 |
| **06115.10 FLOOR SHEATHING** | | |
| Sub-flooring, plywood, CDX | | |
| 1/2" thick | S.F. | 0.010 |
| 5/8" thick | " | 0.011 |
| 3/4" thick | " | 0.013 |
| Structural plywood | | |
| 1/2" thick | S.F. | 0.010 |
| 5/8" thick | " | 0.011 |
| 3/4" thick | " | 0.012 |
| Board type subflooring | | |
| 1x6 | | |
| Minimum | S.F. | 0.018 |
| Maximum | " | 0.020 |
| 1x8 | | |
| Minimum | S.F. | 0.017 |
| Maximum | " | 0.019 |
| 1x10 | | |
| Minimum | S.F. | 0.016 |
| Maximum | " | 0.018 |
| Underlayment | | |
| Hardboard, 1/4" tempered | S.F. | 0.010 |
| Plywood, CDX | | |
| 3/8" thick | S.F. | 0.010 |
| 1/2" thick | " | 0.011 |
| 5/8" thick | " | 0.011 |
| 3/4" thick | " | 0.012 |
| **06115.20 ROOF SHEATHING** | | |
| Sheathing | | |
| Plywood, CDX | | |

# 06 WOOD AND PLASTICS

| ROUGH CARPENTRY | UNIT | MAN/ HOURS |
|---|---|---|
| **06115.20 ROOF SHEATHING** | | |
| 3/8" thick | S.F. | 0.010 |
| 1/2" thick | " | 0.011 |
| 5/8" thick | " | 0.011 |
| 3/4" thick | " | 0.012 |
| Structural plywood | | |
| 3/8" thick | S.F. | 0.010 |
| 1/2" thick | " | 0.011 |
| 5/8" thick | " | 0.011 |
| 3/4" thick | " | 0.012 |
| **06115.30 WALL SHEATHING** | | |
| Sheathing | | |
| Plywood, CDX | | |
| 3/8" thick | S.F. | 0.012 |
| 1/2" thick | " | 0.012 |
| 5/8" thick | " | 0.013 |
| 3/4" thick | " | 0.015 |
| Waferboard | | |
| 3/8" thick | S.F. | 0.012 |
| 1/2" thick | " | 0.012 |
| 5/8" thick | " | 0.013 |
| 3/4" thick | " | 0.015 |
| Structural plywood | | |
| 3/8" thick | S.F. | 0.012 |
| 1/2" thick | " | 0.012 |
| 5/8" thick | " | 0.013 |
| 3/4" thick | " | 0.015 |
| Gypsum, 1/2" thick | " | 0.012 |
| Asphalt impregnated fiberboard, 1/2" thick | " | 0.012 |
| **06125.10 WOOD DECKING** | | |
| Decking, T&G solid | | |
| Cedar | | |
| 3" thick | S.F. | 0.020 |
| 4" thick | " | 0.021 |
| Fir | | |
| 3" thick | S.F. | 0.020 |
| 4" thick | " | 0.021 |
| Southern yellow pine | | |
| 3" thick | S.F. | 0.023 |
| 4" thick | " | 0.025 |
| White pine | | |
| 3" thick | S.F. | 0.020 |
| 4" thick | " | 0.021 |
| **06130.10 HEAVY TIMBER** | | |
| Mill framed structures | | |
| Beams to 20' long | | |
| Douglas fir | | |
| 6x8 | L.F. | 0.080 |
| 6x10 | " | 0.083 |
| 6x12 | " | 0.089 |

| ROUGH CARPENTRY | UNIT | MAN/ HOURS |
|---|---|---|
| **06130.10 HEAVY TIMBER** | | |
| 6x14 | L.F. | 0.092 |
| 6x16 | " | 0.096 |
| 8x10 | " | 0.083 |
| 8x12 | " | 0.089 |
| 8x14 | " | 0.092 |
| 8x16 | " | 0.096 |
| Southern yellow pine | | |
| 6x8 | L.F. | 0.080 |
| 6x10 | " | 0.083 |
| 6x12 | " | 0.089 |
| 6x14 | " | 0.092 |
| 6x16 | " | 0.096 |
| 8x10 | " | 0.083 |
| 8x12 | " | 0.089 |
| 8x14 | " | 0.092 |
| 8x16 | " | 0.096 |
| Columns to 12' high | | |
| Douglas fir | | |
| 6x6 | L.F. | 0.120 |
| 8x8 | " | 0.120 |
| 10x10 | " | 0.133 |
| 12x12 | " | 0.133 |
| Southern yellow pine | | |
| 6x6 | L.F. | 0.120 |
| 8x8 | " | 0.120 |
| 10x10 | " | 0.133 |
| 12x12 | " | 0.133 |
| Posts, treated | | |
| 4x4 | L.F. | 0.032 |
| 6x6 | " | 0.040 |
| **06190.20 WOOD TRUSSES** | | |
| Truss, fink, 2x4 members | | |
| 3-in-12 slope | | |
| 24' span | EA. | 0.686 |
| 26' span | " | 0.686 |
| 28' span | " | 0.727 |
| 30' span | " | 0.727 |
| 34' span | " | 0.774 |
| 38' span | " | 0.774 |
| 5-in-12 slope | | |
| 24' span | EA. | 0.706 |
| 28' span | " | 0.727 |
| 30' span | " | 0.750 |
| 32' span | " | 0.750 |
| 40' span | " | 0.800 |
| Gable, 2x4 members | | |
| 5-in-12 slope | | |
| 24' span | EA. | 0.706 |
| 26' span | " | 0.706 |
| 28' span | " | 0.727 |
| 30' span | " | 0.750 |
| 32' span | " | 0.750 |
| 36' span | " | 0.774 |

# 06 WOOD AND PLASTICS

## ROUGH CARPENTRY

| 06190.20 WOOD TRUSSES | UNIT | MAN/ HOURS |
|---|---|---|
| 40' span | EA. | 0.800 |
| King post type, 2x4 members | | |
| 4-in-12 slope | | |
| 16' span | EA. | 0.649 |
| 18' span | " | 0.667 |
| 24' span | " | 0.706 |
| 26' span | " | 0.706 |
| 30' span | " | 0.750 |
| 34' span | " | 0.750 |
| 38' span | " | 0.774 |
| 42' span | " | 0.828 |

## FINISH CARPENTRY

| 06200.10 FINISH CARPENTRY | UNIT | MAN/ HOURS |
|---|---|---|
| Mouldings and trim | | |
| Apron, flat | | |
| 9/16 x 2 | L.F. | 0.040 |
| 9/16 x 3-1/2 | " | 0.042 |
| Base | | |
| Colonial | | |
| 7/16 x 2-1/4 | L.F. | 0.040 |
| 7/16 x 3 | " | 0.040 |
| 7/16 x 3-1/4 | " | 0.040 |
| 9/16 x 3 | " | 0.042 |
| 9/16 x 3-1/4 | " | 0.042 |
| 11/16 x 2-1/4 | " | 0.044 |
| Ranch | | |
| 7/16 x 2-1/4 | L.F. | 0.040 |
| 7/16 x 3-1/4 | " | 0.040 |
| 9/16 x 2-1/4 | " | 0.042 |
| 9/16 x 3 | " | 0.042 |
| 9/16 x 3-1/4 | " | 0.042 |
| Casing | | |
| 11/16 x 2-1/2 | L.F. | 0.036 |
| 11/16 x 3-1/2 | " | 0.038 |
| Chair rail | | |
| 9/16 x 2-1/2 | L.F. | 0.040 |
| 9/16 x 3-1/2 | " | 0.040 |
| Closet pole | | |
| 1-1/8" dia. | L.F. | 0.053 |
| 1-5/8" dia. | " | 0.053 |
| Cove | | |
| 9/16 x 1-3/4 | L.F. | 0.040 |
| 11/16 x 2-3/4 | " | 0.040 |
| Crown | | |

## FINISH CARPENTRY

| 06200.10 FINISH CARPENTRY | UNIT | MAN/ HOURS |
|---|---|---|
| 9/16 x 1-5/8 | L.F. | 0.053 |
| 9/16 x 2-5/8 | " | 0.062 |
| 11/16 x 3-5/8 | " | 0.067 |
| 11/16 x 4-1/4 | " | 0.073 |
| 11/16 x 5-1/4 | " | 0.080 |
| Drip cap | | |
| 1-1/16 x 1-5/8 | L.F. | 0.040 |
| Glass bead | | |
| 3/8 x 3/8 | L.F. | 0.050 |
| 1/2 x 9/16 | " | 0.050 |
| 5/8 x 5/8 | " | 0.050 |
| 3/4 x 3/4 | " | 0.050 |
| Half round | | |
| 1/2 | L.F. | 0.032 |
| 5/8 | " | 0.032 |
| 3/4 | " | 0.032 |
| Lattice | | |
| 1/4 x 7/8 | L.F. | 0.032 |
| 1/4 x 1-1/8 | " | 0.032 |
| 1/4 x 1-3/8 | " | 0.032 |
| 1/4 x 1-3/4 | " | 0.032 |
| 1/4 x 2 | " | 0.032 |
| Ogee molding | | |
| 5/8 x 3/4 | L.F. | 0.040 |
| 11/16 x 1-1/8 | " | 0.040 |
| 11/16 x 1-3/8 | " | 0.040 |
| Parting bead | | |
| 3/8 x 7/8 | L.F. | 0.050 |
| Quarter round | | |
| 1/4 x 1/4 | L.F. | 0.032 |
| 3/8 x 3/8 | " | 0.032 |
| 1/2 x 1/2 | " | 0.032 |
| 11/16 x 11/16 | " | 0.035 |
| 3/4 x 3/4 | " | 0.035 |
| 1-1/16 x 1-1/16 | " | 0.036 |
| Railings, balusters | | |
| 1-1/8 x 1-1/8 | L.F. | 0.080 |
| 1-1/2 x 1-1/2 | " | 0.073 |
| Screen moldings | | |
| 1/4 x 3/4 | L.F. | 0.067 |
| 5/8 x 5/16 | " | 0.067 |
| Shoe | | |
| 7/16 x 11/16 | L.F. | 0.032 |
| Sash beads | | |
| 1/2 x 3/4 | L.F. | 0.067 |
| 1/2 x 7/8 | " | 0.067 |
| 1/2 x 1-1/8 | " | 0.073 |
| 5/8 x 7/8 | " | 0.073 |
| Stop | | |
| 5/8 x 1-5/8 | | |
| Colonial | L.F. | 0.050 |
| Ranch | " | 0.050 |
| Stools | | |
| 11/16 x 2-1/4 | L.F. | 0.089 |

# 06 WOOD AND PLASTICS

| FINISH CARPENTRY | UNIT | MAN/ HOURS |
|---|---|---|
| **06200.10 FINISH CARPENTRY** | | |
| 11/16 x 2-1/2 | L.F. | 0.089 |
| 11/16 x 5-1/4 | " | 0.100 |
| Exterior trim, casing, select pine, 1x3 | " | 0.040 |
| Douglas fir | | |
| 1x3 | L.F. | 0.040 |
| 1x4 | " | 0.040 |
| 1x6 | " | 0.044 |
| 1x8 | " | 0.050 |
| Cornices, white pine, #2 or better | | |
| 1x2 | L.F. | 0.040 |
| 1x4 | " | 0.040 |
| 1x6 | " | 0.044 |
| 1x8 | " | 0.047 |
| 1x10 | " | 0.050 |
| 1x12 | " | 0.053 |
| Shelving, pine | | |
| 1x8 | L.F. | 0.062 |
| 1x10 | " | 0.064 |
| 1x12 | " | 0.067 |
| Plywood shelf, 3/4", with edge band, 12" wide | " | 0.080 |
| Adjustable shelf, and rod, 12" wide | | |
| 3' to 4' long | EA. | 0.200 |
| 5' to 8' long | " | 0.267 |
| Prefinished wood shelves with brackets and supports | | |
| 8" wide | | |
| 3' long | EA. | 0.200 |
| 4' long | " | 0.200 |
| 6' long | " | 0.200 |
| 10" wide | | |
| 3' long | EA. | 0.200 |
| 4' long | " | 0.200 |
| 6' long | " | 0.200 |
| **06220.10 MILLWORK** | | |
| Countertop, laminated plastic | | |
| 25" x 7/8" thick | | |
| Minimum | L.F. | 0.200 |
| Average | " | 0.267 |
| Maximum | " | 0.320 |
| 25" x 1-1/4" thick | | |
| Minimum | L.F. | 0.267 |
| Average | " | 0.320 |
| Maximum | " | 0.400 |
| Add for cutouts | EA. | 0.500 |
| Backsplash, 4" high, 7/8" thick | L.F. | 0.160 |
| Plywood, sanded, A-C | | |
| 1/4" thick | S.F. | 0.027 |
| 3/8" thick | " | 0.029 |
| 1/2" thick | " | 0.031 |
| A-D | | |
| 1/4" thick | S.F. | 0.027 |
| 3/8" thick | " | 0.029 |
| 1/2" thick | " | 0.031 |
| Base cabinets, 34-1/2" high, 24" deep, hardwood, no tops | | |

| FINISH CARPENTRY | UNIT | MAN/ HOURS |
|---|---|---|
| **06220.10 MILLWORK** | | |
| Minimum | L.F. | 0.320 |
| Average | " | 0.400 |
| Maximum | " | 0.533 |
| Wall cabinets | | |
| Minimum | L.F. | 0.267 |
| Average | " | 0.320 |
| Maximum | " | 0.400 |
| Oil borne | | |
| Water borne | | |

| ARCHITECTURAL WOODWORK | UNIT | MAN/ HOURS |
|---|---|---|
| **06420.10 PANEL WORK** | | |
| Hardboard, tempered, 1/4" thick | | |
| Natural faced | S.F. | 0.020 |
| Plastic faced | " | 0.023 |
| Pegboard, natural | " | 0.020 |
| Plastic faced | " | 0.023 |
| Untempered, 1/4" thick | | |
| Natural faced | S.F. | 0.020 |
| Plastic faced | " | 0.023 |
| Pegboard, natural | " | 0.020 |
| Plastic faced | " | 0.023 |
| Plywood unfinished, 1/4" thick | | |
| Birch | | |
| Natural | S.F. | 0.027 |
| Select | " | 0.027 |
| Knotty pine | " | 0.027 |
| Cedar (closet lining) | | |
| Standard boards T&G | S.F. | 0.027 |
| Particle board | " | 0.027 |
| Plywood, prefinished, 1/4" thick, premium grade | | |
| Birch veneer | S.F. | 0.032 |
| Cherry veneer | " | 0.032 |
| Chestnut veneer | " | 0.032 |
| Lauan veneer | " | 0.032 |
| Mahogany veneer | " | 0.032 |
| Oak veneer (red) | " | 0.032 |
| Pecan veneer | " | 0.032 |
| Rosewood veneer | " | 0.032 |
| Teak veneer | " | 0.032 |
| Walnut veneer | " | 0.032 |

# 06 WOOD AND PLASTICS

| ARCHITECTURAL WOODWORK | UNIT | MAN/ HOURS |
|---|---|---|
| **06430.10 STAIRWORK** | | |
| Risers, 1x8, 42" wide | | |
| White oak | EA. | 0.400 |
| Pine | " | 0.400 |
| Treads, 1-1/16" x 9-1/2" x 42" | | |
| White oak | EA. | 0.500 |
| **06440.10 COLUMNS** | | |
| Column, hollow, round wood | | |
| 12" diameter | | |
| 10' high | EA. | 0.800 |
| 12' high | " | 0.857 |
| 14' high | " | 0.960 |
| 16' high | " | 1.200 |
| 24" diameter | | |
| 16' high | EA. | 1.200 |
| 18' high | " | 1.263 |
| 20' high | " | 1.263 |
| 22' high | " | 1.333 |
| 24' high | " | 1.333 |

# 07 THERMAL AND MOISTURE

| MOISTURE PROTECTION | UNIT | MAN/HOURS |
|---|---|---|
| **07100.10 WATERPROOFING** | | |
| Membrane waterproofing, elastomeric | | |
| Butyl | | |
| 1/32" thick | S.F. | 0.032 |
| 1/16" thick | " | 0.033 |
| Butyl with nylon | | |
| 1/32" thick | S.F. | 0.032 |
| 1/16" thick | " | 0.033 |
| Neoprene | | |
| 1/32" thick | S.F. | 0.032 |
| 1/16" thick | " | 0.033 |
| Neoprene with nylon | | |
| 1/32" thick | S.F. | 0.032 |
| 1/16" thick | " | 0.033 |
| Bituminous membrane waterproofing, asphalt felt, 15 lb. | | |
| One ply | S.F. | 0.020 |
| Two ply | " | 0.024 |
| Three ply | " | 0.029 |
| Four ply | " | 0.033 |
| Five ply | " | 0.042 |
| Modified asphalt membrane waterproofing, fibrous asphalt | | |
| One ply | S.F. | 0.033 |
| Two ply | " | 0.040 |
| Three ply | " | 0.044 |
| Four ply | " | 0.053 |
| Five ply | " | 0.064 |
| Asphalt coated protective board | | |
| 1/8" thick | S.F. | 0.020 |
| 1/4" thick | " | 0.020 |
| 3/8" thick | " | 0.020 |
| 1/2" thick | " | 0.021 |
| Cement protective board | | |
| 3/8" thick | S.F. | 0.027 |
| 1/2" thick | " | 0.027 |
| Fluid applied, neoprene | | |
| 50 mil | S.F. | 0.027 |
| 90 mil | " | 0.027 |
| Tab extended polyurethane | | |
| .050" thick | S.F. | 0.020 |
| Fluid applied rubber based polyurethane | | |
| 6 mil | S.F. | 0.025 |
| 15 mil | " | 0.020 |
| Bentonite waterproofing, panels | | |
| 3/16" thick | S.F. | 0.020 |
| 1/4" thick | " | 0.020 |
| 5/8" thick | " | 0.021 |
| Granular admixtures, trowel on, 3/8" thick | " | 0.020 |
| Metallic oxide waterproofing, iron compound, troweled | | |
| 5/8" thick | S.F. | 0.020 |
| 3/4" thick | " | 0.023 |
| **07150.10 DAMPROOFING** | | |
| Silicone dampproofing, sprayed on | | |
| Concrete surface | | |

| MOISTURE PROTECTION | UNIT | MAN/HOURS |
|---|---|---|
| **07150.10 DAMPROOFING** | | |
| 1 coat | S.F. | 0.004 |
| 2 coats | " | 0.006 |
| Concrete block | | |
| 1 coat | S.F. | 0.005 |
| 2 coats | " | 0.007 |
| Brick | | |
| 1 coat | S.F. | 0.006 |
| 2 coats | " | 0.008 |
| **07160.10 BITUMINOUS DAMPPROOFING** | | |
| Building paper, asphalt felt | | |
| 15 lb | S.F. | 0.032 |
| 30 lb | " | 0.033 |
| Asphalt dampproofing, troweled, cold, primer plus | | |
| 1 coat | S.F. | 0.027 |
| 2 coats | " | 0.040 |
| 3 coats | " | 0.050 |
| Fibrous asphalt dampproofing, hot troweled, primer plus | | |
| 1 coat | S.F. | 0.032 |
| 2 coats | " | 0.044 |
| 3 coats | " | 0.057 |
| Asphaltic paint dampproofing, per coat | | |
| Brush on | S.F. | 0.011 |
| Spray on | " | 0.009 |
| **07190.10 VAPOR BARRIERS** | | |
| Vapor barrier, polyethylene | | |
| 2 mil | S.F. | 0.004 |
| 6 mil | " | 0.004 |
| 8 mil | " | 0.004 |
| 10 mil | " | 0.004 |

| INSULATION | UNIT | MAN/HOURS |
|---|---|---|
| **07210.10 BATT INSULATION** | | |
| Ceiling, fiberglass, unfaced | | |
| 3-1/2" thick, R11 | S.F. | 0.009 |
| 6" thick, R19 | " | 0.011 |
| 9" thick, R30 | " | 0.012 |
| Suspended ceiling, unfaced | | |
| 3-1/2" thick, R11 | S.F. | 0.009 |
| 6" thick, R19 | " | 0.010 |
| 9" thick, R30 | " | 0.011 |
| Crawl space, unfaced | | |
| 3-1/2" thick, R11 | S.F. | 0.012 |
| 6" thick, R19 | " | 0.013 |

# 07 THERMAL AND MOISTURE

| INSULATION | UNIT | MAN/HOURS |
|---|---|---|
| **07210.10 BATT INSULATION** | | |
| 9" thick, R30 | S.F. | 0.015 |
| Wall, fiberglass | | |
| Paper backed | | |
| 2" thick, R7 | S.F. | 0.008 |
| 3" thick, R8 | " | 0.009 |
| 4" thick, R11 | " | 0.009 |
| 6" thick, R19 | " | 0.010 |
| Foil backed, 1 side | | |
| 2" thick, R7 | S.F. | 0.008 |
| 3" thick, R11 | " | 0.009 |
| 4" thick, R14 | " | 0.009 |
| 6" thick, R21 | " | 0.010 |
| Foil backed, 2 sides | | |
| 2" thick, R7 | S.F. | 0.009 |
| 3" thick, R11 | " | 0.010 |
| 4" thick, R14 | " | 0.011 |
| 6" thick, R21 | " | 0.011 |
| Unfaced | | |
| 2" thick, R7 | S.F. | 0.008 |
| 3" thick, R9 | " | 0.009 |
| 4" thick, R11 | " | 0.009 |
| 6" thick, R19 | " | 0.010 |
| Mineral wool batts | | |
| Paper backed | | |
| 2" thick, R6 | S.F. | 0.008 |
| 4" thick, R12 | " | 0.009 |
| 6" thick, R19 | " | 0.010 |
| Fasteners, self adhering, attached to ceiling deck | | |
| 2-1/2" long | EA. | 0.013 |
| 4-1/2" long | " | 0.015 |
| Capped, self-locking washers for fastening insulation | " | 0.008 |
| **07210.20 BOARD INSULATION** | | |
| Insulation, rigid | | |
| Fiberglass, roof | | |
| 0.75" thick, R2.78 | S.F. | 0.007 |
| 1.06" thick, R4.17 | " | 0.008 |
| 1.31" thick, R5.26 | " | 0.008 |
| 1.63" thick, R6.67 | " | 0.008 |
| 2.25" thick, R8.33 | " | 0.009 |
| Composite board, roof | | |
| 1-1/2" thick, R6.67 | S.F. | 0.008 |
| 1-5/8" thick, R7.69 | " | 0.008 |
| 2" thick, R10.0 | " | 0.009 |
| 2-1/4" thick, R12.50 | " | 0.009 |
| 2-1/2" thick, R14.29 | " | 0.010 |
| 2-3/4" thick, R16.67 | " | 0.011 |
| 3-1/4" thick, R20.00 | " | 0.011 |
| Perlite board, roof | | |
| 1.00" thick, R2.78 | S.F. | 0.007 |
| 1.50" thick, R4.17 | " | 0.007 |
| 2.00" thick, R5.92 | " | 0.007 |
| 2.50" thick, R6.67 | " | 0.008 |

| INSULATION | UNIT | MAN/HOURS |
|---|---|---|
| **07210.20 BOARD INSULATION** | | |
| 3.00" thick, R8.33 | S.F. | 0.008 |
| 4.00" thick, R10.00 | " | 0.008 |
| 5.25" thick, R14.29 | " | 0.009 |
| Rigid urethane | | |
| Roof | | |
| 1" thick, R6.67 | S.F. | 0.007 |
| 1.20" thick, R8.33 | " | 0.007 |
| 1.50" thick, R11.11 | " | 0.007 |
| 2" thick, R14.29 | " | 0.007 |
| 2.25" thick, R16.67 | " | 0.008 |
| Wall | | |
| 1" thick, R6.67 | S.F. | 0.008 |
| 1.5" thick, R11.11 | " | 0.009 |
| 2" thick, R14.29 | " | 0.009 |
| Polystyrene | | |
| Roof | | |
| 1.0" thick, R4.17 | S.F. | 0.007 |
| 1.5" thick, R6.26 | " | 0.007 |
| 2.0" thick, R8.33 | " | 0.007 |
| Wall | | |
| 1.0" thick, R4.17 | S.F. | 0.008 |
| 1.5" thick, R6.26 | " | 0.009 |
| 2.0" thick, R8.33 | " | 0.009 |
| Rigid board insulation, deck | | |
| Mineral fiberboard | | |
| 1" thick, R3.0 | S.F. | 0.007 |
| 2" thick, R5.26 | " | 0.007 |
| Fiberglass | | |
| 1" thick, R4.3 | S.F. | 0.007 |
| 2" thick, R8.5 | " | 0.007 |
| Polystyrene | | |
| 1" thick, R5.4 | S.F. | 0.007 |
| 2" thick, R10.8 | " | 0.007 |
| Urethane | | |
| .75" thick, R5.4 | S.F. | 0.007 |
| 1" thick, R6.4 | " | 0.007 |
| 1.5" thick, R10.7 | " | 0.007 |
| 2" thick, R14.3 | " | 0.007 |
| Foamglass | | |
| 1" thick, R1.8 | S.F. | 0.007 |
| 2" thick, R5.26 | " | 0.007 |
| Wood fiber | | |
| 1" thick, R3.85 | S.F. | 0.007 |
| 2" thick, R7.7 | " | 0.007 |
| Particle board | | |
| 3/4" thick, R2.08 | S.F. | 0.007 |
| 1" thick, R2.77 | " | 0.007 |
| 2" thick, R5.50 | " | 0.007 |
| **07210.60 LOOSE FILL INSULATION** | | |
| Blown-in type | | |
| Fiberglass | | |
| 5" thick, R11 | S.F. | 0.007 |
| 6" thick, R13 | " | 0.008 |

# 07 THERMAL AND MOISTURE

| INSULATION | UNIT | MAN/ HOURS |
|---|---|---|
| **07210.60 LOOSE FILL INSULATION** | | |
| 9" thick, R19 | S.F. | 0.011 |
| Rockwool, attic application | | |
| 6" thick, R13 | S.F. | 0.008 |
| 8" thick, R19 | " | 0.010 |
| 10" thick, R22 | " | 0.012 |
| 12" thick, R26 | " | 0.013 |
| 15" thick, R30 | " | 0.016 |
| Poured type | | |
| Fiberglass | | |
| 1" thick, R4 | S.F. | 0.005 |
| 2" thick, R8 | " | 0.006 |
| 3" thick, R12 | " | 0.007 |
| 4" thick, R16 | " | 0.008 |
| Mineral wool | | |
| 1" thick, R3 | S.F. | 0.005 |
| 2" thick, R6 | " | 0.006 |
| 3" thick, R9 | " | 0.007 |
| 4" thick, R12 | " | 0.008 |
| Vermiculite or perlite | | |
| 2" thick, R4.8 | S.F. | 0.006 |
| 3" thick, R7.2 | " | 0.007 |
| 4" thick, R9.6 | " | 0.008 |
| Masonry, poured vermiculite or perlite | | |
| 4" block | S.F. | 0.004 |
| 6" block | " | 0.005 |
| 8" block | " | 0.006 |
| 10" block | " | 0.006 |
| 12" block | " | 0.007 |
| **07210.70 SPRAYED INSULATION** | | |
| Foam, sprayed on | | |
| Polystyrene | | |
| 1" thick, R4 | S.F. | 0.008 |
| 2" thick, R8 | " | 0.011 |
| Urethane | | |
| 1" thick, R7.7 | S.F. | 0.008 |
| 2" thick, R15.4 | " | 0.011 |
| **07250.10 FIREPROOFING** | | |
| Sprayed on | | |
| 1" thick | | |
| On beams | S.F. | 0.018 |
| On columns | " | 0.016 |
| On decks | | |
| Flat surface | S.F. | 0.008 |
| Fluted surface | " | 0.010 |
| 1-1/2" thick | | |
| On beams | S.F. | 0.023 |
| On columns | " | 0.020 |
| On decks | | |
| Flat surface | S.F. | 0.010 |
| Fluted surface | " | 0.013 |

| SHINGLES AND TILES | UNIT | MAN/ HOURS |
|---|---|---|
| **07310.10 ASPHALT SHINGLES** | | |
| Standard asphalt shingles, strip shingles | | |
| 210 lb/square | SQ. | 0.800 |
| 235 lb/square | " | 0.889 |
| 240 lb/square | " | 1.000 |
| 260 lb/square | " | 1.143 |
| 300 lb/square | " | 1.333 |
| 385 lb/square | " | 1.600 |
| Roll roofing, mineral surface | | |
| 90 lb | SQ. | 0.571 |
| 110 lb | " | 0.667 |
| 140 lb | " | 0.800 |
| **07310.30 METAL SHINGLES** | | |
| Aluminum, .020" thick | | |
| Plain | SQ. | 1.600 |
| Colors | " | 1.600 |
| Steel, galvanized | | |
| 26 ga. | | |
| Plain | SQ. | 1.600 |
| Colors | " | 1.600 |
| 24 ga. | | |
| Plain | SQ. | 1.600 |
| Colors | " | 1.600 |
| Porcelain enamel, 22 ga. | | |
| Minimum | SQ. | 2.000 |
| Average | " | 2.000 |
| Maximum | " | 2.000 |
| **07310.60 SLATE SHINGLES** | | |
| Slate shingles | | |
| Pennsylvania | | |
| Ribbon | SQ. | 4.000 |
| Clear | " | 4.000 |
| Vermont | | |
| Black | SQ. | 4.000 |
| Grey | " | 4.000 |
| Green | " | 4.000 |
| Red | " | 4.000 |
| Replacement shingles | | |
| Small jobs | EA. | 0.267 |
| Large jobs | S.F. | 0.133 |
| **07310.70 WOOD SHINGLES** | | |
| Wood shingles, on roofs | | |
| White cedar, #1 shingles | | |
| 4" exposure | SQ. | 2.667 |
| 5" exposure | " | 2.000 |
| #2 shingles | | |
| 4" exposure | SQ. | 2.667 |
| 5" exposure | " | 2.000 |
| Resquared and rebutted | | |
| 4" exposure | SQ. | 2.667 |
| 5" exposure | " | 2.000 |

# 07 THERMAL AND MOISTURE

| SHINGLES AND TILES | UNIT | MAN/ HOURS |
|---|---|---|
| **07310.70 WOOD SHINGLES** | | |
| On walls | | |
| White cedar, #1 shingles | | |
| 4" exposure | SQ. | 4.000 |
| 5" exposure | " | 3.200 |
| 6" exposure | " | 2.667 |
| #2 shingles | | |
| 4" exposure | SQ. | 4.000 |
| 5" exposure | " | 3.200 |
| 6" exposure | " | 2.667 |
| **07310.80 WOOD SHAKES** | | |
| Shakes, hand split, 24" red cedar, on roofs | | |
| 5" exposure | SQ. | 4.000 |
| 7" exposure | " | 3.200 |
| 9" exposure | " | 2.667 |
| On walls | | |
| 6" exposure | SQ. | 4.000 |
| 8" exposure | " | 3.200 |
| 10" exposure | " | 2.667 |

| ROOFING AND SIDING | UNIT | MAN/ HOURS |
|---|---|---|
| **07410.10 MANUFACTURED ROOFS** | | |
| Aluminum roof panels, for structural steel framing | | |
| Corrugated | | |
| Unpainted finish | | |
| .024" | S.F. | 0.020 |
| .030" | " | 0.020 |
| Painted finish | | |
| .024" | S.F. | 0.020 |
| .030" | " | 0.020 |
| V-beam | | |
| Unpainted finish | | |
| .032" | S.F. | 0.020 |
| .040" | " | 0.020 |
| .050" | " | 0.020 |
| Painted finish | | |
| .032" | S.F. | 0.020 |
| .040" | " | 0.020 |
| .050" | " | 0.020 |
| Steel roof panels, for structural steel framing | | |
| Corrugated, painted | | |
| 18 ga. | S.F. | 0.020 |
| 20 ga. | " | 0.020 |
| 22 ga. | " | 0.020 |
| Box rib, painted | | |

| ROOFING AND SIDING | UNIT | MAN/ HOURS |
|---|---|---|
| **07410.10 MANUFACTURED ROOFS** | | |
| 18 ga. | S.F. | 0.021 |
| 20 ga. | " | 0.021 |
| 22 ga. | " | 0.021 |
| 4" rib, painted | | |
| 18 ga. | S.F. | 0.022 |
| 20 ga. | " | 0.022 |
| 22 ga. | " | 0.022 |
| Standing seam roof | | |
| 2" high seam, painted | | |
| 22 ga. | S.F. | 0.032 |
| 24 ga. | " | 0.032 |
| 26 ga. | " | 0.032 |
| **07410.30 MANUFACTURED WALLS** | | |
| Sandwich panels with 1-1/2" fiberglass insulation | | |
| Galvanized 18 ga. steel interior panels | | |
| Exterior panels | | |
| 16 ga. aluminum | S.F. | 0.107 |
| 18 ga. galvanized steel | " | 0.107 |
| 20 ga. painted steel | " | 0.107 |
| 20 ga. stainless steel | " | 0.107 |
| Metal liner panels, 1-3/8" thick, 24" wide | | |
| Galvanized | | |
| 22 ga. | S.F. | 0.027 |
| 20 ga. | " | 0.027 |
| 18 ga. | " | 0.027 |
| Primed | | |
| 22 ga. | S.F. | 0.027 |
| 20 ga. | " | 0.027 |
| 18 ga. | " | 0.027 |
| **07440.10 AGGREGATE COATED PANELS** | | |
| Dryvit type system | | |
| 1" thick | S.F. | 0.027 |
| 1-1/2" thick | " | 0.029 |
| 2" thick | " | 0.033 |
| **07460.10 METAL SIDING PANELS** | | |
| Aluminum siding panels | | |
| Corrugated | | |
| Plain finish | | |
| .024" | S.F. | 0.032 |
| .032" | " | 0.032 |
| Painted finish | | |
| .024" | S.F. | 0.032 |
| .032" | " | 0.032 |
| V. beam | | |
| Plain finish | | |
| .032" | S.F. | 0.032 |
| .040" | " | 0.032 |
| .050" | " | 0.032 |
| Painted finish | | |
| .032" | S.F. | 0.032 |
| .040" | " | 0.032 |

# 07 THERMAL AND MOISTURE

| ROOFING AND SIDING | UNIT | MAN/ HOURS |
|---|---|---|
| **07460.10 METAL SIDING PANELS** | | |
| .050" | S.F. | 0.032 |
| 4" rib | | |
| Plain finish | | |
| .032" | S.F. | 0.036 |
| .040" | " | 0.036 |
| .050" | " | 0.036 |
| Painted finish | | |
| .032" | S.F. | 0.036 |
| .040" | " | 0.036 |
| .050" | " | 0.036 |
| Steel siding panels | | |
| Corrugated | | |
| 22 ga. | S.F. | 0.053 |
| 24 ga. | " | 0.053 |
| 26 ga. | " | 0.053 |
| Box rib | | |
| 20 ga. | S.F. | 0.053 |
| 22 ga. | " | 0.053 |
| 24 ga. | " | 0.053 |
| 26 ga. | " | 0.053 |
| **07460.50 PLASTIC SIDING** | | |
| Horizontal vinyl siding, solid | | |
| 8" wide | | |
| Standard | S.F. | 0.031 |
| Insulated | " | 0.031 |
| 10" wide | | |
| Standard | S.F. | 0.029 |
| Insulated | " | 0.029 |
| Vinyl moldings for doors and windows | L.F. | 0.032 |
| **07460.60 PLYWOOD SIDING** | | |
| Rough sawn cedar, 3/8" thick | S.F. | 0.027 |
| Fir, 3/8" thick | " | 0.027 |
| Texture 1-11, 5/8" thick | | |
| Cedar | S.F. | 0.029 |
| Fir | " | 0.029 |
| Redwood | " | 0.029 |
| Southern Yellow Pine | " | 0.029 |
| **07460.70 STEEL SIDING** | | |
| Ribbed, sheets, galvanized | | |
| 22 ga. | S.F. | 0.032 |
| 24 ga. | " | 0.032 |
| 26 ga. | " | 0.032 |
| 28 ga. | " | 0.032 |
| Primed | | |
| 24 ga. | S.F. | 0.032 |
| 26 ga. | " | 0.032 |
| 28 ga. | " | 0.032 |

| ROOFING AND SIDING | UNIT | MAN/ HOURS |
|---|---|---|
| **07460.80 WOOD SIDING** | | |
| Beveled siding, cedar | | |
| A grade | | |
| 1/2 x 6 | S.F. | 0.040 |
| 1/2 x 8 | " | 0.032 |
| 3/4 x 10 | " | 0.027 |
| Clear | | |
| 1/2 x 6 | S.F. | 0.040 |
| 1/2 x 8 | " | 0.032 |
| 3/4 x 10 | " | 0.027 |
| B grade | | |
| 1/2 x 6 | S.F. | 0.040 |
| 1/2 x 8 | " | 0.320 |
| 3/4 x 10 | " | 0.027 |
| Board and batten | | |
| Cedar | | |
| 1x6 | S.F. | 0.040 |
| 1x8 | " | 0.032 |
| 1x10 | " | 0.029 |
| 1x12 | " | 0.026 |
| Pine | | |
| 1x6 | S.F. | 0.040 |
| 1x8 | " | 0.032 |
| 1x10 | " | 0.029 |
| 1x12 | " | 0.026 |
| Redwood | | |
| 1x6 | S.F. | 0.040 |
| 1x8 | " | 0.032 |
| 1x10 | " | 0.029 |
| 1x12 | " | 0.026 |
| Tongue and groove | | |
| Cedar | | |
| 1x4 | S.F. | 0.044 |
| 1x6 | " | 0.042 |
| 1x8 | " | 0.040 |
| 1x10 | " | 0.038 |
| Pine | | |
| 1x4 | S.F. | 0.044 |
| 1x6 | " | 0.042 |
| 1x8 | " | 0.040 |
| 1x10 | " | 0.038 |
| Redwood | | |
| 1x4 | S.F. | 0.044 |
| 1x6 | " | 0.042 |
| 1x8 | " | 0.040 |
| 1x10 | " | 0.038 |

# 07 THERMAL AND MOISTURE

| MEMBRANE ROOFING | UNIT | MAN/HOURS |
|---|---|---|
| **07510.10 BUILT-UP ASPHALT ROOFING** | | |
| Built-up roofing, asphalt felt, including gravel | | |
| 2 ply | SQ. | 2.000 |
| 3 ply | " | 2.667 |
| 4 ply | " | 3.200 |
| Walkway, for built-up roofs | | |
| 3' x 3' x | | |
| 1/2" thick | S.F. | 0.027 |
| 3/4" thick | " | 0.027 |
| 1" thick | " | 0.027 |
| Cant strip, 4" x 4" | | |
| Treated wood | L.F. | 0.023 |
| Foamglass | " | 0.020 |
| Mineral fiber | " | 0.020 |
| New gravel for built-up roofing, 400 lb/sq | SQ. | 1.600 |
| Roof gravel (ballast) | C.Y. | 4.000 |
| Aluminum coating, top surfacing, for built-up roofing | SQ. | 1.333 |
| Remove & replace gravel, includes flood coat | " | 2.667 |
| **07530.10 SINGLE-PLY ROOFING** | | |
| Elastic sheet roofing | | |
| Neoprene, 1/16" thick | S.F. | 0.010 |
| EPDM rubber | | |
| 45 mil | S.F. | 0.010 |
| 60 mil | " | 0.010 |
| PVC | | |
| 45 mil | S.F. | 0.010 |
| 60 mil | " | 0.010 |
| Flashing | | |
| Pipe flashing, 90 mil thick | | |
| 1" pipe | EA. | 0.200 |
| 2" pipe | " | 0.200 |
| 3" pipe | " | 0.211 |
| 4" pipe | " | 0.211 |
| 5" pipe | " | 0.222 |
| 6" pipe | " | 0.222 |
| 8" pipe | " | 0.235 |
| 10" pipe | " | 0.267 |
| 12" pipe | " | 0.267 |
| Neoprene flashing, 60 mil thick strip | | |
| 6" wide | L.F. | 0.067 |
| 12" wide | " | 0.100 |
| 18" wide | " | 0.133 |
| 24" wide | " | 0.200 |
| Adhesives | | |
| Mastic sealer, applied at joints only | | |
| 1/4" bead | L.F. | 0.004 |
| Fluid applied roofing | | |
| Urethane, 2 components, elastomeric top membrane | | |
| 1" thick | S.F. | 0.013 |
| Vinyl liquid roofing, 2 coats, 2 mils per coat | " | 0.011 |
| Silicone roofing, 2 coats sprayed, 16 mil per coat | " | 0.013 |
| Inverted roof system | | |
| Insulated membrane with coarse gravel ballast | | |

| MEMBRANE ROOFING | UNIT | MAN/HOURS |
|---|---|---|
| **07530.10 SINGLE-PLY ROOFING** | | |
| 3 ply with 2" polystyrene | S.F. | 0.013 |
| Ballast, 3/4" through 1-1/2" dia. river gravel, 100lb/sf | " | 0.800 |
| Walkway for membrane roofs, 1/2" thick | " | 0.027 |

| FLASHING AND SHEET METAL | UNIT | MAN/HOURS |
|---|---|---|
| **07610.10 METAL ROOFING** | | |
| Sheet metal roofing, copper, 16 oz, batten seam | SQ. | 5.333 |
| Standing seam | " | 5.000 |
| Aluminum roofing, natural finish | | |
| Corrugated, on steel frame | | |
| .0175" thick | SQ. | 2.286 |
| .0215" thick | " | 2.286 |
| .024" thick | " | 2.286 |
| .032" thick | " | 2.286 |
| V-beam, on steel frame | | |
| .032" thick | SQ. | 2.286 |
| .040" thick | " | 2.286 |
| .050" thick | " | 2.286 |
| Ridge cap | | |
| .019" thick | L.F. | 0.027 |
| Corrugated galvanized steel roofing, on steel frame | | |
| 28 ga. | SQ. | 2.286 |
| 26 ga. | " | 2.286 |
| 24 ga. | " | 2.286 |
| 22 ga. | " | 2.286 |
| 26 ga., factory insulated with 1" polystyrene | " | 3.200 |
| Ridge roll | | |
| 10" wide | L.F. | 0.027 |
| 20" wide | " | 0.032 |
| **07620.10 FLASHING AND TRIM** | | |
| Counter flashing | | |
| Aluminum, .032" | S.F. | 0.080 |
| Stainless steel, .015" | " | 0.080 |
| 16 oz. | " | 0.080 |
| 20 oz. | " | 0.080 |
| 24 oz. | " | 0.080 |
| 32 oz. | " | 0.080 |
| Valley flashing | | |
| Aluminum, .032" | S.F. | 0.060 |
| Stainless steel, .015 | " | 0.050 |
| Copper | | |
| 16 oz. | S.F. | 0.050 |
| 20 oz. | " | 0.067 |

# 07 THERMAL AND MOISTURE

| FLASHING AND SHEET METAL | UNIT | MAN/ HOURS |
|---|---|---|
| **07620.10 FLASHING AND TRIM** | | |
| 24 oz. | S.F. | 0.050 |
| 32 oz. | " | 0.050 |
| Base flashing | | |
| Aluminum, .040" | S.F. | 0.067 |
| Stainless steel, .018" | " | 0.067 |
| 16 oz. | " | 0.067 |
| 20 oz. | " | 0.050 |
| 24 oz. | " | 0.067 |
| 32 oz. | " | 0.067 |
| Waterstop, "T" section, 22 ga. | | |
| 1-1/2" x 3" | L.F. | 0.040 |
| 2" x 2" | " | 0.040 |
| 4" x 3" | " | 0.040 |
| 6" x 4" | " | 0.040 |
| 8" x 4" | " | 0.040 |
| Scupper outlets | | |
| 10" x 10" x 4" | EA. | 0.200 |
| 22" x 4" x 4" | " | 0.200 |
| 8" x 8" x 5" | " | 0.200 |
| Flashing and trim, aluminum | | |
| .019" thick | S.F. | 0.057 |
| .032" thick | " | 0.057 |
| .040" thick | " | 0.062 |
| Neoprene sheet flashing, .060" thick | " | 0.050 |
| Copper, paper backed | | |
| 2 oz. | S.F. | 0.080 |
| Drainage boots, roof, cast iron | | |
| 2 x 3 | L.F. | 0.100 |
| 3 x 4 | " | 0.100 |
| 4 x 5 | " | 0.107 |
| 4 x 6 | " | 0.107 |
| 5 x 7 | " | 0.114 |
| Pitch pocket, copper, 16 oz. | | |
| 4 x 4 | EA. | 0.200 |
| 6 x 6 | " | 0.200 |
| 8 x 8 | " | 0.200 |
| 8 x 10 | " | 0.200 |
| 8 x 12 | " | 0.200 |
| Reglets, copper 10 oz. | L.F. | 0.053 |
| Stainless steel, .020" | " | 0.053 |
| Gravel stop | | |
| Aluminum, .032" | | |
| 4" | L.F. | 0.027 |
| 6" | " | 0.027 |
| 8" | " | 0.031 |
| 10" | " | 0.031 |
| Copper, 16 oz. | | |
| 4" | L.F. | 0.027 |
| 6" | " | 0.027 |
| 8" | " | 0.031 |
| 10" | " | 0.031 |

| FLASHING AND SHEET METAL | UNIT | MAN/ HOURS |
|---|---|---|
| **07620.20 GUTTERS AND DOWNSPOUTS** | | |
| Copper gutter and downspout | | |
| Downspouts, 16 oz. copper | | |
| Round | | |
| 3" dia. | L.F. | 0.053 |
| 4" dia. | " | 0.053 |
| Rectangular, corrugated | | |
| 2" x 3" | L.F. | 0.050 |
| 3" x 4" | " | 0.050 |
| Rectangular, flat surface | | |
| 2" x 3" | L.F. | 0.053 |
| 3" x 4" | " | 0.053 |
| Lead-coated copper downspouts | | |
| Round | | |
| 3" dia. | L.F. | 0.050 |
| 4" dia. | " | 0.057 |
| Rectangular, corrugated | | |
| 2" x 3" | L.F. | 0.053 |
| 3" x 4" | " | 0.053 |
| Rectangular, plain | | |
| 2" x 3" | L.F. | 0.053 |
| 3" x 4" | " | 0.053 |
| Gutters, 16 oz. copper | | |
| Half round | | |
| 4" wide | L.F. | 0.080 |
| 5" wide | " | 0.089 |
| Type K | | |
| 4" wide | L.F. | 0.080 |
| 5" wide | " | 0.089 |
| Lead-coated copper gutters | | |
| Half round | | |
| 4" wide | L.F. | 0.080 |
| 6" wide | " | 0.089 |
| Type K | | |
| 4" wide | L.F. | 0.080 |
| 5" wide | " | 0.089 |
| Aluminum gutter and downspout | | |
| Downspouts | | |
| 2" x 3" | L.F. | 0.053 |
| 3" x 4" | " | 0.057 |
| 4" x 5" | " | 0.062 |
| Round | | |
| 3" dia. | L.F. | 0.053 |
| 4" dia. | " | 0.057 |
| Gutters, stock units | | |
| 4" wide | L.F. | 0.084 |
| 5" wide | " | 0.089 |
| Galvanized steel gutter and downspout | | |
| Downspouts, round corrugated | | |
| 3" dia. | L.F. | 0.053 |
| 4" dia. | " | 0.053 |
| 5" dia. | " | 0.057 |
| 6" dia. | " | 0.057 |
| Rectangular | | |
| 2" x 3" | L.F. | 0.053 |

# 07 THERMAL AND MOISTURE

| FLASHING AND SHEET METAL | UNIT | MAN/ HOURS |
|---|---|---|
| **07620.20 GUTTERS AND DOWNSPOUTS** | | |
| 3" x 4" | L.F. | 0.050 |
| 4" x 4" | " | 0.050 |
| Gutters, stock units | | |
| 5" wide | | |
| Plain | L.F. | 0.089 |
| Painted | " | 0.089 |
| 6" wide | | |
| Plain | L.F. | 0.094 |
| Painted | " | 0.094 |

| ROOFING SPECIALTIES | UNIT | MAN/ HOURS |
|---|---|---|
| **07700.10 MANUFACTURED SPECIALTIES** | | |
| Moisture relief vent | | |
| Aluminum | EA. | 0.114 |
| Copper | " | 0.114 |
| Expansion joint | | |
| Aluminum | | |
| Opening to 2.5" | L.F. | 0.057 |
| Opening to 3.5" | " | 0.062 |
| Copper, 16 oz. | | |
| Opening to 2.5" | L.F. | 0.057 |
| Opening to 3.5" | " | 0.062 |
| Butyl or neoprene | | |
| 4" wide | | |
| 16 oz. copper bellows | L.F. | 0.067 |
| 28 ga. stainless steel bellows | " | 0.067 |
| 6" wide | | |
| Copper bellows | L.F. | 0.073 |
| Stainless steel | | |
| Opening to 2.5" | L.F. | 0.057 |
| Opening to 3.5" | " | 0.062 |
| Smoke vent, 48" x 48" | | |
| Aluminum | EA. | 2.000 |
| Galvanized steel | " | 2.000 |
| Heat/smoke vent, 48" x 96" | | |
| Aluminum | EA. | 2.667 |
| Galvanized steel | " | 2.667 |
| Ridge vent strips | | |
| Mill finish | L.F. | 0.053 |
| Connectors | EA. | 0.200 |
| End cap | " | 0.229 |
| Soffit vents | | |
| Mill finish | | |
| 2-1/2" wide | L.F. | 0.032 |
| 3" wide | " | 0.032 |

| ROOFING SPECIALTIES | UNIT | MAN/ HOURS |
|---|---|---|
| **07700.10 MANUFACTURED SPECIALTIES** | | |
| 6" wide | L.F. | 0.032 |
| Roof hatches | | |
| Steel, plain, primed | | |
| 2'6" x 3'0" | EA. | 2.000 |
| 2'6" x 4'6" | " | 2.667 |
| 2'6" x 8'0" | " | 4.000 |
| Galvanized steel | | |
| 2'6" x 3'0" | EA. | 2.000 |
| 2'6" x 4'6" | " | 2.667 |
| 2'6" x 8'0" | " | 4.000 |
| Aluminum | | |
| 2'6" x 3'0" | EA. | 2.000 |
| 2'6" x 4'6" | " | 2.667 |
| 2'6" x 8'0" | " | 4.000 |
| Ceiling access doors | | |
| Swing up model, metal frame | | |
| Steel door | | |
| 2'6" x 2'6" | EA. | 0.800 |
| 2'6" x 3'0" | " | 0.800 |
| Aluminum door | | |
| 2'6" x 2'6" | EA. | 0.800 |
| 2'6" x 3'0" | " | 0.800 |
| Swing down model, metal frame | | |
| Steel door | | |
| 2'6" x 2'6" | EA. | 0.800 |
| 2'6" x 3'0" | " | 0.800 |
| Aluminum door | | |
| 2'6" x 2'6" | EA. | 0.800 |
| 2'6" x 3'0" | " | 0.800 |
| Gravity ventilators, with curb, base, damper and screen | | |
| Stationary siphon | | |
| 6" dia. | EA. | 0.533 |
| 12" dia. | " | 0.533 |
| 24" dia. | " | 0.800 |
| 36" dia. | " | 0.800 |
| Wind driven spinner | | |
| 6" dia. | EA. | 0.533 |
| 12" dia. | " | 0.533 |
| 24" dia. | " | 0.800 |
| 36" dia. | " | 0.800 |
| Stationary mushroom | | |
| 16" dia. | EA. | 0.800 |
| 30" dia. | " | 1.000 |
| 36" dia. | " | 1.333 |
| 42" dia. | " | 1.600 |

# 07 THERMAL AND MOISTURE

## SKYLIGHTS

| 07810.10 PLASTIC SKYLIGHTS | UNIT | MAN/ HOURS |
|---|---|---|
| Single thickness, not including mounting curb | | |
| 2' x 4' | EA. | 1.000 |
| 4' x 4' | " | 1.333 |
| 5' x 5' | " | 2.000 |
| 6' x 8' | " | 2.667 |
| Double thickness, not including mounting curb | | |
| 2' x 4' | EA. | 1.000 |
| 4' x 4' | " | 1.333 |
| 5' x 5' | " | 2.000 |
| 6' x 8' | " | 2.667 |
| Metal framed skylights | | |
| Translucent panels, 2-1/2" thick | S.F. | 0.080 |
| Continuous vaults, 8' wide | | |
| Single glazed | S.F. | 0.100 |
| Double glazed | " | 0.114 |

## JOINT SEALERS

| 07920.10 CAULKING | UNIT | MAN/ HOURS |
|---|---|---|
| Caulk exterior, two component | | |
| 1/4 x 1/2 | L.F. | 0.040 |
| 3/8 x 1/2 | " | 0.044 |
| 1/2 x 1/2 | " | 0.050 |
| Caulk interior, single component | | |
| 1/4 x 1/2 | L.F. | 0.038 |
| 3/8 x 1/2 | " | 0.042 |
| 1/2 x 1/2 | " | 0.047 |
| Butyl rubber fillers | | |
| 1/4" x 1/4" | L.F. | 0.016 |
| 1/2" x 1/2" | " | 0.027 |
| 1/2" x 3/4" | " | 0.032 |
| 3/4" x 3/4" | " | 0.032 |
| 1" x 1" | " | 0.036 |
| Seals, "O" ring type cord | | |
| 1/4" dia. | L.F. | 0.020 |
| 1/2" dia. | " | 0.021 |
| 1" dia. | " | 0.022 |
| 1-1/4" dia. | " | 0.024 |
| 1-1/2" dia. | " | 0.025 |
| 1-3/4" dia. | " | 0.026 |
| 2" dia. | " | 0.027 |
| Polyvinyl chloride, closed cell | | |
| 1/4" x 2" | L.F. | 0.029 |
| 1/4" x 6" | " | 0.036 |
| Silicon foam penetration seal | | |
| 1/4" x 1/2" | L.F. | 0.010 |

## JOINT SEALERS

| 07920.10 CAULKING | UNIT | MAN/ HOURS |
|---|---|---|
| 1/2" x 1/2" | L.F. | 0.013 |
| 1/2" x 3/4" | " | 0.016 |
| 3/4" x 3/4" | " | 0.020 |
| 1/8" x 1" | " | 0.010 |
| 1/8" x 3" | " | 0.016 |
| 1/4" x 3" | " | 0.020 |
| 1/4" x 6" | " | 0.027 |
| 1/2" x 6" | " | 0.062 |
| 1/2" x 9" | " | 0.100 |
| 1/2" x 12" | " | 0.145 |
| Oil base sealants and caulking | | |
| 1/4" x 1/4" | L.F. | 0.020 |
| 1/4" x 3/8" | " | 0.021 |
| 1/4" x 1/2" | " | 0.022 |
| 3/8" x 3/8" | " | 0.023 |
| 3/8" x 1/2" | " | 0.024 |
| 3/8" x 5/8" | " | 0.026 |
| 3/8" x 3/4" | " | 0.028 |
| 1/2" x 1/2" | " | 0.031 |
| 1/2" x 5/8" | " | 0.035 |
| 1/2" x 3/4" | " | 0.040 |
| 1/2" x 7/8" | " | 0.041 |
| 1/2" x 1" | " | 0.042 |
| 3/4" x 3/4" | " | 0.043 |
| 1" x 1" | " | 0.044 |
| Polyurethane compounds | | |
| 1/4" x 1/4" | L.F. | 0.020 |
| 1/4" x 3/8" | " | 0.021 |
| 1/4" x 1/2" | " | 0.022 |
| 3/8" x 3/8" | " | 0.023 |
| 3/8" x 1/2" | " | 0.024 |
| 3/8" x 5/8" | " | 0.026 |
| 3/8" x 3/4" | " | 0.028 |
| 1/2" x 1/2" | " | 0.031 |
| 1/2" x 5/8" | " | 0.035 |
| 1/2" x 3/4" | " | 0.040 |
| 1/2" x 7/8" | " | 0.041 |
| 1/2" x 1" | " | 0.044 |
| 3/4" x 3/4" | " | 0.043 |
| 3/4" x 1" | " | 0.044 |
| Backer rod, polyethylene | | |
| 1/4" | L.F. | 0.020 |
| 1/2" | " | 0.021 |
| 3/4" | " | 0.022 |
| 1" | " | 0.024 |

# 08 DOORS AND WINDOWS

| METAL | UNIT | MAN/ HOURS |
|---|---|---|
| **08110.10 METAL DOORS** | | |
| Flush hollow metal, standard duty, 20 ga., 1-3/8" | | |
| 2-6 x 6-8 | EA. | 0.889 |
| 2-8 x 6-8 | " | 0.889 |
| 3-0 x 6-8 | " | 0.889 |
| 1-3/4" | | |
| 2-6 x 6-8 | EA. | 0.889 |
| 2-8 x 6-8 | " | 0.889 |
| 3-0 x 6-8 | " | 0.889 |
| 2-6 x 7-0 | " | 0.889 |
| 2-8 x 7-0 | " | 0.889 |
| 3-0 x 7-0 | " | 0.889 |
| Heavy duty, 20 ga., unrated, 1-3/4" | | |
| 2-8 x 6-8 | EA. | 0.889 |
| 3-0 x 6-8 | " | 0.889 |
| 2-8 x 7-0 | " | 0.889 |
| 3-0 x 7-0 | " | 0.889 |
| 3-4 x 7-0 | " | 0.889 |
| 18 ga., 1-3/4", unrated door | | |
| 2-0 x 7-0 | EA. | 0.889 |
| 2-4 x 7-0 | " | 0.889 |
| 2-6 x 7-0 | " | 0.889 |
| 2-8 x 7-0 | " | 0.889 |
| 3-0 x 7-0 | " | 0.889 |
| 3-4 x 7-0 | " | 0.889 |
| 2", unrated door | | |
| 2-0 x 7-0 | EA. | 1.000 |
| 2-4 x 7-0 | " | 1.000 |
| 2-6 x 7-0 | " | 1.000 |
| 2-8 x 7-0 | " | 1.000 |
| 3-0 x 7-0 | " | 1.000 |
| 3-4 x 7-0 | " | 1.000 |
| Galvanized metal door | | |
| 3-0 x 7-0 | EA. | 1.000 |
| **08110.40 METAL DOOR FRAMES** | | |
| Hollow metal, stock, 18 ga., 4-3/4" x 1-3/4" | | |
| 2-0 x 7-0 | EA. | 1.000 |
| 2-4 x 7-0 | " | 1.000 |
| 2-6 x 7-0 | " | 1.000 |
| 2-8 x 7-0 | " | 1.000 |
| 3-0 x 7-0 | " | 1.000 |
| 4-0 x 7-0 | " | 1.333 |
| 5-0 x 7-0 | " | 1.333 |
| 6-0 x 7-0 | " | 1.333 |
| 16 ga., 6-3/4" x 1-3/4" | | |
| 2-0 x 7-0 | EA. | 1.000 |
| 2-4 x 7-0 | " | 1.000 |
| 2-6 x 7-0 | " | 1.000 |
| 2-8 x 7-0 | " | 1.000 |
| 3-0 x 7-0 | " | 1.000 |
| 4-0 x 7-0 | " | 1.333 |
| 6-0 x 7-0 | " | 1.333 |
| Transom frame | | |

| METAL | UNIT | MAN/ HOURS |
|---|---|---|
| **08110.40 METAL DOOR FRAMES** | | |
| 3-8 x 1-6 | EA. | 1.000 |
| 6-4 x 1-6 | " | 1.000 |
| Transom sash | | |
| 3-0 x 1-4 | EA. | 1.000 |
| 3-4 x 1-4 | " | 1.000 |
| 6-0 x 1-4 | " | 1.000 |
| Sidelights, complete | | |
| 1-0 x 7-2 | EA. | 1.000 |
| 1-4 x 7-2 | " | 1.000 |
| 1-0 x 8-8 | " | 1.000 |
| 1-6 x 8-8 | " | 1.000 |
| 16 ga., 4-3/4" x 1-3/4" | | |
| 2-0 x 7-0 | EA. | 1.000 |
| 2-4 x 7-0 | " | 1.000 |
| 2-6 x 7-0 | " | 1.000 |
| 2-8 x 7-0 | " | 1.000 |
| 3-0 x 7-0 | " | 1.000 |
| 4-0 x 7-0 | " | 1.333 |
| 6-0 x 7-0 | " | 1.333 |
| 3-4 x 1-6 | " | 1.000 |
| 3-8 x 1-6 | " | 1.000 |
| 6-4 x 1-6 | " | 1.000 |
| Transom sash | | |
| 3-0 x 1-4 | EA. | 1.000 |
| 3-4 x 1-4 | " | 1.000 |
| 6-0 x 1-4 | " | 1.000 |
| Sidelights, complete | | |
| 1-0 x 7-2 | EA. | 1.000 |
| 1-4 x 7-2 | " | 1.000 |
| 1-0 x 8-8 | " | 1.000 |
| 1-4 x 8-8 | " | 1.000 |
| 16 ga., 5-3/4" x 1-3/4" | | |
| 2-0 x 7-0 | EA. | 1.000 |
| 2-4 x 7-0 | " | 1.000 |
| 2-6 x 7-0 | " | 1.000 |
| 2-8 x 7-0 | " | 1.000 |
| 3-0 x 7-0 | " | 1.000 |
| 4-0 x 7-0 | " | 1.333 |
| 5-0 x 7-0 | " | 1.333 |
| 6-0 x 7-0 | " | 1.333 |
| Mullions, vertical | | |
| 5-1/4" x 1-3/4" | L.F. | 0.100 |
| 5-1/4" x 2" | " | 0.100 |
| Horizontal | | |
| 5-1/4" x 1-3/4" | L.F. | 0.100 |
| 5-1/4" x 2" | " | 0.100 |
| **08120.10 ALUMINUM DOORS** | | |
| Aluminum doors, commercial | | |
| Narrow stile | | |
| 2-6 x 7-0 | EA. | 4.000 |
| 3-0 x 7-0 | " | 4.000 |
| 3-6 x 7-0 | " | 4.000 |
| Pair | | |

# 08 DOORS AND WINDOWS

| METAL | UNIT | MAN/HOURS |
|---|---|---|
| **08120.10 ALUMINUM DOORS** | | |
| 5-0 x 7-0 | EA. | 8.000 |
| 6-0 x 7-0 | " | 8.000 |
| 7-0 x 7-0 | " | 8.000 |
| Wide stile | | |
| 2-6 x 7-0 | EA. | 4.000 |
| 3-0 x 7-0 | " | 4.000 |
| 3-6 x 7-0 | " | 4.000 |
| Pair | | |
| 5-0 x 7-0 | EA. | 8.000 |
| 6-0 x 7-0 | " | 8.000 |
| 7-0 x 7-0 | " | 8.000 |

| WOOD AND PLASTIC | UNIT | MAN/HOURS |
|---|---|---|
| **08210.10 WOOD DOORS** | | |
| Solid core, 1-3/8" thick | | |
| Birch faced | | |
| 2-4 x 7-0 | EA. | 1.000 |
| 2-8 x 7-0 | " | 1.000 |
| 3-0 x 7-0 | " | 1.000 |
| 3-4 x 7-0 | " | 1.000 |
| 2-4 x 6-8 | " | 1.000 |
| 2-6 x 6-8 | " | 1.000 |
| 2-8 x 6-8 | " | 1.000 |
| 3-0 x 6-8 | " | 1.000 |
| Lauan faced | | |
| 2-4 x 6-8 | EA. | 1.000 |
| 2-8 x 6-8 | " | 1.000 |
| 3-0 x 6-8 | " | 1.000 |
| 3-4 x 6-8 | " | 1.000 |
| Tempered hardboard faced | | |
| 2-4 x 7-0 | EA. | 1.000 |
| 2-8 x 7-0 | " | 1.000 |
| 3-0 x 7-0 | " | 1.000 |
| 3-4 x 7-0 | " | 1.000 |
| Hollow core, 1-3/8" thick | | |
| Birch faced | | |
| 2-4 x 7-0 | EA. | 1.000 |
| 2-8 x 7-0 | " | 1.000 |
| 3-0 x 7-0 | " | 1.000 |
| 3-4 x 7-0 | " | 1.000 |
| Lauan faced | | |
| 2-4 x 6-8 | EA. | 1.000 |
| 2-6 x 6-8 | " | 1.000 |
| 2-8 x 6-8 | " | 1.000 |
| 3-0 x 6-8 | " | 1.000 |

| WOOD AND PLASTIC | UNIT | MAN/HOURS |
|---|---|---|
| **08210.10 WOOD DOORS** | | |
| 3-4 x 6-8 | EA. | 1.000 |
| Tempered hardboard faced | | |
| 2-4 x 7-0 | EA. | 1.000 |
| 2-6 x 7-0 | " | 1.000 |
| 2-8 x 7-0 | " | 1.000 |
| 3-0 x 7-0 | " | 1.000 |
| 3-4 x 7-0 | " | 1.000 |
| Solid core, 1-3/4" thick | | |
| Birch faced | | |
| 2-4 x 7-0 | EA. | 1.000 |
| 2-6 x 7-0 | " | 1.000 |
| 2-8 x 7-0 | " | 1.000 |
| 3-0 x 7-0 | " | 1.000 |
| 3-4 x 7-0 | " | 1.000 |
| Lauan faced | | |
| 2-4 x 7-0 | EA. | 1.000 |
| 2-6 x 7-0 | " | 1.000 |
| 2-8 x 7-0 | " | 1.000 |
| 3-0 x 7-0 | " | 1.000 |
| 3-0 x 7-0 | " | 1.000 |
| Tempered hardboard faced | | |
| 2-4 x 7-0 | EA. | 1.000 |
| 2-6 x 7-0 | " | 1.000 |
| 2-8 x 7-0 | " | 1.000 |
| 3-0 x 7-0 | " | 1.000 |
| 3-4 x 7-0 | " | 1.000 |
| Hollow core, 1-3/4" thick | | |
| Birch faced | | |
| 2-4 x 7-0 | EA. | 1.000 |
| 2-6 x 7-0 | " | 1.000 |
| 2-8 x 7-0 | " | 1.000 |
| 3-0 x 7-0 | " | 1.000 |
| 3-4 x 7-0 | " | 1.000 |
| Lauan faced | | |
| 2-4 x 6-8 | EA. | 1.000 |
| 2-6 x 6-8 | " | 1.000 |
| 2-8 x 6-8 | " | 1.000 |
| 3-0 x 6-8 | " | 1.000 |
| 3-4 x 6-8 | " | 1.000 |
| Tempered hardboard | | |
| 2-4 x 7-0 | EA. | 1.000 |
| 2-6 x 7-0 | " | 1.000 |
| 2-8 x 7-0 | " | 1.000 |
| 3-0 x 7-0 | " | 1.000 |
| 3-4 x 7-0 | " | 1.000 |
| Add-on, louver | " | 0.800 |
| Glass | " | 0.800 |
| Exterior doors, 3-0 x 7-0 x 2-1/2", solid core | | |
| Carved | | |
| One face | EA. | 2.000 |
| Two faces | " | 2.000 |
| Closet doors, 1-3/4" thick | | |
| Bi-fold or bi-passing, includes frame and trim | | |
| Paneled | | |

# 08 DOORS AND WINDOWS

| WOOD AND PLASTIC | UNIT | MAN/ HOURS |
|---|---|---|
| **08210.10 WOOD DOORS** | | |
| 4-0 x 6-8 | EA. | 1.333 |
| 6-0 x 6-8 | " | 1.333 |
| Louvered | | |
| 4-0 x 6-8 | EA. | 1.333 |
| 6-0 x 6-8 | " | 1.333 |
| Flush | | |
| 4-0 x 6-8 | EA. | 1.333 |
| 6-0 x 6-8 | " | 1.333 |
| Primed | | |
| 4-0 x 6-8 | EA. | 1.333 |
| 6-0 x 6-8 | " | 1.333 |
| **08210.90 WOOD FRAMES** | | |
| Frame, interior, pine | | |
| 2-6 x 6-8 | EA. | 1.143 |
| 2-8 x 6-8 | " | 1.143 |
| 3-0 x 6-8 | " | 1.143 |
| 5-0 x 6-8 | " | 1.143 |
| 6-0 x 6-8 | " | 1.143 |
| 2-6 x 7-0 | " | 1.143 |
| 2-8 x 7-0 | " | 1.143 |
| 3-0 x 7-0 | " | 1.143 |
| 5-0 x 7-0 | " | 1.600 |
| 6-0 x 7-0 | " | 1.600 |
| Exterior, custom, with threshold, including trim | | |
| Walnut | | |
| 3-0 x 7-0 | EA. | 2.000 |
| 6-0 x 7-0 | " | 2.000 |
| Oak | | |
| 3-0 x 7-0 | EA. | 2.000 |
| 6-0 x 7-0 | " | 2.000 |
| Pine | | |
| 2-4 x 7-0 | EA. | 1.600 |
| 2-6 x 7-0 | " | 1.600 |
| 2-8 x 7-0 | " | 1.600 |
| 3-0 x 7-0 | " | 1.600 |
| 3-4 x 7-0 | " | 1.600 |
| 6-0 x 7-0 | " | 2.667 |
| **08300.10 SPECIAL DOORS** | | |
| Vault door and frame, class 5, steel | EA. | 8.000 |
| Overhead door, coiling insulated | | |
| Chain gear, no frame, 12' x 12' | EA. | 10.000 |
| Aluminum, bronze glass panels, 12-9 x 13-0 | " | 8.000 |
| Garage door, flush insulated metal, primed, 9-0 x 7-0 | " | 2.667 |
| Sliding metal fire doors, motorized, fusible link, 3 hr. | | |
| 3-0 x 6-8 | EA. | 16.000 |
| 3-8 x 6-8 | " | 10.000 |
| 4-0 x 8-0 | " | 16.000 |
| 5-0 x 8-0 | " | 16.000 |
| Metal clad doors, including electric motor | | |
| Light duty | | |

| WOOD AND PLASTIC | UNIT | MAN/ HOURS |
|---|---|---|
| **08300.10 SPECIAL DOORS** | | |
| Minimum | S.F. | 0.133 |
| Maximum | " | 0.320 |
| Heavy duty | | |
| Minimum | S.F. | 0.400 |
| Maximum | " | 0.500 |
| Hangar doors, based on 150' openings | | |
| To 20' high | S.F. | 0.096 |
| 20' to 40' high | " | 0.060 |
| 40' to 60' high | " | 0.040 |
| 60' to 80' high | " | 0.024 |
| Over 80' high | " | 0.019 |
| Counter doors, (roll-up shutters), standard, manual | | |
| Opening, 4' high | | |
| 4' wide | EA. | 6.667 |
| 6' wide | " | 6.667 |
| 8' wide | " | 7.273 |
| 10' wide | " | 10.000 |
| 14' wide | " | 10.000 |
| 6' high | | |
| 4' wide | EA. | 6.667 |
| 6' wide | " | 7.273 |
| 8' wide | " | 8.000 |
| 10' wide | " | 10.000 |
| 14' wide | " | 11.429 |
| Service doors, (roll up shutters), standard, manual | | |
| Opening | | |
| 8' high x 8' wide | EA. | 4.444 |
| 10' high x 10' wide | " | 6.667 |
| 12' high x 12' wide | " | 10.000 |
| 14' high x 14' wide | " | 13.333 |
| 16' high x 14' wide | " | 13.333 |
| 20' high x 14' wide | " | 20.000 |
| 24' high x 16' wide | " | 17.778 |
| Roll-up doors | | |
| 13-0 high x 14-0 wide | EA. | 11.429 |
| 12-0 high x 14-0 wide | " | 11.429 |
| Top coiling grilles, manually operated, steel or aluminum | | |
| Opening, 4' high x | | |
| 4' wide | EA. | 3.200 |
| 6' wide | " | 3.200 |
| 8' wide | " | 4.444 |
| 12' wide | " | 4.444 |
| 16' wide | " | 6.667 |
| 6' high x | | |
| 4' wide | EA. | 6.667 |
| 6' wide | " | 7.273 |
| 8' wide | " | 8.000 |
| 12' wide | " | 8.889 |
| 16' wide | " | 11.429 |
| Side coiling grilles, manually operated, aluminum | | |
| Opening, 8' high x | | |
| 18' wide | EA. | 60.000 |
| 24' wide | " | 68.571 |
| 12' high x | | |

# 08 DOORS AND WINDOWS

| WOOD AND PLASTIC | UNIT | MAN/ HOURS |
|---|---|---|
| **08300.10 SPECIAL DOORS** | | |
| 12' wide | EA. | 60.000 |
| 18' wide | " | 68.571 |
| 24' wide | " | 80.000 |
| Accordion folding doors, tracks and fittings included | | |
| Vinyl covered, 2 layers | S.F. | 0.320 |
| Woven mahogany and vinyl | " | 0.320 |
| Economy vinyl | " | 0.320 |
| Rigid polyvinyl chloride | " | 0.320 |
| Sectional wood overhead doors, frames not included | | |
| Commercial grade, heavy duty, 1-3/4" thick, manual | | |
| 8' x 8' | EA. | 6.667 |
| 10' x 10' | " | 7.273 |
| 12' x 12' | " | 8.000 |
| Chain hoist | | |
| 12' x 16' high | EA. | 13.333 |
| 14' x 14' high | " | 10.000 |
| 20' x 8' high | " | 16.000 |
| 16' high | " | 20.000 |
| Sectional metal overhead doors, complete | | |
| Residential grade, manual | | |
| 9' x 7' | EA. | 3.200 |
| 16' x 7' | " | 4.000 |
| Commercial grade | | |
| 8' x 8' | EA. | 6.667 |
| 10' x 10' | " | 7.273 |
| 12' x 12' | " | 8.000 |
| 20' x 14', with chain hoist | " | 16.000 |
| Sliding glass doors | | |
| Tempered plate glass, 1/4" thick | | |
| 6' wide | | |
| Economy grade | EA. | 2.667 |
| Premium grade | " | 2.667 |
| 12' wide | | |
| Economy grade | EA. | 4.000 |
| Premium grade | " | 4.000 |
| Insulating glass, 5/8" thick | | |
| 6' wide | | |
| Economy grade | EA. | 2.667 |
| Premium grade | " | 2.667 |
| 12' wide | | |
| Economy grade | EA. | 4.000 |
| Premium grade | " | 4.000 |
| 1" thick | | |
| 6' wide | | |
| Economy grade | EA. | 2.667 |
| Premium grade | " | 2.667 |
| 12' wide | | |
| Economy grade | EA. | 4.000 |
| Premium grade | " | 4.000 |
| Vertical lift doors, channel frame construction | | |
| 20' high x | | |
| 10' wide | EA. | 9.600 |
| 15' wide | " | 9.600 |
| 20' wide | " | 17.143 |

| WOOD AND PLASTIC | UNIT | MAN/ HOURS |
|---|---|---|
| **08300.10 SPECIAL DOORS** | | |
| 25' wide | EA. | 17.143 |
| 25' high | | |
| 20' wide | EA. | 17.143 |
| 25' wide | " | 20.000 |
| 30' high | | |
| 25' wide | EA. | 20.000 |
| 30' wide | " | 20.000 |
| 35' wide | " | 20.000 |
| Residential storm door | | |
| Minimum | EA. | 1.333 |
| Average | " | 1.333 |
| Maximum | " | 2.000 |

| STOREFRONTS | UNIT | MAN/ HOURS |
|---|---|---|
| **08410.10 STOREFRONTS** | | |
| Storefront, aluminum and glass | | |
| Minimum | S.F. | 0.100 |
| Average | " | 0.114 |
| Maximum | " | 0.133 |
| Entrance doors | | |
| 1/2" thick glass | | |
| 3' x 7' | EA. | 6.667 |
| 6' x 7' | " | 10.000 |
| 3/4" thick glass | | |
| 3' x 7' | EA. | 6.667 |
| 6' x 7' | " | 10.000 |
| 1" thick glass | | |
| 3' x 7' | EA. | 6.667 |
| 6' x 7' | " | 10.000 |
| Revolving doors | | |
| 7'diameter, 7' high | | |
| Minimum | EA. | 60.000 |
| Average | " | 96.000 |
| Maximum | " | 120 |

## METAL WINDOWS

### 08510.10 STEEL WINDOWS

| METAL WINDOWS | UNIT | MAN/HOURS |
|---|---|---|
| Steel windows, primed | | |
| Casements | | |
| Operable | | |
| Minimum | S.F. | 0.047 |
| Maximum | " | 0.053 |
| Fixed sash | " | 0.040 |
| Double hung | " | 0.044 |
| Industrial windows | | |
| Horizontally pivoted sash | S.F. | 0.053 |
| Fixed sash | " | 0.044 |
| Security sash | | |
| Operable | S.F. | 0.053 |
| Fixed | " | 0.044 |
| Picture window | " | 0.044 |
| Projecting sash | | |
| Minimum | S.F. | 0.050 |
| Maximum | " | 0.050 |
| Mullions | L.F. | 0.040 |

### 08520.10 ALUMINUM WINDOWS

| METAL WINDOWS | UNIT | MAN/HOURS |
|---|---|---|
| Jalousie | | |
| 3-0 x 4-0 | EA. | 1.000 |
| 3-0 x 5-0 | " | 1.000 |
| Fixed window | | |
| 6 sf to 8 sf | S.F. | 0.114 |
| 12 sf to 16 sf | " | 0.089 |
| Projecting window | | |
| 6 sf to 8 sf | S.F. | 0.200 |
| 12 sf to 16 sf | " | 0.133 |
| Horizontal sliding | | |
| 6 sf to 8 sf | S.F. | 0.100 |
| 12 sf to 16 sf | " | 0.080 |
| Double hung | | |
| 6 sf to 8 sf | S.F. | 0.160 |
| 10 sf to 12 sf | " | 0.133 |
| Storm window, 0.5 cfm, up to | | |
| 60 u.i. (united inches) | EA. | 0.400 |
| 70 u.i. | " | 0.400 |
| 80 u.i. | " | 0.400 |
| 90 u.i. | " | 0.444 |
| 100 u.i. | " | 0.444 |
| 2.0 cfm, up to | | |
| 60 u.i. | EA. | 0.400 |
| 70 u.i. | " | 0.400 |
| 80 u.i. | " | 0.400 |
| 90 u.i. | " | 0.444 |
| 100 u.i. | " | 0.444 |

## WOOD AND PLASTIC

### 08600.10 WOOD WINDOWS

| WOOD AND PLASTIC | UNIT | MAN/HOURS |
|---|---|---|
| Double hung | | |
| 24" x 36" | | |
| Minimum | EA. | 0.800 |
| Average | " | 1.000 |
| Maximum | " | 1.333 |
| 24" x 48" | | |
| Minimum | EA. | 0.800 |
| Average | " | 1.000 |
| Maximum | " | 1.333 |
| 30" x 48" | | |
| Minimum | EA. | 0.889 |
| Average | " | 1.143 |
| Maximum | " | 1.600 |
| 30" x 60" | | |
| Minimum | EA. | 0.889 |
| Average | " | 1.143 |
| Maximum | " | 1.600 |
| Casement | | |
| 1 leaf, 22" x 38" high | | |
| Minimum | EA. | 0.800 |
| Average | " | 1.000 |
| Maximum | " | 1.333 |
| 2 leaf, 50" x 50" high | | |
| Minimum | EA. | 1.000 |
| Average | " | 1.333 |
| Maximum | " | 2.000 |
| 3 leaf, 71" x 62" high | | |
| Minimum | EA. | 1.000 |
| Average | " | 1.333 |
| Maximum | " | 2.000 |
| 4 leaf, 95" x 75" high | | |
| Minimum | EA. | 1.143 |
| Average | " | 1.600 |
| Maximum | " | 2.667 |
| 5 leaf, 119" x 75" high | | |
| Minimum | EA. | 1.143 |
| Average | " | 1.600 |
| Maximum | " | 2.667 |
| Picture window, fixed glass, 54" x 54" high | | |
| Minimum | EA. | 1.000 |
| Average | " | 1.143 |
| Maximum | " | 1.333 |
| 68" x 55" high | | |
| Minimum | EA. | 1.000 |
| Average | " | 1.143 |
| Maximum | " | 1.333 |
| Sliding, 40" x 31" high | | |
| Minimum | EA. | 0.800 |
| Average | " | 1.000 |
| Maximum | " | 1.333 |
| 52" x 39" high | | |
| Minimum | EA. | 1.000 |
| Average | " | 1.143 |
| Maximum | " | 1.333 |

# 08 DOORS AND WINDOWS

| WOOD AND PLASTIC | UNIT | MAN/ HOURS |
|---|---|---|
| **08600.10 WOOD WINDOWS** | | |
| 64" x 72" high | | |
| Minimum | EA. | 1.000 |
| Average | " | 1.333 |
| Maximum | " | 1.600 |
| Awning windows | | |
| 34" x 21" high | | |
| Minimum | EA. | 0.800 |
| Average | " | 1.000 |
| Maximum | " | 1.333 |
| 40" x 21" high | | |
| Minimum | EA. | 0.889 |
| Average | " | 1.143 |
| Maximum | " | 1.600 |
| 48" x 27" high | | |
| Minimum | EA. | 0.889 |
| Average | " | 1.143 |
| Maximum | " | 1.600 |
| 60" x 36" high | | |
| Minimum | EA. | 1.000 |
| Average | " | 1.333 |
| Maximum | " | 1.600 |
| Window frame, milled | | |
| Minimum | L.F. | 0.160 |
| Average | " | 0.200 |
| Maximum | " | 0.267 |

| HARDWARE | UNIT | MAN/ HOURS |
|---|---|---|
| **08710.20 LOCKSETS** | | |
| Latchset, heavy duty | | |
| Cylindrical | EA. | 0.500 |
| Mortise | " | 0.800 |
| Lockset, heavy duty | | |
| Cylindrical | EA. | 0.500 |
| Mortise | " | 0.800 |
| Mortise locks and latchsets, chrome | | |
| Latchset passage or closet latch | EA. | 0.667 |
| Privacy (bath or bedroom) | " | 0.667 |
| Entry lockset | " | 0.667 |
| Classroom lockset (outside key operated) | " | 0.667 |
| Storeroom lock | " | 0.667 |
| Front door lock | " | 0.667 |
| Dormitory or exit lock | " | 0.667 |
| Preassembled locks and latches, brass | | |
| Latchset, passage or closet latch | EA. | 0.667 |
| Lockset | | |

| HARDWARE | UNIT | MAN/ HOURS |
|---|---|---|
| **08710.20 LOCKSETS** | | |
| Privacy (bath or bathroom) | EA. | 0.667 |
| Entry lock | " | 0.667 |
| Classroom lock (outside key, operated) | " | 0.667 |
| Storeroom lock | " | 0.667 |
| Bored locks and latches, satin chrome plated | | |
| Latchset passage or closet latch | EA. | 0.667 |
| Lockset | | |
| Privacy (bath or bedroom) | EA. | 0.667 |
| Entry lock | " | 0.667 |
| Classroom lock | " | 0.667 |
| Corridor lock | " | 0.667 |
| Miscellaneous locks | | |
| Exit lock with alarm, single door | EA. | 3.200 |
| Electric strike | | |
| Rim mounted wrought steel | EA. | 2.000 |
| Mortised, wrought steel with bronze plating | " | 3.200 |
| Dead bolt | | |
| Bored, wrought brass, keyed both sides | EA. | 1.333 |
| Mortised, cast brass | " | 1.333 |
| Lockset, cipher, mechanical | " | 0.800 |
| **08710.30 CLOSERS** | | |
| Door closers | | |
| Surface mounted, traditional type, parallel arm | | |
| Standard | EA. | 1.000 |
| Heavy duty | " | 1.000 |
| Modern type, parallel arm, standard duty | " | 1.000 |
| Overhead, concealed, pivot hung, single acting | | |
| Interior | EA. | 1.000 |
| Exterior | " | 1.000 |
| Floor concealed, single acting, offset, pivoted | | |
| Interior | EA. | 2.667 |
| Exterior | " | 2.667 |
| **08710.40 DOOR TRIM** | | |
| Door bumper, bronze, wall type | EA. | 0.160 |
| Wall type, 4" dia. with convex rubber pad, aluminum | " | 0.160 |
| Floor type | | |
| Aluminum | EA. | 0.160 |
| Brass | " | 0.160 |
| Door holders | | |
| Wall type, bronze | EA. | 0.160 |
| Overhead | " | 0.400 |
| Floor type | " | 0.400 |
| Plunger type | " | 0.400 |
| Wall type, aluminum | " | 0.400 |
| Surface bolt | " | 0.160 |
| Panic device | | |
| Rim type with thumb piece | EA. | 2.000 |
| Mortise | " | 2.000 |
| Vertical rod | " | 2.000 |
| Labelled, rim type | " | 2.000 |
| Mortise | " | 2.000 |

# 08 DOORS AND WINDOWS

| HARDWARE | UNIT | MAN/ HOURS |
|---|---|---|
| **08710.40 DOOR TRIM** | | |
| Vertical rod | EA. | 2.000 |
| Silencers, rubber type | " | 0.016 |
| Dust proof strike with plate, brass | " | 0.267 |
| Flush bolt, lever extension, brass, rated | " | 0.160 |
| Surface bolt with strike, brass, 6" long | " | 0.160 |
| Door coordinator, labelled, brass, satin chrome | " | 0.571 |
| Door plates | | |
| Kick plate, aluminum, 3 beveled edges | | |
| 10" x 28" | EA. | 0.400 |
| 10" x 30" | " | 0.400 |
| 10" x 34" | " | 0.400 |
| 10" x 38" | " | 0.400 |
| Push plate, 4" x 16" | | |
| Aluminum | EA. | 0.160 |
| Bronze | " | 0.160 |
| Stainless steel | " | 0.160 |
| Armor plate, 40" x 34" | " | 0.320 |
| Pull handle, 4" x 16" | | |
| Aluminum | EA. | 0.160 |
| Bronze | " | 0.160 |
| Stainless steel | " | 0.160 |
| Hasp assembly | | |
| 3" | EA. | 0.133 |
| 4-1/2" | " | 0.178 |
| 6" | " | 0.229 |
| Electro-magnetic door holder | | |
| Wall mounted | EA. | 2.667 |
| Floor mounted | " | 2.667 |
| Smoke detector door holder | | |
| Photo electric type | EA. | 2.667 |
| Ionization type | " | 2.667 |
| Pneumatic operators, activated by rubber mats | | |
| Swing | | |
| Single | EA. | 6.667 |
| Double | " | 10.000 |
| Sliding | | |
| Single | EA. | 6.667 |
| Double | " | 10.000 |
| **08710.60 WEATHERSTRIPPING** | | |
| Weatherstrip, head and jamb, metal strip, neoprene bulb | | |
| Standard duty | L.F. | 0.044 |
| Heavy duty | " | 0.050 |
| Spring type | | |
| Metal doors | EA. | 2.000 |
| Wood doors | " | 2.667 |
| Sponge type with adhesive backing | " | 0.800 |
| Astragal | | |
| 1-3/4" x 13 ga., aluminum | L.F. | 0.067 |
| 1-3/8" x 5/8", oak | " | 0.053 |
| Thresholds | | |
| Bronze | L.F. | 0.200 |
| Aluminum | | |
| Plain | L.F. | 0.200 |

| HARDWARE | UNIT | MAN/ HOURS |
|---|---|---|
| **08710.60 WEATHERSTRIPPING** | | |
| Vinyl insert | L.F. | 0.200 |
| Aluminum with grit | " | 0.200 |
| Steel | | |
| Plain | L.F. | 0.200 |
| Interlocking | " | 0.667 |

| GLAZING | UNIT | MAN/ HOURS |
|---|---|---|
| **08810.10 GLAZING** | | |
| Sheet glass, 1/8" thick | S.F. | 0.044 |
| Plate glass, bronze or grey, 1/4" thick | " | 0.073 |
| Clear | " | 0.073 |
| Polished | " | 0.073 |
| Plexiglass | | |
| 1/8" thick | S.F. | 0.073 |
| 1/4" thick | " | 0.044 |
| Float glass, clear | | |
| 3/16" thick | S.F. | 0.067 |
| 1/4" thick | " | 0.073 |
| 5/16" thick | " | 0.080 |
| 3/8" thick | " | 0.100 |
| 1/2" thick | " | 0.133 |
| 5/8" thick | " | 0.160 |
| 3/4" thick | " | 0.200 |
| 1" thick | " | 0.267 |
| Tinted glass, polished plate, twin ground | | |
| 3/16" thick | S.F. | 0.067 |
| 1/4" thick | " | 0.073 |
| 3/8" thick | " | 0.100 |
| 1/2" thick | " | 0.133 |
| Total, full vision, all glass window system | | |
| To 10' high | | |
| Minimum | S.F. | 0.200 |
| Average | " | 0.200 |
| Maximum | " | 0.200 |
| 10' to 20' high | | |
| Minimum | S.F. | 0.200 |
| Average | " | 0.200 |
| Maximum | " | 0.200 |
| Insulated glass, bronze or gray | | |
| 1/2" thick | S.F. | 0.133 |
| 1" thick | " | 0.200 |
| Clear | | |
| 1/2" thick | S.F. | 0.133 |
| 1" thick | " | 0.200 |
| Spandrel glass, polished bronze/grey, 1 side, 1/4" thick | " | 0.073 |

# 08 DOORS AND WINDOWS

| GLAZING | UNIT | MAN/HOURS |
|---|---|---|
| **08810.10 GLAZING** | | |
| Tempered glass (safety) | | |
| Clear sheet glass | | |
| 1/8" thick | S.F. | 0.044 |
| 3/16" thick | " | 0.062 |
| Clear float glass | | |
| 1/4" thick | S.F. | 0.067 |
| 5/16" thick | " | 0.080 |
| 3/8" thick | " | 0.100 |
| 1/2" thick | " | 0.133 |
| 5/8" thick | " | 0.160 |
| 3/4" thick | " | 0.267 |
| Tinted float glass | | |
| 3/16" thick | S.F. | 0.062 |
| 1/4" thick | " | 0.067 |
| 3/8" thick | " | 0.100 |
| 1/2" thick | " | 0.133 |
| Laminated glass | | |
| Float safety glass with polyvinyl plastic interlayer | | |
| 1/4", sheet or float | | |
| Two lites, 1/8" thick, clear glass | S.F. | 0.067 |
| 1/2" thick, float glass | | |
| Two lites, 1/4" thick, clear glass | S.F. | 0.133 |
| Tinted glass | " | 0.133 |
| Insulating glass, two lites, clear float glass | | |
| 1/2" thick | S.F. | 0.133 |
| 5/8" thick | " | 0.160 |
| 3/4" thick | " | 0.200 |
| 7/8" thick | " | 0.229 |
| 1" thick | " | 0.267 |
| Glass seal edge | | |
| 3/8" thick | S.F. | 0.133 |
| Tinted glass | | |
| 1/2" thick | S.F. | 0.133 |
| 1" thick | " | 0.267 |
| Tempered, clear | | |
| 1" thick | S.F. | 0.267 |
| Wire reinforced | " | 0.267 |
| Plate mirror glass | | |
| 1/4" thick | | |
| 15 sf | S.F. | 0.080 |
| Over 15 sf | " | 0.073 |
| Transparent, one way vision, 1/4" thick | " | 0.080 |
| Sheet mirror glass | | |
| 3/16" thick | S.F. | 0.080 |
| 1/4" thick | " | 0.067 |
| Wall tiles, 12" x 12" | | |
| Clear glass | S.F. | 0.044 |
| Veined glass | " | 0.044 |
| Wire glass, 1/4" thick | | |
| Clear | S.F. | 0.267 |
| Hammered | " | 0.267 |
| Obscure | " | 0.267 |
| Bullet resistant glass, plate glass with inter-leaved vinyl | | |
| 1-3/16" thick | | |

| GLAZING | UNIT | MAN/HOURS |
|---|---|---|
| **08810.10 GLAZING** | | |
| To 15 sf | S.F. | 0.400 |
| Over 15 sf | " | 0.400 |
| 2" thick | | |
| To 15 sf | S.F. | 0.667 |
| Over 15 sf | " | 0.667 |
| Glazing accessories | | |
| Neoprene glazing gaskets | | |
| 1/4" glass | L.F. | 0.032 |
| 3/8" glass | " | 0.033 |
| 1/2" glass | " | 0.035 |
| 3/4" glass | " | 0.036 |
| 1" glass | " | 0.040 |
| Mullion section | | |
| 1/4" glass | L.F. | 0.016 |
| 3/8" glass | " | 0.020 |
| 1/2" glass | " | 0.023 |
| 3/4" glass | " | 0.027 |
| 1" glass | " | 0.032 |
| Molded corners | EA. | 0.533 |

| GLAZED CURTAIN WALLS | UNIT | MAN/HOURS |
|---|---|---|
| **08910.10 GLAZED CURTAIN WALLS** | | |
| Curtain wall, aluminum system, framing sections | | |
| 2" x 3" | | |
| Jamb | L.F. | 0.067 |
| Horizontal | " | 0.067 |
| Mullion | " | 0.067 |
| 2" x 4" | | |
| Jamb | L.F. | 0.100 |
| Horizontal | " | 0.100 |
| Mullion | " | 0.100 |
| 3" x 5-1/2" | | |
| Jamb | L.F. | 0.100 |
| Horizontal | " | 0.100 |
| Mullion | " | 0.100 |
| 4" corner mullion | " | 0.133 |
| Coping sections | | |
| 1/8" x 8" | L.F. | 0.133 |
| 1/8" x 9" | " | 0.133 |
| 1/8" x 12-1/2" | " | 0.160 |
| Sill section | | |
| 1/8" x 6" | L.F. | 0.080 |
| 1/8" x 7" | " | 0.080 |
| 1/8" x 8-1/2" | " | 0.080 |
| Column covers, aluminum | | |

| GLAZED CURTAIN WALLS | UNIT | MAN/ HOURS |
|---|---|---|
| **08910.10 GLAZED CURTAIN WALLS** | | |
| 1/8" x 26" | L.F. | 0.200 |
| 1/8" x 34" | " | 0.211 |
| 1/8" x 38" | " | 0.211 |
| Doors | | |
| Aluminum framed, standard hardware | | |
| Narrow stile | | |
| 2-6 x 7-0 | EA. | 4.000 |
| 3-0 x 7-0 | " | 4.000 |
| 3-6 x 7-0 | " | 4.000 |
| Wide stile | | |
| 2-6 x 7-0 | EA. | 4.000 |
| 3-0 x 7-0 | " | 4.000 |
| 3-6 x 7-0 | " | 4.000 |
| Flush panel doors, to match adjacent wall panels | | |
| 2-6 x 7-0 | EA. | 5.000 |
| 3-0 x 7-0 | " | 5.000 |
| 3-6 x 7-0 | " | 5.000 |
| Wall panel, insulated | | |
| "U"=.08 | S.F. | 0.067 |
| "U"=.10 | " | 0.067 |
| "U"=.15 | " | 0.067 |
| Window wall system, complete | | |
| Minimum | S.F. | 0.080 |
| Average | " | 0.089 |
| Maximum | " | 0.114 |

| SUPPORT SYSTEMS | UNIT | MAN/HOURS |
|---|---|---|
| **09110.10 METAL STUDS** | | |
| Studs, non load bearing, galvanized | | |
| 2-1/2", 20 ga. | | |
| 12" o.c. | S.F. | 0.017 |
| 16" o.c. | " | 0.013 |
| 25 ga. | | |
| 12" o.c. | S.F. | 0.017 |
| 16" o.c. | " | 0.013 |
| 24" o.c. | " | 0.011 |
| 3-5/8", 20 ga. | | |
| 12" o.c. | S.F. | 0.020 |
| 16" o.c. | " | 0.016 |
| 24" o.c. | " | 0.013 |
| 25 ga. | | |
| 12" o.c. | S.F. | 0.020 |
| 16" o.c. | " | 0.016 |
| 24" o.c. | " | 0.013 |
| 4", 20 ga. | | |
| 12" o.c. | S.F. | 0.020 |
| 16" o.c. | " | 0.016 |
| 24" o.c. | " | 0.013 |
| 25 ga. | | |
| 12" o.c. | S.F. | 0.020 |
| 16" o.c. | " | 0.016 |
| 24" o.c. | " | 0.013 |
| 6", 20 ga. | | |
| 12" o.c. | S.F. | 0.025 |
| 16" o.c. | " | 0.020 |
| 24" o.c. | " | 0.017 |
| 25 ga. | | |
| 12" o.c. | S.F. | 0.025 |
| 16" o.c. | " | 0.020 |
| 24" o.c. | " | 0.017 |
| Load bearing studs, galvanized | | |
| 3-5/8", 16 ga. | | |
| 12" o.c. | S.F. | 0.020 |
| 16" o.c. | " | 0.016 |
| 18 ga. | | |
| 12" o.c. | S.F. | 0.013 |
| 16" o.c. | " | 0.016 |
| 4", 16 ga. | | |
| 12" o.c. | S.F. | 0.020 |
| 16" o.c. | " | 0.016 |
| 6", 16 ga. | | |
| 12" o.c. | S.F. | 0.025 |
| 16" o.c. | " | 0.020 |
| Furring | | |
| On beams and columns | | |
| 7/8" channel | L.F. | 0.053 |
| 1-1/2" channel | " | 0.062 |
| On ceilings | | |
| 3/4" furring channels | | |
| 12" o.c. | S.F. | 0.033 |
| 16" o.c. | " | 0.032 |
| 24" o.c. | " | 0.029 |

| SUPPORT SYSTEMS | UNIT | MAN/HOURS |
|---|---|---|
| **09110.10 METAL STUDS** | | |
| 1-1/2" furring channels | | |
| 12" o.c. | S.F. | 0.036 |
| 16" o.c. | " | 0.033 |
| 24" o.c. | " | 0.031 |
| On walls | | |
| 3/4" furring channels | | |
| 12" o.c. | S.F. | 0.027 |
| 16" o.c. | " | 0.025 |
| 24" o.c. | " | 0.024 |
| 1-1/2" furring channels | | |
| 12" o.c. | S.F. | 0.029 |
| 16" o.c. | " | 0.027 |
| 24" o.c. | " | 0.025 |

| LATH AND PLASTER | UNIT | MAN/HOURS |
|---|---|---|
| **09205.10 GYPSUM LATH** | | |
| Gypsum lath, 1/2" thick | | |
| Clipped | S.Y. | 0.044 |
| Nailed | " | 0.050 |
| **09205.20 METAL LATH** | | |
| Diamond expanded, galvanized | | |
| 2.5 lb., on walls | | |
| Nailed | S.Y. | 0.100 |
| Wired | " | 0.114 |
| On ceilings | | |
| Nailed | S.Y. | 0.114 |
| Wired | " | 0.133 |
| 3.4 lb., on walls | | |
| Nailed | S.Y. | 0.100 |
| Wired | " | 0.114 |
| On ceilings | | |
| Nailed | S.Y. | 0.114 |
| Wired | " | 0.133 |
| Flat rib | | |
| 2.75 lb., on walls | | |
| Nailed | S.Y. | 0.100 |
| Wired | " | 0.114 |
| On ceilings | | |
| Nailed | S.Y. | 0.114 |
| Wired | " | 0.133 |
| 3.4 lb., on walls | | |
| Nailed | S.Y. | 0.100 |
| Wired | " | 0.114 |

# 09 FINISHES

| LATH AND PLASTER | UNIT | MAN/ HOURS |
|---|---|---|
| **09205.20 METAL LATH** | | |
| On ceilings | | |
| Nailed | S.Y. | 0.114 |
| Wired | " | 0.133 |
| Stucco lath | | |
| 1.8 lb. | S.Y. | 0.100 |
| 3.6 lb. | " | 0.100 |
| Paper backed | | |
| Minimum | S.Y. | 0.080 |
| Maximum | " | 0.114 |
| **09205.60 PLASTER ACCESSORIES** | | |
| Expansion joint, 3/4", 26 ga., galvanized, one piece | L.F. | 0.020 |
| Plaster corner beads, 3/4", galvanized | " | 0.023 |
| Casing bead, expanded flange, galvanized | " | 0.020 |
| Expanded wing, 1-1/4" wide, galvanized | " | 0.020 |
| Joint clips for lath | EA. | 0.004 |
| Metal base, galvanized, 2-1/2" high | L.F. | 0.027 |
| Stud clips for gypsum lath | EA. | 0.004 |
| Sound deadening board, 1/4", nailed or clipped | S.F. | 0.013 |
| **09210.10 PLASTER** | | |
| Gypsum plaster, trowel finish, 2 coats | | |
| Ceilings | S.Y. | 0.250 |
| Walls | " | 0.235 |
| 3 coats | | |
| Ceilings | S.Y. | 0.348 |
| Walls | " | 0.308 |
| Vermiculite plaster | | |
| 2 coats | | |
| Ceilings | S.Y. | 0.381 |
| Walls | " | 0.348 |
| 3 coats | | |
| Ceilings | S.Y. | 0.471 |
| Walls | " | 0.421 |
| Keenes cement plaster | | |
| 2 coats | | |
| Ceilings | S.Y. | 0.308 |
| Walls | " | 0.267 |
| 3 coats | | |
| Ceilings | S.Y. | 0.348 |
| Walls | " | 0.308 |
| On columns, add to installation, 50% | " | |
| Chases, fascia, and soffits, add to installation, 50% | " | |
| Beams, add to installation, 50% | " | |
| **09220.10 PORTLAND CEMENT PLASTER** | | |
| Stucco, portland, gray, 3 coat, 1" thick | | |
| Sand finish | S.Y. | 0.348 |
| Trowel finish | " | 0.364 |
| White cement | | |
| Sand finish | S.Y. | 0.364 |
| Trowel finish | " | 0.400 |

| LATH AND PLASTER | UNIT | MAN/ HOURS |
|---|---|---|
| **09220.10 PORTLAND CEMENT PLASTER** | | |
| Scratch coat | | |
| For ceramic tile | S.Y. | 0.080 |
| For quarry tile | " | 0.080 |
| Portland cement plaster | | |
| 2 coats, 1/2" | S.Y. | 0.160 |
| 3 coats, 7/8" | " | 0.200 |
| **09250.10 GYPSUM BOARD** | | |
| Drywall, plasterboard, 3/8" clipped to | | |
| Metal furred ceiling | S.F. | 0.009 |
| Columns and beams | " | 0.020 |
| Walls | " | 0.008 |
| Nailed or screwed to | | |
| Wood framed ceiling | S.F. | 0.008 |
| Columns and beams | " | 0.018 |
| Walls | " | 0.007 |
| 1/2", clipped to | | |
| Metal furred ceiling | S.F. | 0.009 |
| Columns and beams | " | 0.020 |
| Walls | " | 0.008 |
| Nailed or screwed to | | |
| Wood framed ceiling | S.F. | 0.008 |
| Columns and beams | " | 0.018 |
| Walls | " | 0.007 |
| 5/8", clipped to | | |
| Metal furred ceiling | S.F. | 0.010 |
| Columns and beams | " | 0.022 |
| Walls | " | 0.009 |
| Nailed or screwed to | | |
| Wood framed ceiling | S.F. | 0.010 |
| Columns and beams | " | 0.022 |
| Walls | " | 0.009 |
| Vinyl faced, clipped to metal studs | | |
| 1/2" | S.F. | 0.010 |
| 5/8" | " | 0.010 |
| Taping and finishing joints | | |
| Minimum | S.F. | 0.005 |
| Average | " | 0.007 |
| Maximum | " | 0.008 |
| Casing bead | | |
| Minimum | L.F. | 0.023 |
| Average | " | 0.027 |
| Maximum | " | 0.040 |
| Corner bead | | |
| Minimum | L.F. | 0.023 |
| Average | " | 0.027 |
| Maximum | " | 0.040 |

| TILE | UNIT | MAN/HOURS |
|---|---|---|
| **09310.10 CERAMIC TILE** | | |
| Glazed wall tile, 4-1/4" x 4-1/4" | | |
| Minimum | S.F. | 0.057 |
| Average | " | 0.067 |
| Maximum | " | 0.080 |
| Base, 4-1/4" high | | |
| Minimum | L.F. | 0.100 |
| Average | " | 0.100 |
| Maximum | " | 0.100 |
| Unglazed floor tile | | |
| Portland cement bed, cushion edge, face mounted | | |
| 1" x 1" | S.F. | 0.073 |
| 1" x 2" | " | 0.070 |
| 2" x 2" | " | 0.067 |
| Adhesive bed, with white grout | | |
| 1" x 1" | S.F. | 0.073 |
| 1" x 2" | " | 0.070 |
| 2" x 2" | " | 0.067 |
| Organic adhesive bed, thin set, back mounted | | |
| 1" x 1" | S.F. | 0.073 |
| 1" x 2" | " | 0.070 |
| 2" x 2" | " | 0.067 |
| Unglazed wall tile | | |
| Organic adhesive, face mounted cushion edge | | |
| 1" x 1" | | |
| Minimum | S.F. | 0.067 |
| Average | " | 0.073 |
| Maximum | " | 0.080 |
| 1" x 2" | | |
| Minimum | S.F. | 0.064 |
| Average | " | 0.070 |
| Maximum | " | 0.076 |
| 2" x 2" | | |
| Minimum | S.F. | 0.062 |
| Average | " | 0.067 |
| Maximum | " | 0.073 |
| Back mounted | | |
| 1" x 1" | | |
| Minimum | S.F. | 0.067 |
| Average | " | 0.073 |
| Maximum | " | 0.080 |
| 1" x 2" | | |
| Minimum | S.F. | 0.064 |
| Average | " | 0.070 |
| Maximum | " | 0.076 |
| 2" x 2" | | |
| Minimum | S.F. | 0.062 |
| Average | " | 0.067 |
| Maximum | " | 0.073 |
| Conductive floor tile, unglazed square edged | | |
| Portland cement bed | | |
| 1 x 1 | S.F. | 0.100 |
| 1-9/16 x 1-9/16 | " | 0.100 |
| Dry set | | |
| 1 x 1 | S.F. | 0.100 |

| TILE | UNIT | MAN/HOURS |
|---|---|---|
| **09310.10 CERAMIC TILE** | | |
| 1-9/16 x 1-9/16 | S.F. | 0.100 |
| Epoxy bed with epoxy joints | | |
| 1 x 1 | S.F. | 0.100 |
| 1-9/16 x 1-9/16 | " | 0.100 |
| Ceramic accessories | | |
| Towel bar, 24" long | | |
| Minimum | EA. | 0.320 |
| Average | " | 0.400 |
| Maximum | " | 0.533 |
| Soap dish | | |
| Minimum | EA. | 0.533 |
| Average | " | 0.667 |
| Maximum | " | 0.800 |
| **09330.10 QUARRY TILE** | | |
| Floor | | |
| 4 x 4 x 1/2" | S.F. | 0.107 |
| 6 x 6 x 1/2" | " | 0.100 |
| 6 x 6 x 3/4" | " | 0.100 |
| Wall, applied to 3/4" portland cement bed | | |
| 4 x 4 x 1/2" | S.F. | 0.160 |
| 6 x 6 x 3/4" | " | 0.133 |
| Cove base | | |
| 5 x 6 x 1/2" straight top | L.F. | 0.133 |
| 6 x 6 x 3/4" round top | " | 0.133 |
| Stair treads 6 x 6 x 3/4" | " | 0.200 |
| Window sill 6 x 8 x 3/4" | " | 0.160 |
| For abrasive surface, add to material, 25% | | |
| **09410.10 TERRAZZO** | | |
| Floors on concrete, 1-3/4" thick, 5/8" topping | | |
| Gray cement | S.F. | 0.114 |
| White cement | " | 0.114 |
| Sand cushion, 3" thick, 5/8" top, 1/4" | | |
| Gray cement | S.F. | 0.133 |
| White cement | " | 0.133 |
| Monolithic terrazzo, 3-1/2" base slab, 5/8" topping | " | 0.100 |
| Terrazzo wainscot, cast-in-place, 1/2" thick | " | 0.200 |
| Base, cast in place, terrazzo cove type, 6" high | L.F. | 0.114 |
| Curb, cast in place, 6" wide x 6" high, polished top | " | 0.400 |
| Stairs, cast-in-place, topping on concrete or metal | | |
| 1-1/2" thick treads, 12" wide | L.F. | 0.400 |
| Combined tread and riser | " | 1.000 |
| Precast terrazzo, thin set | | |
| Terrazzo tiles, non-slip surface | | |
| 9" x 9" x 1" thick | S.F. | 0.114 |
| 12" x 12" | | |
| 1" thick | S.F. | 0.107 |
| 1-1/2" thick | " | 0.114 |
| 18" x 18" x 1-1/2" thick | " | 0.114 |
| 24" x 24" x 1-1/2" thick | " | 0.094 |
| Terrazzo wainscot | | |

# 09 FINISHES

## TILE

| 09410.10 TERRAZZO | UNIT | MAN/ HOURS |
|---|---|---|
| 12" x 12" x 1" thick | S.F. | 0.200 |
| 18" x 18" x 1-1/2" thick | " | 0.229 |
| Base | | |
| 6" high | | |
| Straight | L.F. | 0.062 |
| Coved | " | 0.062 |
| 8" high | | |
| Straight | L.F. | 0.067 |
| Coved | " | 0.067 |
| Terrazzo curbs | | |
| 8" wide x 8" high | L.F. | 0.320 |
| 6" wide x 6" high | " | 0.267 |
| Precast terrazzo stair treads, 12" wide | | |
| 1-1/2" thick | | |
| Diamond pattern | L.F. | 0.145 |
| Non-slip surface | " | 0.145 |
| 2" thick | | |
| Diamond pattern | L.F. | 0.145 |
| Non-slip surface | " | 0.160 |
| Stair risers, 1" thick to 6" high | | |
| Straight sections | L.F. | 0.080 |
| Cove sections | " | 0.080 |
| Combined tread and riser | | |
| Straight sections | | |
| 1-1/2" tread, 3/4" riser | L.F. | 0.229 |
| 3" tread, 1" riser | " | 0.229 |
| Curved sections | | |
| 2" tread, 1" riser | L.F. | 0.267 |
| 3" tread, 1" riser | " | 0.267 |
| Stair stringers, notched for treads and risers | | |
| 1" thick | L.F. | 0.200 |
| 2" thick | " | 0.267 |
| Landings, structural, nonslip | | |
| 1-1/2" thick | S.F. | 0.133 |
| 3" thick | " | 0.160 |
| Conductive terrazzo, spark proof industrial floor | | |
| Epoxy terrazzo | | |
| Floor | S.F. | 0.050 |
| Base | " | 0.067 |
| Polyacrylate | | |
| Floor | S.F. | 0.050 |
| Base | " | 0.067 |
| Polyester | | |
| Floor | S.F. | 0.032 |
| Base | " | 0.040 |
| Synthetic latex mastic | | |
| Floor | S.F. | 0.050 |
| Base | " | 0.067 |

## ACOUSTICAL TREATMENT

| 09510.10 CEILINGS AND WALLS | UNIT | MAN/ HOURS |
|---|---|---|
| Acoustical panels, suspension system not included | | |
| Fiberglass panels | | |
| 5/8" thick | | |
| 2' x 2' | S.F. | 0.011 |
| 2' x 4' | " | 0.009 |
| 3/4" thick | | |
| 2' x 2' | S.F. | 0.011 |
| 2' x 4' | " | 0.009 |
| Glass cloth faced fiberglass panels | | |
| 3/4" thick | S.F. | 0.013 |
| 1" thick | " | 0.013 |
| Mineral fiber panels | | |
| 5/8" thick | | |
| 2' x 2' | S.F. | 0.011 |
| 2' x 4' | " | 0.009 |
| 3/4" thick | | |
| 2' x 2' | S.F. | 0.011 |
| 2' x 4' | " | 0.009 |
| Wood fiber panels | | |
| 2" thick | | |
| 2' x 2' | S.F. | 0.013 |
| 2' x 4' | " | 0.010 |
| Air distributing panels | | |
| 3/4" thick | S.F. | 0.020 |
| 5/8" thick | " | 0.016 |
| Acoustical tiles, suspension system not included | | |
| Fiberglass tile, 12" x 12" | | |
| 5/8" thick | S.F. | 0.015 |
| 3/4" thick | " | 0.018 |
| Glass cloth faced fiberglass tile | | |
| 3/4" thick | S.F. | 0.018 |
| 3" thick | " | 0.020 |
| Mineral fiber tile, 12" x 12" | | |
| 5/8" thick | | |
| Standard | S.F. | 0.016 |
| Vinyl faced | " | 0.016 |
| 3/4" thick | | |
| Standard | S.F. | 0.016 |
| Vinyl faced | " | 0.016 |
| Fire rated | " | 0.016 |
| Aluminum or mylar faced | " | 0.016 |
| Wood fiber tile, 12" x 12" | | |
| 1/2" thick | S.F. | 0.016 |
| 3/4" thick | " | 0.016 |
| Metal pan units, 24 ga. steel | | |
| 12" x 12" | S.F. | 0.032 |
| 12" x 24" | " | 0.027 |
| Aluminum, .025" thick | | |
| 12" x 12" | S.F. | 0.032 |
| 12" x 24" | " | 0.027 |
| Anodized aluminum, 0.25" thick | | |
| 12" x 12" | S.F. | 0.032 |
| 12" x 24" | " | 0.027 |
| Stainless steel, 24 ga. | | |

# 09 FINISHES

| ACOUSTICAL TREATMENT | UNIT | MAN/ HOURS |
|---|---|---|
| **09510.10 CEILINGS AND WALLS** | | |
| 12" x 12" | S.F. | 0.032 |
| 12" x 24" | " | 0.027 |
| Metal ceiling systems | | |
| .020" thick panels | | |
| 10', 12', and 16' lengths | S.F. | 0.023 |
| Custom lengths, 3' to 20' | " | 0.023 |
| .025" thick panels | | |
| 32 sf, 38 sf, and 52 sf pieces | S.F. | 0.027 |
| Custom lengths, 10 sf to 65 sf | " | 0.027 |
| Sound absorption walls, with fabric cover | | |
| 2-6" x 9' x 3/4" | S.F. | 0.027 |
| 2' x 9' x 1" | " | 0.027 |
| Starter spline | L.F. | 0.020 |
| Internal spline | " | 0.020 |
| Acoustical treatment | | |
| Barriers for plenums | | |
| Leaded vinyl | | |
| 0.48 lb per sf | S.F. | 0.038 |
| 0.87 lb per sf | " | 0.040 |
| Aluminum foil, fiberglass reinforcement | | |
| Minimum | S.F. | 0.027 |
| Maximum | " | 0.040 |
| Aluminum mesh, paper backed | " | 0.027 |
| Fibered cement sheet, 3/16" thick | " | 0.029 |
| Sheet lead, 1/64" thick | " | 0.020 |
| Sound attenuation blanket | | |
| 1" thick | S.F. | 0.080 |
| 1-1/2" thick | " | 0.080 |
| 2" thick | " | 0.080 |
| 3" thick | " | 0.089 |
| Ceiling suspension systems | | |
| T bar system | | |
| 2' x 4' | S.F. | 0.008 |
| 2' x 2' | " | 0.009 |
| Concealed Z bar suspension system, 12" module | " | 0.013 |

| FLOORING | UNIT | MAN/ HOURS |
|---|---|---|
| **09550.10 WOOD FLOORING** | | |
| Wood strip flooring, unfinished | | |
| Fir floor | | |
| C and better | | |
| Vertical grain | S.F. | 0.027 |
| Flat grain | " | 0.027 |
| Oak floor | | |
| Minimum | S.F. | 0.038 |
| Average | S.F. | 0.038 |
| Maximum | " | 0.038 |
| Maple floor | | |
| 25/32" x 2-1/4" | | |
| Minimum | S.F. | 0.038 |
| Maximum | " | 0.038 |
| 33/32" x 3-1/4" | | |
| Minimum | S.F. | 0.038 |
| Maximum | " | 0.038 |
| Wood block industrial flooring | | |
| Creosoted | | |
| 2" thick | S.F. | 0.021 |
| 2-1/2" thick | " | 0.025 |
| 3" thick | " | 0.027 |
| Parquet, 5/16", white oak | | |
| Finished | S.F. | 0.040 |
| Unfinished | " | 0.040 |
| Gym floor, 2 ply felt, 25/32" maple, finished, in mastic | " | 0.044 |
| Over wood sleepers | " | 0.050 |
| Finishing, sand, fill, finish, and wax | " | 0.020 |
| Refinish sand, seal, and 2 coats of polyurethane | " | 0.027 |
| Clean and wax floors | " | 0.004 |
| **09630.10 UNIT MASONRY FLOORING** | | |
| Clay brick | | |
| 9 x 4-1/2 x 3" thick | | |
| Glazed | S.F. | 0.067 |
| Unglazed | " | 0.067 |
| 8 x 4 x 3/4" thick | | |
| Glazed | S.F. | 0.070 |
| Unglazed | " | 0.070 |
| **09660.10 RESILIENT TILE FLOORING** | | |
| Solid vinyl tile, 1/8" thick, 12" x 12" | | |
| Marble patterns | S.F. | 0.020 |
| Solid colors | " | 0.020 |
| Travertine patterns | " | 0.020 |
| Conductive resilient flooring, vinyl tile | | |
| 1/8" thick, 12" x 12" | S.F. | 0.023 |
| **09665.10 RESILIENT SHEET FLOORING** | | |
| Vinyl sheet flooring | | |
| Minimum | S.F. | 0.008 |
| Average | " | 0.010 |
| Maximum | " | 0.013 |
| Cove, to 6" | L.F. | 0.016 |
| Fluid applied resilient flooring | | |
| Polyurethane, poured in place, 3/8" thick | S.F. | 0.067 |
| Vinyl sheet goods, backed | | |
| 0.070" thick | S.F. | 0.010 |
| 0.093" thick | " | 0.010 |
| 0.125" thick | " | 0.010 |
| 0.250" thick | " | 0.010 |

# 09 FINISHES

| FLOORING | UNIT | MAN/HOURS |
|---|---|---|
| **09665.10 RESILIENT SHEET FLOORING** | | |
| Wall base, vinyl | | |
| Group 1 | | |
| 4" high | L.F. | 0.027 |
| 6" high | " | 0.027 |
| Group 2 | | |
| 4" high | L.F. | 0.027 |
| 6" high | " | 0.027 |
| Group 3 | | |
| 4" high | L.F. | 0.027 |
| 6" high | " | 0.027 |
| Stair accessories | | |
| Treads, 1/4" x 12", rubber diamond surface | | |
| Marbled | L.F. | 0.067 |
| Plain | " | 0.067 |
| Grit strip safety tread, 12" wide, colors | | |
| 3/16" thick | L.F. | 0.067 |
| 5/16" thick | " | 0.067 |
| Risers, 7" high, 1/8" thick, colors | | |
| Flat | L.F. | 0.040 |
| Coved | " | 0.040 |
| Nosing, rubber | | |
| 3/16" thick, 3" wide | | |
| Black | L.F. | 0.040 |
| Colors | " | 0.040 |
| 6" wide | | |
| Black | L.F. | 0.067 |
| Colors | " | 0.067 |

| CARPET | UNIT | MAN/HOURS |
|---|---|---|
| **09680.10 FLOOR LEVELING** | | |
| Repair and level floors to receive new flooring | | |
| Minimum | S.Y. | 0.027 |
| Average | " | 0.067 |
| Maximum | " | 0.080 |
| **09682.10 CARPET PADDING** | | |
| Carpet padding | | |
| Foam rubber, waffle type, 0.3" thick | S.Y. | 0.040 |
| Jute padding | | |
| Minimum | S.Y. | 0.036 |
| Average | " | 0.040 |
| Maximum | " | 0.044 |
| Sponge rubber cushion | | |
| Minimum | S.Y. | 0.036 |
| Average | " | 0.040 |

| CARPET | UNIT | MAN/HOURS |
|---|---|---|
| **09682.10 CARPET PADDING** | | |
| Maximum | S.Y. | 0.044 |
| Urethane cushion, 3/8" thick | | |
| Minimum | S.Y. | 0.036 |
| Average | " | 0.040 |
| Maximum | " | 0.044 |
| **09685.10 CARPET** | | |
| Carpet, acrylic | | |
| 24 oz., light traffic | S.Y. | 0.044 |
| 28 oz., medium traffic | " | 0.044 |
| Residential | | |
| Nylon | | |
| 15 oz., light traffic | S.Y. | 0.044 |
| 28 oz., medium traffic | " | 0.044 |
| Commercial | | |
| Nylon | | |
| 28 oz., medium traffic | S.Y. | 0.044 |
| 35 oz., heavy traffic | " | 0.044 |
| Wool | | |
| 30 oz., medium traffic | S.Y. | 0.044 |
| 36 oz., medium traffic | " | 0.044 |
| 42 oz., heavy traffic | " | 0.044 |
| Carpet tile | | |
| Foam backed | | |
| Minimum | S.F. | 0.008 |
| Average | " | 0.009 |
| Maximum | " | 0.010 |
| Tufted loop or shag | | |
| Minimum | S.F. | 0.008 |
| Average | " | 0.009 |
| Maximum | " | 0.010 |
| Clean and vacuum carpet | | |
| Minimum | S.Y. | 0.004 |
| Average | " | 0.005 |
| Maximum | " | 0.008 |
| **09700.10 SPECIAL FLOORING** | | |
| Epoxy flooring, marble chips | | |
| Epoxy with colored quartz chips in 1/4" base | S.F. | 0.044 |
| Heavy duty epoxy topping, 3/16" thick | " | 0.044 |
| Epoxy terrazzo | | |
| 1/4" thick chemical resistant | S.F. | 0.050 |

# 09 FINISHES

| PAINTING | UNIT | MAN/ HOURS |
|---|---|---|
| **09910.10 EXTERIOR PAINTING** | | |
| Exterior painting | | |
| Wood surfaces, 1 coat primer, two coats paint | | |
| Door and frame | EA. | 1.333 |
| Windows | S.F. | 0.020 |
| Wood trim | " | 0.020 |
| Wood siding | " | 0.010 |
| Hardboard surfaces | | |
| One coat primer, two coats paint | S.F. | 0.010 |
| Asbestos cement surfaces | | |
| One coat primer, two coats paint | S.F. | 0.010 |
| Galvanized surfaces, galvanized primer | | |
| One coat primer, two coats paint | S.F. | 0.009 |
| Stucco surfaces, acrylic primer, acrylic latex paint | | |
| One coat primer, two coats paint | S.F. | 0.013 |
| Concrete masonry unit surfaces, brush work | | |
| One coat filler, one coat paint | S.F. | 0.010 |
| Two coats epoxy | " | 0.013 |
| Texture coating | " | 0.008 |
| Concrete surfaces | | |
| One coat filler, one coat paint | S.F. | 0.010 |
| Two coats paint | " | 0.013 |
| Structural steel | | |
| One field coat paint, brush work | | |
| Light framing | S.F. | 0.007 |
| Heavy framing | " | 0.004 |
| One field coat paint, spray work | | |
| Light framing | S.F. | 0.002 |
| Heavy framing | " | 0.002 |
| Miscellaneous steel items, spray work, one coat | | |
| Exposed decking | S.F. | 0.007 |
| Joist | " | 0.010 |
| Columns | " | 0.010 |
| Pipes, one coat primer, one coat paint | | |
| 4" dia. | L.F. | 0.010 |
| 8" dia. | " | 0.013 |
| 12" dia. | " | 0.020 |
| Paint letters on pipe with brush | " | 0.020 |
| Paint pipe insulation cloth cover | S.F. | 0.010 |
| Sprinkler system piping | L.F. | 0.007 |
| Miscellaneous surfaces | | |
| Stair pipe rails | | |
| Two rails | L.F. | 0.027 |
| One rail | " | 0.016 |
| Stair to 4' wide, including rails, per riser | EA. | 0.114 |
| Gratings and frames | S.F. | 0.027 |
| Ladders | L.F. | 0.023 |
| Miscellaneous exposed metal | S.F. | 0.008 |
| **09920.10 INTERIOR PAINTING** | | |
| Walls, concrete and masonry, brush, primer, acrylic | | |
| One coat primer, one coat paint | S.F. | 0.010 |
| Two coats paint | " | 0.013 |
| Plywood, paint | " | 0.004 |

| PAINTING | UNIT | MAN/ HOURS |
|---|---|---|
| **09920.10 INTERIOR PAINTING** | | |
| Natural finish | S.F. | 0.005 |
| Wood, paint | " | 0.005 |
| Natural finish | " | 0.006 |
| Metal | | |
| One coat filler | S.F. | 0.005 |
| One coat primer, one coat paint | " | 0.010 |
| Two coats paint | " | 0.013 |
| Plaster or gypsum board, paint | " | 0.004 |
| Epoxy | " | 0.005 |
| Ceilings, one coat paint, wood | " | 0.006 |
| Concrete | " | 0.005 |
| Plaster | " | 0.004 |
| Miscellaneous metal, brushwork | " | 0.010 |
| **09920.30 DOORS AND MILLWORK** | | |
| Painting, doors | | |
| Minimum | S.F. | 0.027 |
| Average | " | 0.040 |
| Maximum | " | 0.053 |
| Cabinets, shelves, and millwork | | |
| Minimum | S.F. | 0.013 |
| Average | " | 0.023 |
| Maximum | " | 0.040 |
| **09920.60 WINDOWS** | | |
| Painting, windows | | |
| Minimum | S.F. | 0.016 |
| Average | " | 0.020 |
| Maximum | " | 0.032 |
| **09955.10 WALL COVERING** | | |
| Vinyl wall covering | | |
| Medium duty | S.F. | 0.011 |
| Heavy duty | " | 0.013 |
| Over pipes and irregular shapes | | |
| Lightweight, 13 oz. | S.F. | 0.016 |
| Medium weight, 25 oz. | " | 0.018 |
| Heavy weight, 34 oz. | " | 0.020 |
| Cork wall covering | | |
| 1' x 1' squares | | |
| 1/4" thick | S.F. | 0.020 |
| 1/2" thick | " | 0.020 |
| 3/4" thick | " | 0.020 |
| Wall fabrics | | |
| Natural fabrics, grass cloths | | |
| Minimum | S.F. | 0.012 |
| Average | " | 0.013 |
| Maximum | " | 0.016 |
| Flexible gypsum coated wall fabric, fire resistant | " | 0.008 |
| Vinyl corner guards | | |
| 3/4" x 3/4" x 8' | EA. | 0.100 |
| 2-3/4" x 2-3/4" x 4' | " | 0.100 |

| PAINTING | UNIT | MAN/ HOURS |
|---|---|---|
| **09980.10 PAINTING PREPARATION** | | |
| Cleaning, light | | |
| Wood | S.F. | 0.002 |
| Plaster or gypsum wallboard | " | 0.002 |
| Prepare sprinkler piping for painting | L.F. | 0.002 |
| Prepare insulated pipe for painting | S.F. | 0.002 |
| Normal painting prep, masonry and concrete | | |
| Unpainted | S.F. | 0.001 |
| Painted | " | 0.002 |
| Plaster or gypsum | | |
| Unpainted | S.F. | 0.001 |
| Painted | " | 0.002 |
| Wood | | |
| Unpainted | S.F. | 0.001 |
| Painted | " | 0.002 |
| Painted steel, light rusting | " | 0.003 |
| Sandblasting | | |
| Brush off blast | S.F. | 0.005 |
| Commercial blast | " | 0.013 |
| Near white metal blast | " | 0.023 |
| White metal blast | " | 0.027 |

# 10 SPECIALTIES

| SPECIALTIES | UNIT | MAN/ HOURS |
|---|---|---|
| **10110.10 CHALKBOARDS** | | |
| Chalkboard, metal frame, 1/4" thick | | |
| 48"x60" | EA. | 0.800 |
| 48"x96" | " | 0.889 |
| 48"x144" | " | 1.000 |
| 48"x192" | " | 1.143 |
| Liquid chalkboard | | |
| 48"x60" | EA. | 0.800 |
| 48"x96" | " | 0.889 |
| 48"x144" | " | 1.000 |
| 48"x192" | " | 1.143 |
| Map rail, deluxe | L.F. | 0.040 |
| Average | PCT. | |
| **10165.10 TOILET PARTITIONS** | | |
| Toilet partition, plastic laminate | | |
| Ceiling mounted | EA. | 2.667 |
| Floor mounted | " | 2.000 |
| Metal | | |
| Ceiling mounted | EA. | 2.667 |
| Floor mounted | " | 2.000 |
| Wheel chair partition, plastic laminate | | |
| Ceiling mounted | EA. | 2.667 |
| Floor mounted | " | 2.000 |
| Painted metal | | |
| Ceiling mounted | EA. | 2.667 |
| Floor mounted | " | 2.000 |
| Urinal screen, plastic laminate | | |
| Wall hung | EA. | 1.000 |
| Floor mounted | " | 1.000 |
| Porcelain enameled steel, floor mounted | " | 1.000 |
| Painted metal, floor mounted | " | 1.000 |
| Stainless steel, floor mounted | " | 1.000 |
| Metal toilet partitions | | |
| Toilet partitions, front door and side divider, floor mounted | | |
| Porcelain enameled steel | EA. | 2.000 |
| Painted steel | " | 2.000 |
| Stainless steel | " | 2.000 |
| **10185.10 SHOWER STALLS** | | |
| Shower receptors | | |
| Precast, terrazzo | | |
| 32" x 32" | EA. | 0.667 |
| 32" x 48" | " | 0.800 |
| Concrete | | |
| 32" x 32" | EA. | 0.667 |
| 48" x 48" | " | 0.889 |
| Shower door, trim and hardware | | |
| Economy, 24" wide, chrome frame, tempered glass | EA. | 0.800 |
| Porcelain enameled steel, flush | " | 0.800 |
| Baked enameled steel, flush | " | 0.800 |
| Aluminum frame, tempered glass, 48" wide, sliding | " | 1.000 |
| Folding | " | 1.000 |
| Aluminum frame and tempered glass, molded plastic | | |

| SPECIALTIES | UNIT | MAN/ HOURS |
|---|---|---|
| **10185.10 SHOWER STALLS** | | |
| Complete with receptor and door | | |
| 32" x 32" | EA. | 2.000 |
| 36" x 36" | " | 2.000 |
| 40" x 40" | " | 2.286 |
| Shower compartment, precast concrete receptor | | |
| Single entry type | | |
| Porcelain enameled steel | EA. | 8.000 |
| Baked enameled steel | " | 8.000 |
| Stainless steel | " | 8.000 |
| Double entry type | | |
| Porcelain enameled steel | EA. | 10.000 |
| Baked enameled steel | " | 10.000 |
| Stainless steel | " | 10.000 |
| **10190.10 CUBICLES** | | |
| Hospital track | | |
| Ceiling hung | L.F. | 0.089 |
| Suspended | " | 0.114 |
| Hospital metal dividers, galvanized steel | | |
| Baked enamel finish | | |
| 54" high | | |
| 10" glass light | L.F. | 0.400 |
| 14" glass light | " | 0.400 |
| 24" glass light | " | 0.400 |
| 60" high | | |
| 10" glass light | L.F. | 0.444 |
| 14" glass light | " | 0.444 |
| 24" glass light | " | 0.444 |
| Stainless steel | | |
| 54" high | | |
| 10" glass light | L.F. | 0.444 |
| 14" glass light | " | 0.444 |
| 24" glass light | " | 0.444 |
| 60" high | | |
| 14" glass light | L.F. | 0.500 |
| 14" glass light | " | 0.500 |
| 24" glass light | " | 0.500 |
| **10210.10 VENTS AND WALL LOUVERS** | | |
| Block vent, 8"x16"x4" aluminum, w/screen, mill finish | EA. | 0.267 |
| Standard | " | 0.250 |
| Vents w/screen, 4" deep, 8" wide, 5" high | | |
| Modular | EA. | 0.250 |
| Aluminum gable louvers | S.F. | 0.133 |
| Vent screen aluminum, 4" wide, continuous | L.F. | 0.027 |
| Louvers, aluminum, anodized, fixed blade | | |
| Horizontal line | S.F. | 0.200 |
| Vertical line | " | 0.200 |
| Wall louver, aluminum mill finish | | |
| Under, 2 sf | S.F. | 0.100 |
| 2 to 4 sf | " | 0.089 |
| 5 to 10 sf | " | 0.089 |
| Galvanized steel | | |

# 10 SPECIALTIES

| SPECIALTIES | UNIT | MAN/ HOURS |
|---|---|---|
| **10210.10 VENTS AND WALL LOUVERS** | | |
| Under 2 sf | S.F. | 0.100 |
| 2 to 4 sf | " | 0.089 |
| 5 to 10 sf | " | 0.089 |
| **10225.10 DOOR LOUVERS** | | |
| Fixed, 1" thick, enameled steel | | |
| 8"x8" | EA. | 0.100 |
| 12"x8" | " | 0.100 |
| 12"x12" | " | 0.114 |
| 16"x12" | " | 0.123 |
| 18"x12" | " | 0.200 |
| 20"x8" | " | 0.114 |
| 20"x12" | " | 0.229 |
| 20"x16" | " | 0.267 |
| 20"x20" | " | 0.320 |
| 24"x12" | " | 0.267 |
| 24"x16" | " | 0.286 |
| 24"x18" | " | 0.308 |
| 24"x20" | " | 0.333 |
| 24"x24" | " | 0.364 |
| 26"x26" | " | 0.500 |
| **10270.40 ACCESS & PEDESTAL FLOOR** | | |
| Panels, no covering, 2'x2' | | |
| Plain | S.F. | 0.010 |
| Perforated | " | 0.400 |
| Pedestals | | |
| For 6" to 12" clearance | EA. | 0.080 |
| Stringers | | |
| 2' | L.F. | 0.038 |
| 6' | " | 0.027 |
| Accessories | | |
| Ramp assembly | S.F. | 0.032 |
| Elevated floor assembly | " | 0.030 |
| Handrail | L.F. | 0.400 |
| Fascia plate | " | 0.200 |
| RF shielding components, floor liner | | |
| Hot rolled steel sheet | | |
| 14 ga. | S.F. | 0.020 |
| 11 ga. | " | 0.062 |
| **10290.10 PEST CONTROL** | | |
| Termite control | | |
| Under slab spraying | | |
| Minimum | S.F. | 0.002 |
| Average | " | 0.004 |
| Maximum | " | 0.008 |
| **10350.10 FLAGPOLES** | | |
| Installed in concrete base | | |
| Fiberglass | | |

| SPECIALTIES | UNIT | MAN/ HOURS |
|---|---|---|
| **10350.10 FLAGPOLES** | | |
| 25' high | EA. | 5.333 |
| 50' high | " | 13.333 |
| Aluminum | | |
| 25' high | EA. | 5.333 |
| 50' high | " | 13.333 |
| Bonderized steel | | |
| 25' high | EA. | 6.154 |
| 50' high | " | 16.000 |
| Freestanding tapered, fiberglass | | |
| 30' high | EA. | 5.714 |
| 40' high | " | 7.273 |
| 50' high | " | 8.000 |
| 60' high | " | 9.412 |
| Wall mounted, with collar, brushed aluminum finish | | |
| 15' long | EA. | 4.000 |
| 18' long | " | 4.000 |
| 20' long | " | 4.211 |
| 24' long | " | 4.706 |
| Outrigger, wall, including base | | |
| 10' long | EA. | 5.333 |
| 20' long | " | 6.667 |
| **10400.10 IDENTIFYING DEVICES** | | |
| Directory and bulletin boards | | |
| Open face boards | | |
| Chrome plated steel frame | S.F. | 0.400 |
| Aluminum framed | " | 0.400 |
| Bronze framed | " | 0.400 |
| Stainless steel framed | " | 0.400 |
| Tack board, aluminum framed | " | 0.400 |
| Visual aid board, aluminum framed | " | 0.400 |
| Glass encased boards, hinged and keyed | | |
| Aluminum framed | S.F. | 1.000 |
| Bronze framed | " | 1.000 |
| Stainless steel framed | " | 1.000 |
| Chrome plated steel framed | " | 1.000 |
| Metal plaque | | |
| Cast bronze | S.F. | 0.667 |
| Aluminum | " | 0.667 |
| Metal engraved plaque | | |
| Porcelain steel | S.F. | 0.667 |
| Stainless steel | " | 0.667 |
| Brass | " | 0.667 |
| Aluminum | " | 0.667 |
| Metal built-up plaque | | |
| Bronze | S.F. | 0.800 |
| Copper and bronze | " | 0.800 |
| Copper and aluminum | " | 0.800 |
| Metal nameplate plaques | | |
| Cast bronze | S.F. | 0.500 |
| Cast aluminum | " | 0.500 |
| Engraved, 1-1/2" x 6" | | |
| Bronze | EA. | 0.500 |

# 10 SPECIALTIES

| SPECIALTIES | UNIT | MAN/ HOURS |
|---|---|---|
| **10400.10 IDENTIFYING DEVICES** | | |
| Aluminum | EA. | 0.500 |
| Letters, on masonry or concrete, aluminum, satin finish | | |
| 1/2" thick | | |
| 2" high | EA. | 0.320 |
| 4" high | " | 0.400 |
| 6" high | " | 0.444 |
| 3/4" thick | | |
| 8" high | EA. | 0.500 |
| 10" high | " | 0.571 |
| 1" thick | | |
| 12" high | EA. | 0.667 |
| 14" high | " | 0.800 |
| 16" high | " | 1.000 |
| 3/8" thick | | |
| 2" high | EA. | 0.320 |
| 4" high | " | 0.400 |
| 1/2" thick, 6" high | " | 0.444 |
| 5/8" thick, 8" high | " | 0.500 |
| 1" thick | | |
| 10" high | EA. | 0.571 |
| 12" high | " | 0.667 |
| 14" high | " | 0.800 |
| 16" high | " | 1.000 |
| Interior door signs, adhesive, flexible | | |
| 2" x 8" | EA. | 0.200 |
| 4" x 4" | " | 0.200 |
| 6" x 7" | " | 0.200 |
| 6" x 9" | " | 0.200 |
| 10" x 9" | " | 0.200 |
| 10" x 12" | " | 0.200 |
| Hard plastic type, no frame | | |
| 3" x 8" | EA. | 0.200 |
| 4" x 4" | " | 0.200 |
| 4" x 12" | " | 0.200 |
| Hard plastic type, with frame | | |
| 3" x 8" | EA. | 0.200 |
| 4" x 4" | " | 0.200 |
| 4" x 12" | " | 0.200 |
| **10450.10 CONTROL** | | |
| Access control, 7' high, indoor or outdoor impenetrability | | |
| Remote or card control, type B | EA. | 10.667 |
| Free passage, type B | " | 10.667 |
| Remote or card control, type AA | " | 10.667 |
| Free passage, type AA | " | 10.667 |
| **10500.10 LOCKERS** | | |
| Locker bench, floor mounted, laminated maple | | |
| 4' | EA. | 0.667 |
| 6' | " | 0.667 |
| Wardrobe locker, 12" x 60" x 15", baked on enamel | | |
| 1-tier | EA. | 0.400 |
| 2-tier | " | 0.400 |

| SPECIALTIES | UNIT | MAN/ HOURS |
|---|---|---|
| **10500.10 LOCKERS** | | |
| 3-tier | EA. | 0.421 |
| 4-tier | " | 0.421 |
| 12" x 72" x 15", baked on enamel | | |
| 1-tier | EA. | 0.400 |
| 2-tier | " | 0.400 |
| 4-tier | " | 0.421 |
| 5-tier | " | 0.421 |
| 15" x 60" x 15", baked on enamel | | |
| 1-tier | EA. | 0.400 |
| 4-tier | " | 0.421 |
| Wardrobe locker, single tier type | | |
| 12" x 15" x 72" | EA. | 0.800 |
| 18" x 15" x 72" | " | 0.842 |
| 12" x 18" x 72" | " | 0.889 |
| 18" x 18" x 72" | " | 0.941 |
| Double tier type | | |
| 12" x 15" x 36" | EA. | 0.400 |
| 18" x 15" x 36" | " | 0.400 |
| 12" x 18" x 36" | " | 0.400 |
| 18" x 18" x 36" | " | 0.400 |
| Two person unit | | |
| 18" x 15" x 72" | EA. | 1.333 |
| 18" x 18" x 72" | " | 1.600 |
| Duplex unit | | |
| 15" x 15" x 72" | EA. | 0.800 |
| 15" x 21" x 72" | " | 0.800 |
| Basket lockers, basket sets with baskets | | |
| 24 basket set | SET | 4.000 |
| 30 basket set | " | 5.000 |
| 36 basket set | " | 6.667 |
| 42 basket set | " | 8.000 |
| **10520.10 FIRE PROTECTION** | | |
| Portable fire extinguishers | | |
| Water pump tank type | | |
| 2.5 gal. | | |
| Red enameled galvanized | EA. | 0.533 |
| Red enameled copper | " | 0.533 |
| Polished copper | " | 0.533 |
| Carbon dioxide type, red enamel steel | | |
| Squeeze grip with hose and horn | | |
| 2.5 lb | EA. | 0.533 |
| 5 lb | " | 0.615 |
| 10 lb | " | 0.800 |
| 15 lb | " | 1.000 |
| 20 lb | " | 1.000 |
| Wheeled type | | |
| 125 lb | EA. | 1.600 |
| 250 lb | " | 1.600 |
| 500 lb | " | 1.600 |
| Dry chemical, pressurized type | | |
| Red enameled steel | | |
| 2.5 lb | EA. | 0.533 |

| SPECIALTIES | UNIT | MAN/ HOURS |
|---|---|---|
| **10520.10 FIRE PROTECTION** | | |
| 5 lb | EA. | 0.615 |
| 10 lb | " | 0.800 |
| 20 lb | " | 1.000 |
| 30 lb | " | 1.000 |
| Chrome plated steel, 2.5 lb | " | 0.533 |
| Other type extinguishers | | |
| 2.5 gal, stainless steel, pressurized water tanks | EA. | 0.533 |
| Soda and acid type | " | 0.533 |
| Cartridge operated, water type | " | 0.533 |
| Loaded stream, water type | " | 0.533 |
| Foam type | " | 0.533 |
| 40 gal, wheeled foam type | " | 1.600 |
| Fire extinguisher cabinets | | |
| Enameled steel | | |
| 8" x 12" x 27" | EA. | 1.600 |
| 8" x 16" x 38" | " | 1.600 |
| Aluminum | | |
| 8" x 12" x 27" | EA. | 1.600 |
| 8" x 16" x 38" | " | 1.600 |
| 8" x 12" x 27" | " | 1.600 |
| Stainless steel | | |
| 8" x 16" x 38" | EA. | 1.600 |
| **10550.10 POSTAL SPECIALTIES** | | |
| Mail chutes | | |
| Single mail chute | | |
| Finished aluminum | L.F. | 2.000 |
| Bronze | " | 2.000 |
| Single mail chute receiving box | | |
| Finished aluminum | EA. | 4.000 |
| Bronze | " | 4.000 |
| Twin mail chute, double parallel | | |
| Finished aluminum | FLR | 4.000 |
| Bronze | " | 4.000 |
| Receiving box, 36" x 20" x 12" | | |
| Finished aluminum | EA. | 6.667 |
| Bronze | " | 6.667 |
| Locked receiving mail box | | |
| Finished aluminum | EA. | 4.000 |
| Bronze | " | 4.000 |
| Commercial postal accessories for mail chutes | | |
| Letter slot, brass | EA. | 1.333 |
| Bulk mail slot, brass | " | 1.333 |
| Mail boxes | | |
| Residential postal accessories | | |
| Letter slot | EA. | 0.400 |
| Rural letter box | " | 1.000 |
| Apartment house, keyed, 3.5" x 4.5" x 16" | " | 0.267 |
| Ranch style | " | 0.400 |
| Commercial postal accessories | | |
| Letter box, with combination lock | EA. | 0.286 |
| Key lock | " | 0.286 |
| Mail box, aluminum w/glass front, 4x5 | | |
| Horizontal rear load | EA. | 0.229 |

| SPECIALTIES | UNIT | MAN/ HOURS |
|---|---|---|
| **10550.10 POSTAL SPECIALTIES** | | |
| Vertical front load | EA. | 0.229 |
| **10600.10 MOVABLE PARTITIONS** | | |
| Partition, movable office type, 2-1/2" thick, vinyl-gypsum | S.F. | 0.040 |
| Enameled steel frame, with 1/4" thick clear glass | " | 0.040 |
| Door frame and hardware for movable partitions | EA. | 2.667 |
| Add for acoustic movable partition | S.F. | |
| Accordion partition, 12' high | | |
| Vinyl | S.F. | 0.133 |
| Acoustical | " | 0.133 |
| Standard office cubicles, 8' high, steel framed | | |
| Baked enamel finish | | |
| 100% flush | L.F. | 0.200 |
| 75% flush and 25% glass | " | 0.222 |
| 50% flush and 50% glass | " | 0.222 |
| 100% glass | " | 0.267 |
| Natural hardwood panels | | |
| 100% flush | L.F. | 0.267 |
| 50% flush and 50% glass | " | 0.286 |
| Plastic laminated panels | | |
| 100% flush | L.F. | 0.267 |
| 75% flush and 25% glass | " | 0.286 |
| 50% flush and 50% glass | " | 0.286 |
| Vinyl covered panels | | |
| 100% flush | L.F. | 0.276 |
| 75% flush and 25% glass | " | 0.296 |
| 50% and 50% glass | " | 0.296 |
| Aluminum framed | | |
| Enameled or anodized aluminum panels | | |
| 100% flush | L.F. | 0.200 |
| 75% flush and 25% glass | " | 0.222 |
| 50% flush and 50% glass | " | 0.222 |
| Vinyl covered panels | | |
| 100% flush | L.F. | 0.276 |
| 75% flush and 25% glass | " | 0.296 |
| 50% flush and 50% glass | " | 0.296 |
| 60" high partitions, steel framed | | |
| Enameled panels | L.F. | 0.178 |
| Natural hardwood panels, two sides | " | 0.186 |
| Plastic laminated panels | " | 0.186 |
| Vinyl covered panels | " | 0.178 |
| Aluminum framed | | |
| Anodized or baked enamel panels | L.F. | 0.178 |
| Natural hardwood panels | " | 0.186 |
| Plastic laminated panels | " | 0.186 |
| Vinyl covered panels | " | 0.178 |
| Wire mesh partitions | | |
| Wall panels | | |
| 4' x 7' | EA. | 0.500 |
| 4' x 8' | " | 0.533 |
| 4' x 10' | " | 0.615 |
| Wall filler panels | | |

# 10 SPECIALTIES

| SPECIALTIES | UNIT | MAN/ HOURS |
|---|---|---|
| **10600.10 MOVABLE PARTITIONS** | | |
| 1' x 7' | EA. | 0.500 |
| 1' x 8' | " | 0.533 |
| 1' x 10' | " | 0.571 |
| 2' x 7' | " | 0.500 |
| 2' x 8' | " | 0.533 |
| 2' x 10' | " | 0.571 |
| 3' x 7' | " | 0.500 |
| 3' x 8' | " | 0.533 |
| 3' x 10' | " | 0.571 |
| Ceiling panels | | |
| 10' x 2' | EA. | 1.143 |
| 10' x 4' | " | 1.600 |
| Wall panel with service window | | |
| 5' wide | | |
| 7' high | EA. | 0.500 |
| 8' high | " | 0.533 |
| 10' high | " | 0.571 |
| Doors | | |
| Sliding | | |
| 3' x 7' | EA. | 2.000 |
| 3' x 8' | " | 2.286 |
| 3' x 10' | " | 3.200 |
| 4' x 7' | " | 2.286 |
| 4' x 8' | " | 3.200 |
| 4' x 10' | " | 4.000 |
| 5' x 7' | " | 3.200 |
| 5' x 8' | " | 4.000 |
| 5' x 10' | " | 4.000 |
| Swing door | | |
| 3' x 7' | EA. | 2.000 |
| 4' x 7' | " | 2.286 |
| Swing door, with 1' transom | | |
| 3' x 7' | EA. | 2.286 |
| 4' x 7' | " | 3.200 |
| Swing door, with 3' transom | | |
| 3' x 7' | EA. | 3.200 |
| 4' x 7' | " | 4.000 |
| **10670.10 SHELVING** | | |
| Shelving, enamel, closed side and back, 12" x 36" | | |
| 5 shelves | EA. | 1.333 |
| 8 shelves | " | 1.778 |
| Open | | |
| 5 shelves | EA. | 1.333 |
| 8 shelves | " | 1.778 |
| Metal storage shelving, baked enamel | | |
| 7 shelf unit, 72" or 84" high | | |
| 10" shelf | L.F. | 0.800 |
| 12" shelf | " | 0.842 |
| 15" shelf | " | 0.889 |
| 18" shelf | " | 0.941 |
| 24" shelf | " | 1.000 |
| 30" shelf | " | 1.067 |

| SPECIALTIES | UNIT | MAN/ HOURS |
|---|---|---|
| **10670.10 SHELVING** | | |
| 36" shelf | L.F. | 1.143 |
| 4 shelf unit, 40" high | | |
| 10" shelf | L.F. | 0.667 |
| 12" shelf | " | 0.727 |
| 15" shelf | " | 0.800 |
| 18" shelf | " | 0.842 |
| 24" shelf | " | 0.889 |
| 3 shelf unit, 32" high | | |
| 10" shelf | L.F. | 0.400 |
| 12" shelf | " | 0.421 |
| 15" shelf | " | 0.444 |
| 18" shelf | " | 0.471 |
| 24" shelf | " | 0.500 |
| Single shelf unit, attached to masonry | | |
| 10" shelf | L.F. | 0.133 |
| 12" shelf | " | 0.145 |
| 15" shelf | " | 0.154 |
| 18" shelf | " | 0.163 |
| 24" shelf | " | 0.174 |
| Built-in wood shelves | | |
| Posts and trimmed plywood | L.F. | 0.114 |
| Solid clear pine | " | 0.123 |
| Closet shelf, pine with rod | " | 0.123 |
| **10750.10 TELEPHONE ENCLOSURES** | | |
| Telephone enclosure, wall mounted, shelf, 28" x 30" x 15" | EA. | 2.000 |
| Directory shelf, stainless steel, 3 binders | " | 1.333 |
| **10800.10 BATH ACCESSORIES** | | |
| Ash receiver, wall mounted, aluminum | EA. | 0.400 |
| Grab bar, 1-1/2" dia., stainless steel, wall mounted | | |
| 24" long | EA. | 0.400 |
| 36" long | " | 0.421 |
| 42" long | " | 0.444 |
| 48" long | " | 0.471 |
| 52" long | " | 0.500 |
| 1" dia., stainless steel | | |
| 12" long | EA. | 0.348 |
| 18" long | " | 0.364 |
| 24" long | " | 0.400 |
| 30" long | " | 0.421 |
| 36" long | " | 0.444 |
| 48" long | " | 0.471 |
| Hand dryer, surface mounted, 110 volt | " | 1.000 |
| Medicine cabinet, 16 x 22, baked enamel, steel, lighted | " | 0.320 |
| With mirror, lighted | " | 0.533 |
| Mirror, 1/4" plate glass, up to 10 sf | S.F. | 0.080 |
| Mirror, stainless steel frame | | |
| 18"x24" | EA. | 0.267 |
| 18"x32" | " | 0.320 |
| 18"x36" | " | 0.400 |
| 24"x30" | " | 0.400 |
| 24"x36" | " | 0.444 |

# 10 SPECIALTIES

| SPECIALTIES | UNIT | MAN/ HOURS |
|---|---|---|
| **10800.10 BATH ACCESSORIES** | | |
| 24"x48" | EA. | 0.667 |
| 24"x60" | " | 0.800 |
| 30"x30" | " | 0.800 |
| 30"x72" | " | 1.000 |
| 48"x72" | " | 1.333 |
| With shelf, 18"x24" | " | 0.320 |
| Sanitary napkin dispenser, stainless steel, wall mounted | " | 0.533 |
| Shower rod, 1" diameter | | |
| Chrome finish over brass | EA. | 0.400 |
| Stainless steel | " | 0.400 |
| Soap dish, stainless steel, wall mounted | " | 0.533 |
| Toilet tissue dispenser, stainless, wall mounted | | |
| Single roll | EA. | 0.200 |
| Double roll | " | 0.229 |
| Towel dispenser, stainless steel | | |
| Flush mounted | EA. | 0.444 |
| Surface mounted | " | 0.400 |
| Combination towel dispenser and waste receptacle | " | 0.533 |
| Towel bar, stainless steel | | |
| 18" long | EA. | 0.320 |
| 24" long | " | 0.364 |
| 30" long | " | 0.400 |
| 36" long | " | 0.444 |
| Toothbrush and tumbler holder | " | 0.267 |
| Waste receptacle, stainless steel, wall mounted | " | 0.667 |
| **10900.10 WARDROBE SPECIALTIES** | | |
| Hospital wardrobe units, 24" x 24" x 76", with door | | |
| Baked enameled steel | EA. | 4.444 |
| Hardwood | " | 4.444 |
| Stainless steel | " | 4.444 |
| Plastic laminated | " | 4.444 |
| Dormitory wardrobe units, 24" x 76", with door | | |
| Hardwood | EA. | 4.444 |
| Plastic laminated | " | 4.444 |
| Hat and coat rack | | |
| Single tier | | |
| Baked enameled steel | L.F. | 0.200 |
| Stainless steel | " | 0.200 |
| Aluminum | " | 0.200 |
| Double tier | | |
| Baked enameled steel | L.F. | 0.229 |
| Stainless steel | " | 0.229 |
| Aluminum | " | 0.229 |

# 11 EQUIPMENT

| ARCHITECTURAL EQUIPMENT | UNIT | MAN/ HOURS |
|---|---|---|
| **11010.10 MAINTENANCE EQUIPMENT** | | |
| Vacuum cleaning system | | |
| 3 valves | | |
| 1.5 hp | EA. | 8.889 |
| 2.5 hp | " | 11.429 |
| 5 valves | " | 16.000 |
| 7 valves | " | 20.000 |
| **11020.10 SECURITY EQUIPMENT** | | |
| Bulletproof teller window | | |
| 4' x 4' | EA. | 13.333 |
| 5' x 4' | " | 16.000 |
| Bulletproof partitions | | |
| Up to 12' high, 2.5" thick | S.F. | 0.053 |
| Counter for banks | | |
| Minimum | L.F. | 1.600 |
| Maximum | " | 2.667 |
| Drive-up window | | |
| Minimum | EA. | 11.429 |
| Maximum | " | 26.667 |
| Night depository | | |
| Minimum | EA. | 11.429 |
| Maximum | " | 26.667 |
| Office safes, 30" x 20" x 20", 1 hr rating | " | 2.000 |
| 30" x 16" x 15", 2 hr rating | " | 1.600 |
| 30" x 28" x 20", H&G rating | " | 1.000 |
| Service windows, pass through painted steel | | |
| 24" x 36" | EA. | 8.000 |
| 48" x 40" | " | 10.000 |
| 72" x 40" | " | 16.000 |
| Special doors and windows | | |
| 3' x 7' bulletproof door with frame | EA. | 11.429 |
| 12" x 12" vision panel | " | 5.714 |
| Surveillance system | | |
| Minimum | EA. | 16.000 |
| Maximum | " | 80.000 |
| Vault door, 3' wide, 6'6" high | | |
| 3-1/2" thick | EA. | 100 |
| 7" thick | " | 133 |
| 10" thick | " | 160 |
| Insulated vault door | | |
| 2 hr rating | | |
| 32" wide | EA. | 8.000 |
| 40" wide | " | 8.421 |
| 4 hr rating | | |
| 32" wide | EA. | 8.889 |
| 40" wide | " | 10.000 |
| 6 hr rating | | |
| 32" wide | EA. | 8.889 |
| 40" wide | " | 10.000 |
| Insulated file room door | | |
| 1 hr rating | | |
| 32" wide | EA. | 8.000 |
| 40" wide | " | 8.889 |

| ARCHITECTURAL EQUIPMENT | UNIT | MAN/ HOURS |
|---|---|---|
| **11060.10 THEATER EQUIPMENT** | | |
| Roll out stage, steel frame, wood floor | | |
| Manual | S.F. | 0.050 |
| Electric | " | 0.080 |
| Portable stages | | |
| 8" high | S.F. | 0.040 |
| 18" high | " | 0.044 |
| 36" high | " | 0.047 |
| 48" high | " | 0.050 |
| Band risers | | |
| Minimum | S.F. | 0.040 |
| Maximum | " | 0.040 |
| Chairs for risers | | |
| Minimum | EA. | 0.036 |
| Maximum | " | 0.036 |
| **11080.10 POLICE EQUIPMENT** | | |
| Firing range equipment, rifle | | |
| 3 position | EA. | 26.667 |
| 4 position | " | 40.000 |
| 5 position | " | 44.444 |
| 6 position | " | 47.059 |
| **11090.10 CHECKROOM EQUIPMENT** | | |
| Motorized checkroom equipment | | |
| No shelf system, 6'4" height | | |
| 7'6" length | EA. | 8.000 |
| 14'6" length | " | 8.000 |
| 28' length | " | 8.000 |
| One shelf, 6'8" height | | |
| 7'6" length | EA. | 8.000 |
| 14'6" length | " | 8.000 |
| 28' length | " | 8.000 |
| Two shelves, 7'5" height | | |
| 7'6" length | EA. | 8.000 |
| 14'6" length | " | 8.000 |
| 28' length | " | 8.000 |
| Three shelves, 8' height | | |
| 7'6" length | EA. | 16.000 |
| 14'6" length | " | 16.000 |
| 28' length | " | 16.000 |
| Four shelves, 8'7" height | | |
| 7'6" length | EA. | 16.000 |
| 14'6" length | " | 16.000 |
| 28' length | " | 16.000 |
| 200 lb | | |
| **11110.10 LAUNDRY EQUIPMENT** | | |
| High capacity, heavy duty | | |
| Washer extractors | | |
| 135 lb | | |
| Standard | EA. | 6.667 |
| Pass through | " | 6.667 |
| 200 lb | | |

# 11 EQUIPMENT

| ARCHITECTURAL EQUIPMENT | UNIT | MAN/ HOURS |
|---|---|---|
| **11110.10 LAUNDRY EQUIPMENT** | | |
| Standard | EA. | 6.667 |
| Pass through | " | 6.667 |
| 110 lb dryer | " | 6.667 |
| Hand operated presser | " | 8.889 |
| Mushroom press | " | 8.889 |
| Spreader feeders | | |
| 2 station | EA. | 8.889 |
| 4 station | " | 16.000 |
| Delivery carts | | |
| 12 bushel | EA. | 0.100 |
| 16 bushel | " | 0.107 |
| 18 bushel | " | 0.114 |
| 30 bushel | " | 0.133 |
| 40 bushel | " | 0.160 |
| Low capacity | | |
| Pressers | | |
| Air operated | EA. | 3.200 |
| Hand operated | " | 3.200 |
| Extractor, low capacity | " | 3.200 |
| Ironer, 48" | " | 1.600 |
| Coin washers | | |
| 10 lb capacity | EA. | 1.600 |
| 20 lb capacity | " | 1.600 |
| Coin dryer | " | 1.000 |
| Coin dry cleaner, 20 lb | " | 3.200 |
| **11161.10 LOADING DOCK EQUIPMENT** | | |
| Dock leveler, 10 ton capacity | | |
| 6' x 8' | EA. | 8.000 |
| 7' x 8' | " | 8.000 |
| Bumpers, laminated rubber | | |
| 4-1/2" thick | | |
| 6" x 14" | EA. | 0.160 |
| 6" x 36" | " | 0.178 |
| 10" x 14" | " | 0.200 |
| 10" x 24" | " | 0.229 |
| 10" x 36" | " | 0.267 |
| 12" x 14" | " | 0.211 |
| 12" x 24" | " | 0.250 |
| 12" x 36" | " | 0.296 |
| 6" thick | | |
| 10" x 14" | EA. | 0.229 |
| 10" x 24" | " | 0.276 |
| 10" x 36" | " | 0.400 |
| Extruded rubber bumpers | | |
| T-section, 22" x 22" x 3" | EA. | 0.160 |
| Molded rubber bumpers | | |
| 24" x 12" x 3" thick | EA. | 0.400 |
| Door seal, 12" x 12", vinyl covered | L.F. | 0.200 |
| Dock boards, heavy duty, 5' x 5' | | |
| 5000 lb | | |
| Minimum | EA. | 6.667 |
| Maximum | " | 6.667 |

| ARCHITECTURAL EQUIPMENT | UNIT | MAN/ HOURS |
|---|---|---|
| **11161.10 LOADING DOCK EQUIPMENT** | | |
| 9000 lb | | |
| Minimum | EA. | 6.667 |
| Maximum | " | 7.273 |
| 15,000 lb | " | 7.273 |
| Truck shelters | | |
| Minimum | EA. | 6.154 |
| Maximum | " | 11.429 |
| **11170.10 WASTE HANDLING** | | |
| Incinerator, electric | | |
| 100 lb/hr | | |
| Minimum | EA. | 8.000 |
| Maximum | " | 8.000 |
| 400 lb/hr | | |
| Minimum | EA. | 16.000 |
| Maximum | " | 16.000 |
| 1000 lb/hr | | |
| Minimum | EA. | 24.242 |
| Maximum | " | 24.242 |
| Incinerator, medical-waste | | |
| 25 lb/hr, 2-7 x 4-0 | EA. | 16.000 |
| 50 lb/hr, 2-11 x 4-11 | " | 16.000 |
| 75 lb/hr, 3-8 x 5-0 | " | 32.000 |
| 100 lb/hr, 3-8 x 6-0 | " | 32.000 |
| Industrial compactor | | |
| 1 cy | EA. | 8.889 |
| 3 cy | " | 11.429 |
| 5 cy | " | 16.000 |
| Trash chutes steel, including sprinklers | | |
| 18" dia. | L.F. | 4.000 |
| 24" dia. | " | 4.211 |
| 30" dia. | " | 4.444 |
| 36" dia. | " | 4.706 |
| Refuse bottom hopper | EA. | 4.444 |
| **11400.10 FOOD SERVICE EQUIPMENT** | | |
| Unit kitchens | | |
| 30" compact kitchen | | |
| Refrigerator, with range, sink | EA. | 4.000 |
| Sink only | " | 2.667 |
| Range only | " | 2.000 |
| Cabinet for upper wall section | " | 1.143 |
| Stainless shield, for rear wall | " | 0.320 |
| Side wall | " | 0.320 |
| 42" compact kitchen | | |
| Refrigerator with range, sink | EA. | 4.444 |
| Sink only | " | 4.000 |
| Cabinet for upper wall section | " | 1.333 |
| Stainless shield, for rear wall | " | 0.333 |
| Side wall | " | 0.333 |
| 54" compact kitchen | | |
| Refrigerator, oven, range, sink | EA. | 5.714 |

| ARCHITECTURAL EQUIPMENT | UNIT | MAN/ HOURS |
|---|---|---|
| **11400.10 FOOD SERVICE EQUIPMENT** | | |
| Cabinet for upper wall section | EA. | 1.600 |
| Stainless shield, for | | |
| Rear wall | EA. | 0.364 |
| Side wall | " | 0.364 |
| 60" compact kitchen | | |
| Refrigerator, oven, range, sink | EA. | 5.714 |
| Cabinet for upper wall section | " | 1.600 |
| Stainless shield, for | | |
| Rear wall | EA. | 0.364 |
| Side wall | " | 0.364 |
| 72" compact kitchen | | |
| Refrigerator, oven, range, sink | EA. | 6.667 |
| Cabinet for upper wall section | " | 1.600 |
| Stainless shield for | | |
| Rear wall | EA. | 0.400 |
| Side wall | " | 0.400 |
| Bake oven | | |
| Single deck | | |
| Minimum | EA. | 1.000 |
| Maximum | " | 2.000 |
| Double deck | | |
| Minimum | EA. | 1.333 |
| Maximum | " | 2.000 |
| Triple deck | | |
| Minimum | EA. | 1.333 |
| Maximum | " | 2.667 |
| Convection type oven, electric, 40" x 45" x 57" | | |
| Minimum | EA. | 1.000 |
| Maximum | " | 2.000 |
| Broiler, without oven, 69" x 26" x 39" | | |
| Minimum | EA. | 1.000 |
| Maximum | " | 1.333 |
| Coffee urns, 10 gallons | | |
| Minimum | EA. | 2.667 |
| Maximum | " | 4.000 |
| Fryer, with submerger | | |
| Single | | |
| Minimum | EA. | 1.600 |
| Maximum | " | 2.667 |
| Double | | |
| Minimum | EA. | 2.000 |
| Maximum | " | 2.667 |
| Griddle, counter | | |
| 3' long | | |
| Minimum | EA. | 1.333 |
| Maximum | " | 1.600 |
| 5' long | | |
| Minimum | EA. | 2.000 |
| Maximum | " | 2.667 |
| Kettles, steam, jacketed | | |
| 20 gallons | | |
| Minimum | EA. | 2.000 |
| Maximum | " | 4.000 |
| 40 gallons | | |

| ARCHITECTURAL EQUIPMENT | UNIT | MAN/ HOURS |
|---|---|---|
| **11400.10 FOOD SERVICE EQUIPMENT** | | |
| Minimum | EA. | 2.000 |
| Maximum | " | 4.000 |
| 60 gallons | | |
| Minimum | EA. | 2.000 |
| Maximum | " | 4.000 |
| Range | | |
| Heavy duty, single oven, open top | | |
| Minimum | EA. | 1.000 |
| Maximum | " | 2.667 |
| Fry top | | |
| Minimum | EA. | 1.000 |
| Maximum | " | 2.667 |
| Steamers, electric | | |
| 27 kw | | |
| Minimum | EA. | 2.000 |
| Maximum | " | 2.667 |
| 18 kw | | |
| Minimum | EA. | 2.000 |
| Maximum | " | 2.667 |
| Dishwasher, rack type | | |
| Single tank, 190 racks/hr | EA. | 4.000 |
| Double tank | | |
| 234 racks/hr | EA. | 4.444 |
| 265 racks/hr | " | 5.333 |
| Dishwasher, automatic 100 meals/hr | " | 2.667 |
| Disposals | | |
| 100 gal/hr | EA. | 2.667 |
| 120 gal/hr | " | 2.759 |
| 250 gal/hr | " | 2.857 |
| Exhaust hood for dishwasher, gutter 4 sides, s-steel | | |
| 4'x4'x2' | EA. | 2.963 |
| 4'x7'x2' | " | 3.200 |
| Food preparation machines | | |
| Vertical cutter mixers | | |
| 25 quart | EA. | 2.667 |
| 40 quart | " | 2.667 |
| 80 quart | " | 4.000 |
| 130 quart | " | 6.667 |
| Choppers | | |
| 5 lb | EA. | 2.000 |
| 16 lb | " | 2.667 |
| 40 lb | " | 4.000 |
| Mixers, floor models | | |
| 20 quart | EA. | 1.000 |
| 60 quart | " | 1.000 |
| 80 quart | " | 1.143 |
| 140 quart | " | 1.600 |
| Ice cube maker | | |
| 50 lb per day | | |
| Minimum | EA. | 8.000 |
| Maximum | " | 8.000 |
| 500 lb per day | | |
| Minimum | EA. | 13.333 |
| Maximum | " | 13.333 |

# 11 EQUIPMENT

| ARCHITECTURAL EQUIPMENT | UNIT | MAN/ HOURS |
|---|---|---|
| **11400.10 FOOD SERVICE EQUIPMENT** | | |
| Ice flakers | | |
| 300 lb per day | EA. | 8.000 |
| 600 lb per day | " | 13.333 |
| 1000 lb per day | " | 17.778 |
| 2000 lb per day | " | 20.000 |
| Refrigerated cases | | |
| Dairy products | | |
| Multi deck type | L.F. | 0.533 |
| Delicatessen case, service deli | | |
| Single deck | L.F. | 4.000 |
| Multi deck | " | 5.000 |
| Meat case | | |
| Single deck | L.F. | 4.706 |
| Multi deck | " | 5.000 |
| Produce case | | |
| Single deck | L.F. | 4.706 |
| Multi deck | " | 5.000 |
| Bottle coolers | | |
| 6' long | | |
| Minimum | EA. | 16.000 |
| Maximum | " | 16.000 |
| 10' long | | |
| Minimum | EA. | 26.667 |
| Maximum | " | 26.667 |
| Frozen food cases | | |
| Chest type | L.F. | 4.706 |
| Reach-in, glass door | " | 5.000 |
| Island case, single | " | 4.706 |
| Multi deck | " | 5.000 |
| Ice storage bins | | |
| 500 lb capacity | EA. | 11.429 |
| 1000 lb capacity | " | 22.857 |
| **11450.10 RESIDENTIAL EQUIPMENT** | | |
| Compactor, 4 to 1 compaction | EA. | 2.000 |
| Dishwasher, built-in | | |
| 2 cycles | EA. | 4.000 |
| 4 or more cycles | " | 4.000 |
| Disposal | | |
| Garbage disposer | EA. | 2.667 |
| Heaters, electric, built-in | | |
| Ceiling type | EA. | 2.667 |
| Wall type | | |
| Minimum | EA. | 2.000 |
| Maximum | " | 2.667 |
| Hood for range, 2-speed, vented | | |
| 30" wide | EA. | 2.667 |
| 42" wide | " | 2.667 |
| Ice maker, automatic | | |
| 30 lb per day | EA. | 1.143 |
| 50 lb per day | " | 4.000 |
| Folding access stairs, disappearing metal stair | | |
| 8' long | EA. | 1.143 |
| 11' long | " | 1.143 |

| ARCHITECTURAL EQUIPMENT | UNIT | MAN/ HOURS |
|---|---|---|
| **11450.10 RESIDENTIAL EQUIPMENT** | | |
| 12' long | EA. | 1.143 |
| Wood frame, wood stair | | |
| 22" x 54" x 8'9" long | EA. | 0.800 |
| 25" x 54" x 10' long | " | 0.800 |
| Ranges electric | | |
| Built-in, 30", 1 oven | EA. | 2.667 |
| 2 oven | " | 2.667 |
| Counter top, 4 burner, standard | " | 2.000 |
| With grill | " | 2.000 |
| Free standing, 21", 1 oven | " | 2.667 |
| 30", 1 oven | " | 1.600 |
| 2 oven | " | 1.600 |
| Water softener | | |
| 30 grains per gallon | EA. | 2.667 |
| 70 grains per gallon | " | 4.000 |
| **11470.10 DARKROOM EQUIPMENT** | | |
| Dryers | | |
| 36" x 25" x 68" | EA. | 4.000 |
| 48" x 25" x 68" | " | 4.000 |
| Processors, film | | |
| Black and white | EA. | 4.000 |
| Color negatives | " | 4.000 |
| Prints | " | 4.000 |
| Transparencies | " | 4.000 |
| Sinks with cabinet and/or stand | | |
| 5" sink with stand | | |
| 24" x 48" | EA. | 2.000 |
| 32" x 64" | " | 2.667 |
| 38" x 52" | " | 2.667 |
| 42" x 132" | " | 4.000 |
| 48" x 52" | " | 4.000 |
| 5" sink with cabinet | | |
| 24" x 48" | EA. | 2.000 |
| 32" x 64" | " | 2.667 |
| 38" x 52" | " | 2.667 |
| 42" x 132" | " | 4.000 |
| 48" x 52" | " | 4.000 |
| 10" sink with stand | | |
| 24" x 48" | EA. | 2.000 |
| 32" x 64" | " | 2.667 |
| 38" x 52" | " | 2.667 |
| 10" sink with cabinet | | |
| 24" x 48" | EA. | 2.000 |
| 38" x 52" | " | 2.667 |
| **11480.10 ATHLETIC EQUIPMENT** | | |
| Basketball backboard | | |
| Fixed | EA. | 10.000 |
| Swing-up | " | 16.000 |
| Portable, hydraulic | " | 4.000 |
| Suspended type, standard | " | 16.000 |
| Bleacher, telescoping, manual | | |

# 11 EQUIPMENT

## ARCHITECTURAL EQUIPMENT

| 11480.10 ATHLETIC EQUIPMENT | UNIT | MAN/HOURS |
|---|---|---|
| 15 tier, minimum | SEAT | 0.160 |
| Maximum | " | 0.160 |
| 20 tier, minimum | " | 0.178 |
| Maximum | " | 0.178 |
| 30 tier, minimum | " | 0.267 |
| Maximum | " | 0.267 |
| Boxing ring elevated, complete, 22' x 22' | EA. | 80.000 |
| Gym divider curtain | | |
| Minimum | S.F. | 0.011 |
| Maximum | " | 0.011 |
| Scoreboards, single face | | |
| Minimum | EA. | 8.000 |
| Maximum | " | 40.000 |
| Parallel bars, wall mounted | | |
| Minimum | EA. | 8.000 |
| Maximum | " | 13.333 |

| 11500.10 INDUSTRIAL EQUIPMENT | UNIT | MAN/HOURS |
|---|---|---|
| Vehicular paint spray booth, solid back, 14'4" x 9'6" | | |
| 24' deep | EA. | 8.000 |
| 26'6" deep | " | 8.000 |
| 28'6" deep | " | 8.000 |
| Drive through, 14'9" x 9'6" | | |
| 24' deep | EA. | 8.000 |
| 26'6" deep | " | 8.000 |
| 28'6" deep | " | 8.000 |
| Water wash, paint spray booth | | |
| 5' x 11'2" x 10'8" | EA. | 8.000 |
| 6' x 11'2" x 10'8" | " | 8.000 |
| 8' x 11'2" x 10'8" | " | 8.000 |
| 10' x 11'2" x 11'2" | " | 8.000 |
| 12' x 12'2" x 11'2" | " | 8.000 |
| 14' x 12'2" x 11'2" | " | 8.000 |
| 16' x 12'2" x 11'2" | " | 8.000 |
| 20' x 12'2" x 11'2" | " | 8.000 |
| Dry type spray booth, with paint arrestors | | |
| 5'4" x 7'2" x 6'8" | EA. | 8.000 |
| 6'4" x 7'2" x 6'8" | " | 8.000 |
| 8'4" x 7'2" x 9'2" | " | 8.000 |
| 10'4" x 7'2" x 9'2" | " | 8.000 |
| 12'4" x 7'6" x 9'2" | " | 8.000 |
| 14'4" x 7'6" x 9'8" | " | 8.000 |
| 16'4" x 7'7" x 9'8" | " | 8.000 |
| 20'4" x 7'7" x 10'8" | " | 8.000 |
| Air compressor, electric | | |
| 1 hp | | |
| 115 volt | EA. | 5.333 |
| 5 hp | | |
| 115 volt | EA. | 8.000 |
| 230 volt | " | 8.000 |
| Hydraulic lifts | | |
| 8,000 lb capacity | EA. | 20.000 |
| 11,000 lb capacity | " | 32.000 |
| 24,000 lb capacity | " | 53.333 |

## ARCHITECTURAL EQUIPMENT

| 11500.10 INDUSTRIAL EQUIPMENT | UNIT | MAN/HOURS |
|---|---|---|
| Power tools | | |
| Band saws | | |
| 10" | EA. | 0.667 |
| 14" | " | 0.800 |
| Motorized shaper | " | 0.615 |
| Motorized lathe | " | 0.667 |
| Bench saws | | |
| 9" saw | EA. | 0.533 |
| 10" saw | " | 0.571 |
| 12" saw | " | 0.667 |
| Electric grinders | | |
| 1/3 hp | EA. | 0.320 |
| 1/2 hp | " | 0.348 |
| 3/4 hp | " | 0.348 |

| 11600.10 LABORATORY EQUIPMENT | UNIT | MAN/HOURS |
|---|---|---|
| Cabinets, base | | |
| Minimum | L.F. | 0.667 |
| Maximum | " | 0.667 |
| Full storage, 7' high | | |
| Minimum | L.F. | 0.667 |
| Maximum | " | 0.667 |
| Wall | | |
| Minimum | L.F. | 0.800 |
| Maximum | " | 0.800 |
| Counter tops | | |
| Minimum | S.F. | 0.100 |
| Average | " | 0.114 |
| Maximum | " | 0.133 |
| Tables | | |
| Open underneath | S.F. | 0.400 |
| Doors underneath | " | 0.500 |
| Medical laboratory equipment | | |
| Analyzer | | |
| Chloride | EA. | 0.400 |
| Blood | " | 0.667 |
| Bath, water, utility, countertop unit | " | 0.800 |
| Hot plate, lab, countertop | " | 0.727 |
| Stirrer | " | 0.727 |
| Incubator, anaerobic, 23x23x36" | " | 4.000 |
| Dry heat bath | " | 1.333 |
| Incinerator, for sterilizing | " | 0.080 |
| Meter, serum protein | " | 0.100 |
| Ph analog, general purpose | " | 0.114 |
| Refrigerator, blood bank, undercounter type 153 litres | " | 1.333 |
| 5.4 cf, undercounter type | " | 1.333 |
| Refrigerator/freezer, 4.4 cf, undercounter type | " | 1.333 |
| Sealer, impulse, free standing, 20x12x4" | " | 0.267 |
| Timer, electric, 1-60 minutes, bench or wall mounted | " | 0.444 |
| Glassware washer - dryer, undercounter | " | 10.000 |
| Balance, torsion suspension, tabletop, 4.5 lb capacity | " | 0.444 |
| Binocular microscope, with in base illuminator | " | 0.308 |
| Centrifuge, table model, 19x16x13" | " | 0.320 |
| Clinical model, with four place head | " | 0.178 |

| ARCHITECTURAL EQUIPMENT | UNIT | MAN/ HOURS |
|---|---|---|
| **11700.10 MEDICAL EQUIPMENT** | | |
| Hospital equipment, lights | | |
| Examination, portable | EA. | 0.667 |
| Meters | | |
| Air flow meter | EA. | 0.444 |
| Oxygen flow meters | " | 0.333 |
| Racks | | |
| 40 chart, revolving open frame; mobile caddy | EA. | 0.667 |
| Scales. | | |
| Clinical, metric with measure rod, 350 lb | EA. | 0.727 |
| Physical therapy | | |
| Chair, hydrotherapy | EA. | 0.133 |
| Diathermy, shortwave, portable, on casters | " | 0.320 |
| Exercise bicycle, floor standing, 35" x 15" | " | 0.267 |
| Hydrocollator, 4 pack, portable, 129 x 90 x 160" | " | 0.114 |
| Lamp, infrared, mobile with variable heat control | " | 0.615 |
| Ultra violet, base mounted | " | 0.615 |
| Mirror, posture training, 27" wide and 72" high | " | 0.200 |
| Parallel bars, adjustable | " | 1.000 |
| Platform mat 10'x6', 1" thick | " | 0.200 |
| Pulley, duplex, wall mounted | " | 2.667 |
| Rack, crutch, wall mounted, 66 x 16 x 13" | " | 0.800 |
| Stimulator, galvanic-faradic, hand held | " | 0.053 |
| Ultrasound, muscle stimulator, portable, 13x13x8" | " | 0.067 |
| Sandbag set, velcro straps, saddle bag type | " | 0.114 |
| Whirlpool, 85 gallon | " | 4.000 |
| 65 gallon capacity | " | 4.000 |
| Radiology | | |
| Radiographic table, motor driven tilting table | EA. | 80.000 |
| Fluoroscope image/tv system | " | 160 |
| Processor for washing and drying radiographs | | |
| Water filter unit, 30" x 48-1/2" x 37-1/2" | EA. | 13.333 |
| Cassette transfer cabinet | " | 0.667 |
| Base storage cabinets, sectional design | | |
| With back splash, 24" deep and 35" high | L.F. | 0.667 |
| Wall storage cabinets | " | 1.000 |
| Steam sterilizers | | |
| For heat and moisture stable materials | EA. | 0.800 |
| For fast drying after sterilization | " | 1.000 |
| Compact unit | " | 1.000 |
| Semi-automatic | " | 4.000 |
| Floor loading | | |
| Single door | EA. | 6.667 |
| Double door | " | 8.000 |
| Utensil washer, sanitizer | " | 6.154 |
| Automatic washer/sterilizer | " | 16.000 |
| 16 x 16 x 26", including generator & accessories | " | 26.667 |
| Steam generator, elec., 10 kw to 180 kw | " | 16.000 |
| Surgical scrub | | |
| Minimum | EA. | 2.667 |
| Maximum | " | 2.667 |
| Gas sterilizers | | |
| Automatic, free standing, 21x19x29" | EA. | 8.000 |
| Surgical tables | | |
| Minimum | EA. | 11.429 |

| ARCHITECTURAL EQUIPMENT | UNIT | MAN/ HOURS |
|---|---|---|
| **11700.10 MEDICAL EQUIPMENT** | | |
| Maximum | EA. | 16.000 |
| Surgical lights, ceiling mounted | | |
| Minimum | EA. | 13.333 |
| Maximum | " | 16.000 |
| Water stills | | |
| 4 liters/hr | EA. | 2.667 |
| 8 liters/hr | " | 2.667 |
| 19 liters/hr | " | 6.667 |
| X-ray equipment | | |
| Mobile unit | | |
| Minimum | EA. | 4.000 |
| Maximum | " | 8.000 |
| Film viewers | | |
| Minimum | EA. | 1.333 |
| Maximum | " | 2.667 |
| Autopsy table | | |
| Minimum | EA. | 8.000 |
| Maximum | " | 8.000 |
| Incubators | | |
| 15 cf | EA. | 4.000 |
| 29 cf | " | 6.667 |
| Infant transport, portable | " | 4.211 |
| Beds | | |
| Stretcher, with pad, 30" x 78" | EA. | 2.000 |
| Transfer, for patient transport | " | 2.000 |
| Headwall | | |
| Aluminum, with back frame and console | EA. | 4.000 |
| **11700.20 DENTAL EQUIPMENT** | | |
| Dental care equipment | | |
| Drill console with accessories | EA. | 13.333 |
| Amalgamator | " | 0.400 |
| Lathe | " | 0.267 |
| Finish polisher | " | 0.533 |
| Model trimmer | " | 0.364 |
| Motor, wall mounted | " | 0.364 |
| Cleaner, ultrasonic | " | 0.800 |
| Curing unit, bench mounted | " | 1.333 |
| Oral evacuation system, dual pump | " | 1.000 |
| Sterilizer, table top, self contained | " | 0.444 |
| Dental lights | | |
| Light, floor or ceiling mounted | EA. | 4.000 |
| X-ray unit | | |
| Portable | EA. | 2.000 |
| Wall mounted with remote control | " | 6.667 |
| Illuminator, single panel | " | 11.429 |
| X-ray film processor | " | 6.667 |
| Shield, portable x-ray, lead lined | " | 0.533 |

# 12 FURNISHINGS

| INTERIOR | UNIT | MAN/ HOURS |
|---|---|---|
| **12302.10 CASEWORK** | | |
| Kitchen base cabinet, prefinished, 24" deep, 35" high | | |
| 12"wide | EA. | 0.800 |
| 18" wide | " | 0.800 |
| 24" wide | " | 0.889 |
| 27" wide | " | 0.889 |
| 36" wide | " | 1.000 |
| 48" wide | " | 1.000 |
| Corner cabinet, 36" wide | " | 1.000 |
| Wall cabinet, 12" deep, 12" high | | |
| 30" wide | EA. | 0.800 |
| 36" wide | " | 0.800 |
| 15" high | | |
| 30" wide | EA. | 0.889 |
| 36" wide | " | 0.889 |
| 24" high | | |
| 30" wide | EA. | 0.889 |
| 36" wide | " | 0.889 |
| 30" high | | |
| 12" wide | EA. | 1.000 |
| 18" wide | " | 1.000 |
| 24" wide | " | 1.000 |
| 27" wide | " | 1.000 |
| 30" wide | " | 1.143 |
| 36" wide | " | 1.143 |
| Corner cabinet, 30" high | | |
| 24" wide | EA. | 1.333 |
| 30" wide | " | 1.333 |
| 36" wide | " | 1.333 |
| Wardrobe | " | 2.000 |
| Vanity with top, laminated plastic | | |
| 24" wide | EA. | 2.000 |
| 30" wide | " | 2.000 |
| 36" wide | " | 2.667 |
| 48" wide | " | 3.200 |
| **12390.10 COUNTER TOPS** | | |
| Stainless steel, counter top, with backsplash | S.F. | 0.200 |
| Acid-proof, kemrock surface | " | 0.133 |
| **12500.10 WINDOW TREATMENT** | | |
| Drapery tracks, wall or ceiling mounted | | |
| Basic traverse rod | | |
| 50 to 90" | EA. | 0.400 |
| 84 to 156" | " | 0.444 |
| 136 to 250" | " | 0.444 |
| 165 to 312" | " | 0.500 |
| Traverse rod with stationary curtain rod | | |
| 30 to 50" | EA. | 0.400 |
| 50 to 90" | " | 0.400 |
| 84 to 156" | " | 0.444 |
| 136 to 250" | " | 0.500 |
| Double traverse rod | | |
| 30 to 50" | EA. | 0.400 |
| 50 to 84" | " | 0.400 |

| INTERIOR | UNIT | MAN/ HOURS |
|---|---|---|
| **12500.10 WINDOW TREATMENT** | | |
| 84 to 156" | EA. | 0.444 |
| 136 to 250" | " | 0.500 |
| **12510.10 BLINDS** | | |
| Venetian blinds | | |
| 2" slats | S.F. | 0.020 |
| 1" slats | " | 0.020 |
| **12690.40 FLOOR MATS** | | |
| Recessed entrance mat, 3/8" thick, aluminum link | S.F. | 0.400 |
| Steel, flexible | " | 0.400 |

# 13 SPECIAL

| CONSTRUCTION | UNIT | MAN/HOURS |
|---|---|---|
| **13056.10 VAULTS** | | |
| Floor safes | | |
| Class C | | |
| 1.0 cf | EA. | 0.667 |
| 1.3 cf | " | 1.000 |
| 1.9 cf | " | 1.333 |
| 5.2 cf | " | 1.333 |
| **13121.10 PRE-ENGINEERED BUILDINGS** | | |
| Pre-engineered metal building, 40'x100' | | |
| 14' eave height | S.F. | 0.032 |
| 16' eave height | " | 0.037 |
| 20' eave height | " | 0.048 |
| 60'x100' | | |
| 14' eave height | S.F. | 0.032 |
| 16' eave height | " | 0.037 |
| 20' eave height | " | 0.048 |
| 80'x100' | | |
| 14' eave height | S.F. | 0.032 |
| 16' eave height | " | 0.037 |
| 20' eave height | " | 0.048 |
| 100'x100' | | |
| 14' eave height | S.F. | 0.032 |
| 16' eave height | " | 0.037 |
| 20' eave height | " | 0.048 |
| 100'x150' | | |
| 14' eave height | S.F. | 0.032 |
| 16' eave height | " | 0.037 |
| 20' eave height | " | 0.048 |
| 120'x150' | | |
| 14' eave height | S.F. | 0.032 |
| 16' eave height | " | 0.037 |
| 20' eave height | " | 0.048 |
| 140'x150' | | |
| 14' eave height | S.F. | 0.032 |
| 16' eave height | " | 0.037 |
| 20' eave height | " | 0.048 |
| 160'x200' | | |
| 14' eave height | S.F. | 0.032 |
| 16' eave height | " | 0.037 |
| 20' eave height | " | 0.048 |
| 200'x200' | | |
| 14' eave height | S.F. | 0.032 |
| 16' eave height | " | 0.037 |
| 20' eave height | " | 0.048 |
| Liner panel, 26 ga, painted steel | " | 0.020 |
| Wall panel insulated, 26 ga. steel, foam core | " | 0.020 |
| Roof panel, 26 ga. painted steel | " | 0.011 |
| Plastic (sky light) | " | 0.011 |
| Insulation, 3-1/2" thick blanket, R11 | " | 0.005 |
| **13152.10 SWIMMING POOL EQUIPMENT** | | |
| Diving boards | | |
| 14' long | | |

| CONSTRUCTION | UNIT | MAN/HOURS |
|---|---|---|
| **13152.10 SWIMMING POOL EQUIPMENT** | | |
| Aluminum | EA. | 4.444 |
| Fiberglass | " | 4.444 |
| Ladders, heavy duty | | |
| 2 steps | | |
| Minimum | EA. | 1.600 |
| Maximum | " | 1.600 |
| 4 steps | | |
| Minimum | EA. | 2.000 |
| Maximum | " | 2.000 |
| Lifeguard chair | | |
| Minimum | EA. | 8.000 |
| Maximum | " | 8.000 |
| Lights, underwater | | |
| 12 volt, with transformer | EA. | 2.000 |
| 110 volt | | |
| Minimum | EA. | 2.000 |
| Maximum | " | 2.000 |
| Ground fault interrupter for 110 volt, each light | " | 0.667 |
| Pool covers | | |
| Reinforced polyethylene | S.F. | 0.062 |
| Vinyl water tube | | |
| Minimum | S.F. | 0.062 |
| Maximum | " | 0.062 |
| Slides with water tube | | |
| Minimum | EA. | 6.667 |
| Maximum | " | 6.667 |

## 14 CONVEYING

| ELEVATORS | UNIT | MAN/HOURS |
|---|---|---|
| **14210.10 ELEVATORS** | | |
| Passenger elevators, electric, geared | | |
| Based on a shaft of 6 stops and 6 openings | | |
| 50 fpm, 2000 lb | EA. | 24.000 |
| 100 fpm, 2000 lb | " | 26.667 |
| 150 fpm | | |
| 2000 lb | EA. | 30.000 |
| 3000 lb | " | 34.286 |
| 4000 lb | " | 40.000 |
| 200 fpm | | |
| 2500 lb | EA. | 34.286 |
| 3000 lb | " | 36.923 |
| 4000 lb | " | 40.000 |
| 250 fpm | | |
| 2500 lb | EA. | 34.286 |
| 3000 lb | " | 36.923 |
| 4000 lb | " | 40.000 |
| 300 fpm | | |
| 2500 lb | EA. | 34.286 |
| 3000 lb | " | 36.923 |
| 4000 lb | " | 24.000 |
| Based on a shaft of 8 stops and 8 openings | | |
| 300 fpm | | |
| 3000 lb | EA. | 48.000 |
| 3500 lb | " | 48.000 |
| 4000 lb | " | 53.333 |
| 5000 lb | " | 57.143 |
| 400 fpm | | |
| 3000 lb | EA. | 48.000 |
| 3500 lb | " | 48.000 |
| 4000 lb | " | 53.333 |
| 5000 lb | " | 57.143 |
| 600 fpm | | |
| 3000 lb | EA. | 53.333 |
| 3500 lb | " | 57.143 |
| 4000 lb | " | 58.537 |
| 5000 lb | " | 60.000 |
| 800 fpm | | |
| 3000 lb | EA. | 53.333 |
| 3500 lb | " | 57.143 |
| 4000 lb | " | 58.537 |
| 5000 lb | " | 60.000 |
| Hydraulic, based on a shaft of 3 stops, 3 openings | | |
| 50 fpm | | |
| 2000 lb | EA. | 20.000 |
| 2500 lb | " | 20.000 |
| 3000 lb | " | 20.870 |
| 100 fpm | | |
| 2000 lb | EA. | 20.000 |
| 2500 lb | " | 20.870 |
| 3000 lb | " | 21.818 |
| 150 fpm | | |
| 2000 lb | EA. | 20.000 |
| 2500 lb | " | 20.870 |
| 3000 lb | " | 22.857 |

| ELEVATORS | UNIT | MAN/HOURS |
|---|---|---|
| **14210.10 ELEVATORS** | | |
| Small elevators, 4 to 6 passenger capacity | | |
| Electric, push | | |
| 2 stops | EA. | 20.000 |
| 3 stops | " | 21.818 |
| 4 stops | " | 24.000 |
| Freight elevators, electric | | |
| Based on a shaft of 6 stops and 6 openings | | |
| 50 fpm | | |
| 3500 lb | EA. | 26.667 |
| 4000 lb | " | 26.667 |
| 5000 lb | " | 30.000 |
| 100 fpm | | |
| 3500 lb | EA. | 30.000 |
| 4000 lb | " | 30.000 |
| 5000 lb | " | 34.286 |
| 200 fpm | | |
| 3500 lb | EA. | 34.286 |
| 4000 lb | " | 34.286 |
| 5000 lb | " | 40.000 |
| Based on shaft of 8 stops and 8 openings | | |
| 100 fpm | | |
| 4000 lb | EA. | 30.000 |
| 6000 lb | " | 30.769 |
| 8000 lb | " | 32.432 |
| 150 fpm | | |
| 4000 lb | EA. | 34.286 |
| 6000 lb | " | 35.294 |
| 8000 lb | " | 37.500 |
| 200 fpm | | |
| 4000 lb | EA. | 40.000 |
| 6000 lb | " | 41.379 |
| 8000 lb | " | 43.636 |
| Hydraulic, based on 3 stops and 3 openings | | |
| 50 fpm | | |
| 3000 lb | EA. | 17.143 |
| 4000 lb | " | 17.778 |
| 6000 lb | " | 18.462 |
| 100 fpm | | |
| 3000 lb | EA. | 17.143 |
| 4000 lb | " | 17.778 |
| 6000 lb | " | 18.462 |
| 150 fpm | | |
| 3000 lb | EA. | 17.143 |
| 4000 lb | " | 17.778 |
| 6000 lb | " | 18.462 |
| **14300.10 ESCALATORS** | | |
| Escalators | | |
| 32" wide, floor to floor | | |
| 12' high | EA. | 40.000 |
| 15' high | " | 48.000 |
| 18' high | " | 60.000 |
| 22' high | " | 80.000 |

# 14 CONVEYING

## ELEVATORS

| 14300.10 ESCALATORS | UNIT | MAN/HOURS |
|---|---|---|
| 25' high | EA. | 96.000 |
| 48" wide | | |
| 12' high | EA. | 41.379 |
| 15' high | " | 50.000 |
| 18' high | " | 63.158 |
| 22' high | " | 85.714 |
| 25' high | " | 96.000 |

## LIFTS

| 14410.10 PERSONNEL LIFTS | UNIT | MAN/HOURS |
|---|---|---|
| Residential stair climber, per story | EA. | 6.667 |

| 14450.10 VEHICLE LIFTS | UNIT | MAN/HOURS |
|---|---|---|
| Automotive hoist, one post, semi-hydraulic, 8,000 lb | EA. | 24.000 |
| Full hydraulic, 8,000 lb | " | 24.000 |
| 2 post, semi-hydraulic, 10,000 lb | " | 34.286 |
| Full hydraulic | | |
| 10,000 lb | EA. | 34.286 |
| 13,000 lb | " | 60.000 |
| 18,500 lb | " | 60.000 |
| 24,000 lb | " | 60.000 |
| 26,000 lb | " | 60.000 |
| Pneumatic hoist, fully hydraulic | | |
| 11,000 lb | EA. | 80.000 |
| 24,000 lb | " | 80.000 |

## MATERIAL HANDLING

| 14560.10 CHUTES | UNIT | MAN/HOURS |
|---|---|---|
| Linen chutes, stainless steel, with supports | | |
| 18" dia. | L.F. | 0.057 |
| 24" dia. | " | 0.062 |
| 30" dia. | " | 0.067 |
| Hopper | EA. | 0.533 |
| Skylight | " | 0.800 |

## MATERIAL HANDLING

| 14560.10 CHUTES | UNIT | MAN/HOURS |
|---|---|---|
| Sprinkler unit at top | EA. | 0.889 |

| 14580.10 PNEUMATIC SYSTEMS | UNIT | MAN/HOURS |
|---|---|---|
| Pneumatic message tube system | | |
| Average, 20 station job | | |
| 3" round system | E.A. | 72.727 |
| 4" round system | " | 80.000 |
| 6" round system | " | 88.889 |
| 4" x 7" oval system | " | 160 |
| Trash and linen tube system | | |
| 10 stations | EA. | 120 |
| 15 stations | " | 160 |
| 20 stations | " | 185 |
| 30 stations | " | 218 |

## HOISTS AND CRANES

| 14600.10 INDUSTRIAL HOISTS | UNIT | MAN/HOURS |
|---|---|---|
| Industrial hoists, electric, light to medium duty | | |
| 500 lb | EA. | 4.000 |
| 1000 lb | " | 4.211 |
| 2000 lb | " | 4.444 |
| 3000 lb | " | 4.706 |
| 4000 lb | " | 5.000 |
| 5000 lb | " | 5.333 |
| 6000 lb | " | 5.517 |
| 7500 lb | " | 5.714 |
| 10,000 lb | " | 5.926 |
| 15,000 lb | " | 6.154 |
| 20,000 lb | " | 6.667 |
| 25,000 lb | " | 7.273 |
| 30,000 lb | " | 8.000 |
| Heavy duty | | |
| 500 lb | EA. | 4.000 |
| 1000 lb | " | 4.211 |
| 2000 lb | " | 4.444 |
| 3000 lb | " | 4.706 |
| 4000 lb | " | 5.000 |
| 5000 lb | " | 5.333 |
| 6000 lb | " | 5.517 |
| 7500 lb | " | 5.714 |
| 10,000 lb | " | 5.926 |
| 15,000 lb | " | 6.154 |
| 20,000 lb | " | 6.667 |

| HOISTS AND CRANES | UNIT | MAN/ HOURS |
|---|---|---|
| **14600.10 INDUSTRIAL HOISTS** | | |
| 25,000 lb | EA. | 7.273 |
| 30,000 lb | " | 8.000 |
| Air powered hoists | | |
| 500 lb | EA. | 4.000 |
| 1000 lb | " | 4.000 |
| 2000 lb | " | 4.211 |
| 4000 lb | " | 4.706 |
| 6000 lb | " | 6.154 |
| Overhead traveling bridge crane | | |
| Single girder, 20' span | | |
| 3 ton | EA. | 12.000 |
| 5 ton | " | 12.000 |
| 7.5 ton | " | 12.000 |
| 10 ton | " | 15.000 |
| 15 ton | " | 15.000 |
| 30' span | | |
| 3 ton | EA. | 12.000 |
| 5 ton | " | 12.000 |
| 10 ton | " | 15.000 |
| 15 ton | " | 15.000 |
| Double girder, 40' span | | |
| 3 ton | EA. | 26.667 |
| 5 ton | " | 26.667 |
| 7.5 ton | " | 26.667 |
| 10 ton | " | 34.286 |
| 15 ton | " | 34.286 |
| 25 ton | " | 34.286 |
| 50' span | | |
| 3 ton | EA. | 26.667 |
| 5 ton | " | 26.667 |
| 7.5 ton | " | 26.667 |
| 10 ton | " | 34.286 |
| 15 ton | " | 34.286 |
| 25 ton | " | 34.286 |
| **14650.10 JIB CRANES** | | |
| Self supporting, swinging 8' boom, 200 deg rotation | | |
| 1000 lb | EA. | 6.667 |
| 2000 lb | " | 6.667 |
| 3000 lb | " | 13.333 |
| 4000 lb | " | 13.333 |
| 6000 lb | " | 13.333 |
| 10,000 lb | " | 13.333 |
| Wall mounted, 180 deg rotation | | |
| 2000 lb | EA. | 6.667 |
| 3000 lb | " | 6.667 |
| 4000 lb | " | 13.333 |
| 6000 lb | " | 13.333 |
| 10,000 lb | " | 13.333 |

| BASIC MATERIALS | UNIT | MAN/ HOURS |
|---|---|---|
| **15100.10 SPECIALTIES** | | |
| Wall penetration | | |
| Concrete wall, 6" thick | | |
| 2" dia. | EA. | 0.267 |
| 4" dia. | " | 0.400 |
| 8" dia. | " | 0.571 |
| 12" thick | | |
| 2" dia. | EA. | 0.364 |
| 4" dia. | " | 0.571 |
| 8" dia. | " | 0.889 |
| Non-destructive testing, piping systems | | |
| X-ray of welds | | |
| 3" dia. pipe | EA. | 0.800 |
| 4" dia. pipe | " | 0.800 |
| 6" dia. pipe | " | 0.800 |
| 8" dia. pipe | " | 1.000 |
| 10" dia. pipe | " | 1.000 |
| Liquid penetration of welds | | |
| 2" dia. pipe | EA. | 0.500 |
| 3" dia. pipe | " | 0.500 |
| 4" dia. pipe | " | 0.500 |
| 6" dia. pipe | " | 0.500 |
| 8" dia. pipe | " | 0.500 |
| 10" dia. pipe | " | 0.500 |
| **15120.10 BACKFLOW PREVENTERS** | | |
| Backflow preventer, flanged, cast iron, with valves | | |
| 3" pipe | EA. | 4.000 |
| 4" pipe | " | 4.444 |
| 6" pipe | " | 6.667 |
| 8" pipe | " | 8.000 |
| Threaded | | |
| 3/4" pipe | EA. | 0.500 |
| 2" pipe | " | 0.800 |
| Reduced pressure assembly, bronze, threaded | | |
| 3/4" | EA. | 0.500 |
| 1" | " | 0.571 |
| 1-1/4" | " | 0.667 |
| 1-1/2" | " | 0.800 |
| **15140.10 PIPE HANGERS, HEAVY** | | |
| Hangers | | |
| 1/2" pipe, clevis pipe hanger | | |
| Black steel | EA. | 0.267 |
| Galvanized | " | 0.267 |
| 3/4" pipe, clevis pipe hanger | | |
| Black steel | EA. | 0.267 |
| Galvanized | " | 0.267 |
| 1" pipe, clevis pipe hanger | | |
| Black steel | EA. | 0.267 |
| Galvanized | " | 0.267 |
| U bolt | " | 0.080 |
| 2" pipe, clevis pipe hanger | | |

| BASIC MATERIALS | UNIT | MAN/ HOURS |
|---|---|---|
| **15140.10 PIPE HANGERS, HEAVY** | | |
| Black steel | EA. | 0.267 |
| Galvanized | " | 0.267 |
| 3" pipe, clevis pipe hanger | | |
| Black steel | EA. | 0.267 |
| Galvanized | " | 0.267 |
| 4" pipe, clevis pipe hanger | | |
| Black steel | EA. | 0.267 |
| Galvanized | " | 0.267 |
| 6" pipe, clevis pipe hanger | | |
| Black steel | EA. | 0.320 |
| Galvanized | " | 0.320 |
| 12" pipe, clevis pipe hanger | | |
| Black steel | EA. | 0.320 |
| Galvanized | " | 0.320 |
| Threaded rod, galvanized | | |
| C-clamp, steel, with lock nut | | |
| 3/8" | EA. | 0.100 |
| 1/2" | " | 0.100 |
| **15140.11 PIPE HANGERS, LIGHT** | | |
| A band, black iron | | |
| 1/2" | EA. | 0.057 |
| 1" | " | 0.059 |
| 1-1/4" | " | 0.062 |
| 1-1/2" | " | 0.067 |
| 2" | " | 0.073 |
| 2-1/2" | " | 0.080 |
| 3" | " | 0.089 |
| 4" | " | 0.100 |
| 5" | " | 0.107 |
| 6" | " | 0.114 |
| 8" | " | 0.133 |
| Copper | | |
| 1/2" | EA. | 0.057 |
| 3/4" | " | 0.059 |
| 1" | " | 0.059 |
| 1-1/4" | " | 0.062 |
| 1-1/2" | " | 0.067 |
| 2" | " | 0.073 |
| 2-1/2" | " | 0.080 |
| 3" | " | 0.089 |
| 4" | " | 0.100 |
| 2 hole clips, galvanized | | |
| 3/4" | EA. | 0.053 |
| 1" | " | 0.055 |
| 1-1/4" | " | 0.057 |
| 1-1/2" | " | 0.059 |
| 2" | " | 0.062 |
| 2-1/2" | " | 0.064 |
| 3" | " | 0.067 |
| 4" | " | 0.073 |
| Perforated strap | | |
| 3/4" | | |
| Galvanized, 20 ga. | L.F. | 0.040 |

| BASIC MATERIALS | UNIT | MAN/HOURS |
|---|---|---|
| **15140.11 PIPE HANGERS, LIGHT** | | |
| Copper, 22 ga. | L.F. | 0.040 |
| J-Hooks | | |
| 1/2" | EA. | 0.036 |
| 3/4" | " | 0.036 |
| 1" | " | 0.038 |
| 1-1/4" | " | 0.039 |
| 1-1/2" | " | 0.040 |
| 2" | " | 0.040 |
| 3" | " | 0.042 |
| 4" | " | 0.042 |
| PVC coated hangers, galvanized, 28 ga. | | |
| 1-1/2" x 12" | EA. | 0.053 |
| 2" x 12" | " | 0.057 |
| 3" x 12" | " | 0.062 |
| 4" x 12" | " | 0.067 |
| Copper, 30 ga. | | |
| 1-1/2" x 12" | EA. | 0.053 |
| 2" x 12" | " | 0.057 |
| 3" x 12" | " | 0.062 |
| 4" x 12" | " | 0.067 |
| Wire hook hangers | | |
| Black wire, 1/2" x | | |
| 4" | EA. | 0.040 |
| 6" | " | 0.042 |
| Copper wire hooks | | |
| 1/2" x | | |
| 4" | EA. | 0.040 |
| 6" | " | 0.042 |

| INSULATION | UNIT | MAN/HOURS |
|---|---|---|
| **15260.10 FIBERGLASS PIPE INSULATION** | | |
| Fiberglass insulation on 1/2" pipe | | |
| 1" thick | L.F. | 0.027 |
| 1-1/2" thick | " | 0.033 |
| 3/4" pipe | | |
| 1" thick | L.F. | 0.027 |
| 1-1/2" thick | " | 0.033 |
| 1" pipe | | |
| 1" thick | L.F. | 0.027 |
| 1-1/2" thick | " | 0.033 |
| 2" pipe | | |
| 1" thick | L.F. | 0.033 |
| 1-1/2" thick | " | 0.036 |
| 3" pipe | | |
| 1" thick | L.F. | 0.038 |
| 1-1/2" thick | " | 0.040 |

| INSULATION | UNIT | MAN/HOURS |
|---|---|---|
| **15260.10 FIBERGLASS PIPE INSULATION** | | |
| 6" pipe | | |
| 1" thick | L.F. | 0.042 |
| 2" thick | " | 0.044 |
| 10" pipe | | |
| 2" thick | L.F. | 0.042 |
| 3" thick | " | 0.044 |
| **15260.60 EXTERIOR PIPE INSULATION** | | |
| Fiberglass insulation, aluminum jacket | | |
| 1/2" pipe | | |
| 1" thick | L.F. | 0.062 |
| 1-1/2" thick | " | 0.067 |
| 3/4" pipe | | |
| 1" thick | L.F. | 0.062 |
| 1-1/2" thick | " | 0.067 |
| 1" pipe | | |
| 1" thick | L.F. | 0.062 |
| 1-1/2" thick | " | 0.067 |
| 2" pipe | | |
| 1" thick | L.F. | 0.073 |
| 1-1/2" thick | " | 0.076 |
| 3" pipe | | |
| 1" thick | L.F. | 0.080 |
| 1-1/2" thick | " | 0.084 |
| 6" pipe | | |
| 1" thick | L.F. | 0.089 |
| 2" thick | " | 0.094 |
| 10" pipe | | |
| 2" thick | L.F. | 0.089 |
| 3" thick | " | 0.094 |
| **15260.90 PIPE INSULATION FITTINGS** | | |
| Insulation protection saddle | | |
| 1" thick covering | | |
| 1/2" pipe | EA. | 0.320 |
| 3/4" pipe | " | 0.320 |
| 1" pipe | " | 0.320 |
| 2" pipe | " | 0.320 |
| 3" pipe | " | 0.364 |
| 6" pipe | " | 0.500 |
| 1-1/2" thick covering | | |
| 3/4" pipe | EA. | 0.320 |
| 1" pipe | " | 0.320 |
| 2" pipe | " | 0.320 |
| 3" pipe | " | 0.320 |
| 6" pipe | " | 0.500 |
| 10" pipe | " | 0.667 |
| **15280.10 EQUIPMENT INSULATION** | | |
| Equipment insulation, 2" thick, cellular glass | S.F. | 0.050 |
| Urethane, rigid, field applied jacket, plastered finish | " | 0.100 |
| Fiberglass, rigid, with vapor barrier | " | 0.044 |

## INSULATION

| 15290.10 DUCTWORK INSULATION | UNIT | MAN/HOURS |
|---|---|---|
| Fiberglass duct insulation, plain blanket | | |
| 1-1/2" thick | S.F. | 0.010 |
| 2" thick | " | 0.013 |
| With vapor barrier | | |
| 1-1/2" thick | S.F. | 0.010 |
| 2" thick | " | 0.013 |
| Rigid with vapor barrier | | |
| 2" thick | S.F. | 0.027 |
| 3" thick | " | 0.032 |
| 4" thick | " | 0.040 |
| 6" thick | " | 0.053 |
| Weatherproof, polystyrene, 3" thick, w/vapor barrier | " | 0.080 |
| Urethane board with vapor barrier | " | 0.100 |

## FIRE PROTECTION

| 15330.10 WET SPRINKLER SYSTEM | UNIT | MAN/HOURS |
|---|---|---|
| Sprinkler head, 212 deg, brass, exposed piping | EA. | 0.320 |
| Chrome, concealed piping | " | 0.444 |
| Water motor alarm | " | 1.333 |
| Fire department inlet connection | " | 1.600 |
| Wall plate for fire dept connection | " | 0.667 |
| Swing check valve flanged iron body, 4" | " | 2.667 |
| Check valve, 6" | " | 4.000 |
| Wet pipe valve, flange to groove, 4" | " | 0.889 |
| Flange to flange | | |
| 6" | EA. | 1.333 |
| 8" | " | 2.667 |
| Alarm valve, flange to flange, (wet valve) | | |
| 4" | EA. | 0.889 |
| 8" | " | 6.667 |
| Inspector's test connection | " | 0.667 |
| Wall hydrant, polished brass, 2-1/2" x 2-1/2", single | " | 0.571 |
| 2-way | " | 0.571 |
| 3-way | " | 0.571 |
| Wet valve trim, includes retard chamber & gauges, 4"-6" | " | 0.667 |
| Retard pressure switch for wet systems | " | 1.600 |
| Air maintenance device | " | 0.667 |
| Wall hydrant non-freeze, 8" thick wall, vacuum breaker | " | 0.400 |
| 12" thick wall | " | 0.400 |

| 15330.50 DRY SPRINKLER SYSTEM | UNIT | MAN/HOURS |
|---|---|---|
| Dry pipe valve, flange to flange | | |
| 4" | EA. | 1.600 |
| 6" | " | 2.000 |

## FIRE PROTECTION

| 15330.50 DRY SPRINKLER SYSTEM | UNIT | MAN/HOURS |
|---|---|---|
| Trim, 4" and 6", includes gauges | EA. | 0.667 |
| Field testing and flushing | " | 6.667 |
| Disinfection | " | 6.667 |
| Pressure switch double circuit, open/close contacts | " | 2.000 |
| Low air | | |
| Supervisory unit | EA. | 1.333 |
| Pressure switch | " | 0.667 |
| Halon type fire protection system, per computer room | | |

## PLUMBING

| 15410.05 C.I. PIPE, ABOVE GROUND | UNIT | MAN/HOURS |
|---|---|---|
| No hub pipe | | |
| 1-1/2" pipe | L.F. | 0.057 |
| 2" pipe | " | 0.067 |
| 3" pipe | " | 0.080 |
| 4" pipe | " | 0.133 |
| 6" pipe | " | 0.160 |
| 8" pipe | " | 0.267 |
| 10" pipe | " | 0.320 |
| No hub fittings, 1-1/2" pipe | | |
| 1/4 bend | EA. | 0.267 |
| 1/8 bend | " | 0.267 |
| Sanitary tee | " | 0.400 |
| Wye | " | 0.400 |
| 2" pipe | | |
| 1/4 bend | EA. | 0.320 |
| 1/8 bend | " | 0.320 |
| Sanitary tee | " | 0.533 |
| Wye | " | 0.667 |
| 3" pipe | | |
| 1/4 bend | EA. | 0.400 |
| 1/8 bend | " | 0.400 |
| Sanitary tee | " | 0.500 |
| Wye | " | 0.667 |
| 4" pipe | | |
| 1/4 bend | EA. | 0.400 |
| 1/8 bend | " | 0.400 |
| Sanitary tee | " | 0.667 |
| Wye | " | 0.667 |
| 8" deep | " | 0.400 |
| 6" pipe | | |
| 1/4 bend | EA. | 0.667 |
| 1/8 bend | " | 0.667 |
| Sanitary tee | " | 0.800 |
| Wye | " | 0.800 |
| 8" pipe | | |

| PLUMBING | UNIT | MAN/HOURS |
|---|---|---|
| **15410.05 C.I. PIPE, ABOVE GROUND** | | |
| 1/4 bend | EA. | 0.667 |
| 1/8 bend | " | 0.667 |
| Sanitary tee | " | 1.000 |
| Wye | " | 0.800 |
| 10" pipe | | |
| 1/4 bend | EA. | 0.667 |
| 1/8 bend | " | 0.667 |
| Wye | " | 1.333 |
| 10x6" wye | " | 1.333 |
| **15410.06 C.I. PIPE, BELOW GROUND** | | |
| No hub pipe | | |
| 1-1/2" pipe | L.F. | 0.040 |
| 2" pipe | " | 0.044 |
| 3" pipe | " | 0.050 |
| 4" pipe | " | 0.067 |
| 6" pipe | " | 0.073 |
| 8" pipe | " | 0.089 |
| 10" pipe | " | 0.100 |
| Fittings, 1-1/2" | | |
| 1/4 bend | EA. | 0.229 |
| 1/8 bend | " | 0.229 |
| Wye | " | 0.320 |
| 2" | | |
| 1/4 bend | EA. | 0.267 |
| 1/8 bend | " | 0.267 |
| 3" | | |
| 1/4 bend | EA. | 0.320 |
| 1/8 bend | " | 0.320 |
| Wye | " | 0.500 |
| 4" | | |
| 1/4 bend | EA. | 0.320 |
| 1/8 bend | " | 0.320 |
| Wye | " | 0.500 |
| 6" | | |
| 1/4 bend | EA. | 0.500 |
| 1/8 bend | " | 0.500 |
| 8" | | |
| 1/4 bend | EA. | 0.500 |
| 1/8 bend | " | 0.500 |
| Wye | " | 0.667 |
| 10" | | |
| 1/4 bend | EA. | 0.500 |
| 1/8 bend | " | 0.500 |
| Wye | " | 1.000 |
| **15410.09 SERVICE WEIGHT PIPE** | | |
| Service weight pipe, single hub | | |
| 3" x 5' | EA. | 0.170 |
| 4" x 5' | " | 0.178 |
| 6" x 5' | " | 0.200 |
| 1/8 bend | | |
| 3" | EA. | 0.320 |

| PLUMBING | UNIT | MAN/HOURS |
|---|---|---|
| **15410.09 SERVICE WEIGHT PIPE** | | |
| 4" | EA. | 0.364 |
| 6" | " | 0.400 |
| 1/4 bend | | |
| 3" | EA. | 0.320 |
| 4" | " | 0.364 |
| 6" | " | 0.400 |
| Sweep | | |
| 3" | EA. | 0.320 |
| 4" | " | 0.364 |
| 6" | " | 0.400 |
| Sanitary T | | |
| 3" | EA. | 0.571 |
| 4" | " | 0.667 |
| 6" | " | 0.727 |
| Wye | | |
| 3" | EA. | 0.444 |
| 4" | " | 0.471 |
| 6" | " | 0.571 |
| **15410.10 COPPER PIPE** | | |
| Type "K" copper | | |
| 1/2" | L.F. | 0.025 |
| 3/4" | " | 0.027 |
| 1" | " | 0.029 |
| 1-1/4" | " | 0.031 |
| 1-1/2" | " | 0.033 |
| 2" | " | 0.036 |
| 2-1/2" | " | 0.040 |
| 3" | " | 0.042 |
| 4" | " | 0.044 |
| DWV, copper | | |
| 1-1/4" | L.F. | 0.033 |
| 1-1/2" | " | 0.036 |
| 2" | " | 0.040 |
| 3" | " | 0.044 |
| 4" | " | 0.050 |
| 6" | " | 0.057 |
| Refrigeration tubing, copper, sealed | | |
| 1/8" | L.F. | 0.032 |
| 3/16" | " | 0.033 |
| 1/4" | " | 0.035 |
| 5/16" | " | 0.036 |
| 3/8" | " | 0.038 |
| 1/2" | " | 0.040 |
| 7/8" | " | 0.046 |
| 1-1/8" | " | 0.053 |
| 1-3/8" | " | 0.062 |
| Type "L" copper | | |
| 1/4" | L.F. | 0.024 |
| 3/8" | " | 0.024 |
| 1/2" | " | 0.025 |
| 3/4" | " | 0.027 |
| 1" | " | 0.029 |
| 1-1/4" | " | 0.031 |

| PLUMBING | UNIT | MAN/ HOURS |
|---|---|---|
| **15410.10 COPPER PIPE** | | |
| 1-1/2" | L.F. | 0.033 |
| 2" | " | 0.036 |
| Type "M" copper | | |
| 1/2" | L.F. | 0.025 |
| 3/4" | " | 0.027 |
| 1" | " | 0.029 |
| 1-1/4" | " | 0.031 |
| 2" | " | 0.036 |
| **15410.11 COPPER FITTINGS** | | |
| Coupling, with stop | | |
| 1/4" | EA. | 0.267 |
| 3/8" | " | 0.320 |
| 1/2" | " | 0.348 |
| 5/8" | " | 0.400 |
| 3/4" | " | 0.444 |
| 1" | " | 0.471 |
| 3" | " | 0.800 |
| 4" | " | 1.000 |
| Reducing coupling | | |
| 1/4" x 1/8" | EA. | 0.320 |
| 3/8" x 1/4" | " | 0.348 |
| 1/2" x | | |
| 3/8" | EA. | 0.400 |
| 1/4" | " | 0.400 |
| 1/8" | " | 0.400 |
| 3/4" x | | |
| 3/8" | EA. | 0.444 |
| 1/2" | " | 0.444 |
| 1" x | | |
| 3/8" | EA. | 0.500 |
| 1" x 1/2" | " | 0.500 |
| 1" x 3/4" | " | 0.500 |
| 1-1/4" x | | |
| 1/2" | EA. | 0.533 |
| 3/4" | " | 0.533 |
| 1" | " | 0.533 |
| 1-1/2" x | | |
| 1/2" | EA. | 0.571 |
| 3/4" | " | 0.571 |
| 1" | " | 0.571 |
| 1-1/4" | " | 0.571 |
| 2" x | | |
| 1/2" | EA. | 0.667 |
| 3/4" | " | 0.667 |
| 1" | " | 0.667 |
| 1-1/4" | " | 0.667 |
| 1-1/2" | " | 0.667 |
| 2-1/2" x | | |
| 1" | EA. | 0.800 |
| 1-1/4" | " | 0.800 |
| 1-1/2" | " | 0.800 |
| 2" | " | 0.800 |
| 3" x | | |

| PLUMBING | UNIT | MAN/ HOURS |
|---|---|---|
| **15410.11 COPPER FITTINGS** | | |
| 1-1/2" | EA. | 1.000 |
| 2" | " | 1.000 |
| 2-1/2" | " | 1.000 |
| 4" x | | |
| 2" | EA. | 1.143 |
| 2-1/2" | " | 1.143 |
| 3" | " | 1.143 |
| Slip coupling | | |
| 1/4" | EA. | 0.267 |
| 1/2" | " | 0.320 |
| 3/4" | " | 0.400 |
| 1" | " | 0.444 |
| 1-1/4" | " | 0.500 |
| 1-1/2" | " | 0.533 |
| 2" | " | 0.667 |
| 2-1/2" | " | 0.667 |
| 3" | " | 0.800 |
| 4" | " | 1.000 |
| Coupling with drain | | |
| 1/2" | EA. | 0.400 |
| 3/4" | " | 0.444 |
| 1" | " | 0.500 |
| Reducer | | |
| 3/8" x 1/4" | EA. | 0.320 |
| 1/2" x 3/8" | " | 0.320 |
| 3/4" x | | |
| 1/4" | EA. | 0.364 |
| 3/8" | " | 0.364 |
| 1/2" | " | 0.364 |
| 1" x | | |
| 1/2" | EA. | 0.400 |
| 3/4" | " | 0.400 |
| 1-1/4" x | | |
| 1/2" | EA. | 0.444 |
| 3/4" | " | 0.444 |
| 1" | " | 0.444 |
| 1-1/2" x | | |
| 1/2" | EA. | 0.500 |
| 3/4" | " | 0.500 |
| 1" | " | 0.500 |
| 1-1/4" | " | 0.500 |
| 2" x | | |
| 1/2" | EA. | 0.571 |
| 3/4" | " | 0.571 |
| 1" | " | 0.571 |
| 1-1/4" | " | 0.571 |
| 1-1/2" | " | 0.571 |
| 2-1/2" x | | |
| 1" | EA. | 0.667 |
| 1-1/4" | " | 0.667 |
| 1-1/2" | " | 0.667 |
| 2" | " | 0.667 |
| 3" x | | |
| 1-1/4" | EA. | 0.800 |

| PLUMBING | UNIT | MAN/ HOURS |
|---|---|---|
| **15410.11 COPPER FITTINGS** | | |
| 1-1/2" | EA. | 0.800 |
| 2" | " | 0.800 |
| 2-1/2" | " | 0.800 |
| 4" x | | |
| 2" | EA. | 1.000 |
| 3" | " | 1.000 |
| Female adapters | | |
| 1/4" | EA. | 0.320 |
| 3/8" | " | 0.364 |
| 1/2" | " | 0.400 |
| 3/4" | " | 0.444 |
| 1" | " | 0.444 |
| 1-1/4" | " | 0.500 |
| 1-1/2" | " | 0.500 |
| 2" | " | 0.533 |
| 2-1/2" | " | 0.571 |
| 3" | " | 0.667 |
| 4" | " | 0.800 |
| Increasing female adapters | | |
| 1/8" x | | |
| 3/8" | EA. | 0.320 |
| 1/2" | " | 0.320 |
| 1/4" x 1/2" | " | 0.348 |
| 3/8" x 1/2" | " | 0.364 |
| 1/2" X | | |
| 3/4" | EA. | 0.400 |
| 1" | " | 0.400 |
| 3/4" X | | |
| 1" | EA. | 0.444 |
| 1-1/4" | " | 0.444 |
| 1" x | | |
| 1-1/4" | EA. | 0.444 |
| 1-1/2" | " | 0.444 |
| 1-1/4" x | | |
| 1-1/2" | EA. | 0.500 |
| 2" | " | 0.500 |
| 1-1/2" x 2" | " | 0.533 |
| Reducing female adapters | | |
| 3/8" x 1/4" | EA. | 0.364 |
| 1/2" x | | |
| 1/4" | EA. | 0.400 |
| 3/8" | " | 0.400 |
| 3/4" x 1/2" | " | 0.444 |
| 1" x | | |
| 1/2" | EA. | 0.444 |
| 3/4" | " | 0.444 |
| 1-1/4" x | | |
| 1/2" | EA. | 0.500 |
| 3/4" | " | 0.500 |
| 1" | " | 0.500 |
| 1-1/2" x | | |
| 1" | EA. | 0.533 |
| 1-1/4" | " | 0.533 |
| 2" x | | |

| PLUMBING | UNIT | MAN/ HOURS |
|---|---|---|
| **15410.11 COPPER FITTINGS** | | |
| 1" | EA. | 0.571 |
| 1-1/4" | " | 0.571 |
| 1-1/2" | " | 0.571 |
| Female fitting adapters | | |
| 1/2" | EA. | 0.400 |
| 3/4" | " | 0.400 |
| 3/4" x 1/2" | " | 0.421 |
| 1" | " | 0.444 |
| 1-1/4" | " | 0.471 |
| 1-1/2" | " | 0.500 |
| 2" | " | 0.533 |
| Male adapters | | |
| 1/4" | EA. | 0.364 |
| 3/8" | " | 0.364 |
| 3" | " | 0.667 |
| 4" | " | 0.800 |
| Increasing male adapters | | |
| 3/8" x 1/2" | EA. | 0.364 |
| 1/2" x | | |
| 3/4" | EA. | 0.400 |
| 1" | " | 0.400 |
| 3/4" x | | |
| 1" | EA. | 0.421 |
| 1-1/4" | " | 0.421 |
| 1" x 1-1/4" | " | 0.444 |
| 1-1/2" x | | |
| 3/4" | EA. | 0.471 |
| 1" | " | 0.471 |
| 1-1/4" | " | 0.471 |
| 2" x | | |
| 1" | EA. | 0.500 |
| 1-1/4" | " | 0.500 |
| 1-1/2" | " | 0.500 |
| 2" x 2-1/2" | " | 0.533 |
| Copper pipe fittings | | |
| 1/2" | | |
| 90 deg ell | EA. | 0.178 |
| 45 deg ell | " | 0.178 |
| Tee | " | 0.229 |
| Cap | " | 0.089 |
| Coupling | " | 0.178 |
| Union | " | 0.200 |
| 3/4" | | |
| 90 deg ell | EA. | 0.200 |
| 45 deg ell | " | 0.200 |
| Tee | " | 0.267 |
| Cap | " | 0.094 |
| Coupling | " | 0.200 |
| Union | " | 0.229 |
| 1" | | |
| 90 deg ell | EA. | 0.267 |
| 45 deg ell | " | 0.267 |
| Tee | " | 0.320 |
| Cap | " | 0.133 |

| PLUMBING | UNIT | MAN/HOURS |
|---|---|---|
| **15410.11 COPPER FITTINGS** | | |
| Coupling | EA. | 0.267 |
| Union | " | 0.267 |
| 1-1/4" | | |
| 90 deg ell | EA. | 0.229 |
| 45 deg ell | " | 0.229 |
| Tee | " | 0.400 |
| Cap | " | 0.133 |
| Union | " | 0.286 |
| 1-1/2" | | |
| 90 deg ell | EA. | 0.286 |
| 45 deg ell | " | 0.286 |
| Tee | " | 0.444 |
| Cap | " | 0.133 |
| Coupling | " | 0.267 |
| Union | " | 0.364 |
| 2" | | |
| 90 deg ell | EA. | 0.320 |
| 45 deg ell | " | 0.500 |
| Tee | " | 0.500 |
| Cap | " | 0.160 |
| Coupling | " | 0.320 |
| Union | " | 0.400 |
| 2-1/2" | | |
| 90 deg ell | EA. | 0.400 |
| 45 deg ell | " | 0.400 |
| Tee | " | 0.571 |
| Cap | " | 0.200 |
| Coupling | " | 0.400 |
| Union | " | 0.444 |
| **15410.18 GLASS PIPE** | | |
| Glass pipe | | |
| 1-1/2" dia. | L.F. | 0.160 |
| 2" dia. | " | 0.178 |
| 3" dia. | " | 0.200 |
| 4" dia. | " | 0.229 |
| 6" dia. | " | 0.267 |
| **15410.30 PVC/CPVC PIPE** | | |
| PVC schedule 40 | | |
| 1/2" pipe | L.F. | 0.033 |
| 3/4" pipe | " | 0.036 |
| 1" pipe | " | 0.040 |
| 1-1/4" pipe | " | 0.044 |
| 1-1/2" pipe | " | 0.050 |
| 2" pipe | " | 0.057 |
| 2-1/2" pipe | " | 0.067 |
| 3" pipe | " | 0.080 |
| 4" pipe | " | 0.100 |
| 6" pipe | " | 0.200 |
| 8" pipe | " | 0.267 |
| PVC schedule 80 pipe | | |
| 1-1/2" pipe | L.F. | 0.050 |

| PLUMBING | UNIT | MAN/HOURS |
|---|---|---|
| **15410.30 PVC/CPVC PIPE** | | |
| 2" pipe | L.F. | 0.057 |
| 3" pipe | " | 0.080 |
| 4" pipe | " | 0.100 |
| Polypropylene, acid resistant, DWV pipe | | |
| Schedule 40 | | |
| 1-1/2" pipe | L.F. | 0.057 |
| 2" pipe | " | 0.067 |
| 3" pipe | " | 0.080 |
| 4" pipe | " | 0.100 |
| 6" pipe | " | 0.200 |
| Polyethylene pipe and fittings | | |
| SDR-21 | | |
| 3" pipe | L.F. | 0.100 |
| 4" pipe | " | 0.133 |
| 6" pipe | " | 0.200 |
| 8" pipe | " | 0.229 |
| 10" pipe | " | 0.267 |
| **15410.33 ABS DWV PIPE** | | |
| Schedule 40 ABS | | |
| 1-1/2" pipe | L.F. | 0.040 |
| 2" pipe | " | 0.044 |
| 3" pipe | " | 0.057 |
| 4" pipe | " | 0.080 |
| 6" pipe | " | 0.100 |
| **15410.35 PLASTIC PIPE** | | |
| Fiberglass reinforced pipe | | |
| 2" pipe | L.F. | 0.062 |
| 3" pipe | " | 0.067 |
| 4" pipe | " | 0.073 |
| 6" pipe | " | 0.080 |
| 8" pipe | " | 0.133 |
| 10" pipe | " | 0.160 |
| 12" pipe | " | 0.200 |
| **15410.70 STAINLESS STEEL PIPE** | | |
| Stainless steel, schedule 40, threaded | | |
| 1/2" pipe | L.F. | 0.114 |
| 3/4" pipe | " | 0.118 |
| 1" pipe | " | 0.123 |
| 1-1/2" pipe | " | 0.133 |
| 2" pipe | " | 0.145 |
| 2-1/2" pipe | " | 0.160 |
| 3" pipe | " | 0.178 |
| 4" pipe | " | 0.200 |
| **15410.80 STEEL PIPE** | | |
| Black steel, extra heavy pipe, threaded | | |
| 1/2" pipe | L.F. | 0.032 |
| 3/4" pipe | " | 0.032 |
| 1" pipe | " | 0.040 |

| PLUMBING | UNIT | MAN/ HOURS |
|---|---|---|
| **15410.80 STEEL PIPE** | | |
| 1-1/2" pipe | L.F. | 0.044 |
| 2-1/2" pipe | " | 0.100 |
| 3" pipe | " | 0.133 |
| 4" pipe | " | 0.160 |
| 5" pipe | " | 0.200 |
| 6" pipe | " | 0.200 |
| 8" pipe | " | 0.267 |
| 10" pipe | " | 0.320 |
| 12" pipe | " | 0.400 |
| Fittings, malleable iron, threaded, 1/2" pipe | | |
| 90 deg ell | EA. | 0.267 |
| 45 deg ell | " | 0.267 |
| Tee | " | 0.400 |
| 3/4" pipe | | |
| 90 deg ell | EA. | 0.267 |
| 45 deg ell | " | 0.400 |
| Tee | " | 0.400 |
| 1" pipe | | |
| 90 deg ell | EA. | 0.320 |
| 45 deg ell | " | 0.320 |
| Tee | " | 0.444 |
| 1-1/2" pipe | | |
| 90 deg ell | EA. | 0.400 |
| 45 deg ell | " | 0.400 |
| Tee | " | 0.571 |
| 2-1/2" pipe | | |
| 90 deg ell | EA. | 1.000 |
| 45 deg ell | " | 1.000 |
| Tee | " | 1.333 |
| 3" pipe | | |
| 90 deg ell | EA. | 1.333 |
| 45 deg ell | " | 1.333 |
| Tee | " | 2.000 |
| 4" pipe | | |
| 90 deg ell | EA. | 1.600 |
| 45 deg ell | " | 1.600 |
| Tee | " | 2.667 |
| 6" pipe | | |
| 90 deg ell | EA. | 1.600 |
| 45 deg ell | " | 1.600 |
| Tee | " | 2.667 |
| 8" pipe | | |
| 90 deg ell | EA. | 3.200 |
| 45 deg ell | " | 3.200 |
| Tee | " | 5.000 |
| 10" pipe | | |
| 90 deg ell | EA. | 4.000 |
| 45 deg ell | " | 4.000 |
| Tee | " | 5.000 |
| 12" pipe | | |
| 90 deg ell | EA. | 5.000 |
| 45 deg ell | " | 5.000 |
| Tee | " | 6.667 |

| PLUMBING | UNIT | MAN/ HOURS |
|---|---|---|
| **15410.82 GALVANIZED STEEL PIPE** | | |
| Galvanized pipe | | |
| 1/2" pipe | L.F. | 0.080 |
| 3/4" pipe | " | 0.100 |
| 1" pipe | " | 0.114 |
| 1-1/4" pipe | " | 0.133 |
| 1-1/2" pipe | " | 0.160 |
| 2" pipe | " | 0.200 |
| 2-1/2" pipe | " | 0.267 |
| 3" pipe | " | 0.286 |
| 4" pipe | " | 0.333 |
| 6" pipe | " | 0.667 |
| **15430.23 CLEANOUTS** | | |
| Cleanout, wall | | |
| 2" | EA. | 0.533 |
| 3" | " | 0.533 |
| 4" | " | 0.667 |
| 6" | " | 0.800 |
| 8" | " | 1.000 |
| Floor | | |
| 2" | EA. | 0.667 |
| 3" | " | 0.667 |
| 4" | " | 0.800 |
| 6" | " | 1.000 |
| 8" | " | 1.143 |
| **15430.24 GREASE TRAPS** | | |
| Grease traps, cast iron, 3" pipe | | |
| 35 gpm, 70 lb capacity | EA. | 8.000 |
| 50 gpm, 100 lb capacity | " | 10.000 |
| **15430.25 HOSE BIBBS** | | |
| Hose bibb | | |
| 1/2" | EA. | 0.267 |
| 3/4" | " | 0.267 |
| **15430.60 VALVES** | | |
| Gate valve, 125 lb, bronze, soldered | | |
| 1/2" | EA. | 0.200 |
| 3/4" | " | 0.200 |
| 1" | " | 0.267 |
| 1-1/2" | " | 0.320 |
| 2" | " | 0.400 |
| 2-1/2" | " | 0.500 |
| Threaded | | |
| 1/4", 125 lb | EA. | 0.320 |
| 1/2" | | |
| 125 lb | EA. | 0.320 |
| 300 lb | " | 0.320 |
| 3/4" | | |
| 125 lb | EA. | 0.320 |

| PLUMBING | UNIT | MAN/HOURS |
|---|---|---|
| **15430.60 VALVES** | | |
| 300 lb | EA. | 0.320 |
| 1" | | |
| 125 lb | EA. | 0.320 |
| 300 lb | " | 0.400 |
| 1-1/2" | | |
| 125 lb | EA. | 0.400 |
| 300 lb | " | 0.444 |
| 2" | | |
| 125 lb | EA. | 0.571 |
| 300 lb | " | 0.667 |
| Cast iron, flanged | | |
| 2", 150 lb | EA. | 0.667 |
| 2-1/2" | | |
| 125 lb | EA. | 0.667 |
| 250 lb | " | 0.667 |
| 3" | | |
| 125 lb | EA. | 0.800 |
| 250 lb | " | 0.800 |
| 4" | | |
| 125 lb | EA. | 1.143 |
| 250 lb | " | 1.143 |
| 6" | | |
| 125 lb | EA. | 1.600 |
| 250 lb | " | 1.600 |
| 8" | | |
| 125 lb | EA. | 2.000 |
| 250 lb | " | 2.000 |
| OS&Y, flanged | | |
| 2" | | |
| 125 lb | EA. | 0.667 |
| 250 lb | " | 0.667 |
| 2-1/2" | | |
| 125 lb | EA. | 0.667 |
| 250 lb | " | 0.800 |
| 3" | | |
| 125 lb | EA. | 0.800 |
| 250 lb | " | 0.800 |
| 4" | | |
| 125 lb | EA. | 1.333 |
| 250 lb | " | 1.333 |
| 6" | | |
| 125 lb | EA. | 1.600 |
| 250 lb | " | 1.600 |
| Ball valve, bronze, 250 lb, threaded | | |
| 1/2" | EA. | 0.320 |
| 3/4" | " | 0.320 |
| 1" | " | 0.400 |
| 1-1/4" | " | 0.444 |
| 1-1/2" | " | 0.500 |
| 2" | " | 0.571 |
| Angle valve, bronze, 150 lb, threaded | | |
| 1/2" | EA. | 0.286 |
| 3/4" | " | 0.320 |
| 1" | " | 0.320 |

| PLUMBING | UNIT | MAN/HOURS |
|---|---|---|
| **15430.60 VALVES** | | |
| 1-1/4" | EA. | 0.400 |
| 1-1/2" | " | 0.444 |
| Balancing valve, with meter connections, circuit setter | | |
| 1/2" | EA. | 0.320 |
| 3/4" | " | 0.364 |
| 1" | " | 0.400 |
| 1-1/4" | " | 0.444 |
| 1-1/2" | " | 0.533 |
| 2" | " | 0.667 |
| 2-1/2" | " | 0.800 |
| 3" | " | 1.000 |
| 4" | " | 1.333 |
| Pressure reducing valve, bronze, threaded, 250 lb | | |
| 1/2" | EA. | 0.500 |
| 3/4" | " | 0.500 |
| 1" | " | 0.500 |
| 1-1/4" | " | 0.571 |
| 1-1/2" | " | 0.667 |
| Pressure regulating valve, bronze, class 300 | | |
| 1" | EA. | 0.500 |
| 1-1/2" | " | 0.615 |
| 2" | " | 0.800 |
| 3" | " | 1.143 |
| 4" | " | 1.600 |
| 5" | " | 2.000 |
| 6" | " | 2.667 |
| Solar water temperature regulating valve | | |
| 3/4" | EA. | 0.667 |
| 1" | " | 0.800 |
| 1-1/4" | " | 0.889 |
| 1-1/2" | " | 1.000 |
| 2" | " | 1.143 |
| 2-1/2" | " | 2.000 |
| Tempering valve, threaded | | |
| 3/4" | EA. | 0.267 |
| 1" | " | 0.320 |
| 1-1/4" | " | 0.400 |
| 1-1/2" | " | 0.400 |
| 2" | " | 0.500 |
| 2-1/2" | " | 0.667 |
| 3" | " | 0.800 |
| 4" | " | 1.143 |
| Thermostatic mixing valve, threaded | | |
| 1/2" | EA. | 0.286 |
| 3/4" | " | 0.320 |
| 1" | " | 0.348 |
| 1-1/2" | " | 0.400 |
| 2" | " | 0.500 |
| Sweat connection | | |
| 1/2" | EA. | 0.286 |
| 3/4" | " | 0.320 |
| Mixing valve, sweat connection | | |
| 1/2" | EA. | 0.286 |
| 3/4" | " | 0.320 |

| PLUMBING | UNIT | MAN/ HOURS |
|---|---|---|
| **15430.60 VALVES** | | |
| Liquid level gauge, aluminum body | | |
| 3/4" | EA. | 0.320 |
| 4125 psi, pvc body | | |
| 3/4" | EA. | 0.320 |
| 150 psi, crs body | | |
| 3/4" | EA. | 0.320 |
| 1" | " | 0.320 |
| 175 psi, bronze body, 1/2" | " | 0.286 |
| **15430.65 VACUUM BREAKERS** | | |
| Vacuum breaker, atmospheric, threaded connection | | |
| 3/4" | EA. | 0.320 |
| 1" | " | 0.320 |
| Anti-siphon, brass | | |
| 3/4" | EA. | 0.320 |
| 1" | " | 0.320 |
| 1-1/4" | " | 0.400 |
| 1-1/2" | " | 0.444 |
| 2" | " | 0.500 |
| **15430.68 STRAINERS** | | |
| Strainer, Y pattern, 125 psi, cast iron body, threaded | | |
| 3/4" | EA. | 0.286 |
| 1" | " | 0.320 |
| 1-1/4" | " | 0.400 |
| 1-1/2" | " | 0.400 |
| 2" | " | 0.500 |
| 250 psi, brass body, threaded | | |
| 3/4" | EA. | 0.320 |
| 1" | " | 0.320 |
| 1-1/4" | " | 0.400 |
| 1-1/2" | " | 0.400 |
| 2" | " | 0.500 |
| Cast iron body, threaded | | |
| 3/4" | EA. | 0.320 |
| 1" | " | 0.320 |
| 1-1/4" | " | 0.400 |
| 1-1/2" | " | 0.400 |
| 2" | " | 0.500 |
| **15430.70 DRAINS, ROOF & FLOOR** | | |
| Floor drain, cast iron, with cast iron top | | |
| 2" | EA. | 0.667 |
| 3" | " | 0.667 |
| 4" | " | 0.667 |
| 6" | " | 0.800 |
| Roof drain, cast iron | | |
| 2" | EA. | 0.667 |
| 3" | " | 0.667 |
| 4" | " | 0.667 |
| 5" | " | 0.800 |
| 6" | " | 0.800 |

| PLUMBING | UNIT | MAN/ HOURS |
|---|---|---|
| **15430.80 TRAPS** | | |
| Bucket trap, threaded | | |
| 3/4" | EA. | 0.500 |
| 1" | " | 0.533 |
| 1-1/4" | " | 0.615 |
| 1-1/2" | " | 0.727 |
| Inverted bucket steam trap, threaded | | |
| 3/4" | EA. | 0.500 |
| 1" | " | 0.500 |
| 1-1/4" | " | 0.444 |
| 1-1/2" | " | 0.667 |
| Float trap, 15 psi | | |
| 3/4" | EA. | 0.500 |
| 1" | " | 0.533 |
| 1-1/4" | " | 0.571 |
| 1-1/2" | " | 0.667 |
| 2" | " | 0.800 |
| Float and thermostatic trap, 15 psi | | |
| 3/4" | EA. | 0.500 |
| 1" | " | 0.533 |
| 1-1/4" | " | 0.571 |
| 1-1/2" | " | 0.667 |
| 2" | " | 0.800 |
| Steam trap, cast iron body, threaded, 125 psi | | |
| 3/4" | EA. | 0.500 |
| 1" | " | 0.533 |
| 1-1/4" | " | 0.571 |
| 1-1/2" | " | 0.667 |

| PLUMBING FIXTURES | UNIT | MAN/ HOURS |
|---|---|---|
| **15440.10 BATHS** | | |
| Bath tub, 5' long | | |
| Minimum | EA. | 2.667 |
| Average | " | 4.000 |
| Maximum | " | 8.000 |
| 6' long | | |
| Minimum | EA. | 2.667 |
| Average | " | 4.000 |
| Maximum | " | 8.000 |
| Square tub, whirlpool, 4'x4' | | |
| Minimum | EA. | 4.000 |
| Average | " | 8.000 |
| Maximum | " | 10.000 |
| 5'x5' | | |
| Minimum | EA. | 4.000 |
| Average | " | 8.000 |

| PLUMBING FIXTURES | UNIT | MAN/ HOURS |
|---|---|---|
| **15440.10 BATHS** | | |
| Maximum | EA. | 10.000 |
| 6'x6' | | |
| Minimum | EA. | 4.000 |
| Average | " | 8.000 |
| Maximum | " | 10.000 |
| For trim and rough-in | | |
| Minimum | EA. | 2.667 |
| Average | " | 4.000 |
| Maximum | " | 8.000 |
| **15440.15 FAUCETS** | | |
| Kitchen | | |
| Minimum | EA. | 1.333 |
| Average | " | 1.600 |
| Maximum | " | 2.000 |
| Bath | | |
| Minimum | EA. | 1.333 |
| Average | " | 1.600 |
| Maximum | " | 2.000 |
| Lavatory, domestic | | |
| Minimum | EA. | 1.333 |
| Average | " | 1.600 |
| Maximum | " | 2.000 |
| Hospital, patient rooms | | |
| Minimum | EA. | 2.000 |
| Average | " | 2.667 |
| Maximum | " | 4.000 |
| Operating room | | |
| Minimum | EA. | 2.000 |
| Average | " | 2.667 |
| Maximum | " | 4.000 |
| Washroom | | |
| Minimum | EA. | 1.333 |
| Average | " | 1.600 |
| Maximum | " | 2.000 |
| Handicapped | | |
| Minimum | EA. | 1.600 |
| Average | " | 2.000 |
| Maximum | " | 2.667 |
| Shower | | |
| Minimum | EA. | 1.333 |
| Average | " | 1.600 |
| Maximum | " | 2.000 |
| For trim and rough-in | | |
| Minimum | EA. | 1.600 |
| Average | " | 2.000 |
| Maximum | " | 4.000 |
| **15440.18 HYDRANTS** | | |
| Wall hydrant | | |
| 8" thick | EA. | 1.333 |
| 12" thick | " | 1.600 |
| 18" thick | " | 1.778 |

| PLUMBING FIXTURES | UNIT | MAN/ HOURS |
|---|---|---|
| **15440.18 HYDRANTS** | | |
| 24" thick | EA. | 2.000 |
| Ground hydrant | | |
| 2' deep | EA. | 1.000 |
| 4' deep | " | 1.143 |
| 6' deep | " | 1.333 |
| 8' deep | " | 2.000 |
| **15440.20 LAVATORIES** | | |
| Lavatory, counter top, porcelain enamel on cast iron | | |
| Minimum | EA. | 1.600 |
| Average | " | 2.000 |
| Maximum | " | 2.667 |
| Wall hung, china | | |
| Minimum | EA. | 1.600 |
| Average | " | 2.000 |
| Maximum | " | 2.667 |
| Handicapped | | |
| Minimum | EA. | 2.000 |
| Average | " | 2.667 |
| Maximum | " | 4.000 |
| For trim and rough-in | | |
| Minimum | EA. | 2.000 |
| Average | " | 2.667 |
| Maximum | " | 4.000 |
| **15440.30 SHOWERS** | | |
| Shower, fiberglass, 36"x34"x84" | | |
| Minimum | EA. | 5.714 |
| Average | " | 8.000 |
| Maximum | " | 8.000 |
| Steel, 1 piece, 36"x36" | | |
| Minimum | EA. | 5.714 |
| Average | " | 8.000 |
| Maximum | " | 8.000 |
| Receptor, molded stone, 36"x36" | | |
| Minimum | EA. | 2.667 |
| Average | " | 4.000 |
| Maximum | " | 6.667 |
| For trim and rough-in | | |
| Minimum | EA. | 3.636 |
| Average | " | 4.444 |
| Maximum | " | 8.000 |
| **15440.40 SINKS** | | |
| Service sink, 24"x29" | | |
| Minimum | EA. | 2.000 |
| Average | " | 2.667 |
| Maximum | " | 4.000 |
| Kitchen sink, single, stainless steel, single bowl | | |
| Minimum | EA. | 1.600 |
| Average | " | 2.000 |
| Maximum | " | 2.667 |
| Double bowl | | |
| Minimum | EA. | 2.000 |

## PLUMBING FIXTURES

| PLUMBING FIXTURES | UNIT | MAN/HOURS |
|---|---|---|
| **15440.40 SINKS** | | |
| Average | EA. | 2.667 |
| Maximum | " | 4.000 |
| Porcelain enamel, cast iron, single bowl | | |
| Minimum | EA. | 1.600 |
| Average | " | 2.000 |
| Maximum | " | 2.667 |
| Double bowl | | |
| Minimum | EA. | 2.000 |
| Average | " | 2.667 |
| Maximum | " | 4.000 |
| Mop sink, 24"x36"x10" | | |
| Minimum | EA. | 1.600 |
| Average | " | 2.000 |
| Maximum | " | 2.667 |
| Washing machine box | | |
| Minimum | EA. | 2.000 |
| Average | " | 2.667 |
| Maximum | " | 4.000 |
| For trim and rough-in | | |
| Minimum | EA. | 2.667 |
| Average | " | 4.000 |
| Maximum | " | 5.333 |
| **15440.50 URINALS** | | |
| Urinal, flush valve, floor mounted | | |
| Minimum | EA. | 2.000 |
| Average | " | 2.667 |
| Maximum | " | 4.000 |
| Wall mounted | | |
| Minimum | EA. | 2.000 |
| Average | " | 2.667 |
| Maximum | " | 4.000 |
| For trim and rough-in | | |
| Minimum | EA. | 2.000 |
| Average | " | 4.000 |
| Maximum | " | 5.333 |
| **15440.60 WATER CLOSETS** | | |
| Water closet flush tank, floor mounted | | |
| Minimum | EA. | 2.000 |
| Average | " | 2.667 |
| Maximum | " | 4.000 |
| Handicapped | | |
| Minimum | EA. | 2.667 |
| Average | " | 4.000 |
| Maximum | " | 8.000 |
| Bowl, with flush valve, floor mounted | | |
| Minimum | EA. | 2.000 |
| Average | " | 2.667 |
| Maximum | " | 4.000 |
| Wall mounted | | |
| Minimum | EA. | 2.000 |
| Average | " | 2.667 |
| Maximum | " | 4.000 |

| PLUMBING FIXTURES | UNIT | MAN/HOURS |
|---|---|---|
| **15440.60 WATER CLOSETS** | | |
| For trim and rough-in | | |
| Minimum | EA. | 2.000 |
| Average | " | 2.667 |
| Maximum | " | 4.000 |
| **15440.70 WATER HEATERS** | | |
| Water heater, electric | | |
| 6 gal | EA. | 1.333 |
| 10 gal | " | 1.333 |
| 15 gal | " | 1.333 |
| 20 gal | " | 1.600 |
| 30 gal | " | 1.600 |
| 40 gal | " | 1.600 |
| 52 gal | " | 2.000 |
| 66 gal | " | 2.000 |
| 80 gal | " | 2.000 |
| 100 gal | " | 2.667 |
| 120 gal | " | 2.667 |
| Oil fired | | |
| 20 gal | EA. | 4.000 |
| 50 gal | " | 5.714 |
| **15440.90 MISCELLANEOUS FIXTURES** | | |
| Electric water cooler | | |
| Floor mounted | EA. | 2.667 |
| Wall mounted | " | 2.667 |
| Wash fountain | | |
| Wall mounted | EA. | 4.000 |
| Circular, floor supported | " | 8.000 |
| Deluge shower and eye wash | " | 4.000 |
| **15440.95 FIXTURE CARRIERS** | | |
| Water fountain, wall carrier | | |
| Minimum | EA. | 0.800 |
| Average | " | 1.000 |
| Maximum | " | 1.333 |
| Lavatory, wall carrier | | |
| Minimum | EA. | 0.800 |
| Average | " | 1.000 |
| Maximum | " | 1.333 |
| Sink, industrial, wall carrier | | |
| Minimum | EA. | 0.800 |
| Average | " | 1.000 |
| Maximum | " | 1.333 |
| Toilets, water closets, wall carrier | | |
| Minimum | EA. | 0.800 |
| Average | " | 1.000 |
| Maximum | " | 1.333 |
| Floor support | | |
| Minimum | EA. | 0.667 |
| Average | " | 0.800 |
| Maximum | " | 1.000 |

| PLUMBING FIXTURES | UNIT | MAN/ HOURS |
|---|---|---|
| **15440.95 FIXTURE CARRIERS** | | |
| Urinals, wall carrier | | |
| Minimum | EA. | 0.800 |
| Average | " | 1.000 |
| Maximum | " | 1.333 |
| Floor support | | |
| Minimum | EA. | 0.667 |
| Average | " | 0.800 |
| Maximum | " | 1.000 |
| **15450.30 PUMPS** | | |
| In-line pump, bronze, centrifugal | | |
| 5 gpm, 20' head | EA. | 0.500 |
| 20 gpm, 40' head | " | 0.500 |
| 50 gpm | | |
| 50' head | EA. | 1.000 |
| Cast iron, centrifugal | | |
| 50 gpm, 200' head | EA. | 1.000 |
| 100 gpm | | |
| 100' head | EA. | 1.333 |
| Centrifugal, close coupled, c.i., single stage | | |
| 50 gpm, 100' head | EA. | 1.000 |
| 100 gpm, 100' head | " | 1.333 |
| Base mounted | | |
| 50 gpm, 100' head | EA. | 1.000 |
| 100 gpm, 50' head | " | 1.333 |
| 200 gpm, 100' head | " | 2.000 |
| 300 gpm, 175' head | " | 2.000 |
| Condensate pump, simplex | | |
| 1000 sf EDR, 2 gpm | EA. | 6.667 |
| 2000 sf EDR, 3 gpm | " | 6.667 |
| 4000 sf EDR, 6 gpm | " | 7.273 |
| 6000 sf EDR, 9 gpm | " | 7.273 |
| Duplex, bronze | | |
| 8000 sf EDR, 12 gpm | EA. | 7.273 |
| 10,000 sf EDR, 15 gpm | " | 10.000 |
| 15,000 sf EDR, 23 gpm | " | 11.429 |
| 20,000 sf EDR, 30 gpm | " | 16.000 |
| 25,000 sf EDR, 38 gpm | " | 16.000 |
| **15450.40 STORAGE TANKS** | | |
| Hot water storage tank, cement lined | | |
| 10 gallon | EA. | 2.667 |
| 70 gallon | " | 4.000 |
| 200 gallon | " | 5.714 |
| 900 gallon | " | 10.000 |
| 1100 gallon | " | 10.000 |
| 2000 gallon | " | 10.000 |
| **15480.10 SPECIAL SYSTEMS** | | |
| Air compressor, air cooled, two stage | | |
| 5.0 cfm, 175 psi | EA. | 16.000 |
| 10 cfm, 175 psi | " | 17.778 |
| 20 cfm, 175 psi | " | 19.048 |

| PLUMBING FIXTURES | UNIT | MAN/ HOURS |
|---|---|---|
| **15480.10 SPECIAL SYSTEMS** | | |
| 50 cfm, 125 psi | EA. | 21.053 |
| 80 cfm, 125 psi | " | 22.857 |
| Single stage, 125 psi | | |
| 1.0 cfm | EA. | 11.429 |
| 1.5 cfm | " | 11.429 |
| 2.0 cfm | " | 11.429 |
| Automotive compressor, hose reel, air and water, 50' hose | " | 6.667 |
| Lube equipment, 3 reel, with pumps | " | 32.000 |
| Tire changer | | |
| Truck | EA. | 11.429 |
| Passenger car | " | 6.154 |
| Air hose reel, includes, 50' hose | " | 6.154 |
| Hose reel, 5 reel, motor oil, gear oil, lube, air & water | " | 32.000 |
| Water hose reel, 50' hose | " | 6.154 |
| Pump, air operated, for motor or gear oil, fits 55 gal drum | " | 0.800 |
| For chassis lube | " | 0.800 |
| Fuel dispensing pump, lighted dial, one product | | |
| One hose | EA. | 6.667 |
| Two hose | " | 6.667 |
| Two products, two hose | " | 6.667 |

| HEATING & VENTILATING | UNIT | MAN/ HOURS |
|---|---|---|
| **15555.10 BOILERS** | | |
| Cast iron, gas fired, hot water | | |
| 115 mbh | EA. | 20.000 |
| 175 mbh | " | 21.818 |
| 235 mbh | " | 24.000 |
| 940 mbh | " | 48.000 |
| 1600 mbh | " | 60.000 |
| 3000 mbh | " | 80.000 |
| 6000 mbh | " | 120 |
| Steam | | |
| 115 mbh | EA. | 20.000 |
| 175 mbh | " | 21.818 |
| 235 mbh | " | 24.000 |
| 940 mbh | " | 48.000 |
| 1600 mbh | " | 60.000 |
| 3000 mbh | " | 80.000 |
| 6000 mbh | " | 120 |
| Electric, hot water | | |
| 115 mbh | EA. | 12.000 |
| 175 mbh | " | 12.000 |
| 235 mbh | " | 12.000 |
| 940 mbh | " | 24.000 |
| 1600 mbh | " | 48.000 |
| 3000 mbh | " | 60.000 |

| HEATING & VENTILATING | UNIT | MAN/ HOURS |
|---|---|---|
| **15555.10 BOILERS** | | |
| 6000 mbh | EA. | 80.000 |
| Steam | | |
| 115 mbh | EA. | 12.000 |
| 175 mbh | " | 12.000 |
| 235 mbh | " | 12.000 |
| 940 mbh | " | 24.000 |
| 1600 mbh | " | 48.000 |
| 3000 mbh | " | 60.000 |
| 6000 mbh | " | 80.000 |
| Oil fired, hot water | | |
| 115 mbh | EA. | 16.000 |
| 175 mbh | " | 18.462 |
| 235 mbh | " | 21.818 |
| 940 mbh | " | 40.000 |
| 1600 mbh | " | 48.000 |
| 3000 mbh | " | 60.000 |
| 6000 mbh | " | 120 |
| Steam | | |
| 115 mbh | EA. | 16.000 |
| 175 mbh | " | 18.462 |
| 235 mbh | " | 21.818 |
| 940 mbh | " | 40.000 |
| 1600 mbh | " | 48.000 |
| 3000 mbh | " | 60.000 |
| 6000 mbh | " | 120 |
| **15610.10 FURNACES** | | |
| Electric, hot air | | |
| 40 mbh | EA. | 4.000 |
| 60 mbh | " | 4.211 |
| 80 mbh | " | 4.444 |
| 100 mbh | " | 4.706 |
| 125 mbh | " | 4.848 |
| 160 mbh | " | 5.000 |
| 200 mbh | " | 5.161 |
| 400 mbh | " | 5.333 |
| Gas fired hot air | | |
| 40 mbh | EA. | 4.000 |
| 60 mbh | " | 4.211 |
| 80 mbh | " | 4.444 |
| 100 mbh | " | 4.706 |
| 125 mbh | " | 4.848 |
| 160 mbh | " | 5.000 |
| 200 mbh | " | 5.161 |
| 400 mbh | " | 5.333 |
| Oil fired hot air | | |
| 40 mbh | EA. | 4.000 |
| 60 mbh | " | 4.211 |
| 80 mbh | " | 4.444 |
| 100 mbh | " | 4.706 |
| 125 mbh | " | 4.848 |
| 160 mbh | " | 5.000 |
| 200 mbh | " | 5.161 |

| HEATING & VENTILATING | UNIT | MAN/ HOURS |
|---|---|---|
| **15610.10 FURNACES** | | |
| 400 mbh | EA. | 5.333 |

| REFRIGERATION | UNIT | MAN/ HOURS |
|---|---|---|
| **15670.10 CONDENSING UNITS** | | |
| Air cooled condenser, single circuit | | |
| 3 ton | EA. | 1.333 |
| 5 ton | " | 1.333 |
| 7.5 ton | " | 3.810 |
| 20 ton | " | 4.000 |
| 25 ton | " | 4.000 |
| 30 ton | " | 4.000 |
| With low ambient dampers | | |
| 3 ton | EA. | 2.000 |
| 5 ton | " | 2.000 |
| 7.5 ton | " | 4.000 |
| 20 ton | " | 5.333 |
| 25 ton | " | 5.333 |
| 30 ton | " | 5.333 |
| Dual circuit | | |
| 10 ton | EA. | 4.000 |
| 15 ton | " | 5.714 |
| 20 ton | " | 5.714 |
| 25 ton | " | 5.714 |
| 30 ton | " | 5.714 |
| With low ambient dampers | | |
| 15 ton | EA. | 5.714 |
| 20 ton | " | 5.714 |
| 25 ton | " | 5.714 |
| 30 ton | " | 5.714 |
| **15680.10 CHILLERS** | | |
| Chiller, reciprocal | | |
| Air cooled, remote condenser, starter | | |
| 20 ton | EA. | 8.000 |
| 25 ton | " | 8.000 |
| 30 ton | " | 8.000 |
| 40 ton | " | 12.000 |
| Water cooled, with starter | | |
| 20 ton | EA. | 8.000 |
| 25 ton | " | 8.000 |
| 30 ton | " | 12.000 |
| 40 ton | " | 12.000 |
| Packaged, air cooled, with starter | | |

# 15 MECHANICAL

| REFRIGERATION | UNIT | MAN/ HOURS |
|---|---|---|
| **15680.10 CHILLERS** | | |
| 20 ton | EA. | 6.000 |
| 25 ton | " | 6.000 |
| 30 ton | " | 6.000 |
| 40 ton | " | 6.000 |

| HEAT TRANSFER | UNIT | MAN/ HOURS |
|---|---|---|
| **15780.10 COMPUTER ROOM A/C** | | |
| Air cooled, alarm, high efficiency filter, elec. heat | | |
| 3 ton | EA. | 6.154 |
| 5 ton | " | 6.667 |
| 7.5 ton | " | 8.000 |
| 10 ton | " | 10.000 |
| 15 ton | " | 11.429 |
| Steam heat | | |
| 3 ton | EA. | 6.154 |
| 5 ton | " | 6.667 |
| 7.5 ton | " | 8.000 |
| 10 ton | " | 10.000 |
| 15 ton | " | 11.429 |
| Hot water heat | | |
| 3 ton | EA. | 6.154 |
| 5 ton | " | 6.667 |
| 7.5 ton | " | 8.000 |
| 10 ton | " | 10.000 |
| 15 ton | " | 11.429 |
| Air cooled condenser, low ambient damper | | |
| 3 ton | EA. | 1.600 |
| 5 ton | " | 2.000 |
| 7.5 ton | " | 4.000 |
| 10 ton | " | 5.714 |
| 15 ton | " | 4.706 |
| Water cooled, high efficiency filter, alarm, elec. heat | | |
| 3 ton | EA. | 5.714 |
| 5 ton | " | 6.667 |
| 7.5 ton | " | 10.000 |
| 10 ton | " | 11.429 |
| 15 ton | " | 13.333 |
| Steam heat | | |
| 3 ton | EA. | 5.714 |
| 5 ton | " | 6.667 |
| 7.5 ton | " | 10.000 |
| 10 ton | " | 11.429 |
| 15 ton | " | 13.333 |
| Hot water heat | | |
| 3 ton | EA. | 5.714 |
| 5 ton | " | 6.667 |

| HEAT TRANSFER | UNIT | MAN/ HOURS |
|---|---|---|
| **15780.10 COMPUTER ROOM A/C** | | |
| 7.5 ton | EA. | 10.000 |
| 10 ton | " | 11.429 |
| 15 ton | " | 13.333 |
| **15780.20 ROOFTOP UNITS** | | |
| Packaged, single zone rooftop unit, with roof curb | | |
| 2 ton | EA. | 8.000 |
| 3 ton | " | 8.000 |
| 4 ton | " | 10.000 |
| 5 ton | " | 13.333 |
| 7.5 ton | " | 16.000 |
| **15830.20 FAN COIL UNITS** | | |
| Fan coil unit, 2 pipe, complete | | |
| 200 cfm ceiling hung | EA. | 2.667 |
| Floor mounted | " | 2.000 |
| 300 cfm, ceiling hung | " | 3.200 |
| Floor mounted | " | 2.667 |
| 400 cfm, ceiling hung | " | 3.810 |
| Floor mounted | " | 2.667 |
| 500 cfm, ceiling hung | " | 4.000 |
| Floor mounted | " | 3.077 |
| 600 cfm, ceiling hung | " | 4.420 |
| Floor mounted | " | 3.636 |
| **15830.70 UNIT HEATERS** | | |
| Steam unit heater, horizontal | | |
| 12,500 btuh, 200 cfm | EA. | 1.333 |
| 17,000 btuh, 300 cfm | " | 1.333 |
| 40,000 btuh, 500 cfm | " | 1.333 |
| 60,000 btuh, 700 cfm | " | 1.333 |
| 70,000 btuh, 1000 cfm | " | 2.000 |
| Vertical | | |
| 12,500 btuh, 200 cfm | EA. | 1.333 |
| 17,000 btuh, 300 cfm | " | 1.333 |
| 40,000 btuh, 500 cfm | " | 1.333 |
| 60,000 btuh, 700 cfm | " | 1.333 |
| 70,000 btuh, 1000 cfm | " | 1.333 |
| Gas unit heater, horizontal | | |
| 27,400 btuh | EA. | 3.200 |
| 38,000 btuh | " | 3.200 |
| 56,000 btuh | " | 3.200 |
| 82,200 btuh | " | 3.200 |
| 103,900 btuh | " | 5.000 |
| 125,700 btuh | " | 5.000 |
| 133,200 btuh | " | 5.000 |
| 149,000 btuh | " | 5.000 |
| 172,000 btuh | " | 5.000 |
| 190,000 btuh | " | 5.000 |
| 225,000 btuh | " | 5.000 |
| Hot water unit heater, horizontal | | |
| 12,500 btuh, 200 cfm | EA. | 1.333 |

## HEAT TRANSFER

| 15830.70 UNIT HEATERS | UNIT | MAN/HOURS |
|---|---|---|
| 17,000 btuh, 300 cfm | EA. | 1.333 |
| 25,000 btuh, 500 cfm | " | 1.333 |
| 30,000 btuh, 700 cfm | " | 1.333 |
| 50,000 btuh, 1000 cfm | " | 2.000 |
| 60,000 btuh, 1300 cfm | " | 2.000 |
| Vertical | | |
| 12,500 btuh, 200 cfm | EA. | 1.333 |
| 17,000 btuh, 300 cfm | " | 1.333 |
| 25,000 btuh, 500 cfm | " | 1.333 |
| 30,000 btuh, 700 cfm | " | 1.333 |
| 50,000 btuh, 1000 cfm | " | 1.333 |
| 60,000 btuh, 1300 cfm | " | 1.333 |
| Cabinet unit heaters, ceiling, exposed, hot water | | |
| 200 cfm | EA. | 2.667 |
| 300 cfm | " | 3.200 |
| 400 cfm | " | 3.810 |
| 600 cfm | " | 4.211 |
| 800 cfm | " | 5.000 |
| 1000 cfm | " | 5.714 |
| 1200 cfm | " | 6.667 |
| 2000 cfm | " | 8.889 |

## AIR HANDLING

| 15855.10 AIR HANDLING UNITS | UNIT | MAN/HOURS |
|---|---|---|
| Air handling unit, medium pressure, single zone | | |
| 1500 cfm | EA. | 5.000 |
| 3000 cfm | " | 8.889 |
| 4000 cfm | " | 10.000 |
| 5000 cfm | " | 10.667 |
| 6000 cfm | " | 11.429 |
| 7000 cfm | " | 12.308 |
| 8500 cfm | " | 13.333 |
| 10,500 cfm | " | 16.000 |
| 12,500 cfm | " | 17.778 |
| Rooftop air handling units | | |
| 4950 cfm | EA. | 8.889 |
| 7370 cfm | " | 11.429 |
| 9790 cfm | " | 13.333 |
| 14,300 cfm | " | 11.429 |
| 21,725 cfm | " | 11.429 |
| 33,000 cfm | " | 13.333 |

## AIR HANDLING

| 15870.20 EXHAUST FANS | UNIT | MAN/HOURS |
|---|---|---|
| Belt drive roof exhaust fans | | |
| 640 cfm, 2618 fpm | EA. | 1.000 |
| 940 cfm, 2604 fpm | " | 1.000 |
| 1050 cfm, 3325 fpm | " | 1.000 |
| 1170 cfm, 2373 fpm | " | 1.000 |
| 2440 cfm, 4501 fpm | " | 1.000 |
| 2760 cfm, 4950 fpm | " | 1.000 |
| 3890 cfm, 6769 fpm | " | 1.000 |
| 2380 cfm, 3382 fpm | " | 1.000 |
| 2880 cfm, 3859 fpm | " | 1.000 |
| 3200 cfm, 4173 fpm | " | 1.333 |
| 3660 cfm, 3437 fpm | " | 1.333 |
| Direct drive fans | | |
| 60 to 390 cfm | EA. | 1.000 |
| 145 to 590 cfm | " | 1.000 |
| 295 to 860 cfm | " | 1.000 |
| 235 to 1300 cfm | " | 1.000 |
| 415 to 1630 cfm | " | 1.000 |
| 590 to 2045 cfm | " | 1.000 |

## AIR DISTRIBUTION

| 15890.10 METAL DUCTWORK | UNIT | MAN/HOURS |
|---|---|---|
| Rectangular duct | | |
| Galvanized steel | | |
| Minimum | Lb. | 0.073 |
| Average | " | 0.089 |
| Maximum | " | 0.133 |
| Aluminum | | |
| Minimum | Lb. | 0.160 |
| Average | " | 0.200 |
| Maximum | " | 0.267 |
| Fittings | | |
| Minimum | EA. | 0.267 |
| Average | " | 0.400 |
| Maximum | " | 0.800 |

| 15890.30 FLEXIBLE DUCTWORK | UNIT | MAN/HOURS |
|---|---|---|
| Flexible duct, 1.25" fiberglass | | |
| 5" dia. | L.F. | 0.040 |
| 6" dia. | " | 0.044 |
| 7" dia. | " | 0.047 |
| 8" dia. | " | 0.050 |
| 10" dia. | " | 0.057 |
| 12" dia. | " | 0.062 |

| AIR DISTRIBUTION | UNIT | MAN/HOURS |
|---|---|---|
| **15890.30 FLEXIBLE DUCTWORK** | | |
| 14" dia. | L.F. | 0.067 |
| 16" dia. | " | 0.073 |
| Flexible duct connector, 3" wide fabric | " | 0.133 |
| **15895.10 ROOF CURBS** | | |
| 8" high, insulated, with liner and raised can | | |
| 15" x 15" | EA. | 0.400 |
| 21" x 21" | " | 0.400 |
| 36" x 36" | " | 0.571 |
| 48" x 48" | " | 0.615 |
| 60" x 60" | " | 0.800 |
| 72" x 72" | " | 1.000 |
| **15910.10 DAMPERS** | | |
| Horizontal parallel aluminum backdraft damper | | |
| 12" x 12" | EA. | 0.200 |
| 16" x 16" | " | 0.229 |
| 24" x 24" | " | 0.400 |
| 36" x 36" | " | 0.571 |
| 48" x 48" | " | 0.800 |
| "Up", parallel dampers | | |
| 12" x 12" | EA. | 0.200 |
| 16" x 16" | " | 0.229 |
| 24" x 24" | " | 0.400 |
| 36" x 36" | " | 0.571 |
| 48" x 48" | " | 0.800 |
| "Down", parallel dampers | | |
| 12" x 12" | EA. | 0.200 |
| 16" x 16" | " | 0.229 |
| 24" x 24" | " | 0.400 |
| 36" x 36" | " | 0.571 |
| 48" x 48" | " | 0.800 |
| Fire damper, 1.5 hr rating | | |
| 12" x 12" | EA. | 0.400 |
| 16" x 16" | " | 0.400 |
| 24" x 24" | " | 0.400 |
| 36" x 36" | " | 0.800 |
| 48" x 48" | " | 1.143 |
| **15940.10 DIFFUSERS** | | |
| Ceiling diffusers, round, baked enamel finish | | |
| 6" dia. | EA. | 0.267 |
| 12" dia. | " | 0.333 |
| 16" dia. | " | 0.364 |
| 20" dia. | " | 0.400 |
| Rectangular | | |
| 6x6" | EA. | 0.267 |
| 12x12" | " | 0.400 |
| 18x18" | " | 0.400 |
| 24x24" | " | 0.500 |
| Lay in, flush mounted, perforated face, with grid | | |
| 6x6/24x24 | EA. | 0.320 |

| AIR DISTRIBUTION | UNIT | MAN/HOURS |
|---|---|---|
| **15940.10 DIFFUSERS** | | |
| 12x12/24x24 | EA. | 0.320 |
| 18x18/24x24 | " | 0.320 |
| Two-way slot diffuser with balancing damper, 4' | " | 0.800 |
| **15940.20 RELIEF VENTILATORS** | | |
| Intake ventilator, aluminum, with screen, no curbs | | |
| 12" x 12" | EA. | 0.667 |
| 16" x 16" | " | 0.800 |
| 36" x 36" | " | 1.333 |
| 48" x 48" | " | 1.600 |
| **15940.40 REGISTERS AND GRILLES** | | |
| Lay in flush mounted, perforated face, return | | |
| 6x6/24x24 | EA. | 0.320 |
| 8x8/24x24 | " | 0.320 |
| 9x9/24x24 | " | 0.320 |
| 10x10/24x24 | " | 0.320 |
| 12x12/24x24 | " | 0.320 |
| Rectangular, ceiling return, single deflection | | |
| 10x10 | EA. | 0.400 |
| 12x12 | " | 0.400 |
| 14x14 | " | 0.400 |
| 16x8 | " | 0.400 |
| 16x16 | " | 0.400 |
| 24x12 | " | 0.400 |
| 24x18 | " | 0.400 |
| 36x24 | " | 0.444 |
| 36x30 | " | 0.444 |
| Wall, return air register | | |
| 12x12 | EA. | 0.200 |
| 16x16 | " | 0.200 |
| 18x18 | " | 0.200 |
| 20x20 | " | 0.200 |
| 24x24 | " | 0.200 |
| Ceiling, return air grille | | |
| 6x6 | EA. | 0.267 |
| 8x8 | " | 0.320 |
| 10x10 | " | 0.320 |
| Ceiling, exhaust grille, aluminum egg crate | | |
| 6x6 | EA. | 0.267 |
| 8x8 | " | 0.320 |
| 10x10 | " | 0.320 |
| 12x12 | " | 0.400 |
| 14x14 | " | 0.400 |
| 16x16 | " | 0.400 |
| 18x18 | " | 0.400 |
| **15940.80 PENTHOUSE LOUVERS** | | |
| Penthouse louvers | | |
| 12" high, extruded aluminum, 4" louver | | |
| 6' perimeter | EA. | 2.000 |
| 12' perimeter | " | 2.000 |

| AIR DISTRIBUTION | UNIT | MAN/ HOURS |
|---|---|---|
| **15940.80 PENTHOUSE LOUVERS** | | |
| 20' perimeter | EA. | 5.333 |
| 16" high x 4' perimeter | " | 2.000 |
| 6' perimeter | " | 2.000 |
| 12' perimeter | " | 2.000 |
| 20' perimeter | " | 5.333 |
| 20" high x 4' perimeter | " | 2.000 |
| 6' perimeter | " | 2.000 |
| 12' perimeter | " | 2.000 |
| 20' perimeter | " | 5.333 |
| 24" high x 4' perimeter | " | 2.000 |
| 6' perimeter | " | 2.000 |
| 12' perimeter | " | 2.000 |
| 20' perimeter | " | 5.333 |

| CONTROLS | UNIT | MAN/ HOURS |
|---|---|---|
| **15950.10 HVAC CONTROLS** | | |
| Pressure gauge, direct reading gage cock and siphon | EA. | 0.500 |
| Control valve, 1", modulating | | |
| 2-way | EA. | 0.667 |
| 3-way | " | 1.000 |
| Self contained control valve with sensing element, 3/4" | " | 0.500 |
| Control dampers, round | | |
| 6" dia. | EA. | 0.320 |
| 8" dia | " | 0.320 |
| 10" dia | " | 0.320 |
| 12" dia | " | 0.320 |
| 12" dia | " | 0.400 |
| 18" dia | " | 0.400 |
| 20" dia | " | 0.400 |
| Rectangular, parallel blade standard leakage | | |
| 12" x 12" | EA. | 0.400 |
| 16" x 16" | " | 0.400 |
| 20" x 20" | " | 0.400 |
| 48" x 48" | " | 1.143 |
| 48" x 60" | " | 1.333 |
| 48" x 72" | " | 1.333 |
| Low leakage | | |
| 12" x 12" | EA. | 0.400 |
| 16" x 16" | " | 0.400 |
| 36" x 36" | " | 0.667 |
| 48" x 48" | " | 1.143 |
| 48" x 72" | " | 1.333 |
| Rectangular, opposed horizontal blade | | |
| 12" x 12" | EA. | 0.400 |
| 16" x 16" | " | 0.400 |

| CONTROLS | UNIT | MAN/ HOURS |
|---|---|---|
| **15950.10 HVAC CONTROLS** | | |
| 24" x 24" | EA. | 0.400 |
| 36" x 36" | " | 0.667 |
| 48" x 72" | " | 1.333 |

## BASIC MATERIALS

### 16110.12 CABLE TRAY

| BASIC MATERIALS | UNIT | MAN/HOURS |
|---|---|---|
| Cable tray, 6" | L.F. | 0.059 |
| Ventilated cover | " | 0.030 |
| Solid cover | " | 0.030 |

### 16110.20 CONDUIT SPECIALTIES

| BASIC MATERIALS | UNIT | MAN/HOURS |
|---|---|---|
| Rod beam clamp, 1/2" | EA. | 0.050 |
| Hanger rod | | |
| 3/8" | L.F. | 0.040 |
| 1/2" | " | 0.050 |
| All thread rod | | |
| 1/4" | L.F. | 0.030 |
| 3/8" | " | 0.040 |
| 1/2" | " | 0.050 |
| 5/8" | " | 0.080 |
| Hanger channel, 1-1/2" | | |
| No holes | EA. | 0.030 |
| Holes | " | 0.030 |
| Channel strap | | |
| 1/2" | EA. | 0.050 |
| 3/4" | " | 0.050 |
| 1" | " | 0.050 |
| 1-1/4" | " | 0.080 |
| 1-1/2" | " | 0.080 |
| 2" | " | 0.080 |
| 2-1/2" | " | 0.123 |
| 3" | " | 0.123 |
| 3-1/2" | " | 0.123 |
| 4" | " | 0.145 |
| 5" | " | 0.145 |
| 6" | " | 0.145 |
| Conduit penetrations, roof and wall, 8" thick | | |
| 1/2" | EA. | 0.615 |
| 3/4" | " | 0.615 |
| 1" | " | 0.800 |
| 1-1/4" | " | 0.800 |
| 1-1/2" | " | 0.800 |
| 2" | " | 1.600 |
| 2-1/2" | " | 1.600 |
| 3" | " | 1.600 |
| 3-1/2" | " | 2.000 |
| 4" | " | 2.000 |
| Fireproofing, for conduit penetrations | | |
| 1/2" | EA. | 0.500 |
| 3/4" | " | 0.500 |
| 1" | " | 0.500 |
| 1-1/4" | " | 0.727 |
| 1-1/2" | " | 0.727 |
| 2" | " | 0.727 |
| 2-1/2" | " | 0.899 |
| 3" | " | 0.899 |
| 3-1/2" | " | 1.250 |
| 4" | " | 1.509 |

### 16110.21 ALUMINUM CONDUIT

| BASIC MATERIALS | UNIT | MAN/HOURS |
|---|---|---|
| Aluminum conduit | | |
| 1/2" | L.F. | 0.030 |
| 3/4" | " | 0.040 |
| 1" | " | 0.050 |
| 1-1/4" | " | 0.059 |
| 1-1/2" | " | 0.080 |
| 2" | " | 0.089 |
| 2-1/2" | " | 0.100 |
| 3" | " | 0.107 |
| 3-1/2" | " | 0.123 |
| 4" | " | 0.145 |
| 5" | " | 0.182 |
| 6" | " | 0.200 |

### 16110.22 EMT CONDUIT

| BASIC MATERIALS | UNIT | MAN/HOURS |
|---|---|---|
| EMT conduit | | |
| 1/2" | L.F. | 0.030 |
| 3/4" | " | 0.040 |
| 1" | " | 0.050 |
| 1-1/4" | " | 0.059 |
| 1-1/2" | " | 0.080 |
| 2" | " | 0.089 |
| 2-1/2" | " | 0.100 |
| 3" | " | 0.123 |
| 3-1/2" | " | 0.145 |
| 4" | " | 0.182 |

### 16110.23 FLEXIBLE CONDUIT

| BASIC MATERIALS | UNIT | MAN/HOURS |
|---|---|---|
| Flexible conduit, steel | | |
| 3/8" | L.F. | 0.030 |
| 1/2 | " | 0.030 |
| 3/4" | " | 0.040 |
| 1" | " | 0.040 |
| 1-1/4" | " | 0.050 |
| 1-1/2" | " | 0.059 |
| 2" | " | 0.080 |
| 2-1/2" | " | 0.089 |
| 3" | " | 0.107 |

### 16110.24 GALVANIZED CONDUIT

| BASIC MATERIALS | UNIT | MAN/HOURS |
|---|---|---|
| Galvanized rigid steel conduit | | |
| 1/2" | L.F. | 0.040 |
| 3/4" | " | 0.050 |
| 1" | " | 0.059 |
| 1-1/4" | " | 0.080 |
| 1-1/2" | " | 0.089 |
| 2" | " | 0.100 |
| 2-1/2" | " | 0.145 |
| 3" | " | 0.182 |
| 3-1/2" | " | 0.190 |
| 4" | " | 0.211 |
| 5" | " | 0.286 |

| BASIC MATERIALS | UNIT | MAN/HOURS |
|---|---|---|
| **16110.24 GALVANIZED CONDUIT** | | |
| 6" | L.F. | 0.381 |
| **16110.27 PLASTIC COATED CONDUIT** | | |
| Rigid steel conduit, plastic coated | | |
| 1/2" | L.F. | 0.050 |
| 3/4" | " | 0.059 |
| 1" | " | 0.080 |
| 1-1/4" | " | 0.100 |
| 1-1/2" | " | 0.123 |
| 2" | " | 0.145 |
| 2-1/2" | " | 0.190 |
| 3" | " | 0.222 |
| 3-1/2" | " | 0.250 |
| 4" | " | 0.308 |
| 5" | " | 0.381 |
| **16110.28 STEEL CONDUIT** | | |
| Intermediate metal conduit (IMC) | | |
| 1/2" | L.F. | 0.030 |
| 3/4" | " | 0.040 |
| 1" | " | 0.050 |
| 1-1/4" | " | 0.059 |
| 1-1/2" | " | 0.080 |
| 2" | " | 0.089 |
| 2-1/2" | " | 0.119 |
| 3" | " | 0.145 |
| 3-1/2" | " | 0.182 |
| 4" | " | 0.190 |
| **16110.35 WIREMOLD** | | |
| Wiremold raceway with fittings, surface mounted | | |
| #200 | L.F. | 0.030 |
| #500 | " | 0.030 |
| #700 | " | 0.040 |
| #800 | " | 0.040 |
| Fittings, #200, 90 degree flat elbow | EA. | 0.050 |
| Internal elbow | " | 0.050 |
| Extension adapter | " | 0.059 |
| #200, #500, #700 | | |
| Single pole switch and box | EA. | 0.400 |
| Duplex receptacle with box | " | 0.308 |
| #500, #700 | | |
| 90 deg. flat elbow | EA. | 0.080 |
| Internal elbow | " | 0.080 |
| Junction box | " | 0.133 |
| Fixture box | " | 0.133 |
| Shallow switch and receptacle box | " | 0.080 |
| #800 | | |
| 90 deg. flat elbow | EA. | 0.080 |
| Internal elbow | " | 0.080 |
| Junction box | " | 0.123 |

| BASIC MATERIALS | UNIT | MAN/HOURS |
|---|---|---|
| **16110.80 WIREWAYS** | | |
| Wireway, hinge cover type | | |
| 2-1/2" x 2-1/2" | | |
| 1' section | EA. | 0.154 |
| 2' | " | 0.190 |
| 3' | " | 0.250 |
| 5' | " | 0.381 |
| 10' | " | 0.667 |
| 4" x 4" | | |
| 1' | EA. | 0.250 |
| 2' | " | 0.250 |
| 3' | " | 0.308 |
| 4' | " | 0.308 |
| 10' | " | 0.800 |
| **16120.43 COPPER CONDUCTORS** | | |
| Copper conductors, type THW, solid | | |
| #14 | L.F. | 0.004 |
| #12 | " | 0.005 |
| #10 | " | 0.006 |
| Stranded | | |
| #14 | L.F. | 0.004 |
| #12 | " | 0.005 |
| #10 | " | 0.006 |
| #8 | " | 0.008 |
| #6 | " | 0.009 |
| #4 | " | 0.010 |
| #3 | " | 0.010 |
| #2 | " | 0.012 |
| #1 | " | 0.014 |
| 1/0 | " | 0.016 |
| 2/0 | " | 0.020 |
| 3/0 | " | 0.025 |
| 4/0 | " | 0.028 |
| 250 MCM | " | 0.030 |
| 300 MCM | " | 0.033 |
| 350 MCM | " | 0.040 |
| 400 MCM | " | 0.044 |
| 500 MCM | " | 0.052 |
| 600 MCM | " | 0.059 |
| 750 MCM | " | 0.067 |
| 1000 MCM | " | 0.076 |
| THHN-THWN, solid | | |
| #14 | L.F. | 0.004 |
| #12 | " | 0.005 |
| #10 | " | 0.006 |
| Stranded | | |
| #14 | L.F. | 0.004 |
| #12 | " | 0.005 |
| #10 | " | 0.006 |
| #8 | " | 0.008 |
| #6 | " | 0.009 |
| #4 | " | 0.010 |
| #2 | " | 0.012 |

# 16 ELECTRICAL

| BASIC MATERIALS | UNIT | MAN/ HOURS |
|---|---|---|
| **16120.43 COPPER CONDUCTORS** | | |
| #1 | L.F. | 0.014 |
| 1/0 | " | 0.016 |
| 2/0 | " | 0.020 |
| 3/0 | " | 0.025 |
| 4/0 | " | 0.028 |
| 250 MCM | " | 0.030 |
| 350 MCM | " | 0.040 |
| **16120.47 SHEATHED CABLE** | | |
| Non-metallic sheathed cable | | |
| Type NM cable with ground | | |
| #14/2 | L.F. | 0.015 |
| #12/2 | " | 0.016 |
| #10/2 | " | 0.018 |
| #8/2 | " | 0.020 |
| #6/2 | " | 0.025 |
| #14/3 | " | 0.026 |
| #12/3 | " | 0.027 |
| #10/3 | " | 0.027 |
| #8/3 | " | 0.028 |
| #6/3 | " | 0.028 |
| #4/3 | " | 0.032 |
| #2/3 | " | 0.035 |
| Type U.F. cable with ground | | |
| #14/2 | L.F. | 0.016 |
| #12/2 | " | 0.019 |
| #10/2 | " | 0.020 |
| #8/2 | " | 0.023 |
| #6/2 | " | 0.027 |
| #14/3 | " | 0.020 |
| #12/3 | " | 0.022 |
| #10/3 | " | 0.025 |
| #8/3 | " | 0.026 |
| #6/3 | " | 0.032 |
| **16130.60 PULL AND JUNCTION BOXES** | | |
| 4" | | |
| Octagon box | EA. | 0.114 |
| Box extension | " | 0.059 |
| Plaster ring | " | 0.059 |
| Cover blank | " | 0.059 |
| Square box | " | 0.114 |
| Box extension | " | 0.059 |
| Plaster ring | " | 0.059 |
| Cover blank | " | 0.059 |
| 4-11/16" | | |
| Square box | EA. | 0.114 |
| Box extension | " | 0.059 |
| Plaster ring | " | 0.059 |
| Cover blank | " | 0.059 |
| Switch and device boxes | | |
| 2 gang | EA. | 0.114 |

| BASIC MATERIALS | UNIT | MAN/ HOURS |
|---|---|---|
| **16130.60 PULL AND JUNCTION BOXES** | | |
| 3 gang | EA. | 0.114 |
| 4 gang | " | 0.160 |
| Device covers | | |
| 2 gang | EA. | 0.059 |
| 3 gang | " | 0.059 |
| 4 gang | " | 0.059 |
| Handy box | " | 0.114 |
| Extension | " | 0.059 |
| Switch cover | " | 0.059 |
| Switch box with knockout | " | 0.145 |
| Weatherproof cover, spring type | " | 0.080 |
| Cover plate, dryer receptacle 1 gang plastic | " | 0.100 |
| For 4" receptacle, 2 gang | " | 0.100 |
| Duplex receptacle cover plate, plastic | " | 0.059 |
| 4", vertical bracket box, 1-1/2" with | | |
| RMX clamps | EA. | 0.145 |
| BX clamps | " | 0.145 |
| 4", octagon device cover | | |
| 1 switch | EA. | 0.059 |
| 1 duplex recept | " | 0.059 |
| 4", square face bracket boxes, 1-1/2" | | |
| RMX | EA. | 0.145 |
| BX | " | 0.145 |
| **16130.65 PULL BOXES AND CABINETS** | | |
| Galvanized pull boxes, screw cover | | |
| 4x4x4 | EA. | 0.190 |
| 4x6x4 | " | 0.190 |
| 6x6x4 | " | 0.190 |
| 6x8x4 | " | 0.190 |
| 8x8x4 | " | 0.250 |
| 8x10x4 | " | 0.242 |
| 8x12x4 | " | 0.250 |
| Screw cover | | |
| 10x10x4 | EA. | 0.308 |
| 12x12x6 | " | 0.444 |
| 12x15x6 | " | 0.444 |
| 12x18x6 | " | 0.500 |
| 15x18x6 | " | 0.571 |
| 18x24x6 | " | 0.615 |
| 18x30x6 | " | 0.727 |
| 24x36x6 | " | 0.727 |
| **16130.80 RECEPTACLES** | | |
| 125 volt, 20a, duplex, grounding type, standard grade | EA. | 0.200 |
| Ground fault interrupter type | " | 0.296 |
| 250 volt, 20a, 2 pole, single receptacle, ground type | " | 0.200 |
| 120/208v, 4 pole, single receptacle, twist lock | | |
| 20a | EA. | 0.348 |
| 50a | " | 0.348 |
| 125/250v, 3 pole, flush receptacle | | |
| 30a | EA. | 0.296 |

# 16 ELECTRICAL

| BASIC MATERIALS | UNIT | MAN/ HOURS |
|---|---|---|
| **16130.80 RECEPTACLES** | | |
| 50a | EA. | 0.296 |
| 60a | " | 0.348 |
| 277 v, 20a, 2 pole, grounding type, twist lock | " | 0.200 |
| Dryer receptacle, 250v, 30a/50a, 3 wire | " | 0.296 |
| Clock receptacle, 2 pole, grounding type | " | 0.200 |
| 125v, 20a single recept. grounding type | | |
| Standard grade | EA. | 0.200 |
| Specification | " | 0.200 |
| Hospital | " | 0.200 |
| Isolated ground orange | " | 0.250 |
| Duplex | | |
| Specification grade | EA. | 0.200 |
| Hospital | " | 0.200 |
| Isolated ground orange | " | 0.250 |
| GFI hospital grade recepts, 20a, 125v, duplex | " | 0.296 |
| **16198.10 ELECTRIC MANHOLES** | | |
| Precast, handhole, 4' deep | | |
| 2'x2' | EA. | 3.478 |
| 3'x3' | " | 5.556 |
| 4'x4' | " | 10.256 |
| Power manhole, complete, precast, 8' deep | | |
| 4'x4' | EA. | 14.035 |
| 6'x6' | " | 20.000 |
| 8'x8' | " | 21.053 |
| 6' deep, 9' x 12' | " | 25.000 |
| Cast in place, power manhole, 8' deep | | |
| 4'x4' | EA. | 14.035 |
| 6'x6' | " | 20.000 |
| 8'x8' | " | 21.053 |
| **16199.10 UTILITY POLES & FITTINGS** | | |
| Wood pole, creosoted | | |
| 25' | EA. | 2.353 |
| 40' | " | 3.791 |
| 55' | " | 7.547 |
| Treated, wood preservative, 6"x6" | | |
| 8' | EA. | 0.500 |
| 16' | " | 1.600 |
| 20' | " | 2.000 |
| Aluminum, brushed, no base | | |
| 8' | EA. | 2.000 |
| 20' | " | 3.200 |
| 40' | " | 6.250 |
| Steel, no base | | |
| 10' | EA. | 2.500 |
| 20' | " | 3.810 |
| 35' | " | 6.250 |
| Concrete, no base | | |
| 13' | EA. | 5.517 |
| 30' | " | 12.121 |
| 50' | " | 18.182 |

| BASIC MATERIALS | UNIT | MAN/ HOURS |
|---|---|---|
| **16199.10 UTILITY POLES & FITTINGS** | | |
| 60' | EA. | 20.000 |

| POWER GENERATION | UNIT | MAN/ HOURS |
|---|---|---|
| **16210.10 GENERATORS** | | |
| Diesel generator, with auto transfer switch | | |
| 50kw | EA. | 30.769 |
| 125kw | " | 50.000 |
| 300kw | " | 100 |
| 750kw | " | 200 |
| **16320.10 TRANSFORMERS** | | |
| Floor mounted, single phase, int. dry, 480v-120/240v | | |
| 3 kva | EA. | 1.818 |
| 5 kva | " | 3.077 |
| 7.5 kva | " | 3.478 |
| 10 kva | " | 3.810 |
| 15 kva | " | 4.301 |
| 100 kva | " | 11.594 |
| Three phase, 480v-120/280v | | |
| 15 kva | EA. | 6.015 |
| 30 kva | " | 9.412 |
| 45 kva | " | 10.811 |
| 225 kva | " | 15.385 |
| **16350.10 CIRCUIT BREAKERS** | | |
| Molded case, 240v, 15-60a, bolt-on | | |
| 1 pole | EA. | 0.250 |
| 2 pole | " | 0.348 |
| 70-100a, 2 pole | " | 0.533 |
| 15-60a, 3 pole | " | 0.400 |
| 70-100a, 3 pole | " | 0.615 |
| 480v, 2 pole | | |
| 15-60a | EA. | 0.296 |
| 70-100a | " | 0.400 |
| 3 pole | | |
| 15-60a | EA. | 0.400 |
| 70-100a | " | 0.444 |
| 70-225a | " | 0.615 |
| Load center circuit breakers, 240v | | |
| 1 pole, 10-60a | EA. | 0.250 |
| 2 pole | | |
| 10-60a | EA. | 0.400 |
| 70-100a | " | 0.667 |

| POWER GENERATION | UNIT | MAN/HOURS |
|---|---|---|
| **16350.10 CIRCUIT BREAKERS** | | |
| 110-150a | EA. | 0.727 |
| 3 pole | | |
| 10-60a | EA. | 0.500 |
| 70-100a | " | 0.727 |
| Load center, G.F.I. breakers, 240v | | |
| 1 pole, 15-30a | EA. | 0.296 |
| 2 pole, 15-30a | " | 0.400 |
| Key operated breakers, 240v, 1 pole, 10-30a | " | 0.296 |
| Tandem breakers, 240v | | |
| 1 pole, 15-30a | EA. | 0.400 |
| 2 pole, 15-30a | " | 0.533 |
| Bolt-on, G.F.I. breakers, 240v, 1 pole, 15-30a | " | 0.348 |
| **16360.10 SAFETY SWITCHES** | | |
| Fused, 3 phase, 30 amp, 600v, heavy duty | | |
| NEMA 1 | EA. | 1.143 |
| NEMA 3r | " | 1.143 |
| NEMA 4 | " | 1.600 |
| NEMA 12 | " | 1.739 |
| 60a | | |
| NEMA 1 | EA. | 1.143 |
| NEMA 3r | " | 1.143 |
| NEMA 4 | " | 1.600 |
| NEMA 12 | " | 1.739 |
| 100a | | |
| NEMA 1 | EA. | 1.739 |
| NEMA 3r | " | 1.739 |
| NEMA 4 | " | 2.000 |
| NEMA 12 | " | 2.500 |
| 200a | | |
| NEMA 1 | EA. | 2.500 |
| NEMA 3r | " | 2.500 |
| NEMA 4 | " | 2.759 |
| NEMA 12 | " | 3.478 |
| Non-fused, 240-600v, heavy duty, 3 phase, 30 amp | | |
| NEMA 1 | EA. | 1.143 |
| NEMA 3r | " | 1.143 |
| NEMA 4 | " | 1.739 |
| NEMA 12 | " | 1.739 |
| 60a | | |
| NEMA1 | EA. | 1.143 |
| NEMA 3r | " | 1.143 |
| NEMA 4 | " | 1.739 |
| NEMA 12 | " | 1.739 |
| 100a | | |
| NEMA 1 | EA. | 1.739 |
| NEMA 3r | " | 1.739 |
| NEMA 4 | " | 2.500 |
| NEMA 12 | " | 2.500 |
| 200a, NEMA 1 | " | 2.500 |
| 600a, NEMA 12 | " | 12.308 |

| POWER GENERATION | UNIT | MAN/HOURS |
|---|---|---|
| **16365.10 FUSES** | | |
| Fuse, one-time, 250v | | |
| 30a | EA. | 0.050 |
| 60a | " | 0.050 |
| 100a | " | 0.050 |
| 200a | " | 0.050 |
| 400a | " | 0.050 |
| 600a | " | 0.050 |
| 600v | | |
| 30a | EA. | 0.050 |
| 60a | " | 0.050 |
| 100a | " | 0.050 |
| 200a | " | 0.050 |
| 400a | " | 0.050 |
| **16395.10 GROUNDING** | | |
| Ground rods, copper clad, 1/2" x | | |
| 6' | EA. | 0.667 |
| 8' | " | 0.727 |
| 10' | " | 1.000 |
| 5/8" x | | |
| 5' | EA. | 0.615 |
| 6' | " | 0.727 |
| 8' | " | 1.000 |
| 10' | " | 1.250 |
| 3/4" x | | |
| 8' | EA. | 0.727 |
| 10' | " | 0.800 |
| Ground rod clamp | | |
| 5/8" | EA. | 0.123 |
| 3/4" | " | 0.123 |

| SERVICE AND DISTRIBUTION | UNIT | MAN/HOURS |
|---|---|---|
| **16425.10 SWITCHBOARDS** | | |
| Switchboard, 90" high, no main disconnect, 208/120v | | |
| 400a | EA. | 7.921 |
| 600a | " | 8.000 |
| 1000a | " | 8.000 |
| 1200a | " | 10.000 |
| 1600a | " | 11.940 |
| 2000a | " | 14.035 |
| 2500a | " | 16.000 |
| 277/480v | | |
| 600a | EA. | 8.163 |
| 800a | " | 8.163 |

# 16 ELECTRICAL

## SERVICE AND DISTRIBUTION

| 16425.10 SWITCHBOARDS | UNIT | MAN/ HOURS |
|---|---|---|
| 1600a | EA. | 11.940 |
| 2000a | " | 14.035 |
| 2500a | " | 16.000 |
| 3000a | " | 27.586 |
| 4000a | " | 29.630 |

| 16470.10 PANELBOARDS | UNIT | MAN/ HOURS |
|---|---|---|
| 3 phase, 480/277v, main lugs only, 120a, 30 circuits | EA. | 3.478 |
| 277/480v, 4 wire, flush surface | | |
| 225a, 30 circuits | EA. | 4.000 |
| 400a, 30 circuits | " | 5.000 |
| 600a, 42 circuits | " | 6.015 |
| 208/120v, main circuit breaker, 3 phase, 4 wire | | |
| 100a | | |
| 12 circuits | EA. | 5.096 |
| 20 circuits | " | 6.299 |
| 30 circuits | " | 7.018 |
| 400a | | |
| 30 circuits | EA. | 14.815 |
| 42 circuits | " | 16.000 |
| 600a, 42 circuits | " | 18.182 |
| 120/208v, flush, 3 ph., 4 wire, main only | | |
| 100a | | |
| 12 circuits | EA. | 5.096 |
| 20 circuits | " | 6.299 |
| 30 circuits | " | 7.018 |
| 400a | | |
| 30 circuits | EA. | 14.815 |
| 42 circuits | " | 16.000 |
| 600a, 42 circuits | " | 18.182 |

| 16480.10 MOTOR CONTROLS | UNIT | MAN/ HOURS |
|---|---|---|
| Motor generator set, 3 phase, 480/277v, w/controls | | |
| 10kw | EA. | 27.586 |
| 15kw | " | 30.769 |
| 20kw | " | 32.000 |
| 40kw | " | 38.095 |
| 100kw | " | 61.538 |
| 200kw | " | 72.727 |
| 300kw | " | 80.000 |
| 2 pole, 230 volt starter, w/NEMA-1 | | |
| 1 hp, 9 amp, size 00 | EA. | 1.000 |
| 2 hp, 18amp, size 0 | " | 1.000 |
| 3 hp, 27amp, size 1 | " | 1.000 |
| 5 hp, 45amp, size 1p | " | 1.000 |
| 7-1/2 hp, 45a, size 2 | " | 1.000 |
| 15 hp, 90a, size 3 | " | 1.000 |

| 16490.10 SWITCHES | UNIT | MAN/ HOURS |
|---|---|---|
| Fused interrupter load, 35kv | | |
| 20A | | |

## SERVICE AND DISTRIBUTION

| 16490.10 SWITCHES | UNIT | MAN/ HOURS |
|---|---|---|
| 1 pole | EA. | 16.000 |
| 2 pole | " | 17.021 |
| 3 way | " | 17.021 |
| 4 way | " | 18.182 |
| 30a, 1 pole | " | 16.000 |
| 3 way | " | 17.021 |
| 4 way | " | 18.182 |
| Weatherproof switch, including box & cover, 20a | | |
| 1 pole | EA. | 16.000 |
| 2 pole | " | 17.021 |
| 3 way | " | 18.182 |
| 4 way | " | 18.182 |
| Specification grade toggle switches, 20a, 120-277v | | |
| Single pole | EA. | 0.200 |
| Double pole | " | 0.296 |
| 3 way | " | 0.250 |
| 4 way | " | 0.296 |
| Switch plates, plastic ivory | | |
| 1 gang | EA. | 0.080 |
| 2 gang | " | 0.100 |
| 3 gang | " | 0.119 |
| 4 gang | " | 0.145 |
| 5 gang | " | 0.160 |
| 6 gang | " | 0.182 |
| Stainless steel | | |
| 1 gang | EA. | 0.080 |
| 2 gang | " | 0.100 |
| 3 gang | " | 0.123 |
| 4 gang | " | 0.145 |
| 5 gang | " | 0.160 |
| 6 gang | " | 0.182 |
| Brass | | |
| 1 gang | EA. | 0.080 |
| 2 gang | " | 0.100 |
| 3 gang | " | 0.123 |
| 4 gang | " | 0.145 |
| 5 gang | " | 0.160 |
| 6 gang | " | 0.182 |

| 16490.20 TRANSFER SWITCHES | UNIT | MAN/ HOURS |
|---|---|---|
| Automatic transfer switch 600v, 3 pole | | |
| 30a | EA. | 3.478 |
| 100a | " | 4.762 |
| 400a | " | 10.000 |
| 800a | " | 18.182 |
| 1200a | " | 22.857 |
| 2600a | " | 42.105 |

| 16490.80 SAFETY SWITCHES | UNIT | MAN/ HOURS |
|---|---|---|
| Safety switch, 600v, 3 pole, heavy duty, NEMA-1 | | |
| 30a | EA. | 1.000 |
| 60a | " | 1.143 |
| 100a | " | 1.600 |

# 16 ELECTRICAL

| SERVICE AND DISTRIBUTION | UNIT | MAN/ HOURS |
|---|---|---|
| **16490.80 SAFETY SWITCHES** | | |
| 200a | EA. | 2.500 |
| 400a | " | 5.517 |
| 600a | " | 8.000 |
| 800a | " | 10.526 |
| 1200a | " | 14.286 |

| LIGHTING | UNIT | MAN/ HOURS |
|---|---|---|
| **16510.05 INTERIOR LIGHTING** | | |
| Recessed fluorescent fixtures, 2'x2' | | |
| 2 lamp | EA. | 0.727 |
| 4 lamp | " | 0.727 |
| 2 lamp w/flange | " | 1.000 |
| 4 lamp w/flange | " | 1.000 |
| 1'x4' | | |
| 2 lamp | EA. | 0.667 |
| 3 lamp | " | 0.667 |
| 2 lamp w/flange | " | 0.727 |
| 3 lamp w/flange | " | 0.727 |
| 2'x4' | | |
| 2 lamp | EA. | 0.727 |
| 3 lamp | " | 0.727 |
| 4 lamp | " | 0.727 |
| 2 lamp w/flange | " | 1.000 |
| 3 lamp w/flange | " | 1.000 |
| 4 lamp w/flange | " | 1.000 |
| 4'x4' | | |
| 4 lamp | EA. | 1.000 |
| 6 lamp | " | 1.000 |
| 8 lamp | " | 1.000 |
| 4 lamp w/flange | " | 1.509 |
| 6 lamp w/flange | " | 1.509 |
| 8 lamp, w/flange | " | 1.509 |
| Surface mounted incandescent fixtures | | |
| 40w | EA. | 0.667 |
| 75w | " | 0.667 |
| 100w | " | 0.667 |
| 150w | " | 0.667 |
| Pendant | | |
| 40w | EA. | 0.800 |
| 75w | " | 0.800 |
| 100w | " | 0.800 |
| 150w | " | 0.800 |
| Recessed incandescent fixtures | | |
| 40w | EA. | 1.509 |
| 75w | " | 1.509 |

| LIGHTING | UNIT | MAN/ HOURS |
|---|---|---|
| **16510.05 INTERIOR LIGHTING** | | |
| 100w | EA. | 1.509 |
| 150w | " | 1.509 |
| Exit lights, 120v | | |
| Recessed | EA. | 1.250 |
| Back mount | " | 0.727 |
| Universal mount | " | 0.727 |
| Emergency battery units, 6v-120v, 50 unit | " | 1.509 |
| With 1 head | " | 1.509 |
| With 2 heads | " | 1.509 |
| Light track single circuit | | |
| 2' | EA. | 0.500 |
| 4' | " | 0.500 |
| 8' | " | 1.000 |
| 12' | " | 1.509 |
| Fixtures, square | | |
| R-20 | EA. | 0.145 |
| R-30 | " | 0.145 |
| Mini spot | " | 0.145 |
| Surface mounted fluorescent, wrap around lens | | |
| 1 lamp | EA. | 0.800 |
| 2 lamps | " | 0.889 |
| 4 lamps | " | 1.000 |
| Wall mounted fluorescent | | |
| 2-20w lamps | EA. | 0.500 |
| 2-30w lamps | " | 0.500 |
| 2-40w lamps | " | 0.667 |
| Indirect, with wood shielding, 2049w lamps | | |
| 4' | EA. | 1.000 |
| 8' | " | 1.600 |
| Industrial fluorescent, 2 lamp | | |
| 4' | EA. | 0.727 |
| 8' | " | 1.333 |
| Strip fluorescent | | |
| 4' | | |
| 1 lamp | EA. | 0.667 |
| 2 lamps | " | 0.667 |
| 8' | | |
| 1 lamp | EA. | 0.727 |
| 2 lamps | " | 0.889 |
| Wire guard for strip fixture, 4' long | " | 0.348 |
| Strip fluorescent, 8' long, two 4' lamps | " | 1.333 |
| With four 4' lamps | " | 1.600 |
| Wet location fluorescent, plastic housing | | |
| 4' long | | |
| 1 lamp | EA. | 1.000 |
| 2 lamps | " | 1.333 |
| 8' long | | |
| 2 lamps | EA. | 1.600 |
| 4 lamps | " | 1.739 |
| Parabolic troffer, 2'x2' | | |
| With 2 "U" lamps | EA. | 1.000 |
| With 3 "U" lamps | " | 1.143 |
| 2'x4' | | |
| With 2 40w lamps | EA. | 1.143 |

| LIGHTING | UNIT | MAN/ HOURS |
|---|---|---|
| **16510.05 INTERIOR LIGHTING** | | |
| With 3 40w lamps | EA. | 1.333 |
| With 4 40w lamps | " | 1.333 |
| 1'x4' | | |
| With 1 T-12 lamp, 9 cell | EA. | 0.727 |
| With 2 T-12 lamps | " | 0.889 |
| With 1 T-12 lamp, 20 cell | " | 0.727 |
| With 2 T-12 lamps | " | 0.889 |
| Steel sided surface fluorescent, 2'x4' | | |
| 3 lamps | EA. | 1.333 |
| 4 lamps | " | 1.333 |
| Outdoor sign fluor., 1 lamp, remote ballast | | |
| 4' long | EA. | 6.015 |
| 6' long | " | 8.000 |
| Recess mounted, commercial, 2'x2', 13" high | | |
| 100w | EA. | 4.000 |
| 250w | " | 4.494 |
| High pressure sodium, hi-bay open | | |
| 400w | EA. | 1.739 |
| 1000w | " | 2.424 |
| Enclosed | | |
| 400w | EA. | 2.424 |
| 1000w | " | 2.963 |
| Metal halide hi-bay, open | | |
| 400w | EA. | 1.739 |
| 1000w | " | 2.424 |
| Enclosed | | |
| 400w | EA. | 2.424 |
| 1000w | " | 2.963 |
| High pressure sodium, low bay, surface mounted | | |
| 100w | EA. | 1.000 |
| 150w | " | 1.143 |
| 250w | " | 1.333 |
| 400w | " | 1.600 |
| Metal halide, low bay, pendant mounted | | |
| 175w | EA. | 1.333 |
| 250w | " | 1.600 |
| 400w | " | 2.222 |
| Indirect luminare, square, metal halide, freestanding | | |
| 175w | EA. | 1.000 |
| 250w | " | 1.000 |
| 400w | " | 1.000 |
| High pressure sodium | | |
| 150w | EA. | 1.000 |
| 250w | " | 1.000 |
| 400w | " | 1.000 |
| Round, metal halide | | |
| 175w | EA. | 1.000 |
| 250w | " | 1.000 |
| 400w | " | 1.000 |
| High pressure sodium | | |
| 150w | EA. | 1.000 |
| 250w | " | 1.000 |
| 400w | " | 1.000 |
| Wall mounted, metal halide | | |

| LIGHTING | UNIT | MAN/ HOURS |
|---|---|---|
| **16510.05 INTERIOR LIGHTING** | | |
| 175w | EA. | 2.500 |
| 250w | " | 2.500 |
| 400w | " | 3.200 |
| High pressure sodium | | |
| 150w | EA. | 2.500 |
| 250w | " | 2.500 |
| 400w | " | 3.200 |
| Wall pack lithonia, high pressure sodium | | |
| 35w | EA. | 0.889 |
| 55w | " | 1.000 |
| 150w | " | 1.600 |
| 250w | " | 1.739 |
| Low pressure sodium | | |
| 35w | EA. | 1.739 |
| 55w | " | 2.000 |
| Wall pack hubbell, high pressure sodium | | |
| 35w | EA. | 0.889 |
| 150w | " | 1.600 |
| 250w | " | 1.739 |
| Compact fluorescent | | |
| 2-7w | EA. | 1.000 |
| 2-13w | " | 1.333 |
| 1-18w | " | 1.333 |
| Handball & racquet ball court, 2'x2', metal halide | | |
| 250w | EA. | 2.500 |
| 400w | " | 2.759 |
| High pressure sodium | | |
| 250w | EA. | 2.500 |
| 400w | " | 2.759 |
| Bollard light, 42" w/found., high pressure sodium | | |
| 70w | EA. | 2.581 |
| 100w | " | 2.581 |
| 150w | " | 2.581 |
| Light fixture lamps | | |
| Lamp | | |
| 20w med. bipin base, cool white, 24" | EA. | 0.145 |
| 30w cool white, rapid start, 36" | " | 0.145 |
| 40w cool white "U", 3" | " | 0.145 |
| 40w cool white, rapid start, 48" | " | 0.145 |
| 70w high pressure sodium, mogul base | " | 0.200 |
| 75w slimline, 96" | " | 0.200 |
| 100w | | |
| Incandescent, 100a, inside frost | EA. | 0.100 |
| Mercury vapor, clear, mogul base | " | 0.200 |
| High pressure sodium, mogul base | " | 0.200 |
| 150w | | |
| Par 38 flood or spot, incandescent | EA. | 0.100 |
| High pressure sodium, 1/2 mogul base | " | 0.200 |
| 175w | | |
| Mercury vapor, clear, mogul base | EA. | 0.200 |
| Metal halide, clear, mogul base | " | 0.200 |
| 250w | | |
| High pressure sodium, mogul base | EA. | 0.200 |
| Mercury vapor, clear, mogul base | " | 0.200 |

# 16 ELECTRICAL

| LIGHTING | UNIT | MAN/ HOURS |
|---|---|---|
| **16510.05 INTERIOR LIGHTING** | | |
| Metal halide, clear, mogul base | EA. | 0.200 |
| High pressure sodium, mogul base | " | 0.200 |
| 400w | | |
| Mercury vapor, clear, mogul base | EA. | 0.200 |
| Metal halide, clear, mogul base | " | 0.200 |
| High pressure sodium, mogul base | " | 0.200 |
| 1000w | | |
| Mercury vapor, clear, mogul base | EA. | 0.250 |
| High pressure sodium, mogul base | " | 0.250 |
| **16510.30 EXTERIOR LIGHTING** | | |
| Exterior light fixtures | | |
| Rectangle, high pressure sodium | | |
| 70w | EA. | 2.500 |
| 100w | " | 2.581 |
| 150w | " | 2.581 |
| 250w | " | 2.759 |
| 400w | " | 3.478 |
| Flood, rectangular, high pressure sodium | | |
| 70w | EA. | 2.500 |
| 100w | " | 2.581 |
| 150w | " | 2.581 |
| 400w | " | 3.478 |
| 1000w | " | 4.494 |
| Round | | |
| 400w | EA. | 3.478 |
| 1000w | " | 4.494 |
| Round, metal halide | | |
| 400w | EA. | 3.478 |
| 1000w | " | 4.494 |
| Light fixture arms, cobra head, 6', high press. sodium | | |
| 100w | EA. | 2.000 |
| 150w | " | 2.500 |
| 250w | " | 2.500 |
| 400w | " | 2.963 |
| Flood, metal halide | | |
| 400w | EA. | 3.478 |
| 1000w | " | 4.494 |
| 1500w | " | 6.015 |
| Mercury vapor | | |
| 250w | EA. | 2.759 |
| 400w | " | 3.478 |
| Incandescent | | |
| 300w | EA. | 1.739 |
| 500w | " | 2.000 |
| 1000w | " | 3.200 |
| **16610.30 UNINTERRUPTIBLE POWER** | | |
| Uninterruptible power systems, (U.P.S.) | EA. | 8.000 |
| 5 kva | " | 11.004 |
| 7.5 kva | " | 16.000 |
| 10 kva | " | 21.978 |

| LIGHTING | UNIT | MAN/ HOURS |
|---|---|---|
| **16610.30 UNINTERRUPTIBLE POWER** | | |
| 15 kva | EA. | 22.857 |
| 20 kva | " | 24.024 |
| 25 kva | " | 25.000 |
| 30 kva | " | 25.974 |
| 35 kva | " | 27.027 |
| 40 kva | " | 27.972 |
| 45 kva | " | 28.986 |
| 50 kva | " | 29.963 |
| 62.5 kva | " | 32.000 |
| 75 kva | " | 34.934 |
| 100 kva | " | 36.036 |
| 150 kva | " | 50.000 |
| 200 kva | " | 55.172 |
| 300 kva | " | 74.766 |
| 400 kva | " | 89.888 |
| 500 kva | " | 110 |
| **16670.10 LIGHTNING PROTECTION** | | |
| Lightning protection | | |
| Copper point, nickel plated, 12' | | |
| 1/2" dia. | EA. | 1.000 |
| 5/8" dia. | " | 1.000 |

| COMMUNICATIONS | UNIT | MAN/ HOURS |
|---|---|---|
| **16720.10 FIRE ALARM SYSTEMS** | | |
| Master fire alarm box, pedestal mounted | EA. | 16.000 |
| Master fire alarm box | " | 6.015 |
| Box light | " | 0.500 |
| Ground assembly for box | " | 0.667 |
| Bracket for pole type box | " | 0.727 |
| Pull station | | |
| Waterproof | EA. | 0.500 |
| Manual | " | 0.400 |
| Horn, waterproof | " | 1.000 |
| Interior alarm | " | 0.727 |
| Coded transmitter, automatic | " | 2.000 |
| Control panel, 8 zone | " | 8.000 |
| Battery charger and cabinet | " | 2.000 |
| Batteries, nickel cadmium or lead calcium | " | 5.000 |
| CO2 pressure switch connection | " | 0.727 |
| Annunciator panels | | |
| Fire detection annunciator, remote type, 8 zone | EA. | 1.818 |
| 12 zone | " | 2.000 |
| 16 zone | " | 2.500 |
| Fire alarm systems | | |
| Bell | EA. | 0.615 |

| COMMUNICATIONS | UNIT | MAN/ HOURS |
|---|---|---|
| **16720.10 FIRE ALARM SYSTEMS** | | |
| Weatherproof bell | EA. | 0.667 |
| Horn | " | 0.727 |
| Siren | " | 2.000 |
| Chime | " | 0.615 |
| Audio/visual | " | 0.727 |
| Strobe light | " | 0.727 |
| Smoke detector | " | 0.667 |
| Heat detection | " | 0.500 |
| Thermal detector | " | 0.500 |
| Ionization detector | " | 0.533 |
| Duct detector | " | 2.759 |
| Test switch | " | 0.500 |
| Remote indicator | " | 0.571 |
| Door holder | " | 0.727 |
| Telephone jack | " | 0.296 |
| Fireman phone | " | 1.000 |
| Speaker | " | 0.800 |
| Remote fire alarm annunciator panel | | |
| 24 zone | EA. | 6.667 |
| 48 zone | " | 13.008 |
| Control panel | | |
| 12 zone | EA. | 2.963 |
| 16 zone | " | 4.444 |
| 24 zone | " | 6.667 |
| 48 zone | " | 16.000 |
| Power supply | " | 1.509 |
| Status command | " | 5.000 |
| Printer | " | 1.509 |
| Transponder | " | 0.899 |
| Transformer | " | 0.667 |
| Transceiver | " | 0.727 |
| Relays | " | 0.500 |
| Flow switch | " | 2.000 |
| Tamper switch | " | 2.963 |
| End of line resistor | " | 0.348 |
| Printed ckt. card | " | 0.500 |
| Central processing unit | " | 6.154 |
| UPS backup to c.p.u. | " | 8.999 |
| Smoke detector, fixed temp. & rate of rise comb. | " | 1.600 |
| **16720.50 SECURITY SYSTEMS** | | |
| Sensors | EA. | |
| Balanced magnetic door switch, surface mounted | " | 0.500 |
| With remote test | " | 1.000 |
| Flush mounted | " | 1.860 |
| Mounted bracket | " | 0.348 |
| Mounted bracket spacer | " | 0.348 |
| Photoelectric sensor, for fence | " | |
| 6 beam | " | 2.759 |
| 9 beam | " | 4.255 |
| Photoelectric sensor, 12 volt dc | " | |
| 500' range | " | 1.600 |
| 800' range | " | 2.000 |

| COMMUNICATIONS | UNIT | MAN/ HOURS |
|---|---|---|
| **16720.50 SECURITY SYSTEMS** | | |
| Capacitance wire grid kit | EA. | |
| Surface | " | 1.000 |
| Duct | " | 1.600 |
| Tube grid kit | " | 0.500 |
| Vibration sensor, 30 max per zone | " | 0.500 |
| Audio sensor, 30 max per zone | " | 0.500 |
| Inertia sensor | " | |
| Outdoor | " | 0.727 |
| Indoor | " | 0.500 |
| Ultrasonic transmitter, 20 max per zone | | |
| Omni-directional | EA. | 1.600 |
| Directional | " | 1.333 |
| Transceiver | | |
| Omni-directional | EA. | 1.000 |
| Directional | " | 1.000 |
| Passive infra-red sensor, 20 max per zone | " | 1.600 |
| Access/secure control unit, balanced magnetic switch | " | 1.600 |
| Photoelectric sensor | " | 1.600 |
| Photoelectric fence sensor | " | 1.600 |
| Capacitance sensor | " | 1.739 |
| Audio and vibration sensor | " | 1.600 |
| Inertia sensor | " | 1.600 |
| Ultrasonic sensor | " | 1.739 |
| Infra-red sensor | " | 2.000 |
| Monitor panel, with access/secure tone, standard | " | 1.739 |
| High security | " | 2.000 |
| Emergency power indicator | " | 0.500 |
| Monitor rack with 115v power supply | | |
| 1 zone | EA. | 1.000 |
| 10 zone | " | 2.500 |
| Monitor cabinet, wall mounted | | |
| 1 zone | EA. | 1.000 |
| 5 zone | " | 1.600 |
| 10 zone | " | 1.739 |
| 20 zone | " | 2.000 |
| Floor mounted, 50 zone | " | 4.000 |
| Security system accessories | | |
| Tamper assembly for monitor cabinet | EA. | 0.444 |
| Monitor panel blank | " | 0.348 |
| Audible alarm | " | 0.500 |
| Audible alarm control | " | 0.348 |
| Termination screw, terminal cabinet | | |
| 25 pair | EA. | 1.600 |
| 50 pair | " | 2.500 |
| 150 pair | " | 5.000 |
| Universal termination, cabinets & panel | | |
| Remote test | EA. | 1.739 |
| No remote test | " | 0.727 |
| High security line supervision termination | " | 1.000 |
| Door cord for capacitance sensor, 12" | " | 0.500 |
| Insulation block kit for capacitance sensor | " | 0.348 |
| Termination block for capacitance sensor | " | 0.348 |
| Guard alert display | " | 0.615 |
| Uninterrupted power supply | | |

| COMMUNICATIONS | UNIT | MAN/ HOURS |
|---|---|---|
| **16720.50 SECURITY SYSTEMS** | | |
| Plug-in 40kva transformer | | |
| 12 volt | EA. | 0.348 |
| 18 volt | " | 0.348 |
| 24 volt | " | 0.348 |
| Test relay | " | 0.348 |
| Coaxial cable, 50 ohm | L.F. | 0.006 |
| Door openers | EA. | 0.500 |
| Push buttons | | |
| Standard | EA. | 0.348 |
| Weatherproof | " | 0.444 |
| Bells | " | 0.727 |
| Horns | | |
| Standard | EA. | 1.000 |
| Weatherproof | " | 1.250 |
| Chimes | " | 0.667 |
| Flasher | " | 0.615 |
| Motion detectors | " | 1.509 |
| Intercom units | " | 0.727 |
| Remote annunciator | " | 5.000 |
| **16740.10 TELEPHONE SYSTEMS** | | |
| Communication cable | | |
| 25 pair | L.F. | 0.026 |
| 100 pair | " | 0.029 |
| 400 pair | " | 0.044 |
| Cable tap in manhole or junction box | | |
| 25 pair cable | EA. | 3.810 |
| 100 pair cable | " | 15.094 |
| 400 pair cable | " | 61.538 |
| Cable terminations, manhole or junction box | | |
| 25 pair cable | EA. | 3.756 |
| 100 pair cable | " | 15.094 |
| 400 pair cable | " | 61.538 |
| **16780.50 TELEVISION SYSTEMS** | | |
| TV outlet, self terminating, w/cover plate | EA. | 0.308 |
| Thru splitter | " | 1.600 |
| End of line | " | 1.333 |
| In line splitter multitap | | |
| 4 way | EA. | 1.818 |
| 2 way | " | 1.702 |
| Equipment cabinet | " | 1.600 |
| Antenna | | |
| Broad band uhf | EA. | 3.478 |
| Lightning arrester | " | 0.727 |
| TV cable | L.F. | 0.005 |
| **16850.10 ELECTRIC HEATING** | | |
| Baseboard heater | | |
| 2', 375w | EA. | 1.000 |
| 3', 500w | " | 1.000 |
| 4', 750w | " | 1.143 |

| COMMUNICATIONS | UNIT | MAN/ HOURS |
|---|---|---|
| **16850.10 ELECTRIC HEATING** | | |
| 5', 935w | EA. | 1.333 |
| 6', 1125w | " | 1.600 |
| 7', 1310w | " | 1.818 |
| 8', 1500w | " | 2.000 |
| 9', 1680w | " | 2.222 |
| 10', 1875w | " | 2.286 |
| Unit heater wall mounted | | |
| 1500w | EA. | 1.600 |
| 2500w | " | 1.818 |
| 4000w | " | 2.286 |
| Thermostat | | |
| Integral | EA. | 0.500 |
| Line voltage | " | 0.500 |
| Electric heater connection | " | 0.250 |

# Supporting Construction Reference Data

This section contains information, text, charts and tables on various aspects of construction. The intent is to provide the user with a better understanding of unfamiliar areas in order to be able to estimate better. This information includes actual takeoff data for some areas and also selected explanations of common construction materials, methods and common practices.

# GENERAL / ALLOWANCES | 01024

## TYPICAL BUILDING COST BROKEN DOWN BY CSI FORMAT

*(Commercial Construction)*

| Division | | New Construction | Remodeling Construction |
|---|---|---|---|
| 1. | General Requirements | 6 to 8% | Up to 30% |
| 2. | Sitework | 4 to 6% | |
| 3. | Concrete | 15 to 20% | |
| 4. | Masonry | 8 to 12% | |
| 5. | Metals | 5 to 7% | |
| 6. | Wood And Plastics | 1 to 5% | |
| 7. | Thermal And Moisture Protection | 4 to 6% | |
| 8. | Doors And Windows | 5 to 7% | Up to 30% (Divisions 8–9) |
| 9. | Finishes | 8 to 12% | |
| 10. | Specialties | 6 to 10% (Divisions 10–14) | |
| 11. | Architectural Equipment | | |
| 12. | Furnishings | | |
| 13. | Special Construction | | |
| 14. | Conveying Systems | | |
| 15. | Mechanical | 15 to 25% | Up to 40% (Divisions 15–16) |
| 16. | Electrical | 8 to 12% | |
| | **Total Cost** | 100% | |

## CONVERSION FACTORS

| Change | To | Multiply By |
|---|---|---|
| Atmospheres | Pounds per square inch | 14.696 |
| Atmospheres | Inches of mercury | 29.92 |
| Atmospheres | Feet of water | 34 |
| Barrels, oil | Gallons, of oil | 42 |
| Barrels, cement | Pounds of cement | 376 |
| Bags or sacks, cement | Pounds of cement | 94 |
| Btu/min. | Foot-pounds/sec. | 12.96 |
| Btu/min. | Horse-power | 0.02356 |
| Btu/min. | Kilowatts | 0.01757 |
| Btu/min. | Watts | 17.57 |
| Centimeters | Inches | 0.3937 |
| Centimeters of mercury | Atmospheres | 0.01316 |
| Centimeters of mercury | Feet of water | 0.4461 |
| Cubic inches | Cubic feet | 0.00058 |
| Cubic feet | Cubic inches | 1728 |
| cubic feet | Cubic yards | 0.03703 |
| Cubic yards | Cubic feet | 27 |
| Cubic inches | Gallons | 0.00433 |
| Cubic feet | Gallons | 7.48 |
| Feet | Inches | 12 |
| Feet | Yards | 0.3333 |
| Yards | Feet | 3 |
| Feet of water | Atmospheres | 0.02950 |
| Feet of water | Inches of mercury | 0.8826 |
| Gallons | Cubic Inches | 231 |
| Gallons | Cubic feet | 0.1337 |
| Gallons | Pounds of water | 8.33 |
| Gallons | Quarts | 4 |
| Gallons per min. | Cubic feet sec. | 0.002228 |
| Gallons per min. | Cubic feet hour | 8.0208 |
| Gallons water per min. | Tons water/24 hours | 6.0086 |
| Horse-power | Foot-lbs./sec. | 550 |
| Inches | Centimeters | 2.540 |
| Inches | Feet | 0.0833 |
| Inches | Millimeters | 25.4 |
| Inches of water | Pounds per Sq. inch | 0.0361 |
| Inches of water | Inches of mercury | 0.0735 |
| Inches of water | Ounces per square inch | 0.578 |
| Inches of water | Ounces per square foot | 5.2 |
| Inches of mercury | Inches of water | 13.6 |
| Inches of mercury | Feet of water | 1.1333 |
| Inches of mercury | Pounds per square inch | 0.4914 |
| Kilometers | Miles | 0.6214 |
| Meters | Inches | 39.37 |
| Miles | Feet | 5280 |
| Millimeters | Centimeters | 0.1 |
| Millimeters | Inches | 0.03937 |
| Ounces (fluid) | Cubic inches | 1.805 |
| Ounces | Pounds | 0.0625 |
| Pounds | Ounces | 16 |
| Pounds per square inch | Inches of water | 27.72 |
| Pounds per square inch | Feet of water | 2.310 |
| Pounds per square inch | Inches of mercury | 2.04 |
| Pounds per square inch | Atmospheres | 0.0681 |
| Quarts | Cubic Inches | 67.20 |
| Square Inches | Square feet | 0.00694 |
| Square Feet | Square inches | 144 |
| Square Feet | Square yards | 0.11111 |
| Square yards | Square feet | 9 |
| Square miles | Acres | 640 |
| Short tons | Pounds | 2000 |
| Short tons | Long tons | 0.89285 |
| Tons of water/24 hours | Gallons per minute | 0.16643 |
| Yards | Feet | 3 |
| Yards | Centimeters | 91.44 |
| Yards | Inches | 36 |

## CONVERSION CALCULATIONS

### Commercial Measure

| | |
|---|---|
| 16 drams | =1 ounce |
| 16 ounces | =1 pound |
| 2,000 pounds | =1 ton |

### Long Measure

| | |
|---|---|
| 12 inches | =1 foot |
| 3 feet | =1 yard |
| 16½ feet | =1 rod |
| 40 rods | =1 furlong |
| 8 furlongs (5,280 ft) = | =1 mile |
| 3 miles | =1 league |

### Square Measure

| | |
|---|---|
| 144 square inches | =1 square foot |
| 9 square feet | =1 square yard |
| 30¼ square yards | =1 square rod |
| 160 square rods | =1 acre |
| 4840 square yards | =1 acre |
| 640 acres | =1 square mile |
| 36 square miles | =1 township |

### Surveyors Measure

| | |
|---|---|
| 7.92 inches | =1 link |
| 25 links | =1 rod |
| 4 rods (66 ft.) | =1 chain |
| 10 chains | =1 furlong |
| 8 furlongs | =1 mile |
| 1 square mile | =1 section |

### Cubic Measure

| | |
|---|---|
| 1728 cubic inches | =1 cubic foot |
| 27 cubic feet | =1 cubic yard |
| 128 cubic feet | =1 cord (wood/stone) |
| 231 cubic inches | =1 U.S. gallon |
| 7.48 U.S. Gallons | =1 cubic foot |
| 2150.4 cubic inches | =1 U.S. bushel |

### Liquid Measure

| | |
|---|---|
| 4 fluid ounces | =1 gill |
| 4 gills | =1 pint |
| 2 pints | =1 quart |
| 4 quarts | =1 gallon |
| 9 gallons | =1 firkin |
| 31½ gallons | =1 barrel |
| 2 barrels | =1 hogshead |

### Dry Measure

| | |
|---|---|
| 2 pints | =1 quart |
| 8 quarts | =1 peck |
| 4 pecks | =1 bushel |
| 2150.42 cubic inches | =1 bushel |

**SQUARE**

$A = a^2$

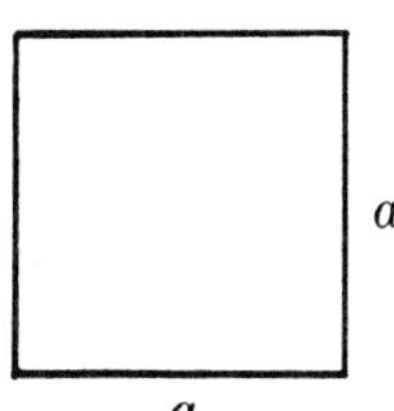


**RECTANGLE**

$A = bh$

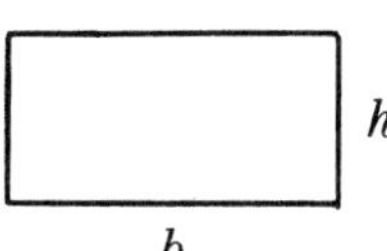


**TRIANGLE**

$A = \frac{1}{2} bh$

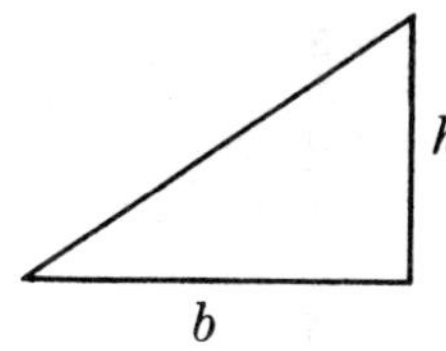


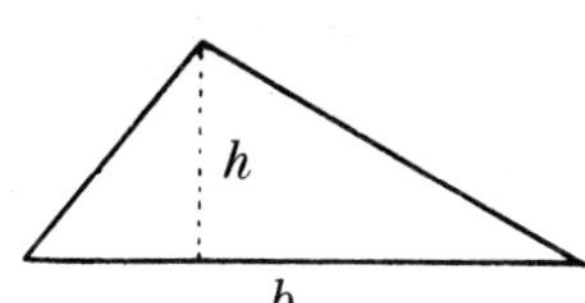


**PARALLELOGRAM**

$A = bh = ab \text{ Sin}\phi$

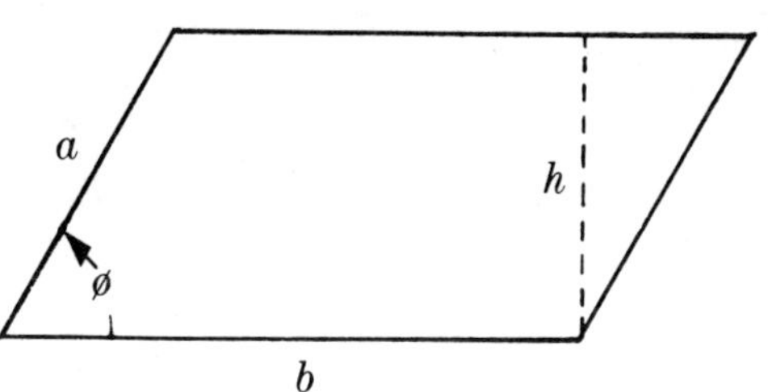


**TRAPEZOID**

$A = \left(\frac{a+b}{2}\right)h$

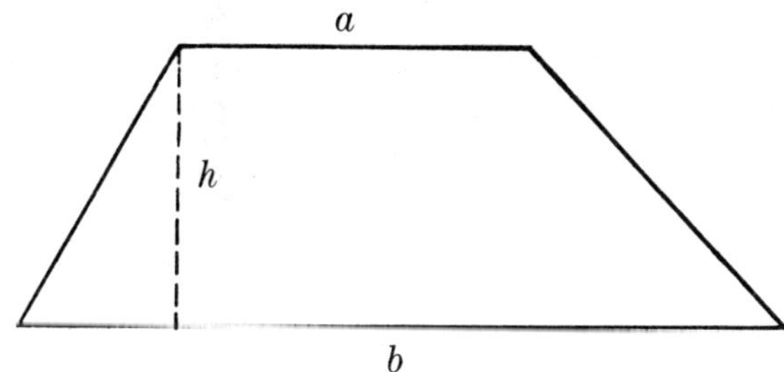


**CIRCLE**

$A = \pi r^2 = \frac{\pi d^2}{4}$

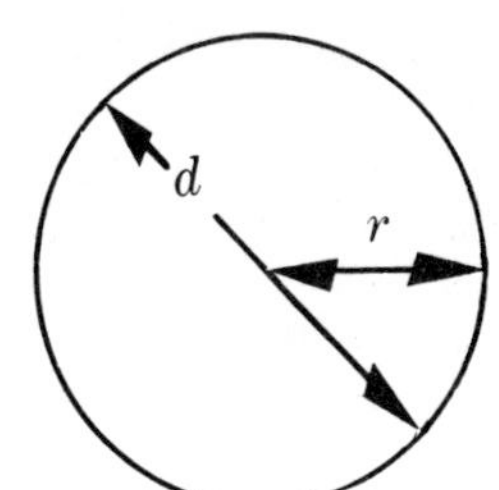


$Circumference = C = 2\pi r = \pi d$

**ELLIPSE**

$A = 0.7854\ ab$

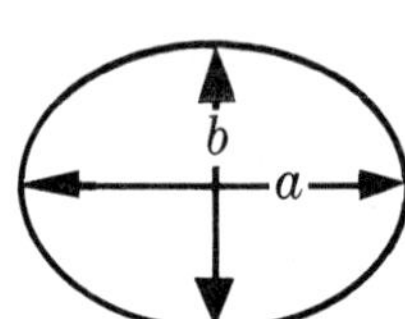


**PARABOLA**

$A = \frac{2}{3} bh$

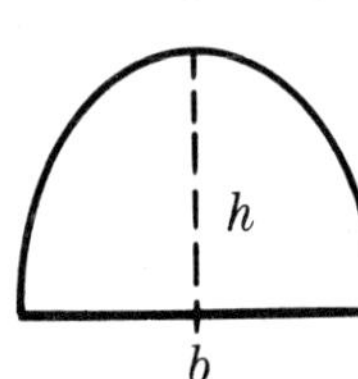


*VOLUMES*

**CUBE**

$V = a^3$

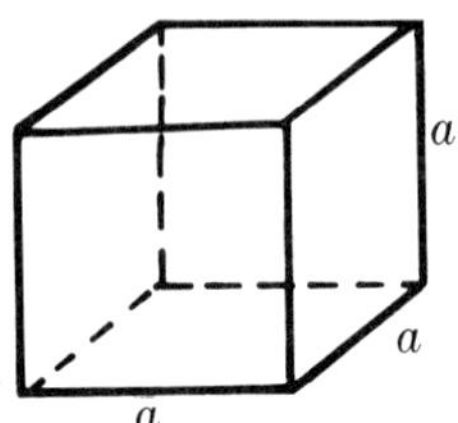


**CYLINDER**

$V = \pi r^3 h$

h
r

**PYRAMID**

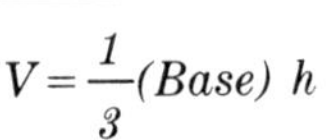

$V = \frac{1}{3} (Base)\ h$

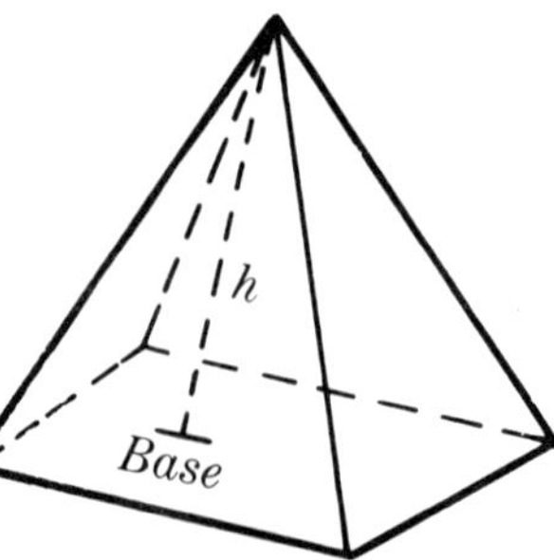


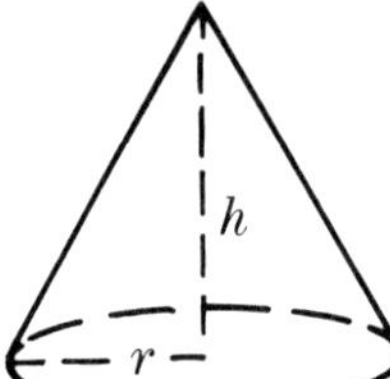


**CONE**

$V = \frac{1}{3} \pi r^2 h$

**SPHERE**

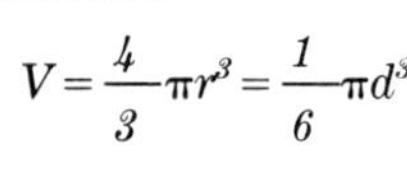

$V = \frac{4}{3} \pi r^3 = \frac{1}{6} \pi d^3$

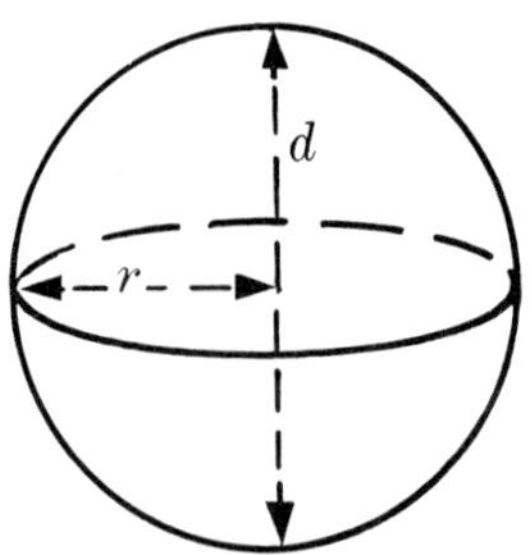

**WEDGE**

$V = \frac{1}{2} abc$

# CONVERSION FACTORS
## ENGLISH TO SI (SYSTEM INTERNATIONAL)

| To Convert from | To | Multiply by |
|---|---|---|
| **LENGTH** | | |
| Inches | Millimetres | 25.4[a] |
| Feet | Metres | 0.3048[a] |
| Yards | Metres | 0.9144[a] |
| Miles (statute) | Kilometres | 1.609 |
| **AREA** | | |
| Square inches | Square millimetres | 645.2 |
| Square feet | Square metres | 0.0929 |
| Square yards | Square metres | 0.8361 |
| **VOLUME** | | |
| Cubic inches | Cubic millimetres | 16.387 |
| Cubic feet | Cubic metres | 0.02832 |
| Cubic yards | Cubic metres | 0.7646 |
| Gallons (U.S. liquid)[b] | Cubic metres[c] | 0.003785 |
| Gallons (Canadian liquid)[b] | Cubic metres[c] | 0.004546 |
| Ounces (U.S. liquid)[b] | Millilitres[c, d] | 29.57 |
| Quarts (U.S. liquid)[b] | Litres[c, d] | 0.9464 |
| Gallons (U.S. liquid)[b] | Litres[c] | 3.785 |
| **FORCE** | | |
| Kilograms force | Newtons | 9.807 |
| Pounds force | Newtons | 4.448 |
| Pounds force | Kilograms force[d] | 0.4536 |
| Kips | Newtons | 4448 |
| Kips | Kilograms force[d] | 453.6 |
| **PRESSURE, STRESS, STRENGTH (FORCE PER UNIT AREA)** | | |
| Kilograms force per sq. centimetre | Megapascals | 0.09807 |
| Pounds force per square inch (psi) | Megapascals | 6895 |
| Kips per square inch | Megapascals | 6.895 |
| Pounds force per square inch (psi) | Kilograms force per square centimetre[d] | 0.07031 |
| Pounds force per square foot | Pascals | 47.88 |
| Pounds force per square foot | Kilograms force per square metre[d] | 4.882 |
| **BENDING MOMENT OR TORQUE** | | |
| Inch-pounds force | Metre-kilog. force[d] | 0.01152 |
| Inch-pounds force | Newton-metres | 0.1130 |
| Foot-pounds force | Metre-kilog. force[d] | 0.1383 |
| Foot-pounds force | Newton-metres | 1.356 |
| Metre-kilograms force | Newton-metres | 9.807 |
| **MASS** | | |
| Ounce (avoirdupois) | Grams | 28.35 |
| Pounds (avoirdupois) | Kilograms | 0.4536 |
| Tons (metric) | Kilograms | 1000[a] |
| Tons, short (2000 pounds) | Kilograms | 907.2 |
| Tons, short (2000 pounds) | Megagrams[e] | 0.9072 |
| **MASS PER UNIT VOLUME** | | |
| Pounds mass per cubic foot | Kilog. per cubic metre | 16.02 |
| Pounds mass per cubic yard | Kilog. per cubic metre | 0.5933 |
| Pds. mass per gallon (U.S. liquid)[b] | Kilog. per cubic metre | 119.8 |
| Pds. mass p/gal. (Canadian liquid)[b] | Kilog. per cubic metre | 99.78 |
| **TEMPERATURE** | | |
| Degrees Fahrenheit | Degrees Celsius | tK = (1F − 32)/1.8 |
| Degrees Fahrenheit | Degrees Kelvin | tK = (1F + 459.67)/1.8 |
| Degree Celsius | Degree Kelvin | tK = 1C + 273.15 |

[a] The factor given is exact.
[b] One U.S. gallon equals 0.8327 Canadian gallon.
[c] 1 litre = 1000 millilitres = 10,000 cubic centimetres = 1 cubic decimetre = 0.001 cubic metre.
[d] Metric but not SI unit.
[e] Called "tonne" in England. Called "metric ton" in other metric systems.

## TRENCH BRACING

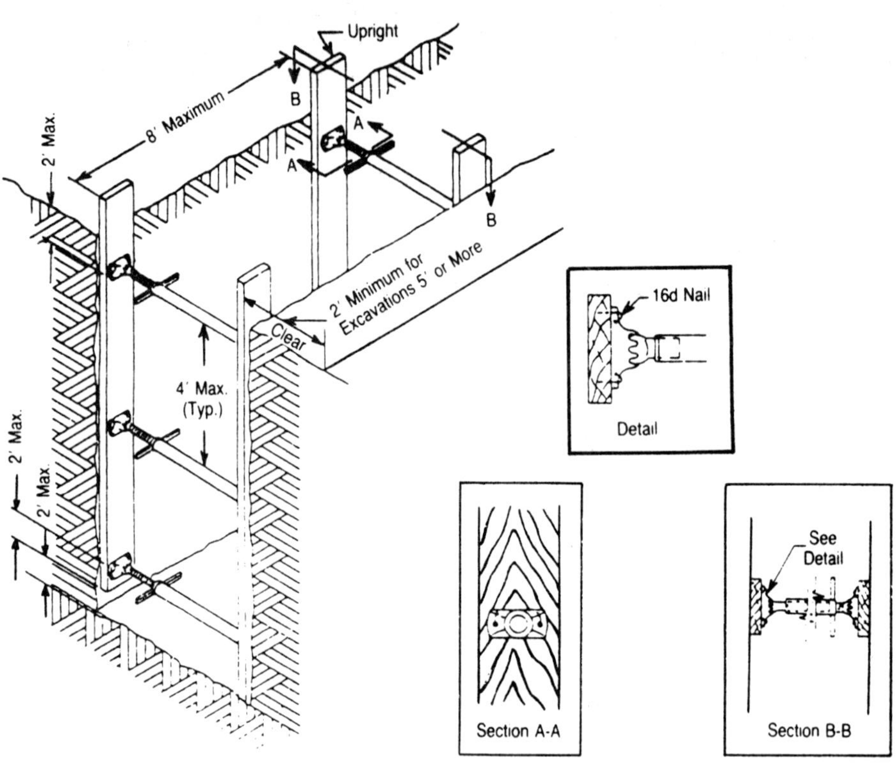


## CLOSED VERTICAL SHEETING

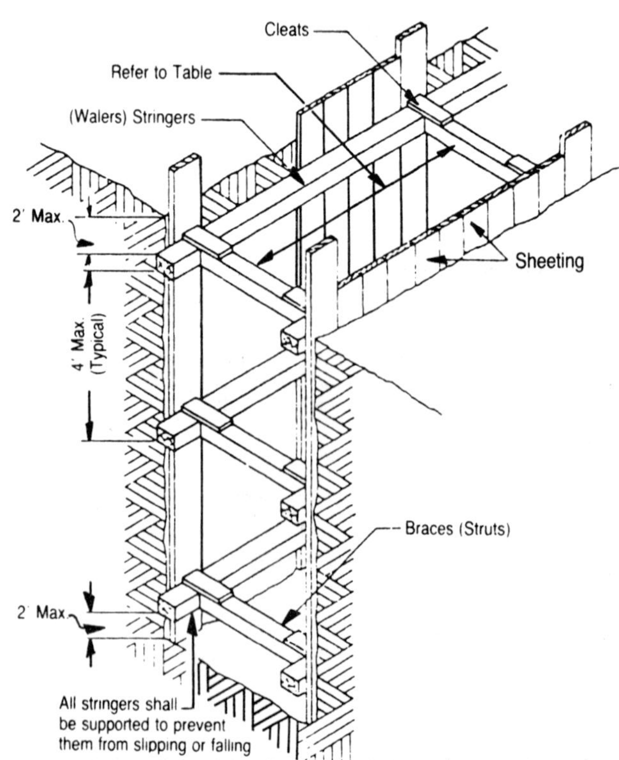

**Soil Identification.** For most purposes, soils can usually be identified visually and by texture, as described in the chart that follows. For design purposes, however, soils must be formally identified and their performance characteristics determined in a laboratory by skilled soil mechanics.

| Classification | Identifying Characteristics |
|---|---|
| Gravel | Rounded or water-worn pebbles or bulk rock grains. No cohesion or plasticity. Gritty, granular and crunchy underfoot. |
| Sand | Granular, gritty, loose grains, passing a No. 4 sieve and between .002 and .079 inches in diameter. Individual grains readily seen and felt. No plasticity or cohesion. When dry, it cannot be molded but will crumble when touched. The coarse grains are rounded; the fine grains are visible and angular. |
| Silt | Fine, barely visible grains passing a No. 200 sieve and between .0002 and .002 inches in diameter. Little or no plasticity and no cohesion. A dried cast is easily crushed. Is permeable and movement of water through the voids occurs easily and is visible. Feels gritty when bitten and will not form a thread. |
| Clay | Invisible particles under .0002 inches in diameter. Cohesive and highly plastic when moist. Will form a long, thin, flexible thread when rolled between the hands. Does not feel gritty when bitten. Will form hard lumps or clods when dry which resist crushing. Impermeable, with no apparent movement of water through voids. |
| Muck and Organic Silt | Thoroughly decomposed organic material often mixed with other soils of mineral origin. Usually black with fibrous remains. Odorous when dried and burnt. Found as deposits in swamps, peat bogs and muskeg flats. |
| Peat | Partly decayed plant material. Mostly organic. Highly fibrous with visible plant remains. Spongy and easily identified. |

**Classification by Particle Size.** Soils can be classified in general terms by the nature of their predominant particle size or the grading of the particle sizes. These particle sizes are usually grouped into gravel, coarse sand, medium sand, fine sand, silt, clay and colloids.

The major divisions of soils are:

| Coarse-Grained (Granular) | | Fine-Grained | | Organic | |
|---|---|---|---|---|---|
| Gravel | Sand | Silt | Clay | Muck | Peat |

Soils comprised primarily of sand particles are referred to as "granular soils," while fine-grained soils are commonly called "heavy soils." It is accepted practice in the field to refer to a particular soil as a coarse sand, or silt, or by any of the particle size groupings which describe the soil generally from a visual examination.

**Classification of Soil Mixtures.**

| Class | % Sand | % Silt | % Clay |
|---|---|---|---|
| Sand | 80-100 | 0- 20 | 0- 20 |
| Sandy clay loam | 50- 80 | 0- 30 | 20- 30 |
| Sandy loam | 50- 80 | 0- 50 | 0- 20 |
| Loam | 30- 50 | 30- 50 | 0- 20 |
| Silty loam | 0- 50 | 50- 80 | 0- 20 |
| Silt | 0- 20 | 80-100 | 0- 20 |
| Silty clay loam | 0- 30 | 50- 80 | 20- 30 |
| Silty clay | 0- 20 | 50- 70 | 30- 50 |
| Clay loam | 20- 50 | 20- 50 | 20- 30 |
| Sandy clay | 50- 70 | 0- 20 | 30- 50 |
| Clay | 0- 50 | 0- 50 | 30-100 |

| Material | Approx. In-Bank Weight (lbs. per cu. yd.) | Percent Swell |
|---|---|---|
| Clay, dry | 2300 | 40 |
| Clay, wet | 3000 | 40 |
| Granite, decomposed | 4500 | 65 |
| Gravel, dry | 3250 | 10-15 |
| Gravel, wet | 3600 | 10-15 |
| Loam, dry | 2800 | 15-35 |
| Loam, wet | 3370 | 25 |
| Rock, well blasted | 4200 | 65 |
| Sand, dry | 3250 | 10-15 |
| Sand, wet | 3600 | 10-15 |
| Shale and soft rock | 3000 | 65 |
| Slate | 4700 | 65 |

## PILES AND PILE DRIVING

**General.** A pile is a column driven or jetted into the ground which derives its supporting capabilities from end-bearing on the underlying strata, skin friction between the pile surface and the soil, or from a combination of end-bearing and skin friction.

Piles can be divided into two major classes: **Sheet piles** and **load-bearing piles.** Sheet piling is used primarily to restrain lateral forces as in trench sheeting and bulkheads, or to resist the flow of water as in cofferdams. It is prefabricated and is available in steel, wood or concrete. Load-bearing piles are used primarily to transmit loads through soil formations of low bearing values to formations that are capable of supporting the designed loads. If the load is supported predominantly by the action of soil friction on the surface of the pile, it is called a **friction pile.** If the load is transmitted to the soil primarily through the lower tip, it is called an **end-bearing pile.**

There are several load-bearing pile types, which can be classified according to the material from which they are fabricated:

- Timber (Treated and untreated)
- Concrete (Precast and cast in place)
- Steel (H-Section and steel pipe)
- Composite (A combination of two or more materials)

Some of the additional uses of piling are to: eliminate or control settlement of structures, support bridge piers and abutments and protect them from scour, anchor structures against uplift or overturning, and for numerous marine structures such as docks, wharves, fenders, anchorages, piers, trestles and jetties.

**Timber Piles.** Timber piles, treated or untreated, are the piles most commonly used thoroughout the world, primarily because they are readily available, economical, easily handled, can be easily cut off to any desired length after driving and can be easily removed if necessary. On the other hand, they have some serious disadvantages which include: difficulty in securing straight piles of long length, problems in driving them into hard formations and difficulty in splicing to increase their length. They are generally not suitable for use as end-bearing piles under heavy load and they are subject to decay and insect attack. Timber piles are resilient and particularly adaptable for use in waterfront structures such as wharves, docks and piers for anchorages since they will bend or give under load or impact where other materials may break. The ease with which they can be worked and their economy makes them popular for trestle construction and for temporary structures such as falsework or centering. Where timber piles can be driven and cut off below the permanent ground-water level, they will last indefinitely; but above this level in the soil, a timber pile will rot or will be attacked by insects and eventually destroyed. In sea water, marine borers and fungus will act to deteriorate timber piles. Treatment of timber piles increases their life but does not protect them indefinitely.

**Concrete Piles.** Concrete piles are of two general types, precast and cast-in-place. The advantages in the use of concrete piles are that they can be fabricated to meet the most exacting conditions of design, can be cast in any desired shape or length, possess high strength and have excellent resistance to chemical and biological attack. Certain disadvantages are encountered in the use of precast piles, such as:

(a) Their heavy weight and bulk (which introduces problems in handling and driving).

(b) Problems with hair cracks which often develop in the concrete as a result of shrinkage after curing (which may expose the steel reinforcement to deterioration).

(c) Difficulty encountered in cut-off or splicing.

(d) Susceptibility to damage or breakage in handling and driving.

(e) They are more expensive to fabricate, transport and drive.

Precast piles are fabricated in casting yards. Centrifugally spun piles (or piles with square or octagonal cross-sections) are cast in horizontal forms, while round piles are usually cast in vertical forms. With the exception of relatively short lengths, precast piles must be reinforced to provide the designed column strengths and to resist damage or breakage while being transported or driven.

Precast piles can be tapered or have parallel sides. The reinforcement can be of deformed bars or be prestressed or poststressed with high strength steel tendons. Prestressing or prestressing eliminates the problem of open shrinkage cracks in the concrete. Otherwise, the pile must be protected by coating it with a bituminous or plastic material to prevent ultimate deterioration of the reinforcement. Proper curing of the precast concrete in piles is essential.

Cast-in-place pile types are numerous and vary according to the manufacturer of the shell or inventor of the method. In general, they can be classified into two groups: shell-less types and the shell types. The shell-less type is constructed by driving a steel shell into the ground and filling it with concrete as the shell is pulled from the ground. The shell type is constructed by driving a steel shell into the ground and filling it in place with concrete. Some of the advantages of cast-in-place concrete piles are: lightweight shells are handled and driven easily, lengths of the shell may be increased or decreased easily, shells may be transported in short lengths and quickly assembled, the problem of breakage is eliminated and a driven shell may be inspected for shell damage or an uncased hole for "pinching off." Among the disadvantages are problems encountered in the proper centering of the reinforcement cages, in placing and consolidating the concrete without displacement of the reinforcement steel or segregation of the concrete, and shell damage or "pinching-off" of uncased holes.

Shell type piles are fabricated of heavy gage metal or are fluted, corrugated or spirally reinforced with heavy wire to make them strong enough to be driven without a mandrel. Other thin-shell types are driven with a collapsible steel mandrel or core inside the casing. In addition to making the driving of a long thin shell possible, the mandrel prevents or minimizes damage to the shell from tearing, buckling, collapsing or from hard objects encountered in driving.

Some shell type piles are fabricated of heavy gauge metal with enlargement at the lower end to increase the end bearing.

These enlargements are formed by withdrawing the casing two to three feet after placing concrete in the lower end of the shell. This wet concrete is then struck by a blow of the pile hammer on a core in the casing and the enlargement is formed. As the shell is withdrawn, the core is used to consolidate the concrete after each batch is placed in the shell. The procedure results in completely filling the hole left by the withdrawal of the shell.

**Steel Piles.** A steel pile is any pile fabricated entirely of steel. They are usually formed of rolled steel H sections, but heavy steel pipe or box piles (fabricated from sections of steel sheet piles welded together) are also used. The advantages of steel piles are that they are readily available, have a thin uniform section and high strength, will take hard driving, will develop high load-bearing values, are easily cut off or extended, are easily adapted to the structure they are to support, and breakage is eliminated. Some disadvantages are: they will rust and deteriorate unless protected from the elements; acid, soils or water will result in corrosion of the pile; and greater lengths may be required than for other types of piles to achieve the same bearing value unless bearing on rock strata. Pipe pile can either be driven open-end or closed-end and can be unfilled, sand filled or concrete filled. After open-end pipe piles are driven, the material from inside can be removed by an earth auger, air or water jets, or other means, inspected, and then filled with concrete. Concrete filled pipe piles are subject to corrosion on the outside surface only.

**Composite Piles.** Any pile that is fabricated of two or more materials is called a composite pile. There are three general classes of composite piles: wood with concrete, steel with concrete, and wood with steel. Composite piles are usually used for a special purpose or for reasons of economy.

Where a permanent ground-water table exists and a composite pile is to be used, it will generally be of concrete and wood. The wood portion is driven to below the water table level and the concrete upper portion eliminates problems of decay and insect infestation above the water table. Composite piles of steel and concrete are used where high bearing loads are desired or where driving in hard or rocky soils is expected. Composite wood and steel piles are relatively uncommon.

It is important that the pile design provides for a permanent joint between the two materials used, so constructed that the parts do not separate or shift out of axial alignment during driving operations.

**Sheet Piles.** Sheet piles are made from the same basic materials as other piling: wood, steel and concrete. They are ordinarily designed so as to interlock along the edges of adjacent piles.

Sheet piles are used where support of a vertical wall of earth is required, such as trench walls, bulkheads, waterfront structures or cofferdams. Wood sheet piling is generally used in temporary installations, but is seldom used where water-tightness is required or hard driving expected. Concrete sheet piling has the capability of resisting much larger lateral loads than wood sheet piling, but considerable difficulty is experienced in securing water-tight joints. The type referred to as "fishmouth" type is designed to permit jetting out the joint and filling with grout, but a seal is not always effected unless the adjacent piles are wedged tightly together. Concrete sheet piling has the advantage that it is the most permanent of all types of sheet piling.

Steel sheet piling is manufactured with a tension-type interlock along its edges. Several different shapes are available to permit versatility in its use. It has the advantages that it can take hard driving, has reasonably water-tight joints and can be easily cut, patched, lengthened or reinforced. It can also be easily extracted and reused. Its principal disadvantage is its vulnerability to corrosion.

**Types of Pile Driving Hammers.** A pile-driving hammer is used to drive load-bearing or sheet piles. The commonly used types are: drop, single-acting, double-acting, differential acting and diesel hammers. The most recent development is a type of hammer that utilizes high-frequency sound and a dead load as the principal sources of driving energy.

**Drop Hammers.** These hammers employ the principle of lifting a heavy weight by a cable and releasing it to fall on top of the pile. This type of hammer is rapidly disappearing from use, primarily because other types of pile driving hammers are more efficient. Its disadvantages are that it has a slow rate of driving (four to eight blows per minute), that there is some risk of damaging the pile from excessive impact, that damage may occur in adjacent structures from heavy vibration and that it cannot be used directly for driving piles under water. Drop hammers have the advantages of simplicity of operation, ability to vary the energy by changing the height of fall and they represent a small investment in equipment.

**Single-Acting Hammers.** These hammers can be operated either on steam or compressed air. The driving energy is provided by a free-falling weight (called a ram) which is raised after each stroke by the action of steam or air on a piston. They are manufactured as either open or closed types. Single-acting hammers are best suited for jobs where dense or elastic soil materials must be penetrated or where long heavy timber or precast concrete piles must be driven. The closed type can be used for underwater pile driving. Its advantages include: faster driving (50 blows or more per minute), reduction in skin friction as a result of more frequent blows, lower velocity of the ram which transmits a greater proportion of its energy to the pile and minimizes piles damage during driving, and it has underwater driving capability. Some of its disadvantages are: requires higher investment in equipment (i.e. steam boiler, air compressor, etc.), higher maintenance costs, greater set-up and moving time required, and a larger operating crew.

**Double-Acting Hammers.** These hammers are similar to the single-acting hammers except that steam or compressed air is used both to lift the ram and to impart energy to the falling ram. While the action is approximately twice as fast as the single-acting hammer (100 blows per minute or more), the ram is much lighter and operates at a greater velocity, thereby making it particularly useful in high production driving of light or medium-weight piles of moderate lengths in granular soils. The hammer is nearly always fully encased by a steel housing which also permits direct driving of piles under water.

Some of its advantages are: faster driving rate, less static skin friction develops between blows, has underwater driving capability and piles can be driven more easily without leads.

Among its disadvantages are: it is less suitable for driving heavy piles in high-friction soils and the more complicated mechanism results in higher maintenance costs.

**Differential-Acting Hammers.** This type of hammer is, in effect, a modified double-acting hammer with the actuating mechanism having two different diameters. A large-diameter piston operates in an upper cylinder to accelerate the ram on the downstroke and a small-diameter piston operates in a lower cylinder to raise the ram. The additional energy added to the falling ram is the difference in areas of the two pistons multiplied by the unit pressure of the steam or air used. This hammer is a short-stroke, fast-acting hammer with a cycle rate approximately that of the double-acting hammer. Its advantages are that it has the speed and characteristics of the double-acting hammer with a ram weight comparable to the single-acting type, and it uses from 25 to 35 percent less steam or air. It is also more suitable for driving heavy piles under more difficult driving conditions than is the double-acting hammer. It is available in the open or closed-type cases, the latter permitting direct underwater pile driving. Its principal disadvantage is higher maintenance costs.

**Diesel Hammers.** This hammer is a self-contained driving unit which does not require an auxilliary steam boiler or air compressor. It consists essentially of a ram operating as a piston in a cylinder. When the ram is lifted and allowed to fall in the cylinder, diesel fuel is injected in the compression space between the ram and an anvil placed on top of the pile. The continued downstroke of the ram compresses the air and fuel to ignition heat and the resultant explosion drives the pile downward and the ram upward to start another cycle. This hammer is capable of driving at a rate of from 80 to 100 blows per minute. Its advantages are that it has a low equipment investment cost, is easily moved, requires a small crew, has a high driving rate, does not require a steam boiler or air compressor and can be used with or without leads for most work. Its disadvantages are that it is not self-starting (the ram must be mechanically lifted to start the action) and it does not deliver a uniform blow. The latter disadvantge arises from the fact that as the reaction of the pile to driving increases, the reaction to the ram increases correspondingly. That is, when the pile encounters considerable resistance, the rebound of the ram is higher and the energy is increased automatically. The operator is required to observe the driving operations closely to identify changing driving conditions and compensate for such changes with his controls to avoid damaging the pile.

Diesel hammers can be used on all types of piles and they are best suited to jobs where mobility or frequent relocation of the pile driving equipment is necessary.

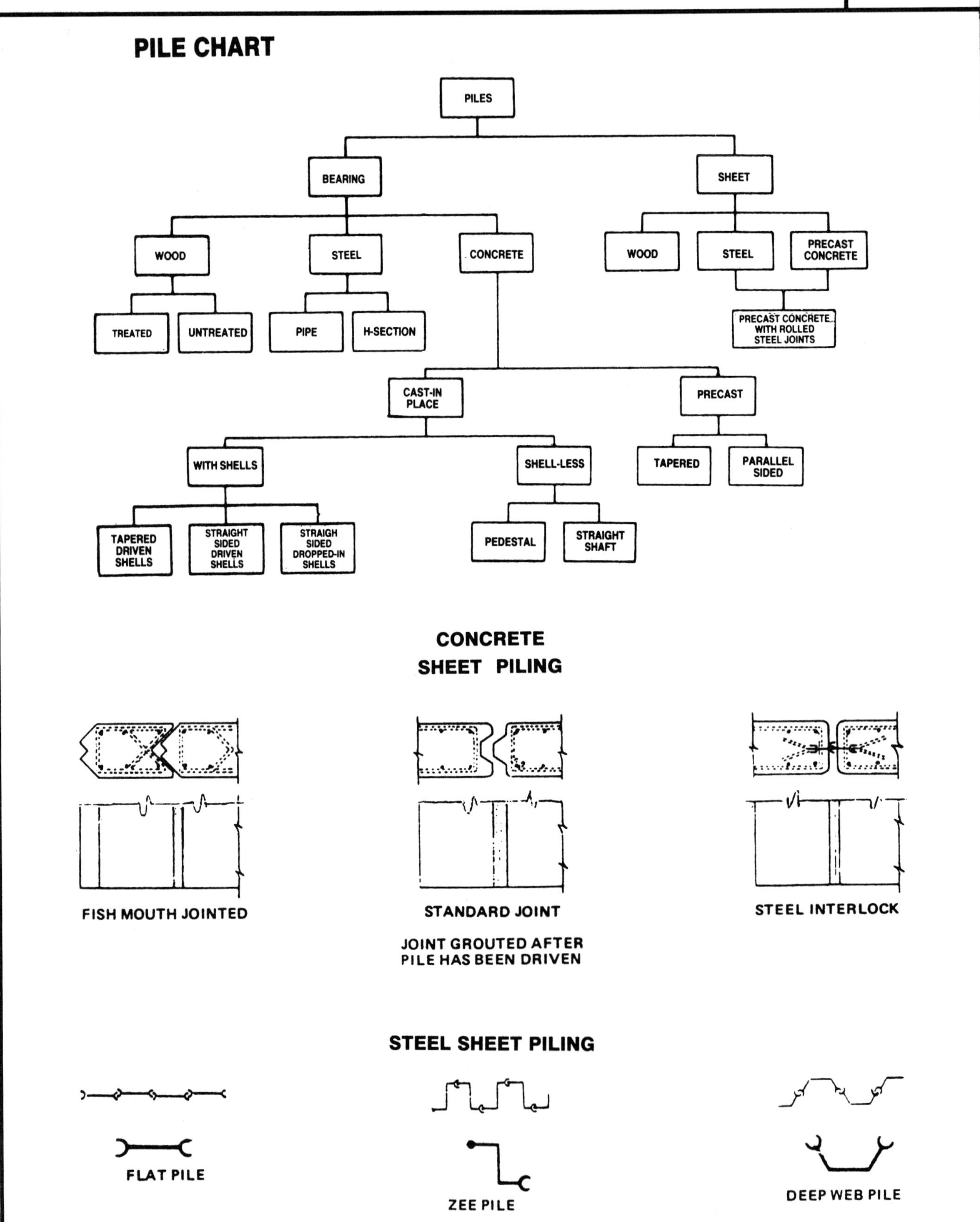
PILE CHART
PILES
BEARING
SHEET
WOOD
STEEL
CONCRETE
WOOD
STEEL
PRECAST CONCRETE
TREATED
UNTREATED
PIPE
H-SECTION
PRECAST CONCRETE WITH ROLLED STEEL JOINTS
CAST-IN PLACE
PRECAST
WITH SHELLS
SHELL-LESS
TAPERED
PARALLEL SIDED
TAPERED DRIVEN SHELLS
STRAIGHT SIDED DRIVEN SHELLS
STRAIGH SIDED DROPPED-IN SHELLS
PEDESTAL
STRAIGHT SHAFT
CONCRETE SHEET PILING
FISH MOUTH JOINTED
STANDARD JOINT
JOINT GROUTED AFTER PILE HAS BEEN DRIVEN
STEEL INTERLOCK
STEEL SHEET PILING
FLAT PILE
ZEE PILE
DEEP WEB PILE

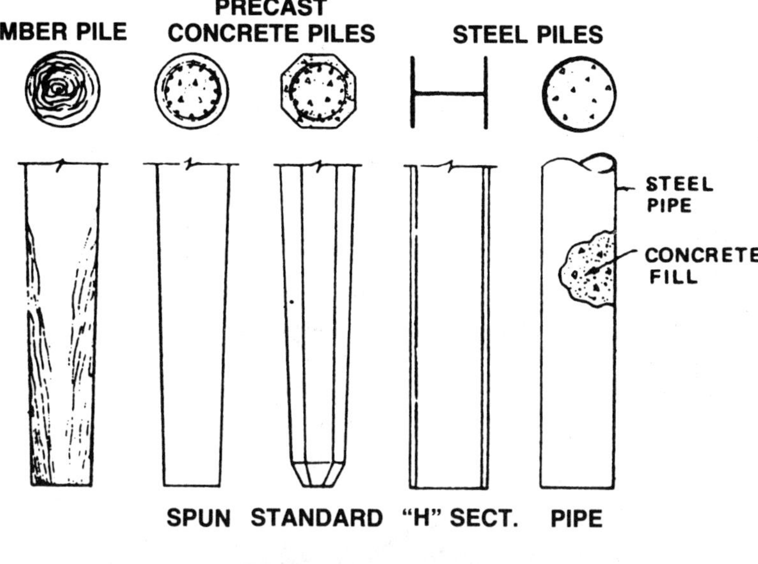


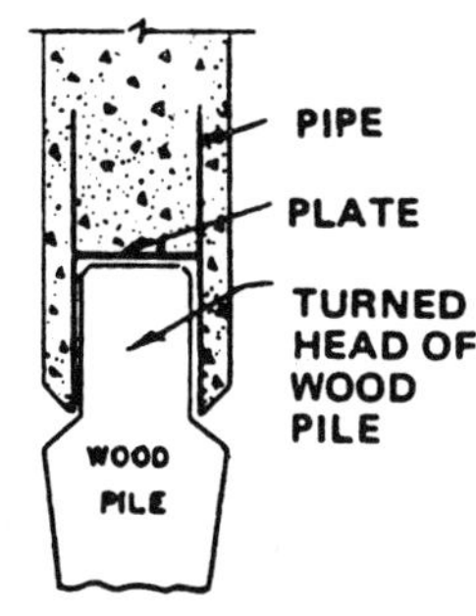


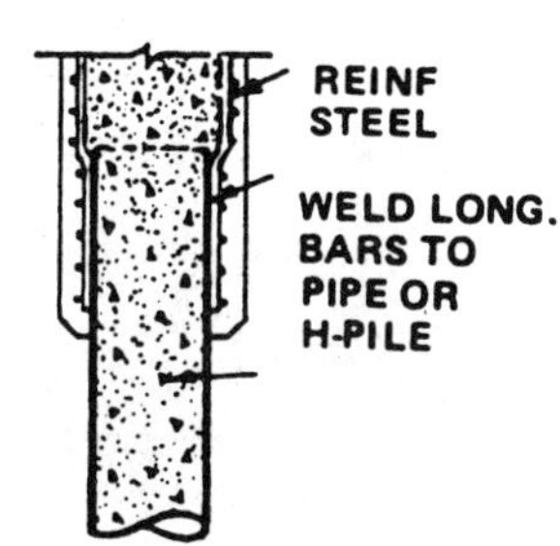


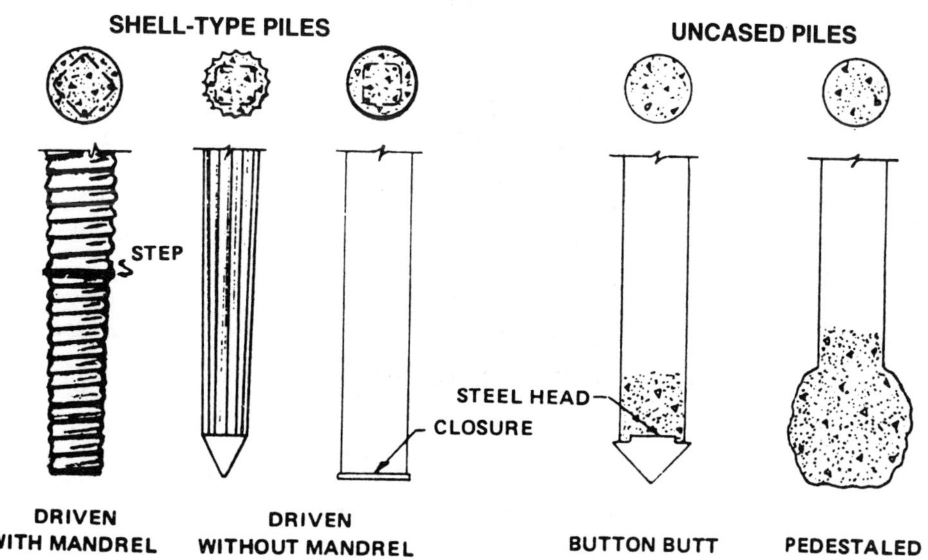


***CAST-IN-PLACE PILES***

## STANDARD NOMENCLATURE FOR STREET CONSTRUCTION

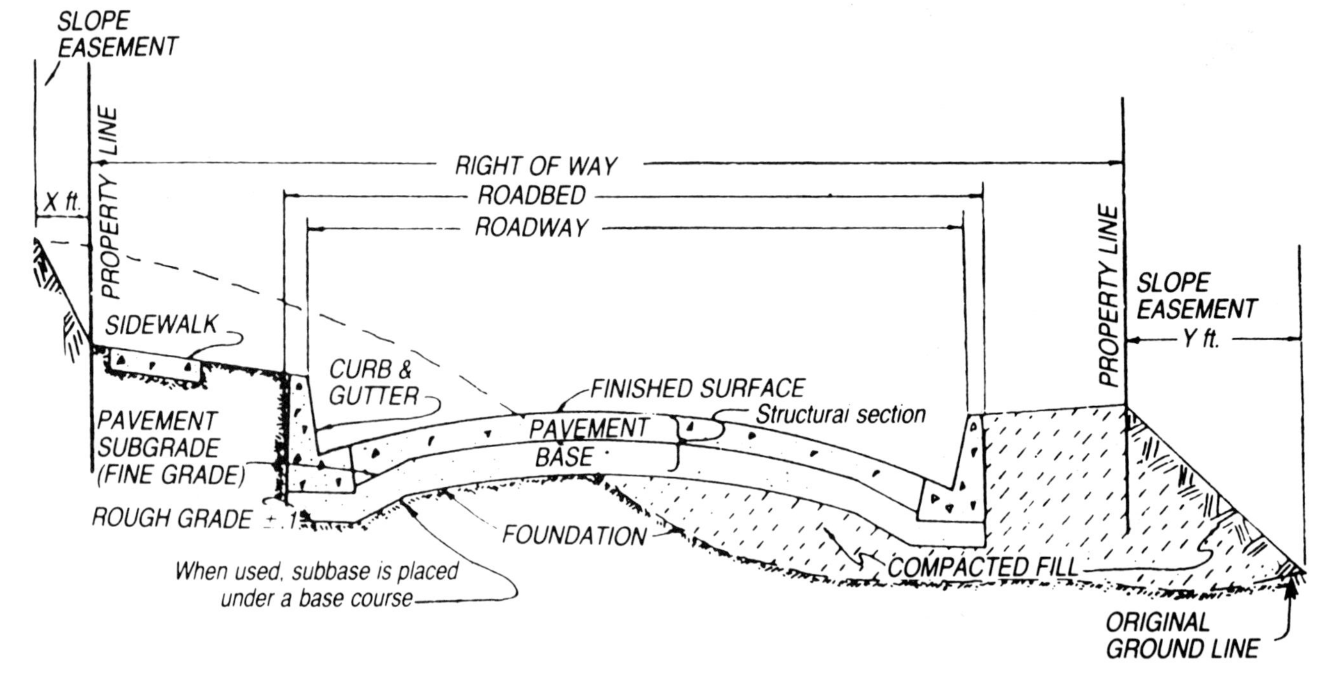

## CONCRETE MASONRY PAVING UNITS

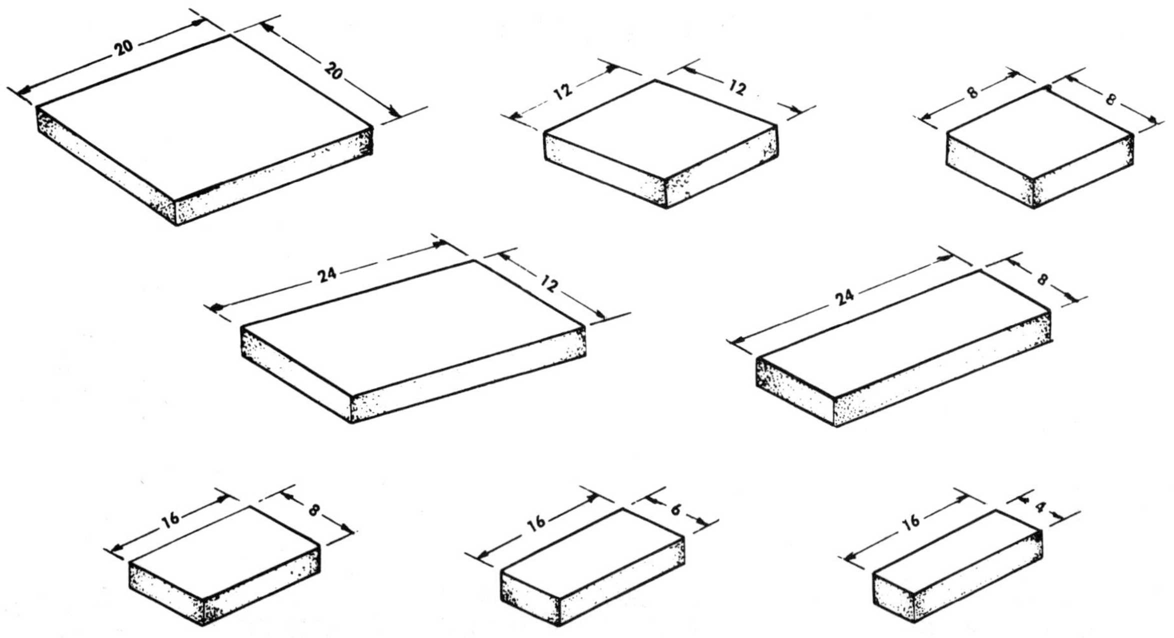


**NOTE:** Sizes are nominal and will vary by manufacturer.

**HEXAGON PAVER UNITS**
**Various Sizes Available**

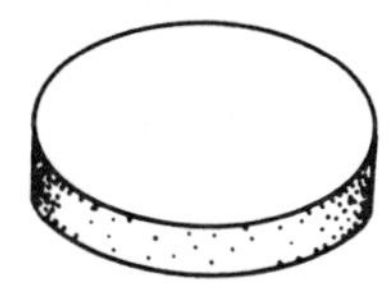

**ROUND PAVING UNITS**
**Various Sizes Available**

## VEHICULAR PAVING UNITS

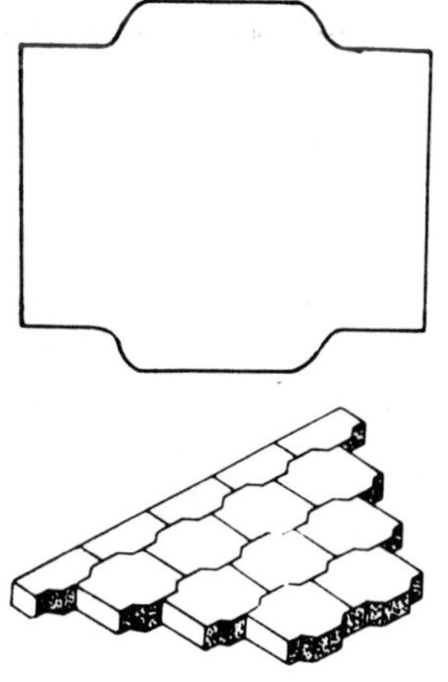

**INTERLOCKING PAVER**
**7¼" x 3" x 8½"**

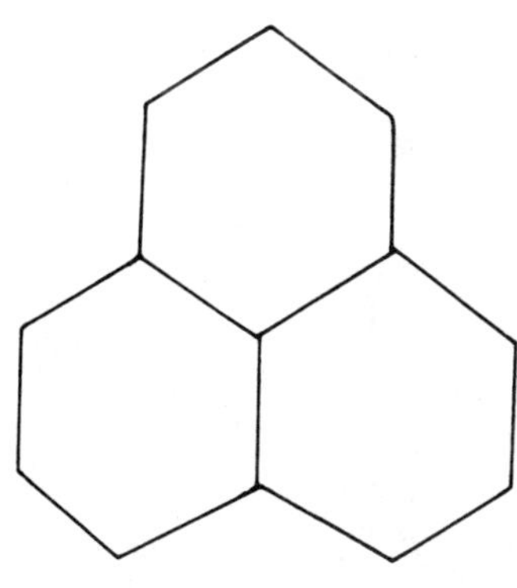

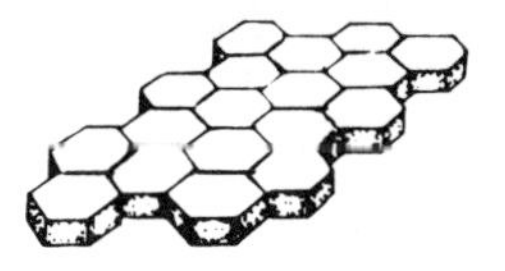

**INTERLOCKING PAVER**
**12" x 3⅝" x 12"**

**TURF PAVER**
**24" x 3⅝" x 24"**

## PIPE

**Clay Pipe.** Clay pipe is manufactured by blending various clays together, milling, mixing, extruding and firing in a kiln to obtain vitrification. The physical properties of the pipe can be changed by varying the proportions of the several clays used. The pipe is supplied in two basic styles: spigot and socket; and plain end.

**Spigot and Socket Pipe** has a spigot on one end and a socket on the other, and is commonly referred to as "bell and spigot" pipe. The plans generally specify the type of joint to be used from the several types of jointing methods available. This type of pipe is manufactured with matching polyurethane gaskets molded on the spigot and socket which form a tight seal when the pipe is jointed.

**Plain End Pipe** is without a socket on either end and is joined with special couplings. The coupling consists of a circular rubber sleeve, two stainless steel compression bands with tightening devices and a corrosion resistant shear ring. Sometimes this joint is supplied with a cardboard form, open at the top, which is filled with portland cement mortar to resist shear and prevent future corrosion of the bands.

**Concrete Pipe.** Unreinforced and reinforced concrete pipe is manufactured by casting in stationary or revolving metal molds. At the present time, the design practice is to specify reinforced concrete pipe for all purposes.

**Unreinforced Concrete Pipe** is cast in vertical steel molds, usually in pipe sizes of 21 inches or less, and is of the spigot and socket type. No steel reinforcement is used and the pipe is usually intended for use in irrigation systems and under light loading conditions.

**Reinforced Concrete Pipe (RCP)** is made in a number of different manufacturing processes and for a wide variety of pressure and non-pressure classes. It is available in standard sizes or it can be made to order to any diameter desired. Some of the larger diameters include diameters of 12 and 14 feet. A large variety of joint details are used with RCP. Tongue and groove joints are used for storm drain pipelines.

Reinforced concrete pipe for wastewater pipeline projects is supplied with gasketed joints and a polyvinyl chloride (PVC) plastic liner cast into the pipe.

(a) **Cast Pipe** is cast vertically in steel forms with the reinforcing cage securely held in place. The reinforcement is generally elliptical in shape to provide the maximum structural strength to resist the loads imposed on the pipe by the backfill and other stresses. Consolidation of the concrete is obtained by the use of external form vibrators.

(b) **Centrifugally Spun Pipe** is manufactured by introducing concrete into a spinning horizontal steel cylinder into which the reinforcement cage has been previously installed and which is equipped with end dams to provide the proper pipe wall thickness. The speed of rotation of the mold is increased and the centrifugal force produces a smooth, dense concrete pipe.

(c) **Pressure Pipe** may be cast or centrifugally spun pipe but it usually has a circular steel reinforcement cage (or cages) designed not only to resist the trench loading, but also the internal pressures exerted on the pipe from the fluid under pressure in the line.

**Concrete Cylinder Pipe.** This class of pipe is generally used for high pressure water lines and sewer force mains and is available in sizes ranging from 10 inches to 60 inches and larger in special cases.

A sheet steel cylinder is wrapped with the designed steel reinforcement and a concrete lining is centrifugally spun in the interior of the steel cylinder. An exterior coating of concrete is applied generally by the gunite process, while the cylinder is slowly rotated. These coatings vary in thickness from ½ to ¾ of an inch. The joints are commonly of the steel ring and rubber gasket type, but are generally designed for the special purpose for which the pipe line is intended.

**Definition of Terms.** In general, the terms used to designate types of reinforced concrete pipe refer to the process used in manufacture.

**Cast RCP** (Cast Reinforced Concrete Pipe). A concrete pipe having one or more cylindrical or elliptical cages of reinforcement steel embedded in it, the concrete for which is cast with the forms in a vertical position.

**CSRCP** (Centrifugally Spun Reinforced Concrete Pipe). A concrete pipe having one or more cylindrical or elliptical cages of reinforcement steel embedded in it, and cast in a horizontal position while the forms are spinning rapidly. This type of pipe may be designated as Spun RCP or as CCP (Centrifugal Concrete Pipe).

**RCP** (Reinforced Concrete Pipe). A reinforced concrete pipe manufactured by either the casting or spinning method.

**Steel Reinforcement.** Steel for reinforcing concrete pipe is generally furnished in large coils which will permit the use of machines to fabricate the "cages." The continuous steel rod is wound spirally at a prescribed pitch on a drum of the proper diameter. Where the rod crosses a longitudinal spacer rod, it is electrically welded to it so that the complete cage is relatively rigid.

Reinforcement cages for pipe designed for external loading are generally elliptical in shape to take full advantage of the steel in tension. Pipe to be used with relatively small external loads or pipe designed for pressure lines will have circular cages.

Reinforcement cages must be rigidly fixed in the forms so that the placement of concrete or the effects of centrifugal spinning will not result in distortion or displacement of the steel. The orientation of an elliptical cage must be marked on the forms to assure that the minor axis can be located after the concrete is placed.

## CAPACITIES FOR SEPTIC TANKS SERVING AN INDIVIDUAL DWELLING

| No. of bedrooms | Capacity of tank (gals.) |
|---|---|
| 2 or less | 750 |
| 3 | 900 |
| 4 | 1,000 |

# CONCRETE / FORMWORK

03110

## FORM NOMENCLAURE

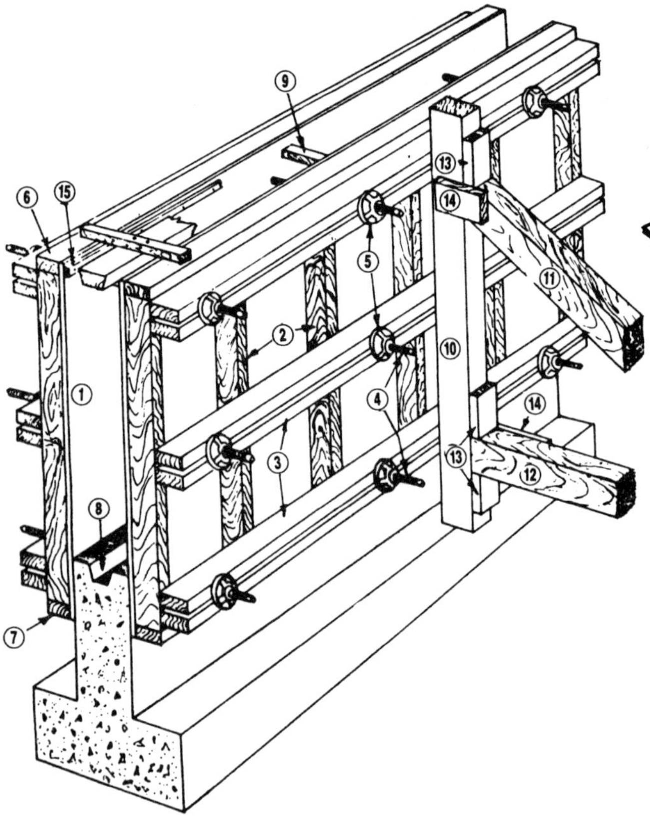

1. SHEATHING
2. STUDS
3. WALES
4. FORM BOLTS
5. NUT WASHER
6. TOP PLATE
7. BOTTOM PLATE
8. KEY-WAY
9. SPREADER
10. STRONGBACK
11. BRACE
12. STRUT
13. CLEATS
14. SCAB
14. POUR STRIP

## FALSEWORK NOMENCLATURE

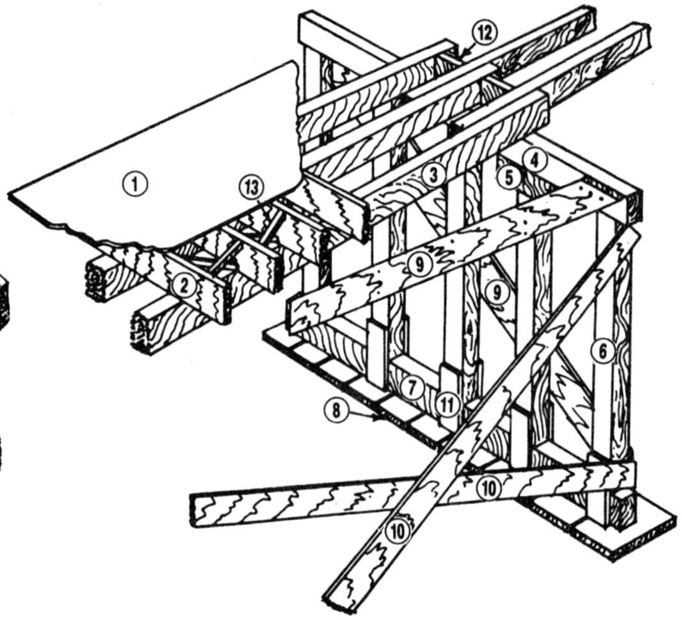

1. SHEATHING
2. JOIST
3. STRINGER
4. CAP
5. CORBEL
6. POST
7. SILL
8. FOOTING
9. SWAY BRACE
10. LONGITUDINAL BRACE
11. SCAB
12. BLOCKING
13. BRIDGING

## TYPICAL PAN-JOIST FORM CONSTRUCTION

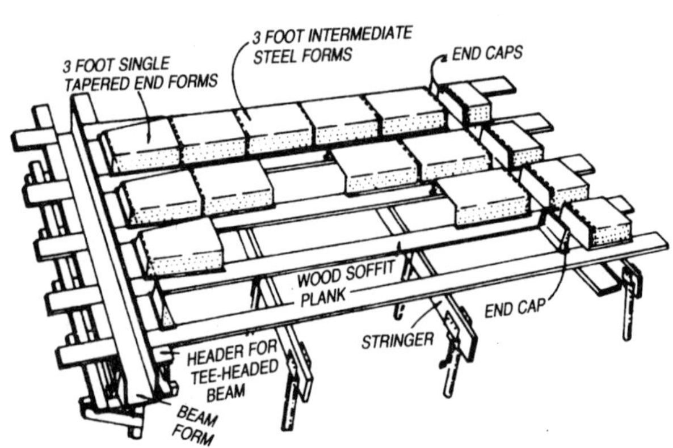


## TYPICAL WAFFLE SLAB FORM CONSTRUCTION

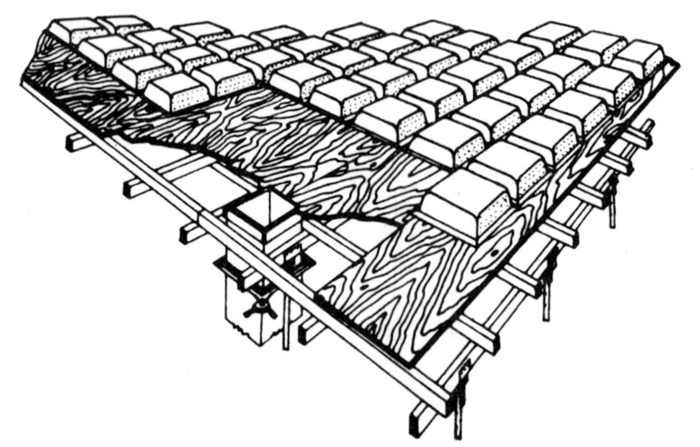

# CONCRETE / REINFORCING STEEL 03210

## STANDARD SIZES OF STEEL REINFORCEMENT BARS

| STANDARD REINFORCEMENT BARS | | | | |
|---|---|---|---|---|
| | | Nominal Dimensions | | |
| Bar Designation Number* | Nominal Weight, lb. per ft. | Diameter, in. | Cross Sectional Area, sq. in. | Perimeter, in. |
| 3 | 0.376 | 0.375 | 0.11 | 1.178 |
| 4 | 0.668 | 0.500 | 0.20 | 1.571 |
| 5 | 1.043 | 0.625 | 0.31 | 1.963 |
| 6 | 1.502 | 0.750 | 0.44 | 2.356 |
| 7 | 2.044 | 0.875 | 0.60 | 2.749 |
| 8 | 2.670 | 1.000 | 0.79 | 3.142 |
| 9 | 3.400 | 1.128 | 1.00 | 3.544 |
| 10 | 4.303 | 1.270 | 1.27 | 3.990 |
| 11 | 5.313 | 1.410 | 1.56 | 4.430 |
| 14 | 7.65 | 1.693 | 2.25 | 5.32 |
| 18 | 13.60 | 2.257 | 4.00 | 7.09 |

*The bar numbers are based on the number of ⅛ inches included in the nominal diameter of the bar.

| Type of Steel and ASTM Specification No. | Size Nos. Inclusive | Grade | Tensile Strength Min., psi | Yield (a) Min., psi |
|---|---|---|---|---|
| Billet Steel A 615 | 3-11 | 40 | 70,000 | 40,000 |
| | 3-11<br>14, 18 | 60 | 90,000 | 60,000 |
| | 11, 14, 18 | 75 | 100,000 | 75,000 |

# CONCRETE / WELDED WIRE FABRIC 03220

## WELDED WIRE FABRIC – COMMON STOCK STYLES OF WELDED WIRE FABRIC

| Style Designation | Steel Area sq. in. per ft. | | Weight Approx. lbs. per 100 sq. ft. |
|---|---|---|---|
| | Longit. | Transv. | |
| **Rolls** | | | |
| 6x6—W1.4xW1.4 | .03 | .03 | 21 |
| 6x6—W2xW2 | .04 | .04 | 29 |
| 6x6—W2.9xW2.9 | .06 | .06 | 42 |
| 6x6—W4xW4 | .08 | .08 | 58 |
| 4x4—W1.4xW1.4 | .04 | .04 | 31 |
| 4x4—W2xW2 | .06 | .06 | 43 |
| 4x4—W2.9xW2.9 | .09 | .09 | 62 |
| 4x4—W4xW4 | .12 | .12 | 86 |
| **Sheets** | | | |
| 6x6—W2.9xW2.9 | .06 | .06 | 42 |
| 6x6—W4xW4 | .08 | .08 | 58 |
| 6x6—W5.5xW5.5 | .11 | .11 | 80 |
| 4x4—W4xW4 | .12 | .12 | 86 |

## COMMON TYPES OF STEEL REINFORCEMENT BARS

ASTM specifications for billet steel reinforcing bars (A 615) require identification marks to be rolled into the surface of one side of the bar to denote the producer's mill designation, bar size and type of steel. For Grade 60 and Grade 75 bars, grade marks indicating yield strength must be show. Grade 40 bars show only three marks (no grade mark) in the following order:

1st — Producing Mill (usually an initial)
2nd — Bar Size Number (#3 through #18)
3rd — Type (N for New Billet)

### NUMBER SYSTEM — GRADE MARKS

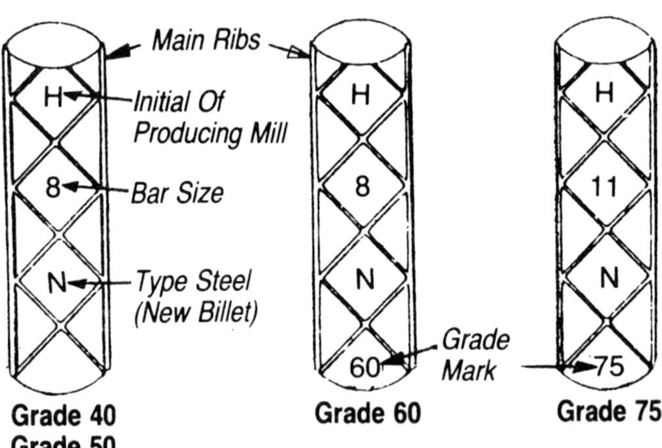


### LINE SYSTEM — GRADE MARKS

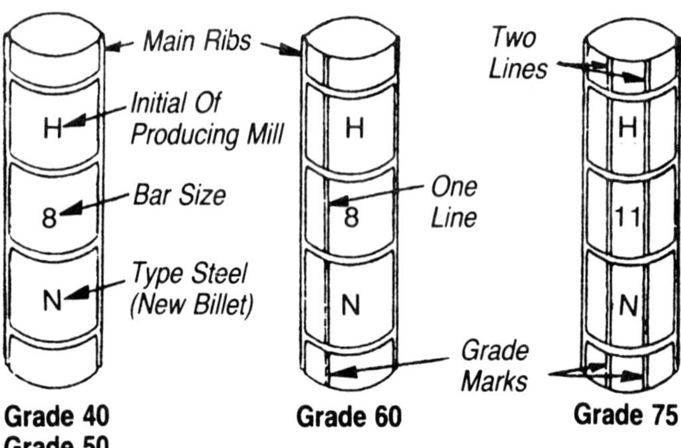

Insofar as is possible, the moisture content should be kept uniform to avoid problems in determining the proper amount of water to be added for mixing. Mixing water must be reduced to compensate for moisture in the aggregate in order to control the slump of the concrete and avoid exceeding the specified water-cement ratio.

**Handling Concrete by Pumping Methods.** Transportation and placement of concrete by pumping is another method gaining increased popularity. Pumps have several advantages, the primary one being that a pump will high-lift concrete without the need for an expensive crane and bucket. Since the concrete is delivered through pipe and hoses, concrete can be conveyed to remote locations in buildings, in tunnels, to locations otherwise inaccessible on steep hillside slopes for anchor walls, pipe bedding or encasement, or for placing concrete for chain link fence post bases. Concrete pumps have been found to be economical and expedient in the placement of concrete, and this has promoted the use and acceptance of this development. The essence of proper concrete pumping is the placement of the concrete in its final location without segregation.

Modern concrete pumps, depending on the mix design and size of line, can pump to a height of 200 feet or a horizontal distance of 1,000 feet. They can handle, economically, structural mixes, standard mixes, low slump mixes, mixes with two-inch maximum size aggregate and light weight concrete. When a special pump mix is required for structural concrete in a major structure, the mix design must be approved by the Engineer and checked and confirmed by the Supervisor of the Materials Control Group. The Inspector should obtain the pump manufacturer's printed information and evaluate its characteristics and ability to handle the concrete mixture specified for the project.

If concrete is being placed for a major reinforced structure, it is important that the placement continue without interruption. The Inspector should be sure that the contractor has ready access to a back-up pump to be used in the event of a breakdown. In order to further insure the success of the concrete placement by the pumping method, the user should be aware of the following points:

(a) A protective grating over the receiving hopper of the pump is necessary to exclude large pieces of aggregate or foreign material.

(b) The pump and lines require lubrication with a grout of cement and water. All of the excess grout is to be wasted prior to pumping the concrete.

(c) All changes in direction must be made by a large radius bend with a maximum bend of 90 degrees. Wye connections induce segregation and shall not be used.

(d) Pump lines should be made of a material capable of resisting abrasion and with a smooth interior surface having a low coefficient of friction. Steel is commonly used for pump lines, ance. Aluminum pipe should not be used for pumping concrete and some of the new plastic or rubber tubing is gaining accept- because a chemical reaction occurs between the concrete and the aluminum. Hydrogen is generated which results in a swelling of the concrete, causing a significant reduction in compressive strength. This reaction is aggravated by any of the following: abrasive coarse aggregate, non-uniformly graded sand, low-slump concrete, low sand-aggregate ratio, high-alkali cement or when no air-entraining agent is used.

(e) During temporary interruptions in pumping, the hopper must remain nearly full, with an occasional turning and pumping to avoid developing a hard slug of concrete in the lines.

(f) Excessive line pressures must be avoided. When this occurs, check these points as the probable cause: segregation caused by too low a slump or too high a slump; large particle contamination caused by large pieces of aggregate or frozen lumps not eliminated by the grating; poor gradation of aggregates or particle shape; rich or lean spots caused by improper mixing.

(g) Corrections must be made to correct excessive slump loss as measured at the transit-mixed concrete truck and as measured at the hose outlet. This may be attributable to porous aggregate, high temperature or rapid setting mixes.

(h) Two transit-mix concrete trucks must be used simultaneously to deliver concrete into the pump hopper. These trucks must be discharged alternately to assure a continuous flow of concrete as trucks are replaced.

(i) Samples of concrete for test specimens prepared to determine the acceptance of the concrete quality are to be taken as required for conventional concrete.

Sampling is done before the concrete is deposited in the pump hopper. However, it is suggested that, where possible, the effect of pumping on the compressive strength be checked by taking companion samples, so identified, from the end of the pump line at the same time. The Record of Test must be properly noted as being a special mix used for pumping purposes. This will enable the Materials Control Group to compile a complete history of mix designs and their respective compressive strengths.

The prudent use of pumped concrete can result in economy and improved quality. However, only the control exercised by the operator will assure continued high standards of quality concrete.

Pump lines must be properly fastened to supports to eliminate excessive vibration. Couplings must be easily and securely fastened in a manner that will prevent mortar leakage. It is preferable to use the flexible hose only at the discharge point. This hose must be moved in such a manner as to avoid kinks or sharp bends. The pump line should be protected from excessive heat during hot weather by water sprinkling or shade.

## COMPRESSIVE STRENGTH FOR VARIOUS WATER-CEMENT RATIOS

*(The strengths listed are based on the use of normal portland cement)*

| WATER/CEMENT RATIO | | PROBABLE 28-DAY STRENGTH | |
|---|---|---|---|
| WEIGHT | GALS./100# | PSI | MEGAPASCALS* |
| .40 | 4.8 | 5000 | 34 |
| .45 | 5.4 | 4500 | 31 |
| .50 | 6.0 | 4000 | 28 |
| .55 | 6.6 | 3500 | 24 |
| .60 | 7.2 | 3000 | 21 |
| .65 | 7.8 | 2500 | 17 |
| .70 | 8.4 | 2000 | 14 |

* International system equivalent.

## APPROXIMATE CONTENT OF SAND, CEMENT AND WATER PER CUBIC YARD OF CONCRETE

Based on aggregates of average grading and physical characteristics in concrete mixes having a water-cement ratio (W/C) of about .65 by weight (or 7.8 gallons) per sack of cement; 3-in. slump; and a medium natural sand having a fineness modulus of about 2.75.

| COURSE AGGREGATE MAX. SIZE | WATER | | CEMENT | % SAND |
|---|---|---|---|---|
| | POUNDS | GALLONS | | |
| 3/8 | 385 | 46 | 590 | 57 |
| 1/2 | 365 | 44 | 560 | 50 |
| 3/4 | 340 | 41 | 525 | 43 |
| 1 | 325 | 39 | 500 | 39 |
| 1 1/2 | 300 | 36 | 460 | 37 |

It can be noted from the above chart that, for a given slump, the amount of mixing water increases as the size of the course aggregate decreases. The size of the course aggregate controls the sand content in the same way; that is, the amount of sand required in the mix increases as the size of the course aggregate decreases.

Other typical examples are contained in the pamphlet published by the Portland Cement Association entitled "Design and Control of Concrete Mixtures."

**Effects of Temperature on Concrete.** Concrete mixtures gain strength rapidly in the first few days after placement. While the rate of gain in strength diminishes, concrete continues to become stronger with time over a period of many years, so long as drying of the concrete is prevented. Its strength at 28 days is considered to be the compressive strength upon which the Engineer bases his calculations. The temperature of the atmosphere has a significant effect upon the development of strength in concrete. Lower temperatures retard and higher temperatures accelerate the gain in strength.

Most destructive of the natural forces is freezing and thawing action. While the concrete is still wet or moist, expansion of the water as it is converted into ice results in severe damage to the fresh concrete. In situations where freezing may be encountered, high early strength cement may be used. Also, the mixing water or the aggregate (or both) may be preheated before mixing. Covering the concrete, and using steam or salamanders to heat the concrete under the covering, will help prevent freezing. Air-entraining agents help to diminish the effects of freezing of fresh concrete as well as in subsequent freezing and thawing cycles throughout the life of the concrete.

Hot weather will present problems of a different nature in placing concrete. Concrete will set up faster and tend to shrink and crack at the surface. To minimize this problem, the concrete should be placed without delay after mixing. Avoid the use of accelerators (perhaps even use a retarding agent), dampen all subgrade and forms, protect the freshly placed concrete from hot dry winds, and provide for adequate curing. Crushed ice or chilled water can be used as part of the mixing water to reduce the temperature of the mix in extremely hot areas.

| Admixture | Purpose | Effects on Concrete | Advantages | Disadvantages |
|---|---|---|---|---|
| Accelerator | Hasten setting. | Improves cement dispersion and increases early strength. | Permits earlier finishing, form removal, and use of the structure. | Increases shrinkage, decreases sulfate resistance, tends to clog mixing and handling equipment. |
| Air-Entraining Agent | Increase workability and reduce mixing water. | Reduces segregation, bleeding and increases freeze-thaw resistance. Increases strength | Increases workability and reduces finishing time. | Excess will reduce strength and increase slump. Bulks concrete volume. |
| Bonding Agent | Increase bond to old concrete. | Produces a non-dusting, slip-resistant finish. | Permits a thin topping without roughening old concrete, self-curing, ready in one day. | Quick setting and susceptible to damage from fats, oils and solvents. |
| Densifier | To obtain dense concrete. | Increased workability and strength. | Increases workability and increases waterproofing characteristics, more impermeable. | Care must be used to reduce mixing water in proportion to amount used. |
| Foaming Agent | Reduce weight. | Increases insulating properties. | Produces a more plastic mix, reduces dead weight loads. | Its use must be very carefully regulated — following instructions explicitly. |
| Retarder | Retard setting. | Increases control of setting. | Provides more time to work and finish concrete. | Performance varies with cement used — adds to slump. Requires stronger forms. |
| Water Reducer and Retarder | Increase compressive and flexural strength. | Reduces segregation, bleeding, absorption, shrinkage, and increases cement dispersion. | Easier to place work, provides better control. | Performance varies with cement. Of no use in cold weather. |
| Water Reducer, Retarder and Air-Entraining Agent | Increases workability. | Improves cohesiveness. Reduces bleeding and segregation. | Easier to place and work. | Care must be taken to avoid excessive air entrainment. |

## ARCHITECTURAL WALL PATTERNS (BONDS)

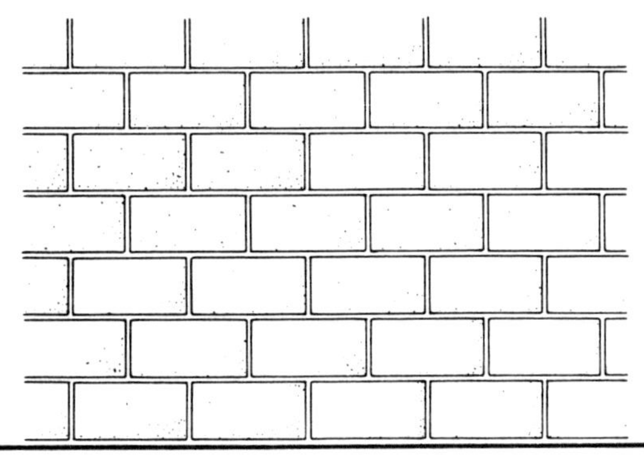

COMMON BOND

8"x 16" UNITS

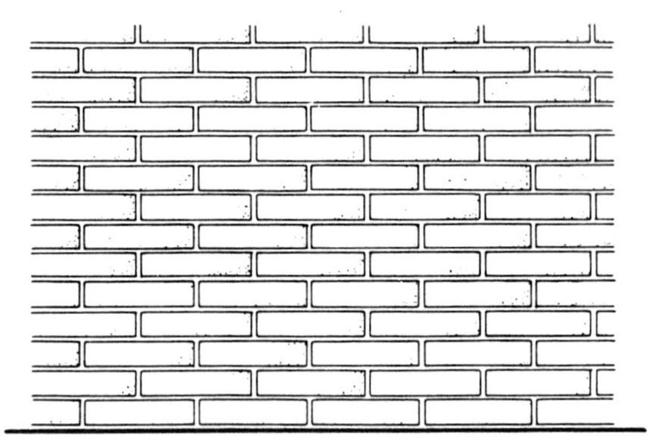

COMMON BOND

4"x 16" UNITS

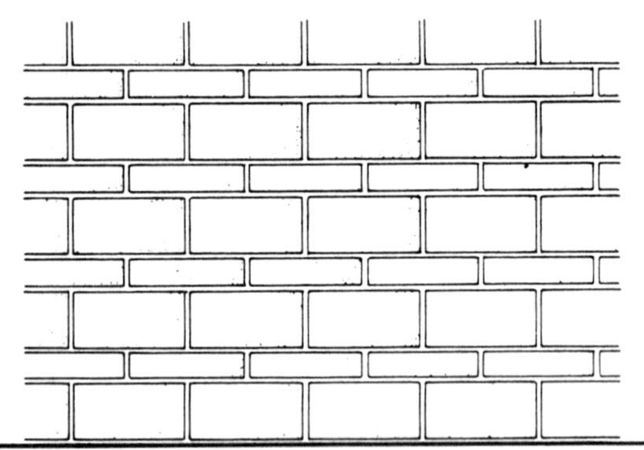

COURSED ASHLAR

8"x 16" 8 4"x 16" UNITS

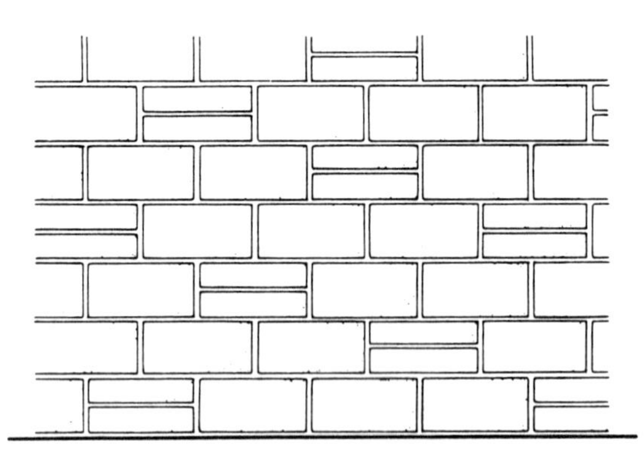

COURSED ASHLAR

8"x 16" 8 4"x 16" UNITS

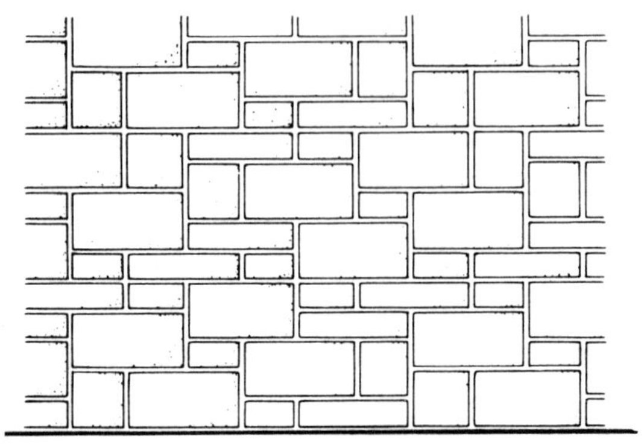

RANDOM ASHLAR

8"x 16", 8"x 8", 4"x 16" AND 4"x 8" UNITS

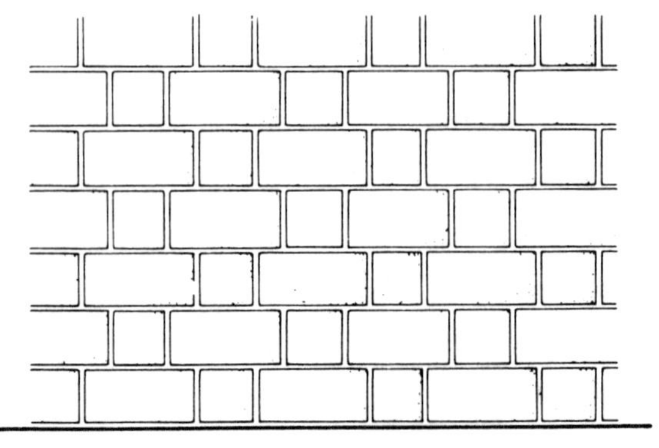

COURSED ASHLAR

8"x 16" 8 8"x 8" UNITS

## ARCHITECTURAL WALL PATTERNS (BONDS)—(Continued)

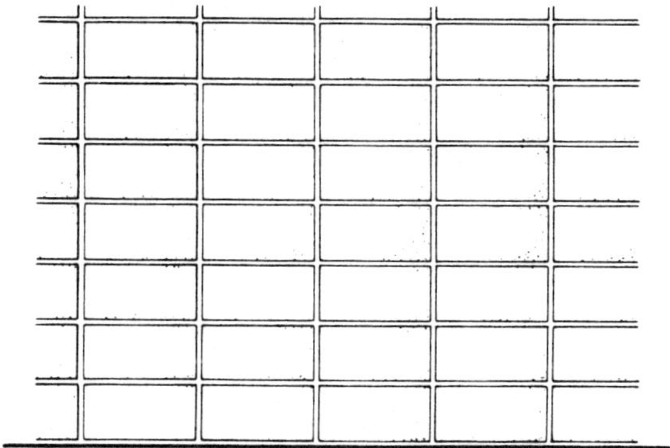

STACKED BOND
8" x 16" UNITS

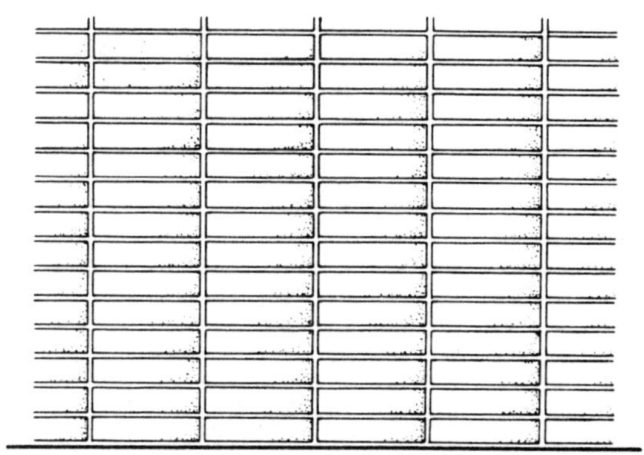

STACKED BOND
4" x 16" UNITS

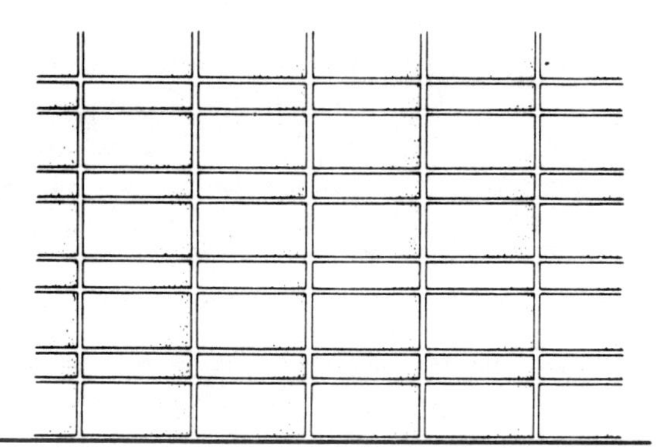

STACKED BOND
8" x 16" & 4" x 16" UNITS

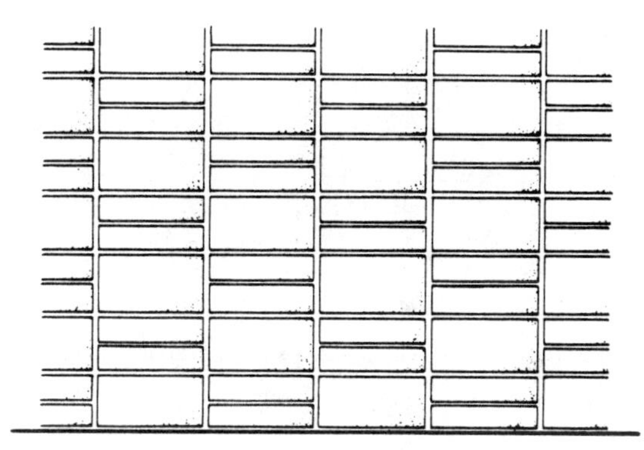

STACKED BOND
8" x 16" & 4" x 16" UNITS

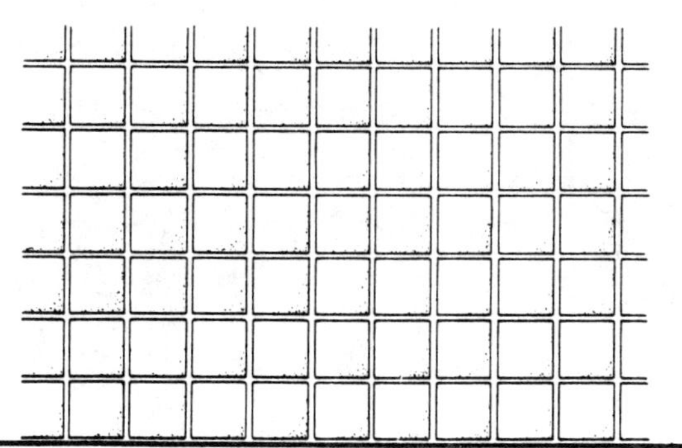

STACKED BOND VERTICAL SCORED UNITS

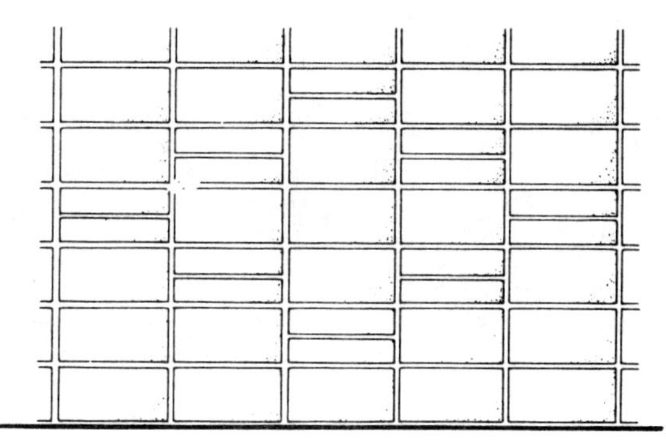

USE OF BLOCK DESIGN IN STACKED BOND

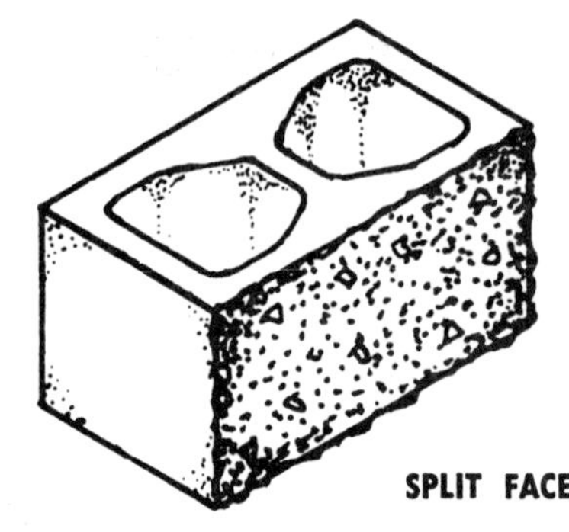
SPLIT FACE

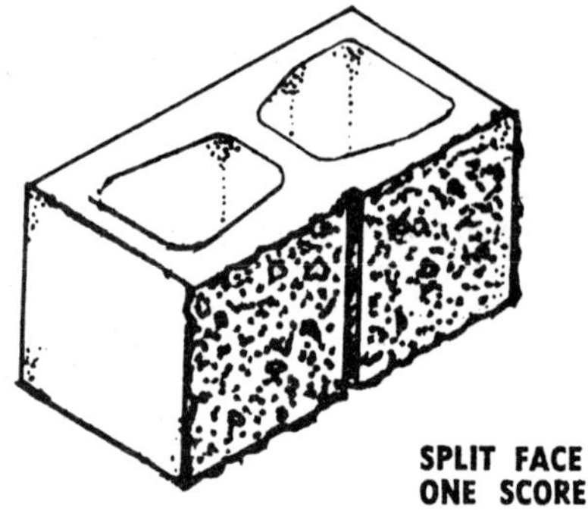
SPLIT FACE ONE SCORE

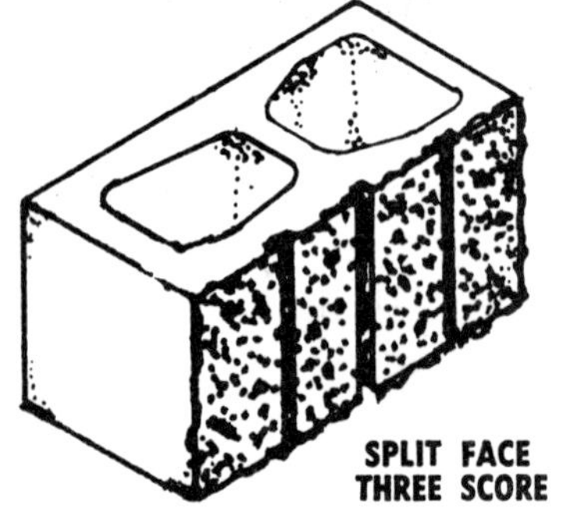
SPLIT FACE THREE SCORE

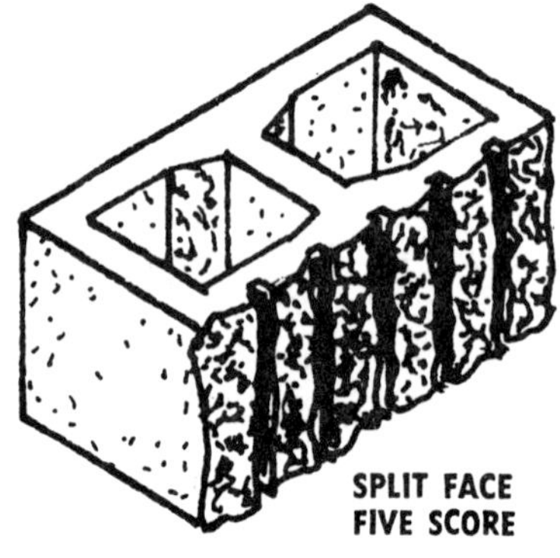
SPLIT FACE FIVE SCORE

## SPLIT FACE BLOCK

Split face block is manufactured as a unit that is normally made double and is literally split apart on a splitter; a machine which resembles a guillotine. The splitter has blades at the top and bottom (and sometimes at the sides) which exert pressure on the blocks, breaking them apart.

Many factors determine the look of the split face, both as to size variances and the amount of aggregate exposure. Split face block is intended to have a rougher texture than precision block. Various configurations of block such as fluted, scored, etc., will split in a different manner than a full split face. The vertical perpendicularity of scored and fluted split face block is subject to variation.

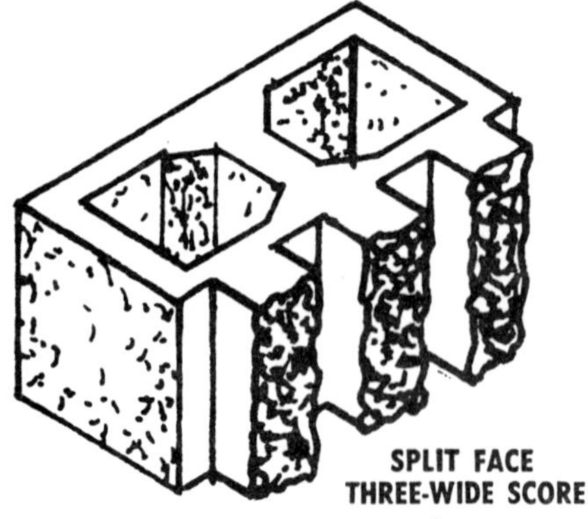
SPLIT FACE THREE-WIDE SCORE

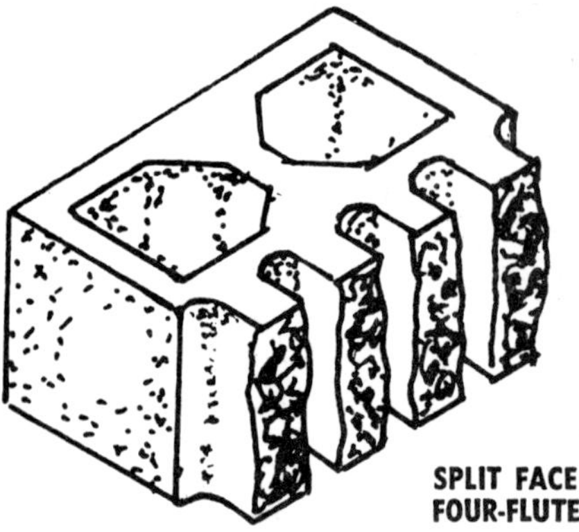
SPLIT FACE FOUR-FLUTE

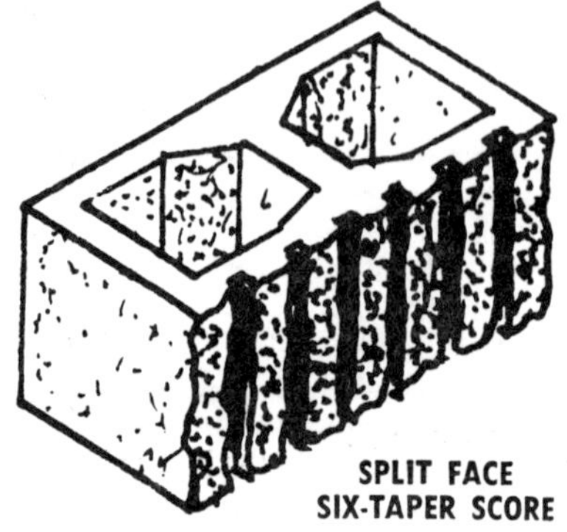
SPLIT FACE SIX-TAPER SCORE

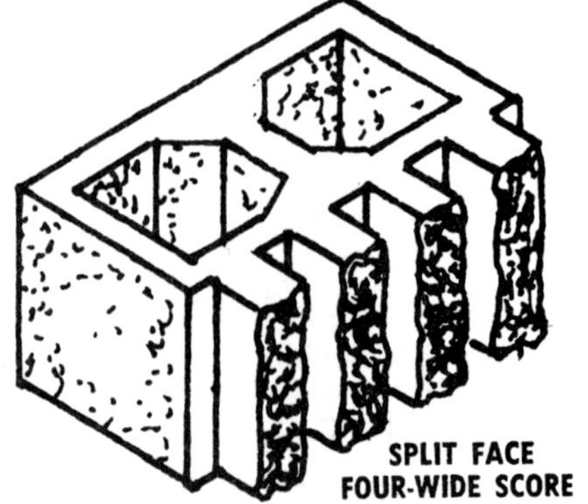
SPLIT FACE FOUR-WIDE SCORE

**NOTE:** Split face units shown in this manual are a small sampling of the broad range of concrete masonry architectural units available from the industry on special order. Depths and widths of scores vary. Consult a local manufacturer for specific information.

## TYPICAL DETAILS—LINTELS AND BOND BEAMS

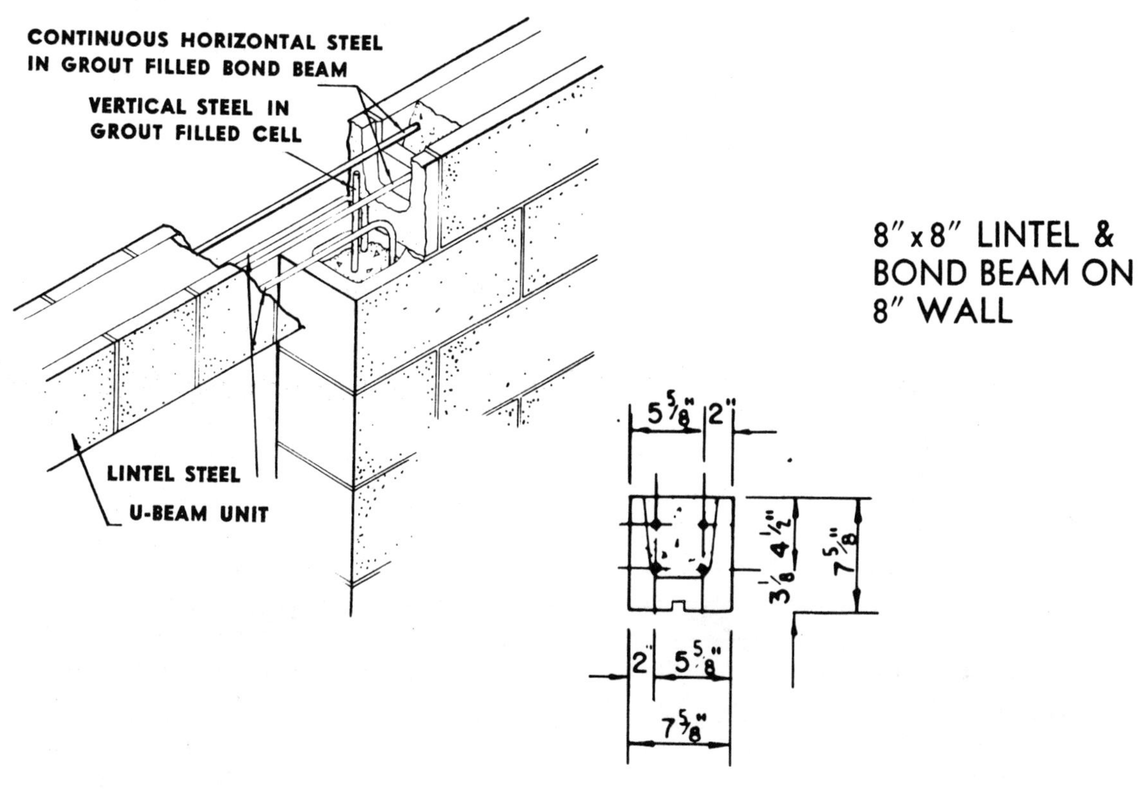


8″ x 8″ LINTEL & BOND BEAM ON 8″ WALL

8″ x 16″ BOND BEAM ON 8″ WALL

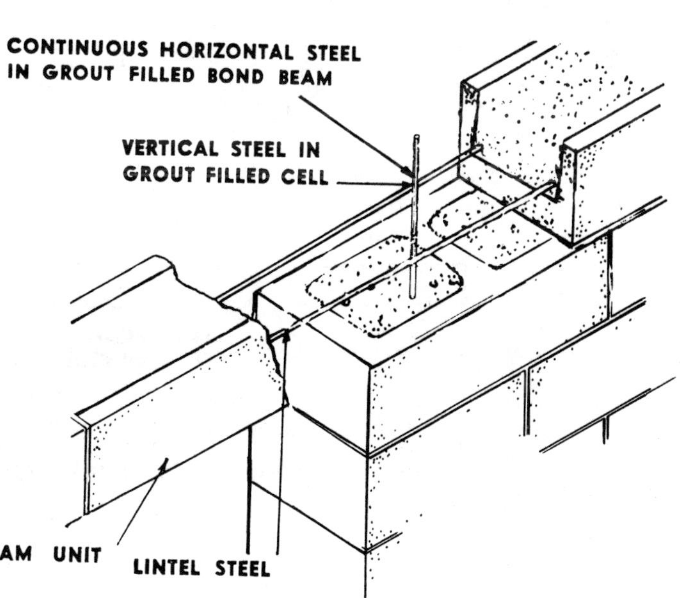

# METALS / WELDING

05050

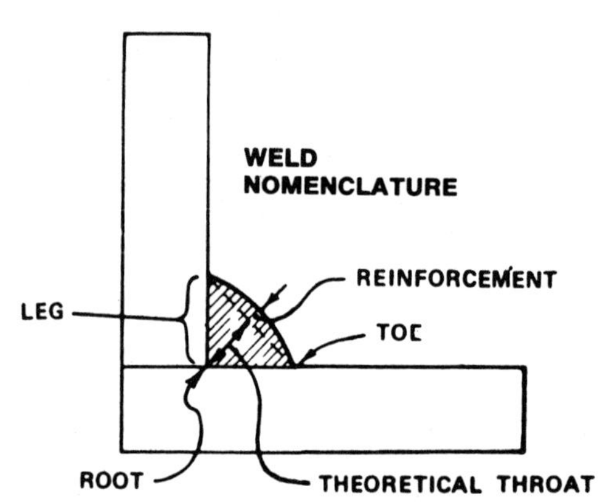


## WELDED JOINTS

SQUARE BUTT

SINGLE VEE BUTT

DOUBLE VEE BUTT

SINGLE U BUTT

DOUBLE U BUTT

SINGLE FILLET LAP

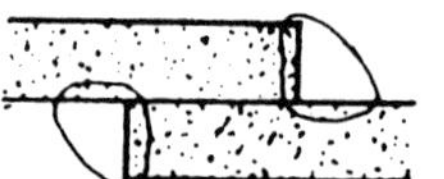
DOUBLE FILLET LAP

STRAP JOINT

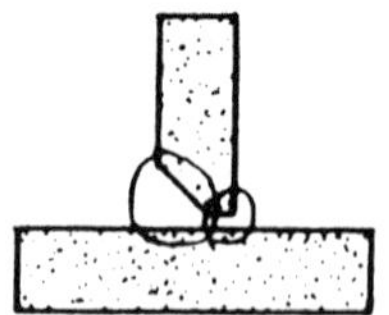
SINGLE BEVEL TEE

DOUBLE BEVEL TEE

SINGLE J TEE

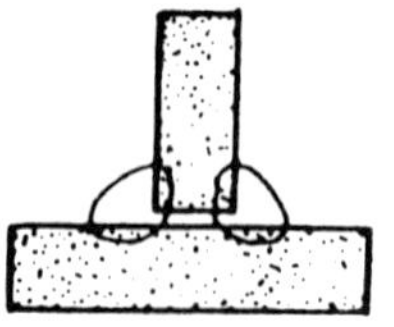
SQUARE TEE

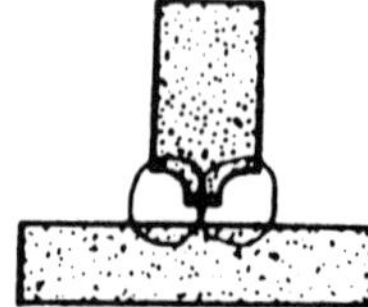
DOUBLE J TEE

CLOSED CORNER (FLUSH) JOINT

HALF OPEN CORNER JOINT

## WELDING POSITIONS

FLAT (F)

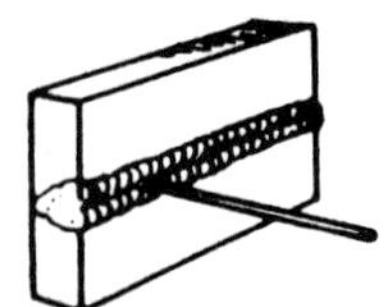
HORIZONTAL (H)

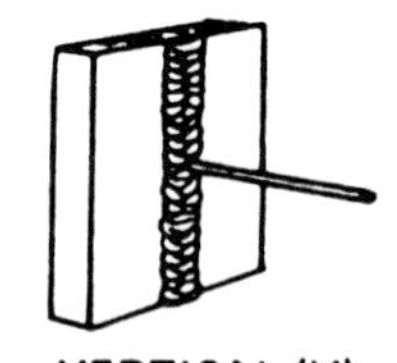
VERTICAL (V)

OVERHEAD (OH)

BOLTS IN COMMON USAGE

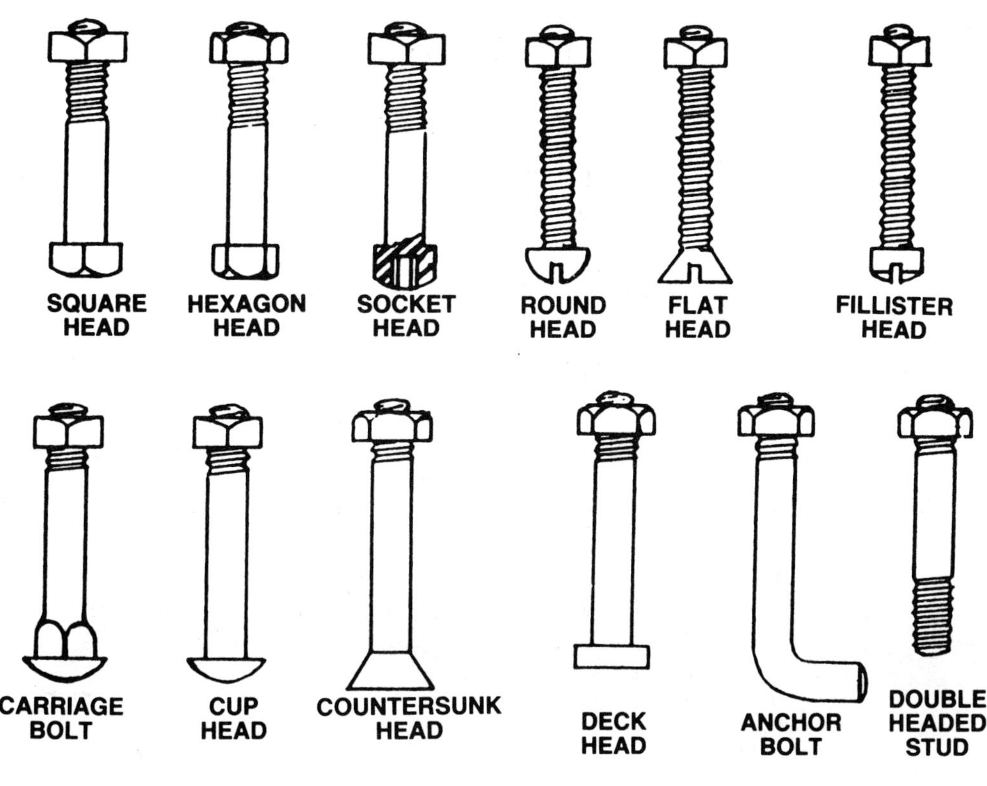


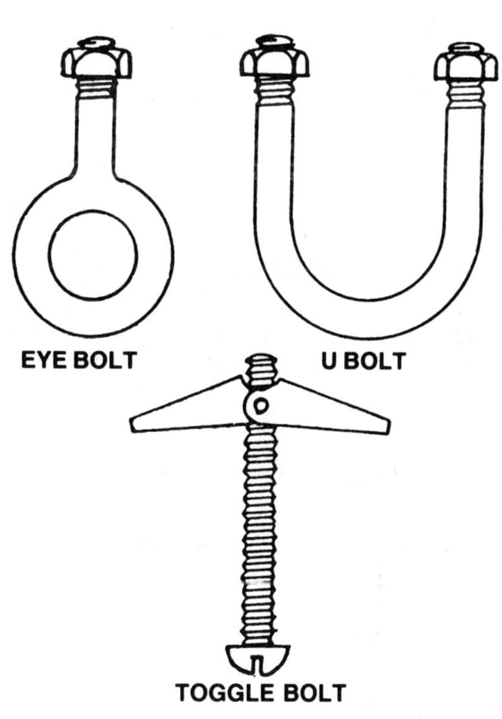

# EXAMPLE OF SIMPLIFIED STRUCTURAL STEEL TAKEOFF METHOD

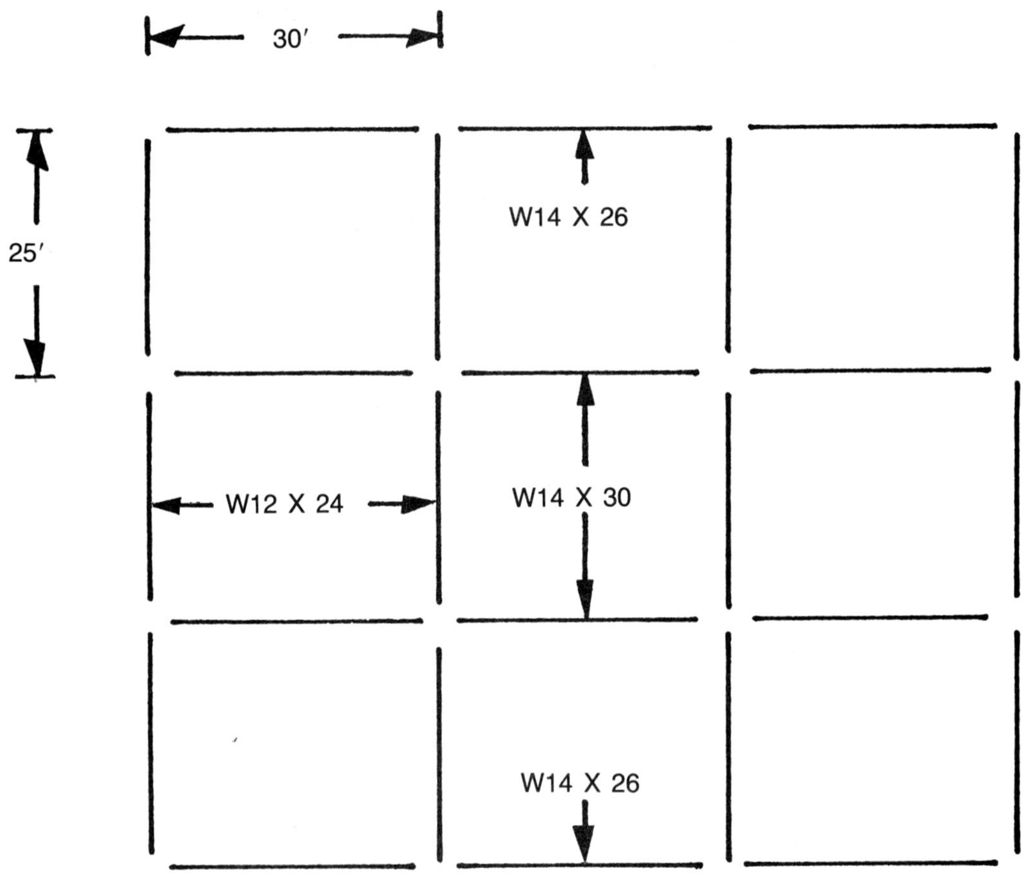


| | # | | LF. EA. | | LBS./LF. | | | |
|---|---|---|---|---|---|---|---|---|
| W14 X 26 | 6 | × | 30 | × | 26 | = | 4,680 | LBS. |
| W14 X 30 | 6 | × | 30 | × | 30 | = | 5,400 | |
| W12 X 24 | 12 | × | 25 | × | 24 | = | 7,200 | |
| | | | | | | | 17,280 | LBS. |
| | | | | | | | OR | |
| | | | | | | | 9 ± | TONS |

**AFTER MAIN MEMBERS ARE ESTIMATED, ADD:**

2 TO 3% FOR BASE PLATES
4 TO 5% FOR COLUMN SPLICES
4 TO 5% FOR MISCELLANEOUS COSTS

## COMMON WIRE NAILS (ACTUAL SIZE)

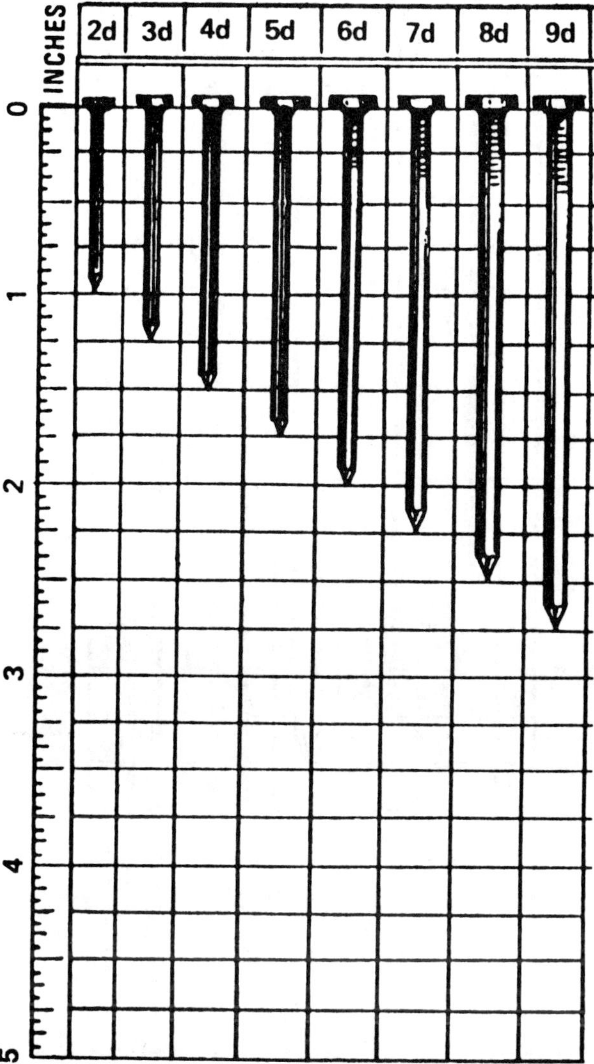


**Cut Nails.** Cut nails are angular-sided, wedge-shaped with a blunt point.

**Wire Nails.** Wire nails are round shafted, straight, pointed nails, and are used more generally than cut nails. They are stronger than cut nails and do not buckle as easily when driven into hard wood, but usually split wood more easily than cut nails. Wire nails are available in a variety of sizes varying from two penny to sixty penny.

**Nail Finishes.** Nails are available with special finishes. Some are galvanized or cadmium plated to resist rust. To increase the resistance to withdrawal, nails are coated with resins or asphalt cement (called cement coated). Nails which are small, sharp-pointed, and often placed in the craftsman's mouth (such as lath or plaster board nails) are generally blued and sterilized.

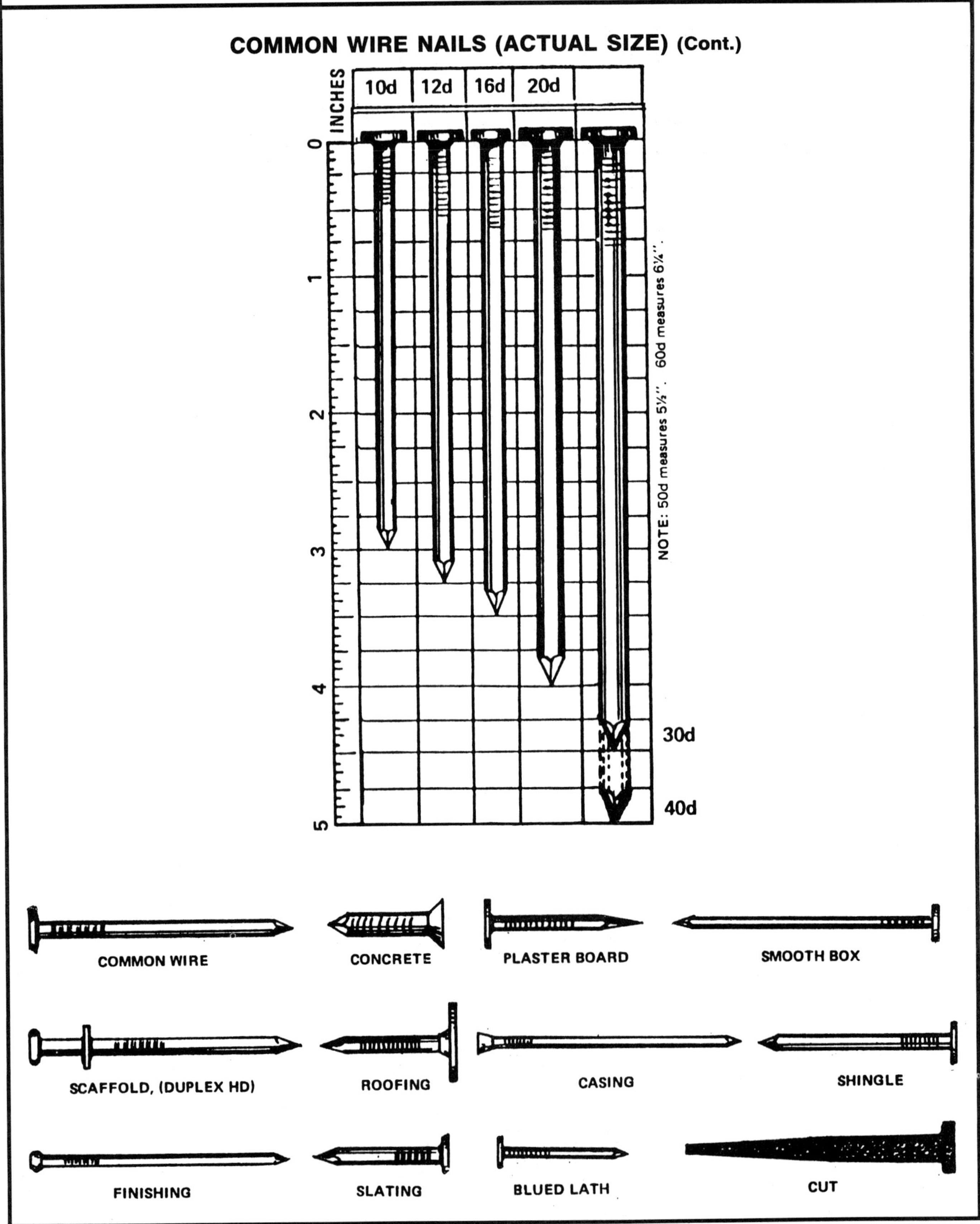
COMMON WIRE NAILS (ACTUAL SIZE) (Cont.)
INCHES
10d
12d
16d
20d
0
1
2
3
4
5
30d
40d
NOTE: 50d measures 5½". 60d measures 6¼".
COMMON WIRE
CONCRETE
PLASTER BOARD
SMOOTH BOX
SCAFFOLD, (DUPLEX HD)
ROOFING
CASING
SHINGLE
FINISHING
SLATING
BLUED LATH
CUT

## PLYWOOD – BASIC GRADE MARKS

### AMERICAN PLYWOOD ASSOCIATION (APA)

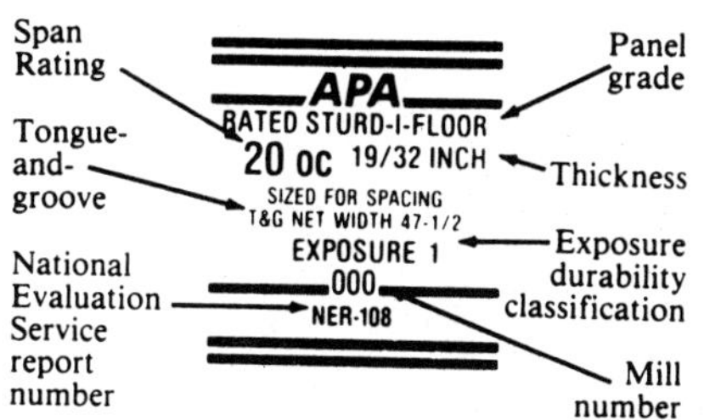


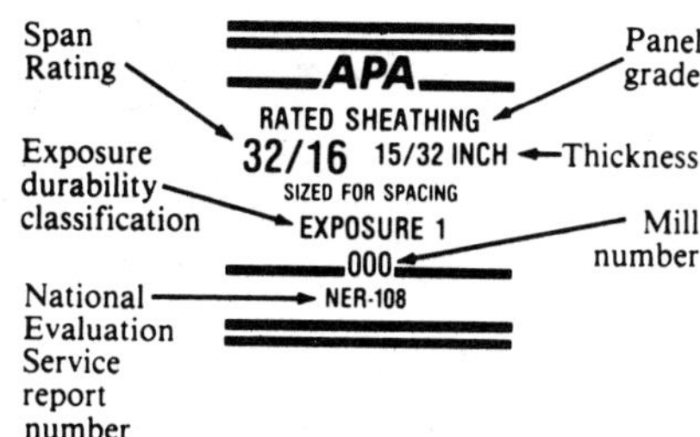


The American Plywood Association's trademarks appear only on products manufactured by APA member mills. The marks signify that the product is manufactured in conformance with APA performance standards and/or U.S. Product Standard PS 1-83 for Construction and Industrial Plywood.

APA
A-C GROUP 1
EXTERIOR
000
PS 1-83

**APA A-C**

For use where appearance of one side is important in exterior applications such as soffits, fences, structural uses, boxcar and truck linings, farm buildings, tanks, trays, commercial refrigerators, etc. ***Exposure Durability Classification:*** Exterior. ***Common Thicknesses:*** ¼, 11/32, ⅜, 15/32, ½, 19/32, ⅝, 23/32, ¾.

APA
A-D GROUP 1
EXPOSURE 1
000
PS 1-83

**APA A-D**

For use where appearance of only one side is important in interior applications, such as paneling, built-ins, shelving, partitions, flow racks, etc. ***Exposure Durability Classifications:*** Interior, Exposure 1. ***Common Thicknesses:*** ¼, 11/32, ⅜, 15/32, ½, 19/32, ⅝, 23/32, ¾.

APA
B-C GROUP 1
EXTERIOR
000
PS 1-83

**APA B-C**

Utility panel for farm service and work buildings, boxcar and truck linings, containers, tanks, agricultural equipment, as a base for exterior coatings and other exterior uses. ***Exposure Durability Classification:*** Exterior. ***Common Thicknesses:*** ¼, 11/32, ⅜, 15/32, ½, 19/32, ⅝, 23/32, ¾.

APA
B-D GROUP 2
INTERIOR
000
PS 1-83

**APA B-D**

Utility panel for backing, sides or builtins, industry shelving, slip sheets, separator boards, bins and other interior or protected applications. ***Exposure Durability Classifications:*** Interior, Exposure 1. ***Common Thicknesses:*** ¼, 11/32, ⅜, 15/32, ½, 19/32, ⅝, 23/32, ¾.

APA
PLYFORM
B-B CLASS I
EXTERIOR
000
PS 1-83

APA proprietary concrete form panels designed for high reuse. Sanded both sides and mill-oiled unless otherwise specified. Class I, the strongest, stiffest and more commonly available, is limited to Group 1 faces, Group 1 or 2 crossbands, and Group 1, 2, 3 or 4 inner plies. Class II is limited to Group 1 or 2 faces (Group 3 under certain conditions) and Group 1, 2, 3 or 4 inner plies. Also available in HDO for very smooth concrete finish, in Structural I, and with special overlays. ***Exposure Durability Classification:*** Exterior. ***Common Thicknesses:*** 19/32, ⅝, 23/32, ¾.

APA
M. D. OVERLAY
GROUP 1
EXTERIOR
000
PS 1-83

Plywood panel manufactured with smooth, opaque, resin-treated fiber overlay providing ideal base for paint on one or both sides. Excellent material choice for shelving, factory work surfaces, paneling, built-ins, signs and numerous other construction and industrial applications. Also available as a 303 Siding with texture-embossed or smooth surface on one side only and Structural I. ***Exposure Durability Classification:*** Exterior. ***Common Thicknesses:*** 11/32, ⅜, 15/32, ½, 19/32, ⅝, 23/32, ¾.

## SPECIALTY PANELS

HDO · A-A · G-1 · EXT-APA · 000 · PS1-83

Plywood panel manufactured with a hard, semi-opaque resin-fiber overlay on both sides. Extremely abrasion resistant and ideally suited to scores of punishing construction and industrial applications, such as concrete forms, industrial tanks, work surfaces, signs, agricultural bins, exhaust ducts, etc. Also available with skid-resistant screen-grid surface and in Structural I. ***Exposure Durability Classification:*** **Exterior.** ***Common Thicknesses:*** 3/8, 1/2, 5/8, 3/4.

MARINE · A-A · EXT-APA · 000 · PS1-83

Specialty designed plywood panel made only with Douglas fir or western larch, solid jointed cores, and highly restrictive limitations on core gaps and faces repairs. Ideal for both hulls and other marine applications. Also available with HDO or MDO faces. ***Exposure Durability Classification:*** Exterior. ***Common Thicknesses:*** 1/4, 3/8, 1/2, 5/8, 3/4.

Unsanded and touch-sanded panels, and panels with "B" or better veneer on one side only, usually carry the APA trademark on the panel back. Panels with both sides of "B" or better veneer, or with special overlaid surfaces (such as Medium Density Overlay), carry the APA trademark on the panel edge, like this:

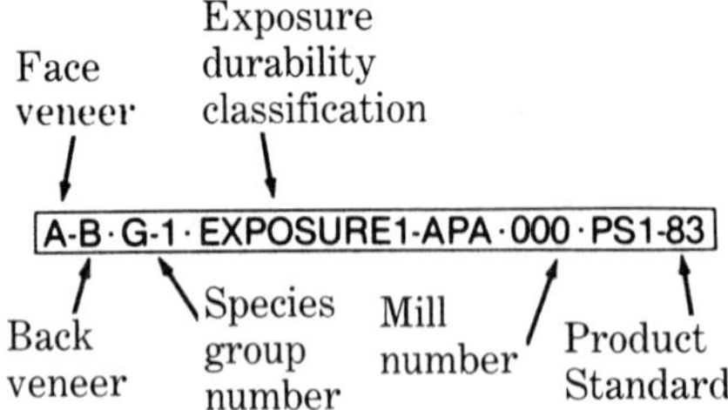


## GLOSSARY OF LUMBER TERMS

Some of the words and terms used in the grading of lumber follow:

***Bow.*** A deviation flatwise from a straight line drawn from end to end of the piece. It is measured at the point of greatest distance from the straight line.

***Checks.*** A separation of the wood which normally occurs across the annual rings and usually as a result of seasoning.

***Crook.*** A deviation edgewise from a straight line drawn from end to end of the piece. It is measured at the point of greatest distance from the straight line.

***Cup.*** A deviation from a straight line drawn across the piece from edge to edge. It is measured at the point of greatest distance from the straight line.

***Flat Grain.*** The annual growth rings pass through the piece at an angle of less than 45 degrees with the flat surface of the piece.

***Mixed Grain.*** The piece may have vertical grain, flat grain or a combination of both vertical and flat grain.

***Pitch.*** An accumulation of resin which occurs in separations in the wood or in the wood cells themselves.

***Shake.*** A separation of the wood which usually occurs between the rings of annual growth.

***Splits.*** A separation of the wood due to tearing apart of the wood cells.

***Vertical Grain.*** The annual growth rings pass through the piece at an angle of 45 degrees or more with the flat surface of the piece.

***Wane.*** Bark or lack of wood from any cause, except eased edges (rounded) on the edge or corner of a piece of lumber.

***Warp.*** Any deviation from a true or plane surface, including crook, cup, bow or any combination thereof.

# LUMBER GRADING

## GRADING-MARK ABBREVIATIONS

### GRADES

*(Listed alphabetically — not by quality)*

| | |
|---|---|
| COM | Common |
| CONST | Construction |
| ECON | Economy |
| No. 1 | Number One |
| SEL-MER | Select Merchantable |
| SEL-STR | Select Structural |
| STAN | Standard |
| UTIL | Utility |

### ALSC TRADEMARKS

| | |
|---|---|
| CLIS | California Lumber Inspection Service |
| NELMA | Northeastern Lumber Manufacturers Association, Inc. |
| NH&PMA | Northern Hardwood & Pine Manufacturers Association, Inc. |
| PLIB | Pacific Lumber Inspection Bureau |
| RIS | Redwood Inspection Service |
| SPIB | Southern Pine Inspection Bureau |
| TP | Timber Products Inspection |
| WCLB | West Coast Lumber Inspection Bureau |
| WWP | Western Wood Products Association |

### SPECIES GROUPINGS

| | |
|---|---|
| AF | Alpine Fir |
| DF | Douglas Fir |
| HF | Hem Fir |
| SP | Sugar Pine |
| PP | Ponderosa Pipe |
| LP | Lodgepole Pine |
| IWP | Idaho White Pine |
| ES | Engelmann Spruce |
| WRC | Western Red Cedar |
| INC CDR | Incense Cedar |
| L | Larch |
| LP | Lodgepole Pine |
| MH | Mountain Hemlock |
| WW | White Wood |

### MOISTURE CONTENT

| | |
|---|---|
| S-GRN | Surfaced at a moisture content of more than 19% |
| S-DRY | Surfaced at a moisture content of 19% or less |
| MC-15 | Surfaced at a moisture content of 15% or less. |

### Framing Estimating Rules of Thumb

For 16″ O.C. stud partitions figure 1 stud for every L.F. of wall; add for top and bottom plates.

For any type of framing, the quantity of basic framing members (in L.F.) can be determined based on spacing and surface area (S.F.):

| | |
|---|---|
| 12″ O.C. | 1.2 L.F./S.F. |
| 16″ O.C. | 1.0 L.F./S.F. |
| 24″ O.C. | 0.8 L.F./S.F. |

*(Doubled-up members, bands, plates, framed openings, etc., must be added.)*

Framing accessories, nails, joist hangers, connectors, etc., should be estimated as separate material costs. Installation should be included with framing. Rule of thumb allowance is 0.5 to 1.5% of lumber cost for rough hardware. Another is 30 to 40 pounds of nails per M.B.F.

## BOARD FEET/LINEAR FEET FOR LUMBER

| Nominal Size | Actual Size | Board Feet Per Linear Foot | Linear Feet Per 1000 Board Feet |
|---|---|---|---|
| 1 × 2 | ¾ × 1½ | .167 | 6000 |
| 1 × 3 | ¾ × 2½ | .250 | 4000 |
| 1 × 4 | ¾ × 3½ | .333 | 3000 |
| 1 × 6 | ¾ × 5½ | .500 | 2000 |
| 1 × 8 | ¾ × 7¼ | .666 | 1500 |
| 1 × 10 | ¾ × 9¼ | .833 | 1200 |
| 1 × 12 | ¾ × 11¼ | 1.0 | 1000 |
| 2 × 2 | 1½ × 1½ | .333 | 3000 |
| 2 × 3 | 1½ × 2½ | .500 | 2000 |
| 2 × 4 | 1½ × 3½ | .666 | 1500 |
| 2 × 6 | 1½ × 5½ | 1.0 | 1000 |
| 2 × 8 | 1½ × 7¼ | 1.333 | 750 |
| 2 × 10 | 1½ × 9¼ | 1.666 | 600 |
| 2 × 12 | 1½ × 11¼ | 2.0 | 500 |

**Redwood.** Redwood is a fairly strong and moderately lightweight material. The heartwood is red but the sapwood is white. One of the principal advantages of redwood is that the heartwood is highly resistant (but not entirely immune) to decay, fungus and insects. Standard Specifications require that all redwood used in permanent installations shall be "select heart." Grade marking shall be in accordance with the standards established in the California Redwood Association. Grade marking shall be done by, or under the supervision of the Redwood Inspection Service. (See Plate 21, Appendix.)

Redwood is graded for specific uses as indicated in the following table:

**REDWOOD GRADING**

| Type of Lumber | Grade | Typical Use |
|---|---|---|
| Grades for Dimension Only Listed Here | Clear All Heart | Exceptionally fine, knot free, straight-grained timbers. This grade is used primarily for stain finish work of high quality. |
| | Clear | Same as Clear All Heart except that this grade may contain sound sapwood and medium stain. |
| | Select Heart | **This grade only is to be used in Agency work, unless otherwise specified in the plans or specifications.** It is sound, live heartwood free from splits or streaks with sound knots. It is generally used where the timber is in contact with the ground, as in posts, mudsills, etc. |
| | Select | |
| | Construction Heart | Slightly less quality than Select Heart. It may have some sapwood in the piece. Used for general construction purposes when redwood is needed. |
| | Construction Common | Same requirement as Construction Heart except that it will contain sapwood and medium stain. Its resistance to decay and insect attack is reduced. |
| | Merchantable | Used for fence posts, garden stakes, etc. |
| | Economy | Suitable for crating, bracing and temporary construction. |

**DOUGLAS FIR GRADING**

| Type of Lumber | Grade | Typical Use |
|---|---|---|
| Select Structural Joists and Planks | Select Structural | Used where strength is the primary consideration, with appearance desirable. |
| | No. 1 | Used where strength is less critical and appearance not a major consideration. |
| | No. 2 | Used for framing elements that will be covered by subsequent construction. |
| | No. 3 | Used for structural framing where strength is required but appearance is not a factor. |
| Finish Lumber | Superior<br>Prime<br>E | For all types of uses as casings, cabinet, exposed members, etc., where a fine appearance is desired. |
| Boards (WCLIB)*<br>*Grading is by West Coast Lumber Inspection Bureau rules, but sizes conform to Western Wood Products Assn. rules. These boards are still manufactured by some mills. | Select Merchantable | Intended for use in housing and light construction where a knotty type of lumber with finest appearance is required. |
| | Construction | Used for sub-flooring, roof and wall sheathing, concrete forms, etc. Has a high degree of serviceability. |
| | Standard | Used widely for general construction purposes, including subfloors, roof and wall sheathing, concrete forms, etc. Seldom used in exposed construction because appearance. |
| | Utility | Used in general construction where low cost is a factor and appearance is not important. (Storage shelving, crates, bracing, temporary scaffolding, etc.) |

## BOARD FEET CONVERSION TABLE

| Nominal Size (In.) | ACTUAL LENGTH IN FEET | | | | | | | | |
|---|---|---|---|---|---|---|---|---|---|
| | 8 | 10 | 12 | 14 | 16 | 18 | 20 | 22 | 24 |
| 1 x 2 | | 1⅔ | 2 | 2⅓ | 2⅔ | 3 | 3½ | 3⅔ | 4 |
| 1 x 3 | | 2½ | 3 | 3½ | 4 | 4½ | 5 | 5½ | 6 |
| 1 x 4 | 2¾ | 3⅓ | 4 | 4⅔ | 5⅓ | 6 | 6⅔ | 7⅓ | 8 |
| 1 x 5 | | 4⅙ | 5 | 5⅚ | 6⅔ | 7½ | 8⅓ | 9⅙ | 10 |
| 1 x 6 | 4 | 5 | 6 | 7 | 8 | 9 | 10 | 11 | 12 |
| 1 x 7 | | 5⅝ | 7 | 8⅙ | 9⅓ | 10½ | 11⅔ | 12⅚ | 14 |
| 1 x 8 | 5⅓ | 6⅔ | 8 | 9⅓ | 10⅔ | 12 | 13⅓ | 14⅔ | 16 |
| 1 x 10 | 6⅔ | 8⅓ | 10 | 11⅔ | 13⅓ | 15 | 16⅔ | 18⅓ | 20 |
| 1 x 12 | 8 | 10 | 12 | 14 | 16 | 18 | 20 | 22 | 24 |
| 1¼ x 4 | | 4⅙ | 5 | 5⅚ | 6⅔ | 7½ | 8⅓ | 9⅙ | 10 |
| 1¼ x 6 | | 6¼ | 7½ | 8¾ | 10 | 11¼ | 12½ | 13¾ | 15 |
| 1¼ x 8 | | 8⅓ | 10 | 11⅔ | 13⅓ | 15 | 16⅔ | 18⅓ | 20 |
| 1¼ x 10 | | 10 5/12 | 12½ | 14 7/12 | 16⅔ | 18¾ | 20⅚ | 22 11/12 | 25 |
| 1¼ x 12 | | 12½ | 15 | 17½ | 20 | 22½ | 25 | 27½ | 30 |
| 1½ x 4 | 4 | 5 | 6 | 7 | 8 | 9 | 10 | 11 | 12 |
| 1½ x 6 | 6 | 7½ | 9 | 10½ | 12 | 13½ | 15 | 16½ | 18 |
| 1½ x 8 | 8 | 10 | 12 | 14 | 16 | 18 | 20 | 22 | 24 |
| 1½ x 10 | 10 | 12½ | 15 | 17½ | 20 | 22½ | 25 | 27½ | 30 |
| 1½ x 12 | 12 | 15 | 18 | 21 | 24 | 27 | 30 | 33 | 36 |
| 2 x 4 | 5⅓ | 6⅔ | 8 | 9⅓ | 10⅓ | 12 | 13⅓ | 14⅔ | 16 |
| 2 x 6 | 8 | 10 | 12 | 14 | 16 | 18 | 20 | 22 | 24 |
| 2 x 8 | 10⅔ | 13⅓ | 16 | 18⅔ | 21⅓ | 24 | 26⅔ | 29⅓ | 32 |
| 2 x 10 | 13⅓ | 16⅔ | 20 | 23⅓ | 26⅔ | 30 | 33⅓ | 36⅔ | 40 |
| 2 x 12 | 16 | 20 | 24 | 28 | 32 | 36 | 40 | 44 | 48 |
| 3 x 6 | 12 | 15 | 18 | 21 | 24 | 27 | 30 | 33 | 36 |
| 3 x 8 | 16 | 20 | 24 | 28 | 32 | 36 | 40 | 44 | 48 |
| 3 x 10 | 20 | 25 | 30 | 35 | 40 | 45 | 50 | 55 | 60 |
| 3 x 12 | 24 | 30 | 36 | 42 | 48 | 54 | 60 | 66 | 72 |
| 4 x 4 | 10⅔ | 13⅓ | 16 | 18⅔ | 21⅓ | 24 | 26⅔ | 29⅓ | 32 |
| 4 x 6 | 16 | 20 | 24 | 28 | 32 | 36 | 40 | 44 | 48 |
| 4 x 8 | 21⅓ | 26⅔ | 32 | 37⅓ | 42⅔ | 48 | 53⅓ | 58⅔ | 64 |
| 4 x 10 | 26⅔ | 33⅓ | 40 | 46⅔ | 53⅓ | 60 | 66⅔ | 73⅓ | 80 |
| 4 x 12 | 32 | 40 | 48 | 56 | 64 | 72 | 80 | 88 | 96 |

# MOISTURE PROTECTION / ROOFING 07000

## SLOPE AREA CALCULATIONS

| Rise and Run | Multiply Flat Area by | LF of Hips or Valleys per LF of Common Run |
|---|---|---|
| 2 in 12 | 1.014 | 1.424 |
| 3 in 12 | 1.031 | 1.436 |
| 4 in 12 | 1.054 | 1.453 |
| 5 in 12 | 1.083 | 1.474 |
| 6 in 12 | 1.118 | 1.500 |
| 7 in 12 | 1.158 | 1.530 |
| 8 in 12 | 1.202 | 1.564 |
| 9 in 12 | 1.250 | 1.600 |
| 10 in 12 | 1.302 | 1.641 |
| 11 in 12 | 1.357 | 1.685 |
| 12 in 12 | 1.413 | 1.732 |

# MOISTURE PROTECTION / DOWNSPOUTS 07600

## DOWNSPOUT/VERTICAL LEADER CALCULATIONS

| Roof Type | Slope | S.F. Roof/ Sq. In. Leader |
|---|---|---|
| Gravel | Less than ¼″ per foot | 300 |
| Gravel | Greater than ¼″ per foot | 250 |
| Metal or Shingle | Any | 200 |

Alternate calculations:

$$\text{Diameter of downspout/leader} = 1.128\sqrt{\frac{\text{Area of drainage}}{\text{SF Roof/Sq. Inch}}}$$

## TYPICAL MINIMUM SIZE OF VERTICAL CONDUCTORS AND LEADERS

| Size of leader or conductor (Inches) | Maximum projected roof area (Square feet) |
|---|---|
| 2 | 544 |
| 2½ | 987 |
| 3 | 1,610 |
| 4 | 3,460 |
| 5 | 6,280 |
| 6 | 10,200 |
| 8 | 22,000 |

## TYPICAL MINIMUM SIZE OF ROOF GUTTERS

| | Maximum projected roof area for gutters of various slopes | | | |
|---|---|---|---|---|
| Diameter gutter (Inches) | 1/16 in. per Ft. slope (Sq. ft.) | 1/8 in. per Ft. slope (Sq. ft.) | 1/4 in. per Ft. slope (Sq. ft.) | 1/2 in. per Ft. slope (Sq. ft.) |
| 3 | 170 | 240 | 340 | 480 |
| 4 | 360 | 510 | 720 | 1020 |
| 5 | 625 | 880 | 1250 | 1770 |
| 6 | 960 | 1360 | 1920 | 2770 |
| 7 | 1380 | 1950 | 2760 | 3900 |
| 8 | 1990 | 2800 | 3980 | 5600 |
| 10 | 3600 | 5100 | 7200 | 10000 |

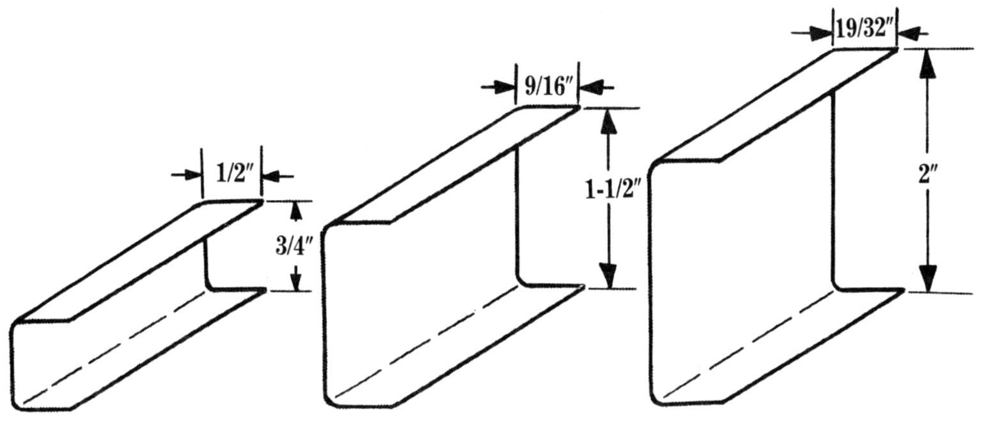


COLD ROLLED CHANNELS

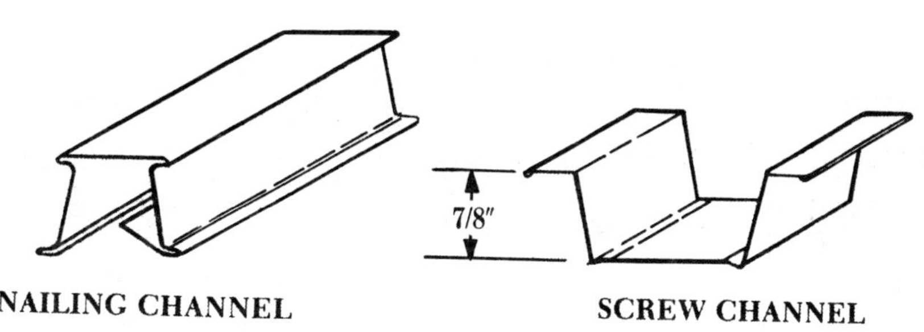


NAILING CHANNEL

SCREW CHANNEL

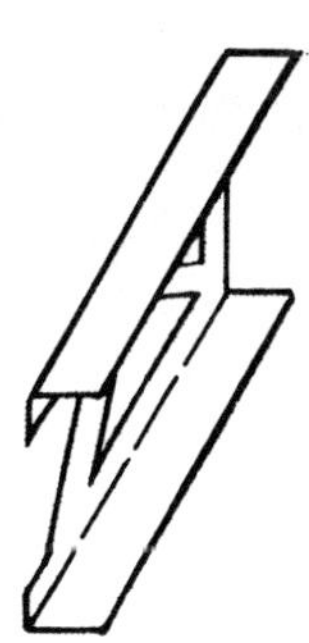

CHANNEL STUD

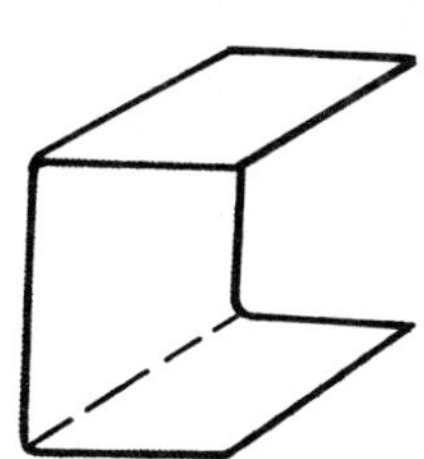

WIDE FLANGE CHANNEL

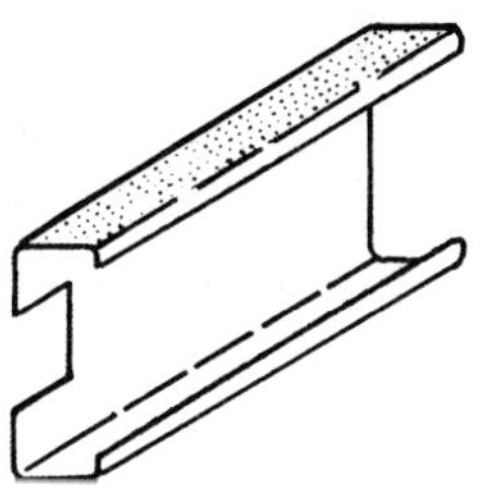

CEE STUD

# STUDLESS SOLID PARTITION

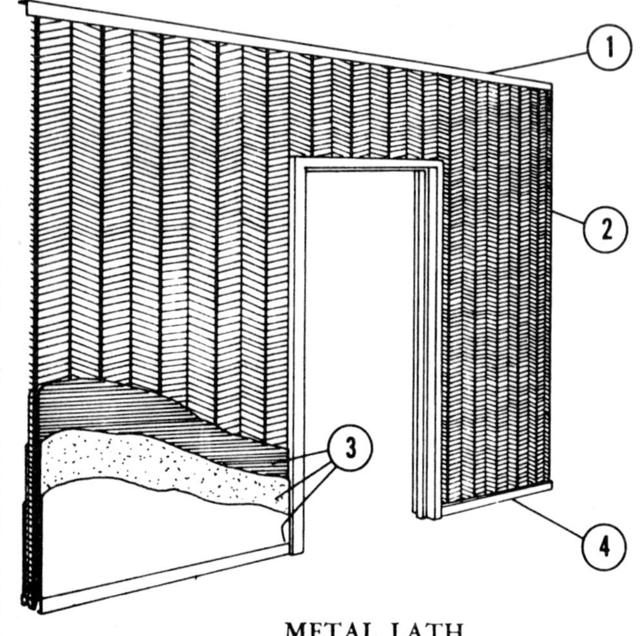


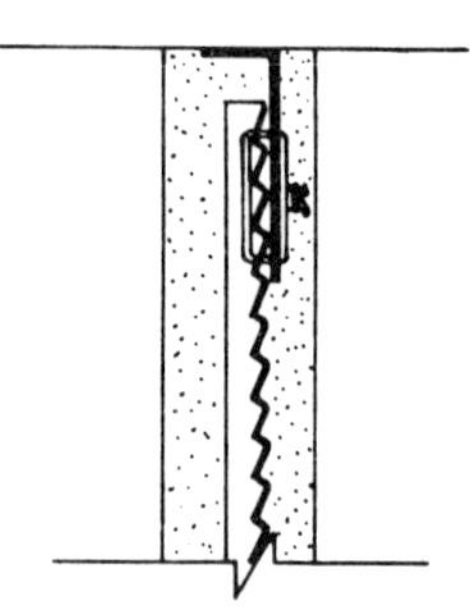

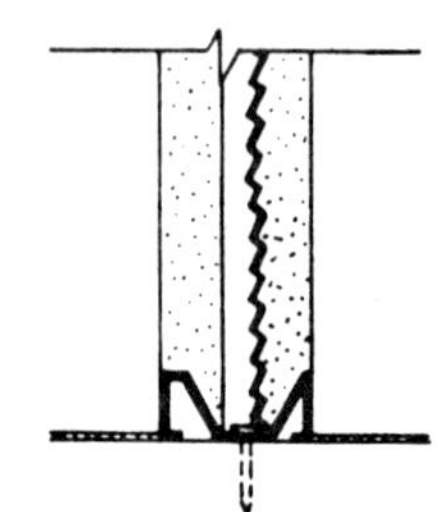

(1) Ceiling Runner
(2) Rib Metal Lath
(3) Plaster
(4) Combination Floor Runner and Screed

METAL LATH

# STUDLESS SOLID PARTITION

(1) Ceiling Runner
(2) Long Length Gypsum Lath
(3) Plaster
(4) Combination Floor Runner and Screed

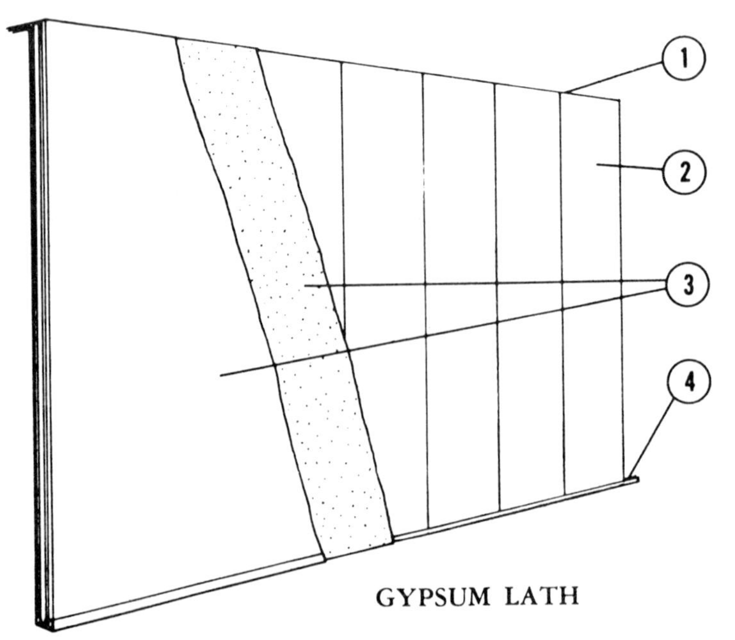


GYPSUM LATH

# SUSPENDED CEILINGS

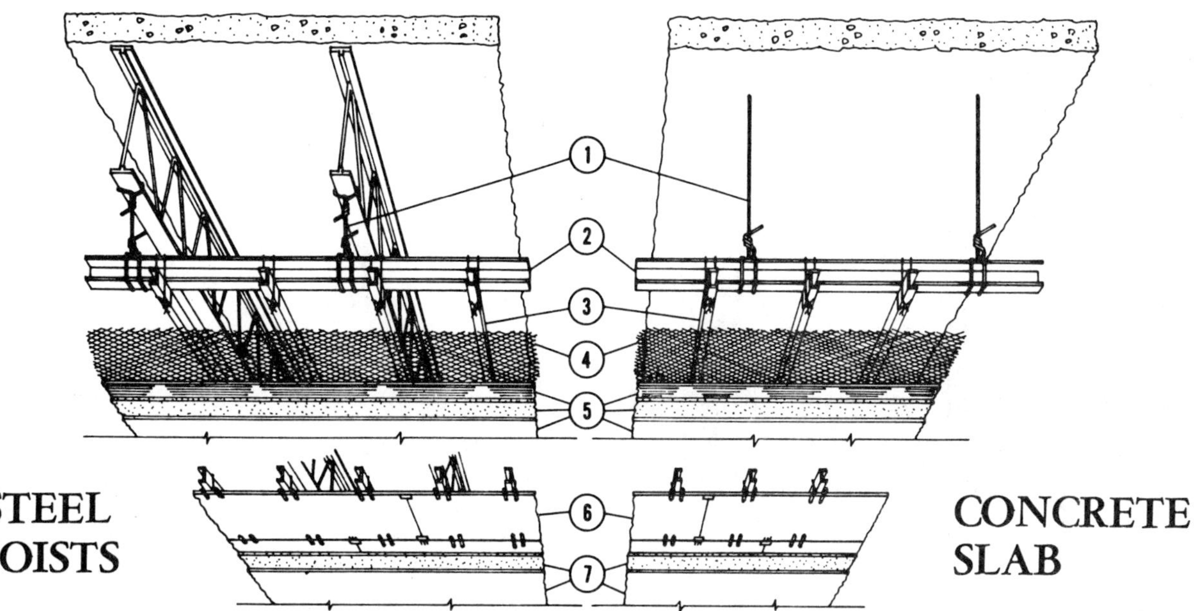


(1) Hanger
(2) Main Runner Channel
(3) Furring Channel
(4) Metal or Wire Fabric Lath
(5) Plaster
(6) Gypsum Lath
(7) Plaster

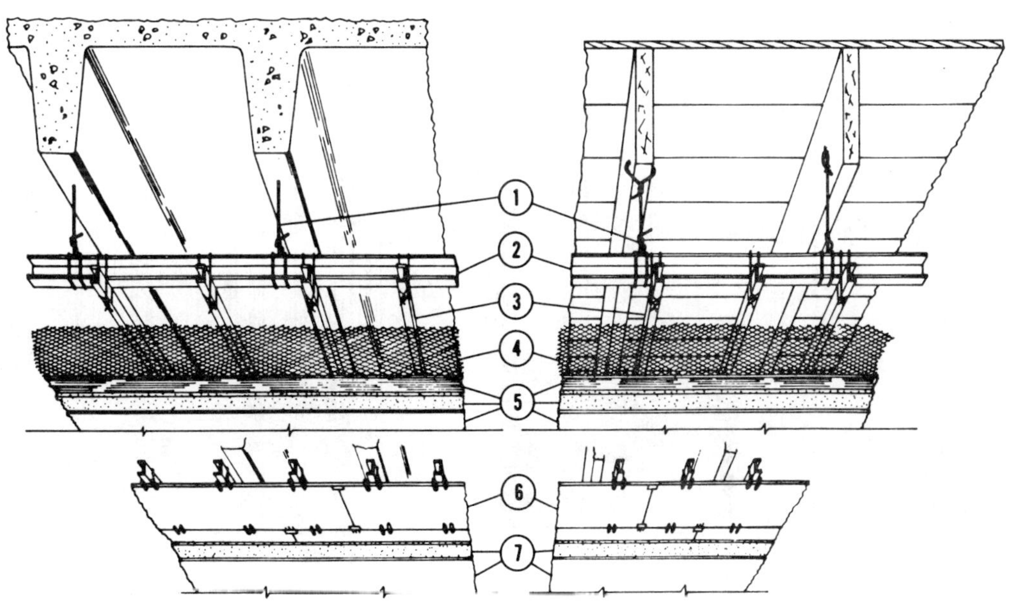


CONCRETE JOISTS

WOOD JOISTS

# STEEL STUD
## Hollow Partition

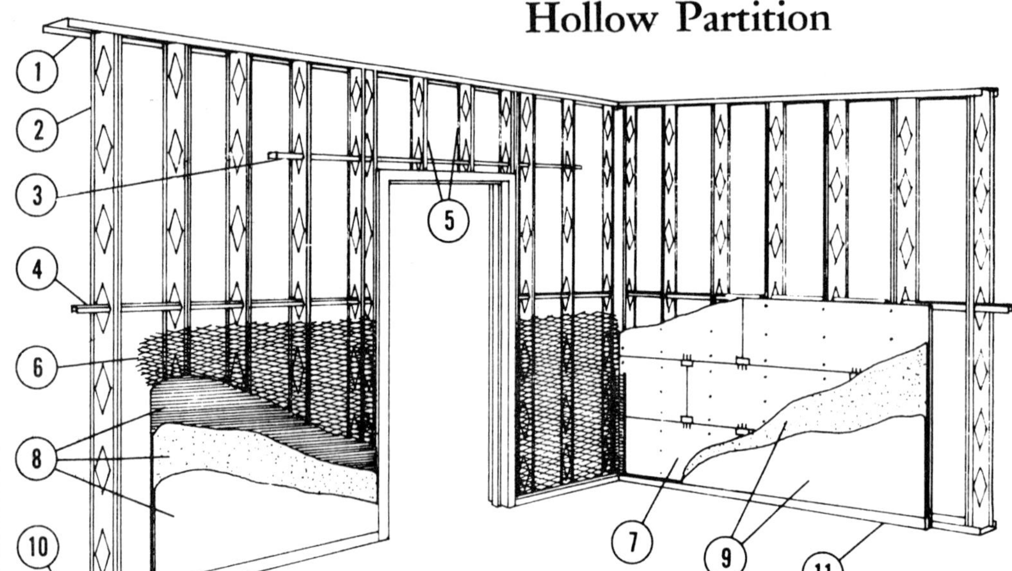


(1) Ceiling Runner Track
(2) Nailable Stud
(3) Door Opening Stiffener
(4) Partition Stiffener
(5) Jack Studs
(6) Metal or Wire Fabric Lath (screwed or wire tied)
(7) Gypsum Lath (nailed, clipped or screwed)
(8) Three Coats of Plaster (Scratch, Brown, Finish)
(9) Two Coats of Plaster (Brown, Finish)
(10) Floor Runner Track
(11) Flush Metal Base

# SCREW STUD
## Hollow Partition

(1) Ceiling Runner Track
(2) Screw Stud
(3) Door Opening Stiffener
(4) Partition Stiffener
(5) Jack Studs
(6) Metal or Wire Fabric Lath (screwed on)
(7) Gypsum Lath (screwed on)
(8) Three Coats of Plaster (Scratch, Brown, Finish)
(9) Two Coats of Plaster (Brown, Finish)
(10) Floor Runner Track
(11) Flush Metal Base

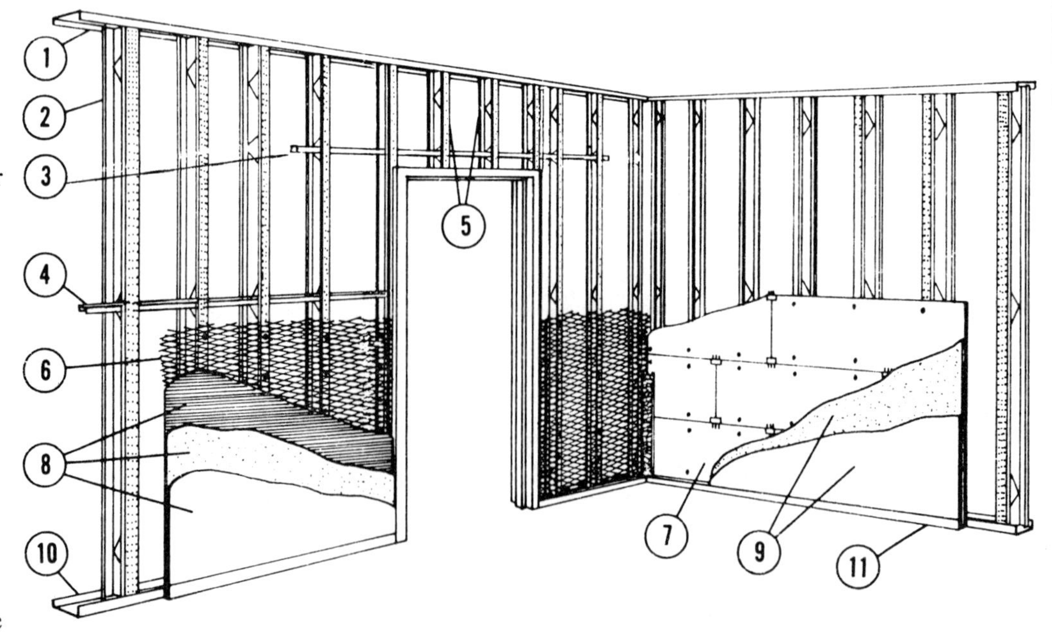

# LOAD-BEARING HOLLOW PARTITION
## Structural Stud

(1) Ceiling Runner Track
(2) Structural Stud (prefabricated)
(3) Structural Stud (nailable)
(4) Jack Studs
(5) Partition Stiffener (bridging)
(6) Metal or Wire Fabric Lath (wire-tied, nailed or stapled)
(7) Gypsum Lath (nailed or stapled)
(8) Three Coats of Plaster (Scratch, Brown, Finish)
(9) Two Coats of Plaster (Brown, Finish)
(10) Floor Runner Track
(11) Flush Metal Base

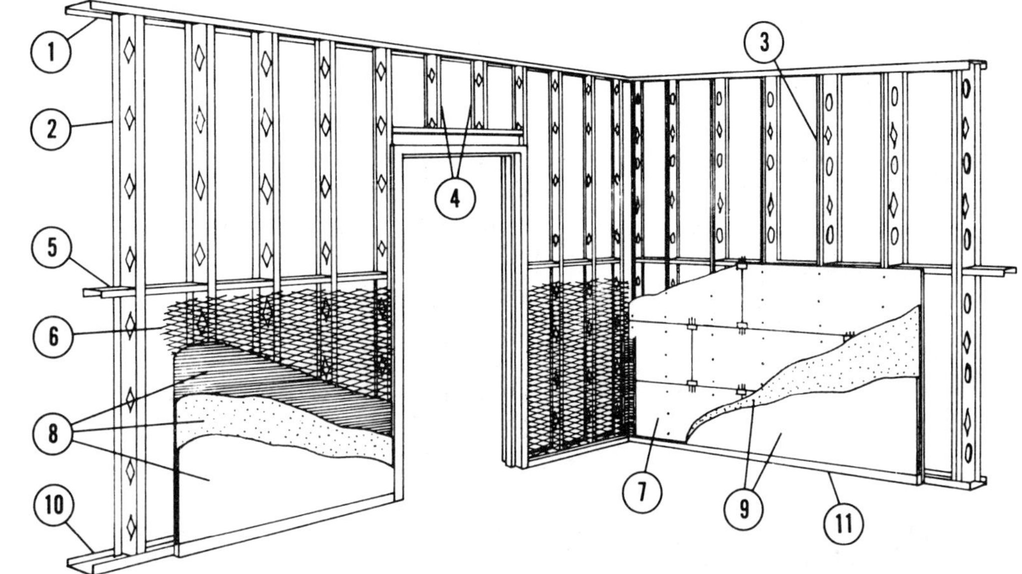


# VERTICAL FURRING
## With Studs

(1) Ceiling Runner
(2) Channel Stud
(3) Horizontal Stiffener
(4) Floor Runner
(5) Metal or Wire Fabric Lath
(6) Gypsum Lath
(7) Bracing
(8) Three Coats of Plaster (Scratch, Brown, Finish)
(9) Two Coats of Plaster (Brown, Finish)
(10) Screw Channel Studs

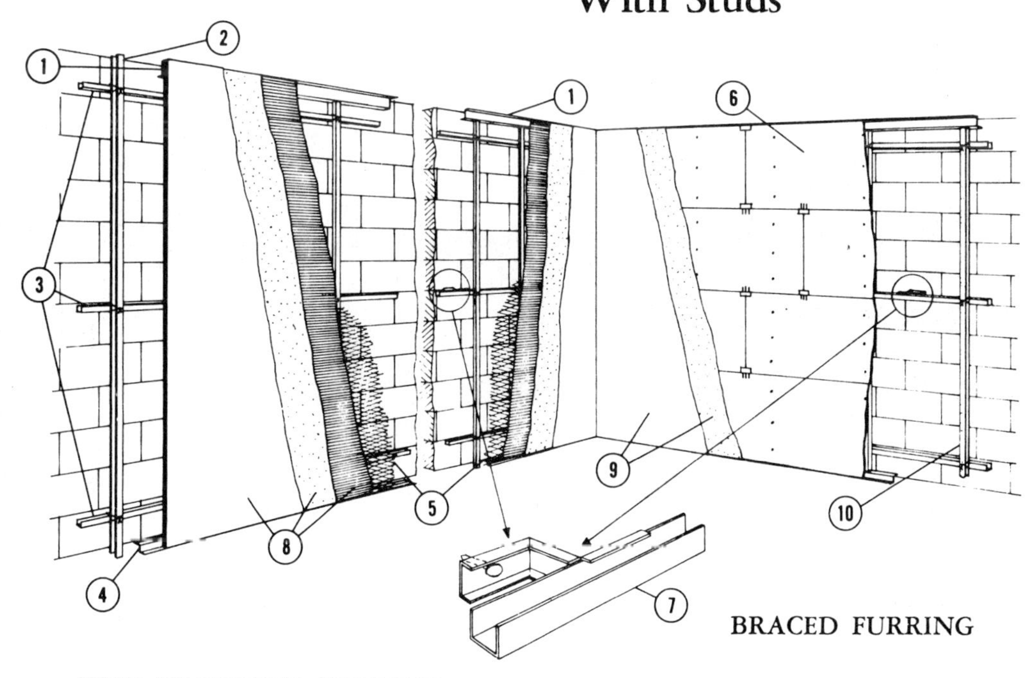


BRACED FURRING

FREE STANDING FURRING

# DOUBLE CHANNEL STUD

## Hollow Partition

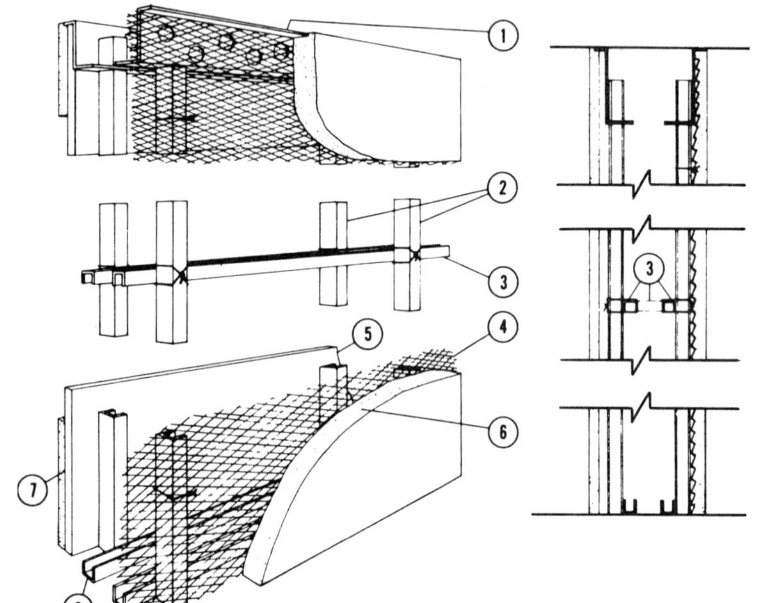


(1) Ceiling Runner
(2) Channel Studs
(3) Partition Stiffener and Channel Spacer
(4) Metal or Wire Fabric Lath (wire-tied)
(5) Gypsum Lath (clipped on)
(6) Three Coats of Plaster (Scratch, Brown, Finish)
(7) Two Coats of Plaster (Brown, Finish)
(8) Floor Runners (channel)

# SINGLE CHANNEL STUD

## Solid Partition

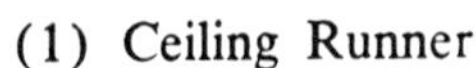

(1) Ceiling Runner
(2) Channel Stud
(3) Metal or Wire Fabric Lath (wire-tied)
(4) Plaster
(5) Combination Floor Runner and Screed

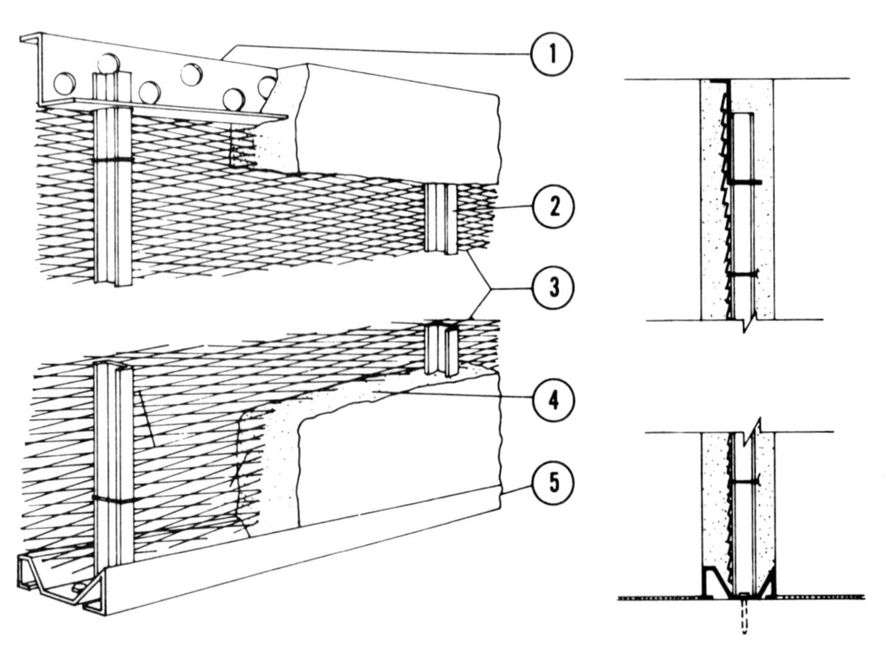

# FURRED CEILINGS

## STEEL JOISTS

## CONCRETE JOISTS

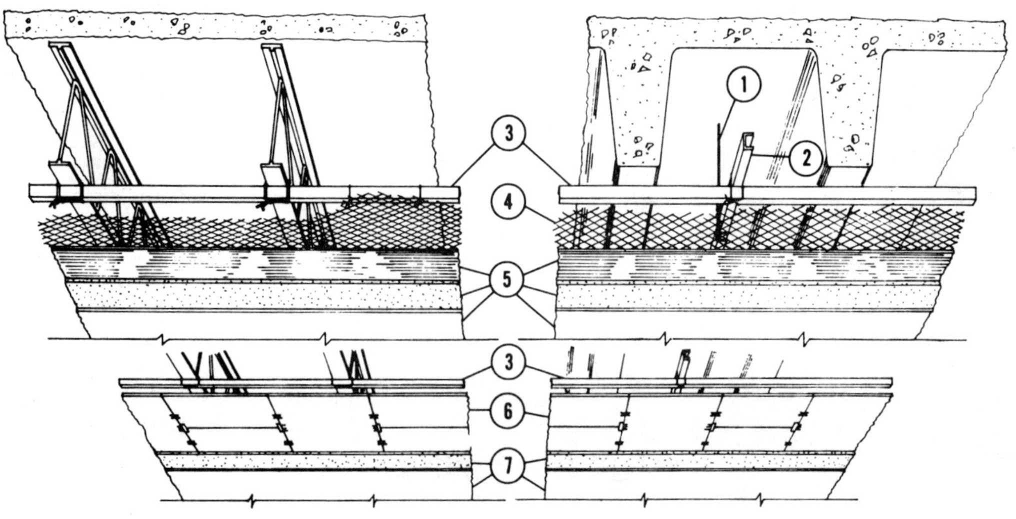


## WOOD JOISTS

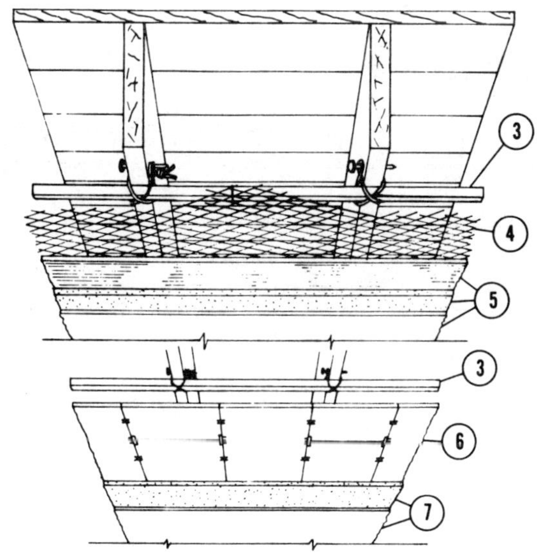


(1) Hanger
(2) ¾-inch Channel
(3) Cross Furring
(4) Metal or Wire Fabric Lath
(5) Plaster
(6) Gypsum Lath
(7) Plaster

# CEILINGS

## STEEL JOISTS

## CONCRETE JOISTS

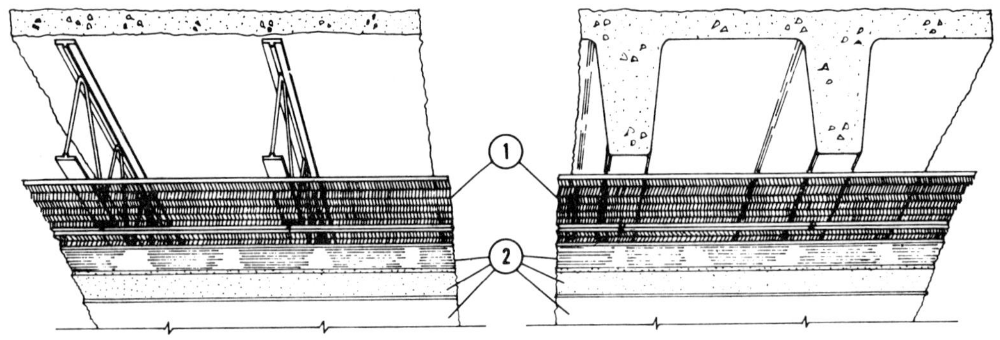


## WOOD JOISTS

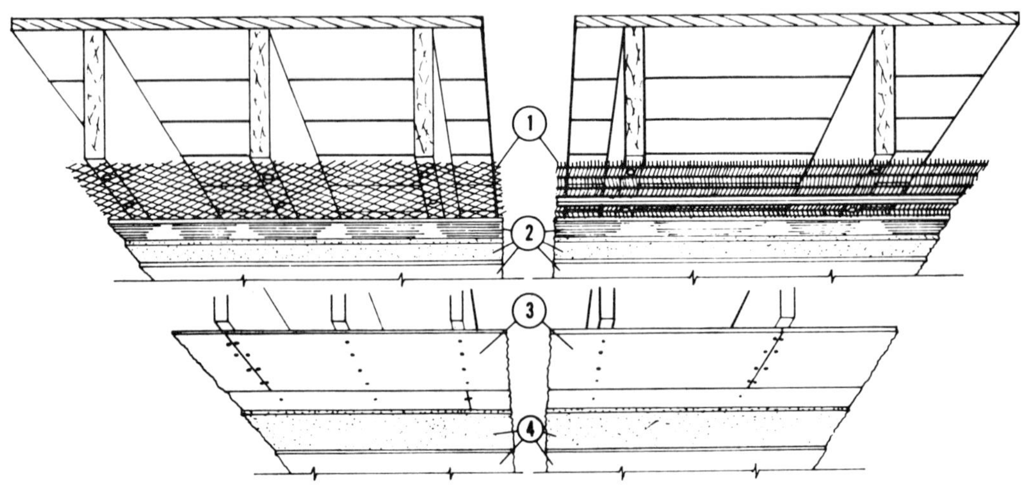


(1) Metal or Wire Fabric Lath

(2) Plaster

(3) Gypsum Lath

(4) Plaster

# FINISHES / PLASTER ACCESSORIES

# 09205

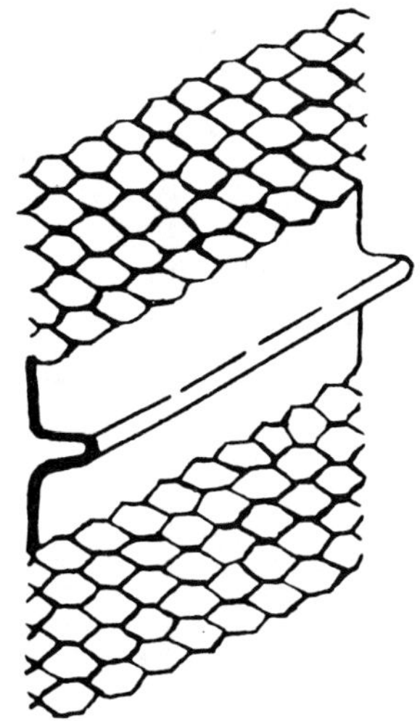
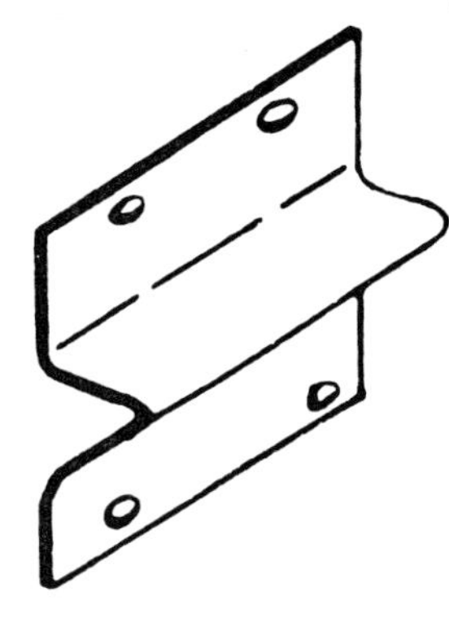

BASE OR PARTING SCREEDS

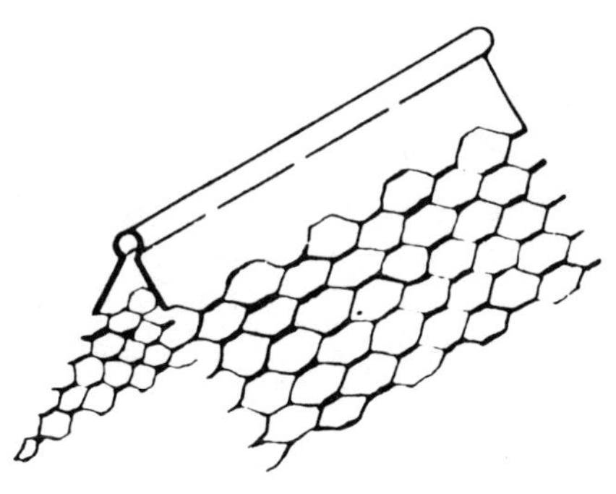
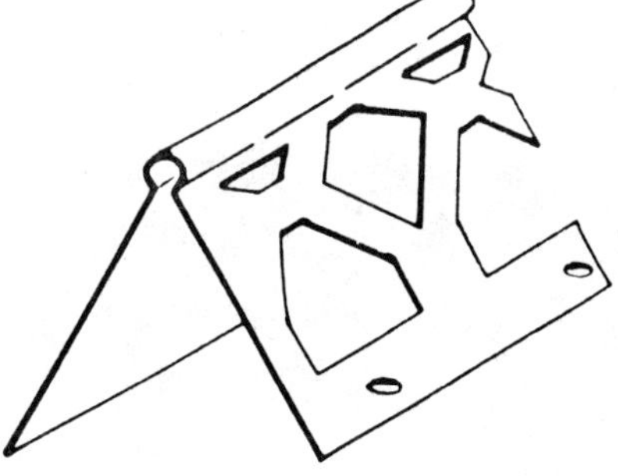

SMALL NOSE CORNER BEADS

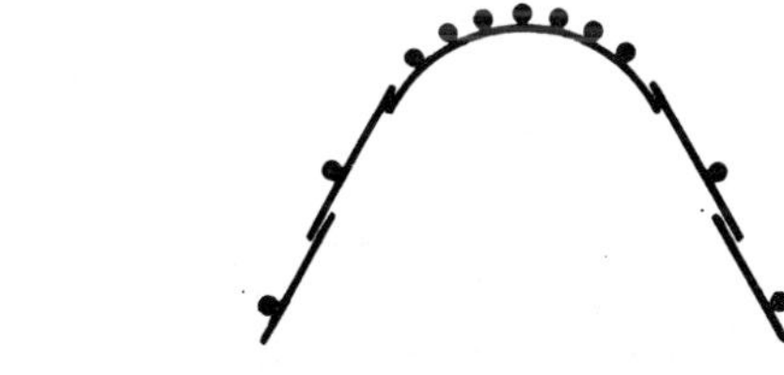

WIRE BULL NOSE
CORNER BEADS

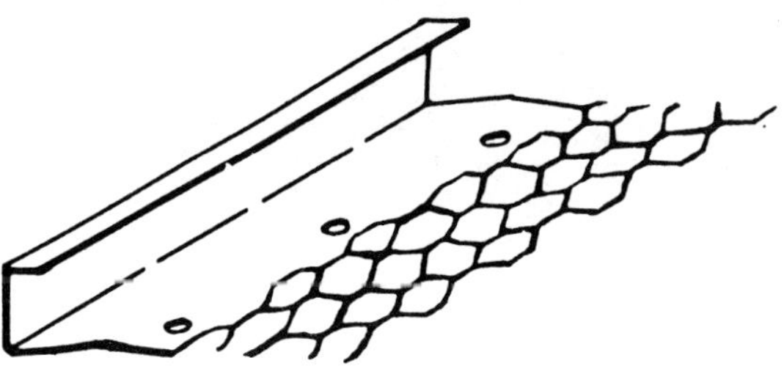
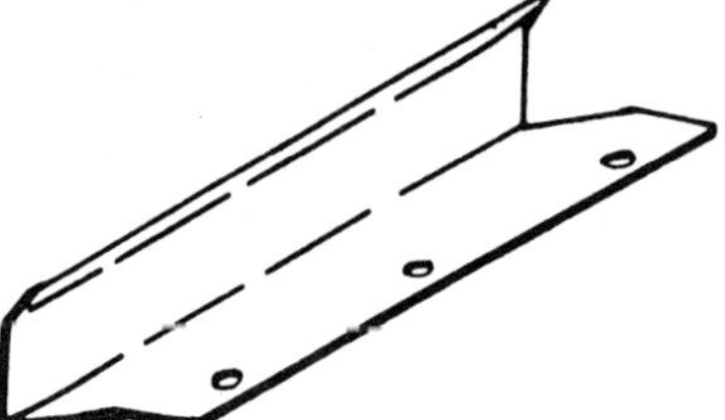

SQUARE CASING BEADS

# FINISHES / METAL LATH

09205

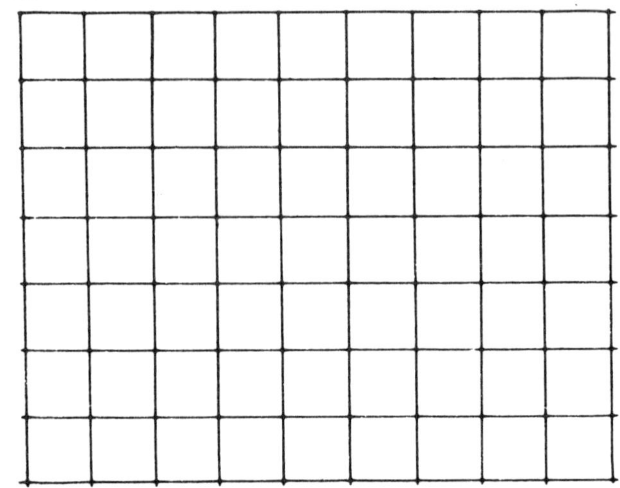

PLAIN WIRE
FABRIC LATH

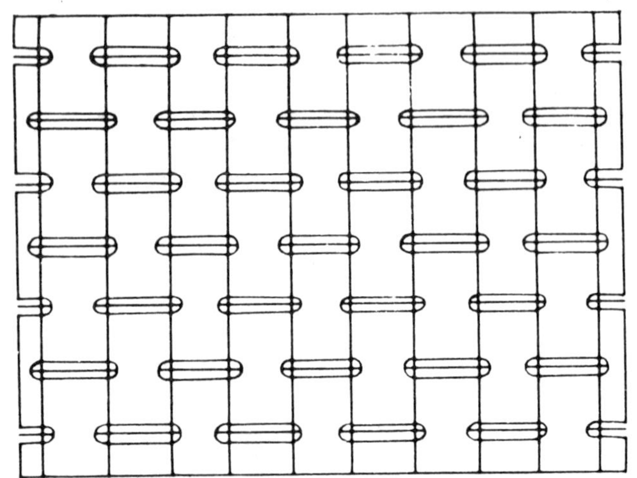

SELF-FURRING
WIRE FABRIC LATH

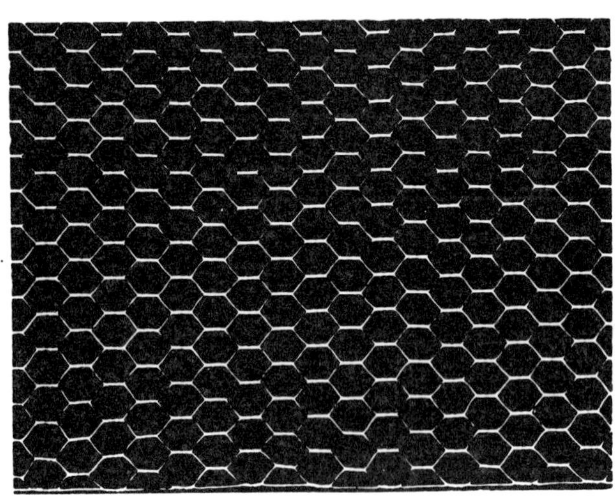

PAPER BACKED
WOVEN WIRE
FABRIC LATH

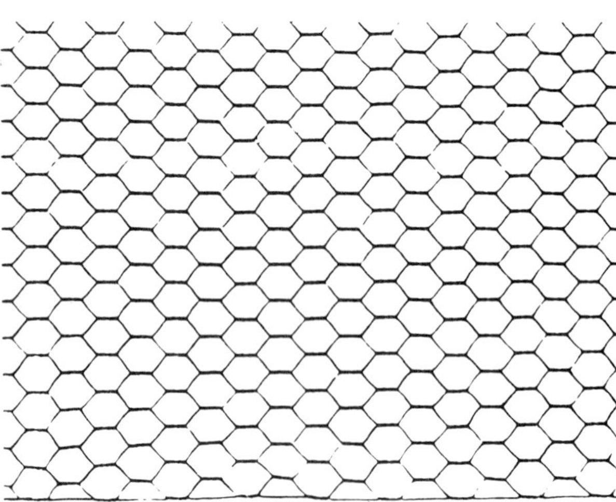

WOVEN WIRE
FABRIC LATH
(Also Available Self-Furred)

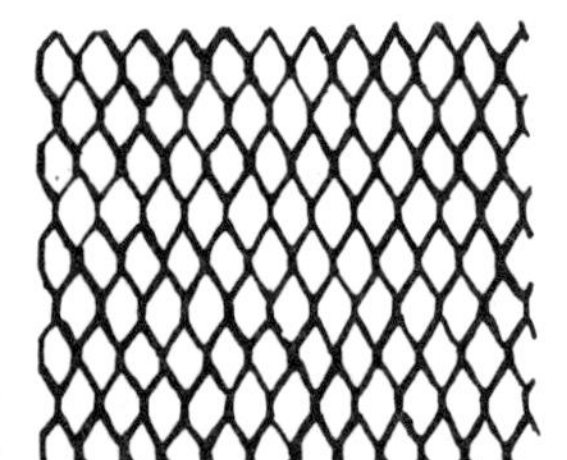

FLAT
DIAMOND MESH
METAL LATH

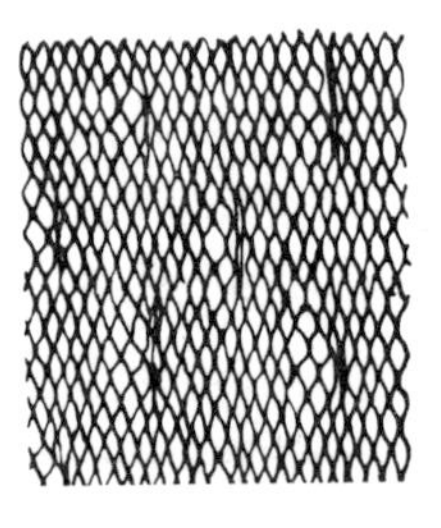

SELF-
FURRING
METAL LATH

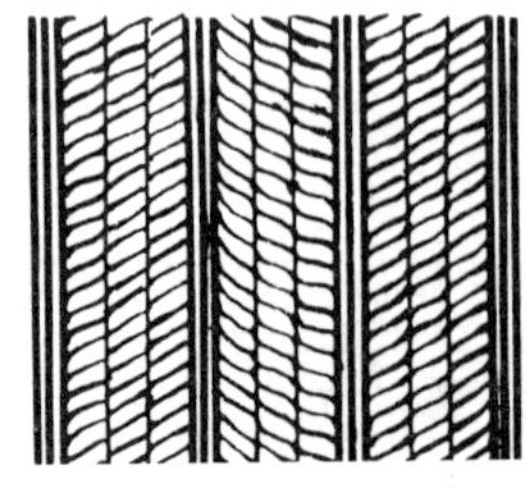

FLAT RIB
METAL LATH

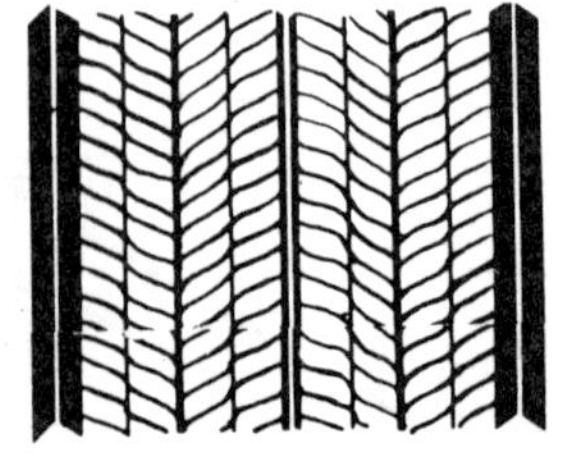

RIB
METAL LATH

RIB
METAL LATH

# VERTICAL FURRING
## Studless

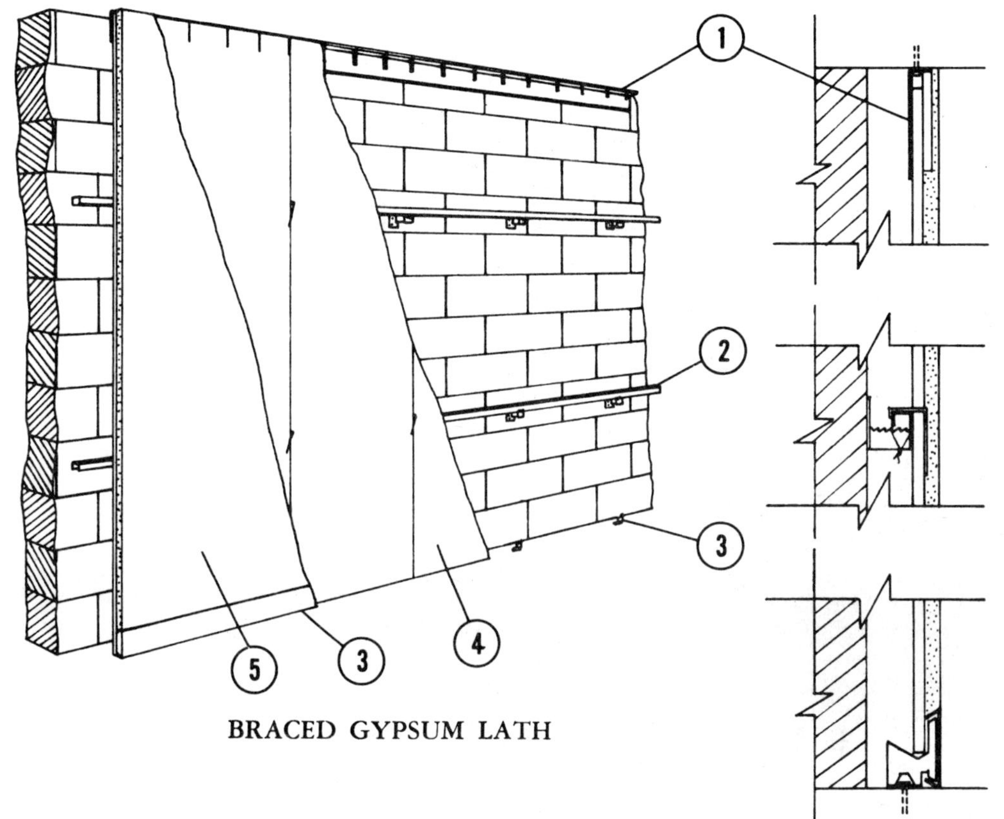


BRACED GYPSUM LATH

(1) Ceiling Runner
(2) Horizontal Stiffener (secured to bracing attachment)
(3) Metal Base and Clips
(4) Gypsum Lath
(5) Three Coats of Plaster (Scratch, Brown, Finish) (Minimum plaster thickness is ¾ inch.)

# COLUMN FURRING

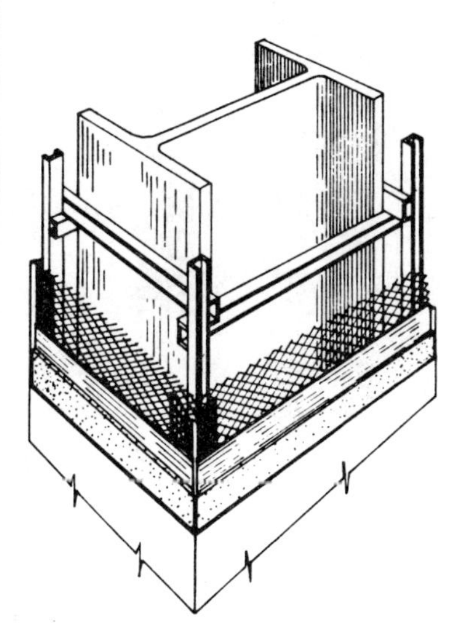

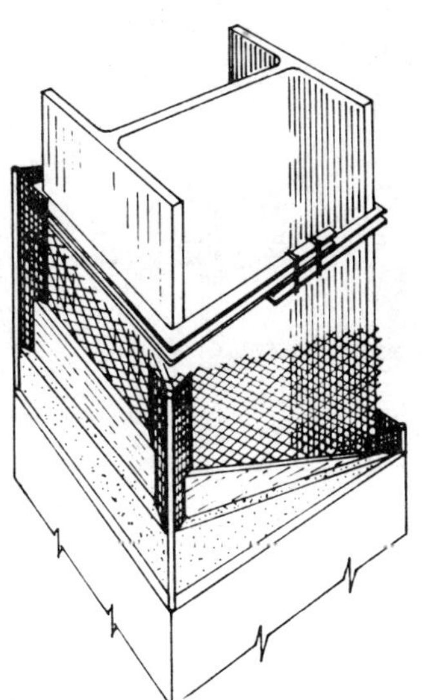

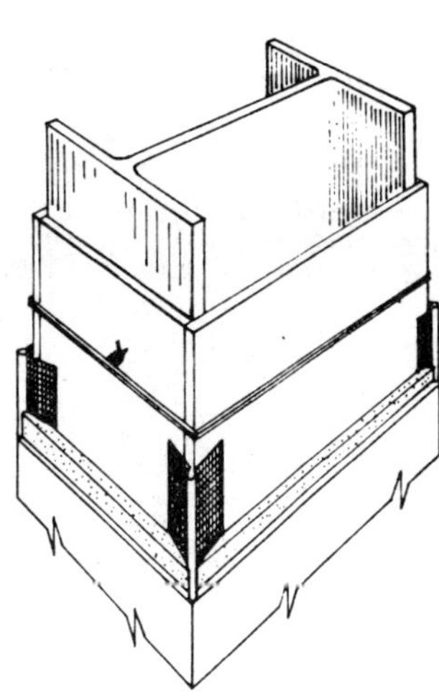

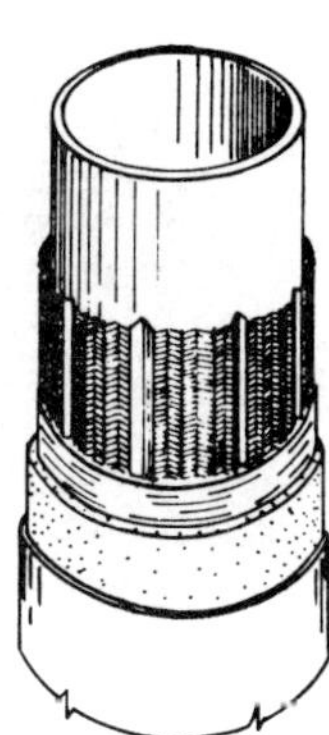

# CONTACT FURRING

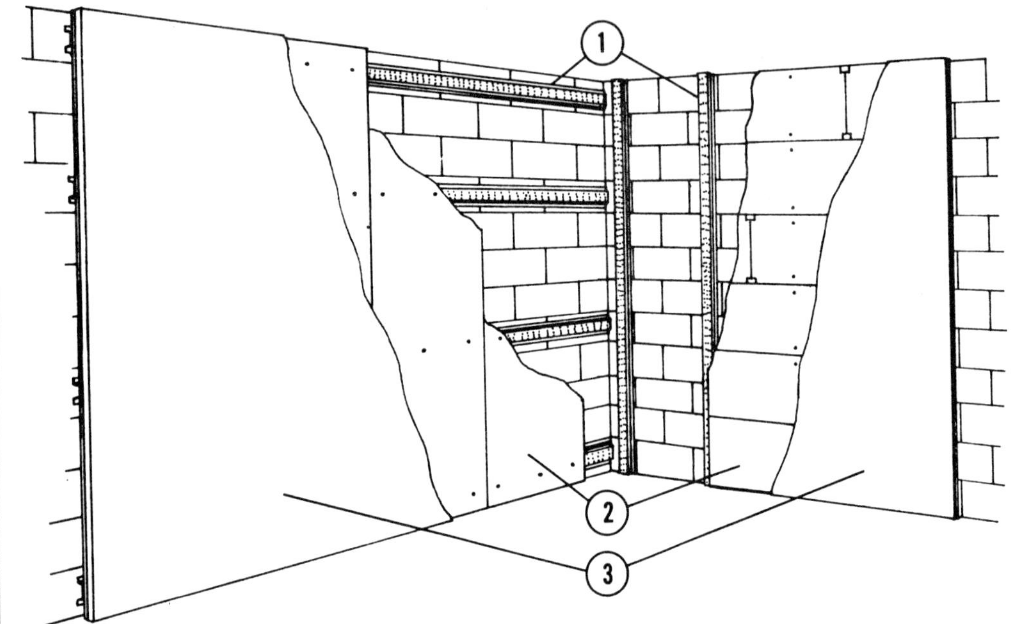


(1) Screw Channel
(2) Gypsum Lath (screwed on)
(3) Two Coats of Plaster (Brown, Finish)

# FALSE BEAMS

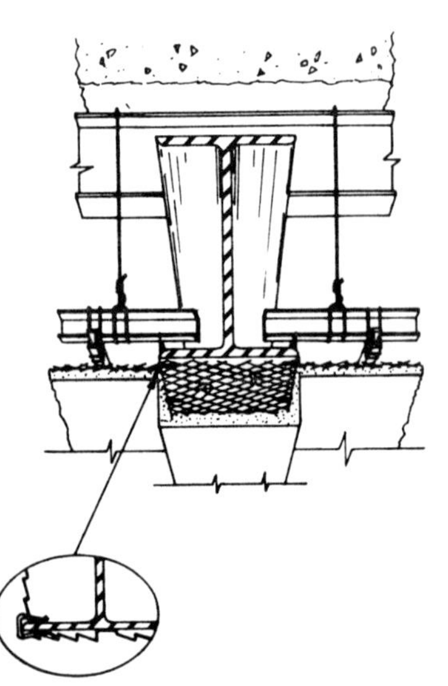

LATH DIRECT TO UNDERSIDE OF BEAM

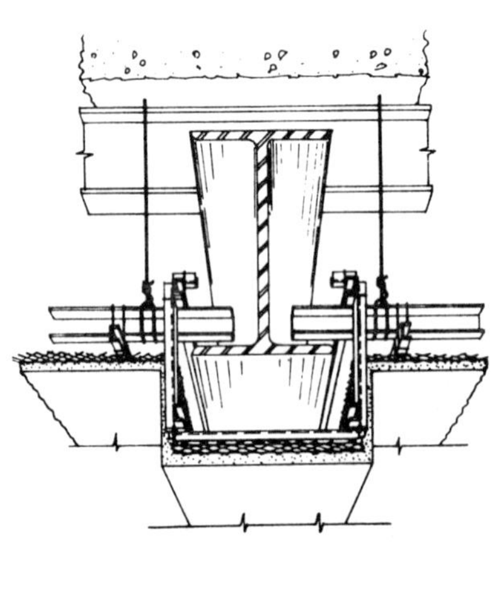

CHANNEL BRACKETS TO MAIN RUNNERS

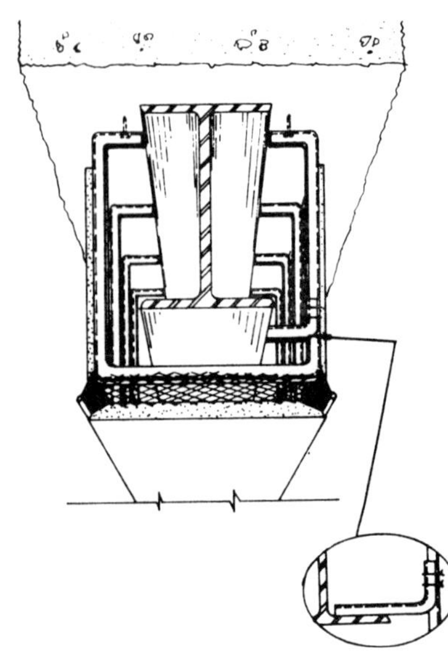

CHANNEL BRACKETS TO SLAB

| MAXIMUM SPACING OF SUPPORTS FOR METAL LATH (Inches) | | | | | | |
|---|---|---|---|---|---|---|
| **Type of Lath** | **Weight of Lath Lb. Per Sq. Yd.** | **WALLS AND PARTITIONS** | | | **CEILINGS** | |
| | | **Wood Studs** | **Solid Partitions** | **Steel Studs, Wall Furring, etc.** | **Wood or Concrete** | **Metal** |
| Diamond Mesh (flat expanded) | 2.5 | 16 | 16 | 13½ | 12 | 12 |
| | 3.4 | 16 | 16 | 16 | 16 | |
| Flat Rib | 2.75 | 16 | 16 | 16 | 16 | 16 |
| | 3.4 | 19 | 24(3) | 19 | 19 | 19 |
| ⅜″ Rib (1) (2) | 3.4 | 24 | (4) | 24 | 24 | 24 |
| | 4.0 | 24 | (4) | 24 | 24 | 24 |
| ¾″ Rib | 5.4 | — | (4) | 24(5) | 36(6) | 36(6) |
| Sheet Lath (2) | 4.5 | 24 | (4) | 24 | 24 | 24 |

**NOTE:** Weights are exclusive of paper, fiber or other backing.

(1) 3.4 lb., ⅜″ Rib Lath is permissible under Concrete Joists at 27″ c.c.

(2) These spacings are based on a narrow bearing surface for the lath. When supports with a relatively wide bearing surface are used, these spacings may be increased accordingly, and still assure satisfactory work.

(3) This spacing permissible for Solid Partitions not exceeding 16′ in height. For greater heights, permanent horizontal stiffener channels or rods must be provided on channel side of partitions, every 6′ vertically, or else spacing shall be reduced 25%.

(4) For studless solid partitions, lath erected vertically.

(5) For interior wall furring or for application over solid surfaces for stucco.

(6) For contact or ceilings only.

# TYPES OF LATH-ATTACHMENT TO WOOD AND METAL SUPPORTS

| TYPE OF LATH | NAILS | MAXIMUM SPACING | | SCREWS MAXIMUM SPACING | | STAPLES Round or Flattened Wire | | | | |
|---|---|---|---|---|---|---|---|---|---|---|
| | | | | | | | | | MAXIMUM SPACING | |
| | Type and Size | Vertical | Horizontal | Vertical | Horizontal | Wire Gauge No. | Crown | Leg | Vertical | Horizontal |
| | | (In Inches) | | (In Inches) | | | (In Inches) | | | |
| 1. Diamond Mesh Expanded Metal Lath and Flat Rib Metal Lath | 4d blued smooth box 1½ No. 14 gauge 7/32" head (clinched)<br>1" No. 11 gauge 7/16" head, barbed<br>1½" No. 11 gauge 7/16" head, barbed | 6<br>6<br>6 | —<br>—<br>6 | 6 | 6 | 16 | 3/4 | 7/8 | 6 | 6 |
| 2. 3/8" Rib Metal Lath and Sheet Lath | 1½" No. 11 ga. 7/16" head, barbed | 6 | 6 | 6 | 6 | 16 | 3/4 | 1½ | At Ribs | At Ribs |
| 3. 3/4" Rib Metal Lath | 4d common 1½" No. 12½ gauge 1/4" head<br>2" No. 11 gauge 7/16" head, barbed | At Ribs | —<br>At Ribs | At Ribs | At Ribs | 16 | 3/4 | 1⅝ | At Ribs | At Ribs |
| 4. Wire Fabric Lath | 4d blued smooth box (clinched)<br>1" No. 11 gauge 7/16" head, barbed | 6<br>6 | —<br>— | 6 | 6 | 16 | 3/4 | 7/8 | 6 | 6 |
| | 1½" No. 11 gauge 7/16" head, barbed<br>1¼" No. 12 ga. 3/8" head, furring<br>1" no. 12 gauge 3/8" head | 6<br>6<br>6 | 6<br>6 | | | 16 | 7/16 | 7/8 | 6 | 6 |
| 5. 3/8" Gypsum Lath | 1⅛" No. 13 gauge 12/61" head, blued | 8 | 8 | 8 | 8 | 16 | 3/4 | 7/8 | 8 | 8 |
| 6. ½" Gypsum Lath | 1¼" No. 13 gauge 12/61" head, blued | 8 | 8<br>6 | 8 | 8<br>6 | 16 | 3/4 | 1⅛ | 8 | 8<br>6 |

## STRESS RELIEF (CONTROL JOINTS)

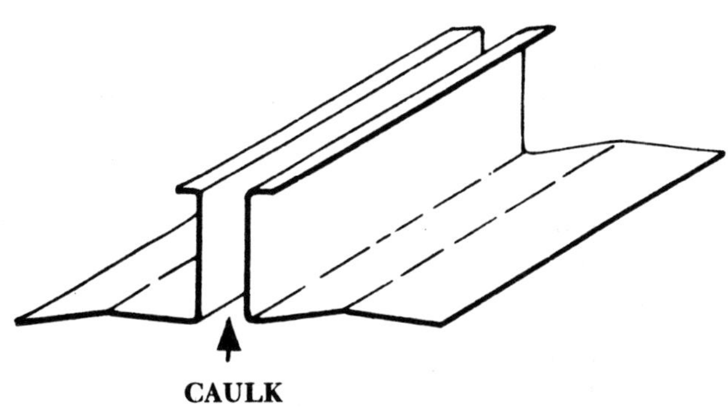


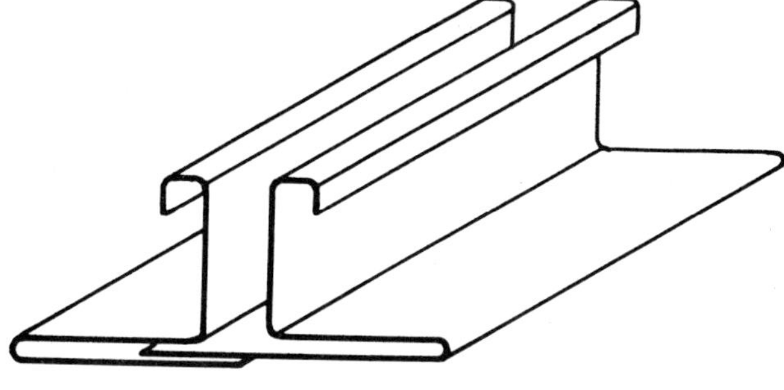

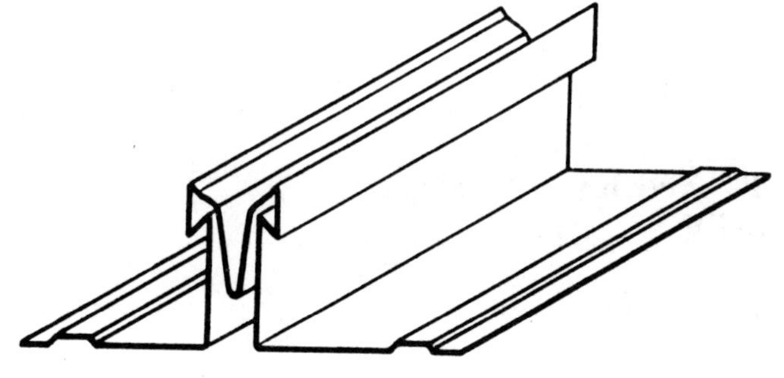

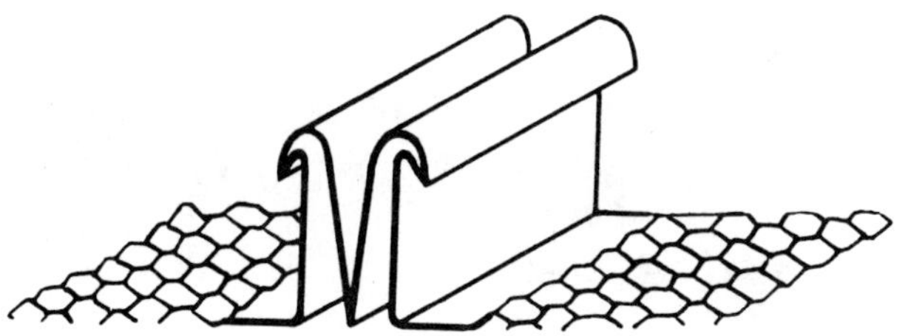

## REVEALS

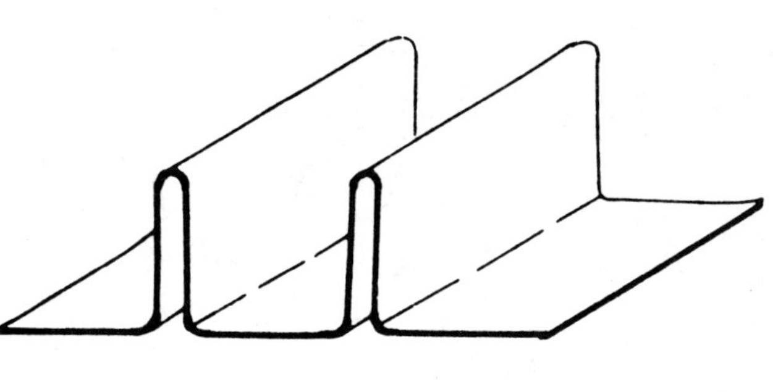

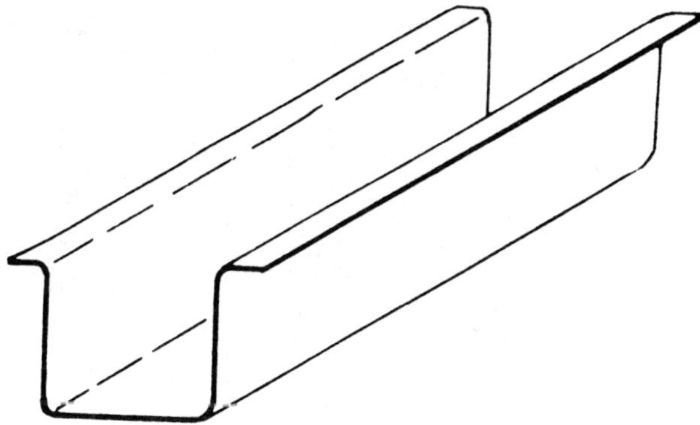

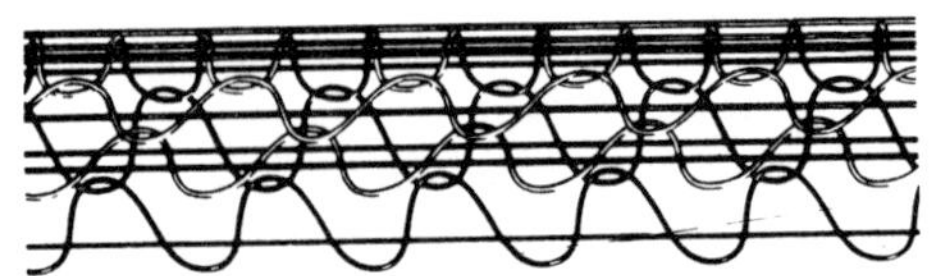

CORNER REINFORCEMENT
(EXTERIOR) WIRE

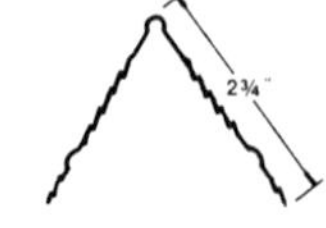


CORNER REINFORCEMENT
(EXTERIOR) EXPANDED METAL

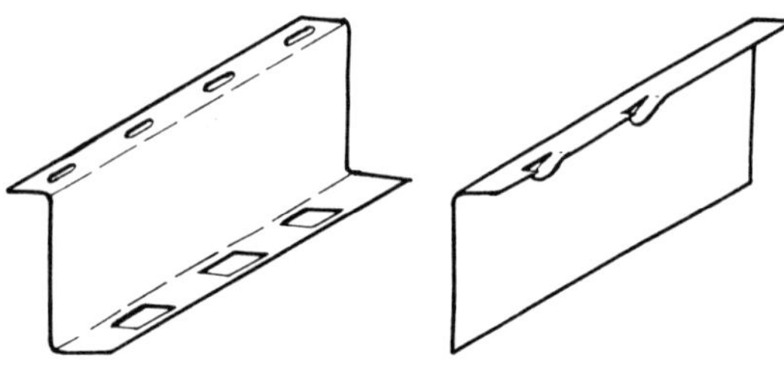

PARTITION RUNNERS
(Z AND L SHAPE)

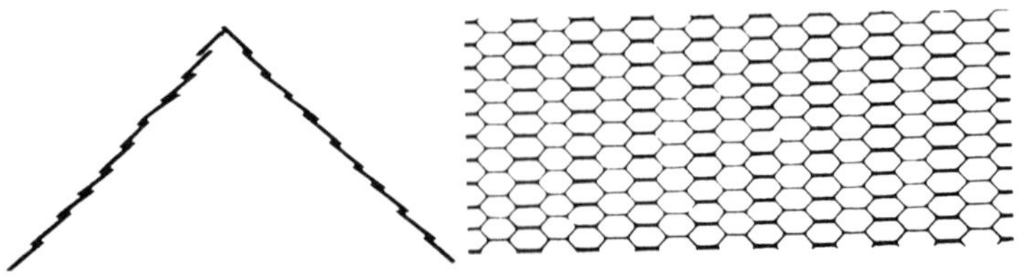

EXPANDED METAL CORNERITE

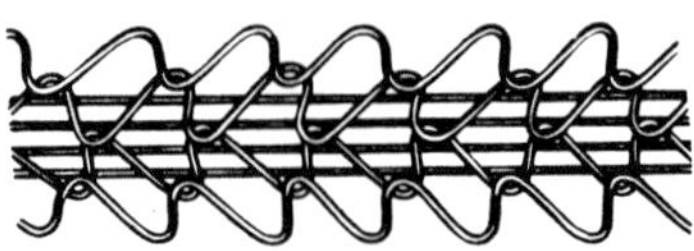

WIRE CORNERITE

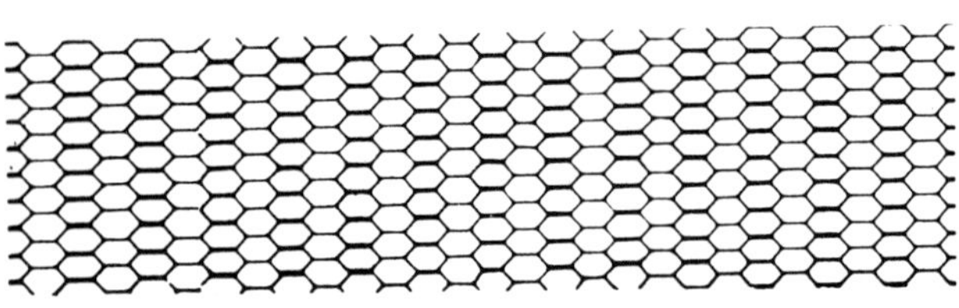

STRIP REINFORCEMENT
(EXPANDED METAL)

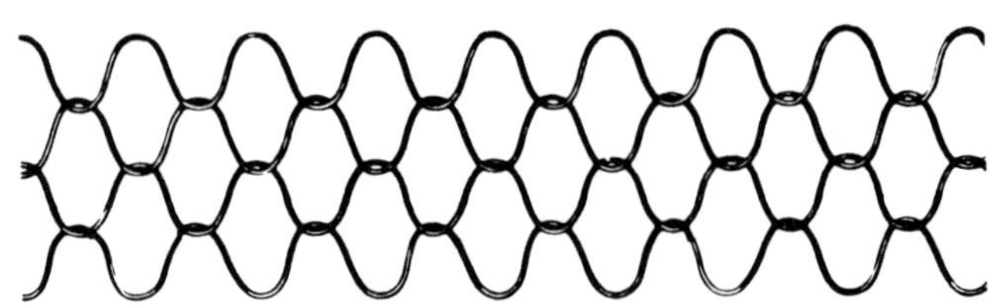

STRIP REINFORCEMENT (WIRE)

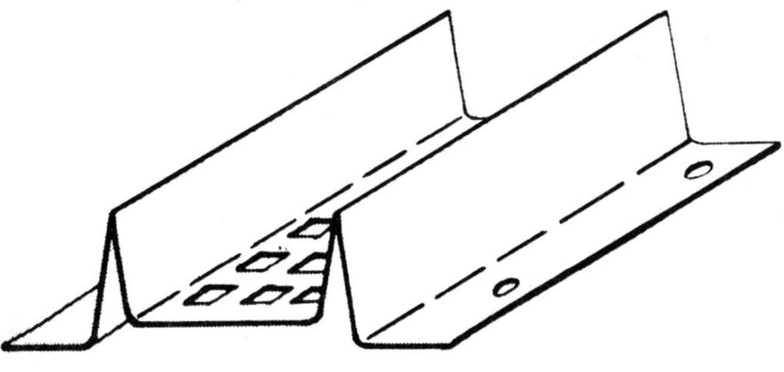

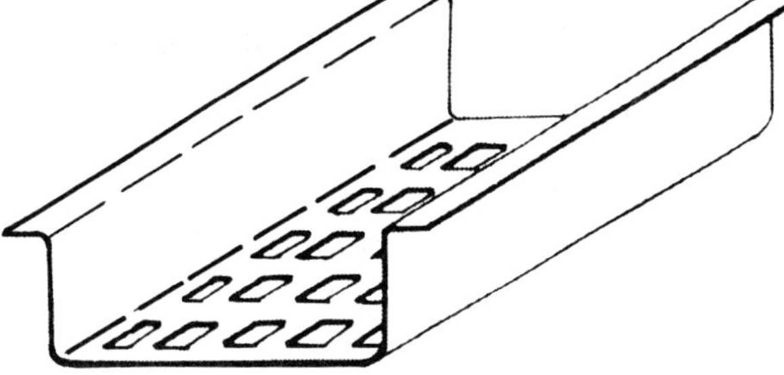

VENTILATING SCREEDS

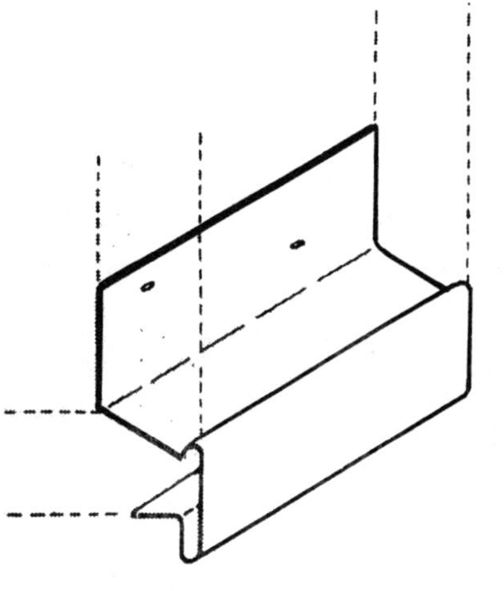

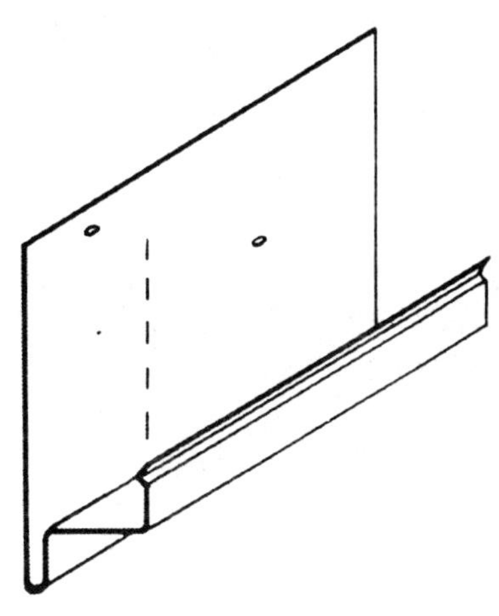

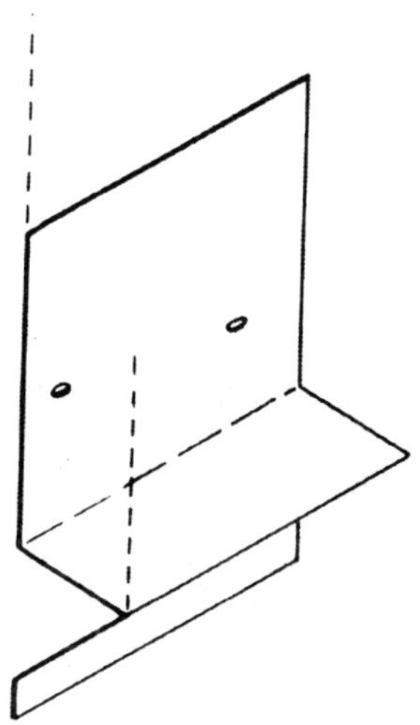

DRIP SCREEDS

WEEP SCREED
(Also available with perforations)

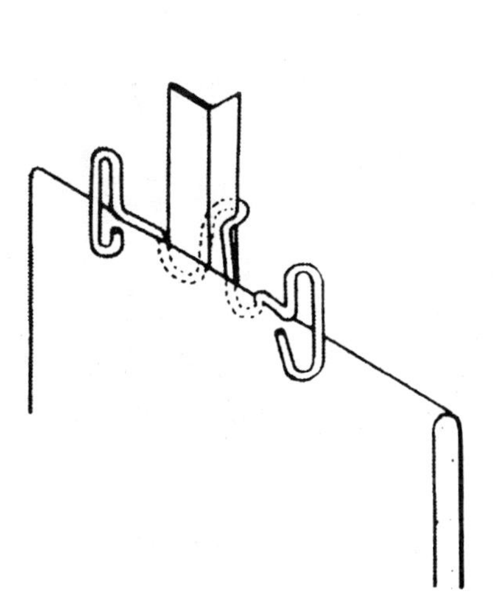

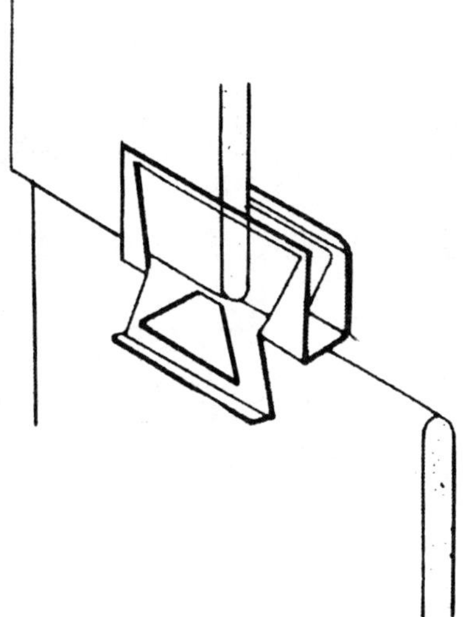

GYPSUM LATH ATTACHMENT CLIPS

## VERTICAL FURRING

| VERTICAL FURRING MEMBER | UNBRACED | | | | BRACED | | | |
|---|---|---|---|---|---|---|---|---|
| | STUD SPACING | | | | STUD SPACING | | | |
| | 24″ | 19″ | 16″ | 12″ | 24″ | 19″ | 16″ | 12″ |
| | Maximum Furring Heights | | | | Maximum Vertical Distance Between Braces | | | |
| ¾″ Channel | 6′ | 7′ | 8′ | 9′ | 5′ | 5′ | 6′ | 7′ |
| 1½″ Channel | 8′ | 9′ | 10′ | 12′ | 6′ | 7′ | 8′ | 9′ |
| 2″ Channel | 9′ | 10′ | 11′ | 13′ | 7′ | 8′ | 9′ | 10′ |
| 2″ Prefab. Stud | 8′ | 9′ | 10′ | 11′ | 6′ | 7′ | 8′ | 9′ |
| 2½″ Prefab. Stud | 10′ | 11′ | 12′ | 14′ | 8′ | 9′ | 10′ | 11′ |
| 3¼″ Prefab. Stud | 14′ | 16′ | 17′ | 20′ | 11′ | 13′ | 14′ | 16′ |

## TYPES OF LATH—MAXIMUM SPACING OF SUPPORTS

| TYPE OF LATH | Minimum Weight (psy), Gauge & Mesh Size | VERTICAL | | | HORIZONTAL | |
|---|---|---|---|---|---|---|
| | | WOOD | METAL | | | |
| | | | Solid Plaster Partitions | Other | Wood or Concrete | Metal |
| Expanded Metal Lath (Diamond Mesh) | 2.5<br>3.4 | 16″<br>16″ | 16″<br>16″ | 12″<br>16″ | 12″<br>16″ | 12″<br>16″ |
| Flat rib Expanded Metal Lath | 2.75<br>3.4 | 16″<br>19″ | 16″<br>24″ | 16″<br>19″ | 16″<br>19″ | 16″<br>19″ |
| Stucco Mesh Expanded Metal Lath | 1.8 and<br>3.6 | 16″ | — | — | — | — |
| ⅜″ Rib Expanded Metal Lath | 3.4<br>4.0 | 24″<br>24″ | — | 24″<br>24″ | 24″<br>24″ | 24″<br>24″ |
| Sheet Lath | 4.5 | 24″ | — | 24″ | 24″ | 24″ |
| ¾″ Rib Expanded Metal Lath (Not manufactured in West) | 5.4 | — | — | — | 36″ | 36″ |
| Wire Fabric Lath — Welded | 1.95 lbs., 11ga., 2″ x 2″<br>1.4 lbs., 16 ga., 2″ x 2″<br>1.4 lbs., 18 ga., 1″ x 1″ | 24″<br>16″<br>16″ | 24″<br>16″<br>— | 24″<br>16″<br>— | 24″<br>16″<br>— | 24″<br>16″<br>— |
| Wire Fabric Lath — Woven | 1.4 lbs., 17 ga., 1½″ Hex.<br>1.4 lbs., 18 ga., 1″ Hex. | 24″<br>24″ | 16″<br>16″ | 16″<br>16″ | 24″<br>24″ | 16″<br>16″ |
| ⅜″ Gypsum Lath (plain) | — | 16″ | — | 16″ | 16″ | 16″ |
| (Large Size) | — | 16″ | — | 16″ | 16″ | 16″ |
| ½″ Gypsum Lath (plain) | — | 24″ | — | 24″ | 24″ | 24″ |
| (Large Size) | — | 24″ | No supports; Erected vertically | 24″ | 24″ | 16″ |
| ⅝″ Gypsum Lath (Large Size) | — | 24″ | No supports; Erected vertically | 24″ | 24″ | 16″ |

# PLASTERING TABLES

## THICKNESS OF PLASTER

| PLASTER BASE | FINISHED THICKNESS OF PLASTER FROM FACE OF LATH, MASONRY, CONCRETE | |
|---|---|---|
| | Gypsum Plaster | Portland Cement Plaster |
| Expanded Metal Lath | ⅝″ minimum | ⅝″ minimum |
| Wire Fabric Lath | ⅝″ minimum | ¾″ minimum (interior)<br>⅞″ minimum (exterior) |
| Gypsum Lath | ½″ minimum | |
| Gypsum Veneer Base | 1⁄16″ minimum | |
| Masonry Walls | ½″ minimum | ½″ minimum |
| Monolithic Concrete Walls | ⅝″ maximum | ⅞″ maximum |
| Monolithic Concrete Ceilings | ⅜″ maximum | ½″ maximum |

## GYPSUM PLASTER PROPORTIONS

| NUMBER OF COATS | COAT | PLASTER BASE OR LATH | MAXIMUM VOLUME AGGREGATE PER 100# NEAT PLASTER (CUBIC FEET) | |
|---|---|---|---|---|
| | | | Damp Loose Sand | Perlite or Vermiculite |
| Two-Coat Work | Basecoat | Gypsum Lath | 2½ | 2½ |
| | Basecoat | Masonry | 3 | 3 |
| Three-Coat Work | First Coat | Lath | 2 | 2 |
| | Second Coat | Lath | 3 | 3 |
| | First & Second Coat | Masonry | 3 | 3 |

## PORTLAND CEMENT PLASTER

| COAT | VOLUME CEMENT | MAXIMUM WEIGHT (OR VOLUME) LIME PER VOLUME CEMENT | MAXIMUM VOLUME SAND PER VOLUME CEMENT | APPROXIMATE MINIMUM THICKNESS | MINIMUM PERIOD MOIST CURING | MINIMUM INTERVAL BETWEEN COATS |
|---|---|---|---|---|---|---|
| First | 1 | 20 lbs. | 4 | ⅜″ | 48 Hours | 48 Hours |
| Second | 1 | 20 lbs. | 5 | 1st and 2nd Coats total ¾″ | 48 Hours | 7 Days |
| Finish | 1 | 1 | 3 | 1st, 2nd and Finish Coats ⅞″ | — | |

## PORTLAND CEMENT-LIME PLASTER

| COAT | VOLUME CEMENT | MAXIMUM VOLUME LIME PER VOLUME CEMENT | MAXIMUM VOLUME SAND PER COMBINED VOLUMES CEMENT AND LIME | APPROXIMATE MINIMUM THICKNESS | MINIMUM PERIOD MOIST CURING | MINIMUM INTERVAL BETWEEN COATS |
|---|---|---|---|---|---|---|
| First | 1 | 1 | 4 | ⅜″ | 48 Hours | 48 Hours |
| Second | 1 | 1 | 4½ | 1st and 2nd Coats total ¾″ | 48 Hours | 7 Days |
| Finish | 1 | 1 | 3 | 1st, 2nd and Finish Coats ⅞″ | — | |

# FINISHES / VENEER PLASTER — 09215

## METAL STUD CONSTRUCTION

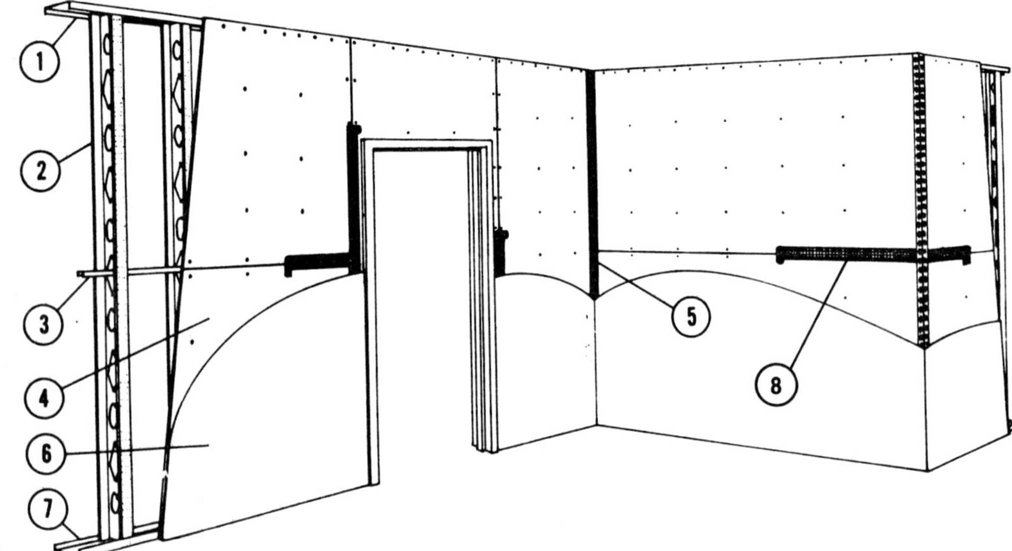


(1) Ceiling Runner Track
(2) Metal Stud (nailable or screw)
(3) Horizontal Stiffener
(4) Large Size Lath
(5) Angle Reinforcement
(6) Veneer Plaster 1/16 to 1/8 inch thick)
(7) Floor Runner Track
(8) Joint Reinforcement

## WOOD STUD CONSTRUCTION

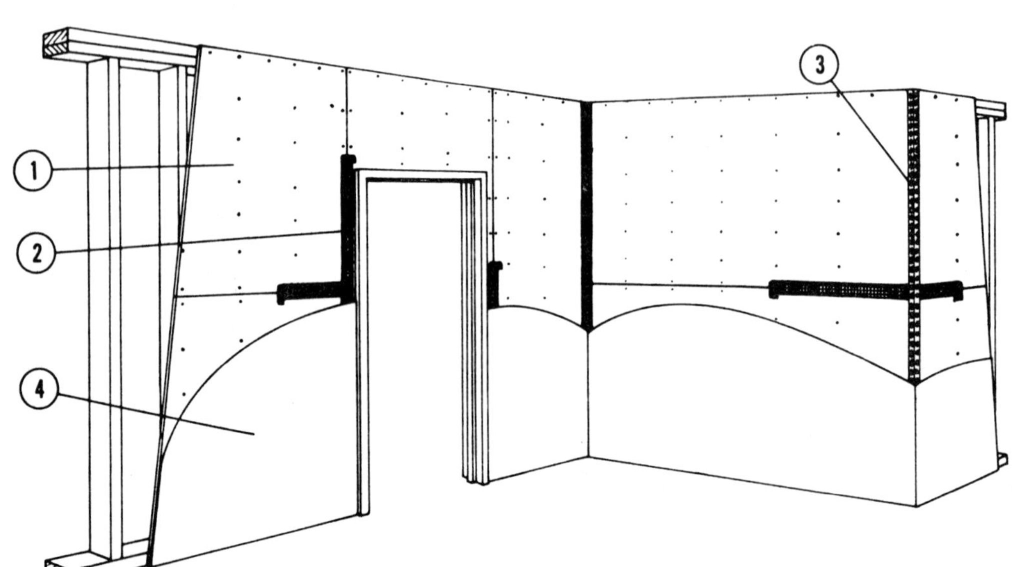


(1) Large Size Lath
(2) Joint Reinforcement
(3) Corner Bead
(4) Veneer Plaster (1/16 to 1/8 inch thick)

# EXTERIOR LATH AND PLASTER

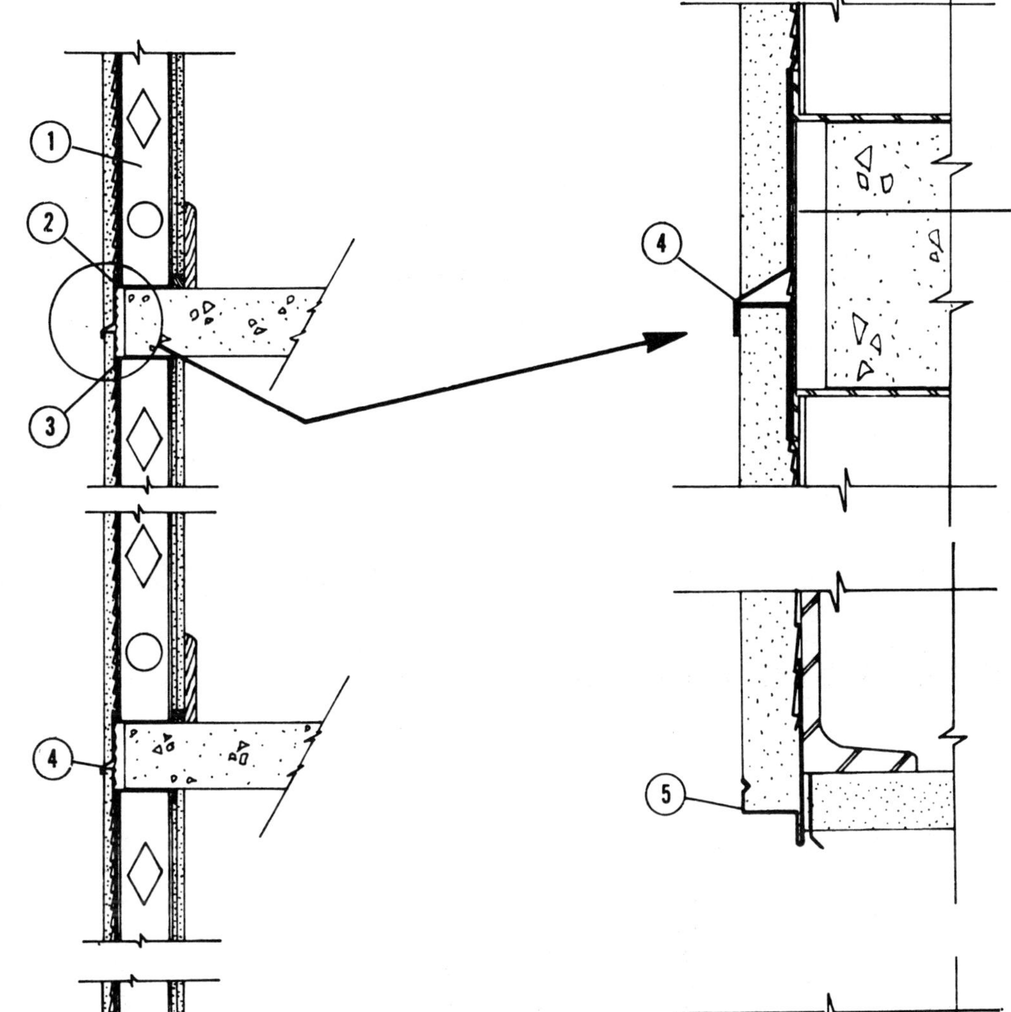


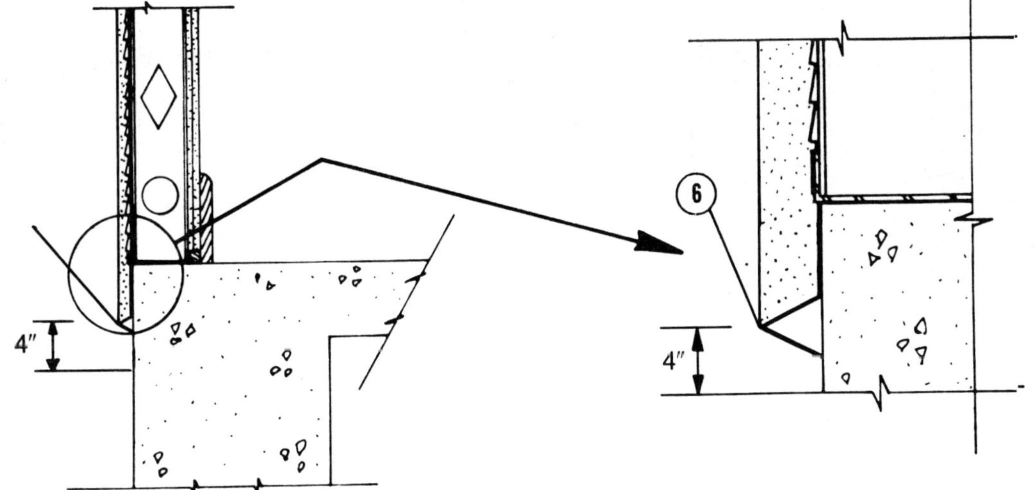


(1) Metal Stud
(2) Floor Runner Track
(3) Ceiling Runner Track
(4) Control Joint
(5) Drip Screed (at soffit)
(6) Weep Screed (at slab or foundation)
(7) Building Paper, Continuous

# EXTERIOR LATH AND PLASTER

## OPEN WOOD FRAME CONSTRUCTION

(1) Wire Backing
(2) Building Paper
(3) Wire Fabric Lath
(4) Approved Fasteners
(5) Weep Screed
(6) Three Coats of Plaster (Scratch, Brown, Finish)

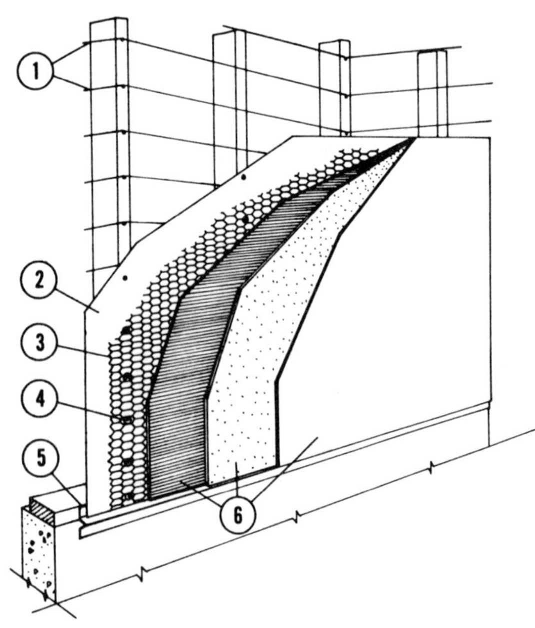


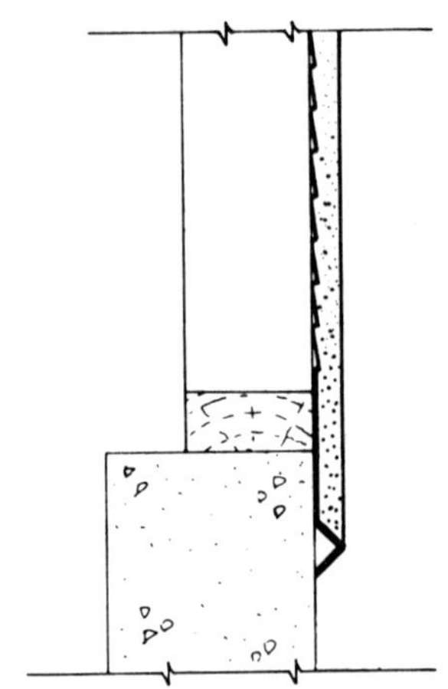

## SHEATHED WOOD FRAME CONSTRUCTION

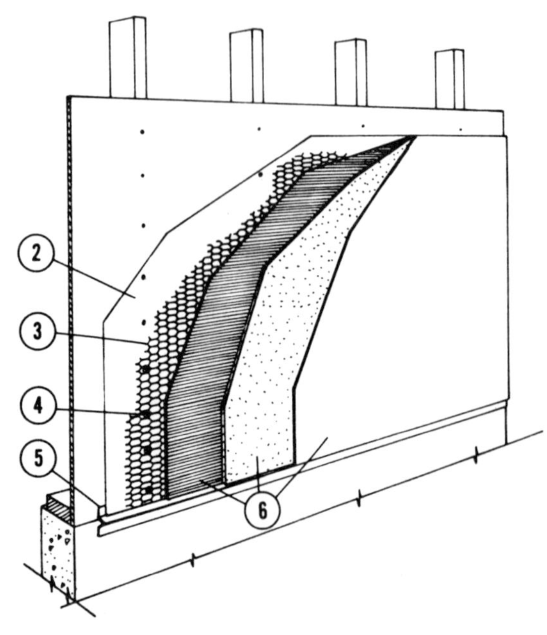


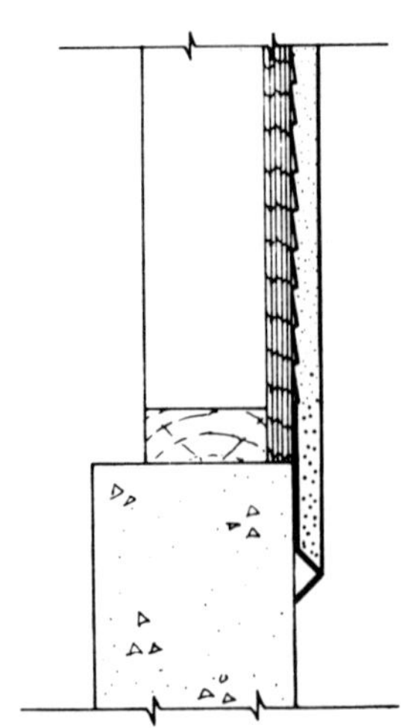

# PIPE WEIGHTS

## CAST IRON PIPE

| SERVICE WEIGHT | | | |
|---|---|---|---|
| Size, Inches | Weight of Pipe | Weight of Water | Total Weight-Lbs. |
| 2″ | 3.8 | 1.45 | 5.3 |
| 3″ | 5.6 | 3.2 | 8.8 |
| 4″ | 7.5 | 5.5 | 13.0 |
| 5″ | 9.8 | 8.7 | 18.5 |
| 6″ | 12.4 | 12.5 | 24.9 |
| 8″ | 18.5 | 21.7 | 40.2 |

| EXTRA HEAVY | | | |
|---|---|---|---|
| Size, Inches | Weight of Pipe | Weight of Water | Total Weight-Lbs. |
| 2″ | 4.3 | 1.45 | 5.8 |
| 3″ | 8.3 | 3.2 | 11.5 |
| 4″ | 10.8 | 5.5 | 16.3 |
| 5″ | 13.3 | 8.7 | 22.0 |
| 6″ | 16.0 | 12.5 | 28.5 |
| 8″ | 26.5 | 21.7 | 48.2 |

## STEEL PIPE

| Pipe Size | W/40 | $H_2O$/Lbs. | Total Lbs./L.F. |
|---|---|---|---|
| 2″ | 3.65 | 1.45 | 5.1 |
| 2½″ | 5.79 | 2.07 | 7.86 |
| 3″ | 7.57 | 3.2 | 10.77 |
| 3½″ | 9.11 | 4.28 | 13.39 |
| 4″ | 10.8 | 5.51 | 16.31 |
| 5″ | 14.6 | 8.66 | 23.26 |
| 6″ | 18.0 | 12.5 | 30.5 |
| 8″ | 28.6 | 21.66 | 50.26 |
| 10″ | 40.5 | 34.15 | 74.65 |
| 12″ | | | |

## SPRINKLER AREA CALCULATIONS

Typical maximum floor area allowed per system riser:

| Light Hazard | Ordinary Hazard | Extra Hazard |
|---|---|---|
| 52,000 S.F. | 40,000-52,000 S.F. | 25,000 S.F. |

Typical maximum floor area coverage allowed per sprinkler head:

| Light Hazard | Ordinary Hazard | Extra Hazard |
|---|---|---|
| 130-200 S.F. | 100-130 S.F. | 90 S.F. |

Typical maximum spacing between lines and sprinkler heads:

| Light Hazard | Ordinary Hazard | Extra Hazard |
|---|---|---|
| 12-15 feet | 12-15 feet | 12 feet |

**Note:** This data is for estimating purposes only. Check all applicable codes and regulations for specific requirements.

## SPRINKLER HEAD CALCULATIONS

*Typical maximum quantity of sprinkler heads allowed by pipe size.*

**Light Hazard:**

For sprinklers below ceiling:

| Steel | | Copper | |
|---|---|---|---|
| 1 in. pipe | 2 sprinklers | 1 in. tube | 2 sprinklers |
| 1¼ in. pipe | 3 sprinklers | 1¼ in. tube | 3 sprinklers |
| 1½ in. pipe | 5 sprinklers | 1½ in. tube | 5 sprinklers |
| 2 in. pipe | 10 sprinklers | 2 in. tube | 12 sprinklers |
| 2½ in. pipe | 30 sprinklers | 2½ in. tube | 40 sprinklers |
| 3 in. pipe | 60 sprinklers | 3 in. tube | 65 sprinklers |
| 3½ in. pipe | 100 sprinklers | 3½ in. tube | 115 sprinklers |

For sprinklers above and below ceiling:

| Steel | | Copper | |
|---|---|---|---|
| 1 in. | 2 sprinklers | 1 in. | 2 sprinklers |
| 1¼ in. | 4 sprinklers | 1¼ in. | 4 sprinklers |
| 1½ in. | 7 sprinklers | 1½ in. | 7 sprinklers |
| 2 in. | 15 sprinklers | 2 in. | 18 sprinklers |
| 2½ in. | 50 sprinklers | 2½ in. | 65 sprinklers |

**Ordinary Hazard:**

For sprinklers above ceiling:

| Steel | | Copper | |
|---|---|---|---|
| 1 in. pipe | 2 sprinklers | 1 in. tube | 2 sprinklers |
| 1¼ in. pipe | 3 sprinklers | 1¼ in. tube | 3 sprinklers |
| 1½ in. pipe | 5 sprinklers | 1½ in. tube | 5 sprinklers |
| 2 in. pipe | 10 sprinklers | 2 in. tube | 12 sprinklers |
| 2½ in. pipe | 20 sprinklers | 2½ in. tube | 25 sprinklers |
| 3 in. pipe | 40 sprinklers | 3 in. tube | 45 sprinklers |
| 3½ in. pipe | 65 sprinklers | 3½ in. tube | 75 sprinklers |
| 4 in. pipe | 100 sprinklers | 4 in. tube | 115 sprinklers |
| 5 in. pipe | 160 sprinklers | 5 in. tube | 180 sprinklers |
| 6 in. pipe | 275 sprinklers | 6 in. tube | 300 sprinklers |

For sprinklers above and below ceiling:

| Steel | | Copper | |
|---|---|---|---|
| 1 in. | 2 sprinklers | 1 in. | 2 sprinklers |
| 1¼ in. | 4 sprinklers | 1¼ in. | 4 sprinklers |
| 1½ in. | 7 sprinklers | 1½ in. | 7 sprinklers |
| 2 in. | 15 sprinklers | 2 in. | 18 sprinklers |
| 2½ in. | 30 sprinklers | 2½ in. | 40 sprinklers |
| 3 in. | 60 sprinklers | 3 in. | 65 sprinklers |

**Extra Hazard:**

For sprinklers below ceiling:

| Steel | | Copper | |
|---|---|---|---|
| 1 in. pipe | 1 sprinkler | 1 in. tube | 1 sprinkler |
| 1¼ in. pipe | 2 sprinklers | 1¼ in. tube | 2 sprinklers |
| 1½ in. pipe | 5 sprinklers | 1½ in. tube | 5 sprinklers |
| 2 in. pipe | 8 sprinklers | 2 in. tube | 8 sprinklers |
| 2½ in. pipe | 15 sprinklers | 2½ in. tube | 20 sprinklers |
| 3 in. pipe | 27 sprinklers | 3 in. tube | 30 sprinklers |
| 3½ in. pipe | 40 sprinklers | 3½ in. tube | 45 sprinklers |
| 4 in. pipe | 55 sprinklers | 4 in. tube | 65 sprinklers |
| 5 in. pipe | 90 sprinklers | 5 in. tube | 100 sprinklers |
| 6 in. pipe | 150 sprinklers | 6 in. tube | 170 sprinklers |

**Note:** This data is for estimating purposes only. Check all applicable codes and regulations for specific requirements.

## SPRINKLER HAZARD OCCUPANCIES

### Typical Light Hazard Occupancies:

Churches
Clubs
Eaves and overhangs, if combustible construction with no combustible beneath
Educational
Hospitals
Institutional
Libraries, except large stack rooms
Museums
Nursing or Convalescent Homes
Office, including Data Processing
Residential
Restaurant seating areas
Theaters seating areas
Theaters and Auditoriums excluding stages and prosceniums
Unused attics

### Typical Ordinary Hazard Occupancies (Group 1):

Automobile parking garages
Bakeries
Beverage manufacturing
Canneries
Dairy products manufacturing and processing
Electronic plants
Glass and glass products manufacturing
Laundries
Restaurant service areas

### Typical Ordinary Hazard Occupancies (Group 2):

Cereal mills
Chemical plants — Ordinary
Cold Storage warehouses
Confectionery products
Distilleries
Leather goods mfg.
Libraries-large stack room areas
Mercantiles
Machine shops
Metal working
Printing and publishing
Textile mfg.
Tobacco Products mfg.
Wood product assembly

### Typical Ordinary Hazard Occupancies (Group 3):

Feed mills
Paper and pulp mills
Paper process plants
Piers and wharves
Repair garages
Tire manufacturing
Warehouses (having moderate to higher combustibility of content, such as paper, household furniture, paint, general storage, whiskey, etc.)[1]
Wood machining

### Typical Extra Hazard Occupancies (Group 1):

Combustible Hydraulic Fluid use areas
Die Casting
Metal Extruding
Plywood and particle board manufacting
Printing (using inks with below 100°F [37.8°C] flash points)
Rubber reclaiming, compounding, drying, milling, vulcanizing
Saw Mills
Textile picking, opening, blending, garnetting, carding, combining of cotton, synthetics, wool shoddy or burlap
Upholstering with plastic foams

### Typical Extra Hazard Occupancies (Group 2):

Asphalt saturating
Flammable liquids spraying
Flow coating
Mobile Home or Modular Building assemblies (where finished enclosure is present and has combustible interiors)
Open Oil quenching
Solvent cleaning
Varnish and paint dipping

...............

## TYPICAL MINIMUM SIZE OF HORIZONTAL BUILDING STORM DRAINS AND BUILDING STORM SEWERS

| Diameter of drain | Maximum projected area in square feet for various slopes | | |
|---|---|---|---|
| | 1/8 inch per feet slope | 1/4 inch per feet slope | 1/2 inch per feet slope |
| 3 | 822 | 1160 | 1644 |
| 4 | 1880 | 2650 | 3760 |
| 5 | 3340 | 4720 | 6680 |
| 6 | 5350 | 7550 | 10700 |
| 8 | 11500 | 16300 | 23000 |
| 10 | 20700 | 29200 | 41400 |
| 12 | 33300 | 47000 | 66600 |
| 15 | 59500 | 84000 | 119000 |

# MECHANICAL / PLUMBING — 15400

## PLUMBING BUDGETS

Budget estimates for plumbing can be determined as a percentage of total building costs depending on building type. For example:

- Apartments ............................9 to 12 percent
- Assembly ..............................4 to 7 percent
- Banks ...................................3 to 6 percent
- Dormitories ...........................7 to 10 percent
- Factories ...............................4 to 8 percent
- Hospitals ...............................8 to 12 percent
- Motels ..................................9 to 12 percent
- Office Buildings ......................4 to 7 percent
- Retail (small) .........................4 to 7 percent
- Retail (large) ..........................3 to 6 percent
- Schools .................................3 to 6 percent
- Warehouses ............................3 to 7 percent

# MECHANICAL / PLUMBING VENTS — 15410

## TYPICAL SIZE AND LENGTH OF PLUMBING VENTS

| Diameter of soil or waste stack (in.) | Total fixture units connected to stack (dfu) | | DIAMETER OF VENT PIPE | | | | | | | | | |
|---|---|---|---|---|---|---|---|---|---|---|---|---|
| | | 1¼ | 1 | 2 | 2½ | 3 | 4 | 5 | 6 | 8 | 10 | 12 |
| 1¼ | 2 | 30 | | | | | | | | | | |
| 1½ | 8 | 50 | 150 | | | | | | | | | |
| 1½ | 10 | 30 | 100 | | | | | | | | | |
| 2 | 12 | 30 | 75 | 200 | | | | | | | | |
| 2 | 20 | 26 | 50 | 150 | | | | | | | | |
| 2½ | 42 | | 30 | 100 | 300 | | | | | | | |
| 3 | 10 | | 42 | 150 | 360 | 1040 | | | | | | |
| 3 | 21 | | 32 | 110 | 270 | 810 | | | | | | |
| 3 | 53 | | 27 | 94 | 230 | 680 | | | | | | |
| 3 | 102 | | 25 | 86 | 210 | 620 | | | | | | |
| 4 | 43 | | | 35 | 85 | 250 | 980 | | | | | |
| 4 | 140 | | | 27 | 65 | 200 | 750 | | | | | |
| 4 | 320 | | | 23 | 55 | 170 | 640 | | | | | |
| 4 | 540 | | | 21 | 50 | 150 | 580 | | | | | |
| 5 | 190 | | | | 28 | 82 | 320 | 990 | | | | |
| 5 | 490 | | | | 21 | 63 | 250 | 760 | | | | |
| 5 | 940 | | | | 18 | 53 | 210 | 670 | | | | |
| 5 | 1400 | | | | 16 | 49 | 190 | 590 | | | | |
| 6 | 500 | | | | | 33 | 130 | 400 | 1000 | | | |
| 6 | 1100 | | | | | 26 | 100 | 310 | 780 | | | |
| 6 | 2000 | | | | | 22 | 84 | 260 | 660 | | | |
| 6 | 2900 | | | | | 20 | 77 | 240 | 600 | | | |
| 8 | 1800 | | | | | | 31 | 95 | 240 | 940 | | |
| 8 | 3400 | | | | | | 24 | 73 | 190 | 720 | | |
| 8 | 5600 | | | | | | 20 | 62 | 160 | 610 | | |
| 8 | 7600 | | | | | | 18 | 56 | 140 | 560 | | |

## TYPICAL SIZES OF FIXTURE WATER SUPPLY PIPES

| Fixture | Nominal pipe size (inches) |
|---|---|
| Bath tubs | ½ |
| Combination sink and tray | ½ |
| Drinking fountain | ⅜ |
| Dishwasher (domestic) | ½ |
| Kitchen sink, residential | ½ |
| Kitchen sink, commercial | ¾ |
| Lavatory | ⅜ |
| Laundry tray, 1, 2 or 3 compartments | ½ |
| Shower (single head) | ½ |
| Sinks (service, slop) | ½ |
| Sinks flushing rim | ¾ |
| Urinal (flash tank) | ½ |
| Urinal (direct flush valve) | ¾ |
| Water closet (tank type) | ⅜ |
| Water closet (flush valve type) | 1 |
| Hose bibs | ½ |
| Wall hydrant | ½ |

## TYPICAL VENTILATION AIR REQUIREMENTS FOR SPECIAL USES

| Occupancy Classification | Required ventilation air in cfm per human occupant |
|---|---|
| Special areas | |
| Lockers | 2* (or 30 per locker) |
| Wardrobes | 2* |
| Public bathrooms | 40** |
| Private bathrooms | 25** |
| Swimming pools | 15 (per occupant) |
| Exitways and corridors | 1½* |

*Per square foot floor area.
**Per water closet or urinal.

## TYPICAL VENTILATION AIR REQUIREMENTS FOR RETAIL USES

| Occupancy Classification | Required ventilation air in cfm per human occupant |
|---|---|
| Mercantile | |
| Sales floors and showrooms (basement & grade floors) | 7 |
| Sales and showrooms (upper floors) | 7 |
| Storage areas | 5 |
| Dressing rooms | 7 |
| Malls | 7 |
| Shipping areas | 15 |
| Elevators | 7 |
| Supermarkets | |
| Meat processing rooms | 5 |
| Drugs stores | |
| Pharmacists' work rooms | 20 |
| Specialty shops | |
| Pet shops | 1.0* |
| Florists | 5 |
| Greenhouses | 5 |

*cfm per sq. ft. floor area.

## TYPICAL VENTILATION AIR REQUIREMENTS FOR RESIDENTIAL USES

| Occupancy Classification | Required ventilation air in cfm per human occupant |
|---|---|
| Residential | |
| General living areas | 5 |
| Bedrooms | 5 |
| Kitchens | 20 |
| Basements, utility rooms | 5 |
| Mobile homes | 5 |
| Hotels, motels | |
| Bedrooms (single, double) | 7 |
| Living rooms (suites) | 10 |
| Corridors | 5 |
| Lobbies | 7 |
| Conference rooms (small) | 20 |
| Assembly rooms (large) | 15 |

## TYPICAL VENTILATION AIR REQUIREMENTS FOR STORAGE USES

| Occupancy Classification | Required ventilation air in cfm per human occupant |
|---|---|
| Storage | |
| Garages, service stations, parking garages (enclosed) | 1.5* |
| Auto repair shops | 1.5** |
| Warehouses | |
| General | 7 |

*cfm/s.f. floor area.
**Must have positive engine exhaust system.

## TYPICAL VENTILATION AIR REQUIREMENTS FOR FACTORY AND INDUSTRIAL USES

| Occupancy Classification | Required ventilation air in cfm per human occupant |
|---|---|
| Factory and industrial | |
| Metalworking & finishing | 35 |
| Automotive engine test | Require |
| Paint spray booths | Special |
| Picking, etching and plating lines | Exhaust |
| Degreasing booths | Systems |
| Sandblasting booths | |
| Chemicals and pharmaceuticals | |
| Dusty operations | 30 |
| Rooms containing potential gas emitters | 20 |
| Drying oven rooms | 15 |
| Fermentation rooms | 15 |
| Pillmaking booths | 10 |
| Packaging areas | 10 |
| Utility rooms | 7 |
| Computer rooms | 7 |
| Textiles-clothes manufacturer | 15 |
| Electronics and aerospace circuit board and soldering rooms | 20 |
| Wood products, papermaking | 20 |
| Brewing, distilling, wineries, bottling | 20* |
| Food processing | 20 |
| Tobacco processing | 20 |
| Power plants | |
| Control rooms | 10 |
| Boiler rooms | 35 |
| Generator rooms | 20 |
| Sewage treatment plants | |
| Control rooms | 10 |
| Compressor/blower motor rooms | 20 |
| Glass and ceramic manufacturer | 20 |
| Agricultural | 20 |

## TYPICAL VENTILATION AIR REQUIREMENTS FOR BUSINESS USES

| Occupancy Classification | Required ventilation air in cfm per human occupant |
|---|---|
| Business | |
| Banks | |
| (see offices) | |
| Vaults | 5 |
| Barber, beauty and health services | |
| Beauty shops (hair dressers) | 25 |
| Reducing salons | 25 |
| Sauna baths, steam rooms | 5 |
| barber shops | 7 |
| Photo studios | |
| Camera rooms, stages | 5 |
| Dark rooms | |
| Shoe repair shops | |
| Workrooms/trade areas | 10 |
| Offices | |
| General office space and showrooms | 15 |
| Conference rooms | 25 |
| Drafting/art rooms | 7 |
| Doctor's consultation rooms | 10 |
| Waiting rooms | 10 |
| Lithographing rooms | 7 |
| Diazo printing rooms | 7 |
| Computer rooms | 5 |
| Keypunch rooms | 7 |
| Communication | |
| TV/radio broadcasting booths, studios | 30 |
| Motion picture and TV stages | 30 |
| Pressrooms | 15 |
| Composing rooms | 7 |
| Engraving rooms | 7 |
| Telephone switchboard rooms (manual) | 7 |
| Telephone switchgear rooms (automatic) | 7 |
| Teletypewriter/facsimile rooms | 5 |
| Research institutes | |
| Laboratories: | |
| Light duty; non-chemical | 15 |
| Chemical | 15 |
| Heavy-duty | 15 |
| Radioisotope, chemical & biologically toxic | 15 |
| Machine shops | 15 |
| Dark rooms, spectroscopy rooms | 10 |
| Animal rooms | 40 |
| Veterinary hospitals | |
| Kennels, stalls | 25 |
| Operating rooms | 25 |
| Reception rooms | 10 |

## TYPICAL VENTILATION AIR REQUIREMENTS FOR INSTITUTIONAL USES

| Occupancy Classification | Required ventilation air in cfm per human occupant |
|---|---|
| Institutional | |
| Prisons | |
| Cell blocks | 7 |
| Eating halls | 15 |
| Guard stations | 7 |

# AIR CONDITIONING

## RECOMMENDED SHEET METAL GAUGES AND CONSTRUCTION FOR RECTANGULAR DUCT

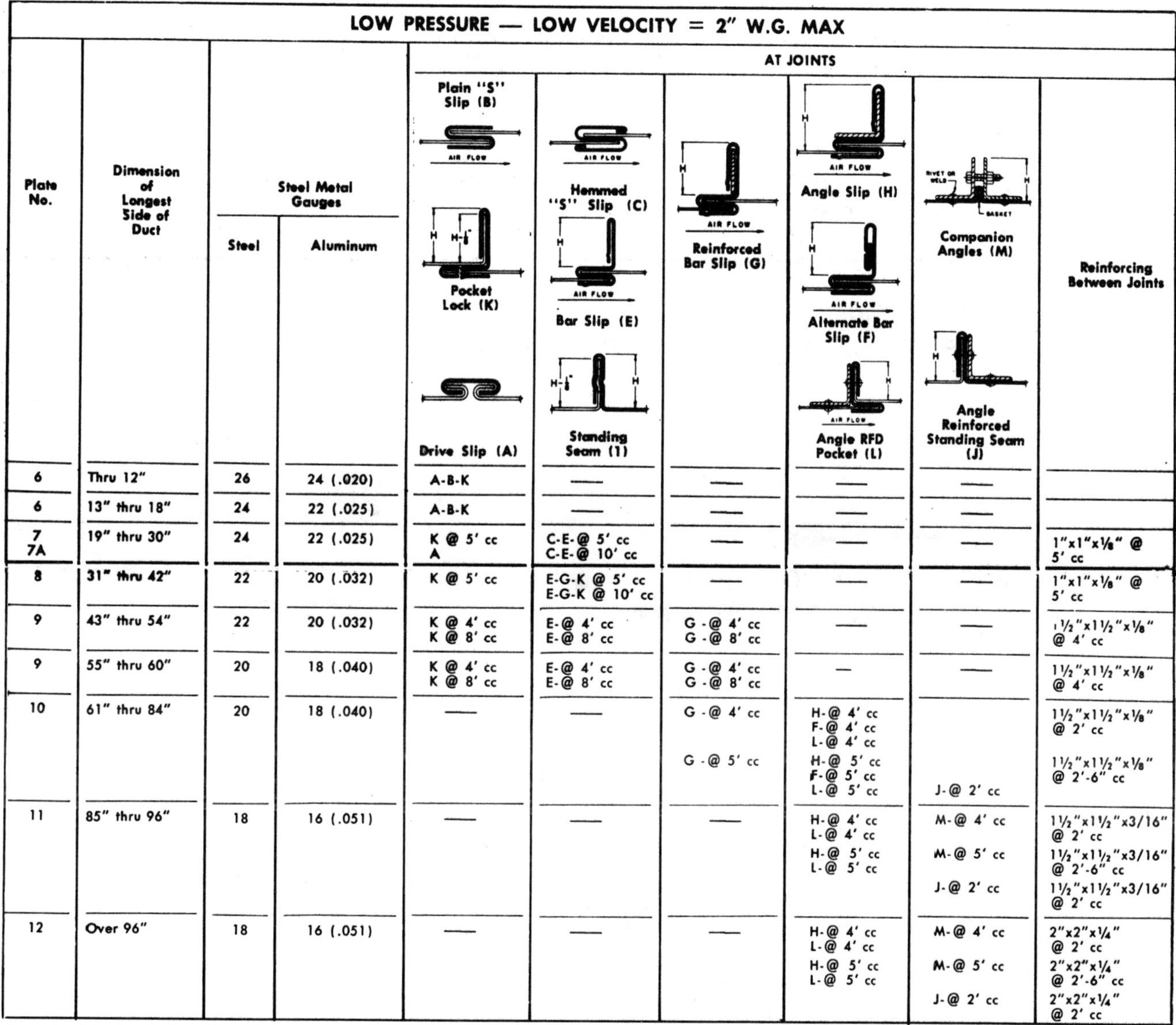

| LOW PRESSURE — LOW VELOCITY = 2" W.G. MAX | | | | | | | | | |
|---|---|---|---|---|---|---|---|---|---|
| | | | | AT JOINTS | | | | | |
| Plate No. | Dimension of Longest Side of Duct | Steel Metal Gauges: Steel | Steel Metal Gauges: Aluminum | Plain "S" Slip (B); Pocket Lock (K); Drive Slip (A) | Hemmed "S" Slip (C); Bar Slip (E); Standing Seam (1) | Reinforced Bar Slip (G) | Angle Slip (H); Alternate Bar Slip (F); Angle RFD Pocket (L) | Companion Angles (M); Angle Reinforced Standing Seam (J) | Reinforcing Between Joints |
| 6 | Thru 12" | 26 | 24 (.020) | A-B-K | — | — | — | — | |
| 6 | 13" thru 18" | 24 | 22 (.025) | A-B-K | — | — | — | — | |
| 7<br>7A | 19" thru 30" | 24 | 22 (.025) | K @ 5' cc<br>A | C-E-@ 5' cc<br>C-E-@ 10' cc | — | — | — | 1"x1"x1/8" @ 5' cc |
| 8 | 31" thru 42" | 22 | 20 (.032) | K @ 5' cc | E-G-K @ 5' cc<br>E-G-K @ 10' cc | — | — | — | 1"x1"x1/8" @ 5' cc |
| 9 | 43" thru 54" | 22 | 20 (.032) | K @ 4' cc<br>K @ 8' cc | E-@ 4' cc<br>E-@ 8' cc | G -@ 4' cc<br>G -@ 8' cc | — | — | 1½"x1½"x1/8" @ 4' cc |
| 9 | 55" thru 60" | 20 | 18 (.040) | K @ 4' cc<br>K @ 8' cc | E-@ 4' cc<br>E-@ 8' cc | G -@ 4' cc<br>G -@ 8' cc | — | — | 1½"x1½"x1/8" @ 4' cc |
| 10 | 61" thru 84" | 20 | 18 (.040) | — | — | G -@ 4' cc<br>G -@ 5' cc | H-@ 4' cc<br>F-@ 4' cc<br>L-@ 4' cc<br>H-@ 5' cc<br>F-@ 5' cc<br>L-@ 5' cc | J-@ 2' cc | 1½"x1½"x1/8" @ 2' cc<br>1½"x1½"x1/8" @ 2'-6" cc |
| 11 | 85" thru 96" | 18 | 16 (.051) | — | — | — | H-@ 4' cc<br>L-@ 4' cc<br>H-@ 5' cc<br>L-@ 5' cc | M-@ 4' cc<br>M-@ 5' cc<br>J-@ 2' cc | 1½"x1½"x3/16" @ 2' cc<br>1½"x1½"x3/16" @ 2'-6" cc<br>1½"x1½"x3/16" @ 2' cc |
| 12 | Over 96" | 18 | 16 (.051) | — | — | — | H-@ 4' cc<br>L-@ 4' cc<br>H-@ 5' cc<br>L-@ 5' cc | M-@ 4' cc<br>M-@ 5' cc<br>J-@ 2' cc | 2"x2"x¼" @ 2' cc<br>2"x2"x¼" @ 2'-6" cc<br>2"x2"x¼" @ 2' cc |

H (height dimension)—up to 42" = 1"
H (height dimension)—43" to 96" = 1½"
H (height dimension)—over 96" = 2"

## AIR CONDITIONING (Cont.)

### LONGITUDINAL SEAMS FOR SHEET METAL DUCTWORK

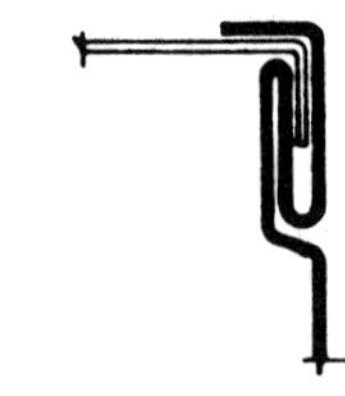

Fig. "N"
PITTSBURGH LOCK

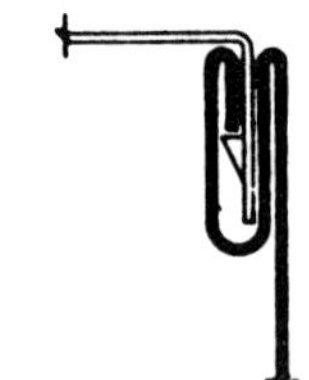

Fig. "Z"
BUTTON PUNCH SNAP LOCK

Fig. "O"
ACME LOCK-GROOVED SEAM

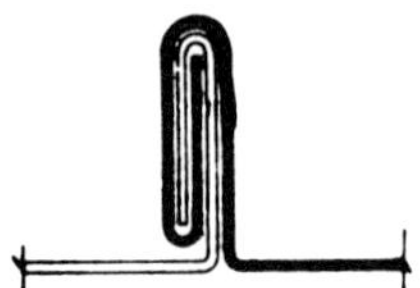

Fig. "T"
DOUBLE SEAM

Approximately 2″ Spacing Between "Buttons"

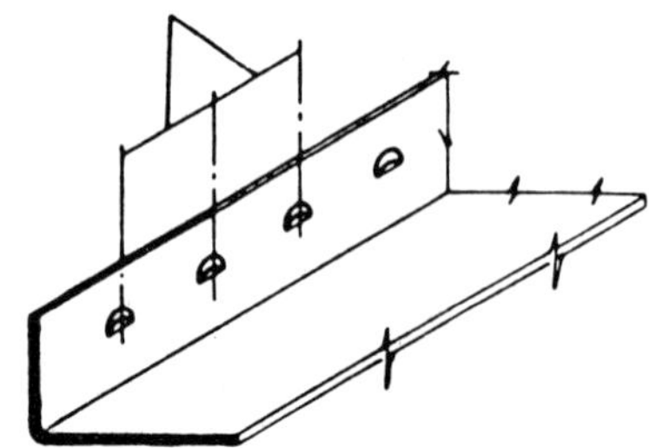

DETAIL NO. 1
MALE PIECE-SNAP LOCK

# AIR CONDITIONING

## TYPICAL DUCT CONNECTIONS

## CROSS JOINTS FOR SHEET METAL DUCTWORK

(NOT TO SCALE)

H=HEIGHT REFERRED TO IN DIMENSIONS

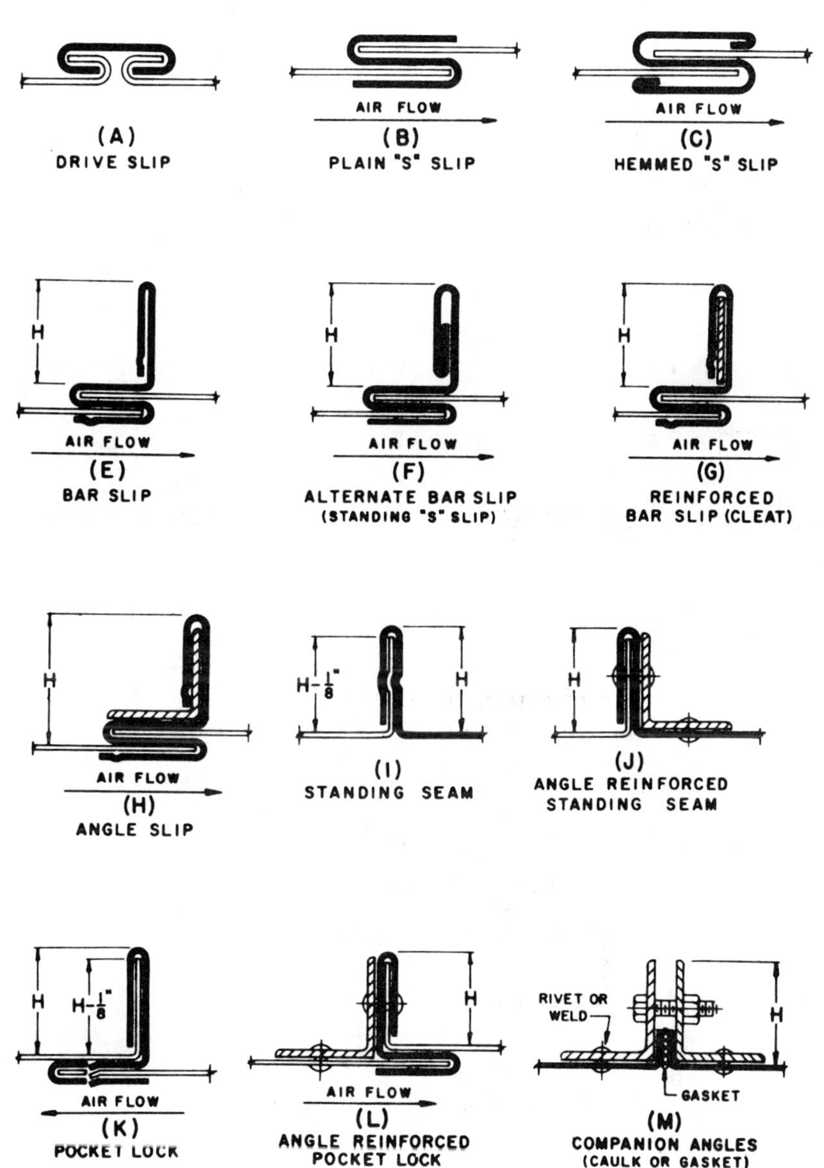

## TYPICAL DUCT CONSTRUCTION SHEET METAL GAGES IN ONE- AND TWO-FAMILY DWELLINGS

| Metal Gauges (duct not enclosed in partitions) | | |
|---|---|---|
| **ROUND DUCTS** | | |
| Diameter, inches | Minimum thickness galvanized sheet gage | Minimum thickness aluminum B&S gage |
| Less than 12 | 30 | 26 |
| 12-14 | 28 | 26 |
| 15-18 | 26 | 24 |
| Over 18 | 24 | 22 |
| **RECTANGULAR DUCTS** | | |
| Width, inches | Minimum thickness galvanized sheet gage | Minimum thickness aluminum B&S gage |
| Less than 14 | 28 | 24 |
| 14-24 | 26 | 22 |
| 25-30 | 24 | 22 |
| Over 30 | 22 | 20 |
| **Metal Gauges (duct enclosed in partitions)** | | |
| Width, inches | Minimum thickness galvanized sheet gage | Minimum thickness aluminum B&S gage |
| 14 or less | 30 | 26 |
| Over 14 | 28 | 24 |

## TYPICAL DUCT CONSTRUCTION SHEET METAL GAGES

*(All uses except 1- and 2-family dwellings)*

| RECTANGULAR DUCTS | | |
|---|---|---|
| Maximum side inches | Steel min. Galv. Sheet Gage | Aluminum Min. B&S Gage |
| Through 12 | 26 (0.022 in.) | 24 (0.020 in.) |
| 13 through 30 | 24 (0.028 in.) | 22 (0.025 in.) |
| 31 through 54 | 22 (0.034 in.) | 20 (0.032 in.) |
| 55 through 84 | 20 (0.040 in.) | 18 (0.040 in.) |
| Over 84 | 18 (0.052 in.) | 16 (0.051 in.) |

| ROUND DUCTS | | | |
|---|---|---|---|
| | Spiral seam duct | Longitudinal seam duct | Fittings |
| Diameter inches | Steel min. Galv. Sht. Gage | Steel min. Galv. Sht. Gage | Steel min. Galv. Sht. Gage |
| Through 12 | 28.(0.019 in.) | 26 (0.022 in.) | 26 (0.022 in.) |
| 13 through 18 | 26 (0.022 in.) | 24 (0.028 in.) | 24 (0.028 in.) |
| 19 through 28 | 24 (0.028 in.) | 22 (0.034 in.) | 22 (0.034 in.) |
| 29 through 36 | 22 (0.034 in.) | 20 (0.040 in.) | 20 (0.040 in.) |
| 37 through 52 | 20 (0.040 in.) | 18 (0.052 in.) | 18 (0.052 in.) |

# APPLICATIONS FOR CONDUCTORS USED FOR GENERAL WIRING

| | AMBIENT TEMPERATURE | | | | | | | | |
|---|---|---|---|---|---|---|---|---|---|
| | 60°C 140°F | 75°C 167°F | 85°C 185°F | 90°C 194°F | 110°C 230°F | 200°C 392°F | Dry | Dry or Wet | FEATURES |
| R | X | | | | | | X | | Code Rubber |
| RH | | X | | | | | X | | Heat Resistant |
| RHH | | | | X | | | X | | More Heat Resistant |
| RW | X | | | | | | | X | Moisture Resistant |
| RH-RW | X | | | | | | | X | Moisture and Heat Resistant |
| | | X | | | | | X | | Moisture and Heat Resistant |
| RHW | | X | | | | | | X | Moisture and Heat Resistant |
| RU | X | | | | | | X | | Latex Rubber |
| RUH | | X | | | | | X | | Heat Resistant |
| RUW | X | | | | | | | X | Moisture Resistant |
| T | X | | | | | | X | | Thermoplastic |
| TW | X | | | | | | | X | Moisture Resistant |
| THHN | | | | X | | | X | | Heat Resistant |
| THW | | X | | | | | | X | Moisture and Heat Resistant |
| THWN | | X | | | | | | X | Moisture and Heat Resistant |
| MI | | | X | | | | | X | Mineral Insulated Metal Sheathed |
| V | | | X | | | | X | | Varnished Cambric |
| AVA | | | | | X | | X | | With Asbestos |
| AVB | | | | X | | | X | | With Asbestos |
| AVL | | | | | X | | | X | With Asbestos |

This table does not include special condition conductors, thickness of conductor insulation, or reference to all outer protective coverings.

**GENERAL CLASSIFICATION OF INSULATIONS:**

A ............... Asbestos
H ............... Heat Resistant
MI ............. Mineral Insulation
R ............... Rubber
RU .............Latex Rubber
V ................Varnished Cambric
T ................Thermoplastic
W ...............(Water) Moisture Resistant

# WIRE AND SHEET METAL GAGES

(In Decimals of an Inch)

| Name of Gage | American Wire Gage (A.W.G.) (Corresponds to Brown & Sharpe Gage) | Birmingham Iron Wire Gage (B.W.G.) | United States Standard Gage (U.S.S.G.) | |
|---|---|---|---|---|
| Principal Use | Electrical Wire & Non-Ferrous Sheet Metal | Iron or Steel Wire | Ferrous Sheet Metal | |
| Gage No. | | | | Gage No. |
| 00 00000 | | | | 00 00000 |
| 0 00000 | .5800 | | | 0 00000 |
| 00000 | .5165 | .500 | | 00000 |
| 0000 | .4600 | .454 | | 0000 |
| 000 | .4096 | .425 | | 000 |
| 00 | .3648 | .380 | | 00 |
| 0 | .3249 | .340 | | 0 |
| 1 | .2893 | .300 | | 1 |
| 2 | .2576 | .284 | | 2 |
| 3 | .2294 | .259 | 23.91 | 3 |
| 4 | .2043 | .238 | .2242 | 4 |
| 5 | .1819 | .220 | .2092 | 5 |
| 6 | .1620 | .203 | .1943 | 6 |
| 7 | .1443 | .180 | .1793 | 7 |
| 8 | .1285 | .165 | .1644 | 8 |
| 9 | .1144 | .148 | .1495 | 9 |
| 10 | .1019 | .134 | .1345 | 10 |
| 11 | .0907 | .120 | .1196 | 11 |
| 12 | .0808 | .109 | .1046 | 12 |
| 13 | .0720 | .095 | .0897 | 13 |
| 14 | .0641 | .083 | .0747 | 14 |
| 15 | .0571 | .072 | .0673 | 15 |
| 16 | .0508 | .065 | .0598 | 16 |
| 17 | .0453 | .058 | .0538 | 17 |
| 18 | .0403 | .049 | .0478 | 18 |
| 19 | .0359 | .042 | .0418 | 19 |
| 20 | .0320 | .035 | .0359 | 20 |
| 21 | .0285 | .032 | .0329 | 21 |
| 22 | .0253 | .028 | .0299 | 22 |
| 23 | .0226 | .025 | .0269 | 23 |
| 24 | .0201 | .022 | .0239 | 24 |
| 25 | .0179 | .020 | .0209 | 25 |
| 26 | .0159 | .018 | .0179 | 26 |
| 27 | .0142 | .016 | .0164 | 27 |
| 28 | .0126 | .014 | .0149 | 28 |
| 29 | .0113 | .013 | .0135 | 29 |
| 30 | .0100 | .012 | .0120 | 30 |
| 31 | .0089 | .010 | .0105 | 31 |
| 32 | .0080 | .009 | .0097 | 32 |
| 33 | .0071 | .008 | .0090 | 33 |
| 34 | .0063 | .007 | .0082 | 34 |
| 35 | .0056 | .005 | .0075 | 35 |
| 36 | .0050 | .004 | .0067 | 36 |
| 37 | .0045 | | .0064 | 37 |
| 38 | .0040 | | .0060 | 38 |
| 39 | .0035 | | | 39 |
| 40 | .0031 | | | 40 |

# Geographic Cost Modifiers

The costs as presented in this book attempt to represent national averages. Costs, however, vary among regions, states and even between adjacent localities.

In order to more closely approximate the probable costs for specific locations throughout the U.S., this table of Geographic Cost Modifiers is provided. These adjustment factors are used to modify costs obtained from this book to help account for regional variations of construction costs and to provide a more accurate estimate for specific areas. The factors are formulated by comparing costs in a specific area to the costs as presented in the Costbook pages. An example of how to use these factors is shown below. Whenever local current costs are known, whether material prices or labor rates, they should be used when more accuracy is required.

| Cost Obtained from Costbook Pages | X | Location Cost Adjustment Factor | = | Adjusted Cost |
|---|---|---|---|---|

For example, a project estimated to cost $125,000 using the Costbook pages can be adjusted to more closely approximate the cost in Los Angeles:

$125,000 X 1.07 = $133,750.

# Geographic Cost Modifiers

|  |  |
|---|---|
| **ALABAMA** | |
| BIRMINGHAM | 0.79 |
| HUNTSVILLE | 0.77 |
| MOBILE | 0.81 |
| MONTGOMERY | 0.75 |
| TUSCALOOSA | 0.75 |
| **ALASKA** | |
| ANCHORAGE | 1.30 |
| JUNEAU | 1.33 |
| FAIRBANKS | 1.37 |
| NOME | 1.43 |
| **ARIZONA** | |
| FLAGSTAFF | 0.91 |
| PHOENIX | 0.89 |
| PRESCOTT | 0.91 |
| TUCSON | 0.89 |
| YUMA | 0.87 |
| **ARKANSAS** | |
| FAYETTEVILLE | 0.73 |
| FORT SMITH | 0.74 |
| LITTLE ROCK | 0.78 |
| PINE BLUFF | 0.75 |
| **CALIFORNIA** | |
| ANAHEIM | 1.02 |
| BAKERSFIELD | 0.97 |
| LOS ANGELES | 1.07 |
| REDDING | 0.92 |
| RIVERSIDE | 0.97 |
| SACRAMENTO | 1.00 |
| SAN DIEGO | 1.02 |
| SAN JOSE | 1.07 |
| SAN FRANCISCO | 1.12 |
| SANTA BARBARA | 1.07 |
| **COLORADO** | |
| BOULDER | 0.92 |
| COLORADO SPRINGS | 0.92 |
| DENVER | 0.95 |
| GRAND JUNCTION | 0.90 |
| PUEBLO | 0.90 |
| **CONNECTICUT** | |
| BRIDGEPORT | 1.03 |
| HARTFORD | 1.01 |
| NEW LONDON | 0.99 |
| STAMFORD | 1.06 |
| WATERBURY | 0.96 |
| **DELAWARE** | |
| DOVER | 0.89 |
| WILMINGTON | 0.91 |
| **FLORIDA** | |
| JACKSONVILLE | 0.80 |
| MIAMI | 0.86 |
| ORLANDO | 0.80 |
| TAMPA | 0.82 |
| WEST PALM BEACH | 0.84 |
| **GEORGIA** | |
| ATLANTA | 0.83 |
| AUGUSTA | 0.75 |
| COLUMBUS | 0.75 |
| MACON | 0.77 |
| SAVANNAH | 0.79 |
| **HAWAII** | |
| HILO | 1.31 |
| HONOLULU | 1.25 |
| MAUI | 1.28 |
| **IDAHO** | |
| BOISE | 0.92 |
| LEWISTON | 0.90 |
| POCATELLO | 0.88 |
| TWIN FALLS | 0.88 |
| **ILLINOIS** | |
| CHICAGO | 1.03 |
| MOLINE | 0.89 |
| PEORIA | 0.92 |
| ROCKFORD | 0.92 |
| SPRINGFIELD | 0.89 |
| **INDIANA** | |
| FORT WAYNE | 0.88 |
| EVANSVILLE | 0.88 |
| GARY | 0.99 |
| INDIANAPOLIS | 0.95 |
| TERRE HAUTE | 0.86 |
| **IOWA** | |
| COUNCIL BLUFFS | 0.85 |
| DAVENPORT | 0.87 |
| DES MOINES | 0.89 |
| SIOUX CITY | 0.85 |
| WATERLOO | 0.87 |
| **KANSAS** | |
| DODGE CITY | 0.77 |
| SALINA | 0.79 |
| TOPEKA | 0.83 |
| WICHITA | 0.81 |
| **KENTUCKY** | |
| BOWLING GREEN | 0.83 |
| LEXINGTON | 0.85 |
| LOUISVILLE | 0.89 |
| PADUCAH | 0.87 |
| **LOUISIANA** | |
| BATON ROUGE | 0.84 |
| LAKE CHARLES | 0.82 |
| MONROE | 0.78 |
| NEW ORLEANS | 0.86 |
| SHREVEPORT | 0.78 |
| **MAINE** | |
| AUGUSTA | 0.85 |
| BANGOR | 0.83 |
| LEWISTON | 0.87 |
| PORTLAND | 0.89 |
| **MARYLAND** | |
| ANNAPOLIS | 0.92 |
| BALTIMORE | 0.90 |
| HAGERSTOWN | 0.88 |
| ROCKVILLE | 0.95 |
| **MASSACHUSETTS** | |
| BOSTON | 1.01 |
| LOWELL | 0.94 |
| FALL RIVER | 0.91 |
| SPRINGFIELD | 0.94 |
| WORCESTER | 0.96 |
| **MICHIGAN** | |
| DETROIT | 0.95 |
| GRAND RAPIDS | 0.88 |
| LANSING | 0.88 |
| MARQUETTE | 0.86 |
| SAGINAW | 0.90 |
| **MINNESOTA** | |
| DULUTH | 0.93 |
| MINNEAPOLIS | 0.98 |
| ROCHESTER | 0.93 |
| SAINT PAUL | 0.98 |
| **MISSISSIPPI** | |
| COLUMBUS | 0.77 |
| GULFPORT | 0.79 |
| JACKSON | 0.81 |
| VICKSBURG | 0.79 |
| **MISSOURI** | |
| JOPLIN | 0.84 |
| KANSAS CITY | 0.86 |
| SAINT LOUIS | 0.90 |
| SAINT JOSEPH | 0.84 |
| SPRINGFIELD | 0.84 |
| **MONTANA** | |
| BILLINGS | 0.87 |
| BUTTE | 0.85 |
| GREAT FALLS | 0.87 |
| HELENA | 0.85 |
| MISSOULA | 0.85 |

# Geographic Cost Modifiers

| NEBRASKA | |
|---|---|
| GRAND ISLAND | 0.81 |
| LINCOLN | 0.85 |
| NORTH PLATTE | 0.81 |
| OMAHA | 0.87 |
| **NEVADA** | |
| CARSON CITY | 0.95 |
| LAS VEGAS | 0.97 |
| SPARKS | 0.97 |
| RENO | 1.00 |
| **NEW HAMPSHIRE** | |
| CONCORD | 0.86 |
| MANCHESTER | 0.88 |
| NASHUA | 0.90 |
| **NEW JERSEY** | |
| ATLANTIC CITY | 0.97 |
| CAMDEN | 0.95 |
| NEWARK | 1.02 |
| PATERSON | 0.99 |
| TRENTON | 0.97 |
| **NEW MEXICO** | |
| ALBUQUERQUE | 0.86 |
| CARLSBAD | 0.84 |
| LAS CRUCES | 0.82 |
| SANTA FE | 0.92 |
| **NEW YORK** | |
| ALBANY | 0.87 |
| BINGHAMTON | 0.83 |
| BUFFALO | 0.90 |
| LONG ISLAND | 1.01 |
| NEW YORK CITY | 1.15 |
| ROCHESTER | 0.90 |
| SYRACUSE | 0.87 |
| WATERTOWN | 0.83 |
| WHITE PLAINS | 1.01 |
| **NORTH CAROLINA** | |
| ASHEVILLE | 0.75 |
| CHARLOTTE | 0.79 |
| GREENSBORO | 0.75 |
| RALEIGH | 0.77 |
| WILMINGTON | 0.74 |
| **NORTH DAKOTA** | |
| BISMARK | 0.86 |
| FARGO | 0.90 |
| GRAND FORKS | 0.88 |
| MINOT | 0.84 |

| OHIO | |
|---|---|
| CINCINNATI | 0.83 |
| CLEVELAND | 0.91 |
| COLUMBUS | 0.89 |
| TOLEDO | 0.89 |
| YOUNGSTOWN | 0.85 |
| **OKLAHOMA** | |
| BARTLESVILLE | 0.77 |
| ENID | 0.77 |
| LAWTON | 0.77 |
| OKLAHOMA CITY | 0.83 |
| TULSA | 0.81 |
| **OREGON** | |
| EUGENE | 0.90 |
| MEDFORD | 0.90 |
| PORTLAND | 0.95 |
| SALEM | 0.92 |
| **PENNSYLVANIA** | |
| ALLENTOWN | 0.91 |
| HARRISBURG | 0.85 |
| PHILADELPHIA | 0.98 |
| PITTSBURGH | 0.89 |
| SCRANTON | 0.87 |
| **RHODE ISLAND** | |
| PAWTUCKET | 0.95 |
| PROVIDENCE | 0.98 |
| NEWPORT | 0.98 |
| WESTERLY | 0.95 |
| WOONSOCKET | 0.95 |
| **SOUTH CAROLINA** | |
| CHARLESTON | 0.79 |
| COLUMBIA | 0.81 |
| FLORENCE | 0.79 |
| GREENVILLE | 0.81 |
| **SOUTH DAKOTA** | |
| ABERDEEN | 0.79 |
| PIERRE | 0.79 |
| RAPID CITY | 0.81 |
| SIOUX FALLS | 0.83 |
| WATERTOWN | 0.81 |
| **TENNESSEE** | |
| CHATTANOOGA | 0.77 |
| JOHNSON CITY | 0.75 |
| KNOXVILLE | 0.79 |
| MEMPHIS | 0.81 |
| NASHVILLE | 0.83 |

| TEXAS | |
|---|---|
| AUSTIN | 0.80 |
| DALLAS | 0.82 |
| HOUSTON | 0.86 |
| LUBBOCK | 0.78 |
| SAN ANTONIO | 0.80 |
| **UTAH** | |
| LOGAN | 0.83 |
| OGDEN | 0.85 |
| PROVO | 0.83 |
| SALT LAKE | 0.87 |
| **VERMONT** | |
| BRATTLEBORO | 0.85 |
| BURLINGTON | 0.94 |
| RUTLAND | 0.87 |
| **VIRGINIA** | |
| ALEXANDRIA | 0.94 |
| LYNCHBURG | 0.81 |
| NORFOLK | 0.83 |
| RICHMOND | 0.85 |
| ROANOKE | 0.81 |
| **WASHINGTON** | |
| BELLINGHAM | 0.99 |
| SEATTLE | 1.06 |
| SPOKANE | 0.99 |
| TACOMA | 1.03 |
| YAKIMA | 0.96 |
| **WASHINGTON, D.C.** | |
| DISTRICT | 0.95 |
| **WEST VIRGINIA** | |
| CHARLESTON | 0.87 |
| HUNTINGTON | 0.89 |
| MORGANTOWN | 0.83 |
| PARKERSBURG | 0.83 |
| **WISCONSIN** | |
| EAU CLAIRE | 0.90 |
| GREEN BAY | 0.92 |
| MADISON | 0.90 |
| MILWAUKEE | 0.97 |
| WAUSAU | 0.90 |
| **WYOMING** | |
| CASPER | 0.89 |
| CHEYENNE | 0.91 |
| ROCK SPRINGS | 0.87 |
| SHERIDAN | 0.85 |

BNi® Building News

# Square Foot Tables

The following Square Foot Tables list hundreds of actual projects for dozens of building types, each with associated building size, total square foot building cost and percentage of project costs for total mechanical and electrical components. This data provides an overview of construction costs by building type. These costs are for actual projects. The variations within similar building types may be due, among other factors, to size, location, quality and specified components, materials and processes. Depending upon all such factors, specific building costs can vary significantly and may not necessarily fall within the range of costs as presented. The data has been updated to reflect current construction costs.

| BUILDING CATEGORY | PAGE |
|---|---|

# SQUARE FOOT TABLES

## COMMERCIAL

### AUTO DEALERSHIP

| Project Size Gross S.F. | Project Cost $/S.F. | % Cost Mechanical | % Cost Electrical |
|---|---|---|---|
| 7,700 | 82.00 | 16.7 | 9.4 |
| 16,100 | 46.20 | 10.2 | 15.9 |
| 20,000 | 48.40 | 12.9 | 23.4 |
| 26,300 | 49.50 | 12.5 | 22.0 |
| 43,600 | 39.10 | 19.4 | 13.2 |
| 53,600 | 65.40 | 12.5 | 11.5 |

### BUSINESS CENTER

| Project Size Gross S.F. | Project Cost $/S.F. | % Cost Mechanical | % Cost Electrical |
|---|---|---|---|
| 3,900 | 49.50 | 12.0 | 9.2 |
| 9,900 | 47.90 | 9.1 | 7.6 |
| 54,400 | 32.00 | 3.4 | 12.2 |
| 135,000 | 34.80 | 8.2 | 1.5 |

### CINEMA

| Project Size Gross S.F. | Project Cost $/S.F. | % Cost Mechanical | % Cost Electrical |
|---|---|---|---|
| 18,000 | 112.00 | 10.9 | 6.7 |
| 22,500 A | 69.00 | 6.6 | 4.2 |

### MALL/PLAZA

| Project Size Gross S.F. | Project Cost $/S.F. | % Cost Mechanical | % Cost Electrical |
|---|---|---|---|
| 9,700 | 35.00 | 15.0 | 13.3 |
| 10,500 | 56.00 | 8.0 | 13.5 |
| 16,300 | 51.00 | 9.4 | 10.6 |
| 26,900 | 50.80 | 15.0 | 7.0 |
| 36,000 | 43.00 | 10.0 | 11.0 |
| 36,300 | 48.80 | 12.4 | 8.2 |
| 44,720 | 51.00 | 18.3 | 12.4 |
| 59,100 | 50.00 | 9.8 | 9.5 |
| 60,000 R | 51.25 | 10.0 | 9.5 |
| 64,100 | 51.50 | 22.4 | 18.4 |
| 66,000 | 59.00 | 13.5 | 11.5 |
| 67,400 | 42.90 | 21.0 | 15.0 |
| 73,500 | 62.00 | 14.9 | 6.9 |

### MALL/PLAZA (Cont.)

| Project Size Gross S.F. | Project Cost $/S.F. | % Cost Mechanical | % Cost Electrical |
|---|---|---|---|
| 142,000 | 39.00 | 7.1 | 8.0 |
| 220,000 | 103.00 | 11.0 | 6.4 |
| 223,700 | 30.25 | 9.0 | 9.3 |
| 321,200 | 36.00 | 7.3 | 7.4 |
| 379,900 | 47.00 | 11.2 | 6.2 |
| 405,100 | 48.50 | 13.9 | 6.0 |
| 482,000 | 80.60 | 10.5 | 9.4 |
| 630,000 | 53.90 | 12.2 | 12.4 |

### RESTAURANT

| Project Size Gross S.F. | Project Cost $/S.F. | % Cost Mechanical | % Cost Electrical |
|---|---|---|---|
| 4,300 R | 89.00 | 6.9 | 8.7 |
| 4,400 R | 131.00 | 14.6 | 8.5 |
| 5,800 | 106.50 | 28.0 | 10.6 |
| 6,800 A | 124.50 | 7.0 | 11.1 |
| 7,360 R | 134.00 | 16.0 | 6.5 |
| 9,600 | 133.10 | 24.7 | 13.1 |
| 10,000 R | 138.50 | 21.0 | 10.0 |
| 10,100 | 119.50 | 28.6 | 18.4 |
| 10,600 | 220.00 | 20.4 | 6.4 |
| 22,900 R | 143.00 | 15.8 | 16.9 |

### RETAIL STORE

| Project Size Gross S.F. | Project Cost $/S.F. | % Cost Mechanical | % Cost Electrical |
|---|---|---|---|
| 1,000 | 134.00 | 12.8 | 6.7 |
| 3,000 R | 120.00 | 14.3 | 10.5 |
| 12,300 | 164.50 | 14.0 | 10.0 |
| 30,000 | 83.10 | 15.6 | 26.2 |
| 61,300 | 43.00 | 13.3 | 13.0 |
| 115,000 | 59.50 | 14.6 | 11.3 |
| 154,700 | 84.30 | 11.2 | 12.4 |
| 314,700 R | 71.10 | 13.8 | 9.4 |

A = Addition R = Remodel

# SQUARE FOOT TABLES

## RESIDENTIAL

### APARTMENTS

| Project Size Gross S.F. | Project Cost $/S.F. | % Cost Mechanical | % Cost Electrical |
|---|---|---|---|
| 3,700 | 49.80 | 14.9 | 4.4 |
| 13,900 | 68.20 | 7.9 | 7.2 |
| 19,200 R | 108.30 | 45.3 | 7.5 |
| 19,700 | 63.50 | 7.4 | 10.4 |
| 23,700 | 69.70 | 10.6 | 4.3 |
| 26,500 | 66.00 | 25.2 | 12.8 |
| 35,100 | 54.00 | 16.4 | 5.6 |
| 54,000 | 93.40 | 23.3 | 13.1 |
| 62,700 | 67.90 | 17.0 | 9.0 |
| 67,300 | 62.00 | 13.8 | 8.4 |
| 70,600 | 34.90 | 18.1 | 7.8 |
| 75,300 | 77.00 | 13.2 | 8.1 |
| 75,600 | 75.00 | 14.5 | 8.9 |
| 72,200 | 65.00 | 18.4 | 10.9 |
| 77,600 | 84.10 | 26.9 | 14.3 |
| 88,100 | 78.00 | 15.3 | 9.3 |
| 89,500 | 73.20 | 10.7 | 11.1 |
| 94,100 | 44.10 | 8.5 | 6.8 |
| 96,000 | 55.00 | 17.0 | 13.3 |
| 102,000 | 95.50 | 17.8 | 12.5 |
| 103,200 | 43.40 | 12.1 | 8.9 |
| 103,600 | 69.90 | 19.1 | 9.0 |
| 105,200 | 89.00 | 14.9 | 8.9 |
| 106,200 | 64.10 | 12.8 | 9.3 |
| 110,900 | 61.60 | 15.6 | 8.4 |
| 111,800 | 87.50 | 17.3 | 7.4 |
| 115,900 | 65.60 | 12.5 | 8.6 |
| 117,200 | 41.00 | 12.9 | 8.0 |
| 119,000 | 56.30 | 15.6 | 8.4 |

### APARTMENTS (Cont.)

| Project Size Gross S.F. | Project Cost $/S.F. | % Cost Mechanical | % Cost Electrical |
|---|---|---|---|
| 119,400 | 36.90 | 18.6 | 8.7 |
| 144,300 | 82.00 | 15.0 | 8.6 |
| 176,300 | 69.70 | 19.1 | 9.0 |
| 192,300 | 41.00 | 10.1 | 6.0 |
| 210,900 | 80.70 | 19.1 | 9.5 |
| 220,200 | 83.00 | 14.8 | 7.5 |
| 253,900 | 107.90 | 20.6 | 7.7 |
| 369,500 | 78.80 | 15.8 | 9.0 |

### CONDOS/TOWNHOUSES

| Project Size Gross S.F. | Project Cost $/S.F. | % Cost Mechanical | % Cost Electrical |
|---|---|---|---|
| 8,600 | 70.70 | 8.5 | 4.4 |
| 16,700 | 57.00 | 13.2 | 6.5 |
| 18,000 | 120.00 | 14.6 | 9.7 |
| 18,400 | 61.50 | 12.9 | 5.8 |
| 74,800 | 55.80 | 13.8 | 5.2 |
| 111,700 | 70.50 | 9.2 | 7.1 |
| 150,300 | 78.70 | 15.9 | 7.9 |
| 278,800 | 123.00 | 14.1 | 7.9 |
| 1,109,900 | 56.60 | 9.8 | 4.7 |

### SINGLE-FAMILY HOMES

| Project Size Gross S.F. | Project Cost $/S.F. | % Cost Mechanical | % Cost Electrical |
|---|---|---|---|
| 600 R | 49.90 | 16.3 | 2.0 |
| 900 | 65.40 | 33.0 | 3.0 |
| 2,100 | 75.80 | 8.8 | 4.0 |
| 2,200 | 163.20 | 23.8 | 3.4 |
| 2,500 | 79.40 | 6.1 | 7.1 |
| 2,900 | 76.70 | 7.7 | 2.8 |
| 3,000 | 53.60 | 9.5 | 6.8 |
| 3,100 | 120.00 | 15.5 | 4.6 |
| 3,600 | 75.40 | 8.5 | 3.0 |
| 3,700 | 104.80 | 9.8 | 3.4 |

A = Addition    R = Remodel

# SQUARE FOOT TABLES

## RESIDENTIAL (Cont.)

### SINGLE-FAMILY HOMES (Cont.)

| Project Size Gross S.F. | Project Cost $/S.F. | % Cost Mechanical | % Cost Electrical |
|---|---|---|---|
| 4,200 | 92.00 | 15.5 | 5.7 |
| 4,600 | 146.00 | 9.0 | 4.7 |
| 5,200 | 190.50 | 8.3 | 6.0 |
| 5,700 | 110.60 | 7.4 | 3.7 |
| 5,700 | 75.40 | 8.7 | 12.6 |
| 21,300* | 53.90 | 26.0 | 4.0 |
| 22,700* | 50.00 | 27.0 | 4.6 |
| 45,000* | 76.00 | 7.8 | 2.5 |
| 51,458* | 55.20 | 11.0 | 5.0 |

*Townhouses.

## EDUCATIONAL

### ADMINISTRATION (OFFICES)

| Project Size Gross S.F. | Project Cost $/S.F. | % Cost Mechanical | % Cost Electrical |
|---|---|---|---|
| 53,700 | 131.00 | 15.4 | 9.0 |

### ATHLETIC FACILITY

| Project Size Gross S.F. | Project Cost $/S.F. | % Cost Mechanical | % Cost Electrical |
|---|---|---|---|
| 38,100 | 130.00 | 16.8 | 9.8 |
| 44,100 | 120.00 | 16.8 | 8.4 |
| 100,000 | 82.00 | 19.2 | 5.3 |
| 160,000 | 148.50 | 13.7 | 7.9 |
| 247,500 | 124.60 | 11.2 | 9.0 |
| 271,000 | 117.00 | 13.2 | 7.6 |
| 283,100 | 141.00 | 14.6 | 6.0 |

### AUDITORIUM/PERFORMING ARTS

| Project Size Gross S.F. | Project Cost $/S.F. | % Cost Mechanical | % Cost Electrical |
|---|---|---|---|
| 9,900 | 210.00 | 17.5 | 29.2 |
| 17,800 | 192.00 | 11.4 | 16.1 |
| 29,200 | 190.00 | 16.2 | 10.5 |
| 62,700 | 138.20 | 13.0 | 12.9 |

### CLASSROOM

| Project Size Gross S.F. | Project Cost $/S.F. | % Cost Mechanical | % Cost Electrical |
|---|---|---|---|
| 35,400 | 190.50 | 10.2 | 6.8 |
| 70,000 | 86.00 | 24.3 | 13.1 |
| 78,900 | 160.00 | 13.3 | 14.3 |
| 80,100 | 130.00 | 16.9 | 10.5 |
| 100,000 | 129.50 | 21.1 | 9.8 |
| 166,000 | 89.50 | 16.1 | 11.2 |
| 298,400 | 84.00 | 11.3 | 11.9 |

### COMPLETE COLLEGE FACILITIES

| Project Size Gross S.F. | Project Cost $/S.F. | % Cost Mechanical | % Cost Electrical |
|---|---|---|---|
| 95,300 | 135.00 | 19.4 | 11.7 |
| 450,000 | 161.10 | 18.6 | 10.8 |

### ELEMENTARY SCHOOL

| Project Size Gross S.F. | Project Cost $/S.F. | % Cost Mechanical | % Cost Electrical |
|---|---|---|---|
| 18,000 | 97.80 | 19.1 | 13.8 |
| 30,800 | 90.00 | 16.0 | 10.0 |
| 31,600 | 75.50 | 12.1 | 11.3 |
| 35,700 | 110.50 | 17.3 | 8.7 |
| 40,000 | 95.50 | 18.5 | 13.4 |
| 40,500 | 81.00 | 22.1 | 11.3 |
| 57,000 | 72.50 | 22.1 | 7.5 |
| 69,700 | 112.90 | 20.6 | 9.3 |
| 91,400 | 91.20 | 18.2 | 7.6 |

### HIGH SCHOOL

| Project Size Gross S.F. | Project Cost $/S.F. | % Cost Mechanical | % Cost Electrical |
|---|---|---|---|
| 116,400 | 100.20 | 18.1 | 12.9 |
| 133,000 | 85.90 | 17.8 | 10.7 |
| 184,000 | 151.50 | 23.0 | 10.0 |
| 217,200 R | 81.70 | 26.4 | 10.6 |
| 254,000 R | 65.00 | 17.2 | 14.6 |
| 431,700 | 96.00 | 13.1 | 9.6 |

A = Addition R = Remodel

# SQUARE FOOT TABLES

## EDUCATIONAL (Cont.)

### JUNIOR HIGH SCHOOL

| Project Size Gross S.F. | Project Cost $/S.F. | % Cost Mechanical | % Cost Electrical |
|---|---|---|---|
| 26,000 | 125.50 | 9.5 | 9.3 |
| 28,100 | 78.00 | 11.9 | 9.5 |
| 52,800 | 100.40 | 18.3 | 8.6 |
| 91,600 | 124.10 | 21.2 | 11.0 |
| 123,700 | 101.60 | 29.1 | 9.1 |

### LABORATORY/RESEARCH

| Project Size Gross S.F. | Project Cost $/S.F. | % Cost Mechanical | % Cost Electrical |
|---|---|---|---|
| 9,200 | 220.00 | 25.4 | 4.5 |
| 80,300 | 173.00 | 20.2 | 15.5 |

### LIBRARY

| Project Size Gross S.F. | Project Cost $/S.F. | % Cost Mechanical | % Cost Electrical |
|---|---|---|---|
| 6,900 | 115.40 | 17.9 | 10.3 |
| 8,200 | 105.90 | 18.8 | 10.6 |
| 12,000 | 137.30 | 19.4 | 15.9 |
| 15,000 | 120.00 | 15.7 | 18.7 |
| 16,300 | 101.60 | 11.3 | 7.4 |
| 28,600 | 91.00 | 15.7 | 9.0 |
| 30,100 | 123.50 | 13.0 | 11.0 |
| 37,700 | 97.00 | 14.6 | 8.0 |
| 43,500 | 82.00 | 16.6 | 6.7 |
| 47,900 | 143.70 | 29.7 | 9.0 |
| 51,400 | 146.80 | 13.1 | 12.4 |
| 63,400 A | 110.40 | 13.5 | 8.3 |
| 64,000 | 103.30 | 10.9 | 11.4 |
| 74,000 | 117.00 | 17.2 | 8.0 |
| 75,600 | 155.40 | 12.5 | 16.7 |
| 176,000 | 86.10 | 11.2 | 9.8 |

### SPECIAL NEEDS FUNCTION

| Project Size Gross S.F. | Project Cost $/S.F. | % Cost Mechanical | % Cost Electrical |
|---|---|---|---|
| 15,200 | 97.60 | 16.6 | 8.4 |
| 27,900 | 117.60 | 19.4 | 11.2 |

### STUDENT CENTER/MULTIPURPOSE

| Project Size Gross S.F. | Project Cost $/S.F. | % Cost Mechanical | % Cost Electrical |
|---|---|---|---|
| 90,000 | 148.40 | 16.2 | 10.0 |
| 49,600 | 109.60 | 18.2 | 9.2 |
| 187,700 | 165.00 | 15.3 | 8.8 |
| 194,800 | 87.20 | 17.2 | 8.5 |

## HOTEL/MOTEL

### CONVENTION/CONFERENCE CENTER

| Project Size Gross S.F. | Project Cost $/S.F. | % Cost Mechanical | % Cost Electrical |
|---|---|---|---|
| 8,600 A | 130.30 | 20.7 | 16.3 |
| 71,900 | 144.50 | 12.5 | 13.8 |
| 433,800 | 75.50 | 22.1 | 8.3 |

### HOTEL

| Project Size Gross S.F. | Project Cost $/S.F. | % Cost Mechanical | % Cost Electrical |
|---|---|---|---|
| 19,900 A | 74.50 | 16.8 | 5.8 |
| 25,875 R | 66.80 | 11.8 | 10.5 |
| 48,400 A | 133.60 | 23.6 | 8.2 |
| 64,300 R | 180.00 | 20.3 | 10.1 |
| 104,200 A | 83.60 | 15.0 | 8.8 |
| 108,040 | 71.40 | 13.0 | 7.0 |
| 110,100 | 92.40 | 18.4 | 10.5 |
| 132,000 | 172.80 | 16.4 | 5.4 |
| 135,900 A | 106.80 | 15.2 | 7.7 |
| 144,100 A | 120.00 | 19.3 | 11.5 |
| 231,000 | 129.80 | 15.2 | 8.4 |
| 449,800 A | 75.00 | 13.1 | 7.0 |

### HOTEL/INN

| Project Size Gross S.F. | Project Cost $/S.F. | % Cost Mechanical | % Cost Electrical |
|---|---|---|---|
| 57,400 | 86.00 | 13.0 | 10.0 |
| 73,000 | 56.10 | 24.4 | 18.1 |
| 75,900 | 74.80 | 16.5 | 7.6 |
| 162,000 | 90.00 | 17.5 | 8.0 |
| 197,000 | 86.10 | 15.7 | 7.7 |
| 277,900 | 63.30 | 18.8 | 9.4 |

A = Addition R = Remodel

# SQUARE FOOT TABLES

## INDUSTRIAL

### MANUFACTURING

| Project Size Gross S.F. | Project Cost $/S.F. | % Cost Mechanical | % Cost Electrical |
|---|---|---|---|
| 14,300 | 73.80 | 13.0 | 7.0 |
| 18,500 | 111.30 | 20.5 | 13.5 |
| 26,600 | 33.80 | 6.1 | 12.1 |
| 31,400 | 87.90 | 18.9 | 17.7 |
| 33,400 | 50.80 | 23.1 | 15.8 |
| 37,300 | 32.00 | 3.0 | 24.0 |
| 43,400 | 50.50 | 14.7 | 11.7 |
| 45,400 | 82.00 | 15.9 | 15.0 |
| 79,800 | 85.20 | 16.0 | 14.5 |
| 81,100 | 91.00 | 8.2 | 8.4 |
| 137,400 | 57.10 | 41.6 | 13.6 |
| 179,600 | 65.90 | 26.3 | 13.6 |
| 186,000 | 61.60 | 19.5 | 11.7 |

### RESEARCH AND DEVELOPMENT

| Project Size Gross S.F. | Project Cost $/S.F. | % Cost Mechanical | % Cost Electrical |
|---|---|---|---|
| 89,140 | 117.70 | 20.8 | 9.0 |
| 100,400 | 155.90 | 36.1 | 25.1 |
| 114,200 | 152.70 | 21.4 | 9.8 |
| 125,000 | 103.90 | 20.8 | 8.4 |
| 140,000 | 145.10 | 20.1 | 11.2 |

### WAREHOUSE W/OFFICE

| Project Size Gross S.F. | Project Cost $/S.F. | % Cost Mechanical | % Cost Electrical |
|---|---|---|---|
| 14,000 | 33.90 | 6.5 | 9.2 |
| 19,000 | 23.80 | 2.3 | 1.8 |
| 19,700 | 39.90 | 9.5 | 7.0 |
| 31,200 | 34.30 | 7.0 | 7.0 |
| 40,500 | 43.80 | 6.6 | 10.5 |
| 62,000 | 55.60 | 10.8 | 10.0 |
| 96,200 | 28.90 | 2.2 | 6.3 |
| 105,000 | 27.10 | 5.3 | 11.1 |
| 149,800 | 28.80 | 14.5 | 8.6 |

### WAREHOUSE W/OFFICE (Cont.)

| Project Size Gross S.F. | Project Cost $/S.F. | % Cost Mechanical | % Cost Electrical |
|---|---|---|---|
| 168,600 | 35.60 | 11.4 | 6.3 |
| 209,600 | 32.90 | 4.9 | 10.3 |
| 402.400 | 45.10 | 14.9 | 8.3 |

## MEDICAL

### EDUCATION CENTER

| Project Size Gross S.F. | Project Cost $/S.F. | % Cost Mechanical | % Cost Electrical |
|---|---|---|---|
| 35,400 | 202.00 | 10.2 | 6.8 |

### HOSPITALS

| Project Size Gross S.F. | Project Cost $/S.F. | % Cost Mechanical | % Cost Electrical |
|---|---|---|---|
| 9,300 R | 199.50 | 29.5 | 12.0 |
| 15,900 A | 147.20 | 27.4 | 15.3 |
| 16,600 R | 53.00 | 14.7 | 9.6 |
| 22,000 | 335.00 | 31.6 | 17.5 |
| 39,100 | 166.90 | 23.3 | 7.9 |
| 63,800 | 134.50 | 23.4 | 7.8 |
| 98,000 A | 193.50 | 20.5 | 17.0 |
| 100,200 A | 308.70 | 22.2 | 9.6 |
| 103,900 A | 171.10 | 31.1 | 11.7 |
| 109,300 | 169.50 | 20.7 | 15.7 |
| 148,700 | 133.20 | 25.2 | 11.8 |
| 154,700 | 216.30 | 19.3 | 10.5 |
| 165,484 | 188.00 | 21.5 | 14.8 |
| 165,700 A | 183.50 | 26.2 | 17.7 |
| 179,400 A | 111.80 | 28.9 | 20.3 |
| 182,800 | 166.40 | 19.1 | 14.0 |
| 265,000 | 198.50 | 31.2 | 14.1 |
| 281,100 | 190.10 | 24.1 | 14.6 |
| 435,000 | 197.00 | 21.5 | 13.7 |
| 694,300 | 109.10 | 28.9 | 9.7 |
| 772,300 | 201.00 | 31.5 | 13.4 |

A = Addition R = Remodel

# SQUARE FOOT TABLES

## MEDICAL (Cont.)

### MEDICAL OFFICES/CENTERS

| Project Size Gross S.F. | Project Cost $/S.F. | % Cost Mechanical | % Cost Electrical |
|---|---|---|---|
| 3,000 | 84.35 | 13.7 | 11.5 |
| 5,500 | 87.80 | 8.5 | 18.7 |
| 10,000 | 109.40 | 13.9 | 9.4 |
| 10,600 | 134.40 | 13.2 | 8.9 |
| 16,300 | 111.20 | 21.8 | 13.2 |
| 18,300 | 60.70 | 7.1 | 13.4 |
| 20,600 | 87.20 | 14.3 | 6.4 |
| 24,900 | 153.50 | 21.0 | 15.1 |
| 27,000 | 155.00 | 19.7 | 10.1 |
| 28,400 | 108.00 | 13.6 | 9.3 |
| 30,500 | 117.00 | 17.2 | 12.7 |
| 32,000 | 92.80 | 18.1 | 10.5 |
| 44,300 | 93.60 | 24.4 | 16.2 |
| 50,200 | 54.00 | 9.8 | 4.9 |
| 51,200 | 136.70 | 21.0 | 12.0 |
| 64,600 | 65.10 | 13.2 | 6.4 |
| 66,000 | 60.70 | 9.8 | 8.8 |
| 80,000 | 43.00 | 8.2 | 8.3 |
| 137,175 | 87.00 | 12.8 | 6.6 |

### NURSING HOMES

| Project Size Gross S.F. | Project Cost $/S.F. | % Cost Mechanical | % Cost Electrical |
|---|---|---|---|
| 11,600 A | 255.00 | 53.2 | 7.9 |
| 16,800 | 166.00 | 33.3 | 7.8 |
| 31,900 A | 124.80 | 22.0 | 11.0 |
| 64,100 | 107.00 | 20.6 | 11.1 |
| 290,000 | 150.00 | 16.1 | 13.6 |

### RESEARCH

| Project Size Gross S.F. | Project Cost $/S.F. | % Cost Mechanical | % Cost Electrical |
|---|---|---|---|
| 34,600 | 142.00 | 18.1 | 3.0 |

## PUBLIC FACILITIES

### ANIMAL CENTER

| Project Size Gross S.F. | Project Cost $/S.F. | % Cost Mechanical | % Cost Electrical |
|---|---|---|---|
| 20,000 | 180.00 | 20.7 | 4.6 |
| 39,100 | 134.50 | 22.9 | 6.8 |
| 44,300 | 128.10 | 8.1 | 5.8 |

### AUTO DEALERSHIP

| Project Size Gross S.F. | Project Cost $/S.F. | % Cost Mechanical | % Cost Electrical |
|---|---|---|---|
| 7,700 | 83.90 | 16.7 | 9.4 |

### BROADCASTING

| Project Size Gross S.F. | Project Cost $/S.F. | % Cost Mechanical | % Cost Electrical |
|---|---|---|---|
| 20,000 | 245.40 | 20.0 | 13.0 |
| 29,500 | 189.30 | 16.6 | 15.0 |
| 45,000 R | 137.30 | 15.0 | 13.0 |

### CIVIC CENTER

| Project Size Gross S.F. | Project Cost $/S.F. | % Cost Mechanical | % Cost Electrical |
|---|---|---|---|
| 6,000 | 135.20 | 9.2 | 2.8 |
| 23,900 | 169.80 | 12.3 | 10.9 |
| 34,400 | 77.40 | 3.5 | 17.8 |
| 69,800 A | 202.80 | 17.7 | 8.8 |
| 206,500 | 109.70 | 15.0 | 11.0 |

### CORRECTION FACILITIES

| Project Size Gross S.F. | Project Cost $/S.F. | % Cost Mechanical | % Cost Electrical |
|---|---|---|---|
| 44,600 | 162.10 | 20.9 | 13.1 |
| 66,000 | 98.80 | 15.0 | 23.0 |
| 257,800 A | 172.00 | 20.9 | 10.7 |
| 360,000 | 115.50 | 32.7 | 13.2 |

### FIRE STATION

| Project Size Gross S.F. | Project Cost $/S.F. | % Cost Mechanical | % Cost Electrical |
|---|---|---|---|
| 6,900 | 146.30 | 12.4 | 9.7 |
| 7,600 | 112.40 | 16.0 | 9.5 |
| 8,430 | 154.60 | 12.8 | 9.6 |
| 9,600 | 134.80 | 13.5 | 11.8 |

A = Addition R = Remodel

# SQUARE FOOT TABLES

## PUBLIC FACILITIES (Cont.)

### GOVERNMENT BUILDINGS

| Project Size Gross S.F. | Project Cost $/S.F. | % Cost Mechanical | % Cost Electrical |
|---|---|---|---|
| 12,300 | 104.90 | 13.5 | 10.7 |
| 23,500 | 171.90 | 11.1 | 15.6 |
| 27,300 | 135.70 | 19.5 | 9.3 |
| 31,600 | 152.80 | 27.1 | 11.3 |
| 46,600 | 143.80 | 17.4 | 13.8 |
| 72,100 | 156.60 | 24.4 | 10.7 |
| 78,200 | 129.90 | 19.7 | 15.2 |
| 332,900 | 126.10 | 16.2 | 14.2 |
| 364,100 | 165.10 | 14.6 | 12.9 |
| 771,000 | 182.40 | 17.6 | 11.3 |

### MUSEUM

| Project Size Gross S.F. | Project Cost $/S.F. | % Cost Mechanical | % Cost Electrical |
|---|---|---|---|
| 27,600 | 126.00 | 17.8 | 14.1 |
| 30,100 | 144.70 | 18.6 | 7.9 |
| 43,264 | 134.80 | 11.1 | 8.3 |
| 63,000 | 145.90 | 8.8 | 18.1 |

### PARKING GARAGE

| Project Size Gross S.F. | Project Cost $/S.F. | % Cost Mechanical | % Cost Electrical |
|---|---|---|---|
| 66,000 | 38.50 | 2.8 | 3.1 |
| 169,000 | 25.60 | 10.3 | 3.7 |
| 562,700 | 22.50 | 2.4 | 6.2 |

### TRANSPORTATION

| Project Size Gross S.F. | Project Cost $/S.F. | % Cost Mechanical | % Cost Electrical |
|---|---|---|---|
| 7,300 | 231.50 | 15.1 | 3.2 |
| 14,300 | 186.00 | 13.3 | 16.6 |
| 23,000 | 128.50 | 9.6 | 13.7 |
| 35,500 | 107.00 | 1.0 | 19.0 |
| 49,100 | 159.60 | 35.5 | 11.6 |

### TRANSPORTATION (Cont.)

| Project Size Gross S.F. | Project Cost $/S.F. | % Cost Mechanical | % Cost Electrical |
|---|---|---|---|
| 288,100 | 92.00 | 8.3 | 13.3 |
| 2,160,000 | 149.00 | 23.3 | 11.5 |

## OFFICES

### BANKS

| Project Size Gross S.F. | Project Cost $/S.F. | % Cost Mechanical | % Cost Electrical |
|---|---|---|---|
| 2,900 | 214.00 | 5.3 | 4.3 |
| 3,100 | 95.50 | 5.9 | 6.7 |
| 3,300 | 138.90 | 10.3 | 16.4 |
| 3,600 | 104.60 | 10.3 | 12.2 |
| 4,000 | 119.20 | 8.0 | 9.0 |
| 4,100 | 118.00 | 21.4 | 13.0 |
| 4,200 | 127.30 | 8.6 | 14.21 |
| 4,400 | 147.00 | 12.2 | 12.7 |
| 4,500 | 91.30 | 12.4 | 13.5 |
| 4,900 | 144.20 | 11.7 | 11.2 |
| 5,900 | 107.20 | 9.3 | 13.8 |
| 6,000 | 120.50 | 11.6 | 7.3 |
| 6,100 | 162.00 | 11.0 | 8.0 |
| 7,000 | 218.00 | 6.0 | 9.0 |
| 7,300 | 133.30 | 11.9 | 11.3 |
| 7,700 | 149.50 | 8.0 | 7.5 |
| 7,800 | 163.80 | 11.0 | 11.7 |
| 8,000 | 83.20 | 10.0 | 14.0 |
| 9,200 | 132.00 | 9.9 | 12.1 |
| 9,400 | 95.60 | 11.7 | 11.8 |
| 10,200 | 199.40 | 12.6 | 12.6 |
| 12,600 | 71.20 | 7.0 | 18.0 |

A = Addition R = Remodel

# SQUARE FOOT TABLES

## OFFICES (Cont.)

### BANKS (Cont.)

| Project Size Gross S.F. | Project Cost $/S.F. | % Cost Mechanical | % Cost Electrical |
|---|---|---|---|
| 13,300 | 125.60 | 9.5 | 8.3 |
| 13,800 | 112.40 | 10.0 | 9.7 |
| 15,000 | 96.30 | 15.4 | 12.4 |
| 15,200 | 76.90 | 9.2 | 12.9 |
| 15,500 | 98.20 | 9.8 | 10.3 |
| 16,000 | 64.40 | 13.4 | 23.1 |
| 20,100 | 60.60 | 13.0 | 11.0 |
| 21,700 | 128.10 | 8.8 | 11.3 |
| 44,800 | 109.50 | 13.0 | 8.2 |
| 53,200 | 197.70 | 14.9 | 7.2 |
| 62,100 | 113.60 | 10.1 | 7.9 |
| 95,100 | 149.40 | 13.5 | 4.3 |

### OFFICE BUILDINGS

| Project Size Gross S.F. | Project Cost $/S.F. | % Cost Mechanical | % Cost Electrical |
|---|---|---|---|
| 2,600 | 134.30 | 17.2 | 9.4 |
| 3,400 | 109.80 | 10.5 | 11.3 |
| 3,800 | 107.80 | 16.4 | 11.8 |
| 4,400 | 106.50 | 12.8 | 8.5 |
| 4,500 | 91.60 | 13.0 | 7.0 |
| 5,100 | 66.50 | 16.6 | 10.2 |
| 5,200 | 87.30 | 8.0 | 5.7 |
| 6,700 | 130.00 | 16.8 | 10.4 |
| 7,500 | 140.00 | 10.1 | 8.0 |
| 7,900 | 109.00 | 17.4 | 9.0 |
| 8,100 | 168.40 | 10.6 | 11.0 |
| 10,600 | 97.40 | 9.9 | 9.7 |
| 10,900 | 56.10 | 13.8 | 10.8 |

### OFFICE BUILDINGS (Cont.)

| Project Size Gross S.F. | Project Cost $/S.F. | % Cost Mechanical | % Cost Electrical |
|---|---|---|---|
| 11,300 | 110.70 | 17.0 | 6.0 |
| 13,000 | 75.30 | 15.0 | 9.0 |
| 14,400 | 98.80 | 19.8 | 12.9 |
| 14,500 | 75.80 | 17.3 | 12.5 |
| 17,000 | 108.20 | 14.7 | 7.9 |
| 17,800 A | 58.50 | 10.4 | 11.0 |
| 18,100 | 113.40 | 22.5 | 11.6 |
| 19,300 | 59.10 | 11.1 | 8.1 |
| 24,600 | 52.90 | 18.5 | 14.1 |
| 27,700 | 68.40 | 19.6 | 5.5 |
| 27,800 | 124.10 | 12.7 | 5.1 |
| 27,800 | 66.80 | 17.8 | 10.3 |
| 32,500 R | 114.30 | 13.9 | 6.4 |
| 35,400 | 66.70 | 15.0 | 12.0 |
| 36,500 | 53.90 | 10.2 | 10.2 |
| 42,300 | 71.80 | 10.3 | 7.7 |
| 44,400 | 97.70 | 23.5 | 14.7 |
| 44,400 | 51.40 | 11.0 | 5.0 |
| 44,500 | 70.70 | 11.5 | 3.1 |
| 45,400 | 61.10 | 19.1 | 13.2 |
| 47,300 | 62.80 | 18.5 | 8.0 |
| 49,700 | 105.60 | 22.1 | 7.2 |
| 50,000 | 100.60 | 19.4 | 15.6 |
| 50,400 | 109.20 | 23.2 | 7.8 |
| 52,200 | 70.90 | 18.3 | 7.8 |
| 52,900 | 82.70 | 4.4 | 3.9 |
| 53,700 | 126.40 | 15.4 | 9.0 |

A = Additional R = Remodel

# SQUARE FOOT TABLES

## OFFICES (Cont.)

### OFFICE BUILDINGS (Cont.)

| Project Size Gross S.F. | Project Cost $/S.F. | % Cost Mechanical | % Cost Electrical |
|---|---|---|---|
| 54,000 | 51.20 | 14.4 | 2.6 |
| 56,000 | 58.10 | 10.6 | 6.2 |
| 56,500 | 79.90 | 19.1 | 11.0 |
| 72,000 R | 28.10 | 12.9 | 20.2 |
| 74,000 | 55.60 | 13.5 | 6.8 |
| 80,800 | 56.20 | 12.0 | 6.0 |
| 81,800 | 77.40 | 21.1 | 9.5 |
| 81,900 | 65.10 | 19.9 | 9.0 |
| 82,000 | 90.50 | 14.0 | 4.2 |
| 83,100 | 106.70 | 16.0 | 8.6 |
| 85,400 | 93.60 | 18.7 | 8.6 |
| 86,200 | 73.30 | 14.7 | 11.0 |
| 99,900 | 84.80 | 16.6 | 6.6 |
| 100,000 | 94.70 | 10.6 | 12.1 |
| 100,000 | 55.90 | 16.3 | 10.4 |
| 116,400 | 89.50 | 12.6 | 10.0 |
| 134,500 | 182.50 | 14.1 | 12.4 |
| 140,000 | 141.50 | 20.1 | 11.2 |
| 155,700 | 176.20 | 11.1 | 6.1 |
| 171,000 | 82.60 | 24.0 | 7.7 |
| 174,300 | 101.70 | 12.3 | 9.1 |
| 203,300 | 137.50 | 19.0 | 13.2 |
| 265,800 | 190.30 | 11.0 | 10.0 |
| 287,300 | 52.70 | 10.7 | 4.1 |
| 319,800 | 82.60 | 13.4 | 5.6 |
| 350,000 | 122.10 | 20.0 | 13.0 |
| 360,900 | 73.90 | 14.2 | 11.1 |
| 394,000 | 58.70 | 22.5 | 8.7 |

### OFFICE BUILDINGS (Cont.)

| Project Size Gross S.F. | Project Cost $/S.F. | % Cost Mechanical | % Cost Electrical |
|---|---|---|---|
| 430,000 | 87.90 | 21.8 | 9.3 |
| 490,000 | 93.50 | 11.2 | 7.5 |
| 588,400 | 192.20 | 15.0 | 9.0 |
| 606,000 | 76.80 | 9.4 | 8.8 |
| 620,000 | 212.00 | 12.6 | 12.8 |
| 733,500 | 58.70 | 10.8 | 4.2 |

## RECREATIONAL

### ARENA

| Project Size Gross S.F. | Project Cost $/S.F. | % Cost Mechanical | % Cost Electrical |
|---|---|---|---|
| 315,200 | 220.00 | 10.4 | 8.00 |
| 385,800 | 126.00 | 13.2 | 8.80 |
| 727,000 | 113.90 | 14.9 | 7.40 |

### HEALTH CLUB

| Project Size Gross S.F. | Project Cost $/S.F. | % Cost Mechanical | % Cost Electrical |
|---|---|---|---|
| 15,900 | 52.10 | 8.7 | 9.7 |
| 21,800 | 87.10 | 10.3 | 10.3 |
| 30,100 | 134.60 | 11.4 | 19.4 |
| 66,400 | 63.20 | 10.7 | 8.5 |

### RECREATIONAL CENTER

| Project Size Gross S.F. | Project Cost $/S.F. | % Cost Mechanical | % Cost Electrical |
|---|---|---|---|
| 9,900 | 119.60 | 6.6 | 9.8 |
| 14,000 | 78.00 | 11.2 | 5.0 |
| 14,000 | 93.60 | 11.3 | 16.7 |
| 15,700 | 71.20 | 17.8 | 14.3 |
| 20,000 | 161.60 | 19.1 | 8.9 |
| 21,200 | 90.90 | 16.0 | 9.6 |

A = Addition  R = Remodel

# SQUARE FOOT TABLES

## RECREATIONAL (Cont.)

### RECREATIONAL CENTER (Cont.)

| Project Size Gross S.F. | Project Cost $/S.F. | % Cost Mechanical | % Cost Electrical |
|---|---|---|---|
| 26,000 | 105.90 | 9.3 | 7.0 |
| 53,400 A | 126.60 | 11.7 | 6.4 |
| 69,800 | 205.40 | 17.7 | 8.8 |

## RELIGIOUS

### CHURCH

| Project Size Gross S.F. | Project Cost $/S.F. | % Cost Mechanical | % Cost Electrical |
|---|---|---|---|
| 4,100 | 144.20 | 8.8 | 13.5 |
| 10,400 R | 196.20 | 12.8 | 8.1 |
| 11,100 | 86.40 | 10.9 | 12.7 |
| 13,400 | 129.50 | 5.5 | 6.0 |
| 14,500 | 88.40 | 12.8 | 7.4 |
| 15,200 | 114.60 | 18.3 | 8. |
| 15,700 | 99.80 | 14.4 | 8.4 |

### CHURCH (Cont.)

| Project Size Gross S.F. | Project Cost $/S.F. | % Cost Mechanical | % Cost Electrical |
|---|---|---|---|
| 16,000 | 162.30 | 14.1 | 9.6 |
| 20,900 | 101.20 | 16.0 | 14.0 |
| 21,500 | 111.10 | 7.3 | 8.0 |
| 22,900 | 100.30 | 12.1 | 9.5 |
| 30,600 | 75.90 | 15.5 | 8.0 |
| 42,700 | 85.70 | 18.6 | 7.7 |

### MULTI-PURPOSE

| Project Size Gross S.F. | Project Cost $/S.F. | % Cost Mechanical | % Cost Electrical |
|---|---|---|---|
| 4,400 | 88.60 | 11.5 | 17.5 |
| 5,800 A | 112.90 | 8.9 | 7.5 |
| 6,400 | 154.20 | 15.6 | 15.8 |
| 9,000 | 75.50 | 7.7 | 5.9 |
| 9,000 | 107.20 | 16.0 | 6.6 |
| 10,100 | 71.90 | 11.1 | 12.0 |
| 12,000 | 142.10 | 13.1 | 10.7 |
| 18,400 | 77.90 | 10.8 | 10.1 |
| 19,500 | 137.30 | 16.0 | 17.3 |

A = Addition R = Remodel

For more information subscribe to **Design Cost & Data**

BNi® Building News

# INDEX

- C -

- D -

- N -



- O -



- P -

- T -

- U -



- V -

# Notes

# Notes